Microbial Control of Pests and Plant Diseases 1970-1980

Edited by

H. D. BURGES

Glasshouse Crops Research Institute
Rustington, Littlehampton, West Sussex, England.

1981

ACADEMIC PRESS

A Subsidiary of Harcourt Brace Jovanovich, Publishers
London   New York   Toronto   Sydney   San Francisco

ACADEMIC PRESS INC. (LONDON) LTD.
24/28 Oval Road
London NW1

United States Edition published by
ACADEMIC PRESS INC.
111 Fifth Avenue
New York, New York 10003

British Library Cataloguing in Publication Data

Microbial control of pests and plant diseases,
1970–1980.
1. Pest control – Biological control
2. Micro-organisms
I. Burges, H D
632'.6   80-41480

ISBN 0-12-143360-9

Printed in Great Britain at the Alden Press
Oxford London and Northampton

Contributors

H. DE BARJAC Institut Pasteur, 25 rue du Docteur Roux, 75724 Paris, France

G.O. BEDFORD   School of Biological Sciences, Sydney Technical College, Broadway, New South Wales 2007, Australia

C.E. BLAND   Department of Biology, East Carolina University, Greenville, North Carolina 27834, USA

H.G. BOMAN   Department of Microbiology, University of Stockholm, S-106 91 Stockholm, Sweden

L.F. BOUSE   Pest Control Equipment and Methods Research Unit, USDA/SEA-AR, Texas A and M University, College Station, Texas, USA

R.J. BRAND   Department of Biomedical and Environmental Health Sciences, University of California, Berkeley, California 94720, USA

W.M. BROOKS   Department of Entomology, North Carolina State University, Raleigh, North Carolina 27607, USA

G.E. BUCHER   Agriculture Canada, 195 Dafoe Road, Winnipeg, Manitoba R3T 2M9, Canada

H.D. BURGES   Glasshouse Crops Research Institute, Littlehampton, West Sussex BN16 3PU, England

A.T.K. CORKE   Long Ashton Research Station, University of Bristol, Avon BS18 9AF, England

J.N. COUCH Department of Botany, University of North Carolina, Chapel Hill, North Carolina 29407, USA

T.L. COUCH Chemical and Agricultural Products Division, Abbott Laboratories, North Chicago, Illinois 60064, USA

J.C. CUNNINGHAM   Forest Pest Management Institute, Canadian Forestry Service, P.O. Box 490, Sault Ste. Marie, Ontario P6A 5M7, Canada

D.H. DEAN   Department of Microbiology, Ohio State University, 484 West 12th Ave., Columbus, Ohio 43210, USA

H.T. DULMAGE Cotton Insects Research, ARS, USDA, P.O. Box 1033, Browns-ville, Texas 78520, USA

E.A. ELLIS Department of Entomology and Nematology, University of Florida, Gainesville, Florida 32611, USA

P.F. ENTWISTLE Institute of Virology, Natural Environment Research Council, 5 South Parks Road, Oxford, OX1 3UB, England

J. FARKAŠ Institute of Organic Chemistry and Biochemistry, Czechoslovak Academy of Sciences, Prague 6, Czechoslovakia

CONTRIBUTORS

P.G. FAST   Forest Pest Management Institute, Canadian Forestry Service, P.O. Box 490, Sault Ste. Marie, Ontario P6A 5M7, Canada

B.A. FEDERICI   Division of Biological Control, Department of Entomology, University of California, Riverside, California 92521, USA

P. FERRON   Station de Recherches de Lutte Biologique, La Minière, 78280 Guyancourt, Yvelines, France

J.R. FINNEY   Research Unit on Vector Pathology, Memorial University of Newfoundland, St. John's, Newfoundland A1C 5S7, Canada

J.R. FUXA   Department of Entomology, North Carolina State University, Raleigh, North Carolina 27607, USA

R.A. HALL   Glasshouse Crops Research Institute, Littlehampton, West Sussex, BN16 3PU, England

E.I. HAZARD   Insects Affecting Man and Animals Research Laboratory, SEA, USDA, Gainesville, Florida 32604, USA

J.E. HENRY   Rangeland Insect Laboratory, SEA, USDA, Montana State University, Bozeman, Montana 59717, USA

K. HORSKÁ   Institute of Organic Chemistry and Biochemistry, Czechoslovak Academy of Sciences, Prague 6, Czechoslovakia

R.A. HUMBER   New England Plant, Soil and Water Laboratory, University of Maine, Orono, Maine 04473, USA

N.W. HUSSEY   Glasshouse Crops Research Institute, Littlehampton, West Sussex, BN16 3PU, England

C.M. IGNOFFO   Biological Control of Insects Research, AR, SEA, USDA, Columbia, Missouri 65205, USA

R.P. JAQUES   Research Station, Harrow, Ontario N0R 1G0, Canada

D.J. JOSLYN   Department of Biology, Camden College of Arts and Sciences, Rutgers University, Camden, New Jersey, USA

K. KATAGIRI   Forestry and Forest Products Research Institute, P.O. Box 2, Ushiku, Ibaraki, 300-12 Japan

D.C. KELLY   Institute of Virology, 5 South Parks Road, Oxford, OX1 3RB, England

D.S. KING   Boyce Thompson Institute for Plant Research, Tower Road, Ithaca, New York State, USA

M.G. KLEIN   Japanese Beetle Research Laboratory, AR, SEA, USDA, Ohio Agricultural Research and Development Centre, Wooster, Ohio 44691, USA

A. KRIEG   Biologische Bundesanstalt für Land- und Forstwirtschaft, Institut für Biologische Schädlingsbekämpfung, Heinrichstrasse 243, D-6100 Darmstadt, German Federal Republic

M. LAIRD   Research Unit on Vector Pathology, Memorial University of Newfoundland, St. John's, Newfoundland A1C 5S7, Canada

G.A. LANGENBRUCH   Biologische Bundesanstalt für Land- und Forstwirtschaft, Institut für Biologische Schädlingsbekämpfung, Heinrichstrasse 243, D-6100 Darmstadt, German Federal Republic

F.B. LEWIS   Forest Insect and Disease Laboratory, USDA, Hamden, Connecticut 06514, USA

J.V. MADDOX   Illinois Institute of Natural Resources, Section of Economic Entomology, 172 Natural Resources Building, Urbana, Illinois 61801, USA

CONTRIBUTORS

M.E. MARTIGNONI Forestry Sciences Laboratory, USDA, 3200 Jefferson Way, Corvallis, Oregon 97331, USA

P.A.W. MARTIN Department of Microbiology, Ohio State University, 484 West 12th Ave., Columbus, Ohio 43210, USA

C.W. McCOY Agricultural Research and Education Centre, University of Florida, P.O. Box 108A, Lake Alfred, Florida 33850, USA

R.J. MILNER Pastoral Research Laboratory, Division of Entomology, CSIRO, Private Bag, Armidale, New South Wales 2350, Australia

O.N. MORRIS Forest Pest Management Institute, Canadian Forestry Service, P.O. Box 490, Sault Ste. Marie, Ontario P6A 5M7, Canada

S.Y. NEWELL Rosenstiel School of Marine and Atmospheric Sciences, University of Miami, Miami, Florida 33149, USA

E.A. OMA Entomology Research Division, ARS, USDA, Montana State University, Bozeman, Montana 59717, USA

C.C. PAYNE Glasshouse Crops Research Institute, Littlehampton, West Sussex, BN16 3PU, England

D.E. PINNOCK Department of Entomology, University of Adelaide, Waite Agricultural Research Institute, Glen Osmond, South Australia, Australia

R.A.J. PRISTON Sittingbourne Research Centre, Shell Research Ltd., Sittingbourne, Kent ME9 8AG, England

D.K. REED Agricultural Fruit and Vegetable Insects Research Laboratory, SEA, USDA, 1118 Chestnut St., P.O. Box 944, Vincennes, Indiana 47591, USA

J. RISHBETH Botany School, University of Cambridge, Cambridge, England

D.W. ROBERTS Boyce Thompson Institute for Plant Research, Tower Road, Cornell University, Ithaca, NY 14853, USA

R.A. SAMSON Centraalbureau voor Schimmelcultures, P.O. Box 273, Oosterstraat 1, Baarn, Netherlands

K. ŠEBESTA Institute of Organic Chemistry and Biochemistry, Czechoslovak Academy of Sciences, Prague 6, Czechoslovakia

S. SINGER Biological Sciences, Western Illinois University, Macomb, Illinois 61455, USA

D.B. SMITH Bioengineering Research Unit, SEA, USDA, Building T-12, University of Missouri, Columbia, Missouri 65201, USA

H. STOCKDALE Sittingbourne Research Centre, Shell Research Ltd., Sittingbourne, Kent ME9 8AG, England

T.W. TINSLEY Institute of Virology, 5 South Parks Road, Oxford, OX1 3UB, England

J. VAŇKOVÁ Department of Insect Pathology, Institute of Entomology, Czechoslovak Academy of Sciences, Prague 6, Czechoslovakia

N. WILDING Rothamsted Experimental Station, Harpenden, Herts. AL5 2JQ, England

G.C. WILSON Forest Pest Management Institute, Canadian Forestry Service, Sault Ste. Marie, Ontario P6A 5M7, Canada

Preface

There is no one way to control crop pests. The ever increasing resistance of pests to pesticides and concern about environmental pollution require an integrated complex of control measures. One component is biological control, of which the harnessing of pathogens is a rapidly growing aspect. "Microbial Control of Insects and Mites", the 1971-forebear of this book, assessed the subject up to 1970, revealing almost as many problems as solutions, inevitably with some gaps in coverage. The present new book stringently avoids repetition of material from the initial book. It is a sequel, not simply a new edition or revision, but an attempt both to cover new material appearing since 1970 and to fill some of the gaps. In particular, the scope has been widened to include the use of competitors, inhibitors and diseases of plant pathogens as alternatives to chemical fungicides and bactericides.

Although essentially a practical book, it delves deeply into fundamental information when necessary to achieve an understanding of the subject and every chapter probes the future. It is aimed at a wide readership of pest control practitioners, research workers, lecturers and students seeking new information on advanced topics. It will interest pathologists, entomologists, plant pathologists, ecologists, chemists and virologists as well as all types of microbiologists. A concluding chapter caters for the general reader desiring a concise analysis of a decade's strategy and progress, and future prospects.

Pathogens, or groups of pathogens, showing particular progress have been selected from the major taxonomic divisions as subjects for a series of compact chapters about their identification, practical use and toxins. Other chapters investigate the potential of genetical engineering, aspects of technology and integration — e.g. formulation, application machinery, ecology, biostatistical modelling — also safety and the insects' defence mechanisms, as well as impressions of use and research in the People's Republic of China. Each of the 63 authors and co-authors is a specialist, writing closely around his own field. Usually the younger leaders of research have been chosen, to seek verve and enthusiasm blended with experience: all but two of my initial selection of authors accepted the challenge. Many problems are now too extensive for any one organisation to cover alone, so international programs have arisen. For instance, the World Health Organisation has financed a special program for the Biological Control of Vectors of Human Disease. Such programs have been given prominence. I have been amazed at the volume of new information in the

PREFACE

past decade and I wish to thank contributors for their co-operation with the severe editing needed to keep the book to a manageable length.

I particularly owe thanks to my wife, Sheila, and family, Stella and Martin, for their help and patient tolerance of my preoccupation with this book. I am indebted to Mrs. A. Warren, Mrs. J.M. Liddle and Mrs. S. Mitchell for help with details, the indexers A.T. Gillespie, P. Jarrett, P. Normansell, A. West and C.F. Williams and numerous colleagues who reviewed manuscripts.

January, 1981   H.D. BURGES

Contents

CONTENTS

CONTENTS

CONTENTS

THE INSECT'S DEFENCE

PEOPLE'S REPUBLIC OF CHINA

CONCLUSIONS

APPENDICES

to
Dr. A. M. Heimpel
in appreciation of his work with the
microbial control of pests

edited by H.D. Burges
Agricultural Research Council, Pest Infestation Control Laboratory, Slough, England

and N.W. Hussey
Agricultural Research Council, Glasshouse Crops Research Institute, Littlehampton, Sussex, England

1971, xxii + 862pp.   **0.12.143350.1**

The status of microbial control, the present use of microorganisms and their potential for the future are thoroughly discussed. Taking into account the practical aspects of insect pathology, the authors examine the role in integrated control of pathogens ranging from viruses to nematodes and including microbial products such as toxins. This is a major reference work for insect pathologists, entomologists, microbiologists and teachers concerned with pest control and pollution of the environment.

"This is an outstanding publication."
Entomological News

"... a truly outstanding contribution to invertebrate pathology and to the field of pest management utilizing microbial agents."
  J.D. Paschke in **The Quarterly Review of Pathology**

"... a necessary handbook for the practical insect pathologist ..."
    F. Wilson in **The Biologist**

"The most authoritative and thorough coverage of microbial control in the English language ..."
    Dudley E. Pinnock in **Science**

"... interestingly, and often stimulatingly written, and the general student of biological control will find much of value."
      N.H.E. Gibson in **Nature**

Printed in England

CHAPTER 1

Progress in the Microbial Control of Pests, 1970-80

H.D. BURGES

Glasshouse Crops Research Institute, Littlehampton, Sussex, England

The practical use of pathogens for insect and mite control and its scientific background up to 1970 were reviewed in a book edited by Burges and Hussey (1971). Since then progress has been exponential. One pathogen was registered for use about 1950, another in 1960 and five more between 1970 and 1979. An increasing number of other pathogens are under investigation, and this research is growing in intensity. International co-operative programs have been organized. Pest control strategies have been improved and the computer is being harnessed. Even so the use of pathogens is still relatively limited in comparison with that of chemical pesticides.

The present book quantifies progress in the past decade, analyses the socio-economic reasons for the directions that progress has followed, and probes the future of research and development. In so doing it makes minimal reference to material covered in the first book, and covers many subjects in greater depth and some new subjects. The scope is widened to include microbial control of plant pathogens, because the use of chemicals against plant diseases becomes more difficult as more biological agents are integrated into pest control systems. Together the two books provide an up to date coverage of pest control by microbial means.

During the past decade, my personal research has moved to the greenhouse industry. This move has provided remarkable inspiration. In greenhouses, plants are grown in intensive monoculture at near optimal conditions and without natural forces to buffer the effects of invading pests. These pests find equable conditions, breeding with great speed and, if unchecked, with severe effects. Control problems and methods evolve with corresponding speed (review: Hussey and Scopes, 1977).

For a while contact insecticides solved the pest problem – until pesticide-resistant strains became dominant. The amount and frequency of insecticide applications progressively increased until, unnoticed, crop losses due to cumulative phytotoxicity approached, in some cases, 30% (Addington, 1966). The industry was virtually compelled to integrate biological methods into control programs. When losses due to phytotoxicity and plant hardening were avoided with considerable profit on high value crops, biological control came to be adopted for financial reasons. However, reduction in the use of broad spectrum chemicals allowed minor pests to assume economic importance, requiring the integration of more control methods, leading to more sophistication and in turn to further complications. In addition, rapid evolution in crop culture altered pest control problems. The result was a dynamic situation. Realization that events under glass move much faster than outside enables us, to some extent, to predict future changes outdoors. Already such changes are evident in cotton culture. In greenhouses, microbial control of pests has a relatively small role, though this appears to be increasing. Thus an analysis of developments in greenhouses may reveal possible future potential outside.

The year-round chrysanthemum crop, grown in dense beds, will be described as an example. The beds are covered 13 h daily by sheets of black polythene to control bud production and produce year-round flower crops. Cuttings about 10 cm high are planted in the beds and take about 3 months to produce flowers. Each bed takes about a week to harvest. A single large greenhouse often contains crops of all ages in a perpetual monoculture. Until recently, easily applied aldicarb granules killed all pests except caterpillars, which could be controlled with more difficulty by high volume sprays of a suitable chemical. However, the red spider mite and a major aphid pest, *Myzus persicae,* became resistant and control failed. This led to the commercial use of a predaceous mite against the spider mite and *Bacillus thuringiensis* against caterpillars, sometimes with a parasitic wasp against the aphid. The wasp does not attack another resistant aphid, *Aphis gossypii,* which became an important pest. A single spray of the fungus, *Verticillium lecanii,* controls all the aphid species for the duration of the crop (control of a minor species may sometimes be only partial). *V. lecanii* is being developed commercially (Chapter 25). High humidity under the polythene covers aids control by the fungus but control has also been effective without the covers (Hall, personal communication). Other pests attack the chrysanthemum crop because broad spectrum chemicals cannot be used as a blanket spray, so additional control measures must be added to the basic program, thus creating a complex situation. This is illustrated in Table I with the various options. Greenhouse staff welcomed biological systems on grounds of safety. Even in a greenhouse, the absence of insecticides for long periods may encourage natural enemies immigrating from outside. In one trial, a high degree of parasitism of the leaf miner by a hymenopterous parasite, *Diglyphus isaea,* resulted in almost complete eradication of the pest by November (Wardlow and Cross, 1980). Another advantage of a regular prophylactic use of biological agents is that chemicals can be used again at any time if a different pest appears because *resistant* major pests are no longer dominant.

The integrated program includes two pathogens, sufficiently broad-spectrum to control all major caterpillar species and all aphids, yet specific enough not to prevent the activities of predators and parasites (Burges and Jarrett, 1979;

TABLE I

Integrated control of pests and diseases of year-round chrysanthemums in greenhouses

Pest[a]	Control[a, b]
Major. Red Spider mite, *Tetranychus urticae*	Commercially available predaceous mite, *Phytoseiulus persimilis*, 3 weeks after planting or after aldicarb or cyhexatin[b, c]
Major. Aphids, *Myzus persicae*, *Aphis gossypii* Minor. Aphids, *Brachycaudus helichrysi*, *Macrosiphoniella sanborni*	*Verticillium lecanii* being developed commercially controls all, *M. sanborni* sometimes only partially.[d] Options: parasitic wasp, *Aphidius matricariae*, effective only on *M. persicae* in trials;[e] selective aphicide in first two weeks, effective versus nonresistant strains. Spot chemical sprays for local aphid patches late in crop.[a, b]
Major. Lepidoptera, *Autographa gamma, Phlogophora meticulosa, Cacoecimorpha pronubana*	High volume sprays of *Bacillus thuringiensis* var. *kurstaki*.[f, g, h, j]
Minor. Lepidoptera, *Mamestra brassicae* Rare. Lepidoptera, *Noctua pronuba, Agrotis segetum*	Commercial strains of *B. thuringiensis* fail.[f] New strains found in laboratory tests.[k]
Major. Leaf miner, *Phytomyza syngenesiae*	Commercially available wasp, *Opius pallipes*. Option,[l] selective chemicals, 2-weekly, only to tops of plants where adults oviposit, bioagents lower down unharmed.[c]
Major. Leaf miner, *Liriomyza trifolii*	New introduction. Eradication being tried by broad spectrum chemicals.
Minor. *Thrips tabaci* (Thysanoptera)	Pupates in soil; non-systemic soil drenches spatially selective, volatile chemicals, e.g. BHC, harm parasites.[a, b] Fungal pathogens under study.[m]
Rare. Whitefly scale, *Trialeurodes vaporariorum*	Commercially available parasite *Encarsia formosa* (Hymenoptera).[n]
Minor. Slugs, Earwig, *Forficula auricularia*	Methiocarb pellets spread on soil, selective.[a]
Minor. Leaf hoppers, *Zygina pallidifrons*	Susceptible to most chemicals.[a, b] Fungal pathogens under study.[m]

TABLE I continued

Pest[a]	Control[a,b]
Minor. Sciarid flies, e.g. *Bradysia paupera*	Soil drenches: diflubenzuron selective;[a] non-systemics spatially selective; volatile chemicals harm parasites.[a,b] Nematode *Tetradonema plicans* promising in small trials.[P]
Major. Fungal diseases: rust *Puccinia boriana:* mildew, *Oidium chrysanthi;* etc.	Versus rust, 3 major fungicides harmless to *V. lecanii*, which may also control the rust and mildew, many fungicides compatible with *V. lecanii*.[d,q]

[a] Machin and Scopes, 1978; [b] Morgan and Ledieu, 1979; [c] Scopes and Biggerstaff, 1973; [d] Chapter 25; [e] Scopes, 1970; [f] Burges, 1974; [g] Jarrett et al., 1978; [h] Burges and Jarrett, 1978; [j] Burges and Jarrett, 1979; [k] Jarrett, 1980; [l] Scopes, personal communication; [m]Gillespie et al. 1980; [n] Scopes and Ledieu, 1979; [P]Chapter 33; [q]Hall, personal communication.

Kanagaratnam et al., 1980). Only certain strains of both the bacterium and the fungus give satisfactory control (Table I). The use of the fungus slightly reduces the range of fungicides applicable against plant pathogens. The program is sophisticated and intimately dependent on environmental conditions. In such a rapidly evolving pest situation research must anticipate new problems. For instance, insecticides and fungicides are continually being tested for compatibility with biological agents, and more biological agents are being studied (Table I).

The strain of *V. lecanii* that attacks aphids also attacks rust and mildew fungi (Hall, personal communication) and so might be a candidate biocontrol agent for both white rust and mildew on chrysanthemums. Carnation rust was prevented in the laboratory and reduced in a greenhouse by application of another strain isolated from rust uredinia (Spencer, 1980).

Under glass each crop, and to some extent each cultural system, introduces different situations that need different programs.

The potential and problems envisaged from the greenhouse scene contributed to the framework for the organization of the present book. Divisions into major topics, i.e. identification, bacteria, viruses, fungi, protozoans and nematodes are now too large and complicated to be treated comprehensively by single authors, so they are split. Identification is divided into major specialized groups of practical interest. A discussion of general aspects of disease recognition in insects is placed with the chapter on bacteria. It is orientated towards the entomologist with acquired microbiological skills — the commonest type of insect pathologist. To meet the exacting standards needed to develop pathogens for practical pest control, identification is aimed, where possible, beyond the species level. Because the crop situation is so important the major pathogen groups are split by selecting major organisms for detailed analysis, ensuring examples from agriculture, forestry and vector control, the latter involving almost exclusively the aquatic habitat. An attempt is made to define special features of each example. Looking forward to a time when we can construct pathogens to our own specifications, a chapter describes the initiation of studies on the genetics of *B. thuringiensis*, our most used pathogen. Chapters on technology, a subject spanning all major pathogen groups, are orientated around the efficient application of living pathogens and the assessment of the results. This requires formulation, machinery, quantification and statistical forecasting. The integration of pathogens into pest control programs involves their compatibility with other components, the use of microorganisms to control plant diseases and a rationalized assessment of their safety to man and his environment. A chapter describing the insect's response to these assaults improves our ability to use microbial weapons more effectively. The People's Republic of China has developed a unique approach to microbial control, interesting to compare with that in Western countries. Finally, the concluding chapter assesses over-all progress, general principles, socio-economic factors and strategy, offering suggestions for the future. Two appendices are attached covering species tested with *B. thuringiensis* and those suffering virus diseases, two subjects now so extensive as to require a dictionary-type approach. A third appendix describes a repository of safety data — a service to insect pathology.

References

Addington, J. (1966). *Grower* **66**, 726.

Burges, H.D. (1974). *Rep. Glasshouse Crops Res. Inst.* **1973**, 99–100.

Burges, H.D. and Hussey, N.W. eds. (1971). "Microbial Control of Insects and Mites". Academic Press, London. 861 pp.

Burges, H.D. and Jarrett, P. (1978). *Grower* **90**, 589–590, 593–595.

Burges, H.D. and Jarrett, P. (1979). Proc. 1979 Br. Crop Prot. Conf. Pests Dis. pp. 433–439. British Crop Protection Council, Croydon.

Gillespie, A.T., Hall, R.A. and Burges, H.D. (1980). *Rep. Glasshouse Crops Res. Inst.* **1979**, 135.

Hussey, N.W. and Scopes, N.E.A. (1977). *In* "Biological Control by Augmentation of Natural Enemies" (R.L. Ridgway and S.B. Vinson, eds), pp. 349–377. Plenum, New York.

Jarrett, P. (1980). *Rep. Glasshouse Crops Res. Inst.* **1979**, 130.

Jarrett, P., Burges, H.D. and Matthews, G.A. (1978). Proc. Symp. Controlled Drop Application, pp. 75–81. British Crop Protection Council, Croydon.

Kanagaratnam, P., Burges, H.D. and Hall, R.A. (1980). *Rep. Glasshouse Crops Res. Inst.* **1979**, 133–134.

Machin, B. and Scopes, N. (1978). "Chrysanthemum Year Round Growing". Blandford, Poole. 233 pp.

Morgan, W.M. and Ledieu, M.S. (1979). "Pest and Disease Control in Glasshouse Crops". British Crop Protection Council, Croydon.

Scopes, N.E.A. (1970) *Ann. appl. Biol.* **66**, 323–327.

Scopes, N.E.A. and Biggerstaff, S.M. (1973). Proc. 7th Brit. Insecticide and Fungicide Conf. pp. 227–234, British Crop Protection Council, Croydon.

Scopes, N.E.A. and Ledieu, M.S. (1979). "The Biological Control of Tomato Pests". Growers' Bulletin No. 3. Glasshouse Crops Research Institute, Littlehampton, England.

Spencer, D.M. (1980). *Trans. Br. mycol. Soc.* **74**, 191–194.

Wardlow, L.R. and Cross, H. (1980). Min. Agric. Fish. Fd, Agric. Dev. Advis. Serv., Agric. Sci. Serv., S.E. Reg., Wye Sub-Centre. Rep. Reg. Entomologist 1979. p. 18.

CHAPTER 2

Identification of Bacteria Found in Insects

G.E. BUCHER

Research Station, Agriculture Canada,
Winnipeg, Manitoba, Canada

The purpose of this chapter is to help the entomologist, without bacteriological experience, search for a presumptive bacterial pathogen as a possible explanation of excessive mortality in a population of insects. The problem of isolating and identifying a pathogen is complicated by our ignorance of what bacteria constitute the normal flora of insects for neither bacteriologists nor entomologists have shown interest in this field. Most of our knowledge stems from searches for pathogens and is so biased that present concepts may be erroneous.

I. Bacteria in Healthy Insects

Apart from symbiotes in a few insect species, the bacterial flora of the healthy insect is confined to the gut. Its composition is determined by the species and numbers ingested and by their ability to survive and/or multiply while in the gut. Modifying factors are temperature, host food, volume and speed of ingestion and defaecation, efficiency of such gut flushing as a dilution factor, and physiological state of the host. Individual sibling insects often differ widely in the number, kinds and ratios of bacteria they contain. In general the number of species increases with age and the total number increases in rough proportion to gut size. Prior to moulting and metamorphosis the insect ceases to feed and partially empties its gut, and the number of bacteria decreases, often to about 5%. In some Diptera a further decrease occurs in the puparium so that the flora of the emerging adult is sparse; in some Lepidoptera bacteria multiply in the pupal gut and the adult has numerous bacteria but voids most of them with the meconium.

Though insect bacteria in culture may reach densities of $5 \times 10^9 - 2 \times 10^{10}$ per ml, their density in the gut of the healthy insect is only $10^{-6} - 10^{-3}$ times this level. An homeostasis is maintained by the combined action of many factors, but numerically common species presumably have multiplied under gut conditions such as a low oxygen level, a rich supply of organic nitrogen and carbon, and a high or low pH. Thus numerically common bacteria are members of the Enterobacteriaceae, Micrococcaceae or of the genera *Streptococcus* and *Brevibacterium*, i.e. they are facultative anaerobes, heterotrophic and pH-tolerant. Less numerous species have at least the capacity to survive in the gut and include the Bacillaceae, which have resistant spores, and various genera common in soil, water or plants. Strictly aerobic groups such as the pseudomonads, the corynebacteria and the actinomycetes are infrequent, probably because they require more oxygen than the gut provides. Allotrophic or photosynthetic forms and nitrogen fixers are apparently absent, and strict anaerobes are rare and associated with specific insects such as termites; such observations may result from the rare use of the special techniques required for isolation. Thus the gut of the healthy insect is apt to contain $10^1 - 10^6$ bacteria depending on its age, size, species, and food habits; from 1–5 species will comprise 95–99% of this number, 1–5 species will be numerically rare. If the entomologist determines the bacteria in his healthy insects, his search for a pathogen is facilitated.

II. Bacteria in Sick Insects

A sick insect may suffer from infection with a pathogen, but sometimes sickness is associated with an abnormal increase (10–100 times) in the number of bacteria in the gut, the increasing species being those commonly found in healthy siblings. Such observations are difficult to interpret, partly because the insect is killed by making the bacterial count, so its original fate is unknown and inferences must be drawn by sampling sick and healthy insects. Presumably the original excessive multiplication results from a breakdown of the homeostatic system, and the sick insect is further stressed by abnormally high bacterial metabolism and multiplication; this causes increased permeability of the gut to both bacteria and their products, so that death is accompanied or followed by mass invasion and multiplication of bacteria in the haemocoel. Such observations usually have been made on populations where the original bacterial multiplication was triggered by stress of improper rearing conditions. However, entomologists often have no measure of survival even in cultured insects; when known, a survival of 50–70% is commonly considered as normal or acceptable; a pathologist wonders why 30–50% died. The homeostatic system may fail more frequently than we thought, even in populations considered unstressed. There is little doubt that an insect with an excessive bacterial population in its gut is sick, that it will probably die, and that the frequency of such insects increases both in stressed populations and at critical stages of development such as moulting or metamorphosis. The condition may increase the probability of infection by a pathogen but occurs in the absence of demonstrable pathogens.

Though the haemocoel of a healthy insect is normally sterile, it may be invaded from time to time by bacteria through wounds in exoskeleton or gut, or through a gut rendered more permeable by loss of the peritrophic membrane at a moult. If pathogens are absent and if the number of invaders does not swamp the blood defence mechanisms, insects eliminate these bacteria and survive.

With few exceptions, known bacterial diseases of insects involve invasion of the haemocoel by pathogens and their multiplication in it. Though such pathogens vary widely in their ability to multiply in the gut or to penetrate the gut wall, the disease develops as a septicemia and the insect dies with large numbers of the pathogen in its blood. An insect's main defence is to prevent original invasion of its haemocoel or to limit the invaders to the small number that the blood defensive mechanisms can eliminate before multiplication. Once multiplication is underway, any defence crumbles and insects succumb to infection.

Exceptions include the following:

1. *Clostridium brevifaciens* and *C. malacosomae* cause disease in tent caterpillars by multiplying only in the gut lumen; they never invade the haemocoel (Bucher, 1957, 1961; Atger, 1964).

2. *Streptococcus pluton* multiplies in the gut lumen of bee larvae and produces the syndrome known as European foulbrood (Bailey, 1968); associated with this pathogen are *Lactobacillus eurydice*, *Bacillus alvei*, and *Streptococcus faecalis.*

3. *Bacillus cereus* multiplies in the gut lumen of larch sawfly larvae (Heimpel, 1955); death is unrelated to invasion of the haemocoel, though this is frequent in late stages of disease.

4. Motile, yellow strains of *Streptococcus faecalis* are pathogens in the gut lumen of the gypsy moth (Cosenza and Lewis, 1965); invasion of the haemocoel is also common.

5. An unidentified bacterium causes disease in a bagworm by intracellular multiplication in midgut epithelium (Bucher, 1963).

6. *Yersinia pestis*, the cause of plague, multiplies in the gut lumen of carrier fleas and kills them, partly by mechanical blockage (Kramer, 1963); it kills some other arthropod vectors, but apparently not by mechanical blockage.

7. *Francisella tularensis*, the cause of tularemia, multiplies within gut epithelial cells of carrier ticks, bedbugs and lice, but does not usually kill unless tissues of the haemocoel are also invaded (Kramer, 1963).

8. *Salmonella* spp., pathogenic to vertebrates, are reported to persist in the

gut of ticks, fleas and lice; some may die but details of these infections have not been elucidated (Kramer, 1963).

9. Some rickettsias are intracellular pathogens of the gut epithelium of insects: *Enterella culicis* in mosquitoes, *E. stethorae* in beetles, and *Rickettsia prowazeki* and *R. typhi*, the causes of typhus, in carrier lice (Krieg, 1963).

10. Some insects, highly susceptible to ingested preparations of *Bacillus thuringiensis*, are killed by the toxic action of the parasporal crystals in the preparations. Death is unrelated to invasion of the haemocoel, though this may occur after death. In most insects, poisoning by the crystals and/or by other preformed toxins favours invasion, with or without prior multiplication in the gut, and septicemia occurs (Heimpel and Angus, 1963).

11. Another example of toxemia is recorded in mosquitoes that have ingested living or dead cells of *Bacillus sphaericus* (Chapter 14, IIB).

III. Isolation of Bacteria; Recognition of Pathogens

Search for a pathogen should be made where and when it exists in numbers relatively free from other bacteria. In septicemic diseases, blood should be examined after the initial invaders have multiplied but before extensive damage to the gut wall permits mass invasion by gut bacteria. Usually this stage coincides with the first visual signs that the insect is abnormal, but it may occur earlier in apparently healthy insects. Gut bacteria multiply rapidly in the blood of dead or dying insects and numerically overwhelm the pathogen, so that delayed examination makes its isolation difficult. In diseases of the gut correct timing is also important, as a pathogen may dominate numerically for only a short time. When faced with the unknown, a pathologist examines insects in what he thinks are successive stages of disease, and pays special attention to comparing bacteria from presumably healthy insects with those from early stages of disease. His first and perhaps most difficult task is to discriminate between sick and healthy individuals (Bucher, 1973).

Individual insects are examined in two steps, a microscopic examination and an attempt at culture; both should be conducted to yield at least rough estimates of bacterial numbers. Culturing is done by spreading bacteria on the surface of an agar plate so that many will form isolated colonies of further use — the common mistake is a crowded and useless plate. Prior knowledge of bacterial density before plating is such a big advantage that microscopic estimates are worthwhile if made rapidly.

Bacteria are commonly handled with a loop, i.e. a long wire bent to form a small circle at one end. A well-manipulated loop will pick up a relatively constant volume of fluid, and will deposit virtually all of this on an agar plate or some proportion of it on a glass slide. Several loops of known capacity provide a method for measuring small volumes. Larger volumes are measured by a dropping pipette, i.e. a calibrated Pasteur pipette. The outer diameter of the tip of a Pasteur pipette is standardized by inserting it in a hole in a thin metal plate and cutting it at plate level. By using different hole sizes, dropping pipettes to deliver from 20–50 drops/ml are rapidly made; note that they must be held vertically in use, and should be individually calibrated for great accuracy. Larger volumes are measured by conventional means.

Rough but useful estimates of bacterial density in an insect can be made during initial microscopic examination if volumes and areas are known. For example, 1/200 ml of insect fluid is placed on a slide, covered with an 18-mm slip, and examined by dark-field. As the field area of the high dry lens (X540) is ca. 1/8520 of the total area, one bacterium per field indicates 8.5×10^3 bacteria in the sample, 1.7×10^6/ml in the insect fluid, and 1.7×10^4 on a culture plate if transferred by a loop delivering 1/100 ml; a dilution of $10^{-1}-10^{-2}$ is indicated to produce a useful plate with individual colonies. Bacteria are readily seen at lower magnification, so that the medium lens (X300) with a field area 3.3 times larger is often better for counting. I prefer dark-field illumination but many workers use phase contrast. Microscopic examination may reveal pathogens other than bacteria. It shows bacteria of unusual size and shape, sporulating rods, loose spores, chains of streptococci, motile bacteria and bacterial capsules. It may provide an adequate immediate diagnosis if known characteristic pathogens are seen. Microscopic examination should be made rapidly then a known volume of the insect fluid, diluted if necessary, should be spread on a culture plate. The cover slip is then removed, and at leisure the preparation is stained by the Gram method and examined for Gram-positive bacteria.

Ideally the culture plate is inoculated so that an estimate of bacterial density is given by counting individual colonies. This limits the original inoculum to 200–400 bacteria; even so, counts may have to be made when colonies are small. The inoculum can be spread with an L-shaped wire or glass rod, or the inoculating loop itself. If confluent growth around the area of the original inoculum prevents colony counts, then some information is lost; the plate is valuable, however, if it contains some areas with well-isolated colonies.

Different colony types and their numbers are recorded after incubation. Many bacteria form colonies of medium size (2–5 mm), convex, entire and smooth. Though colonies may be readily separable under oblique light, and may be characteristic on differential media, differences on common media are not easily described or used for identification.

An individual worker can learn to recognize colonies of species or groups if he constantly uses a single standard medium for plating, e.g. nutrient agar (NA) = beef extract–peptone. Whatever the choice, it should be cheap, clear and low in nutrients, because luxuriant growth tends to obscure colony differences. Most insect bacteria grow on NA. Those that do not are mostly obligatory pathogens, which are noted in the descriptions, and recommendations for their culture are given in the literature. A fastidious pathogen might be cultured by combinations of the following: 1. Different polypeptide linkages, amino acid ratios, or proteins in tryptones, brain or heart infusions, sera, or specific amino acids (glutamine, tryptophan, cysteine). Makers of bacteriological media supply a wide range of possible supplements. 2. A source of vitamins (yeast extract) or specific vitamins (thiamine), or minerals (manganese). 3. Starch or activated charcoal, which presumably act as adsorbers of growth inhibitors. 4. Buffering of the medium to the pH of the insect. 5. Increasing the K/Na ratio by using chemicals containing potassium not sodium. 6. Adding carbohydrate (trehalose, inositol, glucose). 7. Incubation under reduced oxygen tension. A Brewer jar can provide strict anaerobiosis or various levels of O_2 or CO_2. A pure culture is very difficult to isolate in liquid media but these are useful for maintaining cultures of anaerobic or microaerophilic bacteria, if they contain reducing substances (cooked meat particles, cysteine, thioglycollate, glucose) and are thickened with agar to inhibit absorption of atmospheric oxygen.

A few bacteria from one or more colonies representing each colony type are spread on individual subculture plates and incubated. Within each subculture plate colonies should look alike, indicating a pure culture, and one colony is subcultured to two agar slants. One slant provides bacteria for more tests and the other is a back-up from which another working slant can be made when necessary; after bacterial growth both slants are usually refrigerated at $3-5°C$. Bacteria are subcultured in sugar-free broth and one drop of this broth is pipetted into each tube of a set of primary diagnostic culture tubes. This set should be composed of a small number of tests that are easy and cheap to prepare in advance in large quantity, that withstand storage at $3-5°C$ for a long time before use, that give sharp positive or negative reactions, and precontain any required indicators, so that reactions can be observed for several days without ending the test by adding non-sterile chemicals. A typical set might consist of gelatin plus the carbohydrate tests: glucose, lactose, maltose, sucrose, xylose, mannitol and glycerol. Each carbohydrate tube contains a Durham tube to show gas production, and bromcresol purple, which withstands heat steriliz-ation better than other indicators of acid. For several days the gelatin tube is read as not liquefied (−), or liquefied (+) indicating some proteolytic ability, and the carbohydrate tubes are read as no acid (−), acid (+) or both acid and gas (⊕); any alkaline reaction or reversion from acid to neutral or alkaline is noted.

Meanwhile the original isolation plate and subculture plates are examined at intervals to follow changes in colony appearance with age and to ensure that representatives of all typical colonies were subcultured. Wet mounts of bacteria from plates and broths are examined by dark-field, and size, shape, motility and sporulation recorded; similar records are made on Gram-stained bacteria. Plates are illuminated by a BLB $F_{15}T_8$ fluorescent tube or other source of long-wave ultraviolet light (not a short-wave germicidal tube) to determine if colonies or diffusible products from them fluoresce. Finally, colonies are tested for catalase production with a drop or two of H_2O_2, and for oxidase with Kovac's reagent or an alternative. These tests break asepsis so the plates are discarded.

Common mistakes are superficial examination of the original isolation plate and careless subculturing of colonies. These result in missing some colony types and in failing to obtain pure cultures. Examination and colony subculture should be done under a binocular microscope with a strong oblique light against a black background; the plate is tilted to catch both incident and transmitted light at various angles. Colonies, superficially alike, often show characteristic differences in internal granular reflections, iridescent colours, opacity, surface smoothness and reflectivity, shape, and marginal detail. Colonies with a distinguishable area or sector probably contain two species and are ignored. Colony characters are inherently variable and within a single plate are influenced by age, size, and degree of isolation so good judgement comes from experience; a careful beginner is apt to overemphasize differences and pick as different several isolates that further tests show to be identical, but no harm results from this. Colony characters between plates are even more variable, being influenced by composition of the medium, its depth and moisture content, temperature and aeration during incubation, etc. Thus colonies picked as different on several isolation plates may represent the same strain on further testing.

If 10 healthy and 10–20 sick insects are examined, and 5–10 typical colonies subcultured from each isolation plate, a meaningful recording system is mandatory. I use a two-number, one-letter system; 784-23B indicates a bacterial subculture of the second commonest colony type from insect 784-23, the 23rd insect examined from population 784.

The kinds and numbers of bacteria on the isolation plate are compared with those seen during microscopic examination of the insect; ideally these should correspond, indicating that the bacteria seen in the insect are now represented by subcultures. Unfortunately experience is important in assessing an acceptable degree of non-correspondence. For example, some bacteria in the insect may be dead, producing no colonies; others may occur as pairs, chains, or clumps, forming one colony for several bacteria; spores often germinate poorly on agar; bacteria crowded on a plate may inhibit colony formation; such factors can reduce colony counts to 1/10 of microscopic counts. A more serious discrepancy occurs if a bacterium of characteristic appearance in the insect (e.g. a sporeformer) is not isolated. This probably indicates failure to grow on the medium under the conditions provided, but it may indicate that the growth form in culture differs from that in the insect; e.g. the bacterium may grow but not form spores.

The worker may now possess 100–300 bacterial cultures with certain known characteristics. Comparison will show that most are replicates of several different strains. A typical culture of each strain is preserved under its original label, along with its relative numerical importance in both sick and healthy insects. Comparison of relative importance may point to a presumptive pathogen; e.g. strain 784-1A is dominant or numerous in sick insects and rare or absent in healthy ones. If the combined data indicate repeated failure to isolate a bacterium that is visibly or numerically important in sick insects, the pathogen has probably been missed and culture is reattempted with the medium modifications suggested above.

The worker now possesses 5–20 bacterial strains in culture, one or more of which may be implicated as presumptive pathogens. He now has a number of alternatives depending on the depth and purpose of his investigation. He can identify all strains or only those suspected of pathogenicity. If the latter, he may be content if he identifies one or more strains as a known pathogenic species. For example, if strain 784-1A is *Serratia marcescens*, a common cause of septicemia in many insects, he may assume — with some assurance — that it is the pathogen responsible for mortality in his population and end his investigation.

However, this is an assumption and should be reported as such; isolation of a bacterium from sick insects, even if it has previously been reported as a pathogen of other species, is not *prima facie* evidence that it is responsible for the disease syndrome under investigation. Experimental tests of pathogenicity are required: extensive testing is mandatory, if strain 784-1A is a species with little or no previous history of pathogenicity. Bucher (1973) discussed some common errors in pathogenicity tests but two things are essential, a method of measuring and administering a series of doses (Bucher and Morse, 1963; Burges and Thomson, 1971) and a pathogen-free insect population. This population must be demonstrated to be free of strain 784-1A before pathogenicity tests are conducted.

IV. Steps in Identification of Pathogens and Other Bacteria

In approaching his first identification of a bacterial strain, an entomologist is faced with a series of unfamiliar concepts. Bacteriologists regard a species as a set of strains that are sufficiently alike to be recognizably different from other sets; in well-studied species, one strain is designated as the type and embodies the characters used to define the species; other strains may differ from the type, especially in characters known to be variable. The great variability of bacteria is the basic problem in both taxonomy and identification. Variability can occur even in characters considered as taxonomically important because of their relative constancy (e.g. Gram-positive state or spore formation). Variability usually involves loss of a positive character used to define a species or higher taxon. Thus one might identify a Gram-negative strain or a non-sporulating strain as *Bacillus cereus* if their other characters fitted the species definition, but would not identify Gram-positive or sporulating strains as members of the Enterobacteriaceae. Characters of cell structure (Gram staining, sporulation, motility, flagellar arrangement) have value in helping to define higher taxa but are less valuable in separating species than physiological characters that show what the species can do. Thus museum specimens have no place in identification; their bacterial counterpart is the culture collection. Characters important in one group of bacteria may be of no value in another; e.g. structural characters help to identify species of Bacillaceae but not Enterobacteriaceae, which look alike. In the latter family sugar fermentations are important, whereas in the Streptococcaceae antigenic reactions are. Though the genus and higher taxa are collections of more or less similar species, bacterial nomenclature makes no claim to be phylogenetic. Lastly, bacterial taxonomy is in a state of flux not experienced by insect taxonomy since before the discovery of bacteria.

Therefore, a bacterial strain is identified as a given species because its characters fit the definition of that species better than that of some other species; the greater the number of characters and the greater the similarity, the greater the confidence in an identification. This does not mean that all characters have equal weight; some indicate fundamental differences applicable to higher taxa (e.g. cell structure and distinction between oxidative and fermentative metabolism of carbohydrates); others are important because of their relative invariability (e.g. production of urease by the genus *Proteus*); positive characters give more confidence than negative ones, because variation usually involves loss rather than gain. These observations explain why the dichotomous key, so popular with entomologists, is difficult to prepare and to follow, and frequently leads to wrong identification.

A report of strain 784-1A as *Serratia marcescens* implies that it has all characters of the species and higher taxa given in Bergey's Manual (Add. 5*) or in a reference paper that may supercede it. Characters varying from the type, not measured, or known to be variable (in this case colony colour) should be stated along with the identification.

At this stage there are several sets of clues valuable for identification. The first set is given by literature on pathogens of the host insect or its group (Steinhaus, 1947, 1949, 1963; Krieg, 1961). The second set is given by location of the pathogen, whether it is intracellular or extracellular, and whether it occurs exclusively or principally in the gut or produces septicemia of the haemocoel. A third set is given by its culture on NA or its demand for special requirements, and a fourth by bacterial cell structure particularly within the host insect. Combination of these clues gives at least tentative identification of all known obligatory pathogens (*sensu* Bucher, 1960), and most of those confined to the gut. The residual problem consists of identifying pathogens that produce septicemia in many insect species but that are not visibly separable by characteristic cell structure in the insect host. To identify all bacteria isolated from the population, regardless of their location or pathogenicity, is a larger problem.

Tentative identification of some species or groups is provided by colonial appearance (the fifth set of clues) even though all strains may not possess the unusual character. Such colonies are:

1. Red-pink. *Serratia marcescens*; the pigment prodigiosin is soluble in ethanol and does not diffuse into the medium.

2. Blue-green. *Pseudomonas aeruginosa*; the pigment pyocyanin diffuses into the medium, is soluble in chloroform and red when acidified.

3. Fluorescent in long-wave ultraviolet light. *Pseudomonas* spp. in *P. fluorescens*, *P. aeruginosa*, the pigment fluorescin diffuses into the medium and has a light green fluorescence.

4. Yellow-orange. *Erwinia* spp., *Flavobacterium* spp., *Xanthomonas* spp., some cocci. Pigment carotinoid, non-diffusable.

5. Violet. *Chromobacterium* spp.

6. Spreading, without fixed shape. *Proteus vulgaris, P. mirabilis*. Thin growth may spread over the whole plate if the surface is sufficiently damp.

7. Motile with the colony or microcolonies showing rotational movement across the plate if the surface is sufficiently dry. *Bacillus circulans, B. sphaericus, B. alvei.*

8. Large (5–10 mm) irregularly round, white, like galvanized iron under oblique light. *Bacillus cereus* and its relatives *B. thuringiensis* and *B. anthracis*. Long rods are organized into plates, which reflect the light differently depending

---

* References followed by "Add." and a number refer to the references given in the Addendum.

on their angular orientation to it, hence the galvanized appearance. *B. cereus* is so common that its distinctive colonies should be memorized. Though some other bacteria are plane—granular (i.e. they reflect light from plates of cells rather than from points or clumps), the galvanized appearance is less prominent and fades with age and increasing opacity. The surface of *B. cereus* colonies is undulate, looks rough and dry with age and sporulation; the margin is irregular, frequently hairy where strands of cells project. In *B. cereus* var *mycoides* the strands are long and rhizoidal so the colony resembles a fungus.

9. Small (1 mm or less), round, low convex, entire, smooth, shiny mirror-like surface, translucent, colourless. *Brevibacterium* spp., *Streptococcus* spp.

10. Small (1—2 mm), round, high convex, entire, dense white, opaque. *Staphylococcus* spp., *Micrococcus* spp. Some are pigmented, yellow, orange, pink.

The set of primary diagnostic tests gives further information. If acid and copious gas are formed in several carbohydrates, the bacterium is probably one of the fermentative Enterobacteriaceae; acid without gas is indicative of many Gram-positive cocci and rods; weak acid only in glucose and xylose is common in Pseudomonadaceae. If reactions are alkaline, negative, weak, or reversing, the tests (at least the glucose test) should be repeated using reduced levels of protein in the broth; some bacteria produce sufficient ammonia from organic nitrogen to neutralize acid from carbohydrate metabolism.

The Hugh—Leifson test (Hugh and Leifson, 1953) to distinguish oxidative from fermentative metabolism of carbohydrates must be made on all strains, because this is a fundamental character in modern taxonomy. At the same time it is wise to do other tests that are used in several different groups: nitrate reduction, milk tests, indol production, methyl red test, Vosges—Proskauer (VP) test, citrate utilization, and extra carbohydrate tests on those strains that produced acid (e.g. arabinose, cellobiose, fructose, galactose, mannose, raffinose, rhamnose, trehalose, adonitol, dulcitol, inositol, sorbitol, inulin, salicin, starch). The results may indicate that some strains are minor variants of a single species, perhaps preserved because of differences in colony or cell appearance.

If the strain is motile, the number and position of flagella are determined. Flagella are polar in the Pseudomonadaceae. The staining technique frequently fails for various reasons (Bucher and Lüthy, 1969, 1970), and electron microscopy is helpful and sometimes mandatory. Fortunately the technique usually succeeds with the Pseudomonadaceae.

This accumulated information will permit identification of the strains to families or groups, often to species. For some, other tests are needed for species discrimination.

Most diagnostic tests are simple and require no special knowledge, techniques, or equipment. But some modern techniques involve serological reactions, bacteriophage and bacteriolysin sensitivity, determination of DNA base ratios and of specific enzymes or toxins, fluorescent and electron microscopy, chemical analysis of cell components and of metabolic products from given substrates. For these the entomologist probably lacks the tools and may wish to seek help from a practitioner or from a taxonomic expert of a group. He should submit a pure culture and a tentative diagnosis with the results of all the tests he can do himself, and justify his request. The best and perhaps only justification is to send experimental proof of pathogenicity with evidence suggesting a new strain or species.

Some comments may help interpret certain observations.

1. Gram-positive (G +) bacteria retain the blue stain when treated with decolourizing agent and G − ones lose it and stain red with the counterstain. G + may become G − with age, understaining, overdecolourizing, or on artificial media. Cocci and large rods are usually G +; most insect bacteria are G −. If the results are ambiguous, then repeat using a modification of the technique (I prefer Hucker's), a young, rapidly-multiplying culture, a change of medium, or injection into the host.

2. Most insect bacteria are $0.5-0.8\,\mu$ in diameter and $1.0-2.0\,\mu$ long. Large rods, $0.9-1.2\,\mu \times 2.0-4.0\,\mu$, are probably Bacillaceae. Sporulating *Clostridium brevifaciens* is the largest species ($1.4-1.7 \times 6-18\,\mu$). Staphylococci are $0.9-1.0\,\mu$ and other cocci usually smaller. Bacteria shrink during staining, especially in diameter. Bacterial-like organisms $< 0.3\,\mu$ in diameter are probably rickettsias. Some symbiotes are large but occur in healthy insects, usually in special gut diverticula or intracellularly in mycetomes of the haemocoel; they are unculturable, unclassified, and ignored in this paper.

3. Short rods, rapidly multiplying, sometimes resemble oval cocci, and are described as coccobacillary. If G − they are probably short rods, if G + probably cocci. Streptococci, frequently oval, usually form some chains of 3−5 or more cells and chaining is enhanced in the host and on some media under reduced O_2 tension. Discrimination of shape is often improved by intense staining with a blue or polychrome stain.

4. Motility is maximal in young cells in the host. It is frequently reduced or lost with age, with change of pH, r.h. or osmotic pressure, and on some culture media. It is favoured by growth in broth.

5. Spore formation may fail in culture (especially in obligate pathogens) or require media and conditions different from those supporting vegetative growth. Even *B. cereus*, which sporulates readily, may not sporulate on enriched media with high levels of organic N, or in broth where O_2 tension is low; this may explain its frequent failure to sporulate in the insect.

6. Medical bacteria are usually incubated at human body temperature of $37°C$. Insect bacteria are incubated at about $25°C$ unless special knowledge dictates otherwise.

7. In all tests a pathologist includes, as controls, bacteria known to produce both positive and negative results. An entomologist probably lacks the culture bank to provide the controls, but he can obtain for comparison named cultures from various culture collections.

8. Tests for reduction of nitrate must discriminate between bacteria that reduce nitrate to nitrite (reducers) and those that further reduce nitrites usually to atmospheric N_2 (denitrifyers). Some aerobes with oxidative metabolism can grow poorly anaerobically if they are denitrifyers, by using the process as an acceptor of electrons in place of O_2. Some pseudomonads are denitrifyers and tests for anaerobic growth should use nitrate-free media.

9. A medical bacteriologist, confirming a clinical diagnosis by finding a known human pathogen among a number of known contaminants, uses specific isolation media for primary plating. These enhance growth of the pathogen, inhibit growth of contaminants, and induce the pathogen to form characteristic colonies that distinguish it from uninhibited contaminants. An entomologist also could do this to search for a known pathogen among known contaminants, but this is not his usual purpose and his knowledge of contaminants is usually fragmentary. Thus plating on selective, diagnostic, or differential media is not common practice in insect pathology; the use of enriched media and specialized conditions to isolate obligatory pathogens is an exception to this rule. However, if one desires to select only those bacteria likely to be non-obligatory pathogens, one can use some common properties of insect pathogens to design primary plating media. Virtually all potential pathogens of insects are proteolytic, i.e. they liquefy gelatin, and most hydrolyse casein (Bucher, 1960). On a NA plate made with 50% skim milk or 2–5% skim milk powder, a clear zone of digested casein surrounds each colony of proteolytic bacteria, which are easily selected for identification tests. Many insect pathogens are also lipolytic, so that similar plates can be made by emulsifying vegetable oil or tributyrin in the agar (Davis and Ewing, 1964). Some insect pathogens give a positive egg yolk reaction, suggesting production of lecithinase. On agar containing 3–5% egg yolk (ca. 4 yolks/litre) their colonies are surrounded by an opaque zone of precipitated fat; if the bacterium is also lipolytic, the opaque zone is surrounded or replaced by a clear zone, as in *Pseudomonas aeruginosa*. Some insect pathogens produce chitinase and can be recognized by suspending 1% fine chitin particles in the agar. None of these plates inhibit growth of contaminants and none identify the pathogen, but they can reduce the number of colonies selected for identification tests. Sabouraud maltose or dextrose agar favours pigment production in *Pseudomonas aeruginosa* and *P. fluorescens*, and its low pH inhibits many bacteria, so it is a useful primary plating medium to search for these pathogens. Blood agar, widely used both for enrichment and diagnosis in medical bacteriology, is not used because entomologists rarely have a source of sterile blood.

V. Descriptions of Pathogens and Other Bacteria

The following descriptions of insect bacteria are designed for tentative identification using only characters that an entomologist can readily determine. Descriptions of known pathogens are sufficiently detailed to permit identification to species; non-pathogens can be placed in genera or higher taxa. Tentative identifications should be confirmed by comparing determined characters of the strain with more complete descriptions in papers in the addendum or references.

A. BACTERIA-LIKE ORGANISMS

1. Mycoplasms

Bacteria-like, pleomorphic, coccoid to filamentous, lacking rigid cell wall. Usually G − and non-motile, diam. ca. 0.2 μ. On complex media form minute "fried-egg" colonies penetrating into agar (Gibbs and Shapton, Add. 12). *Spiroplasma*, considered a genus of uncertain affiliation, is rarely isolated from arthropods; pathogenic to bees (Clark, 1977); G +, helical, motile, cells large, $0.7-1.2\mu$ diam. (See Note Added in Proof, p. 33).

2. Spirochetes

Cells helically coiled, usually $< 0.5\mu$ diam. but long, $5-100\mu$ or more. G −, motile by flexion, rotation about long axis or along a helical path, no flagella. Often difficult to stain, seen best in dark-field. A few have been cultured on complex media, usually anaerobic, but genus *Leptospira* is aerobic. Found chiefly in bloodsucking vectors and gut of roaches and aquatic insects; associated with protozoa in termite gut (Steinhaus, 1947). One doubtful report of pathogenicity to pierid larvae exists. Strains from insects (except vectors) not well characterized and their names, if any, are doubtful; probably belong to genus *Spirochaeta*, which contains free-living, anaerobic species from mud.

3. Rickettsiaceae

Bacteria-like pleomorphic rods or cocci, $0.1-0.3\mu \times 0.3-0.9\mu$, G −, acid fast, non-motile; obligate pathogens or commensals of arthropods, some transmitted to vertebrate hosts by vectors; usually intracellular; no growth on culture media. Divided into three tribes based on parasitic relationships (Krieg, 1963).
(a) Tribe Wolbachieae. Confined to arthropods. Genus *Rickettsiella*: in fat and blood cells of haemocoel; pathogens of Scarabaeidae, Tipulidae, Chironomidae. Genus *Enterella*: in gut epithelium; pathogens of Culicidae, beetles (*Stethorus* spp.). Genus *Rickettsoides*: in gut lumen in epicellular position on epithelium; non-pathogenic to lice and Hippoboscidae. Genus *Wolbachia*: in many tissues; non-pathogenic to ticks, lice, bed bugs, Culicidae.
(b) Tribe Ehrlichieae. Attack vertebrates, not man. Genus *Ehrlichia*: in haemocoel tissues, not pathogenic to vector ticks. Genus *Cowdria*: in gut epithelium, not pathogenic to vector ticks.
(c) Tribe Rickettsieae. Attack man and other vertebrates; studied chiefly for their medical importance; most are pathogenic to lice by intrahaemocoelic injection; lice may not be susceptible naturally. Genus *Rickettsia*: in gut epithelium, pathogenic to vector lice, not to vector fleas. Genus *Rochalimaea*: in gut lumen, in epicellular position on epithelium, not pathogenic to vector lice; possibly synonymous with genus *Rickettsoides*. Genus *Zinssera*: in haemocoel tissues, not pathogenic to vector mites. Genus *Dermacentroxenus*: in gut epithelium and haemocoel tissues, invades cell nuclei, not pathogenic to vector ticks. Genus *Coxiella*: in gut epithelium and haemocoel tissues, invades only cell cytoplasm, not pathogenic to vector ticks.

4. Chlamydiaceae

Bacteria-like coccoid organisms, $0.2-1.5\mu$ in diam., G −, non-motile. Obligate pathogens of vertebrates, sometimes found in bloodsucking vectors but not recorded as pathogenic or even growing in them. Have an unique growth cycle in vertebrate hosts; no growth on culture media (Page, Add. 20).

B. SPORE-FORMING BACTERIA: BACILLACEAE

Large rods, produce endospores in hosts or in culture. Usually G+ and motile; flagella numerous, peritrichous.

1. *Genus* Clostridium

Anaerobic, catalase —, oxidase —, fermentative. Non-pathogenic species, rarely isolated from insects (Stevenson, 1966), separated by spore position, proteolytic action and fermentations. Two obligate gut pathogens (Bucher, 1957, 1961) require special growth media, do not sporulate in culture.

(i) *C. brevifaciens*. G —; vegetative and sporulating rods motile, flagella numerous, peritrichous; sporulating rods very large, ca. $1.7 \times 9 \mu$.

(ii) *C. malacosomae*. G —; non-motile; sporulating rods slightly fusiform, very large, ca. $1.6 \times 7 \mu$.

2. *Genus* Bacillus

Aerobic; mostly catalase +, oxidase —, and fermentative; species vary in ability to grow under anaerobic or microaerophilic conditions but sporulation favoured by aerobic conditions. Typically G+, motile rods, ca. $0.7 \times 3{-}5 \mu$, spores oval, central, often larger than rods and thus producing fusiform or spindle-like sporangia. Acid +, gas — from a few sugars. Gordon (Add. 14) uses variations from typical structure to divide genus into three primary sections; a pathogenic arrangement is used below.

(a) Section 1. Obligate pathogens of the haemocoel causing specific diseases; catalase —, oxidase —. Require special growth media, particularly thiamine; isolation, spore germination and vegetative growth favoured by microaerophilic conditions; sporulate poorly in culture, sporulation favoured by aerobic conditions; motility poor or absent; acid +, gas — from glucose, fructose, trehalose; other carbohydrates usually —.

(i) *B. popilliae* group. Causes milky disease of Scarabaeidae; see Chapter 4.

(ii) *B. larvae*. Causes American foulbrood of bees. Giant whips of disengaged, agglutinated flagella often seen; proteolytic, reduces nitrates to nitrites; optimum growth temperature of $35°C$ is unusually high for an obligate insect pathogen. Isolation techniques in Bailey and Lee (1962).

(b) Section 2. Associated with insects but not as obligate pathogens; catalase +, oxidase —; grow on NA. Some are potential or facultative pathogens (*sensu* Bucher, 1960), others cause toxemia; some act in different ways in different hosts.

(i) *B. alvei*. Common in bees with European foulbrood, not pathogenic. Colonies with rotation motion common; rods and even free spores align into palisades in smears. Somewhat fastidious, requiring thiamine; facultative anaerobe, proteolytic; acid from glucose, sucrose, starch; does not reduce nitrates or utilize citrate; indole +, VP +. *B. para-alvei* is considered a synonym.

(ii) *B. apiarius*. From dead bee larvae, non-pathogenic. Colonies without motion; rods do not form palisades; spores appear rectangular by retaining part of sporangium wall (Katznelson, 1955). Reactions as for *B. alvei* but reduces nitrates, utilizes citrate, indole —, VP —.

(iii) *B. laterosporus*. From dead bee larvae, non-pathogenic. Spore cradled by non-refractile canoe-shaped parasporal body that accounts for lateral position of the spore and persists with the free spore. Facultative anaerobe, proteolytic; acid from glucose, sucrose, mannitol, not from starch; reduces nitrates; does not utilize citrate; indol variable, VP —.

(iv) *B. pulvifaciens*. From dead bee larvae, non-pathogenic. Similar to *B. laterosporus* except lacks parasporal body, catalase + but delayed, indol —. Produces orange-red colonies when first isolated (Katznelson, 1950).

(v) *B. circulans*. From several insects, non-pathogenic. Colony interior with circular motion; microcolonies with rotational motion across agar. Facultative anaerobe, usually not proteolytic; acid from glucose, sucrose, arabinose, xylose, mannitol, starch; nitrate reduction and citrate utilization variable; indol —, VP —.

(vi) *B. cereus*. Potential pathogen in haemocoel of many insects; facultative pathogen in larch sawfly gut. Rods wide, $1.0-1.2\mu \times 3-5\mu$, not swollen by spore; colonies with characteristic galvanized appearance. Facultative anaerobe, proteolytic, egg-yolk reaction +; acid from glucose, maltose, sucrose, starch; reduces nitrates, utilizes citrate; indol —, VP variable. *B. anthracis*, *B. mycoides*, *B. thuringiensis* are considered varieties by Gordon (Add. 14).

(vii) *B. thuringiensis*. Sporangium with a refractive parasporal crystal that produces toxemia when ingested particularly by Lepidoptera; action as a potential or facultative pathogen influenced by degree of toxemia. See Chapter 3.

(viii) *B. licheniformis*. Rarely isolated from insects, pathogenicity questionable. Very similar to *B. cereus* but rods $< 0.8\mu$ diam., egg yolk reaction —; produces acid from arabinose, xylose and mannitol.

(ix) *B. sphaericus*. Produces toxemia when ingested by Culicidae. Spores round, terminal; strictly aerobic; usually proteolytic; no acid from carbohydrates; does not reduce nitrates; indol —, VP —.

Other strictly aerobic species, e.g. *B. megaterium*, *B. subtilis*, *B. pumilus*, *B. brevis*, and *B. firmus*, are rare in insects. *B. megaterium* may be mildly pathogenic to corn borer eggs (Lynch *et al.*, 1976). *B. polymyxa* and *B. macerans*, which produce acid and gas in carbohydrates, have not been isolated from insects.

3. *Genus* Arthromitis, *Genus* Coleomitis

Cells, $0.6-1.3 \times 3\mu$, form chains or trichomes $300-500\mu$ long. Previously placed in Caryophanales because cells may show a bright central body considered a nucleoid. Sporulating forms from gut of termites and cockroaches now placed in Bacillaceae as genera of uncertain position; not cultured, presumably anaerobic or microaerophilic.

C. GRAM-NEGATIVE BACTERIA

1. *Enterobacteriaceae and Relatives*

Rods, ca. $0.6 \times 0.8-2.0\mu$, often coccobacillary; G —, usually motile, flagella peritrichous. Grow on NA. Aerobic, facultative anaerobes, catalase +, oxidase —, fermentative. Produce acid and usually gas in many carbohydrates but not in starch (except *Klebsiella* spp.); anaerogenic strains and species occur. Reduce nitrates to nitrites. Classification based on biochemical reactions (Edwards and Ewing, Add. 11).

(a) Section 1. Potential pathogens of haemocoel of many insects; some may also be facultative pathogens but supporting evidence is sparse; similar species, not yet recorded as insect pathogens, are included. Proteolytic, some also lipolytic. Most grow in KCN, on Simmons' citrate, indol —, VP +.

(i) *Serratia marcescens.* Colonies sometimes red. Lipolytic; ferments glucose, maltose, sucrose, fructose, galactose, mannose, trehalose, mannitol, adonitol, glycerol, inositol, salicin, but not lactose, raffinose, rhamnose, dulcitol; produces a small gas bubble in fermentable sugars but not in alcohols; may form weak acid, gas — from xylose, arabinose, cellobiose. Urease usually + but weak and delayed; indol —, VP +; decarboxylates lysine, ornithine, not arginine; hydrolyses chitin (Molise and Drake, 1973); only one species in genus.

(ii) *Enterobacter (= Aerobacter) liquefaciens.* Colonies sometimes pink; lipolytic. Not reported from insects but probably confused with *S. marcescens* in literature; reactions so similar that some would transfer to genus *Serratia*; differs in more gas, ferments xylose, arabinose, raffinose strongly and lactose weakly, urease —, VP — or weak.

(iii) *Enterobacter (= Aerobacter) aerogenes* (Cloaca B). Not lipolytic, some strains not proteolytic; acid and much gas from almost all carbohydrates; dulcitol and starch usually not fermented. Distinguishable from other Enterobacteriaceae, except *Klebsiella*, by acid and much gas from both inositol and glycerol. Urease —, indol —, VP +; decarboxylates lysine, ornithine, not arginine.

(iv) *Enterobacter (= Aerobacter) cloacae* (Cloaca A). Not lipolytic, many strains not proteolytic. Reactions similar to *E. aerogenes* except little or no gas in inositol and glycerol, adonitol rarely fermented, urease usually + but weak and delayed, decarboxylates arginine not lysine. Pathogenicity to insects questionable.

(v) *Pectobacterium (Erwinia) carotovorum.* Contains some species previously placed in genus *Erwinia*; non-lipolytic, some strains weakly proteolytic. Separable from other Enterobacteriaceae by liquefying pectates. Produces acid and little gas from most sugars including arabinose, cellobiose, raffinose and rhamnose, but ferments no alcohols except mannitol; indol and VP variable; does not decarboxylate lysine, arginine, or ornithine; some strains urease +. Not recorded as insect pathogen.

(vi) *Proteus* spp. Separable from other Enterobacteriaceae both by producing much urease and by deaminating phenylalanine. Produce acid and little gas from glucose, fructose, galactose, trehalose and usually glycerol, but not from lactose, arabinose, cellobiose, raffinose, dulcitol, sorbitol; other carbohydrate tests vary with species; some strains are anaerogenic. Methyl red +, VP —; do not decarboxylate lysine or arginine. *P. vulgaris* and *P. mirabilis*, the common insect pathogens, are strongly proteolytic, lipolytic, H_2S +, and form thin spreading growth on wet agar; do not ferment rhamnose, mannitol, adonitol, inositol. *P. vulgaris* ferments maltose, sucrose, mannose and usually salicin; indol +, VP —; does not decarboxylate ornithine. *P. mirabilis* does not ferment maltose, mannose, salicin and rarely sucrose; indole —, VP variable; decarboxylates ornithine.

(vii) *Arizona hinshawii.* Not reported as an insect pathogen but strains, occasionally reported as pathogenic salmonellas, may belong to this modern genus. Proteolytic action weak and delayed, not lipolytic. Acid and much gas from carbohydrates except sucrose, raffinose, adonitol, dulcitol, inositol, salicin and starch; reaction delayed in lactose and cellobiose; no gas in glycerol. Indol —, methyl red $+$, VP $-$, urease $-$; no growth in KCN but grows on Simmons' citrate; $H_2S +$; decarboxylates lysine, arginine, ornithine.

(b) Section 2. Non-pathogenic, non-proteolytic, non-lipolytic, indol variable, VP $-$. Most do not grow in KCN or on Simmons' citrate.

(i) *Salmonella* spp. Medically important, isolated from carrier insects, e.g. adult flies, usually after giving experimental doses; see also (vii) above and section II on ticks, fleas, lice; rare in other insects. Reactions similar to *Arizona* except lactose $-$, dulcitol $+$, inositol $+$ without gas. Indol $-$; some strains do not grow on Simmons' citrate; serological typing used in species identification.

(ii) *Citrobacter (Escherichia) freundii* (Bethesda and Ballerup groups). Common in insects. Similar to *Salmonella* and *Arizona* but separable by combination of following characters: grows in KCN; ferments cellobiose rapidly, lactose and glycerol slowly; does not ferment inositol; variable in sucrose, raffinose, dulcitol; indol $-$; does not decarboxylate lysine, usually not ornithine or arginine; most strains $H_2S +$, weak delayed urease $+$.

(iii) *Klebsiella pneumoniae*. Medically important, uncommon in insects. Non-motile, typically capsulated, often producing mucoid colonies. Reactions similar to *Enterobacter aerogenes* including acid and much gas from carbohydrates, especially from both glycerol and inositol. Differs in: produces some urease, does not decarboxylate ornithine, usually ferments dulcitol, starch.

(iv) *Klebsiella rhinoscleromatis*. Uncommon in insects. Non-motile, typically capsulated. Fermentation reactions similar to *K. pneumoniae* but anaerogenic. Readily separated from anaerogenic *Proteus* spp. by fermentation of most carbohydrates, failure to produce urease or deaminate phenylalanine. Differs from anaerogenic *K. pneumoniae* by: methyl red $+$, VP $-$, Simmons' citrate $-$, lysine decarboxylase $-$.

(v) *Enterobacter hafniae*. Uncommon in insects. Resembles non-proteolytic strains of *E. aerogenes*, but attacks less carbohydrates: does not ferment raffinose, adonitol, inositol, or sorbitol; ferments lactose, sucrose, cellobiose, salicin slowly if at all; VP variable.

(vi) *Proteus rettgeri*. Kills insects when inoculated in large doses though non-proteolytic, non-lipolytic. Otherwise similar to *P. vulgaris*, but colonies do not spread, ferments mannitol, adonitol, inositol, usually rhamnose and sucrose; $H_2S -$. Many strains anaerogenic.

(vii) *Proteus morganii*. Rare in insects. Resembles *P. rettgeri*, but ferments only a few sugars, e.g. glucose, fructose, galactose, mannose, trehalose but no alcohols; forms more gas in glucose; decarboxylates ornithine.

(viii) *Providencia* spp. Rare in insects. Genus related to *Proteus*; deaminates phenylalanine but urease $-$, $H_2S -$, decarboxylates no amino acids, ferments few carbohydrates.

(ix) *Escherichia coli*. Medically important but rare in insects other than carrier flies. Separated from above Enterobacteriaceae by: Simmons' citrate $-$, KCN $-$, urease $-$, $H_2S -$, phenylalanine deaminase $-$; non-proteolytic, non-lipolytic; indol $+$, methyl red $+$, VP $-$. It shares most of the above characters with related genera *Shigella* and *Edwardsiella*. Produces acid and gas from most carbohydrates but does not ferment cellobiose, adonitol or inositol; decarboxylates lysine, usually arginine and ornithine.

(x) *Shigella* spp. Medically important, isolated only from carrier flies, usually after experimental dosing. Similar to *E. coli* except non-motile, anaerogenic, less active in fermenting carbohydrates; does not ferment lactose, sucrose, salicin;

does not decarboxylate lysine, only rarely ornithine and arginine; usually indol —. Serological techniques are used to separate the different species.

(xi) *Edwardsiella tarda*. Not reported in insects though possibly misidentified strains belong here. Resembles *E. coli*, but fails to ferment most carbohydrates; ferments only glucose, maltose and sometimes fructose, galactose, mannose; glycerol fermentation weak and delayed; H_2S +; decarboxylates lysine, ornithine, not arginine.

(xii) *Yersinia (Pasteurella) pestis*. Causes plague; pathogenic to fleas and other carriers, not isolated from other insects. Non-motile, non-capsulated, anaerogenic and thus confused only with *Klebsiella rhinoscleromatis*, *Shigella* spp., or non-motile strains that produce little or no gas from carbohydrates. Differs from *Proteus* spp. and *Providencia* spp. in not deaminating phenylalanine. Differs from *K. rhinoscleromatis* in not fermenting lactose, sucrose, cellobiose, raffinose, adonitol, inositol. Differs from *Shigella* in forming H_2S and usually weak delayed acid in inulin, salicin, starch.

(xiii) *Erwinia* spp. Contains miscellaneous strains grouped mainly because of pathogenicity to or association with plants; sometimes isolated from insects. The weakly proteolytic, pectate-liquefying, weakly aerogenic forms are grouped above in genus *Pectobacterium*. The remainder fall into two groups not producing gas; do not decarboxylate lysine, arginine or ornithine, nor grow in KCN; indol —; usually do not reduce nitrates, so differing from all other Enterobacteriaceae. Differ from *Proteus* spp. and *Providencia* spp. in not producing urease or deaminating phenylalanine. Differ from *K. rhinoscleromatis* in not fermenting adonitol, starch, usually not cellobiose or salicin. The *E. herbicola* group typically forms yellow colonies, some strains reduce nitrates; ferment lactose, xylose, arabinose, raffinose, sometimes cellobiose. Whitcomb *et al.* (1966) found a proteolytic, pathogenic strain in leafhoppers. The *E. amylovora* group may need yeast extract, does not reduce nitrates, nor ferment lactose, maltose, cellobiose and usually not xylose, arabinose or raffinose.

2. Pseudomonadaceae

Rods, ca. $0.7 \times 1.5-3.0\mu$, G —, motile, flagella polar, few; aerobic, catalase +, oxidase usually +, oxidative metabolism. Three genera associated with insects.

(a) Genus *Pseudomonas*. Grow on simple media using nitrate as N source. Reduce nitrates and some denitrify; otherwise strict aerobes. Utilize wide range of single compounds as sources of C and energy, but utilize only few carbohydrates and produce weak acid without gas from even fewer, because acid can be neutralized by NH_3 from amino compounds. Colonies may change colour or produce slight fluorescence under UV light — not to be confused with watersoluble, diffusable, strongly fluorescent pigment produced by "fluorescent" species. Modern taxonomy (Stainier *et al.*, Add. 24) reduced several hundred species to 29 divided primarily on intracellular accumulation of poly-β-hydroxybutyrate, production of arginine dihydrolase, autotrophic growth with H_2, requirements for growth factors. Rare in insects except as potential pathogens of haemocoel.

Species (i) to (viii) produce diffusable, fluorescent, yellow-green pigment; do not accumulate poly-β-hydroxybutyrate; require no growth factors, heterotrophic; known pathogens and species most likely to be confused with them; important pathogens are proteolytic, lipolytic, denitrify nitrates, dihydrolyse arginine, egg-yolk reaction usually +.

(i) *P. aeruginosa*. Pathogenic; egg-yolk reaction + but precipitated fat cleared by lipolytic enzymes. Produces pyocyanin, a blue, diffusable phenazine pigment soluble in chloroform; pigment production favoured on Sabouraud maltose agar and on media of King *et al*. (1954); flagellum single. Grows at 41°C, smells of trimethylamine, no levan slime from sucrose; weak acid only from glucose, xylose, fructose, galactose, mannose, and rarely arabinose.

(ii) *P. fluorescens*. Pathogenic, but some strains not lipolytic, some do not denitrify, in some egg-yolk reaction —. No phenazine pigment; no growth at 41°C or smell of trimethylamine; flagella 2–4; usually levan slime from sucrose; weak acid from glucose, xylose, fructose, mannose, arabinose. Several biotypes recognized by Stanier *et al*. (Add. 24).

(iii) *P. chlororaphis*. Pathogenic; similar to *P. fluorescens*, but produces a green, insoluble, phenazine pigment crystallizing on and around colonies.

(iv) *P. aureofaciens*. Pathogenic; similar to *P. fluorescens* but produces a yellow-orange insoluble phenazine pigment in colonies.

(v) *P. lemonnieri*. Not recorded as pathogenic; similar to *P. fluorescens* but produces a blue, insoluble phenazine pigment.

(vi) *P. putida*. Pathogenicity questionable; similar to *P. fluorescens* but non-proteolytic, non-lipolytic, does not denitrify nitrates, levan —, egg-yolk reaction —.

(vii) *P. syringae*. Plant pathogen. Does not denitrify; egg-yolk reaction —; arginine dihydrolase —; no growth at 41°C; flagella 2–4; some strains produce levan, some proteolytic. Only species of genus that is oxidase —. *P. savastonoi*, carried to olives by *Dacus oleae*, considered a synonym.

(viii) *P. cichorii*. Plant pathogen; resembles *P. syringae*, but oxidase +, non-proteolytic, levan —.

Species (ix) to (xiii) do not produce fluorescent green pigment, are not known insect pathogens, but include some proteolytic species that might be confused with pathogens that had become achromogenic.

(ix) *P. stutzeri*. Non-pathogenic, rare in insects. Does not accumulate poly-β-hydroxybutyrate; non-proteolytic, non-lipolytic; egg-yolk reaction —, arginine dihydrolase —; grows at 41°C; denitrifies; levan —; flagellum single; acid + from glucose, xylose, mannose; colonies wrinkled, adherent, often brownish.

(x) *P. alcaligenes*. Not reported from insects, but may have been confused with strains of genus *Alcaligenes* in literature. Does not accumulate poly-β-hydroxybutyrate; weakly proteolytic, egg-yolk reaction —, arginine dihydrolase +; grows at 41°C; does not denitrify; flagellum single; no acid in carbohydrates.

(xi) *P. cepacia*. Plant pathogen; proteolytic, lipolytic, egg-yolk reaction +, arginine dihydrolase —; does not denitrify; accumulates poly-β-hydroxybutyrate; flagella 1–3; produces a variety of phenazine pigments, some diffusable but not fluorescent; grows at 41°C; utilizes some carbohydrates.

(xii) *P. marginata*. Plant pathogen; similar to *P. cepacia*, but separable by utilization of various carbon compounds.

(xiii) *P. mallei*, *P. pseudomallei*, *P. facilis* and *P. maltophila* are proteolytic, but unlikely to be isolated from insects. First two are vertebrate pathogens and accumulate poly-β-hydroxybutyrate, as does *P. facilis*, a soil allotroph. *P. maltophila* occurs in milk, food products, clinical specimens, does not accumulate poly-β-hydroxybutyrate, requires methionine, colonies yellow.

(b) Genus *Xanthomonas*. Plant pathogens, not insect pathogens. Not fluorescent, separable from similar species of *Pseudomonas* by not reducing or denitrifying nitrates, oxidase — or weak, acid from some carbohydrates including starch.

Flagellum single; colonies usually with yellow carotinoid pigment; may require organic N. Isolated from several insects in small numbers.
(c) Genus *Zoogloea*. Normally in contaminated water; recorded as an epibiont on aquatic insects, where it may form scum or flocs that impair motion. Flagellum single, strict aerobe, non-fluorescent; colonies tough, leathery, straw-coloured, coherent; may require added vitamins.

3. Other Gram-negative Aerobic Bacteria

(a) Genus *Vibrio*, Genus *Aeromonas*. Strains, rare in insects, poorly character-ized, probably misidentified. Rods, motile, one or more polar flagella; aerobic, catalase +, oxidase +, and so resemble *Pseudomonas*. However, metabolism is fermentative and they are facultative anaerobes. Valid species are pathogens of vertebrates.

The abandoned family Achromobacteraceae included G − rods, non-motile or motile by peritrichous flagella, producing little or no acid without gas from carbohydrates. It contained three genera: *Flavobacterium*, yellow colonies with a carotinoid pigment; *Achromobacter*, unpigmented, weak acid from some sugars; *Alcaligenes*, unpigmented, usually alkaline reactions in sugar media by liberating NH_3 from peptones. Members of all genera have been isolated in small numbers from a variety of insects and identified by above characters. Abandon-ment of the family leaves most insect strains in an unknown position because they were too sparsely characterized to be assigned with confidence to present definitions of genera.
(b) Genus *Alacaligenes*. Aerobic, metabolism oxidative, catalase +, oxidase typi-cally + but reaction often delayed and occasionally −. Very similar to *Pseudo-monas*, but separable by flagella number and insertion; typically strains have 1−8 peritrichous flagella, but some are non-motile; it is difficult to discriminate between peritrichous and polar insertion if strains have 1−3 flagella. Non-fluorescent, unpigmented poly-β-hydroxybutyrate −. Thus, they are confused with *P. alcaligenes* and with fluorescent strains that have lost the ability to form fluorescent, unpigmented, poly-β-hydroxybutyrate −. Thus, they are confused with *P. alcaligenes* and with fluorescent strains that have lost the ability to form fluorescent pigment. Though Hendrie *et al.* (Add. 15) place most *Achromobacter* litmus milk, but some strains give weak delayed acid in glucose, xylose, arabinose, galactose, mannose; typically do not decarboxylate lysine, ornithine, or arginine, but lysine + or arginine + strains are reported. Positive reactions are often lost on subculture. *A. faecalis* and its varieties common in insects, often misidentified, apparently not pathogenic. *A. odorans*, considered a variety (Bergey, App. 5), common in grasshoppers (*Pseudomonas* sp., Bucher and Stephens, 1959), has characteristic colony and fruity smell.
(c) Genus *Flavobacterium*. Aerobic, metabolism oxidative, catalase +, oxidase +. Motile or non-motile, flagella peritrichous. Colonies yellow, pigment carotinoid. Little or no acid from carbohydrates. Genus still poorly defined. Occasionally in insects, non pathogenic.
(d) Genus *Achromobacter*. *A. nematophilus*, also known as the organism from the nematode DD 136, carried by parasitic neoaplectanid nematodes into haemocoel of insect hosts. Produces lethal septicemia from small inocula in all insects tested. Growth of bacterium in insect host essential for proper develop-ment of nematode cycle. Large rods ca. $1.1 \times 5\,\mu$, G −, poorly motile, flagella

peritrichous; aerobe, anaerobic growth sparse; metabolism fermentative, catalase —, oxidase —. Growth, sparse on NA, favoured by yeast extract, brain—heart infusions, sera. Described by Poinar *et al.* (1971). Presence should be suspected if host contains many tiny nemas; confirm by plating on rich media with triphenyltetrazolium chloride and bromthymol blue; characteristic colonies have deep red centres, blue margins, purple where the pigments mix; colour results from reduction of chloride to red formazan, production of alkali in medium, absorption of blue indicator. No acid in ordinary carbohydrate tubes but weak acid from many carbohydrates in broth low in organic N. Poinar *et al.* (1977) and Khan and Brooks (1977) isolated similar bacteria from heterorhabditid nematodes, but both bacteria bioluminescent, pigmented, did not accumulate bromthymol blue from agar plates, catalase +. The negative catalase reaction of *A. nematophilus* is unusual for an aerobe that grows poorly anaerobically. Studies of metabolism and respiration of these bacteria needed for assignment to a valid genus. (See Note Added in Proof, p. 33.)

(e) Genus *Chromobacterium*. Aerobic, facultative anaerobe, metabolism fermentative, catalase +, oxidase +. Motile by 1—4 polar and lateral flagella. Colonies violet, pigment soluble in ethanol; non-pigmented strains occur. Acid from some carbohydrates (Sivendra and Lo, Add. 22). Rare in insects, non-pathogenic.

(f) Genus *Franciscella (Pasteurella). F. tularensis.* Strict aerobe, oxidative metabolism, catalase +, oxidase +; tiny coccobacilli, $0.2 \times 0.2 - 0.7 \mu$, non-motile. No growth on NA, usually isolated on blood plates. Causes tularemia in vertebrates; multiplies in gut epithelium of carrier ticks, bed bugs, lice; usually not pathogenic if haemocoel not invaded.

(g) Genus *Sphaerotilus*. Normally in contaminated water; an epibiont on aquatic insects. Long rods, $3 - 10 \mu$, in chains encased in sheath and attached by holdfast. Single cells motile, one or more polar or subpolar flagella. Aerobe, metabolism oxidative, but can grow at low O_2 tension.

(h) Genus *Siderocapsa*. Epibiont on aquatic insects. Cells cocci or coccoid, several embedded in gelatinous matrix containing deposits of Fe or Mg. Aerobe, but can grow at low O_2 tension. Not grown in pure culture.

(i) Genus *Neisseria*, Genus *Acinetobacter (Neisseriaceae).* G— aerobic cocci, frequently paired, non-motile. Rare isolates from insects have been placed here because of structural characters; not fully characterized and identity questionable; some probably micrococci that had lost G+ ability. The modern family contains strict aerobes, metabolism oxidative, catalase +. *Neisseria* is oxidase +. *Acinetobacter* is oxidase —. Family composed of vertebrate pathogens requiring special media for growth, but *Acinetobacter* grows on NA and differs considerably from the other genera (Henriksen, Add. 16).

D. GRAM-POSITIVE COCCI

1. Micrococcaceae

G+ cocci; divide in more than one plane producing regular or irregular clumps of cells; usually $0.9 - 1.0 \mu$; non-motile; aerobic but grow slowly anaerobically; catalase +, oxidase —; grow on NA. Do not produce H_2S or urease, deaminate phenylalanine, or decarboxylate lysine, arginine, or ornithine; indol —, VP —; most grow on Simmons' citrate, in 5% or stronger NaCl, reduce or denitrify nitrates. Acid +, gas — from many carbohydrates, not from starch. Some strains proteolytic and lipolytic. Colonies on NA, 1–2 mm, opaque, often waxy or butyrous, sometimes yellow–orange.

(a) Genus *Micrococcus*. Common in insects, usually non-pathogenic; carbohydrate metabolism oxidative; can use $(NH_4)_3 PO_4$ as source of N. Group or species separation in Baird-Parker (Add. 2) and Bergey (Add. 5).

(b) Genus *Staphylococcus*. Lives on human epithelium; so rarely isolated from insects that one suspects the isolations were contaminants from the entomologist. Carbohydrate metabolism fermentative, cannot utilize $(NH_4)_3 PO_4$.

(c) Genus *Gaffkya*. Some isolations from insects were placed in this genus because cells commonly occurred in tetrads. The name has been rejected and strains placed in *Micrococcus*.

(d) Genus *Sarcina*. Some isolations from insects placed in the genus because cells in regular packets and colonies yellow. Present genus restricted to anaerobes so such strains need reidentification.

2. Streptococcaceae

G + cocci, readily become G — with age; usually non-motile; aerobic, usually grow better under anaerobic or microaerophilic conditions; metabolism fermentative. Usually require complex media and growth factors; of five genera only *Streptococcus* in insects.

(a) Genus *Streptococcus*. Cocci divide in one plane, often producing chains; $0.8 \times 0.9–1.2 \mu$, frequently ovoid; rare strains motile, flagella few, insertion variable. Catalase —, oxidase —; do not produce H_2S or urease, deaminate phenylalanine, reduce nitrates, or decarboxylate lysine or ornithine; indol —, VP —; do not grow on Simmons' citrate or utilize $(NH_4)_3 PO_4$ as N source; acid +, gas — from many carbohydrates. Genus divided into serological groups; Lancefield group D contains species common in insects, but entomologists have rarely used group typing. Group D ferment cellobiose and salicin, grow at 45°C and in media with 40% bile; less nutritionally fastidious than other groups (Deibel, Add. 8).

(i) *S. faecalis*. Tiny translucent colonies (< 1 mm) on NA in air. Grows at 10°C, at pH 9.6, in 6.5% NaCl, in 0.1% methylene blue milk, and dihydrolyses arginine; *S. bovis* and *S. equinus* (also group D) do not have these characters. *S. faecalis* reduces tetrazolium salts, ferments sorbitol, not arabinose; *S. faecium* has opposite characters. Streptococci, often poorly characterized but presumably group D enterococci, have been isolated from numerous healthy and sick insects under various binomials. Some strains proteolytic, some implicated as pathogens without experimental proof; Cosenza and Lewis (1965), Doane and Redys (1970) found non-proteolytic, motile, pathogenic strains of *S. faecalis*.

(ii) *S. lactis*. Lancefield Group N; isolates from insects placed here usually without group typing because they dihydrolyse arginine, grow at 10°C not at 45°C, in 0.1% methylene blue milk, not in 6.5% NaCl or at pH 9.6.

(iii) *S. pluton*. Causes European foulbrood of bees; requires glucose or fructose in special medium and microaerophilic conditions for isolation and good growth (Bailey and Gibbs, 1962); no acid; Lancefield group and other characters not reported.

E. GRAM-POSITIVE, NON-SPORULATING RODS

1. Corynebacteriaceae and Relatives

Typically G + aerobic rods, no spores. As taxonomy is cloudy and genera ill-defined, entomologists have placed uncommon isolates from insects here largely on structural grounds. Veldkamp (Add. 25) states that the group tends to have a growth cycle exemplified best in the genus *Arthrobacter*.

(a) Genus *Arthrobacter*. Cocci or coccoid cells resembling micrococci or strepto-cocci; on transfer to fresh culture swell and elongate into rods of irregular size and shape, often slightly curved, club-shaped, or with rudimentary branching. As culture ages, cocci formed either by fragmentation of rods or by shortening of dividing cells. Strict aerobes, metabolism oxidative, catalase +, oxidase +. Most strains require yeast extract to grow but some grow on NA. Soil bacteria, rare in insects, not pathogenic, but see Lüthy and Soper (1969).

(b) Genus *Brevibacterium*. Insect strains placed here are heterogeneous, most so sparsely characterized as to be useless for establishing a tentative definition. The only characters common to all are short rods or oval cocci, G +, may become G − with age. One of the commonest bacteria from healthy grasshoppers was placed in this genus (Bucher and Stephens, 1959). On NA, cells were oval cocci or very short rods but motile rods formed on media rich in organic N; colonies and some reactions similar to *Streptococcus*; aerobic, facultative anaerobe, metabolism fermentative, catalase +, oxidase −, arginine dihydrolase −; acid +, gas − from some sugars not from alcohols. Lysenko (1959) described from insects two strict aerobes with oxidative metabolism and two facultative anaer-obes with fermentative metabolism. Bergey (Add. 5) considers this a genus of uncertain position and does not define it but genus definition is badly needed, as similar non-pathogenic bacteria are common in the insect gut.

(c) Genus *Corynebacterium*. Typically thin rods, ca. $0.6 \times 3-4\,\mu$, often slightly curved or club-shaped, with irregularly stained sections or granules; snapping division produces V-shaped, angular or palisade arrangements of cells. Aerobic, catalase +, oxidase −, metabolism oxidative but some strains have mixed metabolism. Usually non-motile; acid +, gas − from few if any carbohydrates. Valid species are animal and plant pathogens, usually isolated on special media. Rare isolates from insects placed here because cells stained irregularly, were non-motile, inactive in sugar tests, non-pathogenic.

(d) Genus *Kurthia*. Large rods, ca. $0.8 \times 2-8\,\mu$; usually motile, flagella peri-trichous; may fragment into cocci with age. Strict aerobes, metabolism oxidative, catalase +, oxidase −; no acid from carbohydrates; grow on NA. Rare isolates from insects placed here mainly because of large rod size and inactivity in sugar tests, non-pathogenic.

2. Lactobacillaceae: genus Lactobacillus

Rods, G + may become G − with age, usually non-motile. Anaerobic but some strains grow in air, some need complex organic components. Metabolism fermentative, catalase −, oxidase −; much acid (mostly lactic) from carbo-hydrates, some species produce gas. Common in industrial fermentations; rare non-pathogenic isolates from insects placed here largely because of acid from carbohydrates; includes former genus *Catenabacterium*.

(i) L. (= *Achromobacter*) *eurydice*. Common in gut of bee larvae, especially

those with European foulbrood (Bailey, 1957, 1963), non-pathogenic. Placed in
Lactobacillus (Krieg, 1961) because rods in bee are G+ though G− in culture;
non-motile, may form oval cocci on some media. Grows on NA in air but
isolation and growth favoured by anaerobic conditions, yeast extract, and
glucose, fructose or honey. Metabolism presumably fermentative; acid from
glucose and fructose, one strain gas+; non-proteolytic; anaerobic growth is
catalase − but aerobic growth is weakly catalase +; not fully characterized.

3. Actinomycetes

Filamentous bacteria, often branching and developing into a mycelium
resembling fungi. Colonies usually moderate size, convex, coherent tending to
resist fragmentation or picking with a loop; usually have a characteristic earthy
smell, frequently coloured especially underneath; some produce conidia. However, they are procaryotic, cell wall composition is similar to bacteria, and they
fragment into elements of bacterial diameter $(0.5-1.2\,\mu)$. Widely distributed in
soil and water, common air contaminants; one or more colonies, sometimes seen
on insect isolation plates, normally considered contaminants from air or external
surface of insects; most belong to genus *Streptomyces*, which is G +, aerobic,
metabolism oxidative, produces chains of conidia, little or no acid in carbohydrates, proteolytic.

VI. Addendum

The addendum contains current references to bacteriology and identification.
Numbers 1, 18, 21, 26 are general introductions to bacteriology; 5 is the most
comprehensive publication on bacterial identification; 7, 12, 13 deal with
common or important bacteria; 2, 8, 11, 14, 15, 16, 20, 22, 24, 25 cover specific
bacterial groups. Most contain sections on how to do tests; 4 introduces general
techniques, 17 common biochemical tests, and 23 standard methods. Numbers
9, 10 describe commercially available products and their use, with literature
references; 3 gives chemical explanations of biochemical tests, and 6 is a useful
glossary. Number 19 has good coloured plates of bacterial cells, colonies and even
some test reactions; unfortunately most colonies are pictured on blood agar
plates not commonly available to entomologists.

 1. Anderson, D.A. (1973). "Introduction of Microbiology". C.V. Mosby, Saint
    Louis.
 2. Baird-Parker, A.C. (1965). The classification of staphylococci and micro-
    cocci from world-wide sources. *J. Gen. Microbiol.* **38**, 363−387.
 3. Blazevic, D.J. and Ederer, G.M. (1975). "Principles of Biochemical Tests in
    Diagnostic Microbiology". John Wiley, New York.
 4. Bradshaw, L.J. (1973). "Laboratory Microbiology". 2nd ed. W.B. Saunders,
    Philadelphia.
 5. Buchannon, R.E. and Gibbons, N.E., eds (1974). "Bergey's Manual of
    Determinative Bacteriology". 8th ed. Williams & Wilkins, Baltimore.
 6. Cowan, S.T. (1968). "A Dictionary of Microbial Taxonomic Usage". Oliver
    & Boyd, Edinburgh.
 7. Cowan, S.T. (1974). "Cowan and Steel's Manual for the Identification of
    Medical Bacteria". 2nd ed. Cambridge University Press, London.

8. Deibel, R.H. (1964). The group D streptococci. *Bacteriol. Rev.* **28**, 330–366.
9. Difco Manual. 9th ed. (1953). Difco Laboratories, Detroit.
10. Difco Supplementary Literature. (1968). Difco Laboratories, Detroit.
11. Edwards, P.R. and Ewing, W.H. (1972). "Identification of Enterobacteriaceae". 3rd ed. Burgess, Minneapolis.
12. Gibbs, B.M. and Shapton, D.A. eds (1968). "Identification Methods for Microbiologists". Part B Academic Press, New York and London
13. Gibbs, B.M. and Skinner, F.A., eds (1966). "Identification Methods for Microbiologists". Part A Academic Press, New York and London
14. Gordon, R.E. (1973). "The Genus Bacillus". U.S.D.A. Agriculture Handbook 427.
15. Hendrie, M.S., Holding, A.J. and Shewan, J.M. (1974). Emended descriptions of the genus *Alcaligenes* and of *Alcaligenes faecalis* and proposal that the generic name *Achromobacter* be rejected: Status of the named species of *Alcaligenes* and *Achromobacter*. *Int. J. Syst. Bacteriol.* **24**, 534–550.
16. Henriksen, S.D. (1976). *Moraxella, Neisseria, Branhamella,* and *Acinetobacter. Ann. Rev. Microbiol.* **30**, 63–83.
17. Holding, A.J. and Collee, J.G. (1969). Routine biochemical tests. *In* "Methods in Microbiology"' (J.R. Norris and D.W. Ribbons, eds) Vol. 6A, pp. 1–32. Academic Press, London and New York.
18. Lamanna, C., Mallette, M.F. and Zimmerman, L.N. (1973). "Basic Bacteriology". 4th ed. Williams and Wilkins, Baltimore.
19. Olds, R.J. (1975). "Color Atlas of Microbiology". Year Book Medical Publishers, Chicago.
20. Page, L.A. (1966). Revision of the family Chlamydiaceae Rake (Rickettsiales): Unification of the psittacosis-lymphogranuloma venereum-trachoma group of oranisms in the genus *Chlamydia* Jones, Rake and Stearns, 1945. *Int. J. Syst. Bacteriol.* **16**, 223–252.
21. Salle, A.J. (1973). "Fundamental Principles of Bacteriology". 7th ed. McGraw-Hill, New York.
22. Sivendra, R. and Lo, H.S. (1975). Identification of *Chromobacterium violaceum*: pigmented and non-pigmented strains. *J. Gen. Microbiol.* **90**, 21–31.
23. Society of American Bacteriologists (1957). "Manual of Microbiological Methods". McGraw-Hill, New York.
24. Stanier, R.Y., Palleroni, N.J. and Doudoroff, M. (1966). The aerobic pseudomonads: a taxonomic study. *J. Gen. Microbiol.* **43**, 159–271.
25. Veldkamp, H. (1970). Saprophytic coryneform bacteria. *Ann. Rev. Microbiol.* **24**, 209–240.
26. Wilson, G.S. and Miles, A. (1975). "Topley and Wilson's Principles of Bacteriology, Virology and Immunity". Vol. 1, 6th ed. Williams and Wilkins, Baltimore.

References

Atger, P. (1964). *Entomophaga Mém. hors Sér. No.* **2**, 507–509.
Bailey, L. (1957). *J. Gen. Microbiol.* **17**, 39–48.
Bailey, L. (1963). *J. Gen. Microbiol.* **31**, 147–150.

Bailey, L. (1968). *Ann. Rev. Ent.* **13**, 191–212.
Bailey, L. and Gibbs, A.J. (1962). *J. Gen. Microbiol.* **28**, 385–391.
Bailey, L. and Lee, D.C. (1962). *J. Gen. Microbiol.* **29**, 711–717.
Bucher, G.E. (1957). *Can. J. Microbiol.* **3**, 695–709.
Bucher, G.E. (1960). *J. Insect Path.* **2**, 172–195.
Bucher, G.E. (1961). *Can. J. Microbiol.* **7**, 641–655.
Bucher, G.E. (1963). *In* "Insect Pathology" (E.A. Steinhaus, ed) Vol. 2 pp. 117–147. Academic Press. New York and London.
Bucher, G.E. (1973). *Ann. N.Y. Acad. Sci.* **217**, 8–17.
Bucher, G.E. and Lüthy, P. (1969). *J. Invertebr. Path.* **13**, 305–307.
Bucher, G.E. and Lüthy, P. (1970). *J. Invertebr. Path.* **15**, 292–294.
Bucher, G.E. and Morse, P.M. (1963). *J. Insect Path.* **5**, 289–308.
Bucher, G.E. and Stephens, J.M. (1959). *J. Insect Path.* **1**, 374–390.
Burges, H.D. and Thomson, E.M. (1971). *In* "Microbial Control of Insects and Mites" (H.D. Burges and N.W. Hussey, eds), pp. 591–622. Academic Press, London and New York.
Clark, T.B. (1977). *J. Invertebr. Path.* **29**, 112–113.
Cosenza, B.J. and Lewis, F.B. (1965). *J. Invertebr. Path.* **7**, 86–91.
Davis, B.R. and Ewing, W.H. (1964). *J. Bacteriol.* **88**, 16–19.
Doane, C.C. and Redys, J.J. (1970). *J. Invertebr. Path.* **15**, 420–430.
Heimpel. A.M. (1955). *Can. J. Zool.* **33**, 311–326.
Heimpel, A.M. and Angus, T.A. (1963). *In* "Insect Pathology" (E.A. Steinhaus, ed) Vol. 2, pp. 21–73. Academic Press, New York and London.
Hugh, R. and Leifson, E. (1953). *J. Bact.* **66**, 24–26.
Katznelson, H. (1950). *J. Bact.* **59**, 153–155.
Katznelson, H. (1955). *J. Bact.* **70**, 635–636.
Khan, A. and Brooks, W.M. (1977). *J. Invertebr. Path.* **29**, 253–261.
King, E.O., Ward, M.K. and Raney, D.E. (1954). *J. Lab. clin. Med.* **44**, 301–307.
Kramer, J.P. (1963). *In* "Insect Pathology" (E.A. Steinhaus, ed) Vol. 1, pp. 251–272, Academic Press, New York and London.
Krieg, A. (1961). "Grundlagen der Insektenpathologie", Steinkopff, Darmstadt.
Krieg, A. (1963). *In* "Insect Pathology" (E.A. Steinhaus, ed) Vol. 1, pp. 577–617. Academic Press, New York and London.
Lüthy, P. and Soper, R.S. (1969) *J. Invertebr. Path.* **14**, 158–164.
Lynch, R.E., Lewis, L.C. and Brindley, T.A. (1976). *J. Invertebr. Path.* **27**, 325–331.
Lysenko, O. (1959). *J. Insect Path.* **1**, 34–42.
Molise, E.M. and Drake, C.H. (1973). *Int. J. syst. Bact.* **23**, 278–280.
Poinar, G.O. Jr., Thomas, G.M. and Hess, R. (1977). *Nematologica* **23**, 97–102.
Poinar, G.O. Jr., Thomas, G.M., Veremtschuk, G.V. and Pinnock, D.E. (1971). *Int. J. syst. Bact.* **21**, 78–82.
Steinhaus, E.A. (1947). "Insect Microbiology", Comstock, Ithaca.
Steinhaus, E.A. (1949). "Principles of Insect Pathology", McGraw-Hill, New York.

Steinhaus, E.A. ed. (1963). "Insect Pathology", Vols. 1 and 2, Academic Press, New York and London.
Stevenson, J.P. (1966). *J. Invertebr. Path.* **8**, 205–211.
Whitcomb, R.F., Shapiro, M. and Richardson, J. (1966). *J. Invertebr. Path.* **8**, 299–307.

Note Added in Proof

Whitcomb, R.F. and Williamson, D.L. (1975) reviewed the work done on *Spiroplasma* in insects (*Ann. N.Y. Acad. Sci.* **266**, 260–275). "The Mycoplasmas", a three-volume series, discusses insect mycoplasmas particularly in vols. I and III (Barile, M.F., Razin, S., Tully, J.G. and Whitcomb, R.F., eds., 1979. Academic Press, New York and London).

Thomas, G.M. and Poinar, G.O. Jr. (1979) proposed a new genus, *Xenorhabdus*, to accommodate *Achromobacter nematophilus* and its relatives (*Int. J. syst. Bacteriol.* **29**, 352–360).

CHAPTER 3

Identification of H-serotypes of
Bacillus thuringiensis

H. de BARJAC

Institut Pasteur, Paris, France.

I. Introduction

Although the discovery of *Bacillus thuringiensis* took place in the early part of the 20th century, its development for biological control of insect pests was slow, partly due to a long-lasting confusion between the different varieties. The first important attempt to differentiate and classify the early isolates (Heimpel and Angus, 1958) used morphological and biochemical characters. These isolates were named *Bacillus finitimus*, *B. entomocidus* var *entomocidus* and var *subtoxicus*, and *B. thuringiensis* var *berliner*, var *sotto* and var *alesti*.

The introduction of serotypes based on flagellar or H-antigens (de Barjac and Bonnefoi, 1962) and extra biochemical studies enabled these isolates to be more clearly distinguished. More strains were soon discovered. These, and all the species mentioned above, are now regarded as H-serotypes of one species, *B. thuringiensis* (*B.t.*). This is because serotyping by flagellar agglutination proved to be the most sensitive, specific, reliable and rapid method of identification. Its value in representing the basic subdivision of the species is emphasized by the great similarity of the subdivisions made on the basis of five other characteristics which are discussed in Section III.

II. Method for H-serotyping

A. PREPARATION OF AGGLUTINABLE H-ANTIGEN SUSPENSIONS

The method depends on antiserum that agglutinates flagellae, the organs of motility of the vegetative cells. The motility of cells grown on conventional media is usually too poor to give strong agglutination reactions and it has to be increased by selection at 30°C. The bacterial isolate is first grown in broth, or in a basal medium (U.G. medium) with 0.7% pancreatic peptone (de Barjac and Lecadet, 1976), continuously shaken to give adequate aeration. During early exponential growth, an aliquot is inoculated at the agar surface of the inner small tube of a Craigie tube filled with soft nutrient agar containing only 0.2 to 0.3% agar. The bacteria migrate down the inner tube and then up the outer container tube in about 20–24 h. Usually the migration can easily be seen by the progress of a turbid ring in the clear medium. This procedure is repeated in fresh Craigie tubes, one to three times, until the migration accelerates to take about 14–16 h; the bacteria are then motile enough to give good agglutination.

A stock suspension of very motile bacteria for agglutination is grown by inoculating cells from the last Craigie tube into 100 ml of nutrient broth in a litre Fourneau or Erlenmeyer flask and shaking for 5–8 h at 30°C, according to the size of the initial inoculum. Growth is checked by turbidity until $10^6 - 10^9$ cells/ml are obtained. At regular intervals samples are examined microscopically for motility. The cells should appear numerous, mostly isolated, very motile and without clumps. At this stage formalin is added to give a final concentration of 0.5% and the culture is stored at 4°C.

B. PREPARATION OF H-ANTISERA

Antisera are prepared by injecting rabbits with the stock formalin-treated cell suspension. The suspension is diluted with 0.95% NaCl solution to contain 1 to 4×10^6 cells/ml and measured by counting or optically, e.g. by comparison with a standard $BaCl_2$ opacity tube. Two albino rabbits (3–3.5 kg) are injected twice weekly for 2.5 weeks. The first injection is 0.5 ml of suspension subcutaneously, the other four intravenously, progressing from 0.5 to 1, 2, and 4 ml.

To check the agglutination potency, 7 days after the last injection a few ml of blood are taken from an ear vein for testing by the technique described below. If the titre is $\geqslant 25\,600$, the rabbits are bled by carotid puncture. The blood is allowed to coagulate, then the serum is collected aseptically, with a short centrifugation (5 min at 3000 R.P.M.) if there is haemolysis. If the titre is too low, two or three extra injections are made until it is high enough. Two rabbits are needed for each bacterial isolate because some rabbits respond poorly and also to allow for accidents.

The collected antiserum is aseptically dispensed into 1 ml vials, sealed and stored at 4°C. Sodium merthiolate added to the serum improves storage. The sera can also be sterilized by filtration in a bougie candle. Trials have shown that this treatment does not remove any antibodies, in this case.

C. AGGLUTINATION TECHNIQUES

An unknown strain is first tested against antisera to the 15 known H-types for positive reactions, then the titres of the positive reactions are measured.

1. Test for positive reactions

The stock cell suspension is diluted to values between 10^5 and 4×10^5 cells/ml (one tenth the concentration used for injection of the rabbits) with 0.95% NaCl solution. Simultaneously the antisera are diluted by 1:10, 1:20 and 1:40. Then 0.1 ml of each dilution is put into a small Kahn tube and 0.9 ml of diluted cell suspension is added. A control with 1:10 diluted antiserum and NaCl solution is included. After 2 h still incubation at $37°C$, a positive reaction is shown by clear contents of a tube, on the bottom of which is a floccular sediment, easily dispersed by shaking.

2. Measurement of titre

The H-antiserum that reacts positively is diluted in 13 steps of 2-fold or less: 1:10, 1:20, 1:40, 1:80, 1:160, 1:320, 1:640, 1:960, 1:1280, 1:1920, 1:2560, 1:3840, 1:5120, and 0.1 ml of each dilution is put in Kahn tubes with precision pipettes. Then 0.9 ml of diluted cell suspension is added to each and a control included with NaCl solution. After incubation as above, the titre is given by the greatest dilution that agglutinates. Most strains frequently have a titre of 1:25 600.

In our original description of the H-agglutination method for *B.t.* strains, we mentioned a quicker technique involving centrifugation instead of incubation of the antiserum—cell suspension mixtures. We no longer use this technique because it quite often gave false reactions, a fault also found with a "drop agglutination" technique on glass plates.

3. Determination of antigenic subfactors

Several *B.t.* strains can agglutinate with the same H-antiserum, yet have different biochemical characters and different pathogenicity spectra in host insects. Usually, these differences go together with differences in the antigenic formula. A cross-agglutination between two strains, i.e. their respective agglutination by the heterologous antiserum, does not prove their antigenic identity. It means only that they share a common antigenic fraction; but they can differ by another one or more fractions. For checking this possibility and finding the real antigenic pattern, one has to use the cross-saturation method.

D. CROSS-SATURATION TECHNIQUE

This method can be most easily explained by an example. Suppose that strains X and Y cross-agglutinate. Now saturate antiserum X by the strain Y, antiserum Y by the strain X, and retest the saturated antisera towards the two strains. Different possibilities can occur.

If the Y-saturated antiserum X and the X-saturated antiserum Y no longer react with both strains, these strains have the same H-antigen which will be then designated by a number, without subdivision. If the Y-saturated antiserum X still reacts with X, this strain has another distinct antigenic subfactor, say "b", in its H-antigen, in addition to the factor which is common to both strains, say "a". If the X-saturated antiserum Y does not react with Y, this strain bears only "a". On the contrary, if the strain Y still reacts with the X-saturated antiserum Y, this strain has another different antigenic subfactor, say "c". For simplification, the example has been chosen with only two different antigenic subfactors. The number of subfactors could be higher and its demonstration depends on the study of appropriate strains.

Saturation of an H-antiserum with an heterologous strain is achieved as follows: this strain is grown at 37°C for 24 h, on nutrient agar in Roux flasks inoculated with a young, motile broth culture of *B.t.* After 24 h, the cells are dislodged by rolling sterile glass balls on the medium surface with 8 to 10 ml of NaCl solution. This heavy suspension can be used for saturating directly a pure antiserum, by mixing one part of undiluted serum with nine parts of suspension. It is more efficient to centrifuge the suspension, discard the supernatant and mix directly the sedimented bacterial cells with pre-diluted 1:10 antiserum. In both cases, the mixtures are incubated for 2 h at 37°C and centrifuged. The serum is collected and tested for agglutination against the strain used for saturation. If there is a positive reaction, the serum is re-absorbed by the strain until the neutralization of all the antibodies related to the strain is completed.

III. Classification of *Bacillus thuringiensis* By Serotypes

Application of these serological techniques to the *B.t.* strains has progressively led to the recognition of 15 H-serotypes. The classification of *B.t.* according to these serotypes is given in Table I. This up-to-date classification is the result of the analysis of about 700 *B.t.* isolates. In the eighth edition of Bergey's Manual of Determinative Bacteriology (Buchanan and Gibbons, 1974), H-serotypes have

TABLE I
Classification of Bacillus thuringiensis

H-serotype	Variety name	H-serotype	Variety name
1	thuringiensis	7	aizawai
2	finitimus	8a, 8b	morrisoni
3a	alesti	8a, 8c	ostriniae
3a, 3b	kurstaki	9	tolworthi
4a, 4b	sotto	10	darmstadiensis
4a, 4b	dendrolimus	11a, 11b	toumanoffi
4a, 4c	kenyae	11a, 11c	kyushuensis
5a, 5b	galleriae	12	thompsoni
5a, 5c	canadensis	13	pakistani
6	subtoxicus	14	israelensis
6	entomocidus	15[a]	indiana

[a] Editorial note: at present this number is confused, see Appendix 1, Table I.

been retained as a basis for classification and proposals are reported for giving the serotypes the rank of varieties.

The discovery of distinct antigenic subfactors in five serotypes: H_3 (de Barjac and Lemille, 1970), H_4 (Bonnefoi and de Barjac, 1963), H_5 (de Barjac and Bonnefoi, 1972), H_8 (Gaixin *et al.*, 1975) and H_{11} (Ohba and Aizawa, 1977) has resulted in their subdivision into two subtypes each.

In general, the classification by H-antigens is confirmed by the biochemical characterization of the strains. The esterase pattern as analysed by Norris (1964) is usually distinctive for each serotype. Moreover, each serotype has specific physiological characters. Sometimes, one serotype can be subdivided into biotypes, on the basis of different enzymatic reactions: this is so with two serotypes: H_{4a4b} and H_6. These biotypes have also different pathogenicity spectra.

Finally, taking into account antigenic subfactors and biotypes, the classification of *B. thuringiensis* recognizes 22 varieties or subspecies, as listed in Table I. These varieties are synonymous with serovar or biovar. The first three letters of each variety name can be used as abbreviations.

Table II shows the agreement between the serology and the biochemistry of the *B.t.* varieties and gives an up-to-date determination key.

Two other correlations must be mentioned, one between H-serotypes and the polyacrylamide gel electrophoresis of total cell proteins (Lajudie and de Barjac, unpublished observations) and the other between H-serotypes and crystal-serotypes (Krywienczyk, 1976). Usually each H-serotype has a distinct crystal antigenic pattern (see Chapter 2 for details and a discussion). The key feature of this relationship is the correlation with the distinct pathogenicity spectra of each serotype in host insects.

IV. Comparison with *Bacillus cereus* and Conclusions

The species *B. thuringiensis* is very close to the species *B. cereus* in its cultural and biochemical characters. *A priori*, if one ignores crystals and entomopathogenic power, strains of the two species cannot be distinguished by routine bacteriological tests. However, by H-serotyping, the *B.t.* isolates can be easily and quickly identified. This serological test is very specific at the variety level and has a high agglutination titre. As with *Salmonella*, a low agglutination titre (200 to 400) is not considered significant.

In general, the H-antigens of *B. cereus* are different from those of *B.t.* H-serotyping of 33 *B. cereus* strains has shown the presence of 17 different specific H-antigens (Lemille *et al.*, 1969) and the list will grow with more isolates. Cross-agglutination between *B. thuringiensis* and *B. cereus* with significant high titre occurs very rarely, if at all.

On testing more than 150 *B. cereus* strains against *B. thuringiensis* H-antisera we found real cross-reactions in a few cases only. With these precise cases, the problem relates back to the general problem of the relation between *B.t.* and *B. cereus*. Until now, *B.t.* has often been interpreted as a *B. cereus* having the

Table II

Determination key of Bacillus thuringiensis (AMC = acetyl-methyl-carbinol; ADH = arginine-dehydrolase; TW esterase = tween-esterase; ε = weakly positive reaction)

Biochemical characters	H-antigen	Variety
{ ADH +, Chitin −, Mannose +, Starch ++ }	1	thuringiensis
{ ADH −, Chitin +, Mannose −, Starch − }	2	finitimus
{ Chitin ε, Mannose −, Starch + }	13	pakistani
{ Chitin +, Mannose −, Starch + }	9	tolworthi
{ Chitin ++, Mannose −, Starch ε }	7	aizawai
{ Chitin −, Mannose −, Starch + }	4a4c	kenyae
{ Chitin −, Mannose + }	12	thompsoni
{ Chitin +, Mannose − }	5a5c	canadensis
{ Chitin +, Mannose −, Starch ++ }	3a3b	kurstaki
{ Chitin −, Mannose +, Starch + }	11a11c	kyushuensis

Branching key (AMC +, Lecithinase +):

Salicin +
  Proteolysis +, TW esterase +
    Sucrose +, Pellicle +, Urease −
      Esculin +++ → { thuringiensis (1), finitimus (2) }
    Sucrose −, Pellicle −, Urease +, ADH +
      Esculin ++, ADH + → pakistani (13)
      Esculin +, ADH + → tolworthi (9)
  Proteolysis ++, ADH +, Esculin +++
    Sucrose +, Pellicle −, TW esterase +, Starch +
      Esculin + → aizawai (7)
      Esculin ± → kenyae (4a4c)
    Sucrose −, Pellicle −
      Urease ++ → thompsoni (12)
      Urease − → canadensis (5a5c)

AMC +, Lecithinase +
  TW esterase ++, Urease + → kurstaki (3a3b)
  TW esterase +, Urease − → kyushuensis (11a11c)

Salicin −

{ Proteolysis + +
  ADH +
  Sucrose −
  Chitin −

{ Pellicle +
  Urease −
  TW esterase +
{ Esculin −
{ Mannose −
  Starch +
  14   *israelensis*

{ Pellicle −
  Urease + +
  TW esterase −
{ Esculin + + +
{ Mannose +
  Starch + +
  11a11b   *toumanoffi*

Proteolysis + + +

{ Pellicle −
  Urease −
  Esculin +
  Mannose −
  Starch +
{ ADH −
  TW esterase +
{ Chitin −
  Sucrose +
  Cellobiose −
  4a4b   *sotto*

{ ADH +
  TW esterase + +
{ Chitin + +
  Sucrose −
  Cellobiose +
  4a4b   *dendrolimus*

{ Chitin −
  Sucrose −
  Cellobiose −
  3a   *alesti*

Salicin +

{ Proteolysis +
  ADH +
  Sucrose −

{ Pellicle −
  Urease +
{ Esculin + + +
  TW esterase −
{ Chitin +
  Mannose +
  Starch + +
  5a5b   *galleriae*

{ Proteolysis + +
  ADH +
  Sucrose +

{ Pellicle +
  Urease −
{ Esculin −
  TW esterase +
{ Chitin ε
  Mannose +
  Starch ε
  8a8c   *ostriniae*

AMC +
Lecithinase −

Salicin −

{ Pellicle −
  Urease −
{ Esculin +
  TW esterase +
{ Chitin −
  Mannose +
  Starch −
  15   *indiana*

{ Proteolysis +
  ADH −
{ Sucrose +
  Pellicle +
  Urease −
{ Esculin +
  TW esterase + +
{ Chitin +
  Mannose −
  Starch ε
  8a8b   *morrisoni*

{ Proteolysis + +
  ADH +
{ Sucrose −
  Pellicle −
  Urease −
{ Esculin + +
  TW esterase + +
{ Chitin −
  Mannose −
  Starch + +
  10   *darmstadiensis*

AMC −
Lecithinase −

Salicin −

{ ADH −
  Urease −
  Sucrose +
  Chitin −

{ Proteolysis +
  Pellicle +
{ Esculin + +
  TW esterase −
{ Mannose +
  Starch + +
  6   *subtoxicus*

{ Proteolysis + +
  Pellicle −
{ Esculin +
  TW esterase + +
{ Mannose +
  Starch +
  6   *entomocidus*

special ability to synthesize protein crystals specifically larvicidal for some insects. When a *B.t.* H-antiserum co-agglutinates a *B. cereus*-like strain, i.e. a crystal-free, non pathogenic strain, one wonders if it is a real *B. cereus* strain or a *B.t.* strain that has lost the capacity to produce crystals, since known acrystalliferous mutants of *B. thuringiensis* keep their H agglutinability. The answer to this last question is unknown, as we still do not know which genes and/or plasmids are coding for these crystals. There is no real proof of plasmid-linked crystals and the capacity to form crystals has not yet been transduced to crystal-free strains (see Chapter 15).

The modern trend in the classification of the genus *Bacillus* is against excessive simplification and limitation of species, which previously was the rule. At its session in April, 1979, at Longford, U.K., the Sub-Committee for *Bacillus* taxonomy of the International Committee for Systematic Bacteriology has reviewed the subdivision of some officially recognized species and the addition of new species. Accordingly, the erection of *B. thuringiensis* as a species (already achieved in the 7th edition of Bergey's Manual) is not likely to be questioned again and the old interpretation of *B. thuringiensis* as a variety of *B. cereus* is no longer valid. This status of variety is as meaningless for *B.t.* as it is for *B. anthracis*. This view is strengthened by the results of new techniques in taxonomy, such as for instance pyrolysis—GLC studies (O'Donnell and Norris, 1979). When these methods are applied to *B.t.* strains, they emphasize the difference from *B. cereus* strains, by separating strains into two clear-cut groups.

Thus, *B.t.* remains more than ever a distinct species, having as type strain the neotype *B. thuringiensis* var *thuringiensis*, serotype H$_1$, Mattes strain. The species is clearly subdivided into serotypes, regarded as varieties or subspecies. This is especially important at a time when the discovery of strains pathogenic for mosquito and blackfly larvae brings new prospects for *B. thuringiensis* (Goldberg and Margalit, 1977; de Barjac, 1978 a, b, c, d; de Barjac and Coz, 1979; Guillet and de Barjac, 1979).

References

Barjac, H. de (1978a). *C.r. hebd. Séanc. Acad. Sci. Paris* **286** D, 797—800.

Barjac, H. de (1978b). *C.r. hebd. Séanc. Acad. Sci. Paris* **286** D, 1175—1178.

Barjac, H. de. (1978c). *Entomophaga* **23**, 309—319.

Barjac, H. de. (1978d). *La Recherche* **93**, 911—913.

Barjac, H. de and Bonnefoi. A. (1962). *Entomophaga* **7**, 5—31.

Barjac, H. de and Bonnefoi, A. (1972). *J. Invertebr. Path.* **20**, 212—213.

Barjac, H. de and Coz, J. (1979). *Bull. WHO* **57**, 139—141.

Barjac, H. de and Lecadet, M.-M. (1976). *C.R. hebd. Séanc. Acad. Sci. Paris* **282** D, 2119—2122.

Barjac, H. de and Lemille, F. (1970). *J. Invertebr. Path.* **15**, 139—140.

Bonnefoi, A. and Barjac, H. de. (1963). *Entomophaga* **8**, 223—229.

Buchanan, R.E. and Gibbons, N.E. (eds). (1974). *"Bergey's Manual of Determinative Bacteriology"* Williams and Wilkins Co., Baltimore.

Gaixin, R., Ketian, L., Minghua, Y. and Xingmin, Y. (1975). *Acta Microbiol. Sinica* **15**, 292–301.

Goldberg, L.J. and Margalit, J. (1977). *Mosquito News* **37**, 355–358.

Guillet, P. and Barjac, H. de. (1979). *C.r. hebd. Séanc. Acad. Sci. Paris* **D289**, 549–552.

Heimpel, A.M. and Angus, T.A. (1958). *Can. J. Microbiol.* **4**, 531–541.

Krywienczyk, J. (1976). Communication Working Conference on the spectra of activity of *B. thuringiensis*-δ-endotoxins, Brownsville, Texas, January 1976.

Lemille, F., Barjac, H. de and Bonnefoi, A. (1969). *Ann. Inst. Pasteur* **117**, 31–38.

Norris, J.R. (1964). *J. appl. Bact.* **27**, 439–447.

O'Donnell, A. and Norris, J.R. (1979). *Comm. 85th Meet. Soc. Gen. Microbiol.*, Cambridge, April 1979.

Ohba, M. and Aizawa, K. (1977). *Proc. Sericult. Sci. Kyushu* **8**, 48.

CHAPTER 4

Identification of the *Bacillus popilliae* group of Insect Pathogens

R.J. MILNER

CSIRO, Division of Entomology, Armidale,
N.S.W., Australia

I. Introduction

The family Scarabaeidae (Coleoptera) is very large with $> 19\,000$ described species (Britton, 1970). Most species are beneficial decomposers but many are phytophagous pests feeding on roots as larvae and foliage as adults. The sub-family Melolonthinae (cockchafers) includes some of the world's worst pasture pests such as *Melolontha melolontha* in Europe, *Costelytra zealandica* in New Zealand, and *Popillia japonica* in the USA. Scarabaeids are attacked by a wide range of pathogens of which the milky diseases caused by *Bacillus popilliae* are the best known because of their importance in microbial control, particularly of *P. japonica* in the eastern U.S.A. (Fleming, 1968). After Dutky's (1940) description of milky disease in *P. japonica,* others were found in New Zealand

(Dumbleton, 1945; Fowler, 1972), Australia (Beard, 1956; Milner, 1974a), India (David and Alexander, 1975) and Europe (Wille, 1956). Differences between isolates in terms of morphology, host specificity, and country of origin were immediately apparent. *B. popilliae* has been found only in the Scarabaeidae and only a few varieties are described and named. All varieties of *B. popilliae* studied to date are fastidious in their cultural requirements and only limited production of spores is possible *in vitro*. This has severely inhibited their rational classification, since most taxonomic procedures in bacteriology require easy culture of the organism.

The present chapter complements the reviews of Dutky (1963), Steinkraus and Tashiro (1967) and St. Julian and Bulla (1973).

Bacillus popilliae is defined as a mesophylic, aerobic, spore-forming rod. Vegetative growth is facultatively anaerobic. Vegetative cells are catalase negative. The oval, longitudinally-ridged spore is formed within a swollen sporangium which does not elongate during morphogenesis. One, or more rarely two, parasporal bodies may be formed. The sporangium does not autolyse to release a free spore. All varieties are obligate pathogens that sporulate only in the haemolymph of living scarabaeids. There is no growth in simple bacteriological media, e.g. nutrient agar, but vegetative growth of some varieties occurs in a rich yeast-extract–tryptone medium.

II. Methods

A. FIELD COLLECTION

Larvae, with obvious symptoms of milky disease, are commonest in the field after peak feeding of instar III larvae. Symptoms develop in 3 to 4 weeks but underfed larvae succumb to the disease without exhibiting external symptoms. Milky disease develops best at $20–25°C$ and is inhibited at $< 15°C$ and $> 30°C$. Thus field temperature is often limiting and symptom-free field-collected larvae can develop symptoms after a few weeks incubation at $25°C$. Another method of detecting milky disease in the field is to bioassay soil samples.

Spores are collected aseptically by bleeding surface-sterilized, infected larvae into sterile water. After heating ($70°C$, 20 min) to kill vegetative cells and minimize contamination, spores are washed by centrifugation, and stored in a refrigerator or at room temperature. Spores stored below $0°C$ germinate poorly (Milner, 1977).

B. CULTIVATION *IN VIVO*

At present milky disease spores, infectious *per os*, can be mass-produced only in the living host (Milner, 1974b) though limited production is possible in cell- and tissue-culture (Ebersold, 1976). To infect larvae, aged spores must be used as fresh spores germinate poorly. The most reliable method is to inject at least 10^6 spores into each instar III larvae. After 2 to 4 weeks at $25°C$ typical milky symptoms develop and each larva yields 10^9 to 10^{10} spores (Milner, 1976).

Per os methods are less reliable but infection can result, e.g. from force-feeding, feeding spores on a piece of carrot or on grass roots, or mixing spores into the rearing medium. *Per os*, high doses are required since the LD_{50} is about 10^7 spores/larva and symptoms develop more slowly than with injected spores (Milner, 1974b). *In vitro*-grown vegetative cells suspended in 0.1% tryptone infect when injected into larvae (Pridham *et al.*, 1964), but strains grown *in vitro* may lose their capacity to form spores *in vivo* (St. Julian and Bulla, 1973).

C. *IN VITRO* ISOLATION

The best medium for isolation and growth of vegetative cells is "J" medium (0.5% tryptone, 1.5% yeast extract, 0.3% K_2HPO_4, 0.2% glucose and 2.0% agar, pH 7.5). Dormant spores may not germinate. Methods for breaking this dormancy include ageing of spores in water suspension at room temperature for several months (Lüthy, 1975; Milner, 1977), and washing with chemicals, e.g. $0.15 M$ sodium citrate in $1.5 M$ NaCl (Lüthy, 1975). Heat activation has only limited effect (St. Julian and Bulla, 1973; St. Julian and Hall, 1968). Another problem is that other spore-forming bacteria often outgrow *B. popilliae* (Gordon *et al.*, 1973). During isolation from contaminated sources, including soil, contaminant spores can be germinated in a simple medium in which *B. popilliae* spores will not germinate, then killed by heat. If this procedure is repeated several times most contaminants are killed and *B. popilliae* can then be isolated on the rich solid "J" medium (Milner, 1977). Problems of spore dormancy and contaminants can be avoided by injecting a high dose of spores into a susceptible larva and aseptically plating the haemolymph onto solid "J" medium during the log phase of vegetative growth (Milner, unpublished observations). Many varieties of milky disease bacteria still have not been grown *in vitro* becaue of these difficulties and because "J" medium is far from ideal for their fastidious germination and vegetative growth (Milner, 1977).

D. MAINTENANCE OF STRAINS

Problems of storing *B. popilliae* in culture collections and the failure, in general, to sporulate *in vitro* have seriously inhibited the development of a rational taxonomy for milky disease bacteria. *In vitro* grown rods of var *popilliae* and var *lentimorbus* can be preserved by freeze-drying (Gordon *et al.*, 1973) but not those of var *rhopaea* (Haynes, personal communication) and the method has not been used for varieties not grown *in vitro*. The spore is the obvious stage to hold in collections and is best produced *in vivo* in hosts reared from the eggs in disease-free conditions. Luckily transovum transmission of milky disease bacteria does not occur and it is not known for more than one variety to develop simultaneously in a single host larva (Beard, 1946; Milner, unpublished observations). Spores produced *in vivo* can be stored in water or, to reduce the chance of contamination, in ethyl alcohol.

E. TAXONOMIC METHODS

Most of the tests traditionally used to classify bacteria can be modified for use with milky disease organisms (Gordon *et al.*, 1973), while no special techniques are required for transmission electron microscopy, determining base ratios, or for the usual biochemical tests. Details of spore surface structures and parasporal body shape are best seen by scanning electron microscopy: the spore and parasporal body can be released from the sporangium by ultrasonics, washed with water and air-dried directly onto a microscope stub and suitably coated.

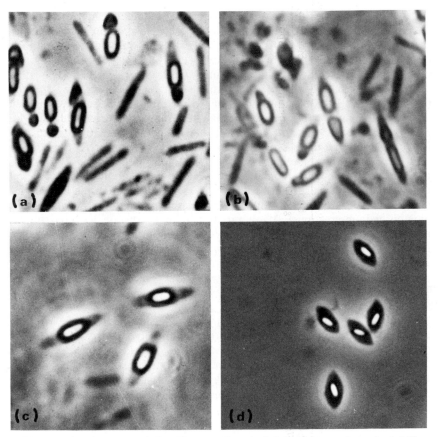

Fig. 1. The 4 types of *Bacillus popilliae* in infected haemolymph. (a) Type A_1, var *rhopaea*, from *Rhopaea verreauxi* ($\times$ 3000). (b) Type A_2 from *Anoplognathus porosus* ($\times$ 3200). (c) Type B_1 from *Antitrogus morbillosus* ($\times$ 3000). (d) Type B_2 from *Ataenius spretulus* ($\times$ 2800).

TABLE I

Comparison of morphology of the sporangium, spore and parasporal body of ten varieties of Bacillus popilliae

Variety of B. popilliae	Host species	Sub-group	Size (μm) Sporangium	Size (μm) Spore	Parasporal body	Shape of parasporal body
popilliae[a]	Popillia japonica	A₁	5.5 × 1.6	1.8 × 0.9	0.5 × 0.5	rhomboid
melolonthae[a]	Melolontha melolontha	A₁	7.1 × 1.4	2.2 × 1.3	0.8 × 0.8	square
rhopaea[a]	Rhopaea verreauxi	A₁	6.0 × 1.5	2.4 × 1.2	1.2 × 0.8	diamond
N.Z. Type I[c]	Costelytra zealandica	A₁	5.0 × 1.5	1.8 × 0.9	0.9 × 0.9	unknown
"RM 12"[b]	Anoplognathus porosus	A₂	8.0 × 1.2	2.6 × 1.1	1.9 × 1.0	diamond
lentimorbus[c]	P. japonica	B₁	3.0 × 1.0	1.8 × 0.9	none	–
"RM 9"[b]	Antitrogus morbillosus	B₁	5.1 × 1.4	2.6 × 1.4[e]	none	–
N.Z. Type II[b]	C. zealandica	B₂	5.2 × 1.8	1.2 × 0.8	none	–
euloomarahae[c]	Heteronychus arator	B₂	3.2 × 0.7	0.4 × 0.2	none	–
unnamed[d]	Ataenius spretulus	B₂	6.4 × 1.4	1.1 × 0.4[f]	none	–

[a]Milner, 1974a; [b]Milner, unpublished observations; [c]Steinkraus and Tashiro, 1967; [d]Splittstoesser and Tashiro, 1977 (off photograph); [e]Spore with unusually thick ridges; [f]Spore with nine ridges in cross section.

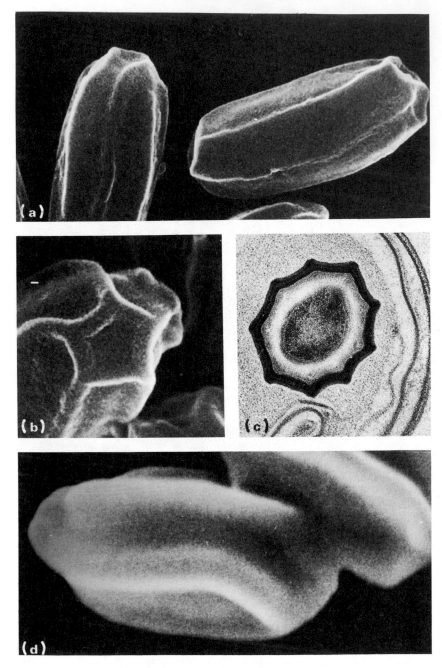

III. Varieties of *B. popilliae*

A. MORPHOLOGY

Morphology of the mature sporangium, with its associated spore and in some varieties parasporal body, is an important and stable taxonomic character as spore morphology is not affected by growth in an abnormal host species (Tashiro and Steinkraus, 1966; Milner, unpubl.). Size, shape and position of the spore and parasporal body are often characteristic for a particular variety. Based on light microscopy, the varieties fall into four subgroups: (i) A_1 — large spore; parasporal body, often small, overlapping the spore; (ii) A_2 — large spore; parasporal body, often large, separated from the spore; (iii) B_1 — large central spore; no parasporal body; (iv) B_2 — small spore in a large sporangium; spore often eccentric; no parasporal body.

Morphological characteristics of some varieties of *B. popilliae* are given in Table I, and Fig. 1. An important spore feature is longitudinal ridges on the surface (Fig. 2a), normally five or six, joined by short cross-ridges near the apex (Fig. 2b). The longitudinal ridges may branch, e.g. in var *popilliae* (Bulla *et al.*, 1969), or may stop before the apex. Individual varieties may be distinguished by the number of longitudinal ridges (Fig. 2c), or by their dimensions (Fig. 2d). The size, shape and position of the parasporal body differs in var *melolonthae* and var *popilliae* (Lüthy, 1968): two other varieties are compared in Figs 3a and 3b. Size of the vegetative cell is normally too variable to be useful, though cells of *B. popilliae*, when young, are often motile and bearing peritrichous flagellae (Wyss, 1971).

The description of two, or occasionally more, morphologically distinct varieties of *B. popilliae* from the same host species, while double infections never occur, is intriguing. Dutky (1940) described a type A_1 variety (which he called *B. popilliae*) and a type B_1 variety (called *B. lentimorbus*) from *P. japonica*. Subsequently both a type A and a type B variety were found in *Amphimallon majalis* (Tashiro *et al.*, 1969), *Costelytra zealandica* (Dumbleton, 1945; Fowler, 1972), *Rhopaea verreauxi* (Milner, 1974a and unpublished observations), *Antitrogus morbillosus* (*Rhopaea morbillosa*) (Milner unpublished observations), and *Sericesthis germinata* (Beard, 1956; Milner, unpublished observations). Splittstoesser and Tashiro (1977) described three, morphologically distinct, type B varieties from *Ataenius spretulus*. In the laboratory, Beard (1946) injected *P. japonica* larvae with var *popilliae* and var *lentimorbus* simultaneously. Only one spore type developed in any one larva, usually var *popilliae* which is commonest in the field. Milner (unpublished observations)

Fig. 2. Morphology of spore ridges of *Bacillus popilliae* varieties under the electron microscope. (a) Spores of *B. popilliae* var *rhopaea* showing typical longitudinal ridges ($\times$ 36 000). (b) Apical cross-ridges of a *B. popilliae* var *rhopaea* spore ($\times$ 72 000). (c) Transverse section through a *B. popilliae* spore from *Ataeneus spretulus* showing an unusually large number of ridges ($\times$ 40 000). (d) Spores of *B. popilliae* from *Antitrogus morbillosus* showing the unusually thick longitudinal ridges ($\times$ 31 000).

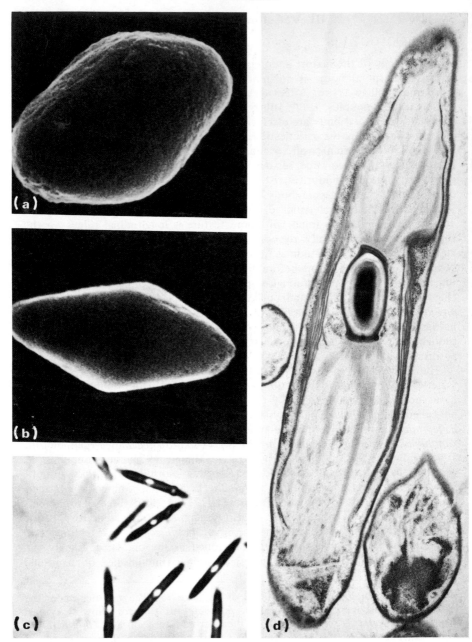

obtained similar results when *R. verreauxi* was injected with its two varieties, except that the proportion of larvae developing the type B disease increased with temperature. *B. popilliae* is not known to produce antibiotics and no explanation for this selective inhibition is known.

B. HOST SPECIFICITY

1. In the Field

Although field data are scant, probably most varieties are each confined to a few host species within a single genus. *B. popilliae* var *rhopaea*, for example, has been found only in *R. verreauxi* (Milner 1974a) while a closely related scarabaeid *An. morbillosus* is infected by a morphologically similar variety. Variety *melolonthae* has been found only in *M. melolontha* in Europe and the morphologically unusual variety from *A. spretulus* (Kawanishi *et al.*, 1974) also is specific, in that it has not been described from either *P. japonica* or *Am. majalis*, two other scarabaeid species common in the same part of the U.S.A. A complex of five genera of pest scarabaeids occurs on the Tablelands of N.S.W., Australia, but no variety of *B. popilliae* is known to cross the generic barrier (Milner, unpublished observations). A possible exception is the var *popilliae* which Dutky (1963) recorded from several genera of scarabaeids in the field. However, caution is needed in the interpretation of these results since more than a single variety may be involved. Thus, many of the *Phyllophaga* spp. found infected in the field could not be infected in the laboratory by feeding soil inoculated with var *popilliae* (Dutky, 1963). Furthermore, "Doom", a commercial product based on var *popilliae* is not recommended for control of any scarabaeid except *P. japonica* (Chittick, personal communication).

2. In the Laboratory

Laboratory data on host specificity are difficult to obtain since erratic results are reported even when scarabaeid species are challenged with their homologous milky disease bacteria. For example Fowler (1974) rarely achieved $> 5\%$ infection of *C. zealandica* with its type I milky disease, despite using high doses and several methods of inoculation. The problems are not fully understood but undoubtedly the correct ageing of spores to promote germination is vital. Other relevant factors are: the nutritional status of the host (Dutky, 1963; Milner, 1974b), larval age (Milner, 1976), temperature, mode of infection, and the dose. The hosts have life cycles of one to 3 years and are not normally laboratory-reared so the presence of inapparent infections in field-collected larvae may also

---

Fig. 3. (a) Parasporal body from *B. popilliae* var *rhopaea* ($\times$ 54 000). (b) Parasporal body from a variety of *B. popilliae* from *Sericesthis geminata* ($\times$ 54 000). (c) Haemolymph from *Aphodius tasmaniae* infected with an unusual *Bacillus* sp., code-named RM 17, showing the long cigar-shaped sporangium with a small central spore ($\times$ 3500). (d) Electron micrograph of longitudinal section through sporangium of RM 17 from *A. tasmaniae* ($\times$ 20 000).

confound the results. Such infected larvae can be detected and eliminated after incubation at 25°C for a few weeks before experimentation.

By injection — the most successful method — most varieties show a relatively wide host range (for example see Table II). However, all attempts to infect

TABLE II
Host range of Bacillus popilliae *var* rhopaea *(Milner, 1974a)*

Host species	10^7 spores/larva by injection			10^9 spores/larva *per os*		
	No. of larvae	Instar	Infection %	No. of larvae	Instar	Infection %
Rhopaea verreauxi	12	III	100	24	III	79
Antitrogus morbillosus	23	III	87	24	III	61
Othnonius batesi	12	III	100	10	III	70
Sericesthis geminata	12	III	100	24	III	0
S. nigrolineata	12	III	75	10	III	0
Anoplognathus porosus	12	III	76	10	III	0
Dasygnathus sp.	12	III	33	12	III	0
Adoryphorus couloni	12	III	0	24	III	0

insects outside the Scarabaeidae have failed (Dutky, 1963; Hurpin, 1973). *Per os*, milky diseases are much more specific (for an example see Table II) but the specificity has not been adequately studied for any variety (Milner, 1974a). Differences in germination rates of different batches of spores are such that all tests should be simultaneous with the same batch of spores and to ensure meaningful results, the original host should be included as a "control". This may be impossible when the original host is found only in a foreign country. Factors such as dose and host nutrition should be standardized as far as possible. Commonly, tests use several local scarabaeid species but few, if any, foreign ones. Since it is permissible to import milky disease bacteria but not scarabaeid larvae into Australia, *R. verreauxi* was extensively tested with imported varieties including var *melolonthae*, var *popilliae* and the two New Zealand strains. However, none was as pathogenic as the native var *rhopaea* (Milner, unpublished observations). Of the three Australian scarabaeids susceptible *per os*, the one natural host was by far the most susceptible (Milner, 1976). Susceptibility may be judged from the proportion infected at a particular dose and also from the yield of spores. Thus var *rhopaea* formed most spores in its natural host (Milner, 1974b).

Three factors, spore germination, host defence and nutrition, all contribute to host specificity. In *Am. majalis* rate of germination and phagocytosis by host haemocytes play a role in determining the fate of the ingested *B. popilliae* spores (Splittstoesser *et al.*, 1978). Also spores of var *melolonthae* were formed in haemocyte cultures but were then destroyed by phagocytosis in cultures of *Pieris brassicae* but not those of *M. melolontha* (Ebersold, 1976). Nutritional factors may be important since var *popilliae* produced abundant vegetative cells but no spores in *M. melolontha* haemocyte cultures.

C. CULTURAL CHARACTERISTICS

Of known *Bacillus* spp., *B. popilliae* has the most specific cultural requirements. No growth occurs on simple media such as nutrient agar but several
varieties grow vegetatively on rich yeast-extract–tryptone media (Steinkraus,
1957). *In vitro* sporulation has been the aim of much recent research, with
limited success for certain strains of var *popilliae* and var *melolonthae* on
specific media (Haynes and Weih, 1972; Wyss, 1971) but spores formed *in vitro*
are not infectious *per os* (Schwartz and Sharpe, 1970). For many varieties,
e.g. New Zealand type 2, even vegetative growth *in vitro* has still to be achieved.
Varieties cultured *in vitro* have a number of features in common: slow growth,
up to 4 days at 25°C for visible colony formation; colonies are small, transparent, round, slightly raised with an undulate margin while older colonies may
develop the "fried-egg" appearance; growth is reduced under anaerobic conditions; vegetative rods, usually non-motile and always catalase negative, die
rapidly in the stationary phase.

Detailed study has been made of only four varieties; var *popilliae*, var
melolonthae, var *lentimorbus* (Gordon *et al.*, 1973; Wyss, 1971), and var
rhopaea (Milner, 1974a). Differences in the main characteristics of these four
varieties are generally insignificant (Table III). They grow only over the relatively narrow pH range of 6.5 to 8.0 within which no differences are apparent.

TABLE III
Some cultural characteristics of four varieties of Bacillus popilliae

Test	Variety			
	popilliae[a]	*lentimorbus*[a]	*melolonthae*[b]	*rhopaea*[c]
Gram stain	—	—	—	±
Motility	±	—	±	±
Catalase	—	—	—	—
Anaerobic growth	+	+	+	+
Max. temp. (°C)	37	37	33	32
Min. temp. (°C)	15	15	18	15
Growth in 2% NaCl	+	—	+	—
Growth at pH 5.7	—	—	—	—
Glucose, sucrose, mannose, maltose	+	+	+	+
Galactose, xylose soluble starch, arabinose	—	—	—	—
Tyrosine	—	+	+	?
Tryptophan, proline	+	—	+	?

+ always positive; ± sometimes; — always negative. [a]Gordon *et al.*, 1973; [b]Wyss, 1971;
[c]Milner, 1974b.

The vitamins, thiamine and biotin, were essential for vegetative growth, as were
such amino acids as arginine and histidine, although the need for tyrosine, tryptophan and proline varied. Varieties also differ in their maximum temperatures

for vegetative growth *in vitro* and sensitivity to NaCl although growth in 2% NaCl is not correlated with presence of a parasporal body as was suggested by Gordon *et al.* (1973). Thus var *rhopaea*, a type A_1 variety, will not grow in 2% NaCl while a type B_1 variety from the same host will grow in 2% NaCl (Milner, unpublished observations). The lethal high temperature for spores of var *rhopaea* was similar to that for var *melolonthae* and var *popilliae*, so this may not be a useful taxonomic character (Milner, 1974a).

D. SEROLOGY

Despite its importance in bacterial taxonomy, serology of milky disease bacteria has received little attention due to problems of *in vitro* growth and because the isolates are host specific. By agglutination var *popilliae* and var *lentimorbus* were distinct from each other and from other *Bacillus* spp., except *Bacillus* larvae (Hrubant and Rhodes, 1968). By double diffusion var *melolonthae* and var *popilliae* were serologically distinct but shared at least one antigen (Lüthy and Krywienczyk, 1972). More detailed studies by electrophoresis technique (Krywienczyk and Lüthy, 1974) on var *lentimorbus, melolonthae* and *popilliae* showed up to 11 precipitin lines, with at least four antigens common to all three varieties. It was concluded that all three were distinct but closely related.

IV. Nomenclature

The early, morphologically distinct milky diseases were each described as species, e.g. *B. popilliae* and *B. lentimorbus* (Dutky, 1940) *B. fribourgensis* (Wille, 1956) and *B. euloomarahae* (Beard, 1956), or as varieties of existing species, e.g. *B. lentimorbus* var *australis* (Beard, 1956). In addition, various strains were described from limited studies of their pathogenicity for a single scarabaeid (Tashiro and Steinkraus, 1966). Other organisms with distinct morphology have been described but not named, e.g. New Zealand type 2 milky disease (Fowler, 1972) and the milky diseases from *Ataenius* (Kawanishi *et al.*, 1974). This situation was rationalized by Wyss (1971) who suggested, on the basis of cultural and serological features, that milky diseases were a homogeneous group, and that most should be varieties of *B. popilliae*, but that *B. euloomarahae* should be retained as a valid species until more was known. Four varieties are now recognized: *popilliae, lentimorbus, melolonthae*, and *rhopaea*. This system, used in this chapter, is not completely accepted as yet (Buchanan and Gibbons, 1974; Gordon *et al.*, 1973).

The "base ratio" of guanine and cytosine to total bases (% GC) in DNA is a measure of the similarity between bacteria (Hill, 1968). Base ratios of both *Bacillus* spp. and *Clostridium* spp. are heterogeneous but predominantly within the range 25—38 for *Clostridium*, and 35—53 for *Bacillus*. The base ratios of var *popilliae* and var *lentimorbus* are 26 to 27 suggesting that *B. popilliae* should be placed in the genus *Clostridium* (Faust and Travers, 1975). In addition *B. popilliae* is catalase negative, a generic feature of *Clostridium* and not of *Bacillus*.

However, the fundamental distinction between the two genera is that sporulation in *Bacillus* is aerobic while *Clostridium* sporulation is anaerobic (Gordon, 1975). Since sporulation in *B. popilliae* is strictly aerobic (Wyss, 1971) it is at present assigned to the correct genus and should be regarded as a microaerobe.

V. Related Diseases: "RM 17"

An organism of doubtful taxonomic position causes a milky disease in Australian Scarabaeidae (Milner, 1979), mainly in *Aphodius tasmaniae*. The symptoms are typical, spores being formed within the haemolymph of the living larva. The unusual features are: the sporangium elongates during spore formation to an exceptional length (10.5 μ); a complex exosporium and feather-like projections within the sporangium (Figs. 3c, 3d); spore wall ultrastructurally simpler than *B. popilliae* without longitudinal ridges; the disease develops at the unusually low temperature of 12°C; the known natural host range is quite wide, although within the Scarabaeidae, though an organism with a similar external appearance has also been found causing a "milky disease" in *Tipula* (Diptera: Nematocera) in England (Sherlock, 1973), and another organism with similar ultrastructure causes a "milky disease" in Isopoda (Yousti, 1976).

VI. Research Needs

Milky diseases offer some of the greatest challenges in insect pathology. Media, suitable for growth of all varieties, are urgently needed and further research is necessary on the problem of sporulation *in vitro*. Developments in the use of cell and tissue cultures for sporulation may provide the basis for solving some of the problems, e.g. the relationship between insect (cell) nutrition and sporulation, and the cause of premature death in sporulating cells on enriched media. The problems of spore dormancy and of the role of the parasporal body need investigation. In describing new varieties authors should deposit ampoules of *in vivo* produced spores in an established culture collection until adequate methods of *in vitro* growth and storage of pure cultures are available. Finally the factors underlying pathogenicity and host specificity are poorly known. A rational taxonomy for *B. popilliae* will not be achieved until many of these problems are resolved. Around these topics, emphasis should be placed on comparative research on the various milky disease bacteria and, in particular, those attacking scarabaeids other than *P. japonica* and *M. melolontha*.

References

Beard, R.L. (1946). *Science* **103**, 371–372.
Beard, R.L. (1956). *Can. Ent.* **88**, 640–647.
Britton, E.G. (1970). *In* "Insects of Australia", 495–621 pp. CSIRO, Melbourne.
Buchanan, R.E. and Gibbons, N.E. (1974). "Bergey's Manual of Determinative Bacteriology" 8th ed., 1246 pp. Williams and Wilkins, Baltimore.

58   R.J. MILNER

Bulla, L.A., St. Julian, G., Rhodes, R.A. and Hesseltine, C.W. (1969). *Appl. Microbiol.* **18**, 490–495.
David, H. and Alexander, K.C. (1975). *Curr. Sci.* **44**, 819–820.
Dumbleton, L.J. (1945). *N.Z. J. Sci. Tech.* **27**, 76–81.
Dutky, S.R. (1940). *J. Agric. Res.* **61**, 57–68.
Dutky, S.R. (1963). In "Insect Pathology, An Advanced Treatise" (E.A. Steinhaus, ed.), Vol. 2, 75–115 pp. Academic Press, New York and London.
Ebersold, H.R. (1976). Eidgenössisohen Technischen Hochschule Dissertation No. 5699, Zürich, 76 pp.
Faust, R.M. and Travers, R.S. (1975). VIIIth Ann. Meet. Soc. Invertebr. Path., Oregon (unpub.).
Fleming, U.E. (1968). *U.S.D.A. Tech. Bull.* 1383, Washington, 30 pp.
Fowler, M. (1972). *J. Invertebr. Path.* **19**, 409–410.
Fowler, M. (1974). *N.Z. J. Zool.* **1**, 97–109.
Gordon, R.E. (1975). In "Biological Control of Vectors" 67–84 pp., U.S. Dep. Hlth Educ. Welf., Washington.
Gordon, R.E., Haynes, W.C. and Pang, C.H.N. (1973). *U.S.D.A. Agric. Handbk* **427**, 283.
Haynes, W.C. and Weih, L.J. (1972). *J. Invertebr. Path.* **19**, 125–130.
Hill, L.R. (1968). In "Identification Methods for Microbiologists" (B.M. Gibbs and D.A. Shapton eds), Pt B, 177–186 pp., Academic Press, London.
Hrubant, G.R. and Rhodes, R.A. (1968). *J. Invertebr. Path.* **11**, 371–376.
Hurpin, B. (1973). *Ann. Zool. Ecol. anim.* **5**, 283–304.
Kawanishi, C.Y., Splittstoesser, C.M. Tashiro, H. and Steinkraus, K.H. (1974). *Environ. Ent.* **3**, 177–181.
Krywienszyk, J. and Lüthy, P. (1974). *J. Invertebr. Path.* **23**, 275–279.
Lüthy, P. (1968). *Zentralbl. Bakteriol. II*, **122**, 671–711.
Lüthy, P. (1975). *Vjschr. Naturforschenden Ges. Zürich* **120**, 81–163.
Lüthy, P. and Krywienczyk, J. (1972). *J. Invertebr. Path.* **19**, 163–165.
Milner, R.J. (1974a). *Aust. J. Biol. Sci.* **27**, 235–247.
Milner, R.J. (1974b). *J. Invertebr. Path.* **23**, 289–296.
Milner, R.J. (1976). *J. Invertebr. Path.* **28**, 185–190.
Milner, R.J. (1977). *J. Invertebr. Path.* **30**, 283–287.
Milner, R.J. (1979). *2nd Int. Coll. Invertebr. Path., Prague*, pp. 139–140.
Pridham, T.E., St. Julian, G., Adams, G.C., Hall, H.H. and Jackson, R.W. (1964). *J. Insect Path.* **6**, 204–213.
St. Julian G. and Bulla, L.A. (1973) In "Current Topics in Comparative Pathobiology" (T.C. Cheng, ed.), Vol. 2. 57–67 pp. Academic Press. New York and London.
St. Julian, G. and Hall, H.H. (1968). *J. Invertebr. Path.* **10**, 48–53.
Schwartz, P.H. and Sharpe, E. (1970). *J. Invertebr. Path.* **15**, 126–128.
Sherlock, P.L. (1973). Ph.D. Thesis, Univ. Newcastle-upon-Tyne, 120 pp.
Splittstoesser, C.M., Kawanishi, C.Y. and Tashiro, H. (1978). *J. Invertebr. Path.* **31**, 84–90.
Splittstoesser, C.M. and Tashiro, H. (1977). *J. Invertebr. Path.* **30**, 436–438.

Steinkraus, K.H. (1957). *J. Bacteriol.* **74**, 625–632.
Steinkraus, K.H. and Tashiro, H. (1967). *Appl. Microbiol.* **15**, 325–333.
Tashiro, H., Gyrisco, G.G., Gambrell, F.L., Fiori, B.J. and Breitfeld, H. (1969). *N.Y. Agric. Expl St. Geneva Bull.* **828**, 71.
Tashiro, H. and Steinkraus, K.H. (1966). *J. Invertebr. Path.* **8**, 382–389.
Wille, H. (1956) *Mitt. Schweitz. Ent. Ges.* **29**, 271–283.
Wyss, C. (1971). *Zentralbl. Bacteriol. II* **126**, 461 –491.
Yousti, A. (1976). Thése de 3ᵉ cycle, Montpellier Univ. 159 pp.

CHAPTER 5

Identification of Insect and Mite Viruses

C.C. PAYNE and D.C. KELLY

Glasshouse Crops Research Institute, Littlehampton, Sussex, England

and

Institute of Virology, 5 South Parks Road, Oxford, England

I. Introduction

It is our intention in this chapter to outline a modern classification of insect viruses and to review the viral properties which can be studied to obtain information on the identity of a particular isolate. Finally, we describe the basic properties of the major insect virus groups and the features most useful for their identification.

II. The Viruses

There are six main groups of viruses recognized as causing disease in insects and mites. These are the baculoviruses, cytoplasmic polyhedrosis viruses, entomopoxviruses, iridoviruses, densoviruses and small RNA viruses. Another virus, *Sigmavirus* (Fenner 1976), has been implicated as the causal agent of CO_2-sensitivity in *Drosophila* spp., but as there is still some doubt as to its authenticity, *Sigmavirus,* although included in Table I, will not be considered further in this chapter. Reviews by Printz (1973) and Sylvester (1977) contain additional information on this virus.

Although six main groups of insect viruses are now recognized, it is likely that as both isolation and diagnostic procedures improve, viruses in other groups will be demonstrated to cause insect disease. For example, *Drosophila* X virus is a recent isolate (Teninges *et al.,* 1979) which resembles a number of unclassified vertebrate viruses, including infectious bursal disease virus of chickens and infectious pancreatic necrosis virus of salmon. In addition, screening insect cell lines for reverse transcriptase activity (an enzyme characteristically associated with C-type viruses or oncornaviruses) showed that some cells possessed this enzyme and that cells harboured these viruses (Heine *et al.,* 1980). This technology, applied directly to insect tissues, should detect tumour viruses if they exist there in nature.

The classification and nomenclature used here for the major groups of insect viruses is that defined by the International Committee on Taxonomy of Viruses (Fenner, 1976). However, many isolates of small RNA viruses as yet remain unclassified by this committee, so that we have adopted a grouping of them similar to that proposed in a recent review by Longworth (1978). Essential properties of the major groups of viruses are shown in Table I, together with the earlier nomenclature used by Weiser and Briggs (1971) in "Microbial Control of Insects and Mites" (Burges and Hussey, 1971), and by Martignoni and Iwai (1977) in their "Catalog of Viral Diseases of Insects and Mites" (see Appendix 2).

III. Properties Used for Virus Identification

Viruses are simply nucleic acids packaged within a coat of protein, glycoprotein, lipoprotein, or combinations of all three. A preliminary identification of a new virus isolate is usually made from morphological criteria followed by an analysis of nucleic acid type. Viruses may contain ribonucleic acid (RNA) or deoxyribonucleic acid (DNA) which may be single- or double-stranded, circular or linear, segmented or intact. Once a virus has been assigned to a group by these criteria, further discrimination is made by analysis of structural components, particularly proteins and nucleic acids. Other structural components of viruses, including lipids, carbohydrates and polyamines, lack sufficiently distinct features to aid identification.

The methods used for, and the approach to, the identification of viruses of insects and mites are governed by the answer required. For certain purposes, assignment of a virus to a major group may be adequate. For other purposes, criteria must be examined which enable closely-related viruses to be distinguished. Although biological properties, including host range, have been used as criteria for identification, these must be supported by some knowledge of the morphological and biochemical properties of the virus. Thus for both purposes, virus characterization by morphological, serological and biochemical methods is the first requirement.

A. MORPHOLOGY

Routine diagnostic techniques for the major virus groups in insects have been available for several years and the virus groups are described in several articles and books (Weiser, 1977; Poinar and Thomas, 1978; Weiser and Briggs 1971; Steinhaus, 1963). Traditionally, invertebrate pathogenic viruses have been divided into two broad groups: (a) those viruses associated with proteinaceous inclusion bodies (IBs), including the "occluded" baculoviruses (nuclear poly-hedrosis and granulosis viruses), cytoplasmic polyhedrosis viruses and entomopoxviruses; and (b) those which are not, i.e iridoviruses, densoviruses, small RNA viruses and "non-occluded" baculoviruses.

This subdivision has helped provide provisional groupings for many new virus isolates by morphological criteria obtained by light and electron microscopy. Virus IBs are readily observed by phase contrast and/or dark field optics, and IBs of some of the different virus groups can be distinguished by shape and size. Some differential staining techniques for IBs from some of the groups are described later under the appropriate virus sections.

Some of the morphological features and terminology used in the description of occluded viruses are shown in diagrammatic representations of IBs of granulosis (GV) and nuclear polyhedrosis viruses (NPV) in Fig. 1. The inclusion

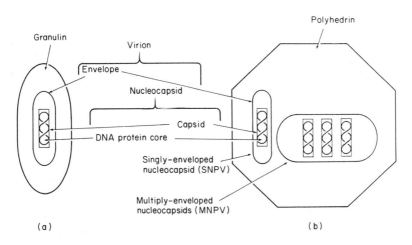

Fig. 1. Diagrams of inclusion bodies of occluded baculoviruses: (a) granulosis virus capsule with virion in longitudinal section; (b) nuclear polyhedrosis virus polyhedron, with virions in longitudinal section.

TABLE I.

Classification, nomenclature and basic properties of viruses pathogenic to insects and mites

Virus Group[a]	Nucleic Acid		Particle Symmetry	Inclusion body
	Type[c]	No. of Segments		
Family BACULOVIRIDAE				
A: Nuclear polyhedrosis ⎱ Genus *Baculo-*	ds DNA	1	Bacilliform	+
B: Granulosis ⎰ *virus*			Bacilliform	+
C: *Oryctes* virus			Bacilliform	−
Family REOVIRIDAE				
Genus Cytoplasmic polyhedrosis	ds RNA	10	Isometric	+
Family POXVIRIDAE				
Genus *Entomopoxvirus* Subgenera, A, B and C	ds DNA	1	Ovoid or Brick-shaped	+
Family IRIDOVIRIDAE				
Genus *Iridovirus*	ds DNA	1	Isometric	−
Family PARVOVIRIDAE				
Genus *Densovirus*	ss DNA	1	Isometric	−
Family PICORNAVIRIDAE				
Genus *Enterovirus*	ss RNA	1	Isometric	−
Unclassified small RNA viruses				
a: Divided genome group	ss RNA	2	Isometric	−
b: *Nudaurelia β* virus group	ss RNA	1	Isometric	−
c: Kelpfly virus group	ss RNA	1	Isometric	−
d: "Group 5" viruses	ss RNA	1	Isometric	−
e: "Minivirus" group	ss RNA	1	Isometric	−
f: Ovoid viruses	ss RNA	1	Ovoid	−
Family RHABDOVIRIDAE				
Genus *Sigmavirus*	Not known		Bullet-shaped	−
UNCLASSIFIED VIRUSES				
Drosophila X virus	ds RNA	2	Isometric	−

[a]Details of nomenclature (Fenner 1976) except small RNA viruses (Longworth, 1978).
[b]Cryptograms of genus or family (Fenner 1976). [c]ds = double-stranded; ss = single-stranded.

TABLE I. (*Contd.*)

Cryptogram[b]	Alternative Nomenclature	
	Weiser and Briggs (1971)	Martignoni and Iwai (1977)
[D/2:80–100/8–15:U or Uo/E:I/O]	Borrelinavirus, Birdiavirus Xerosiavirus	Nucleopolyhedrosis
	Bergoldiavirus	Granulosis, Malaya Disease
[R/2:Σ13–Σ18/25–30:So/S:I/O]	Smithiavirus	Cytoplasmic polyhedrosis
[D/2:140–20-/5–6:X/*:I/O]	Vagoiavirus, Pox-like viruses, Grasshopper inclusion body virus	Spheroidosis
[D/2:130/15–20:S/S:I/I]	Iridescent viruses	Iridescent virosis
[D/1:1.5–2.2/19–32:S/S:I/O]	Densonucleosis virus	Densonucleosis
[R/1:2.5/30:S/S:I, VI, O, R]	—	e.g. Acute paralysis, Sacbrood
	—	—
	—	—
—	—	—
	—	e.g. Flacherie
	e.g. Crystalline array	e.g. Crystalline array
	—	e.g. Chronic paralysis
[*/*/:*/*:Ue/E:I/C]	—	CO_2 sensitivity
—	—	—

bodies of cytoplasmic polyhedrosis viruses (CPV) contain isometric virus particles (virions) without a surrounding envelope, while *Entomopoxvirus* (EPV) IBs contain ovoid or brick-shaped enveloped virions. With the exception of the "non-occluded" baculoviruses, *Sigmavirus* and the "ovoid" small RNA viruses, the major groups of non-occluded viruses have isometric virions with no envelope.

Morphology of virions can be recorded by electron microscopic examination of negatively-stained preparations or thin sections of virus-infected tissues. Morphological characteristics of virions of the groups of invertebrate viruses have recently been well-described (Maramorosch, 1977) and will not be dealt with in detail in this chapter.

B. BIOCHEMICAL PROPERTIES

Morphological criteria alone are probably sufficient to place a new virus isolate within one of the major groups, although they do not easily distinguish densoviruses and different groups of small RNA viruses. Further characterization of virus isolates must be achieved by biochemical and serological analyses. This is largely a comparative exercise that requires access to certain purified virus isolates and antisera prepared against them.

1. Virus Purification and Antiserum Production

No virus purification scheme nor schedule for antiserum production is standardized, but some guidance and examples are given in publications in Table II.

It should be remembered that the successful application of biochemical and serological techniques depends on the use of virus which has been highly purified, using consistent criteria of purity established by the individual worker.

2. Biophysical Properties of Viruses

Although biophysical properties are of little use in the identification of most insect virus groups, the buoyant density and sedimentation coefficient of virions may be useful in placing a small RNA virus within one of the provisional groups of Longworth (1978).

3. Analysis of Virus Structural Proteins

Structural proteins of viruses may represent only a small proportion of the genetic information of the viral genome but they provide readily-available phenotypic characters for use in virus identification. Recent biochemical techniques used to investigate the structural proteins of insect viruses include gel electrophoresis and peptide mapping. Serological methods also enable the antigenic properties of the structural proteins to be compared.

(a) Gel Electrophoresis. Polyacrylamide gel electrophoresis in the presence of the anionic detergent, sodium dodecyl sulphate (SDS) (Maizel, 1966; Laemmli, 1970), is a powerful tool for studies of viral and other polypeptides. Recently it has been used to examine structural proteins of isolates from all the major

TABLE II.

Purification of insect viruses; references for guidance on purification and antiserum production

Virus group	References
Baculoviridae Nuclear polyhedrosis viruses (subgroup a)	Harrap & Longworth, 1974; Harrap *et al.*, 1977; McCarthy & Liu, 1976; Martignoni *et al.*, 1968; Volkman *et al.*, 1976.
Granulosis viruses (subgroup b)	Brown *et al.*, 1977; Tweeten *et al.*, 1977; Yamamoto & Tanada, 1978, 1979.
Non-occluded baculoviruses (subgroup c)	Krell & Stoltz, 1979; Monsarrat *et al.*, 1973; Monsarrat & Veyrunes, 1976; Payne, 1974; Payne *et al.*, 1977a.
Reoviridae Cytoplasmic polyhedrosis viruses	Hayashi & Bird, 1970; Payne, 1976; Payne & Tinsley, 1974; Rubinstein *et al.*, 1975.
Poxviridae Entomopoxviruses	McCarthy *et al.*, 1974; Pogo *et al.*, 1971.
Iridoviridae Iridoviruses	Elliott *et al.*, 1977; Kelly & Tinsley, 1972; Wagner *et al.*, 1973.
Parvoviridae Densoviruses	Kelly *et al.*, 1980; Longworth *et al.*, 1968; Tijssen *et al.*, 1976, 1977.
Small RNA viruses	Longworth & Carey, 1976; Longworth *et al.*, 1973; Morris *et al.*, 1979; Plus *et al.*, 1976; Reinganum *et al.*, 1978; Scotti *et al.*, 1976.

groups of insect viruses. Although providing useful information on the composition and molecular weights (mol. wt) of the polypeptides of a single virus, its greatest use is to analyse the comparative properties of several viruses under identical conditions.

A virus typically contains virus-specific proteins in identical amounts in all virions, so a given virus presents specific and reproducible patterns. Different viruses within some groups, such as baculoviruses (Harrap *et al.*, 1977; Summers and Smith, 1978) and small iridescent viruses (Kelly and Tinsley, 1972; Elliott *et al.*, 1977; Carey *et al.*, 1978; Kelly *et al.*, 1979), present different (yet reproducible) profiles for different isolates. As with many other techniques for virus identification, differences are significant but the detection of proteins with similar electrophoretic mobilities does not imply that they are identical. Additional biochemical and structural evidence is necessary to establish structural similarities.

(b) Peptide Mapping. Probably the greatest use of this technique in virus identification will be in the analysis of viral proteins indistinguishable by other criteria. The method relies on enzymatic cleavage of proteins (e.g. by trypsin or chymotrypsin) and subsequent chromatographic and/or electrophoretic separation of resulting peptides. Co-migration of individual peptides suggests that they are identical. The technique has been applied to the comparative analysis of *Baculovirus* granulins and polyhedrins (Summers and Smith, 1976; Maruniak and Summers, 1978). Proteins from different viruses were related, but contained distinguishing peptides.

(c) Serological Analysis. Serological methods are powerful tools for virus identification and diagnosis. Many techniques can be easily standardized and can often be performed with simple apparatus. For these reasons, a serological technique for virus identification remains the method of choice in field situations or where many samples require processing. Despite this, no concensus has yet been obtained on the relative suitability of the various serological methods available in the study of most insect viruses, although Kelly *et al.* (1979) obtained the greatest specificity with complement fixation and neutralization techniques when comparing several iridoviruses.

Successful use of serological methods depends on production of an avid, specific antiserum. This is influenced, largely, by the purity of the virus or virus component used to raise the antiserum. Much early serology of insect viruses, particularly baculoviruses, is clouded by doubts about the interference of host antigens in tests. Nonetheless, serological techniques can be highly effective in detecting and diagnosing viruses, characterizing their properties and grouping them taxonomically. However, complete reliance on a serological method should come only after supporting evidence of its value for identification from other methods.

Virtually all the common serological techniques have been applied to the study of insect viruses, including tube precipitation, gel immunodiffusion, complement fixation, immunoelectrophoresis, haemagglutination (inhibition), neutralization, latex agglutination, fluorescent antibody and immunoperoxidase, macrophage migration, radio-immunoassay (RIA) and enzyme-linked immunosorbent assay (ELISA). Results obtained with these techniques have been reviewed by Mazzone and Tignor (1976) and Harrap and Payne (1979).

The most promising recent developments have been in such techniques as RIA and ELISA. In the most frequent usage of these methods a specific antibody is adsorbed to a supporting substrate (commonly the wells of polystyrene microtitre plates). A solution containing the viral antigen is then allowed to react with the antibody. Unbound antigen is washed off and replaced by a sample of specific antibody already conjugated with a radioisotope (often ^{125}I) or an enzyme (e.g. alkaline phosphatase or horseradish peroxidase). The extent of the reaction between antigen and conjugated specific antibody is measured by radioactive counting, or by an assessment of enzyme activity after addition of an appropriate substrate (Kalmakoff *et al.* 1977b; Daugharty and Ziegler, 1977; Voller and Bidwell, 1977). Both methods have proved very sensitive in the detection of baculoviruses and other groups (Ohba *et al.*, 1977;

Crawford *et al.*, 1978; Kelly *et al.*, 1978a; Crook and Payne, 1980). They can also detect a virus specifically in impure preparations, including larval extracts, revealing their potential for analysis of field-collected samples (Kelly *et al.*, 1978a; Crawford *et al.*, 1978; Crook and Payne, 1980). The ELISA technique has the advantage of needing simpler equipment and avoiding the use of radio-isotopes. However, more research is still needed to develop the application of both methods for the detection and identification of insect and mite viruses.

4. Analysis of Viral Nucleic Acids

Differences between viruses usually reflect intrinsic differences between the viral genomes. Virus groups may be readily-distinguishable by the nature of the nucleic acid (RNA or DNA) and its strandedness (single or double) while more closely-related viruses may differ only by small regions of distinct base sequences defined only by sophisticated techniques, e.g. restriction endonuclease or homology studies.

(a) Nature and Strandedness. Analysis of the nature of the viral nucleic acid is particularly useful for small isometric viruses with some overlap in size range between the DNA-containing densoviruses and the complex of small RNA viruses (Longworth, 1978). Appropriate tests on viral nucleic acid, e.g. the orcinol test for RNA and diphenylamine colorimetric test for DNA (Shatkin, 1969), should reveal the group to which a new isolate belongs.

Confirmation of the strandedness of insect viral nucleic acid has come from a number of methods, including measurements of hyperchromicity (Summers and Anderson, 1973; Harrap *et al.*, 1977), where increasing temperature produces a sharp rise in absorbance at 260 nm, characteristic of a base-paired (double-stranded) structure. The presence of single (ss) or double-stranded (ds) nucleic acids can also be demonstrated by adding formaldehyde, which interacts with free amino groups on unpaired bases producing characteristic spectral shifts (Miura *et al.*, 1968; Kelly *et al.*, 1977). Ds RNA is relatively resistant to pancreatic ribonuclease at high salt concentrations, while ss RNA is readily digested (Miura *et al.*, 1968). Other enzymes will also differentially digest ss and ds nucleic acids (Marmur *et al.*, 1963).

(b) Base Composition and Size. Other viral nucleic acid properties offering information of rather limited use for virus identification include base composition and genome size. Base composition analysis of insect viral nucleic acids has been made directly by chemical methods, using chromatographic separation, and quantitation of individual bases after hydrolysis of the nucleic acid (Wyatt, 1953; Miura *et al.*, 1968; Cibulsky *et al.*, 1977c). Alternatively, guanosine:cytosine (G:C) compositions for ds nucleic acids have been calculated from melting points (Tm) and buoyant densities, using relationships developed by Mandel and Marmur (1968), Frank-Kamenestkii (1971) and Schildkraut *et al.* (1962) for DNA, and by Kallenbach (1968) for RNA.

Methods for measuring the molecular weights of insect viral nucleic acids include sedimentation analysis, both analytically and in sucrose gradients using standards of known mol. wt (Bellett and Inman, 1967; Summers and Anderson,

1973; Revet and Monsarrat, 1974; Kelly *et al.*, 1977). Perhaps the best physical measurement of nucleic acid size is by electron microscopic examination and several comparative studies of *Baculovirus* DNAs have now been made with this technique (Burgess, 1977; Bud and Kelly, 1977). Electrophoretic analysis on gels has also been used for size estimation whether with naturally segmented ds RNA genomes of CPVs (Kalmakoff *et al.*, 1969; Payne and Rivers, 1976), intact and segmented ss RNA genomes of small RNA viruses (Newman and Brown, 1973; Pudney *et al.*, 1978) or *Baculovirus* DNAs enzymatically fragmented by bacterial restriction endonucleases (see below) (Miller and Dawes, 1978a; Smith and Summers, 1978). The sizes of DNAs from baculoviruses, densoviruses and iridoviruses have also been examined by reassociation kinetics which measure genetic complexity of the genome rather than physical size of the molecule (Kelly and Avery, 1974; Kelly 1977; Kelly *et al.*, 1977).

(c) Restriction Endonuclease Analysis and Segmented Genomes. Restriction endonucleases (RENs) are a range of different bacterial enzymes which each recognize specific base sequences in a DNA molecule and cleave the DNA at or near these recognition sites (Nathans and Smiths, 1975). Thus one enzyme consistently produces specific fragments from a species of DNA molecule. A different enzyme with a different recognition site produces fragments of different sizes. These DNA fragments can then be separated on the basis of their size by electrophoresis on agarose gels. The separation produces a characteristic electrophoretic profile of DNA fragments for each virus. Common patterns from two DNAs suggest genetic relatedness. Completely distinct patterns show that the DNAs are not closely related. Use of several different RENs will improve the identification technique by reducing the chances of obtaining identical profiles with two DNA molecules unless they are very closely related or identical. This method has already shown great promise in the analysis of *Baculovirus* genomes (Miller and Dawes, 1978b, 1979; Lee and Miller, 1978; Smith and Summers, 1978). REN analysis potentially has several advantages over other identification techniques for baculoviruses and other DNA-containing insect viruses: 1. The entire viral genome is screened. 2. Mixtures of viruses can be detected (Miller and Dawes, 1978a; Lee and Miller, 1978; Smith and Summers, 1978). 3. Closely related virus strains can be identified (Miller and Dawes, 1978b, 1979). 4. Genetic changes in viruses can be measured, a feature of particular significance in the quality control of viral insecticides.

Just as fragmentation of *Baculovirus* DNAs by restriction enzymes is a useful technique for future identification of virus isolates, so natural segmentation of the ds RNA from CPVs enables us to provisionally classify the large number of isolates of these viruses (Payne and Rivers, 1976). CPVs contain 10 RNA segments, each representing a single gene. Electrophoretic separation of these segments provides a profile not unlike that obtained by REN analysis of a DNA molecule. However, whereas with REN analysis it is possible to use a range of enzymes (with different base sequence—specific recognition sites) to produce alternative fragment profiles for the same DNA, it is impossible at present to provide alternative segment profiles of the CPV genome. Thus, it should be appreciated that while differences in the sizes of RNA segments imply that the viruses differ in a range of properties, viruses which share the same RNA profile may not necessarily be identical, as the RNA segments could be of the same size but contain base sequence differences (Payne *et al.*, 1977b, 1978). Further techniques, including homology studies, would be needed to compare such viruses.

(d) Homology. Homology studies reveal the extent of genetic relationship between virus isolates by analysing the degree of hybridization between viral nucleic acids. However, an accurate assessment of homology between viral nucleic acids requires some knowledge of the reassociation kinetics of viral double-stranded nucleic acids. Such accurate analyses have been applied only to a few viruses in the *Baculovirus, Iridovirus* and *Densovirus* groups (Kelly and Avery, 1974; Kelly, 1977; Kelly *et al.*, 1977; Wagner and Paschke, 1977; Rohrmann *et al.*, 1978; Jurkovičova *et al.*, 1979). These publications describe the techniques. Other reannealing techniques have been used to study homology between insect viral nucleic acids (Knudson and Tinsley, 1978; Payne *et al.*, 1978) but these methods do not give strictly quantitative data. Undoubtedly, homology studies have a role in the analysis of closely related viruses but the available techniques are time-consuming and so are unsuitable at present for routine identifications. Nonetheless, genomes can be detected in cells by cyto-hybridization techniques, in which a specific radioactive nucleic acid probe is allowed to reanneal with nucleic acids in cells infected with a virus. Where homology exists, the extent of reannealing can be measured by autoradiographic techniques (Tijssen and Kurstak, 1977). Although not yet applied to the diagnosis of insect virus infections, this methodology may prove extremely useful in the future.

C. BIOLOGICAL PROPERTIES

The above techniques enable the definitive characters of a particular virus to be examined. Viruses also differ in a number of biological features including host range (Tompkins *et al.*, 1969), pathogenicity and LD_{50}s in different hosts (Burgerjon *et al.*, 1975; Burgerjon, 1977), symptoms induced in the host, and plaque characteristics in cell cultures (Hink and Vail, 1973). These features, although too poorly defined to aid identification of many viruses or when attempting to identify an isolate, may have a role to play when comparing two or three viruses already identified by other, definitive criteria.

IV. Features of the Major Groups of Insect and Mite Viruses

A. BACULOVIRUSES (BACULOVIRIDAE)

1. Description

Baculoviruses, or *Baculovirus*-like particles (Table I) have been seen in Lepidoptera, Hymenoptera, Diptera, Neuroptera, Coleoptera, Trichoptera, Crustacea and mites (Fenner, 1976). The virions are rod-shaped, 40–70 nm × 250–400 nm, comprising a lipoprotein envelope around a protein capsid containing a DNA—protein core. The capsid and core are known as the nucleo-capsid. In GVs (subgroup b) and *Oryctes*-like viruses (subgroup c), the nucleo-capsids are enveloped singly in virions. In NPVs (subgroup a) a virion may contain more than one nucleocapsid within a single envelope. In subgroups a and b the virions are embedded in IBs.

Virions contained in these three groups are very similar in their basic proper-ties (Harrap *et al.*, 1977; Brown *et al.*, 1977; Payne, 1974). The groups differ in the embedding of virions within IBs. NPVs may contain > 100 virions/IB (poly-hedron), GVs contain one (rarely two) virions per IB (capsule, granule) and *Oryctes*-like viruses have not been seen within inclusion bodies *in vivo* (Huger, 1966; Payne, 1974) or *in vitro* (Kelly, 1976). IBs of NPVs and GVs contain a major polypeptide (polyhedrin or granulin, Summers and Egawa, 1973) which has a mol. wt of ca. 30 000 daltons (Croizier and Croizier, 1977). Most, if not all, occluded baculoviruses contain an alkaline protease enzyme in virions or closely-associated with the IB protein which degrades polyhedrin or granulin to smaller products (Eppstein and Thoma, 1975; Summers and Smith, 1975; McCarthy and Liu, 1976; Payne and Kalmakoff, 1978; Tweeten *et al.*, 1978) but does not dramatically affect the antigenicity of the protein (Norton and DiCapua, 1976; Crawford and Kalmakoff, 1977). The activity of this enzyme can be reduced by low (0°C) and high (70–80°C) temperatures and inhibitory divalent cations (Eppstein and Thoma, 1975; Summers and Smith, 1975, 1978).

Virions of baculoviruses contain 11–25 polypeptides (Payne, 1974; Harrap *et al.*, 1977; Brown *et al.*, 1977; Summers and Smith, 1978; Harrap and Payne, 1979) of which 4–11 are associated with the nucleocapsid and the rest probably with the envelope.

The genome is a single molecule of double-stranded circular supercoiled DNA with a mol. wt of $50-100 \times 10^6$ daltons (Harrap and Payne, 1979). Estimates of genome size vary from laboratory to laboratory, but comparative studies using reliable methods demonstrate that there are real differences in genome size between different isolates (Burgess, 1977; Kelly, 1977; Bud and Kelly, 1977).

Replication of most baculoviruses occurs in the nucleus of cells, though some granulosis viruses and the *Orcytes* virus may be seen apparently replicating within the cytoplasm of infected cells (Kelly, 1976; Summers, 1977).

2. Features for Identification

(a) Gross Pathology. Externally visible, pathological changes in the host occur normally when infection is well-advanced and the insect moribund. Distinguish-ing viral infection from disease caused by other pathogens is possible only on occasion. In some infections behavioural changes may be obvious. Thus the term "Wipfelkrankheit virus" used for NPVs describes how virus-infected larvae (e.g. *Lymantria monacha*) may gather in a typical way at the tops of trees (Aizawa, 1963). In larvae infected with NPV and GV, the integument often changes colour and the insects become flaccid and fragile. Larvae in an advanced disease state may hang in a characteristic inverted position. At death, rupture of the body wall often liberates masses of IBs. In Lepidoptera, infection is frequently generalized in the nucleus (NPVs) and/or cytoplasm (GVs) of cells of most tissues, while in Hymenoptera, NPV replicates only in gut epithelium. *Oryctes rhinoceros* larvae infected with *Oryctes* virus undergo extensive changes in fat-body cells, and the larvae appear almost translucent as infection proceeds. The hindgut may completely evaginate (Huger, 1966). In infected adults, infected gut epithelial cells proliferate (Huger, 1973), leading to the voiding of much virus, readily seen by electron microscopic examination of negatively-stained faeces (Monsarrat and Veyrunes, 1976). The gut also becomes white and swollen with a milky content, and smears of gut contents show many very large cells with abnormally swollen nuclei (Zelazny, 1978).

(b) Light Microscopy. Production of IBs (polyhedra and capsules) is a feature of most NPV and GV infections, but is absent in *Oryctes*-like virus infections. GV capsules can be distinguished from polyhedra by their size (0.5 μm for capsules; 0.5–15 μm for NPV polyhedra; Huger, 1963; Aizawa, 1963). Capsules are usually ellipsoidal in outline while polyhedra may appear angular or almost spherical. Both polyhedra and capsules can be seen in wet mounts of extracts of infected insects by phase-contrast and dark-field microscopy. Capsules appear as small particles in rapid Brownian motion or, in fresh larval extracts, as revolving spherical groups of numerous capsules, first recognized by Paillot (1937) and termed "boules hyalines" (Huger, 1963). Some NPV infections produce polyhedra of distinctive shape, e.g. the cubic polyhedra of *Autographa californica* NPV and the orange segment-shaped inclusions of *Tipula paludosa* NPV.

In unstained preparations, both capsules and polyhedra are seen to best advantage by dark-field microscopy where both "shine" brightly. However, it is possible to confuse these IBs, particularly GV capsules, with a range of small bacteria and granular material from the host, including uric acid crystals (Huger, 1963). A method found particularly useful by C.C. Payne for distinguishing GV capsules from other debris was devised by Etzel and Falcon (1976): a specific antiserum prepared to a granulosis virus is added to a larval extract; the antigen–antibody reaction causes the virus capsules to agglutinate and they can be readily seen as flocculent masses by phase contrast or, preferably, dark-field microscopy.

Many staining methods have been developed for *Baculovirus* IBs. In the authors' experience, the best stain for detecting any large IB in insect smears, is a solution of Buffalo Black (Naphthalene Black or Amido Black) in acetic acid. Sikorowski *et al.* (1971) used 0.1% Buffalo Black in 30% acetic acid, 50% methanol for staining CPV polyhedra. It also stains other IBs. Wigley (1976) used 1.5% Buffalo Black in 40% acetic acid to stain heat-fixed smears of larval extracts containing NPV polyhedra; all proteinaceous inclusions including IBs of baculoviruses, CPVs and entomopoxviruses were stained dark blue by this method. Crystals of *Bacillus thuringiensis* also stain dark blue.

NPV polyhedra can be distinguished from CPV polyhedra by differential staining with Giemsa stain. In a smear of NPV and CPV polyhedra fixed by mild heating and stained with Giemsa solution, NPV remain unstained while CPV stained readily. Provided that the heating during fixation is not too severe, the two types of polyhedra can be differentiated (Smith, 1963), but the method is not always reliable. An alternative method distinguishes EPV, CPV and *Baculovirus* IBs (Wigley, 1976). Air dried smears are heated at 75°C for 60 min and each divided into three zones. Zone A is fixed with ethanol. Zone B is treated with aqueous saturated picric acid. Zone C is also treated with picric acid then stained with hot acidic napthalene black 12B. Finally all three zones are stained with dilute Giemsa. In zone A, EPV spheroids from *Choristoneura fumiferana* stain blue, but other IBs are unstained. In zone B, CPVs and EPV IBs stain blue, while NPVs and GVs are unstained. In zone C, all IBs stain black.

Other diagnostic methods for *Baculovirus* IBs are described by Huger (1963), Weiser and Briggs (1971), Thomas (1974) and Weiser (1977).

(c) Electron Microscopy. A number of distinctive morphological features of baculoviruses are revealed by electron microscopy, using both negative staining and thin section techniques (e.g. Harrap, 1970, 1972). In tissue sections, the fat body of Lepidoptera is usually the most productive tissue, although in Hymenoptera the gut epithelium should be examined. Capsules and polyhedra can be clearly distinguished from other occluded insect viruses at high magnification. NPVs in which the nucleocapsids are enveloped predominantly singly (SNPV) can be distinguished from those with nucleocapsids in groups (MNPV). SNPV virions are ca. 60–70 nm wide by 260–300 nm long. MNPV virions can be much wider, depending on the number of nucleocapsids in the "bundle". This difference is particularly marked in thin sections of infected tissues where the stained electron-dense nucleocapsids stand out clearly in sectioned polyhedra (Summers, 1977).

Oryctes-like viruses can be observed only by electron microscopy, whether by negative staining of faecal extracts (Monsarrat and Veyrunes, 1976), or in sections of infected nuclei (Huger, 1966). In some negatively-stained preparations of *Oryctes* virus a quite distinct "tail-like" structure protrudes from the virion (Monsarrat *et al.*, 1975; Payne *et al.*, 1977a). This is not apparent in thin sections, and has been seen in only one occluded *Baculovirus*, the NPV of *Tipula paludosa* (Bergoin and Guelpa, 1977), and in *Baculovirus*-like particles observed in the parasitoid wasp, *Apanteles melanoscelus* (Krell and Stoltz, 1979).

(d) Biochemical Identification. Comparative biochemical studies of baculoviruses have not yet enabled workers to construct any extensive classification of virus isolates within the group. There is considerable overlap in the properties of GVs, NPVs and *Oryctes*-like viruses, such that the morphologically-different groups cannot be discretely defined biochemically. Even so, studies of viral nucleic acids and proteins have confirmed that many different NPV and GV isolates have quite distinctive properties (Harrap *et al.*, 1977; Kelly, 1977; Cibulsky *et al.*, 1977a,b,c; Summers and Smith, 1978) although in some analyses different virus isolates may appear very similar, if not identical (McCarthy *et al.*, 1978). There is overall similarity in the properties of the IB protein of the NPV and GV groups (Summers and Smith, 1976; Maruniak and Summers, 1978), but even the properties of this protein distinguish several viruses (Croizier and Croizier, 1977; Summers and Smith, 1976). However, it is the features of virions which are most useful for the analysis of virus strains.

Where two very similar viruses have been examined, e.g. *A. californica* NPV and *Trichoplusia ni* NPV (Miller and Dawes, 1978b) or *A. californica* NPV and *Rachiplusia ou* NPV (Smith and Summers, 1978), the techniques that best reveal similarities and minor differences have been REN analysis of viral DNA, and SDS-polyacrylamide gel electrophoresis of virion polypeptides. In contrast, where sucrose gradient profiles of the MNPV virions, SDS-gel electrophoresis and certain DNA studies had shown marked differences between NPVs of different *Spodoptera* spp. (Harrap *et al.*, 1977), DNA homology studies (Kelly, 1977) revealed some genetic similarities between the virus isolates. Similarities were also detected by serology (Harrap *et al.*, 1977). These examples show the importance of using a number of techniques; only thus will it be possible to understand the relationships between the many individual *Baculovirus* isolates.

(e) Serological Detection and Identification. Baculoviruses can be detected with varying degrees of sensitivity and specificity by application of different serological techniques. Specificity is influenced both by the method used as well as the nature of the antiserum and antigen. Assay procedures such as RIA and ELISA, as well as being very sensitive, provide greater specificity than many other serological methods.

Serological analyses of IB proteins of NPVs and GVs led Norton and DiCapua (1975) to propose the existence of a *Baculovirus* group-specific antigen within these proteins, supporting the biochemical similarities observed. However, working on the principle that specificity is a feature of the technique used as well as of the properties of antigen and antiserum, Kalmakoff *et al.* (1977a) distinguished two *Baculovirus* polyhedrins using RIA. Crawford *et al.* (1978) also obtained reasonable specificity between NPV and GV when comparing dissolved IBs of two NPVs and a GV, using RIA methods. The sensitivity and specificity which can be achieved by such RIA techniques and the similar ELISA method (Crook and Payne, 1980) has already been described above, and these methods seem ideal for survey work and the handling of field-collected samples.

Although viruses may differ in the antigenic properties of the IB protein, antigenic differences between virus isolates are clearer using virions or nucleocapsids as antigens. In gel diffusion tests nucleocapsid antigens of three NPVs from different *Spodoptera* spp. gave specific reactions, producing precipitin lines only with the homologous antiserum (Harrap *et al.*, 1977). Virus could also be detected in the presence of much contaminating larval protein (Young *et al.*, 1975; Harrap *et al.*, 1977). The more sensitive ELISA method (Kelly *et al.*, 1978a) distinguished *Heliothis armigera* NPV virions from those of seven other baculoviruses and detected them readily in larval extracts. The fluorescent antibody technique has been used successfully for the detection of *Oryctes* virus in smears of infected tissue (Croizier and Monsarrat, 1974). In summary, serological techniques can offer sensitive diagnostic methods for a specific virus in impure preparations — not yet achieved by any other technique.

B. CYTOPLASMIC POLYHEDROSIS VIRUSES (REOVIRIDAE)

1. Description

CPVs have been found in > 200 insect species, mainly Lepidoptera and Diptera, and morphologically similar virions have been found in a crustacean,

Simocephalus expinosus (Federici and Hazard, 1975). They have properties in common with other animal and plant viruses grouped within the Reoviridae (Joklik, 1974). The virions are icosahedral, 50–65 nm in diameter, with projections or "spikes" at the vertices of the icosahedron (Hosaka and Aizawa, 1964; Payne and Harrap, 1977). These distinct spikes, 20 nm long, appear to be hollow and to originate within the virion core (Miura *et al.*, 1969). Virions may be occluded singly (as in *Anopheles quadrimaculatus* CPV) or more usually in large numbers within a paracrystalline protein lattice to form polyhedra (Payne and Harrap, 1977). Arnott *et al.* (1968) estimated that as many as 10 000 virions were occluded within a single polyhedron of a CPV from *Danaus plexippus*.

The protein lattice of the IB is composed of a single major polypeptide with a mol. wt of 25 000–31 000 daltons (Payne and Rivers, 1976). Virus particles contain three to six polypeptides of which two to three have high mol. wt (> 100 000) (Payne and Rivers, 1976). The genome comprises 10 segments of ds RNA of mol. wt ranging between 0.34–2.59 × 10^6 (Harrap and Payne, 1979) and totalling ca. 15 × 10^6 (Fujii-Kawata *et al.*, 1970). By analogy with other members of the Reoviridae, it may be assumed that each segment represents a single gene. A provisional classification of viruses within the group has been made on the basis of size differences between RNA segments of ca. 40 CPV isolates, involving twelve distinct virus "types" (Payne and Rivers, 1976; Payne *et al.*, 1977b).

Virions contain several enzymes involved in replication. These include a transcriptase, which synthesizes viral messenger RNA (m-RNA) and methylases, nucleotide phosphohydrolase and guanylyl transferase, which modify the 5′-terminal structure of the mRNA (Lewandowski *et al.*, 1969; Storer *et al.*, 1974a,b; Furuichi, 1974, 1978). Ribonuclease activity has also been found in virions (Storer *et al.*, 1974a).

Most CPVs replicate in the cytoplasm of only gut epithelial cells. However, Kawase *et al.* (1973) reported a CPV in *Bombyx mori* with inclusions in the nucleus: this virus strain was indistinguishable by several criteria from a strain producing only cytoplasmic inclusions (Kawase and Yamaguchi, 1974).

2. Features for Identification

(a) Gross Pathology. In larvae infected with a CPV feeding is greatly reduced so the larvae may grow only slowly; the larval stage is prolonged and pupae are small (Bell and Kanavel, 1976). The disease is often chronic rather than lethal. Although the infected gut epithelial cells may lyse and release polyhedra which are voided in the faeces, new epithelial cells may develop from regenerative cells within the midgut tissue (Inoue and Miyagawa, 1978). Late in infection, the IB-loaded cytoplasm gives the intact gut a distinct white or creamy appearance.

(b) Light Microscopy. CPV polyhedra may be observed by techniques similar to those for *Baculovirus* IBs (Section IV, A, 2, a). The IBs in a single infection generally vary more in size than those of NPV. Some CPV strains can be distinguished by polyhedral shape (Aruga *et al.*, 1961) but such features may represent only small genetic differences in the viruses (Payne *et al.*, 1978). CPV polyhedra stain readily with Naphthalene Black, and are distinguished from other IBs by their response to Giemsa stain (Section IV, A, 2, b).

(c) Electron Microscopy. Electron microscope studies readily verify that particular IBs contain CPV-like virions. However, the virions of biochemically distinct CPVs appear morphologically inseparable. Thus, virions of two CPV "types" producing a mixed infection in *Arctia caja* were indistinguishable morphologically (Payne, 1976).

(d) Biochemical Identification. Natural segmentation of the CPV genome provides the best feature for identifying a new virus isolate. RNA can be extracted from virions, polyhedra, or even diseased larvae by a hot phenol-SDS extraction procedure before electrophoresis on polyacrylamide gels (Payne and Tinsley, 1974). The viral ds RNA forms a series of discrete bands on the gel on staining with methylene blue. The profile (preferably obtained in parallel with a standard RNA, such as that from the well-characterized CPV of *B. mori*) can be compared with published examples (Payne and Rivers, 1976; Payne *et al.*, 1977b). It should be possible to recognize major similarities in the sizes of the genome segments of the new isolate and those of the 12 virus "types" accepted to date. Mixed virus infections can also be detected (Payne, 1976). Where RNA profiles appear identical with a previously-described "type", it should not be assumed that the two viruses are, in fact, identical. Base sequence differences may exist within RNA segments of similar size (Payne *et al.*, 1977b, 1978). Further studies of viral proteins by SDS-gel electrophoresis, serological comparisons of viruses and homology studies of viral RNAs should be used to test the validity of the provisional identification.

(e) Serological Detection and Identification. Serological analyses supplement RNA genome "typing" particularly when comparing closely-related viruses. However, during the production of antisera to CPVs, antibodies are produced both to the virus structural proteins and to the virus ds RNA (Payne and Churchill, 1977). The ds RNA antibodies cross-react with ds RNA from quite distinct viruses and thus may produce misleading cross-reactions between viruses whose only common antigen is ds RNA. The problem can be overcome by absorbing out the cross-reacting antibodies (Payne and Churchill, 1977), or by using *intact* virions as antigen (Payne and Kalmakoff, 1974). On taking such precautions, quite specific serological differences can be obtained between distinct virus "types" (defined by RNA analysis) using RIA and ELISA techniques (Payne, 1976; C.C. Payne, R. Rubinstein and N.E. Crook, unpublished observations). CPVs closely-related by RNA segment profile cross-react serologically but can also be distinguished by gel diffusion or ELISA methods (Payne *et al.*, 1978; Payne, Rubinstein and Crook, unpublished observations). In summary, recent serological studies of CPVs closely support and usefully add to the "type" classification based on RNA analysis.

C. ENTOMOPOXVIRUSES (POXVIRIDAE)

1. Description

EPVs occur in Lepidoptera, Orthoptera, Coleoptera and Diptera (Kurstak and Garzon, 1977). They share many morphological and biochemical properties with vertebrate poxviruses. The virions are large, brickshaped or ovoid, with an external coat containing lipid and tubular or globular protein structures, enclosing one or two lateral bodies and a core containing the genome. The viruses have been well-defined morphologically and it is possible to distinguish some isolates on the basis of virion and core shape. Fenner (1976) defines three subgenera. Subgenus A (type species *Melolontha melolontha* EPV) is characterized by ovoid (450 × 250 nm) virions with one lateral body and a unilateral concave core. The virion surface has globular subunits, 22 nm in diameter. Viruses from other Coleoptera are morphologically similar. Subgenus B (type sp. *Amsacta moorei* EPV) is typified by ovoid virions (350 × 250 nm) with a sleeve-shaped lateral body, cylindrical core and globular surface subunits 40 nm in diameter. Morphologically-similar viruses have been found in other Lepidoptera and the orthopteran *Melanoplus sanguinipes*. In subgenus C (type sp. *Chironomus lupidus* EPV), brick-shaped virions (320 × 230 × 110 nm) have two lateral bodies and a biconcave (dumbell-shaped) core. Similar viruses occur in other Diptera.

Virions are occluded within large proteinaceous IBs (spheroids), up to 24 μm in diameter (Kurstak and Garzon, 1977). In subgenera A and B, with few exceptions, smaller fusiform IBs (spindles), devoid of virions, also occur. The spindles themselves may be occluded within the spheroids in some EPVs of Lepidoptera. Virus maturation is completed within the spheroids (Goodwin and Filshie, 1975).

The protein lattice of spheroids is composed of a polypeptide between 90–112 × 10^3 daltons, quite distinct in mol. wt. from the IB proteins of other groups of occluded insect viruses (Langridge and Granados, 1978). EPV virions contain 20–40 structural polypeptides, the number and size varying in different EPVs (Langridge and Granados, 1978).

Virions contain a large linear ds DNA molecule ranging in mean size from 123–200 × 10^6 daltons, with a low G:C content (16–27%). DNAs from lepidopteran and orthopteran EPVs (subgenus B) are smaller (123–136 × 10^6) than DNAs from dipteran or coleopteran EPVs ($\sim$ 200 × 10^6) (Langridge and Roberts, 1977). EPV virions contain at least four enzymes: acidic and neutral deoxyribonucleases, nucleotide phosphohydrolase and DNA–RNA polymerase (Pogo *et al.*, 1971). Virus replication occurs exclusively in the cytoplasm of infected cells, mainly in the fat body (Bergoin and Dales, 1971; Goodwin and Filshie, 1975; Kurstak and Garzon, 1977).

2. Features for Identification

(a) Gross Pathology. At an advanced stage of EPV infection, the fat body often appears white and becomes distended (Goodwin and Roberts, 1975). Haemocytes and hypodermis may also be infected. General torpor and slow development have also been recorded (Henry *et al.*, 1969). Infection may be prolonged, such that larvae may not die until ca. 6 months after the onset of infection (Goodwin and Roberts, 1975).

(b) Light Microscopy. Moore and Milner (1973) state that EPV infections are not always easy to diagnose as the spheroids can be easily confused with fat globules. Also, it is often difficult to see spindles and spheroids in impure preparations. However, when wet tissue smears are stained with concentrated lactophenol cotton blue, spheroids and spindles stain deep blue, distinct from fat globules, uric acid crystals and other types of occluded virus. Lactic—cotton blue staining of air dried smears for 5—10 min in 1 part v/v lactic acid, 3 parts v/v 66% glycerol in distilled H_2O, 0.004 parts w/v, cotton blue distinguishes spindles (dense blue) from spheroids (unaffected or stained only weakly). These techniques were tested with EPVs from both Lepidoptera and Coleoptera (Moore and Milner, 1973). The differential Giemsa stain of Wigley (1976; Section IV, A, 2,b) stained spheroids of the lepidopteran EPV of C. fumiferana blue after alcohol fixation. However, this reaction may not be common to all EPV types. Spindles of Othnonius batesi EPV stained deep blue, but spheroids did not unless pre-treated with acid or alkali (Goodwin and Roberts, 1975). Another useful feature is that spheroids, unlike IBs of most other virus types are insoluble in dilute alkali without a reducing agent (Bergoin et al., 1967). Only the IB of Tipula paludosa NPV reacts similarly (Guelpa et al., 1977).

(c) Electron Microscopy. The morphological differences between the three EPV subgenera, largely in core structure and the number of lateral bodies, can be defined only by electron microscopy of ultra thin sections. Excellent reviews of EPV structure are given by Bergoin and Dales (1971) and Kurstak and Garzon (1977).

(d) Biochemical and Serological Identification. Little is known about the biochemical properties of EPVs and there are no comparative serological studies. Present knowledge of the viral DNAs demonstrates that major differences in DNA size coincide with some of the morphological subgroups. Further studies are needed to determine relationships between EPV isolates.

D. IRIDOVIRUSES (IRIDOVIRIDAE)

1. Description

Insect iridescent viruses (IVs) are large icosahedral cytoplasmic deoxyriboviruses which fall into two size classes, small IVs about 130 nm in diameter and large IVs about 195 nm diameter. Both are similar in structure (Stoltz, 1973). The DNA genome is packaged within a core surrounded by an inner capsid inside an internal lipid membrane and an outer capsid layer (Kelly and Vance, 1973). The viruses contain many structural polypeptides. Strangely, the small IVs contain 20—28 polypeptides (Kelly and Tinsley, 1972; Krell and Lee, 1974; Elliott et al., 1977; Carey et al., 1978; Kelly et al., 1979), whereas the larger IVs contain only 9 (Wagner et al., 1974). The viral DNA is a linear molecule, $114-150 \times 10^6$ daltons for small IVs and $240-288 \times 10^6$ for the large IVs (Bellett and Inman, 1967; Kelly and Avery, 1974; Wagner and Paschke, 1977).

Several enzymes have been detected in IV virions, including DNA—RNA polymerase (Kelly and Tinsley, 1973), nucleotide phosphohydrolase (Kelly and Robertson, 1973; Monnier and Deuvauchelle, 1976) and a protein kinase (Monnier and Devauchelle, 1978; Kelly et al., 1980a).

Virions replicate in the cytoplasm of cells in a wide range of tissues (Lee, 1977) in Lepidoptera, Diptera, Coleoptera, Hemiptera and Hymenoptera.

Tinsley and Kelly (1970) have provisionally classified each isolate as a different virus "type", numbered approximately in the chronological order of isolation. This classification, unlike the "type" groupings proposed for CPVs, is unrelated to any biochemical or morphological property of the virus itself. At the time of writing, 29 IV isolates have been recorded of which 24 are in the smaller particle size class (Kelly et al., 1979).

2. Features for Identification

(a) Gross Pathology. Infection by some IVs is recognizable by a typical iridescence of infected tissues caused by large numbers of virions packed in quasicrystalline array in the cytoplasm of infected cells. Extracts of infected larvae centrifuged at high speed also form brightly-iridescent virus-containing pellets (Carter, 1973). However, other large viruses, including non-occluded baculoviruses, may also produce pellets almost iridescent in appearance.

(b) Microscopic Methods for Virus Detection. The presence of IVs can be confirmed visually only by electron microscopy. The virions are extremely distinctive, being much larger than any other insect virus with icosahedral symmetry. They also tend to pack in arrays within the cytoplasm (Kelly and Robertson, 1973; Lee, 1977). No morphological distinctions have been made between different virus isolates other than the two size classes of virion.

(c) Biochemical Identification. Studies of the viral DNAs have shown that the larger IVs, e.g. "Regular" (RMIV), and "Turquoise" (TMIV) Mosquito Iridescent Viruses (Type 3; Wagner and Paschke, 1977), have a larger genome than the smaller IVs (Kelly and Avery, 1974). Perhaps surprisingly, REN analysis has not yet proved as promising for identification of IV strains as it is in the analysis of baculoviruses. Kelly (unpublished observations) recently used three RENs to analyse the genomes of six IV types, which were serologically different isolates but which yielded similar or even identical DNA fragment profiles. Homology studies of several viral DNAs revealed apparent genetic identity between different isolates (e.g., between types 9 and 18), significant differences in homology (e.g., 26–45% between types 2, 9 and 18) and no homology between DNA from type 6 and other small IVs (Kelly and Avery, 1974).

Analyses of virus structural polypeptides have confirmed that SDS gel electrophoresis can distinguish a number of virus types. None of the comparisons made so far have shown complete identity in the numbers and sizes of polypeptides. The major component (50 000–55 000) may vary in size or the profile of minor polypeptides may be characteristic for each isolate (Kelly and Tinsley, 1972; Elliott et al., 1977; Carey et al., 1978; Kelly et al., 1979).

(d) Serological Detection and Identification. Several serological techniques have been used to detect IVs in larval extracts or infected cells. Carter (1973) used latex agglutination to detect T. paludosa IV. Although sensitive, the method's specificity in relation to other IVs was not tested. Kelly et al. (1978b) used the ELISA technique to detect 10–100 ng virus/ml in extracts containing vast excesses of larval proteins, with no false positive reactions such as seen occasionally in latex agglutination (Carter, 1973).

Of serological methods discussed for IV detection, Kelly et al. (1978b) favoured the ELISA technique for its combination of specificity and sensitivity. Greater specificity was achieved only by immunoneutralization (Carey *et al.*, 1978; Kelly *et al.*, 1979) but this method is impracticable for screening IVs, as bacterial contaminants in larval extracts interfere with culturing the virus in cell culture. Complement fixation and latex agglutination were considered less sensitive, as was the RIA method of Bilimoria *et al.* (1974). The specificity of the ELISA reaction would be particularly useful for studies of the epidemiology and ecology of a particular IV, though, if required, it should be possible to reduce specificity to detect a broader range of IV serotypes.

Recently Kelly *et al.* (1979) summarized the existing knowledge on the serological relationships of the small IVs. Of 24 recorded, 15 have been partially compared. Types 6 (from *Chilo suppressalis*) and 24 (from *Apis cerana*) have no common antigens with each other or any other IV. Types 1, 2, 9, 10, 16, 18, 21, 22, 23, 24, 25 and 28 all share common antigens to some extent and form a broad serogroup, containing at least three groups of closely-related viruses, distinguishable only by the most sensitive serological methods. The only North American isolate examined (Type 29 from *Tenebrio molitor*) is the only isolate to share common antigens with some but not all of the broad serogroup.

E. DENSOVIRUSES (PARVOVIRIDAE)

1. Description

Densoviruses (DNVs) are some of the smallest known insect pathogenic viruses, about 20–22 nm in diameter, with many properties in common with vertebrate parvoviruses. Only two insect isolates have been examined in detail: from *Galleria mellonella* (DNV 1) and *Junonia coenia* (DNV 2). Tijssen *et al.* (1977) reported two virus forms in DNV 1, differing in density and sedimentation rate, a phenomenon not reported by others for this virus or for DNV 2. The isometric virions contain a linear ss DNA (Barwise and Walker, 1970). Complementary DNA strands are packaged in separate virions. When the DNA is extracted the strands reanneal and assume a duplex structure (Barwise and Walker, 1970; Kelly *et al.*, 1977). The viral DNAs contain inverted terminal repetitions of base sequences allowing the DNA to circularize under appropriate conditions (Kelly and Bud, 1978). The ss DNA has a mol. wt. of $1.6–2.2 \times 10^6$ daltons (Barwise and Walker, 1970; Kurstak, 1972; Kelly *et al.*, 1977) and is intimately associated with polyamines in the virions (Kelly and Elliott, 1977).

DNV's 1 and 2 contain four polypeptides although Tijssen *et al.* (1976) suggested that one is a dimer of the smallest protein. Calculated mol. wts of the protein components vary considerably (Tijssen *et al.*, 1976; Longworth, 1978). The viruses replicate in cell nuclei of a wide range of tissues, where they produce characteristic dense, Feulgen-positive nuclear inclusions. Reviews include Kurstak (1972), Kurstak *et al.* (1977) and Longworth (1978).

2. Features for Identification

Virion size and nucleic acid content should distinguish viruses in this group from any small RNA non-occluded virus. The virions also have a higher buoyant density (1.4 g/ml) (Tijssen *et al.*, 1977) than most small RNA viruses (Section

IV,F). Polypeptide components of DNVs 1 and 2 have very similar, but not identical, mobilities in SDS-polyacrylamide gels, although the viruses can be distinguished serologically by gel diffusion tests (Longworth, 1978; Kelly *et al.*, 1980b). Small differences in DNA homology were observed between the two isolates, and REN analysis also indicated a close genetic relationship (Kelly *et al.*, 1977; Kelly, unpublished observations).

The two viruses have different host ranges: DNV 1 has been found in only *G. mellonella*, while DNV 2 infects the original nymphalid host, *J. coenia*, other nymphalids and some noctuids, but not *G. mellonella* (Rivers and Longworth, 1972).

F. SMALL RNA VIRUSES

Many small (35 nm or less) RNA-containing viruses have been isolated from a range of insect and mite species; with continuing improvements in techniques many more should be found. Most have not been classified (Fenner, 1976). Also, although four isolates (Section IV,F,1) are grouped at present in the genus *Enterovirus* (Fenner, 1976), some isolates have properties different from those of vertebrate enteroviruses and re-classification may be necessary (Longworth, 1978). For convenience, we use the grouping of Longworth (1978) to describe properties of some isolates. A number of viruses, known by morphology alone, are omitted here.

All viruses in these groups replicate in the cytoplasm. Gut epithelial cells are often infected and this may lead to vomiting and diarrhoea (Longworth, 1978). Some bee viruses may replicate in the brain and nervous systems, causing paralysis (Bailey, 1976). Others appear to persist as inapparent infections (Bailey, 1976; Morris *et al.*, 1979). Serological detection by gel diffusion is relatively simple as the viruses diffuse readily due to their small size. However, RIA techniques – hence perhaps ELISA – should prove useful in future (Longworth, 1978). Many of the viruses may replicate readily in cell culture, which might be used for the isolation and growth of new virus types (Scotti, 1976; Plus *et al.*, 1978; Longworth, 1978).

1. Enteroviruses (Picornaviridae)

Insect viruses in this group have a diameter of 27–30 nm, a buoyant density of 1.33–1.34 g/ml and a sedimentation coefficient of 153–167 S (Longworth, 1978). *Gonometa* virus, which Fenner (1976) classifies in this group, has been separated into Group "5" below by Longworth (1978). Candidate enteroviruses include *Drosophila* C virus (DCV), cricket paralysis virus (CrPV), Kawino virus, sacbrood and acute bee paralysis. The viruses contain single-stranded RNA with low G:C content, typically 40% or less. Mol. wt estimates for the RNA range from $2.6–3.2 \times 10^6$ (Pudney *et al.*, 1978; Longworth, 1978).

Plus *et al.* (1978) observed three polypeptides in DCV and CrPV. The mol. wts of those of CrPV were significantly higher than those of DCV. Others have detected three major proteins and two minor components in such viruses (Jousset *et al.*, 1977; Pudney *et al.*, 1978). DCV is serologically related to, though distinct from, CrPV. (Plus *et al*., 1978), but unrelated to Kawino virus (Bailey, personal communication, cited in Pudney *et al*., 1978). DCV and CrPV can also be distinguished by a clear-cut host range difference as CrPV but not DCV replicates in the cricket, *Gryllus bimaculatus* (Plus *et al*., 1978).

2. Divided Genome Viruses

Three viruses occur in this group; Nodamura virus, a biochemically-similar virus from *Heteronychus arator*, and Arkansas bee virus (Newman and Brown, 1973; Longworth and Carey, 1976; J.F. Newman and L. Bailey, unpublished observations). They have a diameter of 29—30 nm, a density of 1.33—1.37 g/ml and a sedimentation coefficient of 128—137 S (Newman and Brown, 1973; Bailey, 1976; Longworth and Carey, 1976) and are distinct from any other insect virus in containing two types of single-stranded RNA within a single virion (Newman and Brown, 1978). Nodamura virus is the only insect virus known to infect vertebrates. It kills suckling mice, also bees (Scherer *et al*., 1968; Bailey and Scott, 1973) but not mosquitoes or *G. mellonella,* though it replicates in all these species. The virus also replicates in BHK and mouse L cells in culture. *H. arator* virus does not replicate in these, nor in suckling mice (Longworth, 1978). A major polypeptide of ca. 40 000 daltons occurs in all three isolates, and two minor ones have been described for Nodamura and *H. arator* viruses (Newman and Brown, 1978; Longworth and Carey, 1976; Bailey, 1976). No serological relationships have been detected between the different isolates.

3. Nudaurelia β Virus Group

This group contains a number of serologically-related viruses, with a diameter of 32—35 nm, a buoyant density of 1.275—1.298 S and a sedimentation coefficient of 194—215 S (Reinganum *et al*., 1978; Longworth, 1978; Morris *et al*., 1979). A single capsid polypeptide of $61-67 \times 10^3$ daltons has been found (Struthers and Hendry, 1974; Morris *et al*., 1979; Reinganum *et al*., 1978). Comparative studies of this polypeptide in different isolates have revealed minor differences (Reinganum *et al*., 1978). Theoretical mol. wts of $1.6-1.9 \times 10^6$ have been calculated for the RNA genome and 1.95×10^6 obtained by electrophoretic analysis of one isolate (Reinganum *et al*., 1978; Morris *et al*., 1979).

4. Kelpfly Virus Group

Kelpfly virus is morphologically distinct from other small RNA insect viruses. Virions are 29 nm in diameter with distinct, hollow, surface projections, a high buoyant density (1.425 g/ml) and a sedimentation coefficient of 158 S. The RNA is single stranded, with a mol. wt. of 3.5×10^6. Two structural proteins have been found (Scotti *et al*., 1976).

5. Group "5" Viruses

In this group, Longworth (1978) includes *Gonometa* virus, bee slow paralysis, Kashmir bee virus, *Drosophila* P, *Drosophila* A and Flacherie 1 viruses. Virions are 25–32 nm in diameter, 172–178 S, with a density range of 1.35–1.375 g/ml and typically contain three polypeptides, although the mol. wts. of these vary considerably between different isolates. Some serological relationships between *Drosophila* A, bee slow paralysis and *Gonometa* viruses have been reported (Longworth, 1978).

6. *"Miniviruses"*

This group was formed to include viruses (probably "satellite" viruses) of 13–17 nm diamater and 42–45 S, i.e. *Antheraea* satellite virus, bee chronic paralysis associate virus and crystalline array virus. Little else is known of their properties. However, a virus of similar size (17 nm) has recently been isolated from *T. ni* with a single structural polypeptide (23 000) and an RNA molecule of ca. 3×10^5 (Morris, personal communication).

7. Ovoid Viruses

Only two small RNA viruses, chronic bee paralysis and *Drosophila* RS virus, have typically ellipsoidal or ovoid particles. Chronic bee paralysis virus has 3–4 components, 22 nm wide but differing in length (30–65 nm). It is serologically unrelated to any known bee virus and contains a single polypeptide (Bailey and Woods, 1977). In contrast, RS virus is regular in size and may contain two polypeptides (Plus *et al.*, 1975; Plus unpublished observations, cited in Longworth, 1978).

V. Conclusions and Future Work

Progress has been made in the development of diagnostic techniques for insect viruses, including new differential staining methods to detect virus inclusion bodies and sensitive and specific serological techniques such as RIA and ELISA. The last decade has seen considerable advances in identification and classification of these viruses. Whereas identification was possible only by morphological criteria, recent studies of viral antigens, structural proteins and nucleic acids have initiated an understanding of the relationships between different virus isolates in some of the major groups. This is particularly true of the *Iridovirus* and cytoplasmic polyhedrosis virus groups. However, there is no room for complacency as much work remains to be done. Relationships between many isolates of the largest virus group – the baculoviruses – are still unclear, although analyses of structural proteins and genome studies by REN analyses look promising. A concerted effort is also needed to examine the host range of small RNA viruses and their relationships with other virus isolates from insects, vertebrates and plants.

References

Aizawa, K. (1963). *In* "Insect Pathology. An Advanced Treatise" (E.A. Steinhaus, ed.), Vol. 1, pp. 381–412. Academic Press, New York and London.

Arnott, H.J., Smith, K.M. and Fullilove, S.L. (1968). *J. Ultrastruct. Res.* **24**, 479–507.

Aruga, H., Hukuhara, T., Yoshitake, N. and Ayudhya, I. (1961). *J. Insect Path.* **3**, 81–92.

Bailey, L. (1976). *Adv. Virus Res.* **20**, 271–304.

Bailey, L. and Scott, H.A. (1973). *Nature, Lond.* **241**, 545.

Bailey, L. and Woods, R.D. (1977). *In* "The Atlas of Insect and Plant Viruses" (K. Maramorosch, ed.), pp. 141–156. Academic Press, New York and London.

Barwise, A.H. and Walker, I.O. (1970). *FEBS Lett.* **6**, 13–15.

Bell, M.R. and Kanavel, R.F. (1976). *J. Invertebr. Path.* **28**, 121–126.

Bellett, A.J.D. and Inman, R.B. (1967). *J. molec. Biol.* **25**, 425–432.

Bergoin, M. and Dales, S. (1971). *In* "Comparative Virology" (K. Maramorosch and E. Kurstak, eds), pp. 169–205. Academic Press, New York and London.

Bergoin, M. and Guelpa, B. (1977). *Arch. Virol.* **53**, 243–254.

Bergoin, M., Scalla, R., Duthoit, J.L. and Vago, C. (1967). *In* "Insect Pathology and Microbial Control" (P.A. van der Laan, ed.), pp. 63–68. North-Holland Publ. Co. Amsterdam.

Bilimoria, S.L., Parkinson, A.J. and Kalmakoff, J. (1974). *Appl. Microbiol.* **28**, 133–137.

Brown, D.A., Bud, H.M. and Kelly, D.C. (1977). *Virology* **81**, 317–327.

Bud, H.M. and Kelly, D.C. (1977). *J. gen. Virol.* **37**, 135–144.

Burgerjon, A. (1977). *Entomophaga* **22**, 187–192.

Burgerjon, A., Biache, G. and Chaufaux, J. (1975). *Entomophaga* **20**, 153–160.

Burges, H.D. and Hussey, N.W. (1971). "Microbial Control of Insects and Mites". Academic Press, London and New York.

Burgess, S. (1977). *J. gen. Virol.* **37**, 501–510.

Carey, G.P., Lescott, T., Robertson, J.S., Spencer, L.K. and Kelly, D.C. (1978). *Virology* **85**, 307–309.

Carter, J.B. (1973). *J. gen. Virol.* **21**, 181–185.

Cibulsky, R.J., Harper, J.D. and Gudauskas, R.T. (1977a) *J. Invertebr. Path.* **29**, 182–191.

Cibulsky, R.J., Harper, J.D. and Gudauskas, R.T. (1977b). *J. Invertebr. Path.* **30**, 303–313.

Cibulsky, R.J., Harper, J.D. and Gudauskas, R.T. (1977c). *J. Invertebr. Path.* **30**, 314–317.

Crawford, A.M. and Kalmakoff, J. (1977). *J. Virol.* **24**, 412–415.

Crawford, A.M., Faulkner, P. and Kalmakoff, J. (1978). *Appl. Environ. Microbiol.* **36**, 18–24.

Croizier, G. and Croizier, L. (1977). *Arch. Virol.* **55**, 247–250.

Croizier, G. and Monsarrat, P. (1974). *Entomophaga* **19**, 115–116.

Crook, N.E. and Payne, C.C. (1980). *J. gen. Virol.* **46**, 29–37.

Daugharty, H. and Ziegler, D.W. (1977). *In* "Comparative Diagnosis of Viral Diseases" (E. Kurstak and C. Kurstak, Eds), Vol. 2, pp. 459–487. Academic Press, New York and London.

Elliott, R.M., Lescott, T. and Kelly, D.C. (1977). *Virology* **81**, 309–316.

Eppstein, D.A. and Thoma, J.A. (1975). *Biochem. biophys. Res. Commun.* **62**, 478–484.

Etzel, L.K. and Falcon, L.A. (1976). *J. Invertebr. Path.* **27**, 13–26.

Federici, B.A. and Hazard, E.I. (1975). *Nature, Lond.* **254**, 327–328.

Fenner, F. (1976). *Intervirology* **7**, 1–115.

Frank-Kamenestkii, M.D. (1971). *Biopolymers* **10**, 2623–2624.

Fujii-Kawata, I., Miura, K. and Fuke, M. (1970). *J. molec. Biol.* **51**, 247–253.

Furuichi, Y. (1974). *Nucleic Acids Res.* **1**, 809–822.

Furuichi, Y. (1978). *Proc. natn. Acad. Sci. USA* **75**, 1086–1090.

Goodwin, R.H. and Filshie, B.K. (1975). *J. Invertebr. Path.* **25**, 35–46.

Goodwin, R.H. and Roberts, R.J. (1975). *J. Invertebr. Path.* **25**, 47–57.

Guelpa, B., Bergoin, M. and Croizier, G. (1977). *C.r. hebd. Séanc. Acad. Sci. Paris* **284**, 779–782.

Harrap, K.A. (1970). *Virology* **42**, 311–318.

Harrap, K.A. (1972). *Virology* **50**, 114–123.

Harrap, K.A. and Longworth, J.F. (1974). *J. Invertebr. Path.* **24**, 55–62.

Harrap, K.A. and Payne, C.C. (1979). *Adv. Virus Res.* **25**, 273–355.

Harrap, K.A., Payne, C.C. and Robertson, J.S. (1977). *Virology* **79**, 14–31.

Hayashi, Y. and Bird, F.T. (1970). *Can. J. Microbiol.* **16**, 695–701.

Heine, C.W., Kelly, D.C. and Avery, R.J. (1980). *J. gen. Virol.* **49**, 385–395.

Henry, J.E., Nelson, B.P. and Jutila, J.W. (1969). *J. Virology* **3**, 605–610.

Hink, W.F. and Vail, P.V. (1973). *J. Invertebr. Path.* **22**, 168–174.

Hosaka, Y. and Aizawa, K. (1964). *J. Insect Path.* **6**, 53–77.

Huger, A. (1963). *In* "Insect Pathology. An Advanced Treatise" (E.A. Steinhaus, ed.), Vol. 1, pp. 531–575. Academic Press, New York and London.

Huger, A. (1966). *J. Invertebr. Path.* **8**, 35–51.

Huger, A.M. (1973). *Z. Angew, Ent.* **72**, 309–319.

Inoue, H. and Miyagawa, M. (1978). *J. Invertebr. Path.* **32**, 373–380.

Joklik, W.K. (1974). *In* "Comprehensive Virology" (H. Fraenkel-Conrat and R.R. Wagner, eds), Vol. 2, pp. 231–334. Plenum Press, New York.

Jousset, F-X., Bergoin, M. and Revet, B. (1977). *J. gen. Virol.* **34**, 269–285.

Jurkovičova, M., Van Touw, J.H., Sussenbach, J.S. and Ter Schegget, J. (1979). *Virology* **93**, 8–19.

Kallenbach, N.R. (1968). *J. molec. Biol.* **37**, 445–466.

Kalmakoff, J., Lewandowski, L.J. and Black, D.R. (1969). *J. Virology* **4**, 851–856.

Kalmakoff, J., Crawford, A.M. and Moore, S.G. (1977a). *J. Invertebr. Path.* **29**, 31–35.

Kalmakoff, J., Parkinson, A.J. Crawford, A.M. and Williams, B.R.G. (1977b). *J. Immunol. Meth.* **14**, 73–84.

Kawase, S. and Yamaguchi, K. (1974). *J. Invertebr. Path.* **24**, 106–111.

Kawase, S., Kawamoto, F. and Yamaguchi, K. (1973). *J. Invertebr. Path.* **22**, 266–272.

Kelly, D.C. (1976). *Virology* **69**, 596–606.

Kelly, D.C. (1977). *Virology* **76**, 468–471.

Kelly, D.C. and Avery, R.J. (1974). *Virology* **57**, 425–435.
Kelly, D.C., and Bud, H.M. (1978). *J. gen. Virol.* **40**, 33–43.
Kelly, D.C. and Elliott, R.M. (1977). *J. Virol.* **21**, 408–410.
Kelly, D.C. and Robertson, J.S. (1973). *J. gen. Virol.* (suppl.) **20**, 17–41.
Kelly, D.C. and Tinsley, T.W. (1972). *J. Invertebr. Path.* **19**, 273–275.
Kelly, D.C. and Tinsley, T.W. (1973). *J. Invertebr. Path.* **22**, 199–202.
Kelly, D.C. and Vance, D.E. (1973). *J. gen. Virol.* **21**, 417–423.
Kelly, D.C., Barwise, A.H. and Walker, I.O. (1977). *J. Virol.* **21**, 396–407.
Kelly, D.C., Edwards, M-L., Evans, H.F. and Robertson, J.S. (1978a). *J. gen. Virol.* **40**, 465–469.
Kelly, D.C., Edwards, M-L. and Robertson, J.S. (1978b). *Ann. appl. Biol.* **90**, 369–374.
Kelly, D.C., Ayres, M.D., Lescott, T., Robertson, J.S. and Happ, G.M. (1979). *J. gen. Virol.* **42**, 95–105.
Kelly, D.C., Elliott, R.M. and Blair, G.E. (1980a). *J. gen. Virol.* **48**, 205–211.
Kelly, D.C., Moore, N.F., Spilling, C.R., Barwise, A.H. and Walker, I.O. (1980b). *J. Virol.* **36**, 224–235.
Knudson, D.L. and Tinsley, T.W. (1978). *Virology* **87**, 42–57.
Krell, P. and Lee, P.E. (1974). *Virology* **60**, 315–326.
Krell, P.J. and Stoltz, D.B. (1979). *J. Virol.* **29**, 1118–1130.
Kurstak, E. (1972). *Adv. Virus Res.* **17**, 207–241.
Kurstak, E. and Garzon, S. (1977). *In* "The Atlas of Insect and Plant Viruses" (K. Maramorosch, ed.), pp. 29–66. Academic Press, New York and London.
Kurstak, E., Tijssen, P. and Garzon, S. (1977). *In* "The Atlas of Insect and Plant Viruses" (K. Maramorosch, ed.) pp. 67–91. Academic Press, New York and London.
Laemmli, U.K. (1970). *Nature, Lond.* **227**, 680–685.
Langridge, W.H.R. and Granados, R.R. (1978). Abstr. 4th Int. Congr. Virol., The Hague, p. 548.
Langridge, W.H.R. and Roberts, D.W. (1977). *J. Virol.* **21**, 301–308.
Lee, H.H. and Miller, L.K. (1978). *J. Virol.* **27**, 754–767.
Lee, P.E. (1977). *In* "The Atlas of Insect and Plant Viruses" (K. Maramorosch, ed.), pp. 93–103. Academic Press, New York and London.
Lewandowski, L.J., Kalmakoff, J. and Tanada, Y. (1969). *J. Virol.* **4**, 857–865.
Longworth, J.F. (1978). *Adv. Virus Res.* **23**, 103–157.
Longworth, J.F. and Carey, G.P. (1976). *J. gen. Virol.* **33**, 31–40.
Longworth, J.F., Payne, C.C. and MacLeod, R. (1973). *J. gen. Virol.* **18**, 119–125.
Longworth, J.F., Tinsley, T.W., Barwise, A.H. and Walker, I.O. (1968). *J. gen. Virol.* **3**, 167–174.
McCarthy, W.J. and Liu, S.Y. (1976). *J. Invertebr. Path.* **28**, 57–65.
McCarthy, W.J., Granados, R.R. and Roberts, D.W. (1974). *Virology* **59**, 59–69.
McCarthy, W.J., Mercer, W.E. and Murphy, T.F. (1978). *Virology* **90**, 374–378.
Maizel, J.V. (1966). *Science, N.Y.* **151**, 988–990.

Mandel, M. and Marmur, J. (1968). *In* "Methods in Enzymology". (L. Grossman and K. Moldave, eds), Vol. 12B, pp. 195–206. Academic Press, New York and London.

Maramorosch, K. (1977). "The Atlas of Insect and Plant Viruses". Academic Press, New York and London.

Marmur, J., Schildkraut, C.L. and Rownd, R. (1963). *In* "Progress in Nucleic Acid Research" (J.N. Davidson and W.E. Cohn, eds.) Vol. 1, pp. 231–300. Academic Press, New York and London.

Martignoni, M.E. and Iwai, P.J. (1977). "A Catalog of Viral Diseases of Insects and Mites". USDA Forest Service Gen. Techn. Rep. PNW-40.

Martignoni, M.E., Breillatt, J.P. and Anderson, N.G. (1968). *J. Invertebr. Path.* **11**, 507–510.

Maruniak, J.E. and Summers, M.D. (1978). *J. Invertebr. Path.* **32**, 196–201.

Mazzone, H.M. and Tignor, G.H. (1976). *Adv. Virus Res.* **20**, 237–270.

Miller, L.K. and Dawes, K.P. (1978a). *Appl. Environ. Microbiol.* **35**, 411–421.

Miller, L.K. and Dawes, K.P. (1978b). *Appl. Environ. Microbiol.* **35**, 1206–1210.

Miller, L.K. and Dawes, K.P. (1979). *J. Virol.* **29**, 1044–1045.

Miura, K., Fujii-Kawata, I., Iwata, H. and Kawase, S. (1969). *J. Invertebr. Path.* **14**, 262–265.

Miura, K., Fujii, I., Sakaki, T., Fuke, M. and Kawase, S. (1968). *J. Virol.* **2**, 1211–1222.

Monnier, C. and Devauchelle, G. (1976). *J. Virol.* **19**, *180–186.*

Monnier, C. and Devauchelle, G. (1978). Abstr. 4th Int. Congr. Virol., The Hague, p. 542.

Monsarrat, P. and Veyrunes, J-C. (1976). *J. Invertebr. Path.* **27**, 387–389.

Monsarrat, P., Veyrunes, J-C., Meynadier, G., Croizier, G. and Vago, C. (1973). *C.r. hebd. Séanc. Acad. Sci. Paris* **277**, 1413–1415.

Monsarrat, P., Revet, B. and Gourevitch, I. (1975). *C.r. hebd. Séanc. Acad. Sci. Paris* **281**, 1439–1442.

Moore, S. and Milner, R.J. (1973). *J. Invertebr. Path.* **22**, 467–470.

Morris, T.J., Hess, R.T. and Pinnock, D.E. (1979). *Intervirology* **11**, 238–247.

Nathans, D. and Smith, H.O. (1975). *A. Rev. Biochem.* **44**, 273–293.

Newman, J.F.E. and Brown, F. (1973). *J. gen. Virol.* **21**, 371–384.

Newman, J.F.E. and Brown, F. (1978). *J. gen. Virol.* **38**, 83–95.

Norton, P.W. and DiCapua, R.A. (1975). *J. Invertebr. Path.* **25**, 185–188.

Norton, P.W. and DiCapua, R.A. (1976). Proc. 1st Int. Colloq. Invertebr. Path. Kingston, Ontario, pp. 325–326.

Ohba, M., Summers, M.D., Hoops, P. and Smith, G.E. (1977). *J. Invertebr. Path.* **30**, 362–368.

Paillot, A. (1937). *C.r. hebd. Séanc. Acad. Sci. Paris* **205**, 1264–1266.

Payne, C.C. (1974). *J. gen. Virol.* **25**, 105–116.

Payne, C.C. (1976). *J. gen. Virol.* **30**, 357–369.

Payne, C.C. and Churchill, M.P. (1977). *Virology* **79**, 251–258.

Payne, C.C. and Harrap, K.A. (1977). *In* "The Atlas of Insect and Plant Viruses" (K. Maramorosch, ed.), pp. 105–129. Academic Press, New York and London.

Payne, C.C. and Kalmakoff, J. (1974). *Intervirology* 4, 365–368.

Payne, C.C. and Kalmakoff, J. (1978). *J. Virol.* 26, 84–92.

Payne, C.C. and Rivers, C.F. (1976). *J. gen. Virol.* 33, 71–85.

Payne, C.C. and Tinsley, T.W. (1974). *J. gen. Virol.* 25, 291–302.

Payne, C.C., Compson, D. and de Looze, S.M. (1977a) *Virology* 77, 269–280.

Payne, C.C., Piasecka-Serafin, M. and Pilley, B. (1977b). *Intervirology* 8, 155–163.

Payne, C.C., Mertens, P.P.C. and Katagiri, K. (1978). *J. Invertebr. Path.* 32, 310–318.

Plus, N., Croizier, G., Duthoit, J-L., Anzolabehere, D. and Periquet, G. (1975). *C.r. hebd. Séanc. Acad. Sci. Paris* 280, 1051–1054.

Plus, N., Croizier, G., Veyrunes, J-C. and David, J. (1976). *Intervirology* 7, 346–350.

Plus, N., Croizier, G., Reinganum, C. and Scotti, P.D. (1978). *J. Invertebr. Path.* 31, 296–302.

Pogo, B.G.T., Dales, S., Bergoin, M. and Roberts, D.W. (1971). *Virology* 43, 306–309.

Poinar, G.O. Jr. and Thomas, G.M. (1978). "Diagnostic Manual for the Identification of Insect Pathogens". Plenum Publishing Corp., New York.

Printz, D. (1973). *Adv. Virus. Res.* 18, 143–157.

Pudney, M., Newman, J.F.E. and Brown, F. (1978). *J. gen. Virol.* 40, 433–441.

Reinganum, C., Robertson, J.S. and Tinsley, T.W. (1978). *J. gen. Virol.* 40, 195–202.

Revet, B. and Monsarrat, P. (1974). *C.r. hebd. Séanc. Acad. Sci. Paris* 278, 331–334.

Rivers, C.F. and Longworth, J.F. (1972). *J. Invertebr. Path.* 20, 369–370.

Rohrmann, G.F., McParland, R.H., Martignoni, M.E. and Beaudreau, G.S. (1978). *Virology* 84, 213–217.

Rubinstein, R., Stannard, L. and Polson, A. (1975). *Prep. Biochem.* 5, 79–90.

Scherer, W.F., Verna, J.E. and Richter, G.W. (1968). *Am. J. trop. Med. Hyg.* 17, 120–128.

Schildkraut, C.L., Marmur, J. and Doty, P. (1962) *J. molec. Biol.* 4, 430–443.

Scotti, P.D. (1976). *Intervirology* 6, 333–342.

Scotti, P.D., Gibbs, A.J. and Wrigley, N.C. (1976). *J. gen. Virol.* 30, 1–9.

Shatkin, A.J. (1969). *In* "Fundamental Techniques in Virology" (K. Habel and N.P. Salzman, eds), pp. 231–237. Academic Press, New York and London.

Sikorowski, P.P., Broome, J.R. and Andrews, G.L. (1971). *J. Invertebr. Path.* 17, 451–452.

Smith, G.E. and Summers, M.D. (1978). *Virology* 89, 517–527.

Smith, K.M. (1963). *In* "Insect Pathology. An Advanced Treatise" (E.A. Steinhaus, ed.), Vol. 1, pp. 457–497. Academic Press, New York and London.

Steinhaus, E.A. (1963). "Insect Pathology. An advanced Treatise". Vols. 1 and 2. Academic Press, New York and London.

Stoltz, D.B. (1973). *J. Ultrastruct. Res.* 43, 58–74.

Storer, G.B., Shepherd, M.G. and Kalmakoff, J. (1974a). *Intervirology* 2, 87–94.

Storer, G.B., Shepherd, M.G. and Kalmakoff, J. (1974b). *Intervirology* **2**, 193–199.
Struthers, J.K. and Hendry, D.A. (1974). *J. gen. Virol.* **22**, 355–362.
Summers, M.D. (1977). *In* "The Atlas of Insect and Plant Viruses" (K. Maramorosch, ed.), pp. 3–27. Academic Press, New York and London.
Summers, M.D. and Anderson, D.L. (1973). *J. Virol.* **12**, 1336–1346.
Summers, M.D. and Egawa, K. (1973). *J. Virol.* **12**, 1092–1103.
Summers, M.D. and Smith, G.E. (1975). *J. Virol.* **16**, 1108–1116.
Summers, M.D. and Smith, G.E. (1976). *Intervirology* **6**, 168–180.
Summers, M.D. and Smith, G.E. (1978). *Virology* **84**, 390–402.
Sylvester, E.S. (1977). *In* "The Atlas of Insect and Plant Viruses" (K. Maramorosch, ed.), pp. 131–139. Academic Press, New York and London.
Teninges, D., Ohanessian, A., Richard-Molard, C. and Contamine, D. (1979). *J. gen. Virol.* **42**, 241–254.
Thomas, G.M. (1974). *In* "Insect Diseases" (G.E. Cantwell, ed.), Vol. 1, pp. 1–48, Marcel Dekker, New York.
Tijssen, P. and Kurstak, E. (1977). *In* "Comparative Diagnosis of Viral Diseases" (E. Kurstak and C. Kurstak, eds), Vol. 2, pp. 489–504. Academic Press, New York and London.
Tijssen, P., van der Hurk, J. and Kurstak, E. (1976). *J. Virol.* **17**, 686–691.
Tijssen, P., Tijssen – van der Slikke, T. and Kurstak, E. (1977). *J. Virol.* **21**, 225–231.
Tinsley, T.W. and Kelly, D.C. (1970). *J. Invertebr. Path.* **16**, 470–472.
Tompkins, G.J., Adams, J.R. and Heimpel, A.M. (1969). *J. Invertebr. Path.* **14**, 343–357.
Tweeten, K.A., Bulla, L.A., Jr. and Consigli, R.A. (1977). *Appl. Environ. Microbiol.* **34**, 320–327.
Tweeten, K.A., Bulla, L.A., Jr. and Consigli, R.A. (1978). *J. Virol.* **26**, 702–711.
Voller, A. and Bidwell, D.E. (1977). *In* "Comparative Diagnosis of Viral Diseases" (E. Kurstak and C. Kurstak, eds), Vol. 2, pp. 449–458. Academic Press, New York and London.
Volkman, L.E., Summers, M.D. and Hsieh, C-H. (1976). *J. Virol.* **19**, 820–832.
Wagner, G.W. and Paschke, J.D. (1977). *Virology* **81**, 298–308.
Wagner, G.W., Paschke, J.D., Campbell, W.R. and Webb, S.R. (1973). *Virology* **52**, 72–80.
Wagner, G.W., Paschke, J.D., Campbell, W.R. and Webb, S.R. (1974). *Intervirology* **3**, 97–105.
Weiser, J. (1977). "An Atlas of Insect Diseases". Academia, Praha, C.S.S.R.
Weiser, J. and Briggs, J.D. (1971). *In* "Microbial Control of Insects and Mites" (H.D. Burges and N.W. Hussey, eds), pp. 13–66. Academic Press, London and New York.
Wigley, P.J. (1976). Ph.D. Thesis, Univ. Oxford, 185 pp.
Wyatt, G.R. (1953). *J. gen. Physiol.* **36**, 201–205.
Yamamoto, T. and Tanada, Y. (1978). *J. Invertebr. Path.* **32**, 202–211.

Yamamoto, T. and Tanada, Y. (1979). *Virology* **94**, 71–81.
Young, S.Y., Yearian, W.C. and Scott, H.A. (1975). *J. Invertebr. Path.* **26**, 309–312.
Zelazny, B. (1978). *Pl. Prot. Bull. F.A.O.* **26**, 163–168.

CHAPTER 6

Identification: Entomopathogenic Deuteromycetes

R.A. SAMSON

Centraalbureau voor Schimmelcultures, Baarn, The Netherlands

I. Introduction

Most entomopathogenic fungi belong to the Deuteromycetes (Fungi Imperfecti). About 30 genera have been reported to contain one or more species that infect insects. There has been much confusion about their taxonomy and misapplied or incorrect names still occur. This chapter presents a key to the common and important entomopathogenic genera, followed by a list of genera containing one or more species on insects. Genera with species of unproven pathogenicity to insects (e.g. *Penicillium, Phoma*), or taxa of doubtful identity, are omitted. A glossary of mycological terms is given in Section VII.

II. Macroscopical and Microscopical Features

The Deuteromycetes show different growth patterns on the insect body. Species of *Nomuraea* (Fig. 1a), *Metarhizium* and *Beauveria* usually form a

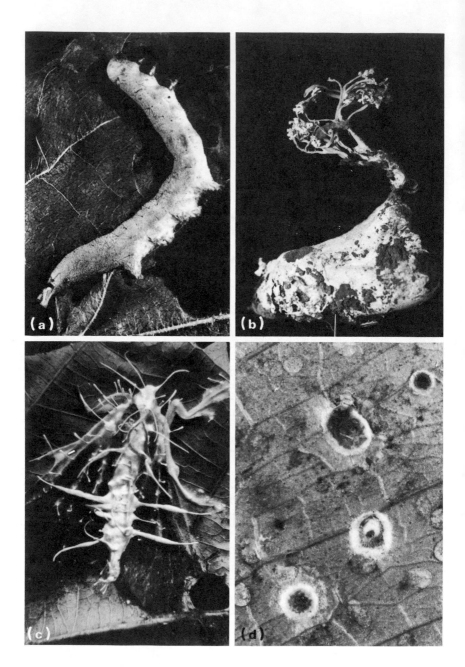

Fig. 1. (a): *Nomuraea rileyi* on *Spodoptera* larva, mycelial mat with coni-diogenous structures. × 3.5. (b): *Paecilomyces tenuipes* on lepidopterous larva, showing loose synnemata with apical conidial blooms. × 1. (c): *Akanthomyces pistillariae-formis* on moth, synnemata arising from different parts of the body. × 5. (d): *Aschersonia aleyrodis* on white flies, sporodochia. × 4.

loose or tough mycelial mat with cushions or areas of conidial structures. Coni-diophores frequently unite into synnemata. In *Paecilomyces* (Fig. 1b) and *Akanthomyces* (Fig. 1c) synnemata are loose, while in *Hymenostilbe* (Fig. 2a), *Hirsutella* and other genera they are tough. The formation of synnemata is probably an ecological necessity on hosts which invariably hide away when infected. Specific light conditions are required and the synnemata usually point towards light. Synnematal production is mostly lost when the fungi grow on artificial media. Species of *Aschersonia* (Fig. 1d) typically form sporodochia and *Synnematium jonesii* produces sclerotia. In insects infected by *Sorosporella* the bodies are filled with characteristic chlamydospores (= resting spores).

Identification requires an examination of conidium ontogeny, the primary character for typifying the different genera in recent taxonomic treatments. Among the entomopathogenic genera two modes of blastic conidiogenesis have to be considered, phialidic and sympodial.

A plurality of conidia are produced in basipetal succession by a phialidic conidiogenous cell (Fig. 3a, b). The shape of the phialide is variable but characteristic for each genus: flask-shaped (e.g. *Paecilomyces*), awl-shaped (e.g. *Verticillium*) or cylindrical (e.g. *Metarhizium*). Conidia can be produced in chains (*Paecilomyces, Nomuraea* and *Metarhizium*), or in slimy heads or droplets (*Verticillium, Fusarium*). Phialides of some *Hirsutella* species form a few conidia held together by a slime sheath, thus appearing to be a single propagule.

Sympodial conidium development is found in *Beauveria* (Fig. 3c), *Sporothrix*, *Pseudogibellula* and *Hymenostilbe*. The conidia are formed solitarily on a

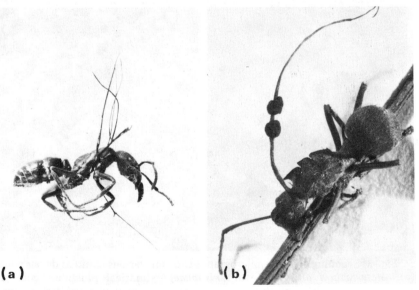

(a)   **(b)**

Fig. 2. (a): *Hymenostilbe longispora* on ant, long tough synnemata. × 2.5. (b): *Cordyceps unilateralis* with *Hirsutella formicarum* on ant, conidial structures at the apical part of the stroma. × 5.

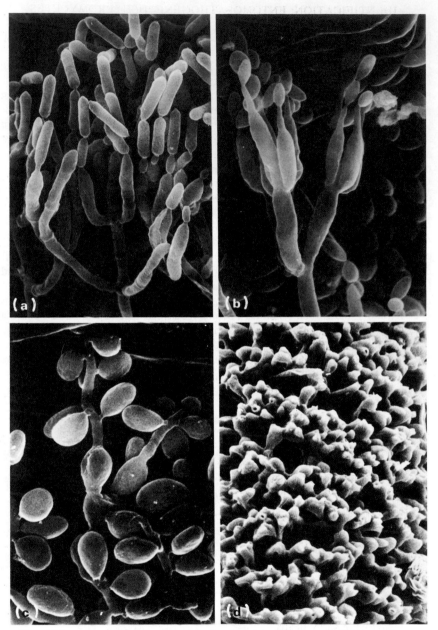

Fig. 3. Scanning electron micrographs showing various conidial developments. (a): *Metarhizium anisopliae* var *anisopliae,* cylindrical phialides, conidia in chains. × 2700. (b): *Paecilomyces farinosus,* flask-shaped phialides, conidia in chains. × 3700. (c): *Beauveria bassiana,* solitary conidia on sympodial conidiogenous cells. × 4300. (d): *Hymenostilbe longispora,* hymenium-like layer of conidiogenous cells with crowded scars. × 1800.

laterally proliferating conidiogenous cell, often showing a geniculate or "zig-zag" type of elongation.

Structure of the conidiophore and arrangement of the conidiogenous cells are often important characters for each genus. Conidiogenous cells occur in compact or divergent whorls (*Paecilomyces*), in a hymenium-like layer along a synnema (Fig. 3d; *Hymenostilbe*), in clusters (*Beauveria*) or on a distinct conidiophore with a vesicle (*Aspergillus*).

III. Perfect States

Many of the entomopathogenic Deuteromycetes are associated with an ascigerous state (*Cordyceps, Torrubiella, Nectria, Hypocrella* and others). The ascigerous and conidial states may occur separately, or on the same stroma (Fig. 2b). It is often difficult to prove a true connection between a perfect and an imperfect fungus unless cultural studies are done. Since some entomopathogenic species do not grow or sporulate on artificial media, their relationships will probably never be clarified.

IV. Microscopical Observations and Cultivation *in vitro*

It is usually possible to identify a fungal pathogen directly from the insect by mounting in lactophenol or lactic acid with some aniline blue. Immature specimens without sporulation should be placed in moist chambers. Old or over-mature specimens, on which no conidial structures can be recognized, must be isolated in pure culture. For this malt or potato-dextrose agars with antibiotics are recommended. Sometimes Sabouraud or mealworm (Samson, 1974) agars can be used for species that grow or sporulate poorly on agar media. Cultures of entomopathogenic species can be maintained on malt or oatmeal agars. *Metarhizium, Beauveria* and *Paecilomyces* sporulate well on sterilized rice. When strains are used for infection or physiological experiments, the cultures should be freeze-dried since regular subculturing reduces physiological and biochemical properties.

V. Key to Common and Important Genera

1a. Conidia produced by phialides in chains or in slimy heads 2

1b. Conidia produced by sympodial conidiogenous cells. 12

2a. Conidia in dry, long chains 3

2b. Conidia in slimy heads 8

3a. Conidiophores consisting of an unbranched stipe terminating in a vesicle
     bearing conidiogenous cells and/or metulae 4

3b. Conidiophores without vesicles 5

4a. Conidiophores united in distinct synnemata; host, spiders (Fig. 4) . . . . . .
. . . .*Gibellula*
4b. Conidiophores arising singly from insect body (Fig. 5) . . . . . . *Aspergillus*
5a. Phialides arranged along synnemata as in a hymenium (Fig. 10) . . . .
. . . . *Akanthomyces*

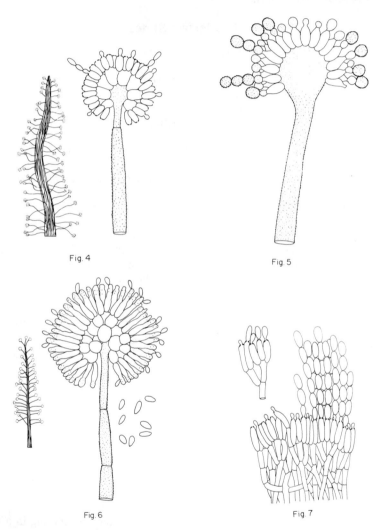

Fig. 4. *Gibellula pulchra:* synnema and conidiophore. Fig. 5. *Aspergillus parasiticus.* Fig. 6. *Pseudogibellula formicarum:* synnema and conidiophore. Fig. 7. *Metarhizium anisopliae* var *anisopliae.*

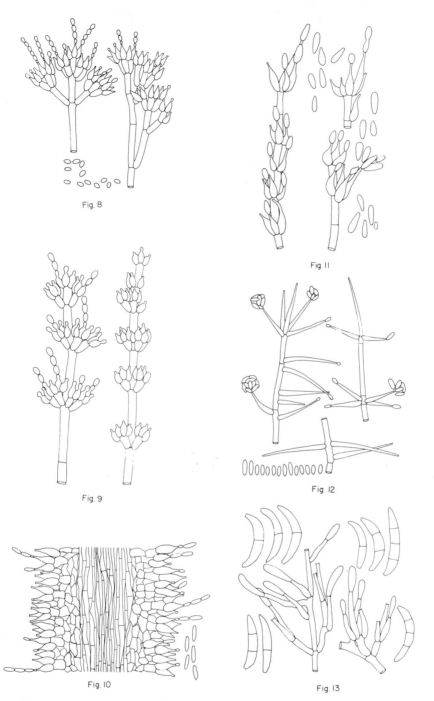

Fig. 8. *Paecilomyces farinosus*. Fig. 9. *Nomuraea rileyi*. Fig. 10. *Akantho-myces pistillariaeformis* part of synnema. Fig. 11. *Culicinomyces clavosporus*. Fig. 12. *Verticillium lecanii*. Fig. 13. *Fusarium larvarum*.

5b. Phialides differently arranged. 6

6a. Conidiophores closely packed in sporodochial structures; conidia in columns (Fig. 7) . . . . Metarhizium

6b. Conidiophores loosely arranged in synnemata or single; conidia in long divergent chains 7

7a. Phialides very short necked; conidiophores bearing dense whorls of branches and phialides (Fig. 9). . . . . .Nomuraea

7b. Phialides with distinct necks; conidiophores are irregular or verticillately branched elements (Fig. 8) . . . . .Paecilomyces

8a. Conidia one- or more-septate, usually curved (Fig. 13) . . . . Fusarium

8b. Conidia non-septate 9

9a. Conidiophores arranged in sporodochia on scale insects or white flies, conidia usually fusiform (Fig. 14). . . . . .Aschersonia

9b. Conidiophores not arranged in sporodochia 10

10a. Phialides mostly single, with a swollen basal part abruptly tapering to a thin, long neck; conidia single or few and covered by a slime sheath (Fig. 15) . . . . Hirsutella

10b. Phialides in whorls on verticillately branched conidiophores 11

11a. Phialides flask-shaped, conidia clavate; on mosquitoes and related Diptera (Fig. 11) . . . . .Culicinomyces

11b. Phialides usually awl-shaped, conidia of various shapes (Fig. 12) . . . . .Verticillium

12a. Conidiogenous cells in a hymenium-like layer along distinct synnemata (Fig. 18) . . . . Hymenostilbe
12b. Conidiogenous cells in cluster or singly 13

13a. Conidiophores with unbranched stipes terminating in a vesicle bearing metulae and conidiogenous cells (Fig. 6) . . . . Pseudogibellula

13b. Conidiophores without swollen vesicles 14

14a. Conidiogenous cells elongate and slender with inconspicuous, terminal or lateral, scars (Fig. 17). . . . . Sporothrix

14b. Conidiogenous cells with a swollen basal part terminating in a "zig-zag" rachis (Fig. 16) . . . . Beauveria

VI. List of Genelra Containing Insect Pathogens

Acremonium Link ex Fr. Type species: *A. alternatum* Link ex S.F. Gray. Gams (1971b) monographed this large genus and placed *Cephalosporium* Corda

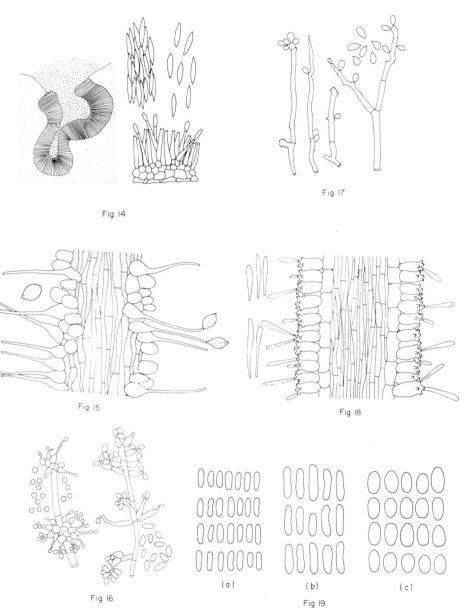

Fig. 14

Fig. 17

Fig. 15

Fig. 18

Fig. 16

Fig. 19

(a)   (b)   (c)

Fig. 14. *Aschersonia aleyrodis·* part of sporodochium and conidiogenous cells. Fig. 15. *Hirsutella citriformis:* part of synnema. Fig. 16. Left *Beauveria bassiana*, right *Beauveria brongniartii*. Fig. 17. *Sporothrix insectorum*. Fig. 18. *Hymenostilbe longispora:* part of synnema. Fig. 19. Conidia of *Metarhizium* spp. (a). *M. anisopliae* var *anisopliae*. (b). *M. anisopliae* var *major*. (c). *M. flavoviride*.

in synonymy with it. The genus contains two species on insects, *A. larvarum* (Petch) W. Gams and *A. zeylanicum* (Petch) Gams and Evans. Other *Cephalosporium* species described on insects are classified under *Verticillium* Nees ex Link.

Acrodontium de Hoog. Type species: *A crateriforme* (van Beyma) de Hoog (1972). *A. crateriforme* on spiders and aphids.

Aegerita Pers. ex Fr. Type species: *A. candida* Pers. Fawcett (1910) reported *A. webberi* as parasitic on *Aleyrodes citri;* see also Petch (1926).

Akanthomyces Lebert. Type species: *A. aculeatus* Lebert. Ca. 10 known species; taxonomy, Mains (1950b) and Samson and Evans (1974). The genus *Insecticola* Mains was erected by Mains (1950b) to separate those species with clavate synnemata and sterile stipes. This generic character is of minor significance and *Insecticola* should be regarded as a synonym of *Akanthomyces.*

Antennopsis Heim. Type species: *A. gallica* Heim and Buchli. Three species (Heim, 1951; Buchli, 1960, 1966; Gouger and Kimbrough, 1969).

Aschersonia Mont. Type species: *A. taitensis* Mont. This genus occurs on scale insects and white flies and contains many species. Taxonomy, Petch (1921) and Mains (1959a, b).

Aspergillus Mich. ex Fr. Type species: *A. flavus* Link ex Fr. *A. flavus, A. parasiticus, A. ochraceus* and some other *Aspergillus* spp. are entomopathogenic (Austwick, 1965). Taxonomy, Raper and Fennell (1965).

Beauveria Vuill. Type species: *B. bassiana* (Bals.) Vuill. Two species on insects: see de Hoog (1972)

Key to the species:

1a. Conidia globose to subglobose ($2-3 \times 2.0-2.5\,\mu m$; conidiogenous structures forming dense clusters . . . . *B. bassiana*

1b. Conidia ellipsoid ($2-3 \times 1.5-2.5\,\mu m$; conidiogenous structures rather slender and rarely clustered . . . . *B. brongniartii* (= *B. tenella*)

Culicinomyces Couch *et al.* Type species: *C. clavosporus* Couch *et al.* Monotypic genus on mosquitoes and related Diptera (Couch *et al.,* 1974; Sweeney, 1975).

Desmidiospora Thaxter. Type species: *D. myrmecophila* Thaxter (1891). Monotypic genus not reported since its original discovery.

Fusarium Link ex Fr. Type species: *F. roseum* Link ex Fr. Some species known on insects but usually wrongly classified as *Atractium, Microcera* or *Sphaerostilbe.* Booth (1971) placed *F. larvarum* Fuckel and *F. coccophilum* (Desm.) Wollenw. and Reink. in a separate section *Coccophilum. Fusarium solani* (Mart.) Sacc. is a weak pathogen of beetles and other invertebrates.

Gibellula Cavara. Type species: *G. pulchra* (Sacc.) Cavara. Four species, all spiders (Mains, 1950a; Samson and Evans, 1973, 1977).

Hirsutella Pat. Type species: *H. entomophila* Pat. Many species known. Mains (1951b) gave a monographic treatment of the synnematous species. However, some species non-synnematous on host also exist, e.g. *H. thompsonii* Fisher, a common pathogen of the citrus rust mite (Speare, 1920b; Mains, 1951b; Fisher, 1950).

Hymenostilbe Petch. Type species: *H. muscaria* Petch. Samson and Evans (1975) accepted nine species, mostly subtropical or tropical.

Isaria Fr. Type species: *I. felina* (D.C. per Fr.) Fr. De Hoog (1972) accepted two species, *I. felina* and *I. amorpha* von Hohnel (= *I. orthopterorum* Petch). Both can be easily confused with *Beauveria.* The species with phialides (e.g. *I. farinosa, I. tenuipes*), formerly described in *Isaria,* are now accommodated in *Paecilomyces* Bain.

Mattirolella Colla. Type species: *M. silvestri* Colla. On termites; second species *M. crustosa* Khan and Kimbrough (Colla, 1929; Khan and Kimbrough, 1974).

Metarhizium Sorok. Type species: *M. anisopliae* (Metsch.) Sorok. Two species, *M. anisopliae* and *M. flavoviride. M. anisopliae* was separated by Tulloch (1976) into the var. *anisopliae* and the var. *major.*

var. *major.*

Key to species

1a. Conidia ellipsoid, either both ends rounded or one slightly truncate. Colonies greyish yellow, green or olivaceous buff. . *M. flavoviride* Gams and Roszypal (Fig. 19c).

1b. Conidia cylindrical to oval, often slightly narrower in the middle; usually truncate at both ends (Fig. 19a, b). Colonies many shades of green, sepia or isabelline 2

2a. Conidia 3.5–9.0 μm long, usually 5.0–8.0 μm. . . . . . *M. anisopliae* var *anisopliae*

2b. Conidia 9.0–18.0 μm long, usually 10–14 μm. . . . . . *M. anisopliae* var *major*

Nomuraea Maublanc Type species: *N. rileyi* (Farlow) Samson = *Spicaria prasina* (Maublanc) Sawada. Detailed description, Samson (1974).

Key to the species:

1a. Colonies green; conidia broadly ellipsoid to cylindrical, 3.5–4.5 × 2– 3.1 μm. . . . . *N. rileyi*

1b. Colonies purple; conidia cylindrical, slightly curved, 4–6 × 1.2–1.5 μm . . . . *N. atypicola*

Paecilomyces Bain. Type species: *P. variotii* Bain. Contains 14 entomopathogenic species. Common species, *P. farinosus, P. tenuipes, P. fumosoroseus.* Detailed descriptions and key, Samson (1974).

Polycephalomyces Kobayasi. Type species: *P. formosus* Kobayasi (1941). Three species (Mains, 1948).

Pseudogibellula Samson and Evans. Type species: *P. formicarum* (Mains) Samson and Evans. Genus monotypic on many arthropod hosts (Samson and Evans, 1973).

Sorosporella Sorok. Type species: *S. uvella* (Krass.) Giard. Speare (1920a) reported *S. uvella* on noctuid larvae in the USA. Petch (1941) considered *Sorosporella* as the resting-spore state of his genus *Syngliocladium.*

Sporothrix Hektoen and Perkins ex Nicot and Mariat. Type species: *S. schenckii* Hektoen and Perkins. De Hoog (1974) described four entomogenous species: *S. isarioides, S. alba, S. insectorum* and *S. ghanensis.*

Stilbella Lindau. Type species: *S. erythrocephala* (Ditm.) Lindau. Most species saprophytic. Four species on insects (Mains, 1948, 1951a; Petch, 1926).

Syngliocladium Petch. Type species: *S. album* Petch. Petch (1932, 1941) accepted three species. *Isaria dubia* Delacroix, a pathogen on *Hepialis* spp. in Great Britain, probably also belongs in this genus.

Synnematium Speare. Type species: *S. jonesii* Speare. One species (Speare, 1920a; Mains, 1951b).

Termitaria Thaxter. Type species: *T. snyderi* Thaxter. A second species is *T. coronata,* both on termites. Khan and Kimbrough (1974) emended the generic diagnosis.

Tilachlidium Preuss. Type species: *T. brachiatum* (Batsch per Fr.) Petch. Genus mostly not associated with insects. However, *Hirsutella liberiana* Mains and some other undescribed taxa do belong in it (Mains, 1949, 1951b; Gams, 1971b).

Tilachlidiopsis Keissler. Type species: *T. racemosa* Keissler. Three species on insects (Mains, 1951a).

Tolypocladium Gams. Type species: *T. inflatum* Gams (1971a). Most species soil-borne, but *T. inflatum* and *T. cylindrosporum* sometimes isolated from insects.

Trichothecium Link ex Fr. Type species: *T. roseum* Link. Madelin (1966) described *T. acridiorum* (Trabut) Madelin on locusts in some detail. *Didymopsis locustanae* Prinsloo is probably the same fungus.

Verticillium Nees per Link. Type species: *V. tenerum* (Nees per Pers.) Link. Most species saprophytes or plant pathogens. Gams (1971b) treated species occurring on insects, mostly known under their *Cephalosporium* name. Ten *Verticillium* species probably entomopathogenic. Balazy (1973) described several new *Cephalosporium* species which are probably only variable forms of the *Verticillium* species of Gams (1971b).

VII. Glossary of Some Mycological Terms

Comprehensive treatments of mycological terminology have been made by Ainsworth *et al.* (1971), Snell and Dick (1971), Kendrick (1971), Cooke (1974) and Cole and Samson (1979).

BASIPETAL (adj.): describes a succession of conidia, the youngest at the base of a chain (e.g. in *Metarhizium*), or closest to the fertile apex of the conidiogenous cell (with conidia in heads, e.g. in *Verticillium*).

BLASTIC (adj.): one of two basic modes of conidium ontogeny — the conidium differentiates from a hypha by budding.

CHLAMYDOSPORE: thick-walled, terminal or intercalary resting propagule; usually non-deciduous.

CONIDIOGENESIS: process of conidium formation.

CONIDIOGENOUS cell: fertile cell from which, or inside which, a conidium (or conidia) is (are) differentiated.

CONIDIOPHORE: specialized hypha, simple or branched, bearing conidiogenous cells.

CONIDIUM: asexual, nonmotile, usually deciduous propagule produced by deuteromycetous fungi. The term "spore" is used for the asexual propagule in the Zygomycetes and other Lower Fungi or the sexual propagules in the Ascomycetes and Basidiomycetes.

METULAE: apical branches of conidiophore, bearing phialides (e.g. *Aspergillus*).

PHIALIDE: conidiogenous cell producing a basipetal succession of conidia.

SCLEROTIUM: resting body produced by aggregation of hyphae, usually forming a hardened mass with or without host tissue.

SPORODOCHIUM: cushion-like mass of conidiogenous cells.

SYMPODIAL (adj.): describes a mechanism of conidiogenous cell proliferation in which each new growing point appears just behind and to one side of the previous apex, often resulting in an elongation of the cell.

SYNNEMA: compact or loose aggregation of erect hyphae, conidiophores and conidiogenous cells.

VESICLE: swollen apical part of conidiophore (e.g. in *Aspergillus*).

VIII. Suggestions for Future Work

Emphasis has been focused on the fungal pathogens of temperate or sub-tropical crops rather than those in natural environments. Evans (1974) has shown natural control of arthropods by Deuteromycetes in the tropical High Forest of Ghana and further such ecological studies will expand knowledge of the number and variation of the taxa. Extensive collections, particularly tropical, are needed for taxonomic work on many little known or doubtful species.

Cultivation of pathogenic Deuteromycetes on artificial media is important for detailed study of the development of the fungi and their life cycles. Isolation experiments are also worthwhile with Ascomycetes and their conidial imperfect states, not only to elucidate the perfect–imperfect association, but also to obtain good (probably mostly conidial) inocula.

References

Ainsworth, G.C.; James, P.W. and Hawksworth, D.L. (1971). "Ainsworth and Bisby's Dictionary of the Fungi". Commonwealth Mycological Institute, Kew.

Austwick, P.K.C. (1965). *In* "The Genus Aspergillus" (K.B. Raper and D.I. Fennell, eds), pp. 82–126. Williams & Wilkins, Baltimore.

Balazy, S. (1973). *Bull. Soc. Sci. Let. Poznan* 14, 101–137.

Booth, C. (1971). "The Genus Fusarium". Commonwealth Mycological Institute, Kew.

Buchli, H.H.R. (1960). *C. r. hebd. Séanc. D Acad. Sci.* 232, 3365–3367.

Buchli, H.H.R. (1966). *Rev. Ecol. Biol. Soc.* 3, 589–610.

Cole, G.T. and Samson, R.A. (1979). "Patterns of Development in Conidial Fungi". Pitman, London.

Colla, S. (1929). *Boll. Lab. Zool. Portici* 22, 39–48.

Cooke, W.B. (1974). *Mycopath. Mycol. appl.* 53, 45–67.

Couch, J.N., Romney, S.V. and Rao, B. (1974). *Mycologia* 66, 374–379.

Evans, H.C. (1974). *J. appl. Ecol.* 11, 37–49.

Fawcett, H.S. (1910). *Mycologia* 2, 164–168.

Fisher, F.E. (1950). *Mycologia* 42, 290–297.

Gams, W. (1971a). *Persoonia* 6, 185–191.

Gams, W. (1971b). *"Cephalosporium* – artige Schimmelpilze (Hyphomycetes)". Gustav Fisher, Stuttgart.

Gouger, R.J. and Kimbrough, J.W. (1969). *J. Invertebr. Path.* 13, 223–228.

Heim, R. (1951). *Bull. trimest. Soc. Mycol. France* 67, 336–364.

Hoog, G.S. de (1972). *Stud. Mycol., Baarn* 1, 1–41.

Hoog, G.S. de (1974). *Stud. Mycol., Baarn* 7, 1–84.

Kendrick, W.B. (1971). "Taxonomy of Fungi Imperfecti". Univ. Toronto Press, Toronto.

Khan, S.R. and Kimbrough, J.W. (1974). *Am. J. Bot.* 61, 395–399.

Kobayasi, Y. (1941). *Sci. Rep. Tokyo Bunrika Daig., Sect. B*, **5**, 53–260.
Madelin, M.F. (1966). *Trans. Brit. mycol. Soc.* **49**, 275–288.
Mains, E.B. (1948). *Mycologia* **40**, 402–416.
Mains, E.B. (1949). *Mycologia* **41**, 303–310.
Mains, E.B. (1950a). *Mycologia* **42**, 306–321.
Mains, E.B. (1950b). *Mycologia* **42**, 566–589.
Mains, E.B. (1951a). *Bull. Torrey bot. Club.* **78**, 122–133.
Mains, E.B. (1951b). *Mycologia* **43**, 691–718.
Mains, E.B. (1959a). *J. Insect Path.* **1**, 43–47.
Mains, E.B. (1959b). *Lloydia* **22**, 215–221.
Petch, T. (1921). *Ann. R. bot. Gdns. Peradeniya* **7**, 167–278.
Petch, T. (1926). *Trans. Brit. mycol. Soc.* **11**, 50–66.
Petch, T. (1932). *Trans. Brit. mycol. Soc.* **17**, 170–178.
Petch, T. (1941). *Trans. Brit. mycol. Soc.* **25**, 250–265.
Raper, K.B. and Fennell, D.I. (1965). "The Genus Aspergillus". Williams & Wilkins, Baltimore.
Samson, R.A. (1974). *Stud. Mycol., Baarn* **6**, 1–119.
Samson, R.A. and Evans, H.C. (1973). *Acta Bot. Neerl.* **22**, 522–528.
Samson, R.A. and Evans, H.C. (1974). *Acta Bot. Neerl.* **23**, 28–35.
Samson, R.A. and Evans, H.C. (1975). *Proc. K. ned. Akad. Wet. Ser. C,* **78**, 73–80.
Samson, R.A. and Evans, H.C. (1977). *Proc. K. ned. Akad. Wet. Ser. C,* **80**, 128–134.
Snell, W.H. and Dick, E.A. (1971). "A Glossary of Mycology". Harvard University Press, Cambridge, Mass.
Speare, A.T. (1920a). *J. agric. Res.* **18**, 399–439.
Speare, A. T. (1920b). *Mycologia* **12**, 62–76.
Sweeney, A.W. (1975). *Aust. J. Zool.* **23**, 49–57.
Thaxter, R. (1891). *Bot. Gaz.* **16**, 201–205.
Tulloch, M. (1976). *Trans. Brit. mycol. Soc.* **66**, 407–411.

CHAPTER 7

Identification of the Entomophthorales

D.S. KING* and R.A. HUMBER**

American Type Culture Collection, Rockville, USA and
*New England Plant, Soil, and Water Laboratory, University of Maine,
Orono, USA*

I. Introduction

A. OBJECTIVE

This chapter characterizes insect pathogens in the fungal order Entomophthorales, illustrated by selected species of genera with potential for use in
biological control of insects: *Entomophthora* Fresenius (1856) and *Conidiobolus*
Brefeld (1884). *Strongwellsea* Batko and Weiser (1965) is also illustrated because

---

* Current address: Boyce Thompson Institute for Plant Research Tower Road, Ithaca,
New York, USA. ** Current address: 12709 Atherton Drive, Wheaton, Maryland, USA.

it is closely related to these genera. Taxonomic difficulties encountered in the Entomophthorales are considered, a key to identification of the major genera that recognizes these difficulties is provided and methods for identification of these fungi are discussed. Their mammalian pathogenicity is discussed in chapters 28 and 40.

B. RELATIONSHIPS OF THE ORDER

The Orders Entomophthorales, Mucorales, and Zoopagales belong to the Class Zygomycetes, which is characterized by absence of flagellate spores and by sexual reproduction through formation of zygospores. The Zygomycetes and Trichomycetes − a group of arthropod-commensal fungi which also lack motile spores and may form zygospores − constitute the Subdivision Zygomycotina, the so-called "terrestrial phycomycetes".

In the Zygomycotina, the primary taxonomic emphasis is on asexual reproductive structures. In the Mucorales, asexual reproduction occurs by means of sporangia containing many spores, few spores, or only one spore, as well as by true conidia. It is believed that conidia are most advanced and have been derived from multispored sporangia by reduction. The precise nature of entomophthoralean asexual spores remains controversial; they are most often called conidia, and are widely considered to be derived by evolutionary reduction, like conidia of the Mucorales. Indications of the sporangial nature of these primary spores are found in a few species of *Basidiobolus*.

Zygomycetes produce sexually derived, thick-walled resting spores called zygospores as well as structurally similar azygospores that arise in the apparent absence of sexual fusions. The mucoralean zygospore wall is formed by modification of the gametangial walls. In the Entomophthorales and Zoopagales, the zygospore (or azygospore) wall is formed *de novo*, and the spore lies free within either the old gametangial wall or a lateral extension that forms after fusion of the gametangia.

With the sole exception of the genus *Massospora* Peck (1879), Entomophthorales are distinguished from all other Zygomycotina by the presence of forcibly discharged conidia. According to the most widely accepted classification, the Entomophthorales includes one family, the Entomophthoraceae, containing < 200 species in nine genera.

C. GENERA OF LIMITED POTENTIAL FOR BIOLOGICAL CONTROL

Fungi of the genus *Massospora* are known only as pathogens of gregarious cicadas. Axenic culture of these fungi has not been reported, and their use for biocontrol is remote despite virulence for cicadas. Our understanding of the taxonomy and biology of this genus has been clarified substantially by Soper (1963, 1974) and by Soper *et al.* (1976a, b); eleven species are recognized currently (Soper, 1974).

The genus *Strongwellsea* contains two relatively rare species parasitic on adult muscoid flies. Both release their conidia through a hole in the host's abdomen, and prevent oviposition by infected females (Humber, 1976).

Although the mycelium of *S. magna* has been cultured (Humber, unpublished observations) the benign and complex host/parasite relationship of these fungi probably excludes their use in biological control.

The evidence for infection of insects by *Basidiobolus* is uncertain. Levisohn (1927) considered passage through beetles part of the life cycle of *Basidiobolus ranarum*. Krejzová (1972) isolated an unidentified strain of *Basidiobolus* from a mosquito, and produced light infections in two termite species with it. However, most attempts to show that *Basidiobolus* invades insects have failed. Greer and Friedman (1966) and Hutchison (personal communication) failed to isolate *Basidiobolus* from field-collected and surface sterilized insects. While confirming these results, the work of Coremans-Pelseneer (1974) and Remaudière *et al.* (1976b) supported Drechsler's (1956) hypothesis that *Basidiobolus* spores are commonly carried on the surface of insects and mites (thus infecting insectivorous amphibians and reptiles), and indicated that dead insects provide a good saprophytic substrate for *Basidiobolus*. There are six species of *Basidiobolus*. A seventh, *B. philippinensis* Josue and Quimio (1976), may be identical with *B. haptosporus* (=*B. meristosporus*), a proven agent of human mycoses.

No evidence has been found to associate several other genera of the Entomophthorales with insects: *Completoria* Lohde (parasitic on fern gametophytes); *Ancylistes* Pfitzer (desmid algae, Waterhouse, 1973); *Meristacrum* Drechsler (nematodes, McCulloch, 1977); *Gonimochaete* Drechsler (nematodes, now considered by Barron, 1977, to be in the Oomycetes); *Ballocephala* Drechsler (tardigrades, Richardson, 1970; Pohlad and Bernard, 1978); and *Zygnemomyces* (nematodes, Miura, 1973).

II. Characterization of Entomogenous Entomophthorales

Thaxter's (1888) was the first comprehensive characterization of fungi now included in the order Entomophthorales. Subsequent studies have concentrated primarily on *Entomophthora* and/or have been very general in their approach (e.g., Bessey, 1950; MacLeod, 1956, 1963; Alexopoulos, 1962; Waterhouse, 1973). In the order, there is greater morphological variation within *Entomophthora* than within any other genus. This variation overlaps the narrower morphological variations within *Conidiobolus* and *Strongwellsea* and complicates their adequate circumscription. Accordingly, the main descriptive characters of these genera must be re-appraised before the taxonomic controversies outlined in Section III can be resolved.

Our concepts necessarily focus on the major entomogenous genera, but have broader applicability to the whole order. The figures exemplify much of the range of morphological variation within the order. The diagnostic characters of the illustrated species figure prominently in recent attempts to stabilize the taxonomy of this group.

A. VEGETATIVE GROWTH

Several diverse modes of vegetative growth occur among the entomopathogenic Entomophthorales, similar to growth patterns in related fungi. Infection

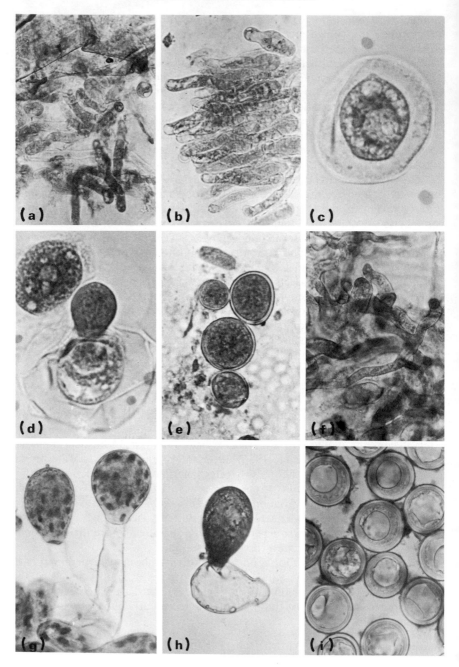

begins by a germ tube from some type of spore penetrating the host's integument, partly mechanically and partly enzymatically (Brobyn and Wilding, 1977). Then, the nature of the vegetative structures changes with time and may vary among hosts of the same or different species: development proceeds by the formation of a mycelium (de Bary, 1887), by fragmentation or budding into relatively short hyphal bodies (Thaxter, 1888), or by release of an amoeboid protoplast (Tyrrell, 1977).

Despite the frequency of hyphal bodies, it is wrong to overemphasize their prevalence (e.g., MacLeod, 1956, 1963; Alexopoulos, 1962; Waterhouse, 1973). It is better to consider hyphal bodies as only one among several interrelated modes of vegetative growth (Brobyn and Wilding, 1977). In fact, the development of many entomogenous species is exclusively mycelial, which may help to explain the confinement of some species to the host's abdomen.

In most species of *Entomophthora*, invasion is initiated by a relatively unbranched mycelium that becomes septate; these multinuclear segments often separate to form characteristic hyphal bodies which then may multiply by budding. These hyphal bodies may be short and irregularly branched as in *E. planchoniana* (Fig. 1a), or long, unbranched, and nonseptate as in *E. aulicae* (Fig. 1f). However, multicellular hyphal segments like those of *E. sphaerosperma* or a well developed mycelium as in the genus *Strongwellsea* (Fig. 2g) may be produced. In short, a continuum of growth forms exists.

Recently, a wall-less amoeboid vegetative protoplast has been discovered in *Entomophthora egressa*. These protoplasts were formed spontaneously when conidia germinated in a liquid insect tissue culture medium (Tyrrell and MacLeod, 1972a). The characteristic disease state, with hyphal bodies and conidiophores, is produced after infection with protoplasts (Otvos *et al.*, 1973), which are detectable in haemolymph of the spruce budworm after infection by conidia (Tyrrell, 1977). Formation of protoplasts is the only character that distinguishes *E. egressa* from *E. aulicae*. Thaxter (1888) reported that the zygospores of *E. fresenii* arise from the conjugation of amoeboid hyphal bodies without a cell wall, and that the resting spores of *F. colorata* (=*E. grylli?*) may arise from peculiar amoeboid bodies (Sorokin *in* Thaxter, 1888).

The vegetative phase in *Strongwellsea* is a coenocytic, infrequently-branched mycelium of hyphae, variable in diameter (Fig. 2g), some of which

---

Fig. 1. (a–b) *Entomophthora planchoniana* from *Acyrthosiphon kondoi:* (a) hyphal bodies, × 330; (b) conidiophores, × 330. (c–e) *Entomophthora muscae* from *Musca domestica:* (c) primary conidium surrounded by cytoplasm expelled from conidiophore, × 1000; (d) replicative conidium with more homogeneous cytoplasm than in adjacent primary conidium, × 1000; (e) developing resting spores, × 650. (f–i) *Entomophthora aulicae:* (f) mycelium from *Choristoneura fumiferana*, × 330; (g) conidiophores and primary conidia from culture with nuclei stained by aceto-orcein, × 450; (h) replicative conidium, × 500; (i) resting spores from *Lymantria dispar*, × 330.

Thanks are due Dr. N. Wilding and Dr. J.P. Kramer for material used in this figure and Fig. 3.

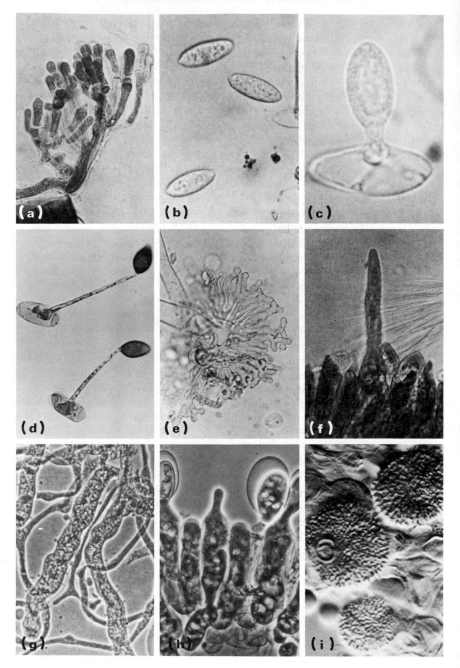

become septate and may form short hyphal bodies in the compact zone just below the cup-like hymenial surface inside the host's abdomen (Humber, 1976).

Most species of *Conidiobolus* form hyphal bodies like those of *Entomophthora*. The mycelium develops septa, forming coenocytic segments, which separate by the decay of intervening lengths of evacuated hyphae (Fig. 4a). Two other less common forms of vegetative growth in this genus are exemplified by *C. lachnodes,* in which hyphal bodies proliferate by budding (King, 1976a), and by *C. adiaeretus* (Fig. 4f), in which a thin coenocytic mycelium forms septa only in delimiting reproductive structures.

B. PRIMARY CONIDIA AND CONIDIOPHORES

The shape, size, colour, and nuclear number of the primary conidia, and the nature of primary conidiophores, are key taxonomic features within the Entomophthorales. These criteria have been used to propose both formal and informal alternative classifications of species usually included in the genus *Entomophthora*. Most taxonomic schemes arbitrarily ignore the diverse mechanisms of conidial discharge. Excepting the non-discharging conidial apparatus in *Massospora*, primary conidia of the Entomophthorales are formed and discharged toward the strongest source of light.

Entomophthora muscae, the type species of the genus, typifies one small group of species. Their primary conidia are broadly ellipsoidal, with an apiculate apex and truncate base (= campanulate), hyaline, and contain from 2 to ca. 10 fairly large and easily seen nuclei (Fig. 1c). Pressure developed within the clavate, unbranched conidiophores (Fig. 1b) causes them to rupture and provides the force for conidial discharge. A liberal coating of cytoplasm from the ruptured conidiophore helps the discharged spore to adhere upon landing (Fig. 1c).

Conidia of *E. aulicae* (Fig. 1g) are broadly pyriform, with an evenly rounded apex and a rounded basal papilla, hyaline, and multinucleate with easily seen nuclei. The conidiophores are simple and clavate but do not participate in conidial discharge, which is by the eversion of the conidial papilla against the conidiophore apex (rounding off of turgid cells).

Conidia of *E. fresenii* (Fig. 3b) are nearly round but have a slightly extended truncate base; these pale grey (smoky) conidia contain four nuclei. The conidiophore is simple (Fig. 3a), and spores discharge by the rounding off of turgid cells.

Fig. 2. (a–d) *Entomophthora sphaerosperma:* (a) developing conidiophores from small fly, × 300; (b) primary conidia, × 330; (c) replicative conidium, × 1650; (d) passively detached conidium atop capillary conidiophore, × 365. (e–f) *Entomophthora aphidis* from *Macrosiphum euphorbiae:* (e) holdfasts at end of rhizoids, × 325; (f) cystidium, × 525. (g–i) *Strongwellsea magna* from *Fannia canicularis:* (g) vegetative hyphae, × 250; (h) conidiophores and primary conidia with outer wall layer detached from conidial surface, × 385; (i) spiny resting spores, showing foramen where originally attached to hypha, × 535.

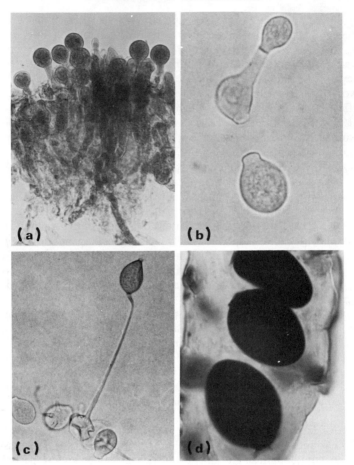

Fig. 3. *Entomophthora fresenii* from *Aphis fabae:* (a) conidiophores, × 350; (b) primary conidia, including one with a replicative conidium, × 720; (c) passively detached conidium atop capillary conidiophore, × 320; (d) resting spores with small apical remnants of gametangia, × 800.

A large group of *Entomophthora* species have ovoidal–cylindrical primary conidia with a single prominent nucleus and a wall whose outer layer readily separates from the spore, except over the papilla. Batko (1964c) considers *E. sphaerosperma* (=*Zoophthora radicans*) to typify these species; its bullet-shaped conidia are borne on highly branched conidiophores (Figs 2a, b). Both species of *Strongwellsea* also have uninucleate conidia with a separable outer wall (Fig. 2h), but their conidiophores are simple, not markedly clavate, and form a cup-shaped hymenium inside a cavity in the host's abdomen. Electron microscope studies of sporogenesis in *S. magna* showed that these spores might be true conidia (Humber, 1975), but are not monosporic sporangia as this conidial type is widely thought to be (Thaxter, 1888; Batko, 1964c, 1974).

Primary conidia of *Entomophthora virulenta* (Fig. 4h) and several related species are globose to pyriform with a more or less prominent basal papilla, hyaline, and contain numerous small nuclei. Conidiophores are simple. Spores are discharged when the papilla everts against the conidiophore. Primary conidia and conidiophores of the 27 species of *Conidiobolus*, e.g. *C. coronatus* (Fig. 4b), are remarkably homogeneous when compared to those of *Entomophthora*, but hardly differ from those of *E. virulenta*. In different species the conidia vary mainly in size, shape of the papilla, and in the ratio of papilla to spore size. The conidiophores, although usually simple, may rarely be bi- or trifurcate. Usually they are not well differentiated from the mycelium but in two species they are larger and are recognizable early in their development (Fig. 4f).

C. REPETITIONAL CONIDIA

In the Entomophthorales, a discharged primary conidium landing on a suitable substrate in appropriate conditions will germinate to form one or more germ tubes which then may develop into vegetative hyphae. In conditions unsuitable for vegetative growth, primary conidia of all species (except in *Massospora* and *Zygnemomyces*) may form one or more new conidia by one of several modes; those conidia are discharged, or, in a few species, are capillospores (= passively detached elongate conidia; Batko, 1974).

Primary conidia discharge secondary conidia, which in turn discharge tertiary conidia, and so on. However, capillospores may also produce subsequent generations of capillospores, or of discharged conidia in *Conidiobolus heterosporus* (Drechsler, 1953). The term repetitional conidia (Drechsler, 1953; King, 1976a) usefully refers to *all* conidia except primary ones, and the term replicative conidia (King, 1976a) refers to all *forcibly* discharged repetitional conidia but not capillospores, regardless of the type of conidia from which they arise. As for primary conidia, formation and discharge of replicative conidia is phototropically oriented.

Entomophthora muscae produces replicative conidia similar to primary conidia but usually lacking an apiculus (Fig. 1d). The spherical replicative conidia of *Strongwellsea* clearly differ from its ellipsoidal to obovoidal primary conidia. Some *Entomophthora* species with elongate primary conidia produce two types of replicative conidia (Thaxter, 1888; Keller, 1977): one resembles the primary conidia, the other is pyriform to spherical. Only replicative conidia similar to the primary conidia are produced by *E. aulicae* (Fig. 1h).

Entomophthora sphaerosperma (Figs. 2c, d) and *E. fresenii* (Figs. 3b, c) each produce both replicative conidia similar to the primary conidia and capillospores that are almond-shaped and borne on long, thin, tapering (capillary) conidiophores.

The only replicative conidia produced by *Entomophthora virulenta* are similar to but smaller than the globose primary conidia, as with *Conidiobolus* (Fig. 4c), except for two species in which both globose and fusiform replicative

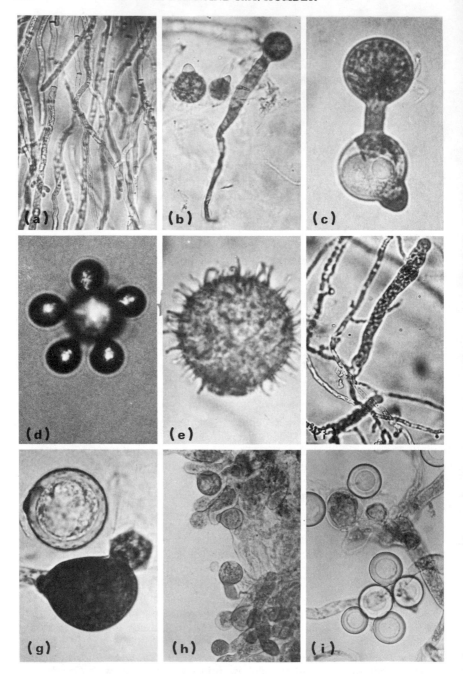

conidia are produced. In addition, four species of *Conidiobolus* also produce capillospores borne on capillary conidiophores, similar to those of *E. sphaerosperma* (Fig. 2d).

D. MICROCONIDIA

Multiplicative conidia (Drechsler, 1952), or microconidia, are small discharged conidia produced from a globose conidium on radially projecting microconidiophores (sterigmata) (Fig. 4d). Their production and discharge are not oriented by light, and moisture relationships appear to be critical for their formation. Microconidia are best known in *Conidiobolus coronatus*, in which five to > 38 may be formed by a primary conidium; unlike repetitional conidia, they either form a germ tube or die (Prasertphon, 1963). However, of 69 *C. coronatus* strains, 20% failed to produce microconidia (King, 1976a). Eight other *Conidiobolus* species (King, 1977) and *Entomophthora apiculata* (Gustafsson, 1965a) also produce microconidia.

A superficially similar type of sporulation occurs in *Basidiobolus microsporus*, in which spores are elongate, passively detached, and arise from uninucleate cells delimited by septa prior to sporulation (Benjamin, 1962). However, in *Conidiobolus* microconidia are globose and discharged, without septation of the precursor conidium (e.g., *C. coronatus*, King and Moss, unpublished observations).

E. RESTING SPORES

There are several types of resting spores in the Entomophthorales. Chlamydospores, as in many other fungi, are formed readily by many species in adverse conditions by the production of a single thick wall in vegetative hyphal elements. When favourable conditions return, chlamydospores readily germinate usually producing germ tubes, but sometimes azygospores.

Zygospores and azygospores are morphologically alike. Each usually has two thick, hyaline walls formed *de novo*, and lies free within a thinner outer wall (episporium) derived from pre-existing elements. The episporium may be hyaline or coloured, and smooth or variously ornamented. Both types are resting spores, and usually require a period of dormancy before germination.

The distinction between zygospores and azygospores is not entirely satisfactory. Zygospores arise from conjugation of two hypha-derived elements,

---

Fig. 4. (a–e) *Conidiobolus coronatus* from culture: (a) vegetative hyphae, × 30; (b) conidiophore with developing conidium and primary conidia, × 260; (c) developing replicative conidium, × 485; (d) microconidia radiating from primary conidium, × 800; (e) villose spore, × 900. (f) *Conidiobolus adiaeretus* from culture: vegetative hyphae and developing conidiophores, × 270. (g) *Conidiobolus osmodes* from culture: mature resting spore and gametangia developing a resting spore, × 725. (h–i) *Entomophthora virulenta* from *Myzus persicae:* (h) conidiophores and primary conidia, × 400; (i) resting spores, × 400.

whereas conjugation is not evident in the production of azygospores. Whether genetic recombination occurs in either of these spore types is unknown. The usual multinucleate nature of the elements that form both types of spores and their similarities in structure suggest their genetic equivalence; the absence of conjugation in the development of azygospores argues against it. Riddle (1906) and Latgé (1976) failed to detect nuclear fusion in resting spores although the number of nuclei may be reduced to one. Contributing to the uncertainty concerning the respective roles of zygospores and azygospores are reports of both occurring in several *Entomophthora* species (e.g., MacLeod and Müller-Kögler, 1973).

It is very difficult to induce germination in zygospores and azygospores of many *Entomophthora* species. However, they are considered by many to be the best fungal stage for field applications in insect control. They survive in dry air and so are easily stored and applied. Methods for their production are described in Chapter 28, Section III, C.

The spherical to oval azygospores of *Entomophthora muscae* (Fig. 1e) are hyaline to brown and contain many oil droplets. They are usually formed at the tips of short hyphae or intercalarily, or sometimes by budding. Azygospores of *E. aulicae* (Fig. 1i) arise as buds; they are multinucleate (Riddle, 1906), spherical, smooth walled, and almost hyaline to dark yellow or brown. Resting spores of *E. sphaerosperma* are spherical, hyaline, smooth, and arise as lateral buds from hyphal segments; whether they are zygospores (Gustafsson, 1965a) or azygospores (Thaxter, 1888; Sawyer, 1931) has been debated for many years.

Zygospores of *E. fresenii* (Fig. 3d) are formed in a distinctive manner. Two hyphal bodies (gametangia) conjugate, and from their point of fusion, the zygospore arises as a bud. The zygospores are elliptical to subovoidal, smooth walled, very dark and opaque, and the empty walls of the gametangia often remain attached to the spore. In *E. virulenta,* smooth, hyaline zygospores (Fig. 4i) are formed within the larger of two gametangia (Latgé, 1976) as in *Conidiobolus.*

It is not known whether the spherical to ovoid resting spores of *Strongwellsea* (Fig. 2i) are zygospores or azygospores. The episporium bears slightly recurved broad spines and is coloured so that the spores appear orange in mass.

Azygospores have been reported for only one species of *Conidiobolus* (Drechsler, 1962), but are now considered to be zygospores (King, 1976a). Zygospores of *C. osmodes* (Fig. 4g) — characteristic of the genus — are formed within one of two gametangia. They have a single large eccentric oil globule and two thick walls, usually marked by fine parallel ridges, though zygospores of most species are smooth. Most *Conidiobolus* zygospores are hyaline, but those in several species, including *C. osmodes,* are yellow in varying degrees.

The villose resting spores of *Conidiobolus coronatus* (Fig. 4e) constitute a fourth type of resting spore formed directly from primary or replicative conidia. These secrete a thick wall with villose appendages of variable length and number. Conidia of *C. coronatus* often acquire thick walls that lack villose appendages, and this occurs in other *Conidiobolus* species (King, 1976a). This, and the production of similar spores (loricoconidia) in *Entomophthora destruens*

(Weiser and Batko, 1966), indicate that this may be a more general phenomenon than heretofore recognized (King, 1976a).

F. RHIZOIDS AND CYSTIDIA

Rhizoids are broad, highly vacuolate, sterile hyphae ending in a more or less complex holdfast or series of short adhesive appendages anchoring an infected insect to the substrate. They emerge shortly before and/or after the host's death and prior to conidiogenesis, but are not formed in culture. They occur in species with nearly every type of conidium described above, but most often among species with digitate conidiophores and uninucleate conidia with a separable outer wall layer.

It is distressingly common for descriptions to state only that rhizoids are present or absent. However, their points of emergence, abundance, dimensions, and the morphology of the terminal holdfast (Fig. 2e) are characteristic for individual species. Their diverse morphologies are best documented by Thaxter (1888) and Gustafsson (1965a).

While the presence of rhizoids is taxonomically significant, their absence is not, e.g. individual strains of *E. sphaerosperma* form rhizoids in some host species but not in others (R. G. Kenneth, personal communication). Also, rhizoids may be absent on hosts whose bodies liquefy on formation of resting spores even though the same fungus produces rhizoids at conidiogenesis on other individuals of the same species. One common species, *E. ignobilis* (= *E. thaxteriana;* Humber, 1978), was recently reported for the first time to form a few delicate rhizoids around mouthparts and forelegs (Brobyn and Wilding, 1977).

Conversely, some species described as producing rhizoids may never do so. Rhizoids described as occasionally present in *E. virulenta* have never been verified and might have belonged to a second (rhizoidal) species in the same individual aphids (Humber *et al.*, 1977). Infected insects are sometimes held by plant exudates or affixed by conidiophores and other hyphae entangled among plant hairs; such fungi should not be regarded as rhizoidal unless differentiated fungal organs of attachment are present.

Cystidia (= pseudocystidia or paraphyses) are more or less differentiated sterile hyphae which extend beyond the level of the conidiophores (Fig. 2f). They occur in some — but not all — rhizoidal *Entomophthora* species; they are unknown from species which never produce rhizoids. Cystidia may be simple or infrequently branched; they may appear like long sterile conidiophores, may taper gently to a point, or may be very stout, tall and easily detected by the unaided eye. When compared to conidiophores and conidia, the cytoplasm of a cystidium is transparent and homogeneous, and the nuclei are often distinct. Unlike rhizoids, cystidia may occur in culture. Cystidia are of little taxonomic value except as ancillary characters. Their function has been debated with little firm evidence, although developing conidiophores of *E. aphidis* surround cystidia which puncture the host cuticle and facilitate emergence of conidiophores (Brobyn and Wilding, 1977).

III. Taxonomic Concepts among Entomogenous Entomophthorales

Entomophthora is the largest genus of the Entomophthorales. There are > 150 species (MacLeod and Müller-Kögler, 1973), almost all from insects, but a few from mites, symphylids, and nematodes. Taxonomic and nomenclatural controversies about them began in the mid 19th century; the early taxonomic history was reviewed thoroughly by MacLeod (1963) and Gustafsson (1965a). Numerous reasons for taxonomic difficulties were recognized as early as Thaxter's (1888) studies, and many of these difficulties remain unresolved today. Relatively few species have been cultured. There is a lack of studies comparing fungi from a common host and of studies about the important implications of host ranges and specificities. Other problems arise (1) from the difficulty of comparing fungi grown in different environments and/or different host species, (2) from the uncertainty of whether two fungi are in comparable physiological – and, therefore, morphological – states when they occur in two individuals, even of the same host species, and (3) the common occurrence of multiple infections.

It is now generally accepted that *Entomophthora* is a heterogeneous group whose species should be divided among several genera if precise and acceptable criteria can be found. The species have been separated into groups on the basis of conidial morphology (Lakon, 1919; Hutchison, 1963; Gustafsson, 1965a), which have no nomenclatural validity but have received a degree of acceptance through usage (MacLeod and Müller-Kögler, 1973; Waterhouse, 1975; King, 1976b; MacLeod *et al.*, 1976). Batko (1964a–e, 1974) offered the most comprehensive scheme. In addition to transferring *E. coronata* to *Conidiobolus*, he proposed that the species of *Entomophthora* be classified in five genera. Despite his apparent correctness in some respects, several weaknesses have prevented general acceptance of his scheme. Some of his criteria must be challenged since the resultant groupings of species into some of his genera seem to be at least as artificial as including all of these species in the single genus *Entomophthora*. Several species resemble *Conidiobolus* and may be eventually assigned to that genus.

The taxonomic history of *Conidiobolus* is considerably simpler than that of *Entomophthora*. Its name has not been disputed since its conception. Among the 27 species currently recognized (King, 1977), only *C. coronatus* and *C. pseudococcus* were not originally described as species of *Conidiobolus*. The complex taxonomic history of *C. coronatus* has been pivotal in the complications of separating *Entomophthora* and *Conidiobolus*. The transfer of this species to *Entomophthora* by Kevorkian (1937) largely because it attacked insects strongly reinforced an unjustifiably persistent criterion for separating *Entomophthora* and *Conidiobolus*, namely entomopathogen or saprobe, as reflected in the keys by MacLeod (1963) and Waterhouse (1973). This concept has persisted even though Gustafsson (1965a) noted that his *Entomophthora* group "c", which corresponds to Hutchison's Apiculata type group of species, bore a strong resemblance to *Conidiobolus*. However, one character is insufficient to separate these genera, and diverse exceptions have made this criterion untenable. Many *Entomophthora* species are now cultured and several with simple nutritional requirements might survive in nature as facultative saprobes (Gustafsson, 1965b). Aside from *C. coronatus* and *C. pseudococcus*, several *Conidiobolus* species originally isolated from plant detritus also have been isolated from insects: *C. osmodes* has been isolated repeatedly from aphids, and its pathogenicity has been proven experimentally (Remaudière *et al.*, 1976b). Both *C. stromoideus* and *C. lamprauges* have been isolated from flies infected by *Entomophthora muscae* (Srinivasan and Thirumalachar, 1962; King, 1977); and an undescribed *Conidiobolus* has been isolated together with *E. arrenoctona* (= *E. caroliniana*) from craneflies, *Tipula paludosa* (Keller, 1977). It is unknown if the above incidences are double infections, or secondary invasions by saprophytic species of *Conidiobolus*.

The criteria used to delimit fungal taxa are predominantly morphological, and the separation of *Entomophthora* and *Conidiobolus* also must be so. The recognition by Tyrrell and MacLeod (1972b) that species forming microconidia belong in *Conidiobolus* was a step in the right direction but offers only a partial solution. Criteria to be used to separate *Entomophthora* as it now stands and *Conidiobolus* are given in the key in Section IV.

Features which may be valuable for separating species in the Entomophthoraceae include shape and size of primary conidia and other structures, host specificities, manner of development in the host and any modifications of host behaviour, culturability (including rate of growth in specified conditions), and features of rhizoids and cystidia, if present. Chemotaxonomy – particularly fatty acid composition (Tyrrell, 1967, 1968, 1971; Tyrrell and Weatherston, 1976) and isozyme analyses (King, 1976c) – has proven valuable in eludicating infrageneric relationships among cultured species. As more species of the Entomophthorales are cultured, chemotaxonomy may aid significantly in clarifying specific relationships, particularly among *Entomophthora* species. Whether a strain is isolated from a saprobic or entomogenous source has no place among the above criteria.

There are many pitfalls in using these criteria, and caution must be used when identifying these fungi, especially when describing a new species. The presence of a given spore type is significant, but its absence may or may not be so. The environmental conditions or the physiological state of the fungal specimen may not be appropriate to produce a specific type of spore. Also, a particular strain may match the description of a particular species to which it should be referred even though it may not produce a type of spore characteristic for that species. For example, the type strain of *Conidiobolus megalotocus* (ATCC 12914) produces microconidia but no zygospores while the type strain (ATCC 14444) of a species synonymized with *C. megalotocus* produces zygospores but no microconidia (King, 1977); other criteria such as overall morphological similarity, fatty acid compositions, and isozyme analyses indicate that these strains are conspecific (King, 1976b, c; Tyrrell, 1971; Tyrrell and Weatherston, 1976).

Little is known of entomophthoralean variability, but some species show remarkable ranges of variation. The size range for primary conidia of *C. coronatus* encompasses those reported for several other *Conidiobolus* species

(King, 1976a). Similarly, size ranges for primary conidia of a single strain of *E. virulenta* (ATCC 14270) from aphids and from cultures were as great as the overall range for 30 strains of this species measured from cultures only (Humber, 1977b).

IV. Key to Genera Associated with Insects

Although considered only briefly, *Basidiobolus* is included in the key because it may infect insects and *Massospora* because it is an obligate entomopathogen. The separation between *Entomophthora* and *Conidiobolus* is less than satisfactory as explained in Section III.

1a. Primary conidiophore with a prominant subapical swelling functional in conidial discharge; zygospores with two more or less closely appressed lateral appendages (beaks). . . . . *Basidiobolus*

1b. Conidiophores and/or resting spores not as above 2

2a. Primary conidia not discharged, enclosed within host; limited to cicada abdomens . . . . *Massospora*

2b. Primary conidia discharged, formed on host's exterior or with external access through a ventrally open cavity 3

3a. Primary conidia elliptical to subovoidal, uninucleate, borne in a hymenium lining a ventrally open abdominal cavity . . . . *Strongwellsea*

3b. Primary conidia variously shaped, uni- to multinucleate, but not in a hymenium lining a ventrally open cavity 4

4a. Primary conidia spherical to subspherical with a smoothly rounded apex, containing > 7 nuclei, conidiophores simple; zygospores (if present) hyaline to yellowish, arising within the larger of two gametangia; villose spores and/or microconidia may be present; culturable on potato dextrose agar . . . .*Conidiobolus**

4b. Primary conidia variable in shape and number of nuclei, but not with the combination of characters in 4a; primary conidiophores simple or branched; zygospores (or azygospores) with hyaline to coloured to very dark outer wall layer, never formed within the larger of two gametangia; villose spores and microconidia absent; most species not culturable on potato dextrose agar . . . .*Entomphthora*

V. Methods of Identification

A. EXAMINATION FROM THE HOST

Usually an entomophthoralean mycosis must be diagnosed from a dead host already bearing conidia or filled with resting spores. A proper identification

---

* Includes some *Entomophthora* species, especially those of Hutchison's (1963) "Apiculata type" and Gustafasson's (1965a) "group c"; see Section III.

usually cannot be given from resting spores and host identity alone since conidia are the most important taxonomic character among these fungi.

Valuable ancillary information is obtained in many ways. Insects dying from *Entomophthora* infections often alter their behaviour to favour the dispersal of the fungus (e.g., Thaxter, 1888; Hutchison, 1962). With living or freshly dead hosts, note nature and morphology of vegetative stages of the fungus, i.e. mycelial or hyphal bodies. Preferably collect an insect on its substrate and examine it under a dissecting microscope or hand lens. A colour reference should be used to identify any colours which seem to be important (Humber, 1977a). Before detaching an insect note rhizoids, and after moving the insect examine the rhizoids and substrate for holdfasts. Cystidia are more easily seen on fresh than on desiccated material.

Collect conidia being discharged from a cadaver onto a slide in a humid chamber and mount in lactophenol (with or without a stain) or aqueous $HgCl_2$ or $CuSO_4$ (King, 1976a; Humber, 1977a). Leave some discharged conidia in wet air for several hours for the development of replicative, micro-, and/or passively detached conidia. While spores can be discharged directly into fungicidal aqueous solutions, lactophenol mounts are advantageous since they can be made semi-permanent by ringing. However, critical dimensions of spores remained unaltered in any of these mounting media. Information on variation of spore dimensions must be obtained from spore populations from several individual hosts (Humber, 1976, 1977b).

Conidiophores, conidia, cystidia, and resting spores from fresh material can be seen best after using fine-pointed needles (e.g., "minuten" insect pins in match sticks) to tease apart very small portions of the fungus in lactophenol. All types of conidia are often observable on wings and legs of the insect as well as on the body.

B. ISOLATION IN CULTURE

Discharge of conidia is exploited by most isolation techniques. Infected material may be stuck inside a Petri dish lid with soft agar, lanolin or petroleum jelly, and suspended over a culture medium in the bottom of the dish, or the fungus may be allowed to discharge conidia upward onto a culture medium (Drechsler, 1952; Hutchison and Nickerson, 1970; King, 1976a; Welton and Tyrrell, 1975). Also, conidia may be allowed to discharge onto a dish lid or a sterile slide from which they can be collected. High humidity is necessary to obtain production of primary conidia (e.g., Wilding, 1969; King, 1976a).

If a fungus has not yet emerged from an insect, the body should be surface sterilized in 70% ethanol for 30 sec followed by ca. 5% sodium hypochlorite solution (commercial bleach) for 1 min and washed in sterile distilled water. The surface-sterilized insect is then placed on the medium where the fungus can grow.

Most axenically cultured species can be isolated by conidial discharge onto coagulated egg yolk (Sawyer, 1929; Müller-Kögler, 1959); a combination of egg yolk and Sabouraud maltose agar (10:3), coagulated at 80°C (Soper *et al.*, 1975); or egg yolk with the addition of milk, Sabouraud maltose agar, or both

(Remaudière *et al.*, 1976a). Growth of some species on egg yolk was stimulated by a complex of rich nitrogen and carbon sources, but not by specific factors (Latgé and Bièvre, 1976). Welton and Tyrrell (1975) failed with egg yolk media due to contamination by faster growing organisms, but succeeded with Grace's insect tissue culture medium supplemented with 5% fetal bovine serum and gentamicin. Naked protoplasts were produced which were readily subcultured in fresh medium of the same kind, or on coagulated egg yolk (Tyrrell and MacLeod, 1972a). For less fastidious species, media such as Sabouraud dextrose agar plus 2% yeast extract or potato dextrose agar suffice.

C. EXAMINATION FROM CULTURE

If axenic growth succeeds, production of primary conidia usually ensues; conidia and conidiophores should be examined as above. Sufficient resting spores for taxonomic purposes often will be formed in culture by fungi with relatively simple nutritional requirements; procedures and media for increasing quantities of resting spores have been reported (Tyrrell, 1970; Gustafsson, 1965b; Gröner, 1975; Soper *et al.*, 1975; Matanmi and Libby, 1976; Latgé *et al.*, 1977; also see Chapter 28, Section III, C). It is usually very difficult to induce resting spores to form in cultures of more fastidious pathogens.

Microconidia or capillospores are often induced to form within 6 h from primary conidia discharged onto 2% water agar (Prasertphon, 1963; King, 1976a; Brobyn and Wilding, 1977). The villose spores unique to *Conidiobolus coronatus* are usually formed freely on a culture surface or a Petri dish lid above a sporulating culture. While induction of villose spores is difficult in a few strains from mammals, a few villose spores can usually be seen among conidia deposited on the upper surface of a cover slip placed directly onto the surface of an actively sporulating culture.

VI. Future Research Requirements

Before a firm basis for the separation of the genera *Conidiobolus* and *Entomophthora* can be developed, comparative morphological studies of these fungi in culture and in their insect hosts will be necessary. These studies should include *at least* the recognized entomogenous species of *Conidiobolus* and those *Entomophthora* species closely resembling *Conidiobolus*. Such *Entomophthora* species, including *virulenta, ignobilis* (= *thaxteriana*), *obscura, major,* and *apiculata,* are included by MacLeod and Müller-Kögler (1973) among those species (1) with pyriform to nearly spherical multinucleate conidia having a broadly rounded apex, (2) which can be cultured on relatively simple myco-logical media, and (3) in which zygospores are formed within the larger of two gametangia (or in which zygospores have not been observed); thus, they conform to our current concept of *Conidiobolus*.

When the circumscription of *Conidiobolus* can be fixed satisfactorily and when *Entomophthora* (in its broad sense) has been purged of its misclassified *Conidiobolus* species, the revisions proposed by Batko should be reappraised.

His *Zoophthora* and *Triplosporium* appear well defined, and it seems likely that emended descriptions modifying the circumscriptions of his remaining segregate genera would clarify and finally stabilize the generic concepts within the entomogenous Entomophthorales.

References

Alexopoulos, C.J. (1962). "Introductory Mycology". 2nd. ed. John Wiley, New York.

Barron, G.L. (1977). "The Nematode Destroying Fungi". Canadian Biological Publications Ltd., Guelph.

Batko, A. (1964a). *Entomophaga Mém. hors sér.* **2**, 129–131.

Batko, A. (1964b). *Bull. Acad. Pol. Sci., Ser. Sci. Biol.* **12**, 319–321.

Batko, A. (1964c). *Bull. Acad. Pol. Sci., Ser. Sci. Biol.* **12**, 323–326.

Batko, A. (1964d). *Bull. Acad. Pol. Sci., Ser. Sci. Biol.* **12**, 399–402.

Batko, A. (1964e). *Bull. Acad. Pol. Sci., Ser. Sci. Biol.* **12**, 403–406.

Batko, A. (1974). *In* "Ewolucja biologiczna: Szkice teoretyczne i metodologiczne" (C. Nowenskiego, ed.), pp. 209–304. Polish Academy of Sciences, Institute of Philosophy and Sociology, Wroclaw.

Batko, A. and Weiser, J. (1965). *J. Invertebr. Path.* **7**, 455–463.

Benjamin, R.K. (1962). *Aliso* **5**, 223–233.

Bessey, E.A. (1950). "Morphology and Taxonomy of Fungi". Hafner, New York and London.

Brefeld, O. (1884). *Unters. Ges. Mykol.* **6**, 35–72; 75–78.

Brobyn, P.J. and Wilding, N. (1977). *Trans. Br. mycol. Soc.* **69**, 349–366.

Coremans-Pelseneer, J. (1974). *Acta Zool. Path.* **60**, 1–143.

De Bary, A. (1887). "Comparative Morphology and Biology of the Fungi, Mycetozoa, and Bacteria". (Trans., H.E.F. Garnsey. Revised, I.B. Balfour). Clarendon, Oxford.

Drechsler, C. (1952). *Science* **115**, 575–576.

Drechsler, C. (1953). *J. Wash. Acad. Sci.* **43**, 29–43.

Drechsler, C. (1956). *Am. J. Bot.* **43**, 778–786.

Drechsler, C. (1962). *Bull. Torrey Botan. Club* **89**, 233–240.

Fresenius, G. (1856). *Bot. Ztg.* (Berlin) **14**, 882–883.

Greer, D.L. and Friedman, L. (1966). *Sabouraudia* **4**, 231–241.

Gröner, A. (1975). *J. Invertebr. Path.* **26**, 393–394.

Gustafsson, M. (1965a). *Lantbrukshögskolans Ann.* **31**, 103–212.

Gustafsson, M. (1965b). *Lantbrukshögskolans Ann.* **31**, 405–457.

Humber, R.A. (1975). Ph. D. Diss., Univ. of Wash., Seattle.

Humber, R.A. (1976). *Mycologia* **68**, 1042–1060.

Humber, R.A. (1977a). *J. Invertebr. Path.* **30**, 1–4.

Humber, R.A. (1977b). Abs. 2nd Int. Mycol. Congr. Vol. A–L, p. 309. Tampa, Florida.

Humber, R.A. (1978) *Mycologia* **70**, 208–210.

Humber, R.A., Soper, R.S., Wilding, N. and Remaudière, G. (1977). *Mycotaxon* **5**, 307–310.

Hutchison, J.A. (1962). *Mycologia* 54, 258–271.

Hutchison, J.A. (1963). *Trans. Kansas Acad. Sci.* 66, 237–254.

Hutchison, J.A. and Nickerson, M.A. (1970). *Mycologia* 62, 585–587.

Josue, A.R. and Quimio, T.H. (1976). *Nova Hedwegia* 27, 483–492.

Keller, S. (1977). *Bull. Soc. Ent. Suisse* 50, 277–284.

Kevorkian, A.G. (1937). *J. Agric. Puerto Rico* 21, 191–200.

King, D.S. (1976a). *Can J. Bot.* 54, 45–65.

King, D.S. (1976b). *Can. J. Bot.* 54, 1285–1296.

King, D.S. (1976c). Proc. 1st Int. Colloq. Invertebr. Path. IXth Ann. Meet. Soc. Invertebr. Path., pp. 277–281. Queen's University, Kingston.

King, D.S. (1977). *Can. J. Bot.* 55, 718–729.

Krejzová, R. (1972). *Vest. Csl. zool. Spol.* 36, 253–255.

Lakon, G. (1919). *Z. angew. Ent.* 5, 191–216.

Latgé, J.-P., (1976). *C.r. hebd. Séanc. Acad. Sci. Paris, Sér. D.* 282, 605–608.

Latgé, J.-P. and de Bièvre, C. (1976). *Annls Microbiol. (Inst. Pasteur)* 127A, 261–274.

Latgé, J.-P., Soper, R.A. and Madore, C.D. (1977). *Biotechnol. Bioengin.* 19, 1269–1284.

Levisohn, I. (1927). *Jahrb. Wiss. Bot.* 66, 513–555.

MacLeod, D.M. (1956). *Can. J. Bot.* 34, 16–26.

MacLeod, D.M. (1963). In "Insect Pathology: An Advanced Treatise" (E.A. Steinhaus, ed.), Vol. 2, pp. 189–231. Academic Press, New York and London.

MacLeod, D.M. and Müller-Kögler, D. (1973). *Mycologia* 65, 823–893.

MacLeod, D.M., Müller-Kögler, E. and Wilding, N. (1976). *Mycologia* 68, 1–29.

Matanmi, B.A. and Libby, J.L. (1976). *J. Invertebr. Path.* 27, 279–285.

McCulloch, J.S. (1977). *Trans. Br. mycol. Soc.* 68, 173–179.

Miura, K. (1973). *Rep. Tottori Mycol. Inst.* 10, 517–522.

Müller-Kögler, E. (1959). *Entomophaga* 4, 261–274.

Otvos, I.S., MacLeod, D.M. and Tyrrell, D. (1973). *Can. Ent.* 105, 1435–1441.

Peck, C.H. (1879). N. Y. St. Mus. Nat. Hist. 31st Ann. Rep., pp. 19–44.

Pohlad, B.R. and Bernard, E.C. (1978). *Mycologia* 70, 130–139.

Prasertphon, S. (1963). *J. Insect. Path.* 5, 318–335.

Remaudière, G., Keller, S., Papierok, B. and Latgé, J.-P. (1976a). *Entomophaga,* 21, 163–177.

Remaudière, G., Latgé, J.-P., Papierok, B. and Coremans-Pelseneer, J. (1976b). *C.r. Hebd. Séanc. Acad. Sci. Paris* 283, sér. D, 1065–1067.

Richardson, M.J. (1970). *Trans. Br. mycol. Soc.* 55, 307–340.

Riddle, L.W. (1906). *Proc. Am. Acad. Arts. Sci.* 42, 177–198.

Sawyer, W.H., Jr. (1929). *Am. J. Bot.* 16, 87–121.

Sawyer, W.H., Jr. (1931). *Mycologia* 23, 411–432.

Soper, R.S. (1963). *Can. J. Bot.* 41, 875–878.

Soper, R.S. (1974). *Mycotaxon* 1, 13–40.

Soper, R.S., Holbrook, F.R., Majchrowicz, I. and Gordon, C.C. (1975). Life Sci. Agri. Exp. Stn, Univ. Maine, Orono, Tech. Bull. No. 76, 1–15.

Soper, R.S., Delyzer, A.J. and Smith, L.F.R. (1976a). *Ann. Ent. Soc. Am.* 69, 89–95.

Soper, R.S., Smith, L.F.R. and Delyzer, A.J. (1976b). *Ann. Ent. Soc. Am.* **69**, 275–283.

Srinivasan, M.C. and Thirumalachar, M.J. (1962). *Sydowia Ann. Mycol.* **16**, 60–66.

Thaxter, R. (1888). *Mem. Boston Soc. Nat. Hist.* **4**, 133–201.

Tyrrell, D. (1967). *Can. J. Microbiol.* **13**, 755–760.

Tyrrell, D. (1968). *Lipids* **3**, 368–372.

Tyrrell, D. (1970). *Can. For. Serv. Dep. Fish. For., Bimon. Res. Notes* **26**, 12–13.

Tyrrell, D. (1971). *Can. J. Microbiol.* **17**, 1115–1118.

Tyrrell, D. (1977). *Experl. Mycol.* **1**, 259–263.

Tyrrell, D. and MacLeod, D.M. (1972a). *J. Invertebr. Path.* **19**, 354–360.

Tyrrell, D. and MacLeod, D.M. (1972b). *J. Invertebr. Path.* **20**, 11–13.

Tyrrell, D. and Weatherston, J. (1976). *Can. J. Microbiol.* **22**, 1058–1060.

Waterhouse, G.M. (1973). *In* "The Fungi" (G.C. Ainsworth, F.K. Sparrow, and A.S. Sussman, eds), Vol IVB, pp. 219–229. Academic Press, New York and London.

Waterhouse, G.M. (1975). *Bull. Brit. Mycol. Soc.* **9**, 15–41.

Weiser, J. and Batko, A. (1966). *Folia Parasit.* (*Praha*) **13**, 144–149.

Welton, M.A. and Tyrrell, D. (1975). *J. Invertebr. Path.* **26**, 405.

Wilding, N. (1969). *Trans. Br. mycol. Soc.* **53**, 126–130.

NOTE ADDED IN PROOF

Several publications on the generic taxonomy of the Entomophthorales have appeared since this chapter went to press. These include G. Remaudière and G.L. Hennebert (1980), *Mycotaxon* **11**, 269–321; G. Remaudière and S. Keller (1980), *Mycotaxon* **11**, 323–338; and R.A. Humber, R.S. Soper, D.M. McLeod, D. Tyrrell, R.G. Kenneth, and I. Ben-Ze'ev (1981), *Taxon*, in press. Though substantial differences still exist concerning generic names and the proper genera for some species, we are now much closer to a generally acceptable scheme of classification for this taxonomically difficult group. Also, *Entomophthora virulenta* is now considered a synonym of *Conidiobolus thromboides* (J.-P. Latgé, D.S. King and B. Papierok, 1980, *Mycotaxon* **11**, 255–268).

These reconsiderations and revisions of the generic taxonomy of the Entomophthorales will figure prominently in the 1980's. It is hoped that by the latter part of this decade a general agreement will exist regarding taxonomically significant criteria and the circumscriptions of the entomopathogenic genera of the Entomophthorales.

CHAPTER 8

Identification of *Coelomomyces, Saprolegniales* and Lagenidiales

C.E. BLAND, J.N. COUCH and S.Y. NEWELL

*Department of Biology, East Carolina University, Greenville, North Carolina, USA,
Department of Botany, University of North Carolina, Chapel Hill, North
Carolina, USA* and *Rosenstiel School of Marine and Atmospheric Science,
University of Miami, Miami, Florida, USA*

I. Introduction

Although grouped here for convenience, the taxa *Coelomomyces*, Saprolegniales, and Lagenidiales include fungi which differ greatly. All, however, contain zooparasitic representatives. Procedures for the identification of forms within each group follow.

II. *Coelomomyces*

Described first by Keilin (1921), it was Couch (1945, 1962) who recognized the true affinities of members of the genus *Coelomomyces* (Class: Chytridiomycetes, Order: Blastocladiales, Family: Coelomomycetaceae). Host specific, endoparasites of larvae of mosquitoes, black flies, chironomids and possibly tabanids, species of *Coelomomyces* are characterized by posteriorly uniflagellate motile zoospores and gametes, branched or lobed, coenocytic hyphae, and variously ornamented, thick-walled, resting sporangia. Recent studies have shown that, at least in some species, the copepod, *Cyclops vernalis*, serves as host for certain stages in the life cycle. (Whisler *et al.*, 1974, 1975; Federici, 1975; Weiser, 1976). The life cycle of such a species, characterized by alternation between copepod and mosquito hosts, is depicted and explained in Figs. 1–11 and in Chapter 29, Section IV, A.

Infection with *Coelomomyces* is most easily detected in larvae containing mature, resting sporangia (Fig. 8). Heavily infected larvae generally appear rust-coloured to the unaided eye and usually remain at or near the water surface. For microscopic examination, larvae may be mounted in water or in lactophenol. Observation at X100–X200 readily discloses numerous, thick-walled, resting sporangia in infected larvae (Fig. 8). In light infections involving small numbers of sporangia or only mycelia (Figs. 6, 7), closer scrutiny is required. Although infection has been observed in larvae as young as instar I, it is most easily detected in instars III and IV. Identification to species generally requires dissection of infected larvae so that surface ornamentation of liberated sporangia may be seen clearly. Although not necessary for species determination, observation of sporangia by scanning electron microscopy graphically illustrates surface features (Anthony *et al.*, 1971; Bland and Couch, 1973; Nolan *et al.*, 1973; Laird *et al.*, 1975).

Figures 12–19 show resting sporangia of representative species from each of the eight main morphological groups described by Bland and Couch (1973). Also important in the recognition of certain species are the extent and branching pattern of mycelia and presence or absence of thin walled sporangia (mitosporangia). The following key and host index with citations of relevant literature are provided to facilitate identification of species. For a more complete annotated survey of the literature see Strand and Roberts (1977), and for description of microbial control potential see Chapter 29, Section IV,B,C.

A. KEY TO SPECIES OF THE GENUS *COELOMOMYCES*

1. Resting sporangia in Culicidae 7

1'. Resting sporangia in Notonectidae, Chironomidae, or Psychotidae 2

2. Notonectid host. Sporangia smooth, oval, thick-walled 18–20 x 30–40 μm . . . . *C. notonectae* Bogoyavlensky (1922) Keilin (1927).

2'. Chironomid or Psychotid host 3

3. Psychotid host (*Lutzomyia*). Sporangia spherical to subsperical, 35–40 μm . . . . *C. ciferrii* Arêa Leão and Pedroso (1965).

3'. Chironomid host 4

4. Sporangial surface smooth 5

4'. Sporangial surface with regular polygonal areas 6

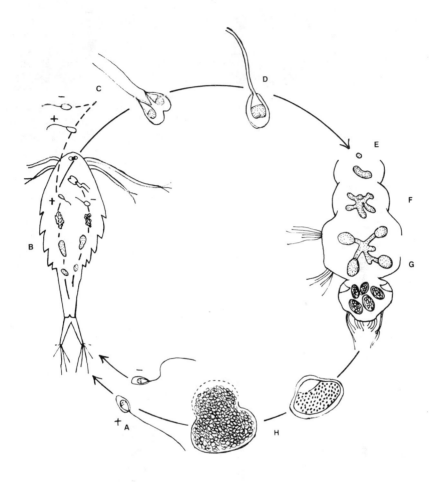

Fig. 1. Life cycle of *Coelomomyces spp.* Zoospores (A) of opposite mating types infect alternate host, *Cyclops vernalis* (B). Each zoospore develops into a thallus and eventually into male and female gametangia. Gametes of opposite mating type (C) fuse either inside or outside the copepod to form the mosquito infecting zygote (D). Thallus development inside the infected mosquito begins with formation of hyphal bodies or hyphagens (E). Hyphagens develop into mycelia (F) and, ultimately, into thick-walled resistant sporangia (G). Under appropriate conditions sporangia release zoospores (H) of opposite mating types. Thin-walled mitosporangia (not figured in drawing, see Fig. 9 for photograph) are formed by some species and give rise to zoospores, presumably capable of infecting other mosquito larvae.

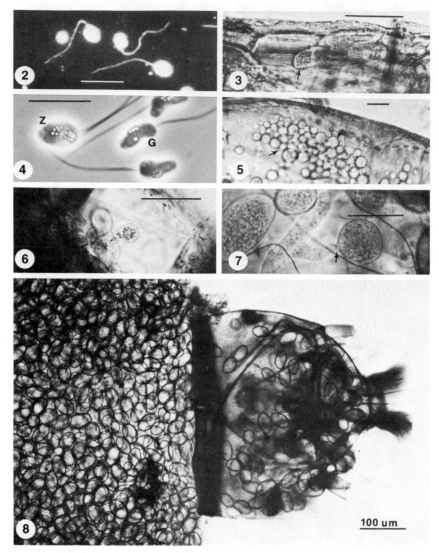

Figs. 2–8. Stages in life cycle of *Coelomomyces* spp. Bar = 10 μm unless otherwise indicated. *Fig. 2* Meiozoospores. Note single whip-lash flagellum. *Fig. 3*. Mycelium (arrow) inside infected copepod. *Fig. 4*. Gametes (G) and biflagellate zygote (Z). *Fig. 5*. Early infection of mosquito evident via "hyphagens" (arrow). *Fig. 6*. Mycelium inside mosquito larva. *Fig. 7*. Formation of sporangium at hyphal tip (arrow). *Fig. 8*. Heavily infected mosquito larva.

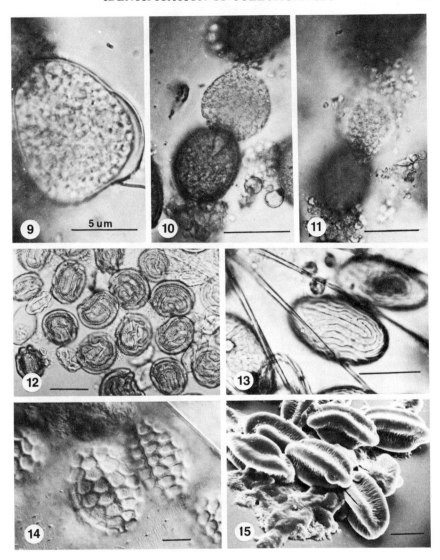

Figs. 9–15. Sporangia of *Coelomomyces* spp. Bar = 10 μm unless indicated otherwise. *Fig. 9.* Thin-walled sporangium. *Fig. 10.* Germination of resting sporangium. Note early stages of cytoplasmic cleavage into zoospores. *Fig. 11.* Zoospore release. *Figs. 12–15.* Resting sporangia of selected species as seen under a variety of conditions. *Figs. 12 and 13.* C. *anophelesicus* (Group V) and *C. lativittatus* (Group II), respectively, as seen via bright-field microscopy. *Fig. 14.* C. n. sp. (Group VIII) as seen via Nomarski Interference — Contrast optics. *Fig. 15.* C. *quadrangulatus* (Group VII) via scanning electron microscopy.

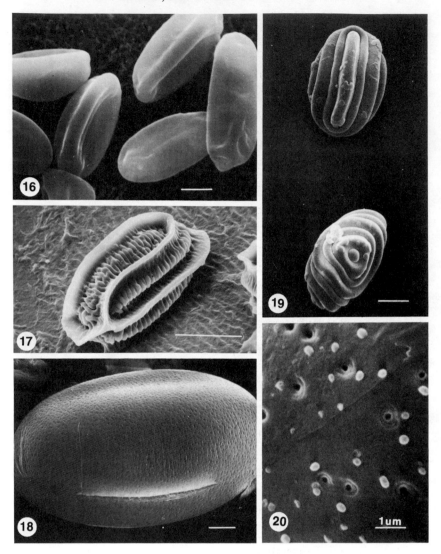

Figs. 16–20. Resting sporangia of selected species of *Coelomomyces* as seen via scanning electron microscopy. Bar = 10 μm unless indicated otherwise. *Fig. 16. C. pentangulatus* (Group I). *Fig. 17. C. uranotaeniae* (Group VI). *Fig. 18. C. psorophorae* (Group III). *Fig. 19. C. bisymmetricus* (Group IV). *Fig. 20. C. stegomyiae* (Group III).

5. Sporangia 8–10 × 21–22 μm. . . . *C. beirnei* Weiser and McCauley (1971).

5'. Sporangia 18–30 × 27–47 μm. . . . .
. . . . *C. tuzetae* Manier, Rioux, Coste and Murand (1970).

6. Sporangial surface with regular polygonal areas containing irregular grooves or pits . . . . *C. chironomi* Răsin (1928).

6'. Sporangial surface with regular polygonal areas containing complex, radiating, deep furrows . . . .
C. canadense (Weiser and McCauley) Nolan stat. et comb. nov. (1978).

7. Sporangial wall smooth. 8

7'. Sporangial wall with pits, striae, bands, ridges or grooves. 9

8. Sporangia oval, 5-angled in end view, 12–18 × 18–40 μm (Fig. 16) . . . . .
. . . . *C. pentangulatus* Couch (1945).

8'. Sporangia oval, rarely flattened on one side . . . . . . Incompletely known species, *C. solomonis* Laird (1956a) or *C. walkeri* van Thiel (1954).

9. Sporangial wall uniformly ornamented with fine, regular pits. 10

9'. Sporangial wall with bands, ridges, grooves, irregular pits, striae or combinations of these features. 15

10. Pits numerous, 35–50/10 μm² (< 1.5 μm apart). Sporangia large, 30–70 × 50–170 μm 11

10'. Pits scattered, 6–15/10 μm² (> 1.5 μm apart). Sporangia medium sized, 20–46 × 30–71 μm. 13

11. Sporangia oval with one side frequently flattened or concave, 37–67 × 46–119 μm. Hyphae 3–10 μ thick 12

11'. Sporangia spherical to oval, 38–70 × 50–177 μm (oval) or 33–60 μm in diam. (spherical). Hyphae 5–15 μm thick. *C. opifexi* Pillai and Smith (1968). . . . .

12. Sporangia oval, always flattened or strongly concave on one side. Hyphae 3–6 μm thick . . . . *C. tasmaniensis* Laird (1956b).

12'. Sporangia oval, frequently flattened on one side. Hyphae 8–10 μm thick (Fig. 18). . . . . *C. psorophorae* Couch (1945).

13. Pits simple 14

13'. Pits star-like (Fig. 20). . . . . *C. keilini* Couch and Dodge (1947).

14. Mean sporangial dimensions 18–24 × 39–44 μm. . . . .
. . . . *C. stegomyiae* var *stegomyiae* Keilin (1921).

14'. Mean sporangial dimensions 25 × 50 μm . . . .
. . . . *C. stegomyiae* var *rotumae* Laird (1959b).

15. Sporangial wall furrowed to form bands and/or irregular pits 16

15'. Sporangial wall with ridges, mounds, and pits or striae or 4-angled in end view 18

16. Wall with irregular, elongate pits. . . . .
. . . . C. punctatus Couch and Dodge (1947).

16'. Wall furrowed to form bands or with irregular, elongate pits and bands . 17

17. Wall with elongate pits and bands . . . . C. dodgei Couch (1945).

17'. Wall with regular to irregular bands (Fig. 13) . . . .
. . . . C. lativittatus Couch and Dodge (1947).

18. Four-angled in end view, transverse striae between angles (Fig. 15) . . .19

18'. Wall with ridges, mounds, and pits or striae 22

19. Sporangia 26–31 × 36–58 μm. . . . .
. . . . C. quadrangulatus var lamborni Couch and Dodge (1947).

19'. Sporangia 8.8–21 × 12.9–40 μm 20

20. Sporangia 8.8–15.6 × 12.9–21.4 μm . . . .
. . . . C. quadrangulatus var parvus Laird (1959b).

20'. Sporangia generally larger than 12 × 19 μm 21

21. Sporangial wall smooth and regular. . . . .
. . . . C. quadrangulatus var quadrangulatus Couch (1945).

21'. Sporangial wall crenate and very irregular. . . . .
. . . . C. quadrangulatus var irregularis Couch and Dodge (1947).

22. Wall with rounded ridges or mounds (dome-like in optical section) separated by lower ridges or mounds or pits or striae. 23

22'. Wall with steep, narrow, anastomosing ridges (spine-like in optical section) separated by depressed areas containing pits or striae 34

23. Wall with rounded ridges or mounds separated by lower ridges or mounds. 24

23'. Wall with rounded ridges separated by depressed areas with pits or striae. 26

24. Ridges forming an irregular network 25

24'. Ridges alternately high and low, originating from a circular dome on one side of the sporangium. Sporangia bisymmetrical (Fig. 19) . . . .
. . . . C. bisymmetricus Couch and Dodge (1947).

25. Network of ridges enclosing a few oblong, elongate (rarely circular) areas. .
. . . . C. sculptosporus Couch and Dodge (1947).

25'. Network of ridges enclosing 3–20 + circular to oblong areas . . . .
. . . . *C. cribrosus* Couch and Dodge (1947).

26. Ridges interconnecting to enclose numerous regular, depressed areas. . . 27

26'. Ridges longitudinal or concentric, connected by pits or striae 28

27. Sporangia small, 12–17 × 23–28 μm. Depressed areas and area adjacent to
dehiscence slit faintly pitted . . . . *C. africanus* Walker (1938).

27'. Sporangia 12–21 × 38–57 μm. Depressed areas devoid of pits or striae . . .
. . . .*C. lairdi* Maffi and Nolan (1977).

28. Ridges anastomosing and usually concentric, connected by striae. Sporangia
flattened on side view (Fig. 12) . . . . .*C. anophelesicus* Iyengar (1935).

28'. Ridges anastomosing or longitudinal to irregular, connected by pits or
striae. Sporangia oval 29

29. Ridges anastomosing, longitudinal to irregular, connected by depressed
areas with striae 30

29'. Ridges longitudinal and parallel to irregular, connected by depressed pitted
areas. 31

30. Sporangia 25–36 × 38–60 μm. Ridges longitudinal, freely anastomosing. .
. . . . *C. indicus* Iyengar (1935).

30'. Sporangia 27–43 × 34–56 μm. Ridges longitudinal, rarely anastomosing. .
. . . .*C. omorii* Laird, Noland and Mogi (1975).

31. Ridges irregular. 32

31'. Ridges longitudinal, parallel. 33

32. Sporangia 18–23 × 35–45 μm, hyaline domes 4–6 μm long × 2–3 μm
high . . . . *C. raffaelei* var *raffaelei* Coluzzi and Rioux (1962).

32'. Sporangia 12.4–20.6 × 21.2–33.6 μm, hyaline domes 3.5 μm long × 1.3–
2.5 μm high . . . .*C. raffaelei* var *parvum* Laird, Nolan and Mogi (1975).

33. Pits arranged in lines to resemble transverse striae . . . .
. . . . *C. grassei* Rioux and Pesch (1960).

33'. Pits scattered . . *C. iliensis* Dubitskii, Dzerzhinskii and Danebekov (1973).

34. Ridges separated by depressed areas containing striae (Fig. 17). . . . .
. . . .*C. uranotaeniae* Couch (1945).

34'. Ridges separated by depressed areas containing pits 35

35. Sporangial faces with same ornamentation . . . . *C. finlayae* Laird (1959b).

35'. Sporangial faces differently ornamented 36

36.   Ridges of one face enclosing hexagonal to irregular areas, inner ridges of other face strongly convoluted . . . .*C. cairnsensis* Laird (1956a).

36'.   Ridges of one face enclosing irregular polygonal areas, some ridges of the other face longitudinal . . . .*C. macleayae* Laird (1959b).

NOTE: *Coelomomyces ascariformis* van Thiel (1954) is considered an inadequately known species and is not included in the key.

B. HOST INDEX FOR SPECIES OF THE GENUS *COELOMOMYCES*

Host	Parasite: *Coelomomyces*	Location	Reference
CULICID (Mosquito) HOSTS			
Aedes aegypti	*C. stegomyiae*	Singapore	Laird, 1959a
Ae. aegypti	*C. stegomyiae*	Ceylon	Rajapaksa, 1964
Ae. albopictus	*C. stegomyiae*	Singapore	Keilin, 1921
Ae. albopictus	*C. stegomyiae*	Ceylon	Rajapaksa 1964
Ae. albopictus	*C. quadrangulatus,* var *lamborni*	Malaya	Couch and Dodge, 1947
Ae. (atropalpus) epactius	*Coelomomyces* sp.	USA (Utah)	Romney *et al.*, 1971
Ae. australis	*C. tasmaniensis*	Tasmania	Laird, 1956b
Ae. australis	*C. opifexi*	New Zealand	Pillai and Smith, 1968; Pillai, 1969
Ae. australis	*C. opifexi*	New Zealand	Pillai and Woo, 1973
Ae. cantans	*Coelomomyces* sp.	England	Service, 1974
Ae. caspius dorsalis	*Coelomomyces* sp.	USSR	Khaliulin and Ivanov, 1973
Ae. cinereus	*C. psorophorae* (?)	France (Strasbourg)	Eckstein, 1922
Ae. cyprius	*Coelomomyces* sp.	USSR	Khaliulin and Ivanov, 1973
Ae. excrucians	*Coelomomyces* sp.	USA (Alaska)	Briggs, 1969
Ae. hebrideus	*Coelomomyces* sp.	Solomon Islands	Genga and Maffi, 1973
Ae. melanimon	*Coelomomyces* sp.	USA (California)	Kellen *et al.*, 1963
Ae. multiformis	*C. stegomyiae*	New Guinea	Huang, 1968
Ae. notoscriptus	*C. finlayae*	Australia	Laird, 1959b
Ae. polynesiensis	*C. macleayae*	Fiji	Pillai and Rakai, 1970
	C. stegomyiae	Tokelau	Laird, 1966
Ae. quadrispinatus	*C. stegomyiae*	New Guinea	Briggs, 1967
Ae scatophagoides	*C. psorophorae* var	Northern Rhodesia	Muspratt, 1946a
Ae. scutellaris	*C. stegomyiae*	Solomon Islands	Laird, 1956a
Ae. simpsoni	*Coelomomyces* sp.	Africa (Bwamba)	McCrae, 1972
Ae. sollicitans	*Coelomomyces* sp.	USA (Louisiana)	Chapman and Woodward, 1966
Ae. taeniorhynchus	*C. psorophorae* var	USA (Florida)	Lum, 1963
Ae. togoi	*Coelomomyces* sp.	USSR	Fedder *et al.*, 1971
Ae. triseriatus	*C. macleayae*	Louisiana	Couch, unpublished
Ae. variabilis	*C. stegomyiae*	New Guinea	Briggs, 1967
Ae. vexans	*C. psorophorae* (?)	France	Eckstein 1922
Ae. vexans	*C. psorophorae* var	USA (Minnesota)	Laird, 1961
Ae. vexans	*C. psorophorae*	USSR	Zharov, 1973
⟍ *des (Macleaya)* sp.	*C. macleayae*	Australia	Laird, 1959b

Host	Parasite: *Coelomomyces*	Location	References
Aedes (Stegomyia) sp.	*C. stegomyiae* var *rotumae*	Rotuma Island	Laird, 1959b
Aedomyia catasticta	*C. indicus*	Australia	Laird, 1956a
Anopheles aconitus	*C. indicus*	India (Bengal)	Iyengar, 1935
An. annularis	*C. indicus*	India (Bengal)	Iyengar, 1935
An. barbirostris	*C. indicus*	India (Bengal)	Iyengar, 1935
An. claviger	*C. raffaelei*	Italy	Coluzzi and Rioux, 1962
An. crucians	*C. bisymmetricus*	USA (Georgia)	Couch and Dodge, 1947
An. crucians	*C. cribrosus*	USA (Georgia)	Couch and Dodge, 1947
An. crucians	*C. dodgei*	USA (Georgia)	Umphlett, 1960, (unpublished)
An. crucians	*C. keilini*	USA (Georgia)	Couch and Dodge, 1947
An. crucians	*C. lativittatus*	USA (Georgia)	Couch and Dodge, 1947
An. crucians	*C. quadrangulatus*	USA (Georgia)	Couch and Dodge, 1947
An. crucians	*C. sculptosporus*	USA (Georgia)	Couch and Dodge, 1947
An. crucians	*C. punctatus*	USA (Louisiana)	Chapman and Glenn, 1972
An. crucians	*C. dodgei*	USA (Louisiana)	Chapman and Glenn, 1972
An. earlei	*C. lativittatus* (var ?)	USA (Minnesota)	Laird, 1961
An. farauti	*C. cairnsensis*	Australia	Laird, 1956a
An. farauti	*Coelomomyces* sp.	British Solomon Is.	Maffi and Genga, 1970
An. funestus	*C. africanus*	Kenya	Haddow, 1942
An. funestus	*C. walkeri* (Walker's type 1)	Sierra Leone	Walker, 1938
An. funestus	*C. indicus*	Northern Rhodesia	Muspratt, 1946a
An. gambiae	*C. africanus*	Kenya	Haddow, 1942
An. gambiae	*C. walkeri*	Sierra Leone	Walker, 1938; (validated by van Thiel, 1962)
An. gambiae	*C. ascariformis*	Africa	Walker, 1938; (validated by van Thiel, 1962)
An. gambiae	*C. indicus*	Northern Rhodesia	Muspratt, 1946a, b
An. gambiae	*C. indicus*	Northern Rhodesia	Muspratt, 1962
An. gambiae	*C. grassei*	North Chad Central Africa	Rioux and Pech, 1960, 1962
An. gambiae	*C. indicus*		Madelin, 1968
An. gambiae	*C. ascariformis*	Volta	Rodhain and Gayral, 1971
An. georgianus	*C. quadrangulatus*	USA (Georgia)	Couch and Dodge, 1947
An. hyrcanus var *nigerrimus*	*C. indicus*	Bengal	Iyengar, 1935
An. jamesi	*C. indicus*	Bengal	Iyengar, 1935
An. minimus	*C. ascariformis*	Philippines	Manalang, 1930
An. pharoensis	*C. indicus*	Egypt	Gad and Sadek, 1968
An. pretoriensis	*C. indicus*	Northern Rhodesia	Muspratt, 1946a
An. pretoriensis	*C. indicus*	Northern Rhodesia	Muspratt, 1962
An. punctipennis	*C. cribrosus*	USA (Georgia)	Couch and Dodge, 1947
An. punctipennis	*C. quadrangulatus*	USA (Georgia)	Couch and Dodge, 1947
An. punctipennis	*C. quadrangulatus* var *irregularis*	USA (Georgia)	Couch and Dodge, 1947
An. punctipennis	*C. sculptosporus*	USA (Georgia)	Couch and Dodge, 1947
An. punctulatus	*C. solomonis*	Solomon Islands and Guadalcanal	Laird, 1956a

Host	Parasite: Coelomomyces	Location	References
An. punctulatus	*C. lairdi*	New Guinea	Maffi and Nolan, 1977
An. quadrimaculatus	*C. punctatus*	USA (Georgia)	Couch and Dodge, 1947
An. quadrimaculatus	*C. quadrangulatus*	USA (Georgia)	Couch and Dodge, 1947
An. ramsayi	*C. indicus*	India (Bengal)	Iyengar, 1935
An. rivulorum	*C. indicus*	Northern Rhodesia	Muspratt, 1946a
An. rivulorum	*C. indicus*	Northern Rhodesia	Muspratt, 1962
An. rufipes	*C. indicus*	Northern Rhodesia	Muspratt, 1946a
An. rufipes	*C. indicus*	Northern Rhodesia	Muspratt, 1962
An. rufipes	*C. africanus*	Upper Volta	Rodhain and Gayral, 1971
An. sinensis	*C. raffaelei* var *parvum*	Japan	Laird *et al.*, 1975
An. squamosus	*C. indicus*	Northern Rhodesia	Muspratt, 1946a
An. squamosus	*C. indicus* (?)	Northern Rhodesia	Muspratt, 1962
An. subpictus	*C. anophelesicus*	India (Bengal)	Iyengar, 1935
An. subpictus	*C. indicus*	Cambodia	Laird, 1959b
An. subpictus	*C. indicus*	India (Bangalore)	Iyengar, 1961, (unpublished)
An. subpictus	*C. indicus*	India	Gugnani *et al.*, 1965
An. tesselatus	*C. walkeri*	Java	van Thiel, 1954
An. vagus	*C. anophelesicus*	India (Bengal)	Iyengar, 1935
An. varuna	*C. indicus*	India (Bengal)	Iyengar, 1935
An. varuna	*C. anophelesicus*	India (Bengal)	Iyengar, 1935
An. walkeri	*C. quadrangulatus* (var ?)	USA (Minnesota)	Laird, 1961
An. walkeri	*C. sculptosporus*	USA (Minnesota)	Laird, 1961
Armigeres obturbans	*C. stegomyiae*	Singapore	Laird, 1959b
Culex annulirostris	*Coelomomyces* sp.	British Solomon Is.	Maffi and Genga, 1970
Cx. erraticus	*C. pentangulatus*	USA (Georgia)	Couch, 1945
Cx. fraudatrix	*C. cribrosus*	North Borneo	Laird, 1959b
Cx. gelidus	*Coelomomyces* n. sp.	Ceylon	Rajapaksa, 1964
Cx. modestus	*C. iliensis*	USSR (Kazakhstan)	Dubitskii *et al.*, 1970, 1973
Cx. orientalis	*Coelomomyces* sp.	USSR (Vladivostock)	Briggs, 1969
Cx. peccator	*Coelomomyces* sp.	USA	Briggs, 1969
Cx. quinquefasciatus	*Coelomomyces* sp.	Ceylon	Rajapaksa, 1964
Cx. quinquefasciatus	*Coelomomyces* sp.	New Caledonia	Lacour and Rageau, 1957
Cx. portesi	*Coelomomyces* sp.	Trinidad	Briggs, 1969
Cx. restuans	*Coelomomyces* sp.	USA (Louisiana)	Chapman and Woodward, 1966
Cx. salinarius	*Coelomomyces* sp.	USA (Louisiana)	Chapman and Woodward, 1966
Cx. simpsoni	*C. indicus*	Northern Rhodesia	Muspratt, 1946a
Cx. tritaeniorhynchus siamensis	*C. quadrangulatus* var. *parvus*	Singapore	Laird, 1959a
Cx. tritaeniorhynchus siamensis	*C. cribrosus*	Singapore	Laird, 1959a
Cx. t. summorosus	*C. cribrosus*	British North Borneo	Laird, 1959b
Cx. t. summorosus	*C. omorii*	Japan	Laird *et al.*, 1975
Culiseta inornata	*C. psorophorae* var	Canada (Alberta)	Shemanchuk, 1959
Opifex fuscus	*C. opifexi*	New Zealand	Pillai and Smith, 1968

Host	Parasite: *Coelomomyces*	Location	References
Psorophora ciliata	*C. psorophorae*	USA (Mississippi)	Laird, 1961
Ps. howardii	*C. psorophorae*	USA (South Carolina)	Couch and Dodge, 1947
Toxorynchites rutilus septentrionalis	*C. macleayae*	USA (Louisiana)	Nolan *et al.*, 1973
Uranotaenia barnesi	*Coelomomyces* sp.	British Solomon Is.	Maffi and Genga, 1970
Ur. sapphirina	*C. uranotaeniae*	USA (Georgia)	Couch, 1945
CHIRONOMID (Midge) HOSTS			
Ablabesymia monilis	*C. beirnei*	Canada	McCauley, 1976
Chironomus modestus	*C. chironomi* var *canadense*	Canada	McCauley, 1976
Chironomus paraplumosus	*C. chironomi*	Czechoslovakia	Weiser and Vávra, 1964
Chironomus plumosus	*C. chironomi*	Czechoslovakia	Răsin, 1928, 1929; Weiser, 1976
Cladotanytarsus sp.	*C. beirnei*	Canada	Weiser and McCauley, 1971; McCauley, 1976
Cricotopus sp.	*C. tuzetae*	France	Manier *et al.*, 1970
Heterotrissocladius grimshawi	*C. beirnei*	Canada	McCauley, 1976
Microtendipes pedellus	*C. beirnei*	Canada	McCauley, 1976
Orthocladius sp.	*C. tuzetae*	France	Manier *et al.*, 1970
Pagastiella sp.	*C. beirnei*	Canada	McCauley, 1976
Parakiefferiella nigra	*C. beirnei*	Canada	McCauley, 1976
Phaenospectra albascens	*C. beirnei*	Canada	McCauley, 1976
Procladius denticulatus	*C. beirnei*	Canada	McCauley, 1976
Psectrocladius semicirculatus	*C. chironomi* var *canadense*	Canada	Weiser and McCauley, 1971, 1972; McCauley, 1976
Psectrocladius sp.	*C. chironomi* var *canadense*	Canada	McCauley, 1976; Nolan, 1978
Tanytarsus spp.	*C. beirnei*	Canada	Weiser and McCauley, 1971; McCauley, 1976
Zavielia sp.	*C. chironomi* var *canadense*	Canada	McCauley, 1976
HEMIPTERAN (True bug) HOSTS			
Notonecta sp.	*C. notonectae*	USSR	Bogoyavlensky, 1922; Keilin, 1927
SIMULID (Black fly) HOSTS			
Simulium sp.	*Coelomomyces* sp.	Morocco	Briggs, 1967
Simulium metallicum	*Coelomomyces* sp.	Honduras	Garnham and Lewis, 1959
TABANID (Horse fly) HOSTS			
Chrysops reflictus	*C. milkoi**	USSR	Andreeva, 1972
Tabanus autumnalis	*C. milkoi**	USSR	Andreeva, 1972

* Recently put in the Entomophthorales, not of *Coelomomyces* by J.N. Couch.

Host	Parasite: Coelomomyces	Location	Reference
PSYCHODID (Sand fly) HOSTS			
Lutzomyia sp. or spp. C. ciferrii		Brazil	Leão and Pedroso, 1965

III. Saprolegniales

The order Saprolegniales (Class: Oomycetes) is characterized by: (1) biflagellate (anteriorly directed tinsel and posteriorly directed whiplash), dimorphic zoospores, always delimited within zoosporangia; (2) generally eucarpic, nonsegmented, aseptate thalli; and (3) uni- or pluriovulate oogonia with oospores possessing an ooplast. Although the order has numerous species, relatively few are zooparasitic. The following is a key to accepted families of the Saprolegniales and to major zooparasitic forms occurring on invertebrates. Figs. 21–29, illustrate diagnostic features of selected zooparasites. The family Haliphthoraceae, considered by some (Vishniac, 1958; Dick, 1973) to belong in the Saprolegniales, is treated (as by Sparrow, 1973b) in the section on Lagenidiales.

A. KEY TO FAMILIES AND SELECTED ZOOPARASITIC MEMBERS OF THE SAPROLEGNIALES

1.   Filamentous, eucarpic, non-septate thallus. Zoosporangia delimited by basal septum. Female gametangia distinct. Oospores with ooplast and granular contents. Zoospore cysts 8–12 μm . . . . *Saprolegniaceae*   2

1′. Filamentous, holocarpic to eucarpic thallus. Morphologically distinct gametangia rare but sometimes present. Parthenospores (oospores) with thick, two-layered, separable endospore membrane. Zoospore cysts 5–10 μm . . . . *Leptolegniellaceae*   9

1″. Saccate to stoutly filamentous, nonseptate, holocarpic thallus. Parasitic in diatoms, fungi, algae or eggs of marine invertebrates. Oospores lacking or, when present, without complex, layered wall. Zoospore cysts 3–6 μm . . . . *Ectrogellaceae*   10

2.   Zoosporangia filamentous, not differentiated from vegetative hyphae, zoospores in one rank, at least distally. Zoospores swimming immediately after discharge from sporangium . . . . *Leptolegnia*   3

2′. Zoosporangia filamentous, not differentiated from vegetative hyphae, zoospores in one rank. Zoospores encysting in cluster at mouth of sporangium . . . . *Aphanomyces* (Fig. 21) After Scott (1961)   4

3.   Marine. Parasitic in *Eurytemora hirundoides* (copepod) . . . .
. . . . *L. baltica* Höhnk and Vallin (1953)

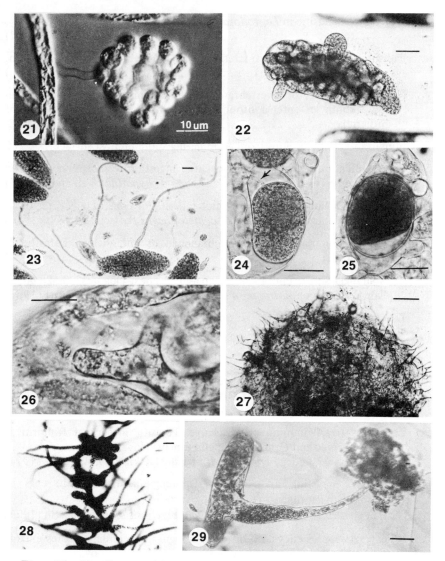

Figs. 21–29. Zooparasitic members of the Saprolegniales. (Bar = 40 μm unless indicated otherwise). *Fig. 21. Aphanomyces* sp. Encystment of primary zoopores at sporangial tip. *Figs. 22–25. Aphanomycopsis sexualis. Fig. 22.* Host egg containing a sporangium. Note lobes arising from sporangium and penetrating chorion. *Fig. 23.* Sporangium with four discharge tubes. *Fig. 24.* Oogonium with attached fertilization tube (arrow) from antheridium. *Fig. 25.* Mature oospores. *Fig. 26. Leptolegniella marina.* Hyphae in gill leaflet of *Pinnotheres. Fig. 27. Atkinsiella dubia.* Discharge tubes penetrating surface of infected egg of *Pinnotheres. Fig. 28–29. Atkinsiella entomophaga. Fig. 28.* Thalli in infected midge eggs. *Fig. 29.* Discharge of primary zoospores.

3'.   Freshwater. Parasitic in *Leptodora kindtii* (cladoceran). . . . .
. . . . *L. caudata* deBary (1887)

4.   Parasitic on crayfish. Characteristics of oogonia unknown (Johnson, personal communication) . . . . *A. astaci* Schikora (1906)

4'.   Parasitic on variety of invertebrates. Outer and inner contours of oogonial wall smooth, or outer contour appearing roughened due to presence of antheridia 5

5.   Fungus infection develops from germinating spore on host surface 6

5'.   Fungus infection develops within host after ingestion of zoospores. Parasitic in *Hydatina* (rotifer) . . . . *A. hydatinae* Valkanov (1931)

6.   Zoosporangia short, unbranched, not tapering or constricted. Parasitic in *Acineto* (protozoan) . . . . . *A. acinetophagus* Bartsch and Wolf (1938)

6'.   Zoosporangia variable in length, branched or unbranched. Parasitic in cladocerans 7

7.   Zoosporangia short, unbranched, tapered at apex. Oogonia 20–28 μm in diam. Parasitic on *Bosmina* or *Cyclops*. . . . . *A. bosminae* Scott (1961)

7'.   Zoosporangia long, branched or unbranched. Oogonia > 28 μm in diam. . .
. . . . 8

8.   Zoosporangia branched, tapered at apex, constricted at base. Parasitic in *Daphnia*. . . . . *A. daphniae* Prowse (1954)

8'.   Zoosporangia unbranched, not tapered at apex. Parasitic in *Daphnia* . . . . .
. . . . *A. patersonii* Scott (1961)

9.   Zoospores free-swimming on discharge. Parasitic in eggs of *Pinnotheres* (crab), *Barnea* and *Cardium* (lamellibranchs) (Fig. 26) . . . .
. . . . *Leptolegniella marina* (Atkins, 1954a) Dick (1971)

9'.   Zoospores encysting at mouth of sporangium. Parasitic in eggs of Chironomidae (midges) (Figs. 22–25). . . . .
. . . . *Aphanomycopsis sexualis* Martin (1975)

10.   Primary zoospores encysting inside sporangium. Secondary spores emerge through irregular discharge tubes. Marine. Parasitic on ova, and zoeae of *Pinnotheres* and other crabs (Fig. 27) . . . .
. . . .*Atkinsiella dubia* (Atkins, 1954b) Vishniac (1958)

10'.   Primary zoospores encysting at mouth of discharge tube. Freshwater. Parasitic in eggs of midges and caddis flies (Figs. 28, 29). . . . .
. . . . *Atkinsiella entomophaga* Martin (1977)

Dick (1971) erected the family Leptolegniellaceae but considered it to have somewhat questionable placement in the Saprolegniales. Inclusion of *Atkinsiella* in the Ectrogellaceae is based on occurrence of dimorphic zoospores in members of this genus and on similarities between it and recognized members of the Ectrogellaceae (Sparrow, 1973b, c).

IV. Lagenidiales

The following key includes all described zooparasitic species of Lagenidiales and allied (lagenidioid) fungi of uncertain affinities. It is amended from that of Newell *et al.* (1977) and updates and/or complements keys of Sparrow (1960; 1973b), Johnson and Sparrow (1961), Cooke and Godfrey (1964), and Barron (1977). The first few dichotomies in the key differentiate the Lagenidiales and lagenidioid fungi from other zooparasitic fungi, and descriptive references to these other groups are given. Other notes regarding use of the key are: (1) host animals are not emphasized because host ranges may be quite wide and overlap, e.g., those of *Haptoglossa heterospora* and *Myzocytium vermicola*; (2) light-microscopic observations of cytology (e.g. presence of vacuoles) are not emphasized because of their relative imprecision and subjectivity, and the possibility that conditions of observation (e.g. salinity, temperature) might affect them; (3) "oospores (?)" signifies absence of clear evidence of antheridia, and that parthenogenetic production of oospores is suspected; (4) if only a single, one-celled infection thallus is observed, the fungus may not be identifiable, for several of the fungi included in the key can produce these; (5) basic taxonomic information is provided as each species is reached in the key, and references follow the key; (6) incompletely described species, possibly lagenidialean, are listed after the references by host organism.

Although this key is constructed along systematic lines, all species of given genera do not lie together in it. This is an indication of the weak taxonomic differentiation of some genera and species, and the questionable generic assignment of some species. The order is in need of critical examination and revision.

A. KEY TO THE LAGENIDIALEAN AND LAGENIDIOID ZOOPARASITES

1. Fungi parasitic (digesting host tissues and causing death), entirely or mostly within bodies of animals, spores formed by simultaneous cleavage of all or part of the sporangial cytoplasm and released by expulsion and/or flagellar motility through a papilla or long discharge tube; or, if without a discharge tube or papilla, then spores formed after successive division of the entire sporangial cytoplasm; never with discharge tube and successive cleavage of the entire sporangial cytoplasm (*Dermocystidium marinum* Mackin *et al.*, 1950: see Perkins and Menzel, 1966; Perkins, 1974a; Jones, 1976); thallus not consisting of massed cells which slowly glide among one another (Labyrinthulales: see Perkins, 1974b; Quick, 1974; Alderman *et al.*, 1974; Jones, 1976). 2

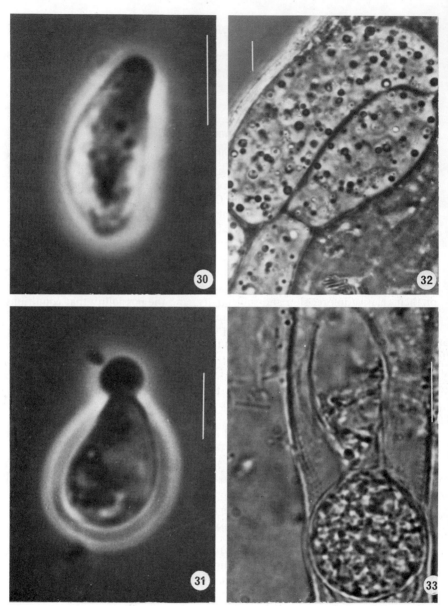

Figs. 30–35. Stages in the life cycle of a lagenidialean fungus, *Myzocytium vermicola.* Bar = 5 µm. *Fig. 30.* Quiescent zoospore, immediately after loss of flagella. *Fig. 31.* Production of an adhesive protuberance by a quiescent zoospore. *Fig. 32.* Sporangial thallus in nematode host; individual sporangia beginning to undergo disjunction at septa. *Fig. 33.* Oogonium (left) and antheridium (right); antheridial cytoplasm is flowing through a septal pore into the oogonium, where the oospore is taking shape. *Fig. 34.* Discharge tube, extending from sporangium through host cuticle; refractile material at tip indicates imminent release of zoospores.

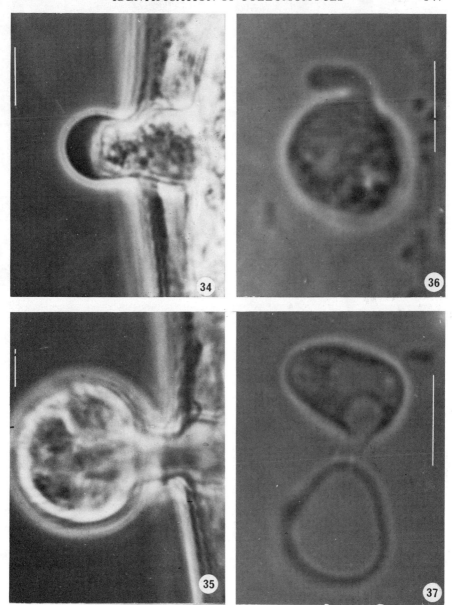

Figs. 30–35 (cont.). *Fig. 35.* Moment of zoospore release; zoospores are tightly gathered in a ball at the mouth of the discharge tube, where they remain momentarily.

Figs. 36 and 37. Formation of a glossoid body by *Haptoglossa heterospora.* Bar = 5 µm. *Fig. 36.* Aplanospore with glossoid body initial. *Fig. 37.* Empty aplanospore wall with attached, maturing glossoid body.

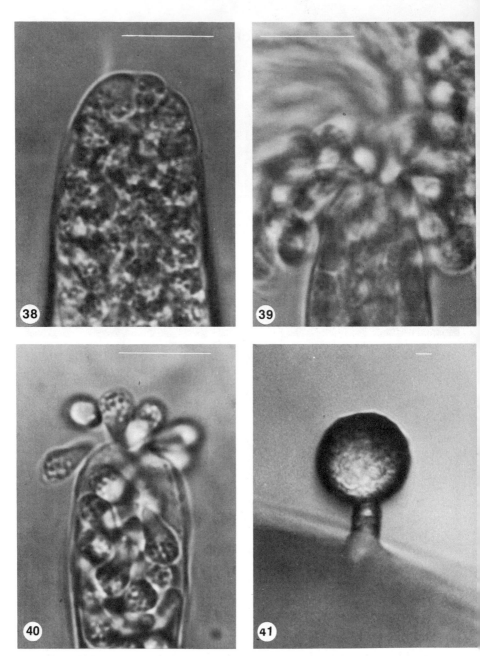

Figs. 38–41. Release of aplanospores by a lagenidioid fungus, *Gonimochaete latitubus*. Bar = 5 µm. *Fig. 38.* Tip of discharge tube immediately prior to aplanospore release. *Fig. 39.* Moment of expulsion of aplanospores. *Fig. 40.* Free aplanospores at resealed tip of discharge tube; aplanospores remaining within the tube are released in subsequent expulsions. *Fig. 41.* Aerial discharge tube with ball of aplanospores characteristic of *Gonimochaete* spp.

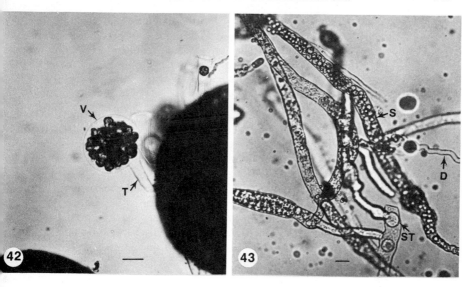

Fig. 42. Sporulation of *Lagenidium callinectes* on ova of the blue crab, *Callinectes sapidus.* Note discharge tube (T) and vesicle (V) containing zoospores. Bar = 10 µm.

Fig. 43. Hyphae of *Haliphthoros milfordensis*. Note sporangia (S), discharge tube (T), and subthalli (ST). Bar = 20 µm.

1'. Parasitic fungi in the bodies of animals, but otherwise not entirely as above: spores formed externally on spore-bearing structures, or, if internally formed, never with flagella and/or never expelled from sporangia Zygomycetes, Hyphomycetes, Basidiomycetes (see Cooke and Godfrey, 1964; Hesseltine and Ellis, Waterhouse, and Duddington in Ainsworth *et al.*, 1973; Kendrick and Carmichael, 1973; Giuma and Cooke, 1972).

2. Sporangiospores immotile (aplanospores), without flagella when released by forcible expulsion from sporangia (Figs. 38–40), or if sporangiospores biflagellate when released then these always encyst and form an attached glossoid body (Figs. 36, 37); entire thallus (all cells) developing directly into sporangia and/or sexual cells (if any). . . . .
. . . . .lagenidioid fungi, 27

2'. Sporangiospores motile (zoospores) upon release and without an attached glossoid body after encystment 3

3. Zoospores with two oppositely directed flagella attached laterally within a groove (the groove is not always easily seen); zoospores never encysting near or within the sporangium and later re-emerging as zoospores of a

second shape (Saprolegniales, Section III), and never producing a glossoid body; sexual spore (if any) formed by fusion of contents of antheridial and oogonial cells, or parthenogenetically, and maturing aplerotically (except in the case of 12 below) within oogonia (oospores); entire thallus (all cells) developing directly into sporangia and/or sexual cells; sporangia not on sporangiophores . . . .
. . . . lagenidialean fungi (Fig. 30–35),   4

3'. With a circumferentially directed flagellum: Ellobiopsidae (Galt and Whisler, 1970); flagella and zoospores otherwise than in (3) and/or thallus eucarpic (with cells which do not develop directly into reproductive structures). . . . .
. . . . .Mastigomycotina other than Lagenidiales (Sparrow, 1973a).
Note: Karling (1944) described a zooparasite of rotifers and nematodes as *Lagenidium parthenosporum*; it, however, has zoospores of two shapes, a primary one which encysts at the discharge tube orifice, and which gives rise subsequently to a secondary one – this feature prevents inclusion of this species in the Lagenidiales as presently defined (Sparrow, 1973b).

4. Zoospores released, either cleaved or uncleaved, into a vesicle (Fig. 35) at the tip of the discharge tube; vesicle may be quickly evanescent (in seconds) and/or difficult to see, but is indicated when zoospores are situated at the discharge tube tip in a spheroid bundle before swimming away (Fig. 35) 14

4'. Zoospores maturing within sporangium, escaping by swimming out the tip of the discharge tube or broken sporangial wall; never gathered tightly together outside the sporangium at the tip of the discharge tube before swimming away. 5

5. Thallus one-celled, spheroid; zoospores formed by successive division of entire sporangial cytoplasm; within bodies of euglenophyte or cryptomonad protozoans. . . . .*Pseudosphaerita*,   6

5'. Thallus tubular or lobed; unknown within bodies of protozoans. 7

6. In *Euglena* spp . . . .*Pseudosphaerita euglenae*

6'. In *Cryptomonas* spp. . . . .*Pseudosphaerita radiata*

7. Sporangial thallus septate (but cytoplasm not regularly spatially discontinuous); disjunction of sporangia may occur at the septa (Fig. 32). . .8

7'. Sporangial thallus rarely septate, converting into one large sporangium in which cytoplasm may be spatially discontinuous without separation by septa (these cytoplasmic units are termed "subthalli": Fig. 43) 11

8. Sporangial thalli not undergoing disjunction at the septa; sporangia obovoid to irregular, to 82 μm diam.; discharge tube 15–142 μm in length; zoospores "more or less" pyriform, 5 × 2 μm; oospores (?)

spherical, 40–90 μm diam.; in clam and oyster larvae (but taxonomically described only as it appears in axenic culture) . . . .
. . . . *Sirolpidium zoophthorum*

8'. Sporangia often formed after disjunction at septa of cells of the thallus . 9

9. Sporangia irregularly globose or saccate (22–60 μm diam.) though an "isthmus" may connect them at the base; discharge tube often contorted, "short or long"; zoospores 8–11 × 3–6 μm; not germinating by formation of adhesive protuberances; oospores (?) spherical, 18–38 μm diam.; in rotifer eggs . .unnamed species (Seymour and Johnson, 1973)

9'. Sporangia spherical to ellipsoid, without interconnecting isthmus; zoospores germinating by formation of adhesive protuberances (Figs. 30–31) 10

10. Oospores with "slightly roughened" walls, and zoosporic adhesive protuberances spherical to ellipsoid (to 2.5 μm long), often in chains; sporangia 20–40 × 15–30 μm; discharge tube 10–12 (to 250 under "dry" conditions) × 3–7 μm; zoospores ellipsoid to cylindric or pyriform, 5–11 × 5 μm; antheridia hypogynous, elongate–cylindrical; oogonia spherical; oospores spherical, 10–20 μm diam.; in bodies of *Rhabditis* sp. (nematode) . . . . *Myzocytium humicola*

10'. Oospores with "echinulate" walls, and zoosporic adhesive protuberances spherical, often in chains; sporangia 18–38 × 14–30 μm; zoospores 8–12 μm long, tapering slightly apically; antheridia hypogynous; oospores spherical, 15–22 μm diam.; in bodies of *Rhabditis terricola* (nematode) . . . . *Myzocytium glutinosporum*

11. Never form subthalli (Fig. 43) 12

11'. Regularly form subthalli, which become sporangial, by spatial discontinuation of cytoplasm (not septal) 13

12. Sporangia unbranched, obovoid to elongate tubular, 22–130 × 8–20 μm; discharge tube (one or two) 6–35 × 4–8 μm; zoospores 6–7 × 4–5 μm; oospores (?) broad to narrowly barrel-shaped, conic, ovoid, subspherical, or angular, 12–22 × 7–14 μm, formed by cleavage and disjunction of oosporic thalli; in rotifer eggs. . . . . *Lagenidium septatum*

12'. Sporangia branched, irregularly lobed; discharge tube "short" zoospores 6–8 μm long; oospores unknown; in eggs of *Cyclops* (copepod) . . . . . .
. . . . *Blastulidiopsis chattoni*

13. Young sporangial thallus with generally cylindrical, branched hyphae; sporangia tubular to irregularly ellipsoid or ovoid, up to 350 × 10–56 μm; discharge tube 100–725 × 5–9 μm; zoospores 7–11 × 5–11 μm; oospores unknown; in *Homarus americanus* (American lobster), in ova and embryos of *Urosalpinx cinerea* (marine snail), and ova of

Pinnotheres pisum (marine crab) and other crustacea; described only from axenic culture . . . . *Haliphthoros milfordensis*

13'. Young sporangial thallus with irregularly lobed branches of varying diameter; sporangial and discharge tube sizes not reported; zoospores 10 × 8 μm; oospores unknown; in crustacean eggs; described only from axenic culture. . . . . *Haliphthoros* sp. (Sparrow, 1975)

14. Vesicle (or zoospore bundle) at tip of discharge tube either completely formed before contents of sporangium flow into it, or persisting (> 5 min) following release of zoospore mass into it (Fig. 42); endozoic thallus extensive and branched (except for 18) 15

14'. Vesicle never preformed, deliquescing and zoospores swimming away from discharge tube quickly (< 5 min) (Figs 34–35) 19

15. Mature vesicle full of cytoplasm or zoospores, or not visible (possibly very thin and tightly appressed to the zoospore mass) 18

15'. Mature vesicle only partially filled, clearly visible (Fig. 42) 16

16. Zoospores 15–20 × 7–10 μm, often with pronounced posterior apiculus; cytoplasm migrating from elongate hyphal zoosporangial segments (3–5 μm diam.) into vesicles at tips of hyphal discharge tubes (to 150 × 3–5 μm); zoospores released by rupture of vesicle wall, and forming spherical cysts primarily at body orifices of host; in bodies of unidentified nematodes . . . . *Lagenidium caudatum*

16'. Zoopores with largest dimension < 14 μm, without a pronounced posterior apiculus. 17

17. Vesicle completely formed before sporangial contents flow into it and deliquescing rapidly (~ 30 sec) after beginning of rapid zoospore motion; sporangia 10–39 μm wide, length "variable"; zoospores 9–10 × 7–9μm; oospores (?) spheroid or ovoid to ellipsoid, 18–27 × 16–23 μm; in ova of *Chthamalus fragilis* and *Chelonibia patula* (barnacles) . . . . *Lagenidium chthamalophilum*

17'. Vesicle forming as sporangial contents migrate into it, and persisting (> 5 min) after zoospores swim away; cytoplasm migrating out of hyphal sporangia and into discharge tubes (25–140 × 11–29 μm) and vesicles (as large as 100 μm diam.); zoospores 9 × 13 μm; oospores (?) spheroid to ovoid, 18–36 μm diam.; in ova of *Callinectes sapidus* (the blue crab), *Chelonibia patula* (barnacle), and *Penaeus setiferus* (shrimp). . . . . *Lagenidium callinectes*

18. Sporangia irregularly tubular to ellipsoid, 30–40 × 6–9 μm; discharge tube 7–25 × 3–4 μm; zoospores 8 × 6 μm, capable of encystment and re-emergence (without changing shape): spheroid to ovoid (8–18 μm diam.) oospores formed after fusion of tubular antheridia (20–30 × 5–8 μm,

androgynous but not hypogynous) and "broadly spindle-shaped" oogonia; in eggs of *Distyla* sp. (rotifer) . . . . *Lagenidium distylae*

18′. Sporangia tubular, 50–300 × 4–6 μm; discharge tube 50–300 × 6–10 μm; zoospores 9–10 × 8–9 μm, not encysting and re-emerging; spheroid (18–30 μm diam.) oospore formed by fusion of contents of hypogynous (or diclinous ?) antheridia with oogonia; in mosquito larvae (species of *Aedes, Anopheles, Culex, Culiseta*), unidentified copepods, and *Daphnia* (cladoceran); see also Chapter 29, Section III . . . .
. . . . *Lagenidium giganteum*

19. Sporangial thalli unicellular throughout developmental stages 20

19′. Sporangial thalli regularly (but not necessarily always) septate in developmental stages 21

20. Zoospores 8 × 6 μm; sporangia irregularly spheroid to ellipsoid, lobed or unlobed, 20–40 × 12–25 μm; discharge tube a short papilla, 4–5 μm wide; oospore not observed; in eggs and embryos of rotifers . . . .
. . . . *Lagenidium oophilum*

20′. Zoospores 4 × 3 μm; sporangia spheroid to ovoid, not lobed, 6–25 × 5–15 μm; discharge tube 5–20 × 4–6 μm; sexual thalli 2-celled, hypogynous antheridia ovoid, 7–9 × 5–7 μm, oogonia ovoid to spheroid, 12–18 μm diam.; oospore spheroid to ovoid, 10–14 × 7–9 μm; in adult *Distyla* sp. (rotifer) . . . . *Myzocytium microsporum*

21. Zoospores encysting, attaching to, and penetrating adult nematode host exclusively at body orifices; sporangia irregularly cylindrical, to 50 μm in length; discharge tubes "short"; zoospores 8–10 × 5–6 μm; hypogynous antheridia and oogonia formed from thallic segments similar to sporangia; oospores smooth, 10–17 μm diam.; in bodies of *Rhabditis* sp. (nematode) . . . . *Myzocytium intermedium*

21′. Zoospores not encysting on and penetrating host primarily at body orifices
. . . . 22

22. Zoospores not more than 7 μm in largest dimension, sporangial thallus not undergoing disjunction at septa, zoospore germinating by production of hyphae (penetration tubes), never adhesive protuberances (Figs. 30–31), or, if zoospores have adhesive protuberances, then mature sporangia remain joined by small-diameter tubules 23

22′. Zoospores up to 10–11 μm in largest dimension, or, if less, then sporangia often (but not always) disjoining from one another at maturity, never joined by small-diameter tubules, and zoospores always terminating by production of adhesive protuberances (Figs 30–31). 24

23. Mature sporangia connected by small-diameter (~ 2 μm) tubules, rather than septa; sporangia oblong to spheroid, not lobed, 11–27 × 11–23 μm; discharge tube 5–24 × 3–4 μm; zoospores 3–6 × 3–4 μm;

oospores (?) pale brown, smooth, ovoid to spheroid, 16–27 μm diam.; in rotifers . . . . *Myzocytium indicum*

23'. Mature sporangia connected by septa, not tubules; sporangia subspheroid to ellipsoid (sometimes curved, sometimes lobed) 19–36 × 14–17 μm; discharge tube 10–53 × 6–8 μm; zoospores 7 × 4–5 μm; antheridia hypogynous; oospores spheroid, 10–16 μm diam., smooth, hyaline; in eggs of *Distyla* sp. (rotifer) . . . .*Myzocytium fijiensis*

24. Zoospores germinating by production of adhesive protuberance (Figs 30–31) or a chain of these, but never a hypha, and sporangia often (but not always) disjoining from one another at maturity; sporangia subspheroid, ovoid, ellipsoid, or irregular, 21–54 × 6–12 μm; discharge tubes 5–37 × 4–6 μm; zoospores 5–10 × 4–6 μm; antheridia hypogynous, 17–23 × 9–11 μm; oogonia 21–28 × 15–20 μm (Fig. 33); oospores slightly tuberculate, spheroid to subspheroid, 17–22 μm diam.; in bodies of several nematode genera . . . . *Myzocytium vermicola*

24'. Zoospore germinating by production of penetration hyphae, not adhesive protuberances. 25

25. Zoospores "lens-shaped" (concave–convex), forming hemispherical cysts on host 26

25'. Zoospores not clearly concave–convex, forming spherical cysts on host; sporangia subspheroid to irregularly lobed, 5–17 μm diam. (length "variable"), often disjoining at maturity; discharge tubes "generally short"; zoospores 10–11 × 6–7 μm; antheridia hypogynous; oospores smooth, spheroid, 12–15 μm diam.; in adults and eggs of rotifers . . . . *Myzocytium zoophthorum*

26. Oospore spherical, thick-walled, smooth, 12–22 μm diam.; sporangia globose, sometimes apiculate or biapiculate, 20–40 × 10–22 μm, often disjoining at maturity; zoospores 8–10 μm in length; antheridia hypogynous; in *Rhabditis terricola* (nematode) . . . *Myzocytium papillatum*

26'. Oospore markedly tuberculate or angular, 14–22 μm diam.; sporangia ellipsoid to ovoid or irregular, 15–50 × 8–16 μm, often disjoining at maturity; zoospores 8–12 × 4–5 × 2–3 μm; antheridia hypogynous; in unidentified nematodes . . . .*Myzocytium lenticulare*

27. All or most of sporangium evacuated prior to aplanospore expulsion, through movement of cytoplasm into discharge tube(s); discharge tubes tend to be aerial (Fig. 41); aplanospore mass often remains at the tips of aerial discharge tubes in a spheroid, cohesive mass . . . .
. . . . *Gonimochaete*,   28

27'. Sporangia not evacuated prior to sporangiospore release; aplanospores not remaining in spheroid cohesive heads at tips of aerial discharge tubes. . .
. 30

28. Sporangia always < 50 μm in length, never branched, thallus often undergoing disjunction at septa; discharge tubes never > 10 μm wide at the base 29

28′. Mature sporangia 30–468 × 7–28 μm, tubular, often branched; thallus aseptate, not undergoing disjunction; discharge tubes broadly pyriform to cylindrical, 14–75 × 5–44 μm; "bubble-like" formations occur at the base of migrating sporangial cytoplasm (Fig. 40); aplanospores conic to ovoid or pyriform, 5–7 × 3–4 μm, germinating by production of a single, apical, spheroid adhesive protuberance (1–2 μm diam.); sexual phenomena unknown; in *Rhabditis marina* (littoral marine nematode) . . . . *Gonimochaete latitubus*

29. Aplanospores truncate–cylindric, 4–10 × 3–4 μm, in single row within discharge tube; sporangial thalli undergoing repeated disjunction at septa, never becoming more than one-septate; mature sporangia tubular, 6–33 × 6–11 μm; discharge tube 10–100 × 3–5 μm; aplanospores germinating by production of a single, spheroid to subspheroid (2–3 × 1–2 μm) adhesive protuberance; sexual phenomena unknown; in *Acrobeloides* sp. (nematode) . . . . *Gonimochaete horridula*

29′. Aplanospores pyriform, 10–18 × 4–6 μm, in double row within discharge tube; sporangial thalli may have several septa and undergo disjunction at septa; mature sporangia tubular, 20–40 × 14–20 μm; discharge tubes up to 110 × 6–8 μm; aplanospores germinating by the production of a single, spheroid, adhesive protuberance (2–3 μm diam); sexual phenomena unknown; in *Diploscapter* sp. (nematode) . . . .
. . . .*Gonimochaete pyriforme*

30. Aplanospores narrowly clavate and slightly arched, 20–25 × 3 μm, germinating by production of one or a chain of small adhesive protuberances at the narrow end, or attaching directly to host's cuticle by the narrow end; sporangial thallus often septate, undergoing disjunction at septa; mature sporangia tubular, 16–110 × 6–7 μm; discharge tube very short to 25 μm; oospores said to be formed after conjugation of separated sporangium-like thalli (antheridium is narrower than oogonium); oospores smooth, very opaque, 15–30 μm diam.; in *Anguillula* sp., *Rhabditis* spp., and *Chromadora ratzeburgensis* (nematodes) . . . .
. . . . . *Protascus subuliformis* and *Protascus subuliformis* var *maupasii*

30′. Aplanospores or zoospores not narrowly clavate, < 15 μm long 31

31. Primary spore an aplanospore which gradually converts to a biflagellate zoospore, and does not form a glossoid body (Figs 36, 37); sporangial thallus septate at maturity, and sporangia may undergo disjunction at septa; sporangia ellipsoid to ovoid, 25–55 × 10–15 μm; discharge tube up to 25 × 8–10 μm; aplanospores 8–10 × 4–6 μm; zoospores concave–convex, 8–10 μm long, forming hemispherical cysts on host cuticle;

oospores (?) spherical, smooth-walled, 12—18 μm diam.; in an unidentified nematode. . . . . *Myzocytium anomalum*

31'. Primary spore either an aplanospore or zoospore, but always forming an attached glossoid body (Figs 36, 37); sporangial thallus aseptate. . . . . .
. . . . *Haptoglossa*,   32

32. Sporangiospores aplanospores; mature sporangia tubular, 7—350 × 4—18 μm; discharge papillae (1—4 per sporangium) 4—7 μm wide; aplanospores spheroid or with rounded angles, 5—6 μm diam., or, in separate sporangia, 8—10 μm diam.; glossoid bodies shaped like "3-cornered cushions" with one curved corner longer than the others, 5—10 × 4—8 × 5—6 μm (occasionally 9—13 × 7—10 × 7—10 μm); tertiary infection spores cylindrical, 5—7 × 2—3 μm; sexual phenomena may include the fusion of cells of parallel sexual thalli; sexual spores unknown; in several species of nematodes . . . . *Haptoglossa heterospora*

32'. Sporangiospores biflagellate zoospores; mature sporangia tubular, 20—450 × 10—60 μm; discharge tubes one to several, occasionally branched, 20—50 × 5—15 μm; zoospores cylindrical to broadly conical 7—9 × 4—7 μm; glossoid bodies broadly conical with rounded base, curved at tip, 5—13 × 3—4 μm; tertiary infection spores cylindrical, 5 × 2—3 μm; sexual phenomena unknown; in *Rhabditis* sp., (nematode) . . . .
. . . . *Haptoglossa zoospora*

B. LITERATURE FOR SPECIES IN KEY

The species of zooparasitic Lagenidiales and lagenidioid fungi are listed below and ordered as in the key above, followed by the major references describing or discussing each. Four basic references which will not be listed below, but which can be consulted for descriptions and/or discussions of several of the species are: Sparrow (1973b — genera of Lagenidiales), Johnson and Sparrow (1961 — genera and species of marine Lagenidiales), Sparrow (1960 — genera and species of aquatic Lagenidiales), and Dollfus (1946 — species of nematophagous Lagenidiales and lagenidioid fungi).

Pseudosphaerita euglenae Dangeard 1894 and *Pseudosphaerita radiata* (Dang.) Sparrow 1943: Dangeard (1933); Karling (1972).

Sirolpidium zoophthorum Vishniac 1955: Davis *et al.* (1954).

Myzocytium humicola Barron and Percy 1975.

Myzocytium glutinosporum, Barron 1976 (c).

Lagenidium septatum Karling 1969.

Blastulidiopsis chattoni Sigot 1931.

Haliphthoros milfordensis Vishniac 1958: Ganaros (1957); Fuller *et al.* (1964); Sparrow (1975); Fisher *et al.* (1975); Tharp and Bland (1977).

Lagenidium caudatum Barron 1976 (b).

Lagenidium chthamalophilum Johnson 1958; Johnson (1960); Fuller *et al.* (1964).

Lagenidium callinectes Couch 1942: Rogers-Talbert (1948); Johnson and Bonner (1960); Fuller *et al.* (1964); Amerson and Bland (1973); Lightner and Fontaine (1973); Bland and Amerson (1973a, b; 1974); Bland (1974); Gotelli (1974).

Lagenidium distylae Karling 1944.

Lagenidium giganteum Couch 1935 *emend.* Couch and Romney 1973: Willoughby (1969); Couch and Romney (1973); McCray *et al.* (1973); Umphlett (1973).

Lagenidium oophilum Sparrow 1939.

Myzocytium microsporum (Karling) Sparrow 1960: Karling (1944, 1966).

Myzocytium intermedium Barron 1976(a).

Myzocytium indicum Singh 1973.

Myzocytium fijiensis Karling 1969.

Myzocytium vermicola (Zopf) Fischer 1892 *emend.* Newell *et al.* 1977: Zopf (1884); Dangeard (1906a).

Myzocytium zoophthorum Sparrow 1936: Sparrow (1952).

Myzocytium papillatum Barron 1976(c).

Myzocytium lenticulare Barron 1976(b).

Gonimochaete latitubus Newell *et al.* 1977.

Gonimochaete horridula Drechsler 1946; Miura (1973).

Gonimochaete pyriforme Barron 1973.

Protascus subuliformis Dangeard 1906(a) and *Protascus subuliformis* var *maupasii* Maire 1915: Dangeard (1906b); Maupas (1915); Micoletzky (1925); Juniper (1957); Sachchidananda and Swarup (1966); Barron, (1977).

Haptoglossa heterospora Drechsler 1940: Shepherd (1956); Juniper (1957); Miura (1970); Giuma and Cooke (1972); Davidson and Barron (1973); Newell *et al.* (1977).

Haptoglossa zoospora Davidson and Barron 1973.

The following incompletely known zooparasitic species, which may be lagenidialean, are listed by host organism: in larval lobsters, unnamed species (Gorham, 1905); in *Calanus finmarchicus* (marine copepod), unnamed (Apstein, 1911b); in *Synchaeta* sp. (marine rotifer), *Synchaetophagous balticus* Apstein 1911 (Apstein, 1911a; Valkanov, 1932); in eggs of unidentified copepods, *Oovorus copepodorum* Entz 1930 (the description is incomplete, but does include Entz's belief that the zoospores were *uniflagellate*); in gills of *Dichelopandalus leptocerus* (marine shrimp), unnamed species (Uzmann and Haynes, 1968); in *Lampanyctus ritteri* (marine midwater fish), *Lagenidium* sp. (Noble and Collard, 1970).

Acknowledgements

The following were reproduced by permission: Figs. 1 and 4, *Proc. Natl. Acad. Sci.;* Figs. 21–23, *Mycologia*; Fig. 25 and 26, Paul Elek Ltd; Figs. 27 and 28, *Am. J. Bot.* Fig. 30–41, *Bull. Marine Sci.*

158   C.E. BLAND, J.N. COUCH AND S.Y. NEWELL

References

Ainsworth, G.C., Sparrow, F.K. and Sussman, A.S. (eds) (1973). "The Fungi: an Advanced Treatise. Vol. 4B. A Taxonomic Review with Keys: Basidiomycetes and Lower Fungi." Academic Press, New York and London.

Alderman, D.J., Harrison, J.L., Bremer, G.B. and Jones E.B.G. (1974). *Mar. Biol.* **25**, 345–357.

Amerson, H.V. and Bland, C.E. (1973). *Mycologia* **65**, 966–970.

Andreeva, R.V. (1972). *Probl. Parasit. Trans. Sci. Conf. Parasit.* USSR, Part 1, 33 pp.

Anthony, D.W., Chapman, H.C. and Hazard, E.I. (1971). *J. Invertebr. Path.* **17**, 395–403.

Apstein, C. (1911a). *Wiss. Meeresunters. Abt. Kiel. (N.S.)* **12**, 163–166.

Apstein, C. (1911b). *Wiss. Meeresunters. Abt. Kiel. (N.S.)* **13**, 205–222.

Atkins, D. (1954a). *J. mar. Biol. Ass. U.K. (N.S.)* **33**, 613–625.

Atkins, D. (1954b). *J. mar. Biol. Ass. U.K. (N.S.)* **33**, 721–732.

Barron, G.L. (1973). *Can. J. Bot.* **51**, 2451–2453.

Barron, G.L. (1976a). *Can. J. Bot.* **54**, 1–4.

Barron, G.L. (1976b). *Antonie van Leeuwenhoek* **42**, 131–139.

Barron, G.L. (1976c). *Can. J. Microbiol.* **22**, 752–762.

Barron, G.L. (1977). *Can. J. Bot.* **55**, 819–824.

Barron, G.L. and Percy, J.G. (1975). *Can. J. Bot.* **53**, 1306–1309.

Bartsch, A.F. and Wolf, F.T. (1938). *Am. J. Bot.* **25**, 392–395.

Bary, A. de. (1887). *Bot. Ztg* **46**, 597–610.

Bland, C.E. (1974). *In* "Gulf Coast Regional Symposium on Diseases of Aquatic Animals". (R.L. Amborski, M.A. Hood and R.R. Miller, eds), pp. 47–54. Center for Wetland Resources, Louisiana State Univ., Baton Rouge.

Bland, C.E. and Amerson, H.V. (1973a). *Mycologia* **65**, 310–320.

Bland, C.E. and Amerson, H.V. (1973b). *Arch. Microbiol.* **94**, 47–64.

Bland, C.E. and Amerson, H.V. (1974). *Chesapeake Sci.* **15**, 232–235.

Bland, C.E. and Couch, J.N. (1973). *Can. J. Bot.* **51**, 1325–1330.

Bogoyavlensky, N. (1922). *Arch. Soc. Russe Protist.* **1**, 113–119.

Briggs, J.D. (1967). WHO/VBC/67.8. 8 pp.

Briggs, J.D. (1969). WHO/VBC/69.171. 6 pp.

Chapman, H.C. and Glenn, F.E., Jr. (1972). *J. Invertebr. Path.*, **19**, 256–261.

Chapman, H.C. and Woodward, D.B. (1966). *Mosq. News* **26** (2), 121–123.

Coluzzi, M. and Rioux, J.A. (1962). *Riv. Malar.* **41**, 29–37.

Cooke, R.C. and Godfrey, B.E.S. (1964). *Trans. Br. mycol. Soc.* **47**, 61–74.

Couch, J.N. (1935). *Mycologia* **27**, 376–387.

Couch, J.N. (1942). *J. Elisha Mitchell scient. Soc.* **58**, 158–162.

Couch, J.N. (1945). *J. Elisha Mitchell scient. Soc.* **61**, 124–136.

Couch, J.N. (1962). *J. Elisha Mitchell scient. Soc.* **78**, 135–138.

Couch, J.N. and Dodge, H.R. (1947). *J. Elisha Mitchell scient. Soc.* **63**, 69–79..

Couch, J.N. and Romney, S.V. (1973). *Mycologia* **65**, 250–252.

Dangeard, P.A. (1894). *Le Botaniste Ser.* **4**, 199–248.

Dangeard, P.A. (1906a). *Le Botaniste Ser.* **9**, 207–215.

Dangeard, P.A. (1906b). *Le Botaniste Ser.* **9**, 253–282.

Dangeard, P.A. (1933). *Le Botaniste Ser.* **25**, 3–46.

Davidson, J.G.N. and Barron, G.L. (1973). *Can. J. Bot.* **51**, 1317–1323.

Davis, H.C., Loosanoff, V.L., Weston, W.H. and Martin, C. (1954). *Science* **120**, 36–38.

Dick, M.W. (1971). *Trans. Br. mycol. Soc.* **57**, 417–425.

Dick, M.W. (1973). In "The Fungi, an Advanced Treatise. A Taxonomic Review with Keys: Basidiomycetes and Lower Fungi." (G.C. Ainsworth, F.K. Sparrow and A.S. Sussman, eds), Vol. 4B, pp. 113–141. Academic Press, New York and London.

Dollfus, R.P. (1946). "Encyclopedie Biologique 27. Parasites (Animaux et Végétaux) des Helminthes". Lechevalier, Paris.

Drechsler, C. (1940). *J. Wash. Acad. Sci.* **30**, 240–253.

Drechsler, C. (1946). *Bull. Torrey Bot. Club* **73**, 1–17.

Dubitskii, A.M., Dzerzhinski, V.A. and Danebekov, A.E. (1970). *Medskaya Parazit.* **39**, 737–738.

Dubitskii, A.M., Dzerzhinski, V.A. and Danebekov, A.E. (1973). *Mikologiya i fitapatologiya* **7**, 136–139.

Eckstein, F. (1922). *Zentbl. Bakt. ParasitKde.* **88**, 128.

Entz, G. (1930). *Arch. Protistenk.* **69**, 175–194.

Fedder, M.L. Danilevskii, M.L. and Reznik, E.P. (1971). *Medskaya Parazitol.* **40** (2), 201–204.

Federici, B.A. (1975). Proc. 43rd Conf. Calif. Mosq. Control Ass. California, USA pp. 172–174.

Fischer, A. (1892). *Rabenhorst. Kryptogamen-Fl.* **1**, 1–490.

Fisher, W.S., Nilson, H. and Shleser, R.A. (1975). *J. Invertebr. Path.* **26**, 41–46.

Fuller, M.S., Fowles, B.E. and McLaughlin, D.J. (1964). *Mycologia* **56**, 746–756.

Gad, A.M. and Sadek, S. (1968). *J. Egypt. Publ. Hlth Ass.* **43**, 387–391.

Galt, J.H. and Whisler, H.C. (1970). *Arch. Mikrobiol.* **71**, 295–303.

Ganaros, A.E. (1957). *Science* **125**, 1194.

Garnham, P.C.C. and Lewis, D.J. (1959). *Trans. R. Soc. trop. Med. Hyg.* **53**, 12–35.

Genga, R. and Maffi, M. (1973). *J. med. Ent.* **10** (4), 413–414.

Giuma, A.Y. and Cooke, R.C. (1972). *Trans. Br. mycol. Soc.* **59**, 213–218.

Gorham, F.P. (1905). *Rep. U.S. Comm. Fish Fisheries* 1903, 175–194.

Gotelli, D. (1974). *Mycologia* **66**, 639–647.

Gugnani, H.C. Wattal, B.L. and Kalva, N.L. (1965). *Bull. Indian Soc. Malaria Commun. Dis.* **2** (4), 333–337.

Haddow, A.J. (1942). *Bull. ent. Res.* **33**, 91–142.

Höhnk, W. and Vallin, S. (1953). *Veröff. Inst. Meeresforsch. Bremerh.* **7**, 63–69.

Huang, Y.M. (1968). *Pacific Insects Monogr.* **17**, 1–74, Bishop Museum, Hawaii.

Iyengar, M.O.T. (1935). *Parasitology* **27**, 440–449.

Iyengar, M.O.T. (1962). *J. Elisha Mitchell scient. Soc.* **78**, 133–134.

Johnson, T.W. (1958). *Biol. Bull.* **114**, 205–214.

Johnson, T.W. (1960). *Am. J. Bot.* **47**, 383–385.

Johnson, T.W. and Bonner, R.R. (1960). *J. Elisha Mitchell scient. Soc.* **76**, 147–149.
Johnson, T.W. and Sparrow, F.K. (1961). "Fungi in Oceans and Estuaries". J. Cramer, New York.
Jones, E.B.G. (1976). "Recent Advances in Aquatic Mycology". Paul Elek, London.
Juniper, A.F. (1957). *Trans. Br. mycol. Soc.* **40**, 346–348.
Karling, J.S. (1944). *Lloydia* **7**, 328–342.
Karling, J.S. (1966). *Sydowia Ann. Mycol.* **20**, 190–199.
Karling, J.S. (1969). *Mycopath. Mycol. appl.* **37**, 161–170.
Karling, J.S. (1972). *Bull. Torrey Bot. Club* **99**, 223–228.
Keilin, D. (1921). *Parasitology* **13**, 225–234.
Keilin, D. (1927). *Parasitology* **19**, 365.
Kellen, W.R., Clark, T.B. and Lindegreni, J.E. (1963). *J. Insect Path.* **5** 167–173.
Kendrick, W.B. and Carmichael, J.W. (1973). *In* "The Fungi. An Advanced Treatise. A Taxonomic Review with Keys: Ascomycetes and Fungi Imperfecti". (G.C. Ainsworth, F.K. Sparrow and A.S. Sussman, eds), Vol. 4A, pp. 323–512. Academic Press, New York and London.
Khaliulin, G.L. and Ivanov, S.L. (1973). *Medskaya Parazitol. Bolezni.* **42** (4), 487. (in Russian).
Lacour, M. and Rageau, J. (1957). *Tech. Pap. South Pacific Comm.* **110**, 1–24.
Laird, M. (1956a). *R. Soc. N.Z. Bull.* **6**, 1–213.
Laird, M. (1956b). *J. Parasit.* **42**, 53–55.
Laird, M. (1959a). *Ecology* **40**, 206–221.
Laird, M. (1959b). *Can. J. Zool.* **37**, 781–791.
Laird, M. (1961). *J. Insect Path.* **3**, 249–253.
Laird, M. (1966). Wld. Hlth. Orgn. EBL/66.69, FIL/66.63, Vector Control/66.204. 9 pp.
Laird, M., Nolan, R.A. and Mogi, M. (1975). *J. Parasit.* **61** (3), 539–544.
Leão, A.E.A. and Pedroso, M.C. (1965). *Mycopath. Mycol. appl.* **26**, 305–307.
Lightner, D.V. and Fontaine, C.T. (1973). *J. Invertebr. Path.* **22**, 94–99.
Lum, P.T.M. (1963). *J. Insect Path.* **5**, 157–166.
Mackin, J.G., Owen, H.M. and Collier, A. (1950). *Science* **111**, 328–329.
Madelin, M.F. (1968). *J. Elisha Mitchell scient. Soc.* **84**, 115–124.
Maffi, M. and Genga, R. (1970). *Parassitologia (Rome)* **12**, 171–178.
Maffi, M. and Nolan, R.A. (1977). *J. med. Ent.* **14**, 29–32.
Manalang, C. (1930). *Philippine J. Sci.* **42**, 279.
Maire, R. (1915). *Bull. Soc. Hist. Nat. Afr. N.* **7**, 50–51.
Manier, J.F., Rioux, J.A., Coste, F. and Murand, J. (1970). *Ann. Parasit. hum. comp.* **45**, 119–128.
Martin, W.W. (1975). *Mycologia* **67**, 923–933.
Martin, W.W. (1977). *Am. J. Bot.* **64**, 760–769.
Maupas, E. (1915). *Bull. Soc. Hist. Nat. Afr. N.* **7**, 34–39.
McCauley, V.J.E. (1976). *Hydrobiologia* **48**, 3–8.
McCrae, A.W.R. (1972). *J. med. Ent.* **9** (6), 545–550.

IDENTIFICATION OF *COELOMOMYCES* 161

McCray, E.M., Umphlett, C.J. and Fay, R.W. (1973). *Mosq. News* 33, 54–60.
Micoletzky, H. (1925). *Mem. Acad. R. Sci. Lettres Danemark. Sect. Sci. Ser.* 8 10, 55–310.
Miura, K. (1970). *J. Jap. Bot.* 45, 233–241.
Miura, K. (1973). *Rep. Tottori Mycol. Inst.* 10, 517–522.
Muspratt, J. (1946a). *Ann. Trop. Med. Parasit.* 40, 10–17.
Muspratt, J. (1946b). *Nature Lond.* 158, 202.
Muspratt, J. (1962). Wld Hlth Orgn EBL/2, Vector Control/2, 2 July 1962.
Newell, S.Y., Cefalu, R. and Fell, J.W. (1977). *Bull. Mar. Sci.* 27, 177–207.
Noble, E.R. and Collard, S.B. (1970). *Am. Fish. Soc. Spec. Publ.* 5, 405–408.
Nolan, R.A. (1978). *Can. J. Bot.* 56, 2303–2306.
Nolan, R.A., Laird, M., Chapman, H.C. and Glenn, Jr. F.E. (1973). *J. Invertebr. Path.* 21, 172–175.
Perkins, F.O. (1974a). *Veröff. Inst. Meeresforsch. Bremerh. Suppl.* 5, 43–63.
Perkins, F.O. (1974b). *Mycologia* 66, 697–702.
Perkins, F.O. and Menzel, R.W. (1966). *Proc. natn. Shellfish. Ass.* 56, 23–30.
Pillai, J.S. (1969). *J. Invertebr. Path.* 14 (1), 93–95.
Pillai, J.S. and Rakai, I. (1970). *J. med. Ent.* 7 (1), 125–126.
Pillai, J.S. and Smith, J.M.B. (1968). *J. Invertebr. Path.* 11, 316–320.
Pillai, J.S. and Woo, A. (1973). *Hydrobiologia* 41, 169–181.
Prowse, G.A. (1954). *Trans. Br. mycol. Soc.* 37, 22–28.
Quick, J.A. (1974). *Trans. Am. microsc. Soc.* 93, 344–365.
Rajapaksa, N. (1964). *Bull. Wld Hlth Org.* 30, 149–151.
Răsin, K. (1928). *Věst. VI Sjezdu ces, Privodozp. Lek. Praze.* 3, 146.
Răsin, K. (1929). *Biol. Spisy vys. Šk. zvěrolék., Brno. B* 8, 1–13.
Rioux, J.A. and Pech, J. (1960). *Acta Trop.* 17, 179–182.
Rioux, J.A. and Pech, J. (1962). *J. Elisha Mitchell scient. Soc.* 78, 134–135.
Rodhain, F. and Gayral, P. (1971). *Annls Parasit. hum. comp.* 46 (3), 295–300.
Rogers-Talbert, R. (1948). *Biol. Bull., Woods Hole* 95, 214–228.
Romney, S.V., Boreham, M.M. and Nielsen, L.T. (1971). *Utah Mosq. Abatement Assoc. Proc.* 24, 18–19.
Sachchidananda, J. and Swarup, G. (1966). *Indian Phytopath.* 19, 279–285.
Schikora, F. (1906). *Fischerei – Leitung* 9, 529–532.
Scott, W.W. (1961). Bull. 51, Virginia Agric. Exp. Stn, Blacksburg, Virginia, 95 pp.
Service, M.W. (1974). *J. med. Ent.* 11 (4), 471–479.
Seymour, R.L. and Johnson, T.W. (1973). *Mycologia* 65, 944–948.
Shemanchuk, J.A. (1959). *Can. Ent.* 91, 743–744.
Shepherd, A.M. (1956). *Friesia* 5, 396–408.
Sigot, A. (1931). *C. R. Soc. Biol. (Strasbourg)* 108, 34–37.
Singh, V.P. (1973). *Hydrobiologia* 42, 445–450.
Sparrow, F.K. (1936). *J. Linn. Soc. Lond. (Bot.)* 50, 417–478.
Sparrow, F.K. (1939). *Mycologia* 31, 527–532.
Sparrow, F.K. (1943). "The Aquatic Phycomycetes, exclusive of the Saprolegniaceae and *Pythium.* Univ. Michigan Press, Ann Arbor.
Sparrow, F.K. (1952). *Rev. Soc. Cubana Bot.* 9, 34–40.

Sparrow, F.K. (1960). "Aquatic Phycomycetes". Univ. Michigan Press, Ann Arbor.

Sparrow, F.K. (1973a). *In* "The Fungi, An Advanced Treatise. A Taxonomic Review with Keys: Basidiomycetes and Lower Fungi". (G.C. Ainsworth, F.K. Sparrow and A.S. Sussman, eds), Vol. 4B, pp. 61–74. Academic Press, New York and London.

Sparrow, F.K. (1973b). *In* "The Fungi. An Advanced Treatise. A Taxonomic Review with Keys: Basidiomycetes and Lower Fungi". (G.C. Ainsworth, F.K. Sparrow and A.S. Sussman, eds), Vol. 4B pp. 159–164. Academic Press, New York and London.

Sparrow, F.K. (1973c). *Arch. Mikrobiol.* **93**, 137–144.

Sparrow, F.K. (1975). *Veröff. Inst. Meeresforsch. Bremerh. Suppl.* **5**, 9–18.

Strand, M.A. and Roberts, D.W. (1977). *In* "Pathogens of Medically Important Arthropods". (D.W. Roberts, and M.A. Strand, eds), pp. 131–145. Supplement No. I., Vol. 55. *Bull. WHO,* Genéve, Switzerland.

Tharp, T.P. and Bland, C.E. (1977). *Can. J. Bot.*, **55**, 2936–2944.

Umphlett, C.J. (1973). *Mycologia* **65**, 970–972.

Uzmann, J.R. and Haynes, E.B. (1968). *J. Invertebr. Path.* **12**, 275–277.

Valkanov, A. (1931). *Arch. Protistenk.* **74**, 5–17.

Valkanov, A. (1932). *Arch. Protistenk.* **78**, 485–496.

van Thiel, P.H. (1954). *J. Parasit.* **40** (3), 271–279.

van Thiel, P.H. (1962). *J. Elisha Mitchell scient. Soc.*, **78**, 135.

Vishniac, H.S. (1955). *Mycologia* **47**, 633–645.

Vishniac, H.S. (1958). *Mycologia* **50**, 75.

Walker, A.J. (1938). *Ann. trop. Med. Parasit.* **32**, 231–244.

Weiser, J. (1976). *J. Invertebr. Path.* **28**, 273–274.

Weiser, J. and McCauley, V.J.E. (1971). *Can. J. Zool.* **49**, 65–68.

Weiser, J. and McCauley, V.J.E. (1972). *Can. J. Zool.* **50**, 365.

Weiser, J. and Vávra, J. (1964). *Z. Tropenmed. Parasit.* **15**, 38–42.

Whisler, H.C., Zebold, S.L. and Shemanchuk, J.A. (1974). *Nature Lond.* **251**, 715–716.

Whisler, H.C., Zebold, S.L. and Shemanchuk. J.A. (1975). *Proc. natn. Acad. Sci. U.S.A.*, **72**, 963–966.

Willoughby, L.G. (1969). *Trans. Br. mycol. Soc.* **52**, 393–410.

Zharov, A.A. (1973). *Medshaya Parazitol.* **42** (4), 485–487.

Zopf, W. (1884). *Nova Acta Der Ksl. Leop. Carol. Deuts. Akad. Naturf.* **46**, 141–236.

CHAPTER 9

Identification of Microsporidia

E.I. HAZARD

*Insects Affecting Man and Animals Research Laboratory,
Science and Education Administration, Agricultural Research,
United States Department of Agriculture, Gainesville, Florida, USA*

and

E. ANN ELLIS and D.J. JOSLYN

*Department of Entomology and Nematology, University of Florida,
Gainesville, Florida USA*

I. Introduction

Microsporidia are common parasites of animals. They occur most frequently in invertebrates, but certainly are not restricted to them. Recently, they were moved from the Protozoa to a new Phylum (Microspora) by Sprague (1977). There has been an increasing awareness of the number of species of micro-sporidia, the large variety of their animal hosts, and of the diversity of their developmental cycles. Microspora are obligate intracellular parasites and are distinguishable from other Protists by having unicellular spores, single tubular polar filaments and being without mitochondria. Hosts become infected when ingested spores both extrude polar filaments and inject sporoplasms into gut epithelium or haemocoel.

It was to be our objective, in this chapter, to fit this large group of organisms into an orderly taxonomic system in which taxonomic keys and illustrations could be used for the identification of families, genera, and species. Recent studies, not yet completed, repudiate those morphological attributes which, defined by light microscopy, served as criteria for past classification systems. These revolutionary findings come from studies of both chromosomes and the ultrastructure of developmental stages, and are expected to result in a drastic change in the present taxonomic status of orders, families, and genera. Therefore, we present a restricted key for the identification of genera only. We shall identify some of the taxonomic problems now facing us and discuss their resolution. Finally, we carefully describe important techniques, report on the status of each described genus found in insects and mites, refer to their occurrence in specific host groups, and group them into similar but not necessarily related types. Most persons will not easily identify microsporidia from the taxonomic information given in this chapter. If nothing else, however, we sincerely hope that many will realize the seriousness of the situation and either delay new descriptions until the matter has been resolved, or consult a specialist for advice before assigning names. Hopefully, further confusion in the literature from the addition of invalid names will be eliminated.

II. Description and Discussion of Taxonomic Methods

Traditional light microscopy techniques used by the classical protozoan taxonomist are slowly being supplemented by more sophisticated and informative methods. These methods are primarily products of the continually expanding fields of molecular and cellular biology yet include principles from the more familiar methods of genetic analysis; they involve biochemistry, cytology, genetics, and immunology. Even so, light microscopy remains an essential tool for the clarification of developmental sequences and for the identification of microsporidian stages recorded in electron micrographs. Without light microscopy to relate to, electron micrographs are difficult to orientate at high magnifications.

Several techniques must be used in order to properly identify microsporidia; therefore, enough material is needed to accommodate each method. They fall into three categories: (1) light microscopy, (2) electron microscopy, and (3) genetic and biochemical analyses. The first two are essential and the beginner must master them if he is seriously interested in identifying microsporidia. The last one, at present, is primarily for the experimental taxonomist, and most likely will be emphasized in any future comprehensive and stable classification system. All of these methods are briefly discussed below.

A. LIGHT MICROSCOPY

1. Fresh smears

These are best examined by phase-contrast microscopy. The microsporidium can be photographed, and spores can be accurately measured without the shrinkage and drastic distortion often caused by fixation and staining. Information about shape, colour, and refractivity of spores and disposition of vacuoles can also be obtained.

To make smears, dissect a small piece of infected tissue from the host, place it in a drop of distilled H_2O on a microscope slide, apply a coverslip and press lightly with the nail of one finger. Next, add several drops of immersion oil to a second slide, enough to cover an area just under the size of a coverslip. Carefully remove the coverslip from the first slide, making sure that some of the infected tissue adheres to the coverslip. If not, remove a piece of the tissue from the slide, with a pair of forceps, and smear on the underside of the coverslip. Gently lower the coverslip to the immersion oil on the second slide and drop it on top of the oil. Do not press it down, as this may cause the spores and vegetative stages to sink into the oil. Let stand a few minutes until the oil is evenly distributed beneath the coverslip. Microsporidia can then be photographed and their spores measured without the interference of Brownian movement.

2. Giemsa stained smears

These are useful for the identification of microsporidia to genus and/or family level. The number of nuclei in meronts and sporonts is easily resolved and, more rarely, the pansporoblastic membrane may be observed if it is thick and somewhat persistent or its presence may be indicated by the frequent grouping of spores (Figs. 1—9). These smears are also important for identification of equivalent stages observed in the electron microscope where their correlation is necessary for proper orientation at high magnifications. The smears should be immediately photographed for permanent records, since they often fade or deteriorate. The preparation procedure is as follows:

(1) Remove small piece of infected tissue, smear on microscope slide.
(2) Allow smear to air dry.
(3) Flood with 95% methanol 5 min.
(4) Pour off methanol.
(5) Flood smear with 10% buffered Giemsa 10 min.
(6) Wash gently with tap water (few sec.).
(7) Blot with absorbent paper.

To prepare buffered Giemsa: Dilute 1:9 with pH 7.4 buffer in distilled H_2O just before use. Not all Giemsa stains are satisfactory. We have found that Giemsa No. SO-G-28 (Fisher Scientific Company), and Fisher Gram-Pac® buffer, pH 7.4 (No. B-82), gives consistent and good results. Never wash stained smears with distilled H_2O as both nuclei and cytoplasm will stain blue. Ideal differentiation of red nuclei and blue cytoplasm will occur when the smears are washed in tap water (pH 5-6).

3. Heidenhain haematoxylin stained wet smears

These are excellent for demonstrating true spore shape because no distortion occurs. Although spore shape is not usually diagnostic among microsporidia, it is important in a few families and for the separation of some genera within a few others, e.g. *Amblyospora* (spores with blunt end, Fig. 7), *Parathelohania* (spores with constricted end, Fig. 8), and *Systenostrema* (pyriform spores, Fig. 9) belonging to the family Thelohaniidae. The family Mrazekiidae has cylindrical spores (Fig. 10) and the genus *Culicospora* has typically elongate, coniform spores (Fig. 11).

Pansporoblastic membranes usually burst when sporonts are removed from the host into water and dried as in Giemsa-stained smears. However, these membranes often remain intact in wet smears as they are rapidly fixed and not dried, especially if the membrane is somewhat persistent as in species of *Vavria* (Fig. 12). Also taxonomically important is the response of different species to Heidenhain stain, as some do not absorb it readily. Finally, these smears are important for type collections, because they do not fade and provide a permanent record for each microsporidian species collected. The preparation procedure follows:

(1) To clean insect, rinse in distilled H_2O; blot dry.
(2) Smear piece of infected insect on coverslip and immediately fix in aqueous Bouin's fixative (Preece, 1965) for at least 2 h. Do not allow smear to dry.
(3) Rinse coverslip in 70% ethanol several times until all picric acid (yellow) is removed.
(4) Rinse coverslip in distilled H_2O, 2–3 min.
(5) Mordant in iron alum solution (Preece, 1965), 4–5 h.
(6) Rinse in distilled H_2O, 3–4 min.
(7) Stain in Heidenhain hematoxylin solution (Preece 1965), overnight.
(8) Rinse in slowly running tap water, 5 min.
(9) Destain in iron alum solution until spore wall can be resolved. Control this under the microscope, rinsing the coverslip in tap water each time before examining.
(10) After destaining, rinse coverslip briefly (10 sec) in tap water containing a few drops (ca. 5 drops/100 ml) of concentrated NH_4OH. Then rinse in slowly running tap water, 30–45 min.
(11) Dehydrate in graded ethanols, clear in 3 changes of xylene and mount.

4. Heidenhain haematoxylin stained sections

These provide limited information for identification, but demonstrate the manifestation of the disease, eg. host tissues attacked. Heidenhain stain is better than other stains because it gives better resolution of stages when photographed in black and white. Specimens may be fixed in Bouin's, Carnoy's, or Zenker's solution, and processed by any paraffin embedding and sectioning procedure (e.g., Preece, 1965).

B. TRANSMISSION ELECTRON MICROSCOPY (TEM)

Currently TEM provides the most meaningful taxonomic information of all of the methods used by the microsporidiologist. Only in ultra-thin sections (Figs. 13 and 14) of infected host tissues can polar filaments, polaroplasts, pansporoblastic membranes, and synaptonemal complexes be resolved. Definition of these structures is now essential in the taxonomy of microsporidia, especially at the family and generic levels.

1. Fixation

Adequate fixation is the most important preparational step, without which the most careful dehydration and infiltration will not yield satisfactory results. There is no single ideal fixative for both vegetative stages and spores because of several inherent problems of mature spores. Specimens are fixed in two steps: primary fixative of glutaraldehyde, alone or in combination with H_2O_2 (Perrachia and Mittler, 1972), acrolein, or formaldehyde and a postfixative of OsO_4. Choice of primary fixative used depends upon the degree of difficulty of the material. Glutaraldehyde (2.5–4%) alone is satisfactory for some easily penetrated microsporidia (e.g., *Amblyospora*). Mixtures of 2.5% glutaraldehyde + 0.2% H_2O_2 can be used for most microsporidia. Spores not easy to penetrate with fixatives (e.g., spores of *Nosema* or *Pleistophora*) can be fixed in hot (70°C) 2.5% glutaraldehyde + 1.0% acrolein and allowed to cool to room temperature (2–3 h). The recommended buffer for all fixatives and washes is 0.1 M sodium cacodylate, pH 7.5.

(1) Dissect specimens in 2.5% glutaraldehyde into small pieces (< 1 mm per side).
(2) Transfer into primary fixative of 2.5% glutaraldehyde–0.2% H_2O_2, overnight at room temperature in the dark.
(3) Wash in 4 changes (15 min each) of buffer.
(4) Postfix in 1.0% OsO_4 for 2 h.
(5) Wash twice in distilled H_2O (15 min each).

When laboratory facilities are unavailable or time is short, specimens dissected and fixed overnight in glutaraldehyde may be held in 5% buffered formalin (5 ml commercial formalin + 95 ml 0.1 M sodium cacodylate buffer, pH 7.5), preferably in the cold, until further processing.

2. Dehydration, infiltration, embedding

(1) Dehydrate pieces at room temperature: 10, 20, 30, 40, 50, 60% ETOH, 15–20 min/change.
(2) *En bloc* stain with 0.5% uranyl acetate in 70% ETOH, 4–6 h or overnight.
(3) Continue dehydration: 80, 90, 95, 100% ETOH, 15–20 min/change.
(4) Make several changes of absolute (100%) ETOH (stored over a molecular sieve or some other drying agent) over 24 h.
(5) Infiltrate in 20–25% steps of resin over 24–48 h. Resin : solvent mixtures should be changed at 4–6 h intervals until pure resin is achieved. Pure resin should be changed 3–4 times over 24 h.
(6) Embed in molds or capsules which have been dried at least 24 h in a 70°C oven over a desiccant.
(7) Polymerize at 70°C.

An alternative dehydrating agent, acidified 2,2-dimethoxypropane (DMP), (Lin et al. 1977) has been used to dehydrate tissues for both transmission and scanning electron microscopy. Acidified DMP (3 drops of $0.1\,N$ HCl to 25 ml of DMP) reacts rapidly with water in an endothermic reaction to produce methanol and acetone. Specimens can be dehydrated in 3 changes (5 min/change) to absolute acetone. This is a rapid, efficient method in that DMP reacts directly with water in the tissue rather than relying on passive exchange of solvents as occurs in conventional dehydration.

3. Embedding media

Although methacrylates and polyester resins have been used for embedding microsporidia-infected specimens, epoxy resins which polymerize uniformly (when thoroughly mixed) with little shrinkage of tissue are better. Mixtures of Epon and Epon-Araldite have been used with varying degrees of success. Their high viscosities and short pot lives often result in poor infiltration and inconsistent, poor sectioning properties. Low viscosity resins have been used to achieve better infiltration and overall improvement in the ultrastructural image of microsporidia. Spurr low viscosity embedding medium (Spurr, 1969) or a modification with Epon 812 at the rate of 10% by weight in standard Spurr mixture is a consistent, reliable embedding medium.

4. Sectioning and staining

Sections showing gold interference colour are easier to stain, show better contrast, and are more stable in the electron beam than thinner sections. Methanolic uranyl acetate (methanol must be acetone free), either at room temperature or hot, followed by lead citrate (Reynolds, 1963), stains microsporidia best. Mordanting with tannic acid, either 0.25% in the primary fixative or 1% aqueous after OsO_4 post fixation, for 30–60 min, greatly increases staining and overall contrast in sections (Wagner, 1976).

C. SCANNING ELECTRON MICROSCOPY (SEM)

Spores of some microsporidia (*Amblyospora*, *Caudospora*, *Parathelohania*, and *Weiseria*) have longitudinal ridges or appendages, which are best resolved in the scanning electron microscope (Fig. 15). These structures may be diagnostic for genera and/or species, but the spore surface of most species is smooth, which limits the use of scanning electron microscopy.

1. Fixation

Spore suspensions should be washed several times with distilled H_2O to remove debris, then fixed and dehydrated by the procedures for transmission electron microscopy up to the final absolute ETOH change. Mordanting with tannic acid before osmicating enhances the uptake of OsO_4 (Sweney and Shapiro, 1977). Although spores can be air dried after absolute ETOH, critical point drying prevents shrinkage and collapse in spores.

2. Mounting and shadowing

A drop of spore suspension in the final solution is placed on coated TEM grids or glass coverslips and attached to SEM stubs. Since most biological specimens are poor conductors of electrons, they need a surface coat of a heavy metal to improve conductivity. Once mounted and dried, spores can be coated with gold, gold palladium, or carbon in a vacuum evaporator or a sputter coater.

D. GENETIC AND BIOCHEMICAL STUDIES

Contributions of genetics and biochemistry to taxonomy have long been recognized for diverse organisms. Recent chromosomal, electrophoretic and serological studies show that these two disciplines are also informative for the systematics of microsporidia.

1. Chromosome analysis

Differences in chromosome structure, number and behaviour between taxa provide valuable cytotaxonomic information because the same genetic mechanism responsible for these differences underscores other taxonomic characters as well.

Chromosome studies of microsporidia show that meiosis in life cycles may be a diagnostic character (Hazard *et al.*, 1978). The meiotic configurations of chromosomes found under the light microscope in species of *Amblyospora* and *Parathelohania*, for example, have not been demonstrated in other genera. The evolution of meiosis in some microsporidia might systematically distinguish them from those without reduction divisions. Furthermore, since fusion of meiotic products has not been observed in microsporidia in primary hosts, the possible existence of alternate hosts in life cycles of some groups as opposed to others may be shown to have taxonomic significance.

Differences in chromosome number and structure also have taxonomic value. Only one report of a definitive number of chromosomes in microsporidia is available and the haploid number was seven (Hazard *et al.*, 1979). No structural differences in chromosomes between species have been noted, but it is expected that, as chromosome studies continue, the degree of relatedness of different microsporidia will be clarified.

The following method of preparing microsporidian chromosomes has been used in our laboratory:

(1) Fix hosts having young sporonts in Carnoy's fluid until hosts turn white (10 min).
(2) Transfer into 45% acetic acid, allow tissue to clear (2–5 min).
(3) Dissect infected tissue from thorax and abdomen in the acid; macerate with sharp minuten pins.
(4) Remove excess acid but do not let tissue dry; place 2 drops of 2% lacto-aceto-orcein stain (French *et al.*, 1962) onto clumped tissue for 10 min.
(5) Place coverslip over tissue and place slide in folded filter paper; squash with even and sustained pressure without moving coverslip.

(6) Specimens with sporonts enclosed in pansporoblastic membranes require considerably more pressure to spread chromosomes. Tapping the slide with a forefinger to disrupt nuclear membranes may help flatten preparations.

(7) Examine chromosomes by phase-contrast microscopy.

2. Identification and separation of gene products

Electrophoresis is currently the most economical method of assaying those gene products which may help to identify taxa. Proteins have been the macromolecules most extensively studied this way and, as *secondary* gene products, they reflect the genic content of a species. Histochemical staining reveals particular enzymes and their structural variants (isoenzymes) separated in the gel medium. Taxonomic estimates rely upon the distances isoenzymes migrate in the gel. Therefore, differences in migration rates of isozymes between two species are indicative of their genetic differences. Electrophoretic procedures, adapted to microsporidia from other organisms, include an inexpensive apparatus (Bush and Huettel, 1972) and the histochemical methods of Steiner and Joslyn (1979).

In microsporidia, horizontal starch gel electrophoresis combined with histochemical staining has revealed variants of twelve enzymes of known function between species (Joslyn et al., 1979). Comparative zymograms (Figs. 16–18) show diagnostic bands among species and, though the genetic basis of these differences has yet to be established, the electrophoretic separation is consistent and reliable as a taxonomic aid.

Other electrophoretic methods used to study microsporidia include disc gel (Fowler, 1971; Fowler and Reeves, 1974a,b, Knell, 1975), isoelectric focusing (Knell, 1975) and two-dimensional electrophoresis (Knell, personal communication). These methods have not yet identified differences for known enzymes; nevertheless, they are taxonomically useful in that they show differences in overall protein content when stained with such general stains as Coomassie brilliant blue or amido black. Histochemical localization of specific gene products with these techniques is being investigated in our laboratory.

Spores have been the stage analysed in all the protein separation techniques discussed. Many spores, ca. 10^9, are needed to obtain sufficient protein for analysis. Isolation of such large numbers of spores has been described (Undeen and Maddox, 1973), and methods of spore purification and fractionation for electrophoresis have also been developed (Conner, 1970). However, to make these techniques more economical, procedures that use less protein and so fewer spores, should be developed. Current volumes of protein required for electrophoretic analyses are ca. $10–15$ mg/ml. Also, other developmental stages of microsporidia may be useful in the future as an alternative source of protein.

3. Serological techniques

Microsporidia have been taxonomically investigated using double immunodiffusion techniques involving spore antigens from six isolates (Knell and Zam, 1978). Spore cultivation, isolation, purification and fractionation with the

Braun MSK cell homogenizer (Conner, 1970) for serological studies are identical to the spore preparation procedures used for electrophoresis. Antigen—antibody reactions involving antimicrosporidian serum from rabbits produced two types of precipitin bands of systematic value. Of six isolates, three were antigenically unrelated, while the other three showed distinct cross-reactions indicating immunological similarities. Technical details include characterization of temperature-sensitive serological responses between microsporidia (Knell and Zam, 1978). Taxonomic separations of microsporidia with these techniques require considerable effort. Though highly sensitive, they may require much additional interest on the part of immunologists before they can be more widely applied.

III. Classification Systems

Prior to 1977 the only classification systems were those of Thélohan (1892), Gurley (1893), Labbé (1899), Stempell (1909) and Léger and Hesse (1922) with modifications incorporated by Kudo (1924), Weiser (1961), and Tuzet et al. (1971). Tuzet et al. stressed the taxonomic importance of the pansporoblastic membrane, now seen clearly in electron micrographs. Recently, Sprague (1977) and Weiser (1977) proposed, almost simultaneously, two new taxonomic systems to accommodate the many newly discovered forms, which could not be incorporated into the older systems. There is a similarity between the two systems at the higher taxonomic scale, but great differences are noted at the family and genus levels.

These two new proposed classification systems will not be analysed and assessed here, because this must wait for additional electron microscopy studies and further analysis of recent reports of meiosis in microsporidia.

IV. Taxonomic Problems

A stable and comprehensive system for classification of microsporidia cannot be established until certain serious problems are resolved. These problems involve the uncertain morphology of existing genera defined by observations made in the light microscope and the ignorance about complete life cycles of many forms. Advanced technologies, e.g. electron microscopy, have given new insight into the nature of microsporidia by revealing a whole new world of morphological structure from which to select new taxonomic attributes. Of 35 genera recognized today in insects and mites, nearly one-half were defined by light microscopy alone. Therefore, to determine their taxonomic status, the type species of these old genera must now be re-examined in all the host's life stages, utilizing electron microscopy and chromosome studies. The problem, however, is compounded by the likelihood that some old types may never be re-collected since old ecosystems may have been destroyed by man's encroachment, particularly in Europe where most of these genera were described near the turn of this century.

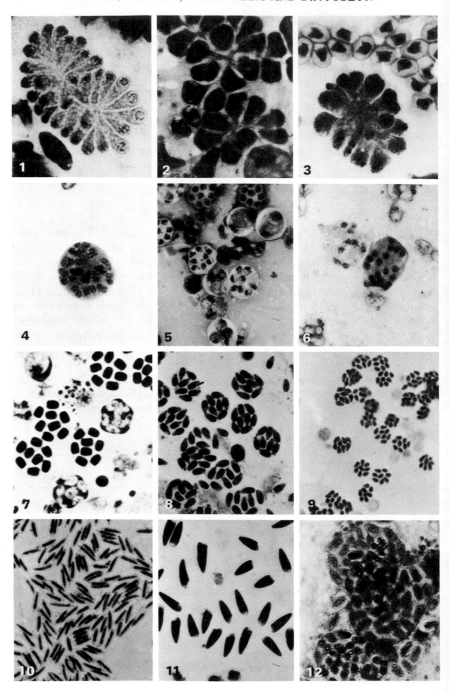

Another important problem concerns the apparent lack of awareness among many protozoologists that many forms have variable and complex life cycles. Too often, the investigator considers only the most apparent part of the life cycle, i.e. stages visible in gross symptomologies, usually in immature forms of aquatic arthropods, where whitish cysts containing sporonts and spores can be readily observed under the host cuticle. Here may be found numerous stages of the parasite, though other equally important developmental sequences may occur in other life stages of the host. For instance, in *Amblyospora* and *Parathelohania* in mosquitoes, stages are involved in transmission of the disease to progeny via the ovaries and eggs, and two morphologically different spore types are produced in host stages. Even though dimorphism has been well documented (Hazard and Weiser, 1968), dual infections continue to be reported, two kinds of spores being called separate species without any awareness that they may constitute only one species.

A third and perhaps most important problem is the controversy over what constitute diagnostic criteria. Here, electron microscopists find themselves opposing classical taxonomists as to suitable taxonomic criteria for families, genera and species. These criteria are briefly discussed next.

V. Taxonomic Criteria

Early protozoologists assigned taxonomic characters as they were seen in the light microscope. These included number of nuclei in vegetative stages (Fig. 1–6), presence or absence of pansporoblastic membranes around groups of spores (Figs. 1–12), spore shape (Figs. 7–11), number of spores formed by sporonts, spore size, and length of polar filaments. Some of these attributes were not always easily resolved in the light microscope, so if certain structures were not observed, they were presumed to be lacking. Electron microscopy produced additional morphological information, particularly internal spore structures (Fig. 14), e.g. the layers of the spore wall, polar filament, polaroplast, polar cap, number of nuclei in spores, vacuole, and specific cell components such as the golgi apparatus and endoplasmic reticulum. Also, it gave more accurate structural detail of diplokarya, pansporoblastic membrane (Fig. 13), and spore surface (Fig. 15).

Some attributes seen by light microscopy have proved to be of limited use in the delineation of taxa. For example, spore size and shape — once important for

Figs. 1–6. Giemsa-stained dry smears; Figs. 7–12. Heidenhain stained wet smears. X600. Fig. 1. *Culicospora magna* from *Culex restuans*. Fig. 2. *Culicosporella lunata* from *Culex pilosus*. Fig. 3. *Tuzetia debaisieuxi* from *Simulium* sp. Fig. 4. *Weiseria* sp. from *Cnephia mutata*. Fig. 5. *Amblyospora callosa* from *Rhyacophila fuscula*. Fig. 6. *Vavraia culicis* from *Anopheles albimanus*. Fig. 7. *Amblyospora* sp. from *Culex salinarius*. Fig. 8. *Parathelohania octolagenella* from *Anopheles pretoriensis*. Fig. 9. *Systenostrema tabani* from *Tabanus lineola*. Fig. 10. *Mrazekia* sp. from *Polypedilum convictum*. Fig. 11. *Culicospora magna* from *Culex restuans*. Fig. 12. *Vavraia culicis* from *Anopheles albimanus*.

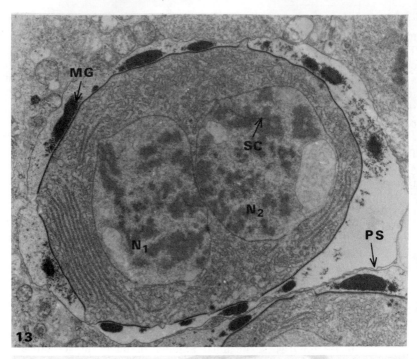

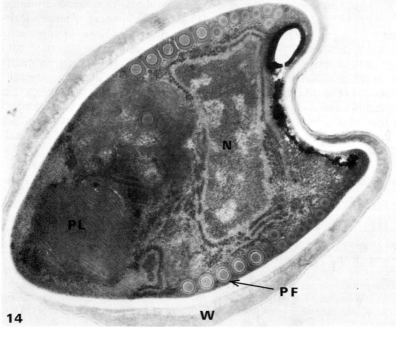

the separation of many species — now are rarely important. Therefore, researchers have not agreed on many of the attributes listed above and now find it difficult to decide what constitute diagnostic characters. The disagreement has become even more apparent today, since some investigators have mistakenly abandoned the light microscope altogether to use more advanced instruments, e.g. the electron microscope.

Current studies and others by Loubés *et al*. (1976), Vávra (1976) and Hazard *et al*. (1979) may change many of our previous concepts of microsporidian taxonomy. Microsporidia, particularly those from aquatic animals, are shown to have very complex life cycles involving polymorphism, meiosis, and the possibility of sexuality through union of gametes in alternate hosts. Phylogenetic relationships can now be demonstrated, in certain taxa, through their similar developmental behaviour when compared with morphological structures defined in light and electron microscopy. Continued studies of these kinds may help establish stable taxonomic groups and lead to a comprehensive classification that will be universally accepted.

VI. Definition of Genera and Host Range

The genera listed here are currently recognized in insects and mites. Keys for separation of classes, orders and families are not presented because the proper taxonomic placements of many genera under these higher taxa are now uncertain. Genera, however, are divided into arbitrary categories based on whether they are monomorphic or dimorphic, the presence or absence of pansporoblastic membranes or cysts, the number of spores they enclose and the number of nuclei in spores. Host groups are listed for each genus to help the generic placement of species. Genera of uncertain status are noted. In the key presented below, (a) refers to a genus not well defined, (b) to a heterogeneous group, and (c) to those forms where horizontal transmission is not known, nor are their complete life cycles known. The key is designed for those having information about species from light and/or electron microscopy. A collective group, *Microsporidium*, is presented at the end of the key for identifiable species which cannot be placed in presently established genera. The reader should refer to Sprague (1977) and Weiser (1977) for complete references to original descriptions of genera and type species.

Key for Separation of Genera of Microsporidia in Insects and Mites

I. Species appearing to have only one spore type, i.e., spores appear to be formed by a single developmental sequence; spores may vary in size; may or may not be transmitted to host progeny via the ovaries.

Figs. 13 and 14. Fig. 13. Sporont of *Amblyospora* sp. in male larva of *Culex salinarius*. X9,400. MG, metabolic granules; N, nucleus; PS, pansporoblast membrane; SC, synaptonemal complex. Fig. 14. Spore of *Amblyospora* sp. in male larva of *Culex salinarius*. X22,500. N, nucleus; PF, polar filament; PL, polaroplast; W, Wall.

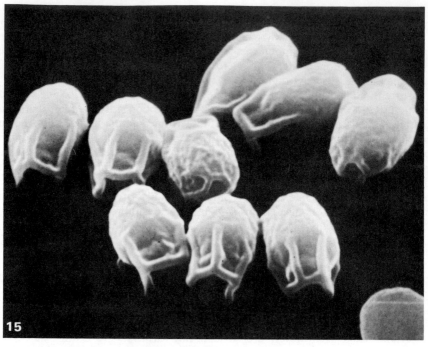

Fig. 15. Surface of spores of *Parathelohania anophelis* in *Anopheles quadri-maculatus.* X9,600.

A. Spores uninucleate, enclosed in a pansporoblastic membrane or cyst wall; many forms transmitted to host progeny via the ovaries.

    (1) Cyst enclosing 1 spore.

        Tuzetia[b, c] Maurand *et al.*, 1971 (Fig. 3). Pansporoblastic membrane resolved only by electron microscopy. Only 1 species reported in insects (*Simulium* spp.). This microsporidium is transmitted to host progeny via the ovaries (Hazard *et al.*, unpublished observations) and has diplokarya and sporonts similar to *Amblyospora* (Fig. 13) including meiotic divisions before sporulation. An examination of infected adult female hosts may predicate its placement close to *Amblyospora* and *Parathelohania* (Fig. 8).

    (2) Cysts enclosing 2 spores.

        Issia[a, c] Weiser, 1977. A monotypic genus in *Trichoptera*.

        Telomyxa[b, c] Léger and Hesse, 1910. In Coleoptera, Diptera and Ephemeroptera. These two genera are not distinguishable through their present definitions.

    (3) Cysts enclosing 4 spores.

        Gurleya[a, b, c] Doflein, 1898. *G. chironomi* Loubés and Maurand, 1975, has sporonts containing synaptonemal complexes as in *Amblyospora* (Fig. 13); therefore, it and other species may be closely related to *Amblyospora* and *Parathelohania*. In Diptera, Ephemeroptera, Lepidoptera, Odonata and Acarina.

    (4) Cysts enclosing 8 spores.

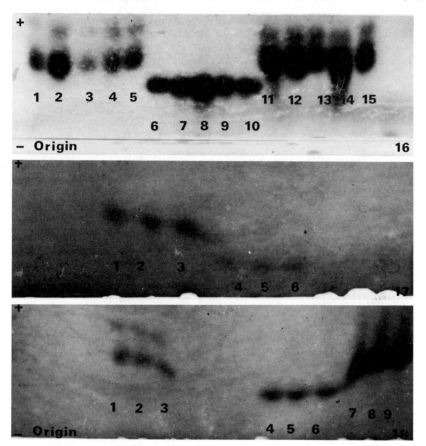

Figs. 16–18. Starch gel zymograms of microsporidia. Fig. 16. Phosphoglucose isomerase from *Nosema heterosporum* (channels 1–5 and 11–15) c.f. host *Heliothis zea* (channels 6–10). Fig. 17. Malate dehydrogenase in *Nosema heterosporum* (channels 1–3) c.f. an undescribed species of *Amblyospora* (channels 4–6). Fig. 18. Phosphoglucomutase diagnostic between *Nosema heterosporum* (channels 1–3) and *Vairimorpha necatrix* (channels 4–6), c.f. host *Heliothis zea* (channels 7–9).

Amblyospora Hazard and Oldacre, 1975 (Figs. 5, 13, 14). Distinguished by spores (found in host larvae) having blunt ends when fixed as wet smears and Heidenhain stained (Fig. 7). Actually represents dimorphic species; therefore, adult female hosts should be examined for presence of a second developmental sequence that produces a second spore type responsible for transmission of the disease to progeny via the ovaries. Common in mosquitoes (Diptera: Culicidae).

Chapmanium[c] H. and O., 1975. Distinguished from other genera by elongate fusiform pansporoblasts. In Diptera: Chaoboridae and in Hemiptera: Nepidae.

Cryptosporina[c] H. and O., 1975. Distinguished by opaque amber pansporoblasts, reminiscent of sporangia in certain fungi. A monotypic genus in an aquatic mite (Acarina).

Hyalincocysta[c] H. and O., 1975. Distinguished by clear pansporoblasts and clearly visible pyriform spores. Transmitted to host progeny via the ovaries. Only in mosquitoes (Diptera: Culicidae).

Parathelohania[c] Codreanu, 1966. Distinguished in host larvae by spores constricted posteriorly when fixed as wet smears (Fig. 8) and in scanning electron micrographs (Fig. 15). Represents dimorphic species; therefore, adult female hosts should be examined for a second developmental sequence and a second spore type that transmits the disease to host progeny via the ovaries. Common in *Anopheles* mosquitoes (Diptera: Culicidae).

Pilosporella[c] H. and O., 1975. Distinguished by subspherical spores. Pansporoblastic membrane resolved only through electron microscopy. Only in mosquitoes (Diptera: Culicidae).

Pegmatheca[c] H. and O., 1975. Distinguished by pansporoblasts formed from plasmodia containing nuclei in diplokaryotic arrangement. Synaptonemal complexes occur in the diplokaryotic nuclei (Hazard, unpublished observations). Transmitted to host progeny via the ovaries (Hazard, unpublished observations). This shows its close relationship to *Amblyospora* and *Parathelohania*. Only in blackflies (Diptera: Simuliidae).

Systenostrema[c] H. and O., 1975. A monotypic genus, its sole species having small pyriform spores (Fig. 9) in oval cysts. In horsefly larvae (Diptera: Tabanidae).

Thelohania[b, c] Henneguy, 1892. Little diagnostic definition has been given to this genus. It is now used as a catch-all genus for forms not having the characteristics of other genera in the Thelohaniidae. Its type species, *T. giardi* Henneguy, was found in a decapod crustacean. However, species occur in insects of the orders Coleoptera, Diptera, Ephemeroptera, Hymenoptera, and Trichoptera.

Toxoglugea[a, b, c] Léger and Hesse, 1924. Distinguished by extremely long, slender, U-shaped spores. In Diptera, Hemiptera, Homoptera and Plecoptera.

(5) Cysts enclosing variable number of spores > 8.

Chytridiopsis[b, c] Schneider, 1884. Distinguished by thick-walled, durable cysts formed in membrane-bound vacuoles in host cell cytoplasm, closely associated with host cell nucleus. In Tenebrionidae and Trogidae (Coleoptera).

Duboscquia[b, c] Pérez, 1908. Distinguished by unornamented cyst containing a consistent and even number (16) of spores. In Diptera and Isoptera.

Hessea[c] Ormiéres and Sprague, 1973. A monotypic genus having thick-walled cysts not closely associated with host cell nucleus. In a fungus gnat (Diptera: Sciaridae).

Mitoplistophora[c] Codreanu, 1966. A monotypic genus distinguished by fusiform or triangular cysts with angles prolonged into flagelliform filaments. In a mayfly (Ephemeroptera).

Pleistophora[b, c] Gurley, 1893. Not well defined. Many species including the type are reported in fish. Pansporoblasts contain many spores. Some species (in both fish and insects) at present placed in this genus are transmitted to host progeny via the ovaries and therefore may be dimorphic (Hazard *et al.*, unpublished observations). In Coleoptera, Dermaptera, Diptera, Ephemeroptera, Isoptera, Lepidoptera, Orthoptera and Plecoptera.

Stempellia[a, c] Léger and Hesse, 1910. Distinguished by having 1, 4 and 8 spores in cysts. A monotypic genus in a mayfly (Ephemeroptera).

Trichoduboscqia[c] Léger, 1926. Distinguished by oval cysts having 3 or 4 long spines and containing a consistent number (16) of spores. A monotypic genus found in a mayfly (Ephemeroptera).

Vavraia[a] Weiser, 1977 (Fig. 6, 12). The type species *V. culicis* (Weiser) transmitted to host larvae *per os*, but not transmitted to host progeny via the ovaries. Not distinguishable by original definition from *Pleistophora*. In Diptera: Culicidae.

B. Spores binucleate and enclosed in a pansporoblastic membrane or cyst wall; none known to be transmitted to host progeny via the ovaries.

(1) Cysts containing 8 spores.

Mrazekia[a, c] Léger and Hesse, 1922. Distinguished by rod-shaped spores (Fig. 10). In midges (Diptera: Chironomidae).

Octosporea[b] Flu, 1911. Distinguished by shorter elongate oval spores. Transmitted to host larvae *per os*. Transovarial transmission not documented. In Diptera, Hemiptera, and Lepidoptera.

(2) Cysts of 32 spores.

Mrazekia[a, c] Léger and Hesse, 1922 (Fig. 10). Distinguished by rod-shaped spores. One species in a chironomid midge (Diptera) (Federici, personal communication).

C. Spores appearing not to be enclosed in a pansporoblastic membrane or cyst wall; some known to be transmitted to host progeny via the ovaries.

(1) Spores uninucleate.

Cougourdella[a, c] Hesse, 1935. Distinguished by lageniform spores. In Trichoptera.

Culicospora[a, c] Weiser, 1977 (Fig. 1, 11). Distinguished by coniform spores (Fig. 11). Sporonts actually have pansporoblastic membranes enclosing variable numbers of spores and species are dimorphic. Pansporoblast membrane is usually resolved only in the electron microscope. Adult female hosts should be examined for second developmental sequence producing a second type of spore that transmits the disease to progeny via the egg. In mosquitoes (Diptera: Culicidae).

Pilosporella[c] H. and O., 1975. Distinguished by subspherical spores. Pansporoblastic membrane actually present and encloses 8 spores, but it can be resolved only through electron microscopy. In mosquito larvae (Diptera: Culicidae).

(2) Spores binucleate.

Caudospora[c] Weiser, 1946. Distinguished by long posterior spore appendage or tail. Only in blackflies (Diptera: Simuliidae).

Culicosporella[a, c] Weiser, 1977 (Fig. 2). Distinguished by coniform spores. A monotypic genus. The type, *C. lunata* (Hazard and Savage), is actually dimorphic; however, it often fails to produce its second sequential stages which give rise to 8 uninucleate (oval) membrane-bound spores. In mosquitoes (Diptera: Culicidae).

Golbergia[c] Weiser, 1977. Distinguished by short nail-like spike posteriorly on the spore. A monotypic genus in mosquitoes (Diptera: Culicidae).

Jirovecia[c] Weiser, 1977. Distinguished by rod-shaped spores with a short caudal appendage. Common parasites of oligochaetes, but a single species in midge (Diptera: Chironomidae).

Nosema[b] Naegeli, 1857. Distinguished by disporous sporonts with nuclei in diplokaryotic arrangement. Species have no special sequence of development, causing transmission to host progeny via the ovaries. However, transmission via the ovaries may occur when meronts invade ovaries from adjacent heavily infected tissues. In mites (Acarina) and numerous insects in the Anoplura, Coleoptera, Diptera, Ephemeroptera, Hemiptera, Hymenoptera, Lepidoptera, Odonata, Orthoptera, and Siphonaptera.

Octosporea[b] Flu, 1911. Spores actually enclosed in a pansporoblast membrane but usually resolved only in the electron microscope. Distinguished here by elongate spores. Spores infectious to host *per os*. Transovarial transmission not documented. In Diptera, Hemiptera and Lepidoptera.

Weiseria[c] Doby and Saguez, 1964. Distinguished by spores with crests or ridges. In blackfly larvae (Diptera: Simuliidae).

II. Species with 2 spore types each formed by a separate sequence of development, one type binucleate, the other uninucleate; uninucleate spores may or may not be bound by pansporoblastic membrane; both spore types formed in host larvae and/or pupae; not transmitted via the ovaries.

Burenella Jouvenaz and Hazard, 1978. Distinguished by the tissue specificity of sporonts of each type, e.g. hypodermis for non-membrane-bound binucleate spores (in larvae) and fat cells for membrane-bound uninucleate spores (in pupae). Spores infectious to host *per os*. Monotypic genus in ants (Hymenoptera: Formicidae).

Culicosporella[a, c] Weiser, 1977. Distinguished by binucleate non-membrane-bound conic spores which greatly predominate (Fig. 11). Sometimes has octosporous sequence with 2–8 spores enclosed in a pansporoblast membrane. These spores often appear aberrant. Monotypic genus in mosquito larvae (Diptera: Culicidae).

Hazardia Weiser, 1977. Distinguished by binucleate oval spores mixed with uninucleate pyriform spores. Oval spores have thick walls and deep surface ridges, and are infectious to their hosts *per os*. Monotypic genus in *Culex* mosquitoes (Diptera: Culicidae).

Vairimorpha Pilley, 1976. Distinguished by binucleate non-membrane-bound elongate oval spores. In addition 8 uninucleate spores (oval) are formed in a pansporoblastic membrane late in the infection. Spores infectious, *per os*, to host larvae. In a large variety of Lepidoptera.

III. *Microsporidium* Balbiani, 1884. A collective group without established attributes, but resurrected by Sprague (1977) for identifiable species which cannot be placed in presently established genera.

VII. Summary and Future Research

Current taxonomic methods for microsporidia are discussed and evaluated. Each procedure is shown to have its own merit; however, certain techniques are essential for identification of species. Studies to determine complete life cycles of species are critical for their proper placement in genera, families, and orders. Biochemical techniques and chromosome studies are in an experimental phase, but preliminary studies show that they may be equally important, especially for microsporidia with the more complex life cycles. Definition of taxa above genera cannot be presented now because they contain at present both species with and without meiotic divisions. We hope, however, that enough taxonomic information about genera is given here to help workers place at least some species. Those studying forest insects and crop pests may find this an easy task since most species they will encounter fall into one of two genera, *Nosema* or *Vairimorpha*. In contrast, those studying aquatic arthropods will have more difficulty placing species because of the diversity of genera in these host groups. Here we suggest that those unfamiliar with microsporidian taxonomy or with the present dubiety of its higher taxonomic structure, either consult with specialists or place new species into the collective group *Microsporidium*, which has been temporarily provided for taxonomically undetermined forms.

References

Bush, G.L. and Huettel, R.N. (1972). Int. Biol. Progrm: Wkg. Grp. Fruit Flies: Popul. Genet.: Project Phase I. 56 pp.

Conner, R. (1970). *J. Invertebr. Path.* **15**, 193–195.

Fowler, J.L. (1971). Ph.D. Diss., Univ. Calif., Riverside. 106 pp + xi.

Fowler, J.L. and Reeves, E.L. (1974a). *J. Invertebr. Path.* **23**, 3–12.

Fowler, J.L. and Reeves, E.L. (1974b). *J. Invertebr. Path.* **23**, 63–69.

French, W., Baker, R.H. and Kitzmiller, J.B. (1962). *Mosq. News* **22**, 377–383.

Gurley, R. (1893). *Bull. U.S. Fish Com. 1891* **11**, 407–420.

Hazard, E.I. and Weiser, J. (1968). *J. Protozool.* **15**, 817–823.

Hazard, E.I., Andreadis, T.G., Joslyn, D.J. and Ellis, E.A. (1979). *J. Parasit.* **65**, 117–122.

Hazard, E.I., Joslyn, D.J., Ellis, E.A. and Andreadis, T.G. (1978). Proc. IVth Cong. Parasitol. p. 11. Warsaw.

Joslyn, D.J., Kelly, J.F., Knell, J.D. and Dillard, C.R. (1979). *Isozyme Bull.* 12:60.

Knell, J.D. (1975). Ph.D. Diss., Univ. Fla., Gainesville. 78 pp. + v.

Knell, J.D. and Zam, S.G. (1978). *J. Invertebr. Path.* **31**, 280–288.

Kudo, R. (1924). *Ill. Biol. Monogr.* **9**, 1–268.

Labbé, A. (1899). *In* "Das Tierreich" 5 Lief. (O. Butschli, ed.), pp. 1–180. Freidlander, Berlin.

Léger, L. and Hesse, E. (1922). *C. r. hebd. Séanc. Acad. Sci. Paris* **174**, 327–330.

Lin, C.H., Falk, R.H. and Stocking, C.R. (1977). *Am. J. Bot.* **64**, 602–605.

Loubés, C., Maurand, J. and Rousset-Galangau, V. (1976). *C. r. hebd. Séanc. Acad. Sci., Paris* **282**, 1025–1027.

Perrachia, C. and Mittler, B.S. (1972). *J. Cell Biol.* **53**, 234–238.

Preece, A. (1965). "A Manual for Histologic Techniques". Little, Brown, and Co. Boston.

Reynolds, E.S. (1963). *J. Cell Biol.* **17**, 208–211.

Sprague, V. (1977). *In* "Comparative Pathobiology Vol. 2, Systematics of the Microsporidia" (L. Bulla and T. Cheng, eds), pp. 1–510. Plenum Press, New York.

Spurr, A.R. (1969). *J. Ultrastruct. Res.* **20**, 346–355.

Steiner, W.W.M. and Joslyn, D.J. (1979). *Mosq. News* **39**, 35–54.

Stempell, W. (1909). *Arch. Protistenk.* **16**, 281–358.

Sweney, L.R. and Shapiro, B.L. (1977). *Stain Technol.* **52**, 221–227.

Thélohan, P. (1892). *Bull. Soc. Philom.* **4**, 165–178.

Tuzet, O., Maurand, J., Fize, A., Michel, R. and Fenwick, B. (1971). *C. r. hebd. Séanc. Acad. Sci. Paris* **272**, 1268–1271.

Undeen, A.H. and Maddox, J. (1973). *J. Invertebr. Path.* **22**, 258–265.

Vávra, J. (1976). *In* "Comparative Pathobiology Vol. I, Biology of the Microsporidia" (L. Bulla and T. Cheng, Eds), pp. 1–371. Plenum Press, New York.

Wagner, R.C. (1976). *J. Ultrastruct. Res.* **57**, 132–139.

Weiser, J. (1961). *Monogr. Angew. Ent.* **17**, 149 pp.

Weiser, J. (1977). *Vest. Csl. Spol. Zool.* **24**, 308–320.

CHAPTER 10

Advances in the Use of *Bacillus popilliae* for Pest Control

M.G. KLEIN

*Japanese Beetle Research Laboratory, USDA, SEA, AR, OARDC,
Wooster, Ohio, USA*

I. Introduction

Milky disease was first noted in 1933 when a few abnormally white grubs of the Japanese beetle, *Popillia japonica,* were found in central New Jersey. The causative agents were soon described as *Bacillus popilliae* or *Bacillus lentimorbus,* two bacteria, either of which grew and sporulated in the blood of living larvae, causing the normally clear haemolymph to appear milky (Fleming, 1968). Production and distribution of the bacteria, first as part of the U.S. Government program and later by private industry, has resulted in one of the classic examples of microbial control.

This chapter will look at recent developments in the use of milky disease bacteria against Japanese beetles, as well as work with closely related *Bacillus* species infecting other scarabaeid larvae in North America, Europe, Asia, Australia and New Zealand.

II. Production and Infectivity of Milky Disease Agents

Two U.S. companies at present produce and sell *B. popilliae* products: Fairfax Biological Laboratory, which markets Doom[R] and Japidemic, and Reuter Laboratories Inc., which markets Milky Spore. Since the products are produced in living larvae (Ignoffo and Hink, 1971), preparations of milky disease bacteria have been expensive and in short supply. Alternative methods of spore production are being explored and will be described in more detail.

A. *IN VITRO* STUDIES

Studies on *in vitro* production of milky disease bacteria have generated much information including nearly $\frac{1}{3}$ of 250 references listed in a recent milky disease bibliography (Klein *et al.*, 1976). Readers interested in details on nutrition, growth, metabolism and sporogenicity should refer to specific articles in that bibliography, and also to recent reviews by St. Julian and Bulla (1973), Lüthy (1975) and Bulla *et al.* (1978).

Limited sporulation of *B. popilliae* var *popilliae* has been achieved on solid media (1–20%) and more recently in a chemostat (1%) (Bulla *et al.*, 1978; Lüthy, 1975; Sharpe and Bulla, 1978). Lüthy (1975) also reports up to 20–30% sporulation of *B. popilliae* var *melolonthae* in liquid media. However, *in vitro* spores of var *melolonthae* weighed less than *in vivo* spores, and tests for pathogenicity by feeding have all been negative for both the *melolonthae* and *popilliae* varieties.

Bulla *et al.* (1978) cited a need for clearer understanding of the metabolism of both host and pathogen, as well as insight into the regulatory mechanisms that control sporogenesis before industrial fermentation will succeed. In addition, Lüthy (1975) concludes that as long as the problem of low infectivity remains the gains in sporulation have no practical value. In view of the remaining problems and the cessation of the extensive USDA research effort on the subject (Beegle, 1979), it appears that *in vitro* production of virulent spores cannot be expected soon.

B. TISSUE CULTURE STUDIES

Lüthy *et al.* (1970) and Lüthy (1975) report growth and sporulation of two strains of var *melolonthae,* and growth but no sporulation with var *popilliae* in a tissue culture medium containing haemocytes of *Phyllophaga anxia.* Later substitution of *Melolontha* haemocytes allowed formation of spores in the European strains with great regularity. If spores produced in tissue culture are infective and if changes in the culture, such as using *P. japonica* blood cells, would stimulate sporulation of var *popilliae*, spores from tissue culture could ease the shortage of milky disease powder, but the cost may still be high.

C. ADULT BEETLES AS HOSTS

Although transmission of milky disease bacteria from one generation to the next by adult beetles is not of importance (Dutky, 1963; Hurpin, 1967), *in vivo* production of spores in adults may be of value. Langford *et al.* (1942) infected adults by inoculation with spores, and diseased adults produced about 5×10^8 spores each. The recent tests at the Japanese Beetle Laboratory (unpublished) have shown that injection rates of 10^4 and 10^5 spores/adult gave 53 and 95% milky adults in about 2 weeks with a spore return similar to that reported by Langford *et al.* (1942). In addition, spores from adults were infective to larvae by injection or *per os* as spores from larvae. Also, Klein and Ladd (unpublished observations) obtained a 25% infection rate of adults fed on artificial diet containing *B. popilliae* spores or milky disease spore powder. Although the yield of spores from adults is less than that from larvae, the ease of obtaining adults in traps with potent lures (Ladd *et al.*, 1976; Tumlinson *et al.*, 1977) and the ability to make spore powder during the summer months when larvae are unavailable could offset the difference.

D. OTHER METHODS

Since vegetative cells of *B. popillae* can be produced in large numbers on artificial media, there has been interest in using them to control Japanese beetles. Lyophilized rods have remained viable under field conditions for at least 1 month when suspended in pellets of tung oil polymer coated with paraffin wax (Cloran and McMahon, 1973; Hepper and McMahon, 1974). Klein and McMahon (unpublished observations) found that rods extracted from pellets gave up to 93% milky larvae when injected. Although larvae readily fed on pellets placed in soil, results from feeding on *B. popilliae* rods were inconclusive. Adequate proof is still lacking for Beard's (1945) contention that larvae can become milky after feeding on rods. However, rods have been shown penetrating the gut wall and initiating infection in the European chafer, *Rhizotrogus* (= *Amphimallon*) *majalis* (Kawanishi *et al.*, 1978; Splittstoesser *et al.*, 1978).

It may be necessary to have additional components such as spore or parasporal material with the rods to aid infection. The role of the parasporal body is not understood. Parasporal bodies of various varieties differ in shape, position, and chemistry (Vago and Delahaye, 1961; Lüthy and Ettlinger, 1967; Milner, 1974a; Lüthy, 1975; Bulla *et al.*, 1978). Lüthy (1975) could not demonstrate any toxic effect from parasporal bodies fed or injected in *Melolontha* larvae, nor find any difference in infectivity of spores with or without parasporal bodies. However, Bulla *et al.* (1978) reported indications of pathogenicity to Japanese beetles associated with the parasporal crystal.

Recently demonstrated toxicity of the parasporal crystal of *Bacillus thuringiensis* to Japanese beetle larvae (Sharpe, 1976) offers the first evidence that the *B. thuringiensis* crystal is toxic to a beetle. Perhaps this crystal may elicit changes in the larval gut that would increase infectivity by milky disease spores or rods. In addition, the toxicity of both *B. popilliae* and *B. thuringiensis* for Japanese beetle larvae is interesting in view of planned studies on genetic manipulations of the bacteria (Beegle, 1979). Faust *et al.* (1979) have studied extrachromosomal DNA from three varieties of *B. thuringiensis* and from *B.*

popilliae for possible work on recombinant DNA. It is hoped that *B. popilliae* organisms that are more fermentable can be obtained (Beegle, 1979).

III. Suppression With Milky Disease Bacteria

A. PERSISTENCE OF BACTERIA

Studies in New Jersey and Delaware showed the presence of milky disease bacteria 25 to 30 years after their original application. Ladd and McCabe (1967) demonstrated that the pathogen was still present at colonization sites after 25 years, and that it had spread into nearby pastures and cultivated fields. Hutton and Burbutis (1974) found the disease still active in 1972 at sites colonized in the early 1940's. They concluded that the pathogen's presence and effectiveness in inoculated areas is important in grub population reductions. However, the disease appeared not to spread rapidly since five of eight untreated sample areas had no milky disease in the grub population.

Beard (1964) reported that larval infection by *B. popilliae* in Connecticut was so impressive that Japanese beetles would be a problem only in areas where the bacterium and other natural control agents were absent. However, about a decade later, heavy infestations of beetle larvae severely damaged turf throughout Connecticut, so the effectiveness of milky disease was seriously questioned (Dunbar and Beard, 1975). Little infection (0–17%) was found at sites that had 40–100% milky grubs in earlier studies. Also, they reported a loss in virulence in spores from Connecticut, a reduced incidence of infection in Connecticut larvae from spores and spore powder, and about 1/4 the normal yield of spores from infected larvae. The presence of an abnormal strain of *B. popilliae* in Connecticut is discussed in Bulla *et al.* (1978). In addition to the loss of virulence, the variant strain is characterized by unusually large parasporal bodies and the presence of multiple parasporal bodies in many cells. Questions are raised about what happened to the highly infective spores present in the 1960's and how a less virulent strain could become dominant. Similar occurrences have been reported before (Fleming, 1968) and linked to strains of bacteria from different hosts.

Laboratory studies indicate a loss in infectivity in stored *B. popilliae* spores (St. Julian *et al.*, 1978). There was almost no infectivity (1.5–3%) in larvae fed spores stored for 7 years in soil or a freezer, and only 20% infectivity from spores stored for 7 years on slides. In contrast, spores stored 2 months in soil gave 50% infectivity. Reductions in infectivity were similar after injection of 7-years old spores. However, strains of bacteria vary considerably in their ability to withstand storage. Dunbar and Beard (1975) found equal or much greater infectivity in spores stored on slides for 17 and 12 years than in spores from freshly prepared slides. Also, in tests during 1975 at the Japanese Beetle Laboratory (unpublished), spores on slides prepared by Dr. S.R. Dutky in 1945 were equal in infectivity to spores 5 years old and less. The 30-year old spores gave infectivity rates of 50 and 85% at 10 000 and 100 000 cells/larva, respectively. In contrast, St. Julian *et al.* (1978) report only 12 and 36% milky at the two rates with their best material.

It was found that *B. popilliae* var *melolonthae* lost much of its pathogenicity after 2 years in soil in pots outside or in the laboratory, a quality that would severely limit its practical effectiveness (Hurpin and Robert, 1976).

B. SPORULATION IN FIELD-INFECTED LARVAE

St. Julian *et al.* (1970) divided the infection process of milky disease in Japanese beetles into four phases with vegetative cells appearing in the haemolymph in Phase II, vegetative cells, prespores and spores present in Phase III, and full sporulation in Phase IV. St. Julian *et al.* (1972) reported that 70% of larvae infected in the field and then reared in the laboratory died before reaching Phase IV, i.e. the massive sporulation phase. They concluded that natural buildup of spores in field plots from diseased larvae occurs less rapidly than previously assumed. Caution must be exercised when laboratory data from stressed larvae (crowded, high temperatures, unknown nutritional state) are extrapolated to make predictions of what is happening in the field. The occurrence of strains too virulent to allow sporulation was discussed by Dutky (1963), who suggested that strain virulence under natural conditions will adjust to the level that produces most spores in the population. Field experience at the Japanese Beetle Laboratory indicates that death of *B. popilliae*-infected larvae before Phase IV is the exception rather than the rule. Klein (unpublished observations) examined individually isolated field-infected larvae from four locations in Ohio and found that 95% died in Phase IV.

Lüthy (1975) reported that *B. popilliae* var. *melolonthae* does not go through the above four phases of development. Rather, sporulation and increase of vegetative cells occur simultaneously. Only some of the *Melolontha* larvae contain bacteria in Phase IV, and produce the maximum number of spores before death. Lüthy believed that larval development and nourishment are important, and that only instar III larvae with well-formed fat bodies support development of the disease up to the end phase.

C. INTERACTIONS BETWEEN HOSTS, ENVIRONMENTAL FACTORS AND BACTERIA

The common cockchafer, *Melolontha melolontha*, and its pathogen, *B. popilliae* var *melolonthae*, have been well studied. Although all three larval instars are susceptible to the disease (Hurpin, 1967, 1968; Lüthy, 1975), Hurpin and Robert (1972) found infections only in instar III at the summer's end in the second year of larval life in field plots. This phenomenon was attributed to interactions of developmental temperatures for the bacteria (15–27°C) and field activity of the larvae (Hurpin, 1967, 1968). Hurpin and Robert (1972) concluded that the bacterium is of little use in the suppression of *M. melolontha* populations. Hurpin and Robert (1970) also found that soil characteristics were of secondary importance in relation to infectivity of grubs by microorganisms. In addition, no interactions between the bacterium and three other organisms (mermithids, rickettsiae, and a virus) that could aid infectivity have been noted (Hurpin and Robert, 1968, 1975). A mutual stimulation between *B. popilliae* var *melolonthae* and *Beauveria tenella* was related to the physiological state of the insect (Ferron *et al.*, 1969).

Milky disease bacteria associated with the European chafer, *R. majalis*, have been extensively studied in New York. Tashiro *et al.* (1969) concluded that milky disease was not a dominant factor in reducing European chafer populations. The limitations did not relate to strain virulence but were similar to those just discussed for *Melolontha*. Although soil temperatures were high enough (above 15–20°C) during instar I to expect substantial infections, dry soils during this period forced grubs to seek deeper levels out of range of maximum spore concentrations. Late summer rains made instar II susceptible, but soil temperatures were then too low to permit consistent infections of instar III. By the time spring soil temperatures averaged 15°C, the chafers had stopped feeding and had become prepupae and pupae. The European chafer has recently been used to study the infectivity of *B. popilliae* (Splittstoesser *et al.*, 1973, 1978; Kawanishi *et al.*, 1978), but no developments that would improve the performance of milky disease have been reported.

Temperature is also important in the activity of milky disease organisms for control of the grass grub, *Costelytra zealandica*. Two distinct strains of bacteria were observed killing grubs in New Zealand (Farrell, 1972; Fowler, 1972, 1974). Farrell (1972) reported milky disease (strain I) was of minor importance as a mortality factor near Nelson. Although incidences up to 58% milky were seen, the incidence is usually much lower, seldom exceeding 5% (Fowler, 1972, 1974). Fowler (1974) concluded that there was little prospect of inducing a high infection rate by field application of spores. While summer temperatures are within the range for development of milky disease, lower temperatures during the winter would preclude continued development of the infection.

Ataenius spretulus, the black turfgrass ataenius, has caused serious damage to turfgrass in the USA and Canada (Niemczyk, 1977). Kawanishi *et al.* (1974) first reported an associated milky disease organism. Splittstoesser and Tashiro (1977) found two more bacilli in the milky disease complex and noted that always when larvae were numerous enough to destroy turf, epizootics occurred in grubs remaining after mid-July. Studies in Ohio have shown a buildup of milky disease (25–28%) in second generation larvae, but the value of the disease in substantially reducing populations is questionable (G. Wegner, personal communication)

The association of milky disease bacteria with the northern masked chafer, *Cyclocephala borealis*, has been well documented (White, 1947; Adams, 1949; Harris, 1959; Dutky, 1963). The strain was atypical of *B. popilliae* and up to 17% of the larvae were milky. Klein and Scoles (1978) found up to 20% milky larvae from five Ohio locations. All of the spores from these infected larvae were characterized by abnormally large parasporal bodies and multiple parasporal bodies and were thus similar to the spores from Connecticut Japanese beetle larvae described by Bulla *et al.* (1978). The Ohio study found virtually no cross-infectivity between spores from *Cyclocephala* and Japanese beetles. Lack of cross-infection indicates that commercial preparations of spore powder for use against Japanese beetles must be made from Japanese beetles and explains the lack of infectivity to *Cyclocephala* from standard spore powder reported by Adams (1949), and Klein and Scoles (1978).

Laboratory studies have shown that milky disease bacteria may have promise for suppression of several different larvae. Milner (1976) found that *B. popilliae* var *rhopaea* lacked sufficient pathogenicity for *Rhopaea morbillosa* and *Othnonius batesi* in Australia to be of practical value. However, its potential to control *R. verreauxi* was high. Milner suggested that a small inoculum could infect a small proportion of larvae in the field, and the inoculum would build up naturally to effect population suppression. Spores and spore powder of *B. popilliae* were sufficiently infective in the laboratory against *Phyllophaga* (= *Holotrichia*) *serrata* to suggest that milky disease could be used to advantage in problem areas in India (David *et al.*, 1973, 1976; David and Alexander, 1975). Tests in China (Chang *et al.*, 1977) with *Bacillus* sp. showed high mortality (100% when injected and 60% when fed) with *Popillia quadriguttata*. This is the first indication of infectivity of a *Bacillus* for another member of the genus *Popillia*.

IV. Future Research Needs

Questions raised by Dunbar and Beard (1975) on the value of milky disease in Connecticut must be answered. Is there milky disease resistance in the larvae, attenuation of spore virulence, or possibly both? What happened to the spores that were so effective just 10 years earlier? Could chlordane resistance or sublethal amounts of chlorinated hydrocarbons in the soil be involved in the problem? Are reductions in the storage capabilities of spores and in spore production in the field (St. Julian *et al.*, 1972, 1978) widespread or restricted phenomena?

In view of the lack of infective spores from *in vitro* production, alternative methods of producing spores and infecting larvae in the field need more study. Milner's (1974b) suggestions on increasing *in vivo* spore yields, by improving the proportion of larvae that survive to reach milky state and inducing all milky larvae to yield the maximum number of spores, need more attention. The specific causes of larval death, and the importance of parasporal bodies and toxic materials exclusive of cellular material discussed by Dutky (1963) also need study.

The importance of host nutrition, and interactions of the disease and fat bodies, have been discussed by many authors (Dutky, 1963; Milner, 1974b; Kawanishi *et al.*, 1974; Lüthy, 1975; Bulla *et al.*, 1978) but there were more questions than answers. Recently, Sharpe and Detroy (1979a) showed that trophocyte cells in milky larvae are depleted of their fat reserves. Dried fat bodies were reduced to a mean of 5.3 mg/milky larva as compared to 22.6 mg/ healthy larva. Sharp and Detroy (1979b) proposed that death of milky larvae results from a lack of energy-rich lipid reserves necessary to physiologically trigger the next step in metamorphosis. Their proposal on the cause of larval death, as well as their suggestion of examining depleted fat body cells for substances that might contribute to sporulation of *B. popilliae* in cultures warrants further consideration. The nutritional status of larvae has caused concern in conducting bioassays of spore powder and field treatments (Dutky, 1963; Milner, 1974b). Milner (1977) has outlined an isolation procedure for spores of *B. popilliae* var *rhopaea* that may prove useful for other scarabs if tests with additional hosts show a direct correlation between number of colonies and percent infection of larvae.

With the disappearance of persistent insecticides, milky disease has an opportunity to become more important against scarabaeid larvae worldwide. Particular emphasis must be placed on selecting strains of bacteria to fit prevailing conditions of temperature limitations and larval populations. New methods of application and formulation should also be explored. The state of North Carolina has developed a jeep-mounted dispenser to help treat large turf areas (E.B. Parker, N.C. Dept. Agric., personal communication), but additional advances are needed. It is possible that olfactory and gustatory stimuli, such as those found by Sutherland (1972a, b) for *C. zealandica* could be added to spore preparations to improve effectiveness.

The expanding host range for *B. popilliae* offers hope for new areas of success. While low soil temperatures are apparently the critical limiting factor in milky disease effectiveness, where the correct conditions of soil temperature, larval development and infective strains of bacteria can be brought together, population suppression with milky disease bacteria offers promise for the present and the future.

References

Adams, J.A. (1949). *J. econ. Ent.* **42**, 626–628.

Beard, R.L. (1945). *Conn. Agric. Stn Bull.* **491**, 503–583.

Beard, R.L. (1964). Proc. 2nd Int. Colloq. Insect Path. Paris, 1962, pp. 47–49.

Beegle, C.C. (1979). *Devel. Indust. Microbiol.* **20**, 97–104.

Bulla, L.A., Jr., Costilow, R.N. and Sharpe, E.S. (1978). *Adv. Appl. Microbiol.* **23**, 1–18.

Chang, S.F., Feng, X.X., Liu, Y.P., Cui, J.Y., Li, K.W. and Li, S.Z. (1977). *Acta ent. sin.* **20**, 355–356.

Cloran, J. and McMahon, K.J. (1973). *Appl. Microbiol.* **26**, 502–504.

David, H. and Alexander, K.C. (1975). *Current Sci.* **44**, 819–820.

David, H. Alexander, K.C. and Ananthanarayana, K. (1973). *Current Sci.* **42**, 695–696.

David, H., Ananthanarayana, K., Alexander, K.C. and Ethirajan, A.S. (1976). *Madras agric. J.* **63**, 537–541.

Dunbar, D.M. and Beard, R.L. (1975). *J. econ. Ent.* **68**, 453–457.

Dutky, S.R. (1963). *In* "Insect Pathology, An Advanced Treatise" (E.A. Steinhaus, ed.), Vol. 2, pp 75–115. Academic Press, New York and London.

Farrell, J.A.K. (1972). *N.Z.J. agric. Res.* **15**, 878–892.

Faust, R.M., Spizizen, J., Gage, V. and Travers, R.S. (1979). *J. Invertebr. Path.* **33**, 233–238.

Ferron, P., Hurpin, B. and Robert, P.H. (1969). *Entomophaga* **14**, 429–437.

Fleming, W.E. (1968). U.S. Dep. Agric. Tech. Bull. 1383. 78 pp.

Fowler, M. (1972). *J. Invertebr. Path.* **19**, 409–410.

Fowler, M. (1974). *N.Z. J. Zool.* **1**, 97–109.

Harris, E.D., Jr. (1959). *Fla. Ent.* **42**, 81–83.

Hepper, K.P. and McMahon, K.J. (1974). N. Dak. Res. Rep. 54. 4 pp.

Hurpin, B. (1967). *Ann. Epiphytes* **18**, 127–173.

Hurpin, B. (1968). *J. Invertebr. Path.* **10**, 252–262.

Hurpin, B. and Robert, P. (1968). *J. Invertebr. Path.* **11**, 203–213.

Hurpin, B. and Robert, P.H. (1970). *Ann. Soc. ent. Fr. (N.S.)* **6**, 825–838.

Hurpin, B. and Robert, P.H. (1972), *J. Invertebr. Path.* **19**, 291–298.

Hurpin, B. and Robert, P.H. (1975), *Ann. Soc. ent. Fr. (N.S.)* **11**, 63–72.

Hurpin, B. and Robert, P.H. (1976). *Entomophaga* **21**, 73–80.

Hutton, P.O., Jr. and Burbutis, P.P. (1974). *J. econ. Ent.* **67**, 247–248.

Ignoffo, C.M. and Hink, W.F. (1971). In "Microbial Control of Insects and Mites" (H.D. Burges and N.W. Hussey, eds) pp. 541–580. Academic Press, London and New York.

Kawanishi, C.Y., Splittstoesser, C.M., Tashiro, H. and Steinkraus, K.H. (1974). *Environ. Ent.* **3**, 177–181.

Kawanishi, C.Y., Splittstoesser, C.M. and Tashiro, H. (1978). *J. Invertebr. Path.* **31**, 91–102.

Klein, M.G. and Scoles, L.E. (1978). *Proc. N.C. Branch Ent. Soc. Am.* **33**, 42. (abstract).

Klein, M.G., Johnson, C.H. and Ladd, T.L. Jr. (1976). *Bull. ent. Soc. Am.* **22**, 305–310.

Ladd, T.L. Jr. and McCabe, P.J. (1967). *J. econ. Ent.* **60**, 493–495.

Ladd, T.L. Jr., McGovern, T.P., Beroza, M., Buriff, C.R. and Klein, M.G. (1976). *J. econ. Ent.* **69**, 468–470.

Langford, G.S., Vincent, R.H. and Cory, E.N. (1942). *J. econ. Ent.* **35**, 165–169.

Lüthy, P. (1975). *Vischt. naturf. Ges. Zurich* **120**, 81–163.

Lüthy, P. and Ettlinger, L. (1967). *In* "Insect Pathology and Microbial control" (P.A. van der Laan, ed.), pp. 54–58. North-Holland Publishing Co., Amsterdam.

Lüthy, P., Wyss, Ch. and Ettlinger, L. (1970). *J. Invertebr. Path.* **16**, 325–330.

Milner, R.J. (1974a). *Aust. J. Biol. Sci.* **27**, 235–247.

Milner, R.J. (1974b). *J. Invertebr. Path.* **23**, 289–296.

Milner, R.J. (1976). *J. Invertebr. Path.* **28**, 185–190.

Milner, R.J. (1977). *J. Invertebr. Path.* **30**, 283–287.

Niemczyk, H.D. (1977). *Ohio Rep.* **62**, 3–5.

St. Julian, G. and Bulla, L.A., Jr. (1973). In "Current Topics in Comparative Pathobiology (T.C. Cheng, ed.), Vol. 2. pp. 57–87. Academic Press, New York and London.

St. Julian, G., Sharpe, G.E. and Rhodes, R.A. (1970). *J. Invertebr. Path.* **15**, 240–246.

St. Julian, G., Bulla, L.A., Jr. and Adams, G.L. (1972). *J. Invertebr. Path.* **20**, 109–113.

St. Julian, G., Bulla, L.A., Jr. and Detroy, R.W. (1978). *J. Invertebr. Path.* **32**, 258–263.

Sharpe, G.E. (1976). *J. Invertebr. Path.* **27**, 421–422.

Sharpe, G.E. and Bulla, L.A., Jr. (1978). *Appl. Environ. Microbiol.* **35**, 601–609.

Sharpe, G.E. and Detroy, R.W. (1979a). *J. Invertebr. Path.* **34**, 90–91.

Sharpe, G.E. and Detroy, R.W. (1979b). *J. Invertebr. Path.* **34**, 92–94.

Splittstoesser, C.M. and Tashiro, H. (1977). *J. Invertebr. Path.* **30**, 436–438.

Splittstoesser, C.M., Tashiro, H., Lin, S.L., Steinkraus, K.H. and Fiori, B.J. (1973). *J. Invertebr. Path.* **22**, 161–167.

Splittstoesser, C.M., Kawanishi, C.Y. and Tashiro, H. (1978). *J. Invertebr. Path.* **31**, 84–90.

Sutherland, O.R.W. (1972a). *Proc. 25th N.Z. Weed and Pest Cont. Conf.* **25**, 248–252.

Sutherland, O.R.W. (1972b). *N.Z.J. Sci.* **15**, 165–172.

Tashiro, H., Gyrisco, G.G., Gambrell, F.L., Fiori, B.J. and Breitfeld, H. (1969). N.Y. Agric. Exp. Stn Geneve Bull. 828. 71 pp.

Tumlinson, J.H., Klein, M.G., Doolittle, R.E., Ladd, T.L. and Proveaux, A.T., (1977). *Science* **197**, 789–792.

Vago, C. and Delahaye, F. (1961). *Mikroskopie* **16**, 198–206.

White, R.T. (1947). *J. econ. Ent.* **40**, 912–914.

CHAPTER 11

Insecticidal Activity of Isolates of *Bacillus thuringiensis* and their Potential for Pest Control

H.T. DULMAGE

*Cotton Insects Research, Agricultural Research,
Science and Education Administration, U. S. Department of Agriculture,
Brownsville, Texas, USA*

and

COOPERATORS*

I. Background of the International Cooperative Program

A. PURPOSE OF THE PROGRAM

Bacillus thuringiensis (*B.t.*) is a complex species divisible into > 20 varieties or H-serotypes by serological and biochemical tests. These produce several insecticidal toxins, two of which are used in agriculture: the δ-endotoxin and the

* This chapter reports the collaborative efforts of the author and the scientists listed in Table I. The senior author emphasizes that this was truly integrated and collaborative work, and that this chapter should be considered as equal contributions from the senior author and each collaborator.

β-exotoxin. The insecticidal actions of isolates of *B.t.* are complex. The relative activity of each isolate against different insect species — i.e. its "spectrum of activity" — arises partly from the combined effects of the potencies of the varying concentrations of the different insecticidal materials that it produces. However, the primary cause of the spectral differences lies in the spectrum of activity of the δ-endotoxin itself: δ-endotoxins of different isolates of *B.t.* can kill different insect species or differ in the degree of their activity toward them. Much of the early work on this subject has been excellently reviewed by Burgerjon and Martouret (1971) and will not be discussed here.

This variation of activity spectrum according to the *B.t.* isolate is very important. Failure to control a pest insect with a particular *B.t.* preparation does not mean that all preparations will fail; it may mean just that the wrong isolate was selected for use against the target insect. Similarly, even though a pest species may be satisfactorily controlled by the toxins in the present commercial preparations, it is possible that a different isolate may be more effective and thus cheaper to use. These observations reveal the value of understanding the spectra of activity of the toxins and the rewards attainable from research into this subject.

The picture is further complicated because different isolates can produce more of the same δ-endotoxin than others (Dulmage, 1970, 1971). Also, maximum toxin production can be achieved only by careful attention to the interaction of fermentation conditions, media and the isolates involved — there is, for example, no one medium best suited to all isolates. In view of present knowledge about fermentations of other organisms, these conclusions are not surprising; even so it is important to remember them while searching for cheaper and better preparations of *B.t.*

Unfortunately, although important, these differences between isolates are also very difficult to determine. To understand activity spectra, we must study many different isolates and test them both serologically and against a wide variety of insect species. No one laboratory is equipped to undertake such a study: it requires the cooperative efforts of many. To achieve this, we organized the "International Cooperative Program on the Spectra of Activity of *Bacillus thuringiensis*", involving 16 scientists from 14 laboratories. Powders derived from about 320 isolates were tested against 23 insect species. Our experiments attempted to ask, and at least partially to answer, the following questions:

1. Is the activity spectrum of a *B.t.* powder dependent on the variety (not just on the isolate) used to produce it?
2. Can the same δ-endotoxin be produced by more than one isolate?
3. Can different isolates of the same variety produce different endotoxins?
4. Can a single isolate produce more than one endotoxin?
5. Can we serologically identify these endotoxins?
6. What other toxins can be present in powders of *B.t.*?
7. Are there isolates that form δ-endotoxins more toxic to a particular insect species than the toxin formed by HD-1, the isolate at present used in most commercial production?

Our program was divided into several stages. In the first stage, now complete, we attempted to survey a large number of isolates so as to answer these questions at least partially. These results are now being collated for publication by the United States Department of Agriculture in the near future, and some are presented in this chapter. Table I lists the cooperators and their roles in our program.

TABLE I

Roles of Cooperators in the International Cooperative Program on the spectrum of activity of Bacillus thuringiensis

Cooperator	Contribution
A. *Microbiological or Supportive*	
H. de Barjac	Identification of isolates
H.T. Dulmage	Production of powders
J. Krywienczyk	Serology of crystals
H. del var Petersen	Biometrical analyses
B. *Insect Bioassays*	
K. Aizawa, N. Fujiyoshi, M. Ohba	*Bombyx mori, Hyphantria cunea, Spodoptera (Prodenia) litura*
C.C. Beegle, H.T. Dulmage, D.S. Needleman	*Agrotis ipsilon*
H.D. Burges, P. Jarrett	*Galleria mellonella*
H.T. Dulmage, C.C. Beegle	*Heliothis virescens, Spodoptera exigua, Trichoplusia ni*
N.R. Dubois	*Lymantria (Porthetria) dispar*
L.P.S. van der Geest, H.J.M. Wassink	*Pieris brassicae*
R.E. Gingrich, M. Haufler, N. Allan	*Haematobia irritans, Bovicola crassipes*
I.M. Hall, K.Y. Arakawa	*Aedes triseriatus, Ae. aegypti, Anopheles albimanus, Culex pipiens, Cx. quinquefasciatus, Cx. tarsalis*
L.C. Lewis	*Ostrinia nubilalis*
W.H. McGaughey, E.B. Dicke	*Ephestia (Cadra) cautella, Plodia interpunctella*
C.G. Thompson	*Choristoneura occidentalis, Orgyia pseudotsugata*

B. DEFINITIONS OF TERMS AND CONCEPTS

1. *Classification of* Bacillus thuringiensis − *the Varietal Concept*

The concept that there are different varieties of *B.t.* is so deeply interwoven into the studies in our program that it is impossible to discuss results without first clearly understanding just what these varieties are.

By the late 1950's, a number of closely-related entomopathogenic Bacilli had variously been named as varieties of *B. cereus* or *B. thuringiensis*. Although alike in shape and size, the key to their close relationship was a crystalline parasporal body in their cells at sporulation, a body so distinctive that they were informally called the "crystalliferous bacteria." However, there were some differences between the various isolates. In 1958 and 1960, Heimpel and Angus constructed a biochemical classification and proposed retention of the so-widely accepted *B. thuringiensis* because the crystal so clearly differentiated the isolates from *B. cereus*. This view predominates today (Chapter 3). In 1962, de Barjac and Bonnefoi differentiated the varieties serologically by comparing antibodies to their flagellar proteins, the "H-antigens". This proved reliable and reproducible and is now virtually universally accepted. Norris (1964) and Norris and Burges (1965) showed that the electrophoretic patterns of the esterases produced in vegetative cells of *B.t.* could also be used to distinguish varieties. Significantly and surprisingly, there was a close relationship between classifications made by esterase patterns and by H-antigens. Electrophoretic patterns are now used to help differentiate H-types *sotto* and *dendrolimus,* both classified as Serotype 4a 4b. Many of the biochemical tests have been retained and are used to distinguish H-types *subtoxicus* and *entomocidus,* which are similar in H-serotype and esterase patterns, but differ in some of their biochemical responses. These different types of character are given together in a determinative key in Chapter 3.

Varieties may sometimes be further subdivided by other tests. For example, working in our program, Krywienczyk found that, in general, division of isolates by serology of the crystals corresponded well with that based on H-antigens. However, there were some exceptions. Crystal serology divides var *kurstaki,* H-serotype 3a 3b, into two large groups (Krywienczyk *et al.*, 1978). Furthermore, as discussed later in this chapter, this subdivision was supported by the spectra of activities of isolates in these sub-varieties, designated "*k-1*" and "*k-73*" after the HD-Nos of the type cultures in which they were first recognized. The importance of this observation for classification of isolates is not yet known; however, it is crucial in understanding the spectra of their activities. The significance of this will be discussed in more detail in Section III; the important fact to remember now is that the varietal concept is not academic: there is a relationship between variety, type of toxin and spectrum of activity. This is of major significance in attempts to develop better insect control with *B.t.* Table II lists varieties of *B.t.* included in our program. Several factors have intensified research on *B.t.* recently, and it is interesting and potentially very important that, since the first phase of our program was completed, at least six new varieties (Table III) have been discovered, including one (var *israelensis*) with considerable promise for control of mosquitos (Culicidae) and blackflies (Simuliidae) and three not yet incorporated in de Barjac's key (Chapter 3). Discovery of these new varieties also emphasizes the cosmopolitan nature of *B.t.*: two are from the People's Republic of China, two from the USA, one from Pakistan, and one from Israel.

2. *Insecticidal Toxins Produced by* Bacillus thuringiensis

a. *Introduction. B.t.* produces several toxins, four of which will be considered here: α-exotoxin (= "heat-labile exotoxin"), β-exotoxin (= "fly factor" or "heat-stable exotoxin"), δ-endotoxin (= "crystalline toxin" or just "crystal") (classification by Heimpel, 1967), and the recently discovered "louse-factor"

TABLE II

Varieties of Bacillus thuringiensis studied in the International Cooperative Screening Program

Variety	H-type	Discoverer	Year	Reference
aizawai	7	Aizawa	1962	Bonnefoi & de Barjac, 1963
alesti	3a	Toumanoff and Vago	1951	Toumanoff & Vago, 1951
canadensis	5a 5c	Morris	1972	de Barjac & Bonnefoi, 1972
darmstadiensis	10	Krieg	1961	Krieg et al., 1968
dendrolimus	4a 4b	Talalaev	1956	Talalaev, 1956
entomocidus	6	Steinhaus	1950	Steinhaus, 1951; Heimpel & Angus, 1958
finitimus	2	MacNamee	1956	Heimpel & Angus, 1958
galleriae	5a 5b	Isakova	1956	Isakova, 1958
kenyae	4a 4c	Norris and Burges	1961	Norris & Burges, 1963
kurstaki	3a 3b	Kurstak	1962	Kurstak, 1964
kurstaki, HD-1	3a 3b	Dulmage	1967	Dulmage, 1970
morrisoni	8	Norris	1963	Norris, 1964
sotto	4a 4b	Ishiwata	1901	Ishiwata, 1901
subtoxicus	6	Steinhaus	1945	Heimpel & Angus, 1958
thuringiensis	1	Berliner	1911	Berliner, 1911
thuringiensis, Mattes strain	1	Mattes	1927	Mattes, 1927
thompsoni	12	Thompson	1969	de Barjac & Thompson, 1970
tolworthi	9	Norris	1963	Norris, 1964
toumanoffi	11	Toumanoff	1956	Krieg, 1969

TABLE III

New varieties of Bacillus thuringiensis not yet tested in the International Cooperative Screening Program

Variety	Discoverer	Year	Reference
dakota	de Lucca and Larson	1978	de Lucca et al., 1979
indiana	de Lucca and Larson	1978	de Lucca et al., 1979
israelensis	Goldberg and Margalit	1976	de Barjac 1978
kyushuensis	Ohba and Aizawa	1977	Ohba et al., 1977
ostriniae	Ren	1975	Ren et. al., 1975
pakistani	Shaikh	1975	de Barjac et al., 1977
wuhanensis	Hubei Institute	1976	Hubei Institute, 1976

(Gingrich et al., 1974). Some knowledge of these toxins, critical to an understanding of our program, is summarized here.

b. α-exotoxin. The little-studied α-exotoxin, identified as lecithinase C by Toumanoff (1953), is water-soluble, heat-labile and toxic to insects. Krieg (1971) reported an exotoxin, toxic per os to mice and to the diamondback moth, Plutella xylostella (= maculipennis), in supernatants of beers of B.t., including some that contained no lecithinase. He believed it to be identical to Toumanoff's factor, but definitely not lecithinase C, so he named it the "mouse factor" or "thermosensitive exotoxin". This definition will be further discussed later in this section.

c. *β-exotoxin*. Beers of *B.t.* were early recognized to contain a water-soluble heat-stable toxin highly toxic to larvae of several species of flies. First named "β-exotoxin" by Heimpel (1967), it has since been defined chemically as an adenine nucleotide and ATP analogue and given the name "thuringiensin". Reviewed in Chapter 13, it is much more toxic by parenteral injection than when given orally in both insects and vertebrates. It is not damaged or absorbed by the gut of cattle and was at one time proposed as an additive to their feed that would pass through the gut and control flies in faeces. However, its teratogenic effect in insects and possible mutagenicity has led regulatory authorities to prevent its use in the USA and Canada, but it is used for insect control in the USSR. Considering the diversity of biological systems, it would not be surprising if some isolates of *B.t.* produced other heat-stable, water-soluble insecticides with different chemical or insecticidal properties arising either as derivatives of thuringiensin or as unrelated chemical entities. This will also be discussed at more length later in this section.

d. *δ-endotoxin*. While the β-exotoxin is a broad-spectrum poison, the δ-endotoxin in the crystal of *B.t.* has an activity spectrum limited, so far as we know, to certain Lepidoptera, mosquitoes, chironomids and blackflies. Its chemistry and mode of action are reviewed in detail in Chapter 12. The crystal is atoxic until dissolved. Solubilizations by alkali, alkaline reducing buffers, and various enzymes produce molecules of a variety of sizes, some toxic, others not. It is thus very wrong to regard the crystal itself as the δ-endotoxin. The toxin is, or resides in, or is attached to, one or more of the molecules produced by solubilization. Fast's opinion (Chapter 12) is divided about these alternatives and about the number of toxic entities in the crystals of the various H-serotypes studied. This is one of the great unanswered questions about *B.t.* Susceptibility of a lepidopterous insect to the toxin may depend partially, and perhaps entirely, on its ability to digest the toxin, and the degree of its susceptibility may reflect the rate at which it brings about this digestion (Lecadet and Martouret, 1964; Lecadet, 1970). We shall accept this definition in this chapter. However, we should not forget that δ-endotoxins from different isolates of *B.t.* can differ, both quantitatively and qualitatively, in their insecticidal activities. Fortunately, the conclusions drawn in our program depend on the insecticidal activities of these toxins and not on their chemistries, but we cannot ignore such questions for long.

e. *The Spore*. Certainly, no discussion of δ-endotoxin can ignore the potential importance of the spore. Originally, *B.t.* was considered only an infective agent, and standardization of commercial products in the USA was made on spore counts. However, along with the realization that δ-endotoxin was the principal factor in the insecticidal activity came the belief that the spore played no part at all. Then close similarity was found between the protein of the crystal and that in the spore coat, part of the spore coat protein being serologically related to the crystal and also being toxic to lepidopterous larvae — the inference being that δ-endotoxin merely represented an excess of spore coat protein (Chapter 12). Interest in the spore revived.

As is so often true in research, there is no single answer. Certainly, the spore plays little or no role in the action of lethal doses of B.t. against certain highly susceptible insects, e.g. *Bombyx mori*, the "Type 1" insects of Heimpel and Angus (1959). There is not enough time for the spore to germinate and infect Type 1 insects before they die from the action of the toxin. However, in most insects, and probably at low dosages in Type 1 insects, death comes more slowly, and the spore can play a role. Sometimes it appears to do so, sometimes not. It is important to the understanding of the work of our program that the spore at times does influence action of δ-endotoxin. Somerville *et al.* (1970) showed that, under certain conditions, a mixture of spores and crystals was more effective than crystals alone. Even more noteworthy was the work of Burges *et al.* (1976), who showed that a mixture of spores and crystals of *B.t.* var *galleriae* was more effective than spores alone, the pure crystals without spores had little effect, also that a mixture of spores of var *galleriae* and crystals of this variety was more effective than a mixture of spores of var *thuringiensis* and crystals of var *galleriae*, indicating that the action of the spores was more than either the addition of more toxin (from the spore coat) or a secondary infection of an insect whose midgut had been damaged by the toxin. Burges' work is particularly important to the understanding of the results obtained in our program and will be discussed later. For now, it is important to note that all our powders were mixtures of spores and δ-endotoxin.

f. *"Louse-Factor"*. In 1974, Gingrich *et al.* reported that four species of mammal-biting lice (*Bovicola bovis, B. crassipes, B. limbata,* and *B. ovis*) were susceptible to powders containing the spore–endotoxin complex of *B.t.* var *kurstaki* (HD-1), an isolate of *B.t.* that does not produce β-exotoxin, a toxin to which these lice are susceptible. In view of past experience with δ-endotoxin, it seemed improbable that the endotoxin could be responsible for this activity, so the authors suggested that the toxicity to lice was due to a new chemical entity, which they named the "louse-factor." *B. crassipes* was therefore included in our screening, although presence of the louse factor could be easily masked if the powder contained significant levels of β-exotoxin.

3. *Toxins of* Bacillus thuringiensis *and the Cooperative Program*

When the program was designed, we expected that the only toxins in our powders would be δ-endotoxins, because our harvest method should effectively eliminate water-soluble toxins. However, Gingrich showed that many powders from isolates that could produce β-exotoxin contained significant levels of this toxin. We believe that this arose from two possible causes. First, the method involved recovering sediment from centrifuged fermentation beers. Some beer is trapped in this sediment, and, since water-soluble toxins are present in this trapped beer, they would be carried over to the final product. This possibility seems to be confirmed by the presence of water-soluble, heat-labile, toxins in some powders. Second, the water at Brownsville, where the powders were prepared, contains much calcium. This could have precipitated some of the β-exotoxin as a water-insoluble Ca-salt, retaining it in the powder.

What began unplanned became an asset, and, in the process, gave a warning about the classification of these toxins. First, workers in our program found that some of the so-called β-exotoxins in these powders appeared to differ among themselves and from thuringiensin because their toxicities towards different insect species were not the same (see Section III, C8 of this chapter). Apparently, there was more than one kind of "β-exotoxin". Perhaps this should not be surprising. Bacteria produce many different and unrelated heat-stable, water-soluble compounds, a few of which might be expected to be insecticidal. Thus, in our discussions of *B.t.,* use of the term "β-exotoxin" could be misleading if this possibility were forgotten.

The same care should be taken in the use of the term "α-exotoxin". Beegle examined the heat-stability of toxins active against *Agrotis ipsilon* and found some evidence of differing heat-stabilities between powders. He also found evidence of a heat-labile exotoxin active against the beet armyworm, *Spodoptera exigua.* There is no reason why these toxins should not differ from each other or why a fermentation product with such broad chemical characteristics should be identical to Krieg's toxin, even though both have some insecticidal attributes, and so the term "α-exotoxin" should be used cautiously.

A similar argument could be made against our use of the term "δ-endotoxin". Most of our isolates were insecticidal, and some of those that were not produced crystals. Our harvest method was designed to recover δ-endotoxin. However, in a screening program such as ours we could not evaluate each powder chemically. While we believe that the insecticidal activities attributed to δ-endotoxin are indeed due to this toxin, some caution must be applied in evaluating this assumption.

II. Protocols

A. PRODUCTION OF *BACILLUS THURINGIENSIS* POWDERS

There are three basic questions that can be asked in any study of production of toxins by *B.t.:* (1) What toxins are produced? (2) In what quantity? (3) How reproducible is the fermentation? In the discussions of our program which follow we will direct most of our attention to asking what toxins are produced. However, our interpretation must not be confused by the quantities of toxin present in our powders.

This is important to remember. We will use several concepts to help analyse what toxins are present: activity ratios, distributions of toxicities, crystal-types, H-types, etc. It is impossible to eliminate quantity as a variable. The quantity of toxin produced in a fermentation can be influenced by the isolate of *B.t.* used and by the medium on which it is grown (Dulmage, 1970, 1971; Dulmage and de Barjac, 1973). Thus, two powders may differ many-fold in toxicity yet still contain the same toxin. All of our analyses will be made in the light of this factor and will attempt to pinpoint *type* as differentiated from *quantity* of toxin.

One cannot, at present, identify individual δ-endotoxins chemically. Serology offers hope for the future. Krywienczyk *et al.* (1978) have shown that serological procedures can distinguish two δ-endotoxins with different spectra of insecticidal activity. However, much more needs to be learned in this area. At present, we can hope only to distinguish these toxins by differences in the insecticidal activities of the various powders.

To accomplish this, and to minimize any effects of changes in media or fermentation conditions, each powder was produced under the same conditions. For each we grew ca. 2 litres of beer, using B-4 medium (Table IV) and fermentation procedures of Dulmage (1971). We harvested the spore-toxin complex by the acetone precipitation method of Dulmage *et al.* (1970). Yields averaged 10.2 g/litre, sufficient for the distributions of a 1-g portion to each participant, who bioassayed it against one or more species of insect. Krywienczyk used her sample for serological analysis of the crystal. Since with a few exceptions all scientists received samples of the same powder, each used exactly the same material, and so their results were directly comparable.

TABLE IV

Medium B-4 used for fermentation of Bacillus thuringiensis (Dulmage, 1971)

Ingredient	Amount (g/litre)	Ingredient	Amount (g/litre)
Proflo[a]	10.0	$MgSO_4 \cdot 7H_2O$	0.3
Dextrose	15.0	$FeSO_4 \cdot 7H_2O$	0.02
Yeast extract (Difco)	2.0	$ZnSO_4 \cdot 7H_2O$	0.02
Bacto-peptone	2.0	$CaCO_3$	1.0

[a] A partially defatted cooked cotton seed flour. Traders Protein Division of Traders Oil Mill Company, Fort Worth, Texas.

It was also important to know the variety (H-type) of *B.t.* used to produce each powder. This was accomplished in two ways. First, a lyophilized stock of the culture used in each fermentation was identified by de Barjac. Second, Aizawa isolated the organism present in the final powder and classified it. A comparison of the two results confirmed the variety grown.

B. BIOASSAYS, ACTIVITY RATIOS AND DISTRIBUTION PATTERNS

1. *Bioassay Procedures*

The types of insects used, their growth rates, eating habits and behaviours varied widely. All these variables influenced the choice of assay procedures. Thus there was no one way that we could use to test a *B.t.* powder against all insect species — procedures had to be adapted to the insect. However, there were unifying principles throughout these assays. (1) In all assays, the powder was administered to the insects by mixing it into their diet, with larvae being allowed to feed *ad libitum* on the powder-diet mixtures. (2) The effect of this exposure was judged by a single criterion: death. A severely retarded or moribund larva was considered alive if it could move. The assays measured only per cent dead. (3) Preliminary assays determined kill at two levels, usually $500 \mu g$ and $50 \mu g$ powder/unit of diet. If activity was sufficient, repeat assays were run with an appropriate series of dilutions to determine the LC_{50}. The amount of replication in the assays, the level of activity judged adequate to warrant determining an LC_{50} and the accuracy of these determinations, all varied according to the insect tested. Details of the procedures will be described in the final publication. A few examples can be found in papers by Burges and Bailey (1968) and Burges (1976) on "artificial food" assays with the greater wax moth, Dulmage *et al.* (1971, 1976) on assays with certain lepidopterous insects, and Hall *et al.* (1977) on assays with mosquitoes.

2. LC_{50} and International Unit

Theoretically, one could rear such a highly inbred and totally homogeneous colony of an insect species that the LC_{50} of a powder determined with that insect would be absolutely reproducible. In actuality, this is impossible. Minor fluctuations in rearing or in assay conditions can influence the response of an insect. To compensate for these variations, it is desirable to include a standard powder in every assay set and to compare the response of the insect to a test powder with its response to this standard. For convenience, a standard is assigned a potency in units of activity, the most common being the "International Unit" or "IU". Then a direct comparison of the LC_{50} of the standard and that of the test powder expresses potency of the test powder in IU's. This unit is much more reproducible over time and between laboratories than the LC_{50} (Dulmage, 1973a).

The first generally accepted standard was prepared in France from a fermentation of H-type *thuringiensis* and called "E-61". E-61 was assigned a potency of 1000 IU/mg and recommended as an international standard in 1966 (Burges, 1967). When the HD-1 strain of H-type *kurstaki* was selected for commercial production of *B.t.* in the USA, a formulation of HD-1, labelled HD-1-S-1971, was selected for use as a primary reference standard in that country (Dulmage, 1973b, 1975). HD-1-S-1971 was assigned a potency of 18,000 IU/mg on the basis of assays against E-61, using *Trichoplusia ni* as test insect. All units presented in this paper are derived from assays against HD-1-S-1971, except those for *G. mellonella*, against which HD-1 has low activity. An H-type *galleriae* powder was used as standard for *G. mellonella* bioassays. The conversion of LC_{50}s into IUs using bioassays against a standard is shown in Table V.

3. Activity Ratios

The IU reflects only the relative activity of a test material against a standard in assays with a specific insect. If the standard and the test material contain the same δ-endotoxin, then the IUs determined in different bioassays will be the same, even if the assays are conducted with different insect species.

In such a case, we are merely comparing the toxin with itself. However, if the test sample contains a different δ-endotoxin — one with a different spectrum of insecticidal activity from that of the standard — then the IUs can vary according to the insect species used.

This leads to an important corollary: *If the IUs determined in assays against two different insect species are not identical, this is evidence that the test sample is not homologous with the standard.* This principle not only allows us to differentiate a sample from the standard, it also offers a convenient and quantitative tool for identifying this difference and thus comparing toxins, as illustrated in Table VI. In the hypothetical example, the δ-endotoxins present in powders A and B were the same as the standard, while that in powder C was different. All three powders were bioassayed against both *T. ni* and *Heliothis virescens*. As expected, the results of bioassays of A and B were the same, regardless of the insect species in the assay. A mathematical way of expressing this is to divide the potency in IUs determined vs *T. ni* by that determined vs *H. virescens*. This yields a ratio, which we call the "Tn/Hv" ratio, or in earlier publications (Dulmage, 1975, 1979; Krywienczyk *et al.*, 1978, 1981), the "T/H" ratio. In our hypothetical example, the Tn/Hv ratio for both A and B was 1.00, indicating that they compared similarly to the standard — and to each other. However, powder B was 5 times more active than A: although the potencies were different, the Tn/Hv ratios were the same. An activity ratio is not changed by potency: it does not measure *quantity* of toxin; it measures only the *character of the toxin relative to a standard*. This is further illustrated by assays with powder C (Table VI). Potency in assays with *T. ni* was 50,000 IU/mg while that in assays with *H. virescens* was only 25,000 IU/mg. Thus the Tn/Hv ratio determined for powder C was 2.00, indicating that the toxin in C was different from that in the standard. Once again, we find that potency alone will not distinguish between toxins as both B and C had a potency of 50,000 IU/mg vs *T. ni*. It was not until a second insect species was used that the difference between the two toxins became apparent.

TABLE V

Calculations of potencies of dry powders of δ-endotoxins of
Bacillus thuringiensis

I. *Basic formula:*

$$\frac{LC_{50}\ \text{Standard}}{LC_{50}\ \text{Test Sample}} \times \text{Potency of Standard, IU/mg} = \text{Potency of Test Sample, IU/mg}$$

II. *When HD-1-S-1971 is used as the standard, the equation becomes:*

$$\frac{LC_{50}\ \text{HD-1-S-1971}}{LC_{50}\ \text{Test Sample}} \times 18\,000\ \text{IU/mg} = \text{Potency of Test Sample, IU/mg}$$

The numerical value of the ratio gives a quantitative evaluation of the difference between a sample and a standard, and this can lead to further

TABLE VI

Hypothetical results of bioassays on powders containing two δ-endotoxins with different spectra of activity (adapted from Dulmage, 1979)

	Powder A	Powder B	Powder C
1. *Assays against* T. ni:			
LC_{50} standard, μg/ml diet	11.2	11.2	11.2
LC_{50} sample, μg/ml diet	20.0	4.0	4.0
Potency of sample, IU/mg	10 000	50 000	50 000
2. *Assays against* H. virescens:			
LC_{50} standard, μg/ml diet	2.8	2.8	2.8
LC_{50} sample, μg/ml diet	5.0	1.0	2.0
Potency of sample, IU/mg	10 000	50 000	25 000
3. *Tn/Hv ratio:*	1.0	1.0	2.0

comparisons e.g. a toxin with a Tn/Hv ratio of 2.00 would not be homologous with a toxin with a Tn/Hv ratio of 5.00.

This type of ratio is not restricted to assays between *T. ni* and *H. virescens.* The relative IU's of any two insect species can form a ratio, even if the assays are performed in different laboratories, as long as the assays are performed on the same powder. We have used several ratios: Tn/Hv, Hc/Bm, Hc/Tn, Bm/Hv, Ec/Pi, and Se/Hi. The letters indicate the insects (Hc = *Hyphantria cunea*; Bm = *Bombyx mori*; Ec = *Ephestia cautella*; Pi = *Plodia interpunctella*; Se = *Spodoptera exigua*; Hi = *Haematobia irritans*). The ratio should be based on IUs, although, in this preliminary work, we used LC_{50} s for the Ec/Pi and Se/Hi ratios.

4. Distribution Patterns

As a tool to uncover the interrelations between toxins in these studies, we have attempted to examine deviations from normal distribution curves. Thus, if there were only one toxin produced by a series of isolates of an H-type, two factors should be evident in the results of our bioassays. (1) Activity ratios determined on powders from these isolates should fluctuate according to a normal distribution curve, with the shape of the curve delineated by the accuracy of the assays. (2) Potencies, whether expressed as IUs or as LC_{50} s, should also fluctuate according to a normal distribution curve, with the shape of the curve, in this case, being delineated both by the accuracy of the assay and the variability of the fermentation efficiencies inherent in the population of isolates being studied. Deviations from both activity ratio curves and potency curves may indicate the presence of a different toxin or a mixture of toxins in the deviating isolates. On the other hand, if the mean and standard deviation of curves from sets of powders of two H-types are quite similar, it may indicate that the toxins present in the two sets are the same.

It would be foolish to regard analyses of these distribution patterns as proof by themselves. It would be even more foolish to do so in these studies, where we frequently have only a few powders with which to judge distribution of results. However, examinations of the distributions of ratios and toxicities within H-types or crystal-types can be useful deductive tools, and this is how we shall now use them.

III. Summary of Results

A. INTRODUCTION

The program uncovered a surprising complexity in the insecticidal activities of *B.t.* powders. We found, as expected, that there were large differences in the quantities of toxin (expressed as IUs or comparative LC_{50}s) produced by the various isolates. However, there were many unexpected interrelationships between variety, crystal, and activity. We cannot present all our data here, but will examine portions in detail to demonstrate these interrelationships. First, we will look at the association between crystal-type and H-serotype. Next, we will explore the interrelationships between crystal-type, H-serotype, and insecticidal activity by examining three selected types of powders: two produced by the same H-serotype, but containing different crystal-types, and a third containing one of these crystal-types, but produced by a different H-serotype. Finally, we will briefly summarize the activities observed in the varieties of *B.t.* studied in our program, hoping that this detailed discussion will warn that the insecticidal activities of *B.t.* are not simple, and that this brief overview, if taken to be all encompassing, could be misleading. Always we will examine the characteristics of the toxins produced; the quantity of toxin will be incidental to the discussion. In a final section, we shall discuss some of the promising isolates that appear to produce high yields of toxin.

B. RELATIONSHIP BETWEEN CRYSTAL-TYPE AND H-SEROTYPE

Krywienczyk and Angus (1967, 1969) and Krywienczyk *et al.* (1978) have already shown the value of serology in differentiating crystals of *B.t.* Krywienczyk broadened this work in our program and developed serological methods to identify most of the crystal-types encountered in our powders. For simplicity, she identified each crystal-type by the first three letters of the varietal name with which it is most commonly associated, although she used "*k-1*" and "*k-73*" for the two crystal-types of H-serotype *kurstaki* differentiated by crystal antigens (see Section I, B, 1). Her data (Table VII) show that apparently identical crystals can appear in different H-serotypes. The frequency of such occurrence depends on the H-type. For example, there is almost a 50—50 division between crystal types *k-1* and *k-73* in H-type *kurstaki* and about a 4:1 division between *thu* and *k-1* crystals in H-type *thuringiensis;* in contrast, 61/65 isolates of H-type *galleriae* produced *gal* crystals, and the *gal* crystal was found in no other serotype. Of particular interest was the apparent presence in crystals of some isolates of more than one type of crystal antigen, e.g. eight isolates of H-type *thuringiensis* appeared to contain both *thu* and *k-1* type antigens.

TABLE VII

Distribution of crystal types among varieties of Bacillus thuringiensis

Variety	H-serotype	Crystal-type	No. of isolates
thuringiensis	1	*thu*	38
		k-1	10
		atypical *k-1*	1
		Mixed *thu* and *k-1*	8
		Weak reaction	8
alesti	3a	*ale*	8
		k-73	1
		Weak reaction	13
kurstaki	3a3b	*k-1*	26
		atypical *k-1*	4
		k-73	21
		Mixed *k-1* and *k-73*	3
		aiz	2
		thu	1
		Weak reaction	6
dendrolimus	4a4b	*den*	5
		sot	2
		Weak reaction	3
kenyae	4a4c	*ken*	9
		atypical *ken*	3
		thu	1
		Mixed thu and *k-1*	1
		Weak reaction	1
galleriae	5a5b	*gal*	61
		k-1	2
		aiz	2
		Weak reaction	7
entomocidus	6	*ent*	1
		thu	1
		Weak reaction	2
subtoxicus	6	*sub*	1
		Weak reaction	1
aizawai	7	*aiz*	16
		Weak reaction	9
tolworthi	9	*tol*	1
		Weak reaction	5

Thus a particular crystal-type may be produced by more than one H-serotype. Still, each H-type can be broadly characterized as containing a typical type of crystal. And, as we shall see, both play an important role in the insecticidal activities of the isolates. Thus, in future, the classification of isolates of *B.t.*

should involve both H-antigen and crystal serology — one will not distinguish an isolate and its activities without the other.

C. INTERRELATIONSHIPS BETWEEN H-SEROTYPE, CRYSTAL-TYPE AND INSECTICIDAL ACTIVITIES IN H-TYPES *THURINGIENSIS* AND *KURSTAKI*

1. *Introduction*

As discussed previously, 38 of the powders of H-type *thuringiensis* contained *thu* crystals while 10 contained, instead, *k-1* crystals. Also, 26 powders from isolates of H-type *kurstaki* produced *k-1* crystals. Bioassays of these three sets of powders offered the opportunity of examining the influences of both H- and crystal-type on insecticidal activity. In this section, we shall analyse our results in this light, using activity ratios and percentage kills. (The eight powders from H-type *thuringiensis* containing crystals with both *thu* and *k-1* antigens will be omitted from consideration, except for a brief discussion when we examine activities vs mosquito larvae.)

TABLE VIII

Grouping of 32 thu *and 10* k-1 *powders of H-type* thuringiensis *and 26* k-1 *powders of H-type* kurstaki *by Tn/Hv activity ratios*[a]

Crystal type	No. in group	Tn/Hv ratios	Crystal type	No. in group	Tn/Hv ratios
thu[b]	4	7.57	*k-1*	3	5.23
	23	4.84		30	2.10
	5	2.50		3	0.755

[a] Ratios based on international units determined in bioassays against *T. ni* (Tn) and *H. virescens* (Hv). [b] Tn/Hv ratios could not be calculated on six isolates containing *thu* crystals.

2. *Tn/Hv Ratios*

Tn/Hv ratios generally distinguished between *thu* and *k-1* powders, but not between the *k-1*s produced by the two H-types (Table VIII). The Tn/Hv ratios of 23/32 isolates of H-type *thuringiensis* with *thu* crystals averaged 4.84. Preparations from four isolates had a much higher Tn/Hv ratio: 7.57. In contrast, *k-1* powders from 30/36 isolates of both H-types had Tn/Hv ratios averaging 2.10. Furthermore, while it is not shown in the table, there was no distinction in the distribution of Tn/Hv ratios between *k-1* powders produced by H-type *thuringiensis* and those produced by H-type *kurstaki*.

There were exceptions to the above observations. Three *k-1* powders (two from H-type *kurstaki* and one from H-type *thuringiensis*) had Tn/Hv ratios averaging 5.23, similar to the ratios of *thu* powders, and five *thu* powders had Tn/Hv ratios averaging 2.50, similar to those of *k-1*. Also, three *k-1* powders from H-type *kurstaki* had Tn/Hv ratios averaging 0.76, similar, as we shall see later, to the ratio usually found only in *k-73* powders. However, these last three materials did not resemble *k-73* powders in their activities vs *B. mori* and *H. cunea:* in contrast, these *k-1* powders were not only active vs *B. mori,* they were more active against this insect than against *H. cunea.*

3. *Hc/Bm Ratios*

Interrelationships between H-type, crystal-type, and spectrum of activity appeared much more complex when Hc/Bm ratios were examined (Fig. 1). Crystal-type influenced ratios: Hc/Bm ratios of *thu* powders were mostly much higher than those of *k-1* — a major difference in ratios between crystal-types that continued the trend seen with Tn/Hv ratios. However, H-type also appeared to influence Hc/Bm ratios: there were distinct differences in the distribution of Hc/Bm ratios of *k-1* powders depending on their H-type, i.e. 7/10 powders of H-type *thuringiensis* had an average Hc/Bm ratio of 0.12 or less, while none of H-type *kurstaki* had a ratio that low. Also, 7/25 powders of H-type *kurstaki* had ratios averaging 2.61, higher than any observed in *k-1* powders from H-type *thuringiensis.* Thus, while there was some overlap in the ratio ranges of the *k-1* powders produced by the two serotypes, there was a definite association of distribution patterns of Hc/Bm ratios with H-serotype.

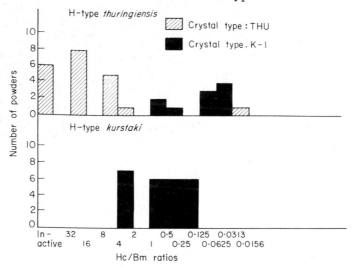

Fig. 1. Distribution patterns of Hc/Bm ratios of powders of H-types *thuringiensis* and *kurstaki* containing *thu* and *k-1* crystals. Hc/Bm ratio = IU/mg vs *Hyphantria cunea* ÷ IU/mg vs *Bombyx mori.*

4. *Ec/Pi Ratios*

Ec/Pi ratios of powders from H-types *thuringiensis* and *kurstaki* appeared to be influenced both by their H-type and their crystal-type (Fig. 2). As with

Hc/Bm ratios, the Ec/Pi ratios of *k-1* powders differed greatly according to H-type: ratios of 8/9 H-type *thuringiensis* powders were < 0.08; 16/22 from *kurstaki* were 0.16 or more. Crystal-type was also important. There was little similarity between ratios of *thu* and *k-1* powders. Ratios of powders with *thu* crystals varied widely, but most fell into two groups: 16/32 averaged 0.55, and 8 others, 0.19. Only 3/32 *thu* powders were inactive vs *P. interpunctella* compared to 7/9 *k-1* powders from H-type *thuringiensis*. As Fig. 2 shows, the patterns of distribution of these powders differed, not only according to H-type, but also according to crystal type.

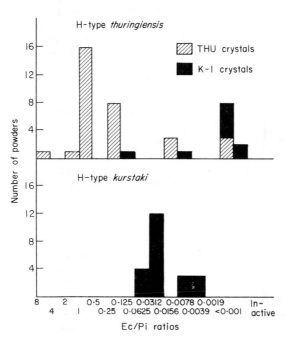

Fig. 2. Distribution patterns of Ec/Pi ratios of powders of H-types *thuringiensis* and *kurstaki* containing *thu* and *k-1* crystals. Ec/Pi ratios = LC_{50} vs *Ephestia cautella* ÷ LC_{50} vs *Plodia interpunctella*.

5. *Activity vs* Ostrinia nubilalis

Most powders killed neonate *O. nubilalis* at 500 µg/ml diet. However, distinct differences between powders appeared at 25 µg/ml, so this was chosen as the level at which to attempt to differentiate *thu* and *k-1* powders produced by the two H-types (Table IX): no H-type *thuringiensis* powder, with either *thu* or *k-1* crystals, killed 70% of the larvae, while 17/26 of the *kurstaki* powders killed > 70%, and the LC_{50}s of 13 of these were < 12 µg/ml.

Clearly activity vs. *O. nubilalis* at the 25 µg/ml level was associated solely with the H-type, since both *thu* and the *k-1* powders from H-type *thuringiensis* were "inactive", while most *k-1* powders from H-type *kurstaki* were "active". However, there were 9 *kurstaki* powders with only low activities, corresponding to those of *thuringiensis* powders. Furthermore, about half of the *kurstaki* powders had *k-73* crystals, and 10/19 of such powders were "inactive". Thus, within the H-type other factors must also play a role. Raun *et al.* (1966) and

TABLE IX.

Groupings of 38 thu *and 10* k-1 *powders of H-serotype* thuringiensis *and 26* k-1 *powders of H-type* kurstaki *by activities against* O. nubilialis *and* H. irritans

H-type	Crystal type	No. active/No. tested[a]	
		O. nubilalis[b]	*H. irritans*[c]
thuringiensis	*thu*	0/37	24/32
	k-1	0/10	2/10
kurstaki	*k-1*	17/26	0/30

[a] Discrepancies in totals in the table arise because not all powders were tested against both insects. [b] Active = > 70% kill at 25 μg powder/ml diet. [c] Active = $LC_{50} < 500$ μg powder/ml diet.

Sutter and Raun (1967) suggested that the activities of *B.t.* formulations were enhanced when vegetative cells of *B.t.*, multiplying in the midgut of 4th and 5th instar larvae of *O. nubilalis*, ultimately penetrated the damaged intestinal wall and infected the blood. Since no antibiotic was used in our bioassay diet, it might be speculated that such multiplication contributed to the activities observed in our results, but this is unlikely because Beegle and coworkers (personal communication) showed that presence or absence of antibiotic had no effect on the susceptibility of neonate larvae.

6. *Activity vs* Galleria mellonella

High toxicity of *B.t.* powders vs *G. mellonella* was, with few exceptions, confined to H-types *galleriae* and *aizawai*.

Powders of H-types *thuringiensis* and *kurstaki* were much less active, although their toxicities varied widely. Distribution patterns of LC_{50}s of H-type *thuringiensis* with *thu* crystals and those of H-type *kurstaki* with *k-1* crystals were nearly the same (Fig. 3) — the former had a mean LC_{50} of 6.2 (variance, 6.0) and the latter 5.9 (4.4) mg powder/g diet. Interestingly, the distribution pattern of H-type *thuringiensis* containing *k-1* crystals differed, with 7/10 powders inactive vs this insect.

The data above — particularly the similarity of different crystal-types from H-types *thuringiensis* and *kurstaki* and the difference between *k-1* powders from the two H-types — indicated no correlation of activity with either H-type or crystal-type. Yet the high activities observed in H-type *galleriae* appeared to be strongly correlated with H-type, or crystal-type or both. Burges *et al.* (1976) proposed that *G. mellonella* was a "Type 3" insect (Heimpel and Angus, 1959) since it was only weakly susceptible to *gal* crystals alone, but highly susceptible to a spore-crystal mixture. H-type *galleriae* spores alone were moderately active. They further noted that the activity of crystals of H-type *galleriae* against this insect was only slightly enhanced by the addition of spores of H-type *thuringiensis*, but was greatly enhanced by the addition of spores of H-type galleriae. This leads one to ask whether the differences in activities observed between the powders from H-types *thuringiensis* and *kurstaki* and the powders from

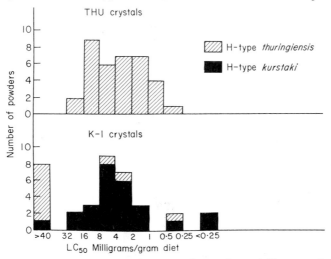

Fig. 3. Distribution patterns of LC_{50} of powders of H-types *thuringiensis* and *kurstaki* containing *thu* and *k-1* crystals vs *Galleria mellonella*.

H-type *galleriae* against *G. mellonella* might not be due to the differences in the H-types of the spores that they contain, i.e. that the same toxic factor might exist in all three types of powders, but be enhanced only in the presence of H-type *galleriae* spores. This question could be answered by mixing H-type *galleriae* spores with *thu* or *k-1* crystals, a very valuable experiment for future research.

7. *Activity vs* Lymantria dispar

Many powders were active vs *L. dispar,* so, for convenience, Dubois divided them into four categories according to the degree of their activity:

Category	Mortality Range	Potency Designation
1	< 80% mortality at 25 µg/ml	nonpotent
2	< 80% mortality at 6.25 µg/ml	weakly potent
3	< 80% mortality at 1.56 µg/ml	moderately potent
4	> 80% mortality at 1.56 µg/ml	very potent

When powders of H-types *thuringiensis* and *kurstaki* were grouped by categories, we found a strong correlation with crystal-type (Fig. 4). Most powders containing *thu* crystals were no more than weakly active: 36/44 fell into Categories 1 and 2. No *thu* powders were in Category 4. On the other hand, 26/35 *kurstaki* powders with *k-1* crystals fell into Categories 3 and 4, with 11 of these in Category 4. The influence of crystal-type predominated in H-type *thuringiensis:* whereas only 8/44 powders with *thu* crystals were moderately potent, 7/10 containing *k-1* crystals were as active.

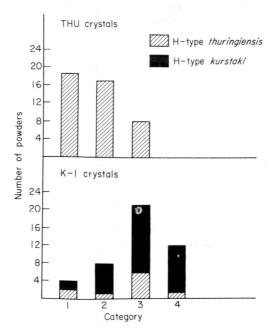

Fig. 4. Distribution patterns of activity categories of powders of H-types *thuringiensis* and *kurstaki* containing *thu* and *k-1* crystals determined against *Lymantria dispar.* The categories are defined in the text.

Dubois further examined his results in terms of Tn/Hv ratios and of IUs determined vs *T. ni* and *H. virescens.* The mean Tn/Hv ratio of Category 4 powders was 1.91 — in good agreement with the presumption that the high activity vs *L. dispar* was associated with the *k-1* crystal-type. Also powders with high activity vs *L. dispar* were very active vs *T. ni,* but the reverse was not necessarily true — some powders highly active vs *T. ni* were not very active vs *L. dispar.*

8. *Activities vs* Haematobia irritans, Spodoptera exigua, Agrotis ipsilon *and* Bovicola crassipes *in Relation to the "β-exotoxin"*

(a) *Introductory Comments.* Activities discussed in this section appeared to be associated with heat-stable, water-soluble toxins active against the horn fly, *H. irritans,* that thus met the definition of β-exotoxin (Heimpel, 1967; Section I, B, 2, c of this chapter). For convenience, we shall call these toxins "β-exotoxins" even though there were discrepancies between the insecticidal activities of different powders indicating that in fact these β-exotoxins were *not* all alike.

(b) H. irritans *and* S. exigua. There was a definite, but not invariable, association of β-exotoxin with H-type (Table X; see also Chapter 13, Section II). However, there also seemed to be some association with crystal-type. Since β-exotoxin is a nucleotide produced during vegetative growth, and δ-endotoxin is a protein formed at sporulation, no correlation would be expected between the production of such unrelated molecules. Yet, although not absolute, the association appeared to exist. While most H-type *thuringiensis* powders with *thu* crystals alone also contained β-exotoxin, less than a third of the powders of this variety with either *k-1* crystals alone or a mixture of *k-1* and *thu* crystals contained the exotoxin. Powders from H-type *kurstaki* were much more uniform: none of 46 powders with either *k-1* or *k-73* crystals alone were toxic to *H. irritans*, although one with *thu* crystals and one with a mixture of crystal-types did contain β-exotoxin.

TABLE X

Production of heat-stable activity vs H. irritans *as a function of H-type and crystal-type*

Crystal-type	H-type *thuringiensis*		H-type *kurstaki*	
	Active powders	Inactive powders	Active powders	Inactive powders
thu	24	8	1	0
k-1	2	8	0	30
thu + *k-1*	2	6	–	–
k-73	–	–	0	23
k-1 + *k-73*	–	–	1	1
aiz	–	–	0	1
unknown	–	–	0	4

All powders of H-type *thuringiensis* and all but one of H-type *kurstaki* that were active against *S. exigua* contained β-exotoxin − i.e., a heat-stable toxin active vs *H. irritans*. Beegle extracted some of these powders and confirmed that the exotoxins themselves were also active vs *S. exigua*, although there appeared to be some differences in the heat-stability of extracts from different powders. To investigate possible differences in the β-exotoxins, we calculated Se/Hi ratios from the LC_{50}s (Table XI). The ratios appeared to group these exotoxins into three categories with ratios of 0.64, 1.92 and 7.16, respectively. Some overlaps between categories may indicate that these exotoxins will eventually be divided into more than these three groups. There was also a possible relationship between these categories and activity vs *B. crassipes*. This will be discussed later in this section. The impact of all these data is to warn that it may be unwise to assume that all β-exotoxins are alike.

Two H-type *kurstaki* powders, including one containing β-exotoxin, also contained a toxin active vs *S. exigua* but not vs *H. irritans*, apparently associated

TABLE XI

Grouping of powders of H-type thuringiensis by Se/Hi ratios[a]

No. of powders	Mean Se/Hi ratio	Coefficient of variation
1	7.16	–
14	1.92	0.285
7	0.639	0.339
5	Low activity vs *S. exigua,* ratio not determined.	

[a] Se/Hi ratio = LC_{50} vs *S. exigua*/LC_{50} vs *H. irritans.*

with δ-endotoxin. Such activity was unusual in our powders, particularly outside of H-type *aizawai:* it occurred in only 18 powders, 14 of which were H-type *aizawai.* Interestingly, while the above two *kurstaki* powders were unique in their action vs *S. exigua,* they were unremarkable in any other way, with activity spectra typical of other powders produced by H-type *kurstaki.*

(c) A. ipsilon. Only 9/319 powders were active vs the black cutworm, *A. ipsilon.* At first, it seemed that this activity was due to β-exotoxin, i.e. an exotoxin with some degree of heat stability and present only in powders active vs *H. irritans* and *S. exigua,* two insects susceptible to β-exotoxin. However, anomalies in these data warn against being too categorical in the assumption that the activities vs *A. ipsilon* and vs the other two insects are the same. First, the activities towards *A. ipsilon* were usually lost on storage, while activities towards *S. exigua* and *H. irritans* were not. Second, very highly active powders vs *H. irritans* were not active vs *A. ipsilon.* In these the *A. ipsilon* activity may never have been present, or it may have been lost in storage before bioassay, but, either way, the toxin vs *A. ipsilon* appears distinct from the activity vs *H. irritans.* Finally, activity ratios of LC_{50}s vs *A. ipsilon* and *H. irritans* failed to reveal any correlation between ratios of the different powders – again an indication of the presence of two unrelated toxins. These data are, however, not conclusive because the samples of the powders used in bioassays against each insect were stored in different locations, and there may have been some unknown factor in the storage of powders tested vs *A. ipsilon* that affected stability of the β-exotoxin. Although this seems unlikely such an occurrence could explain the observed results, even if only one toxin were present.

Thus activity vs *A. ipsilon* is not due to δ-endotoxin, but to a heat-stable exotoxin which may be β-exotoxin, but is more likely an as yet unidentified exotoxin.

(d) B. crassipes. Gingrich *et al.* (1974) reported the presence of a "louse-factor" in certain preparations of *B.t.* var *kurstaki,* HD-1. In our program, Gingrich and his associates found that there were at least four separate toxins against the louse in different powders tested. They grouped these as follows: Group 1, β-exotoxin-like – heat stable, active vs *H. irritans* and *B. crassipes;* Group 2, endotoxin-like – heat labile, not active vs *H. irritans,* active vs *B. crassipes;* Group 3, heat stable, not active vs *H. irritans,* active vs *B. crassipes;* Group 4, like Group 1 in being heat stable and active vs *H. irritans,* but only slightly toxic for *B. crassipes.*

These groups differed in their relative toxicities vs lice. For example, all Group 1 powders had LC_{50}s vs *B. crassipes* of < 8.0 mg/g diet. With two exceptions, LC_{50}s of Groups 2 and 3 powders were > 8 mg/g diet, and some were as high as 72 mg/g diet. There seemed to be some difference between varieties as to which type powder was produced: only 2/19 powders of H-type *thuringiensis* definitely fell into Group 2, while 5/7 powders of H-type *kurstaki* belonged to this group.

In view of the different groups that appeared when we compared activities of β-exotoxin-like powders of H-type *thuringiensis* vs *H. irritans* and *S. exigua,* we determined ratios of LC_{50}s vs *H. irritans* and *B. crassipes*. As seen in Table XII, powders could be divided into four categories by these Hi/Bc ratios. When we carried this grouping one step further, taking into account both Se/Hi and Hi/Bc ratios, we found evidence of five major groups and three minor groups. The significance of this cannot yet be determined, but, once again, we have warnings that all β-exotoxins are not necessarily alike.

TABLE XII

Grouping of powders of H-type thuringiensis *by Se/Hi and Hi/Bc ratios*[a]

No. of powders	Se/Hi ratio		Hi/Bc ratio	
	Mean	CV	Mean	CV
Groupings based on Hi/Bc ratios alone				
2			222.0	—
5			131.0	0.176
7			71.2	0.214
13			34.5	0.242
Groupings based on both ratios				
1	7.08	—	24.0	—
5	2.35	0.211	34.2	0.264
7	1.41	0.287	71.2	0.214
2	1.18	—	184.0	—
3	1.13	0.304	40.3	0.103
3	0.440	0.228	126.0	0.231
4	Low activity vs *S. exigua*		36.5	0.224
2	Se/Hi ratio not determined		176.0	—

[a] Se/Hi = LC_{50} vs *S. exigua*/LC_{50} vs *H. irritans*. Hi/Bc ratio = LC_{50} vs *H. irritans*/LC_{50} vs *B. crassipes*.

(e) *Toxicity of β-exotoxin to other insects in the program.* Since β-exotoxin can cause significant mortality of *S. exigua* and *A. ipsilon* larvae, it is logical to ask what effect this toxin has on other insects in our program, e.g. its effect on activity ratios or on mosquitoes. There is no positive answer, and each insect species must be treated individually. β-exotoxin is a general poison, and highly purified β-exotoxin kills larvae of some of our insect species, e.g. *O. nubilalis, T. ni,* and *H. virescens.* Yet this activity is much weaker than that of δ-endotoxin. Several powders have high toxicity vs *H. irritans* and so appear to contain much β-exotoxin, but these have little or no activity vs most of the insects in the program, including both Lepidoptera and mosquitoes (e.g., compare the activities of powders vs *O. nubilalis* and *H. irritans* in Table IX). Thus we believe that the β-exotoxin in a powder does not materially affect the interpretation of our data on the δ-endotoxin.

9. *Activity vs* Spodoptera litura

Most powders had little or no activity vs *S. litura.* Only 15/297 powders killed > 80% of instar III larvae at 500 μg/g diet. Most active powders were H-type *aizawai.* Two of H-type *kurstaki* with *k-1* crystals (including HD-1 itself) and one of H-type *thuringiensis* with a mixture of *k-1* and *thu* crystals had low activity vs this insect (LC$_{50}$s ca. 250 μg/g diet), but there were too few powders to reveal whether the toxicities were associated with crystal-type or not. While all three powders contained *k-1* crystals, it must also be remembered that most *k-1* powders were inactive vs this insect.

10. *Activity vs Mosquitoes*

B.t. activity was long considered to be restricted to Lepidoptera, until Hall *et al.* (1977) surprisingly reported that many *B.t.* powders killed mosquitoes. It is equally surprising that in the interactions between spores, crystals, and insect activities the most consistent correlations observed were between mosquito activity and crystal-type (Table XIII). All powders with *k-1* crystals, either alone or in a mixture, had some activity vs *Aedes aegypti, Ae. triseriatus,* and *Culex tarsalis.* Activity vs *Cx. pipiens* and *Cx. quinquefasciatus* was either very low or absent. In contrast, powders with *thu*-type crystals had little or no activity against these mosquitoes.

Interestingly, this correlation was not restricted to this one crystal type. For example, the same three mosquito species sensitive to powders with *k-1* crystals were sensitive to powders with *ken* crystals in H-type *kenyae.* Similarly, the two mosquito species resistant to *k-1* crystals were resistant to *ken* crystals.

TABLE XIII

Relationship between mosquito activity and crystal-type within serotypes

Serotype	Crystal-type	No. active/No. tested
1-*thuringiensis*	*thu*	1/34
	k-1	10/10
	thu + k-1	7/7
3a3b-*kurstaki*	*k-1*	28/28
	k-73	1/19
	k-1 + k-73	2/2
	aiz	2/3

11. *Discussion*

Thus far we have examined various aspects of insecticidal activities of powders of H-types *thuringiensis* and *kurstaki* only as isolated independent entities, i.e. the Tn/Hv ratio as independent of the Hc/Bm ratio, and both as independent of activities vs *G. mellonella*, etc. This is, of course, not so. These activities are interdependent, and the total insecticidal activity of an individual powder is a combination of all these — only when the powders are examined in this light can the true complexity of *B.t.* be seen. Our analyses are still incomplete, but the diversity of the insecticidal activities within these two H-types can be illustrated by grouping their powders according to five activity ratios: Tn/Hv, Hc/Bm, Hc/Tn, Bm/Hv, and Ec/Pi (Table XIV; see Section II, B,3 this chapter, for an explanation of the ratios). The first four ratios completely interrelate the results observed in bioassays against four insect species, while the last ratio is independent of the others.

The variety observed in activity spectra of these powders was surprising — 74 powders were tested and 32 different groupings were compiled. Some of these groupings may have been false — particularly those with only one powder — but most appeared valid. For example, the Tn/Hv and Ec/Pi ratios of the first 17 *thu*-type powders of H-type *thuringiensis* listed in the table were quite uniform, with means of ca. 4.8 and 0.55, respectively. Based on these ratios alone, the 17 powders would appear a homogenous group. However, examination of their Hc/Bm and Hc/Tn ratios clearly divides the 17 into three groups of 3, 10, and 4,

TABLE XIV.
Activity ratios of 38 thu *and 10 k-1 powders of H-serotype* thuringiensis *(H1) and 26 k-1 powders of H-type* kurstaki *(H3a3b)*

Crystal type	No. H1	No. H3a3b	Tn/Hv	Hc/Bm	Hc/Tn	Bm/Hv	Ec/Pi	Comments
thu	3	0	4.12	28.3	12.3	1.79	0.57	Ec/Pi ratios varied widely. Only 2 determined: 0.24, 0.89
	10	0	4.79	21.0	4.77	1.09	0.50	Hc/Bm, Hc/Tn, and Bm/Hv ratios based on 5 powders
	4	0	5.51	8.01	1.94	0.688	0.66	
	3	0	2.71	6.43	4.15	1.82	0.26	
	3	0	3.73	---	---	---	0.13	Only 1 Ec/Pi ratio
	3	0	Inact. vs Hv	---	---	---	0.13	Only 1 Ec/Pi ratio
	3	0	Inact vs Hv and Tn	---	---	---	0.23	Only 1 Ec/Pi ratio; 2 powders inactive vs Ec and Pi
	1	0	2.05	0.0727	0.397	11.2	0.078	Low activity vs Pi.
	1	0	2.34	2.24	29.7	31.0	---	Active vs Ec. Inact. vs Pi
	3	0	5.97	---	---	---	1.93	One Ec/Pi ratio. Two powders inact. vs Ec
	2	0	8.08	---	---	---	0.078	Inact. vs Hc and Bm
	1	0	7.10	---	---	---	0.436	Inact. vs Hc and Bm
	1	0	7.01	---	--	---	1.23	Inact. vs Hc and Bm
k-1	2	0	2.08	0.041	0.611	29.1	---	Inact. vs Ec and Pi
	2	0	1.50	0.052	0.254	7.65	---	Active vs Ec. Inact. vs Pi
	0	2	5.25	0.138	0.372	14.2	0.016	Only one Ec/Pi ratio.
	3	0	2.74	0.199	0.991	13.5	---	Active vs Ec. Inact. vs Pi
	0	4	2.73	0.234	0.329	---	---	Wide range Ec/Pi. 2, mean 0.020; 2, mean 0.22.
	0	3	1.93	0.347	0.299	1.58	0.012	2 Pi LC$_{50}$'s estimated. Activity low
	1	0	5.20	0.436	1.23	14.7	---	Active vs Ec. Inact. vs Pi
	0	3	1.43	0.571	0.371	0.920	0.085	
	1	0	2.35	0.782	3.04	9.15	0.085	
	1	0	1.58	0.978	2.98	4.82	0.77	
	0	2	2.81	0.968	1.29	3.79	0.159	Only one Ec/Pi ratio
	0	1	2.06	0.768	1.11	2.78	0.039	
	0	1	1.20	1.31	0.731	0.671	0.19	
	0	1	---	1.78	---	---	0.12	
	0	2	1.43	2.27	2.16	1.13	0.42	
	0	2	2.21	3.45	3.45	5.95	0.52	
	0	1	2.55	2.76	6.63	6.13	0.19	
	0	1	---	2.31	---	---	---	Inact. vs Ec and Pi
	0	3	0.755	0.732	0.602	0.778	0.138	

[a]Activity ratios are based on IU's for the following; *Hyphantria cunea* (HC), *Bombyx mori* (Bm), *Trichoplusia ni* (TN), *Heliothis virescens* (Hv); and on LC$_{50}$'s for; *Ephestia cautella* (Ec), *Plodia interpunctella* (Pi).

with, for example, Hc/Tn ratios of 12.3, 4.77, and 1.94. Similar exercises can be applied to the rest of the Table.

These results also show the value of ratios based on bioassays in different laboratories: the Hc/Tn ratios which played a useful role in deriving these groups were calculated from assays vs *H. cunea* in Japan and vs *T. ni* in Texas.

D. SPECTRA OF ACTIVITIES OF OTHER VARIETIES OF *BACILLUS THURINGIENSIS*

1. *Introduction*

Insecticidal activities within *thu* and *k-1* powders of H-serotypes *thuringiensis* and *kurstaki* were very complex (Section IIIC). In contrast, activities of powders of some H-types were very uniform. Having illustrated the complexity of the problem, I now briefly overview insecticidal patterns in other H-types and in the *k-73* powders of H-type *kurstaki*. I caution that for brevity these reviews ignore some complexities that might have occurred within a particular serotype.

2. *H-type* alesti

Powders of H-type *alesti* with *ale*-type crystals have little or no activity vs *T. ni, H. virescens, E. cautella, P. interpunctella, O. nubilalis, S. exigua, S. litura, A. ipsilon, H. irritans,* and five species of mosquitoes – a discouraging series of negatives. Yet these powders are quite active vs *H. cunea* and *B. mori,* and also compare with H-type *thuringiensis* in activity vs *P. brassicae* and *G. mellonella.* These results show the value of screening a broad spectrum of insect species in evaluating isolates of *B.t.*

3. K-73 *Powders of H-type* kurstaki

A remarkably homogeneous group. Tn/Hv ratios of 18/19 were very close to each other with a mean of 0.590 and a coefficent of variation of 0.25, well within experimental error. Most powders had little or no activity and 14/18 were inactive vs *B. mori.* Ec/Pi ratios were also very consistent, with a mean ratio from 18 powders of 0.297: a relatively high coefficient of variation (0.399) may indicate that the powders should be divided into two nearly equal-sized groups. Some differences appeared vs *O. nubilalis:* 6/18 were as active as *k-1* powders, the rest relatively inactive at 25 μg powder/ml diet. All powders had a relatively low activity vs *G. mellonella* and *P. brassicae,* resembling *k-1* powders in this regard. Against mosquitoes, 17/18 had no activity and one resembled typical *k-1* powders. None of the 18 were active vs *S. litura, S. exigua, A. ipsilon,* or *H. irritans.*

4. *H-type* dendrolimus

With little activity vs *H. virescens,* much more vs *T. ni* and with Tn/Hv ratios similar to those observed with *galleriae* powders, these powders did not share the high activity vs *G. mellonella* typical of *galleriae* powders, but were about the weakest of all powders vs this insect. Five out of seven powders were more active vs *H. cunea* than vs *B. mori* — the other two *vice versa*. Only one powder had a trace of activity vs either *E. cautella* or *P. interpunctella*. Activity vs *O. nubilalis* was low in 6/8 powders. None were active vs *S. litura, S. exigua, A. ipsilon* or five mosquito species and 5/8 had no activity vs *H. irritans*.

5. *H-type* kenyae

All of only six powders had low Tn/Hv ratios, resembling the ratios of *k-1* powders, low and roughly equal activity vs either *H. cunea* or *B. mori* and very low activity vs *G. mellonella*. Powders differed against *E. cautella* and *P. interpunctella*. Interestingly, four powders were active vs three mosquito species, resembling *k-1* powders.

6. *H-types* galleriae *and* aizawai

The two H-types resembled each other by having a remarkably high activity vs *G. mellonella* and a common esterase pattern (Norris, 1964). Also, var *aizawai* and about half the powders of var *galleriae* had an unusually high Tn/Hv ratio (ca. 10.0). Ratios of the other *galleriae* powders were much lower (ca. 1.4). Both H-types were mainly less active vs *H. virescens* than *B. mori*. However, there were differences: vs mosquitoes H-type *galleriae* had little activity but H-type *aizawai* was quite active, resembling k-1 powders; H-type *galleriae*, but not *aizawai*, had a heat-labile activity vs *H. irritans*, which — in view of work by Gingrich and Haufler (personal communication) — may have been due to spore infectivity rather than to a toxin.

IV. Summary

In this brief overview of our International Cooperative Program, we believe the results have given preliminary answers to the questions raised in Section I, A. There is an interrelationship between insecticidal activity spectrum and variety. Our mosquito data alone indicate that the same δ-endotoxin can be produced by more than one isolate, although much of our other data indicate this too. Different isolates of the same variety can produce different endotoxins — this is most clearly seen in the *k-1* and *k-73* toxins produced by H-type *kurstaki*, but is indicated in the many differences of activity spectra in different powders of the same H-type. Spectra and crystal-type are interrelated: progress towards serologically identifying these δ-endotoxins is considerable, but serology does not yet differentiate between some of the groupings in H-type *thuringiensis* and *kurstaki*. Other toxins are present in these powders besides the endotoxin — most noticeably heat-stable toxins, loosely grouped as "β-exotoxins", and, more rarely, a heat-labile exotoxin. We have evidence, based on the work of Burges and of Gingrich and Haufler, that the spore is important.

Finally, and most practically, we have found isolates more active in a particular insect species than HD-1 (Table XV). Some improvements are relatively modest (although even a 50% improvement can markedly affect costs), others are very large. All this encourages us as we proceed in a second stage to retest selected powders to confirm our results and attempt to develop useful formulations.

TABLE XV

Comparison of activities of selected "superior powders" with HD-1 powder

		Characteristics of superior powder			
Insect	Potency[a] HD-1 powder	Potency[a]	H-type	Crystal-type	% increase over HD-1
T. ni	39 800	52 400	*kurstaki*	*k-1*	32
H. virescens	15 400	70 600	*kurstaki*	*k-73*	360
H. cunea	47 500	143 000	*kurstaki*	*k-1*	200
B. mori	50 500	92 800	*aizawai*	*aiz*	83
E. cautella	(2.6)	(1.4)	*kurstaki*	*k-1*	86
P. interpunctella	(16.4)	(10.6)	*tolworthi*	*?*	55
O. nubilalis	29 700	70 900	*kurstaki*	*k-1*	140
G. mellonella	(0.056)	(0.00088)	*galleriae*	*gal*	640
P. brassicae	(0.78)	(0.46)	*kurstaki*	*k-1*	70
S. exigua	Est. (800)	130	*aizawai*	*aiz*	Est. 650
S. litura	22 900	96 000	*aizawai*	*aiz*	420

[a] Potencies expressed in IU/mg or (LC$_{50}$/unit of diet).

References

Barjac, H. de. (1978). *C. r. hebd. Séanc. Acad. Sci. Paris* ser. D, 797–800.

Barjac, H. de and Bonnefoi, A. (1962). *Entomophaga* 7, 5–31.

Barjac, H. de and Bonnefoi, A. (1972). *J. Invertebr. Path.* 20, 212–213.

Barjac, H. de and Thompson, J.V. (1970). *J. Invertebr. Path.* 15, 141–144.

Barjac, H. de, Cosmao-Dumanoir, V., Shaikh, M. R. and Viviani, G. (1977). *C. r. hebd. Séanc. Acad. Sci. Paris* 284, 2051–2053.

Berliner, E. (1911). *Z. ges. Getriedew.* 252, 3160–3162.

Bonnefoi, A. and Barjac, H. de. (1963). *Entomophaga* 8, 223–229.

Burgerjon, A. and Martouret, D. (1971). *In* "Microbial Control of Insects and Mites" (H.D. Burges and N.W. Hussey, eds.), pp. 305–325. Academic Press, London and New York.

Burges, H.D. (1967). *In* "Insect Pathology and Microbial Control" (P.A. van der Laan, ed.), pp. 306–338. North-Holland Publ. Co., Amsterdam.

Burges, H.D. (1976). *Entomologia exp. appl.* 19, 243–254.

Burges, H.D. and Bailey, L. (1968). *J. Invertebr. Path.* 11, 184–195.

Burges, H.D., Thompson, E.M. and Latchford, R.A. (1976). *J. Invertebr. Path.* 27, 87–94.

Delucca, A.J. II, Simonson, J.G. and Larson, A.D. (1979). *J. Invertebr. Path,* 34, 323–324.

Dulmage, H.T. (1970). *J. Invertebr. Path.* 16, 385–389.

Dulmage, H.T. (1971). *J. Invertebr. Path.* 18, 353–358.

Dulmage, H.T. (1973a). *Ann. N. Y. Acad. Sci.* 217, 187–199.

Dulmage, H.T. (1973b). *Bull. ent. Soc. Am.* **19**, 200–202.

Dulmage, H.T. (1975). *J. Invertebr. Path.* **25**, 279–281.

Dulmage, H.T. (1979) *In* "Genetics in Relation to Insect Management" (M.A. Hoy and J.J. McKelvey, Jr., eds.), pp. 116–127. Working Papers — The Rockefeller Foundation, New York.

Dulmage, H.T. and Barjac, H. de. (1973). *J. Invertebr. Path.* **22**, 273–277.

Dulmage, H.T., Correa, J.A. and Martinez, A.J. (1970). *J. Invertebr. Path.* **15**, 15–20.

Dulmage, H.T., Boening, O.P., Rehnborg, C.S. and Hansen, G.D. (1971). *J. Invertebr. Path.* **18**, 240–245.

Dulmage, H.T., Martinez, A.J. and Pena, T. (1976). U.S. Dep. Agric. Tech. Bull. 1528, 1–15.

Gingrich, R.E., Allan, N. and Hopkins, D.E. (1974). *J. Invertebr. Path.* **23**, 232–236.

Hall, I.M., Arakawa, K.Y. Dulmage, H.T. and Correa, J.A. (1977). *Mosq. News* **37**, 246–251.

Heimpel, A.M. (1967). *Ann. Rev. Ent.* **12**, 287–322.

Heimpel, A.M. and Angus, T.A. (1958). *Can. J. Microbiol.* **4**, 531–541.

Heimpel, A.M. and Angus, T.A. (1959). *J. Insect Path.* **1**, 152–170.

Heimpel, A.M. and Angus, T.A. (1960). *J. Insect Path.* **2**, 311–319.

Hubei Institute of Microbiology, Entomogenous Organism Research Group. (1976). *Acta microbiol. Sin.* **16**, 12–16.

Isakova, N.P. (1958). *Dokl. Akad. Sci. Nauk. Selsk.* **23**, 26–27.

Ishiwata, S. (1901). *Dainihon Sanshi Keiho* **9**, 1–5.

Krieg, A. (1969). *J. Invertebr. Path.* **14**, 279-281.

Krieg, A. (1971). *J. Invertebr. Path.* **17**, 134–135.

Krieg, A., Barjac, H. de and Bonnefoi, A. (1968). *J. Invertebr. Path.* **10**, 428–430.

Krywienczyk, J. and Angus, T.A. (1967). *J. Invertebr. Path.* **9**, 126–128.

Krywienczyk, J. and Angus, T.A. (1969). *J. Invertebr. Path.* **14**, 258–261.

Krywienczyk, J., Dulmage, H.T. and Fast, P.G. (1978). *J. Invertebr. Path.* **31**, 372–375.

Krywienczyk, J., Dulmage, H.T., Hall, I.M., Beegle, C.C., Arakawa, K.Y. and Fast, P.G. (1981). *J. Invertebr. Path.* In Press.

Kurstak, E.S. (1964). *Entomophaga Mém. Hors. Sér. No.* **2**, 245–247.

Lecadet, M-M. (1970). *In* "Microbial Toxins" (T.C. Montie, S. Kadis and S.J. Aul, eds.), pp. 437–471. Academic Press, London and New York.

Lecadet, M-M. and Martouret, D. (1964). *Entomophaga Mém. Hors. Sér.* **2**, 205–212.

Mattes, O. (1927). *Gesell. f. Beford. Gesam. Naturw. Sitzber. Marburg* **62**, 381–417.

Norris, J.R. (1964). *J. appl. Bacteriol.* **27**, 439–447.

Norris, J.R. and Burges, H.D. (1963). *J. Insect Path.* **5**, 460–472.

Norris, J.R. and Burges, H.D. (1965). *Entomophaga* **10**, 41–47.

Ohba, M. and Aizawa, K. (1977). *Proc. Sericult. Sci. Kyushu* **8**, 48.

Raun, E.S., Sutter, G.R. and Revelo, M.A. (1966). *J. Invertebr. Path.* 8, 365–375.

Ren, G., Li, K., Yang, M and Yi, X. (1975). *Acta microbiol. Sin.* 15, 292–301.

Somerville, H.J., Tanada, Y. and Omi, E.M. (1970). *J. Invertebr. Path.* 16, 241–248.

Steinhaus, E.A. (1951). *Hilgardia* 20, 359–381.

Sutter, G.R. and Raun, E.S. (1967). *J. Invertebr. Path.* 9, 90–103.

Talalaev, E.V. (1956). *Mikrobiologiya.* 25, 99–102.

Toumanoff, C. (1953). *Ann. Inst. Pasteur* 85, 90–99.

Toumanoff, C. and Vago, C. (1951). *C. r. hebd. Séanc. Acad. Sci. Paris* 233, 1504–1506.

CHAPTER 12

The Crystal Toxin of *Bacillus thuringiensis*

P.G. FAST

Forest Pest Management Institute,
Sault St. Marie, Ontario, Canada

I. Introduction

The principal insecticidal component of present commercial preparations of *Bacillus thuringiensis* (*B.t.*) is the crystalline protein body formed during sporulation; it is variously called the crystal, parasporal body, or delta endotoxin (Heimpel, 1967). Each bacterial cell forms a spore at one end and a crystal (Fig. 1) at the other. When cells lyse after sporulation crystals and spores are set free into the medium and sediment together. Data from electron microscopic and X-ray powder diffraction studies indicate that the readily visible subunit comprising the crystal is rod or dumb-bell shaped with dimensions of the order of 5 x 15 nm (Norris, 1971) to 9 x 13.5 nm (Holmes and Monro, 1965) and has a calculated molecular weight (mol. wt.) of 230 000 daltons.

This review encompasses the information available to the end of July, 1978 on biogenesis, chemistry and mode of action of the parasporal toxin. Unpublished information and theses not generally available are used to provide as comprehensive a discussion as is possible. Each section will be organized to present a coherent picture rather than a chronological or historical perspective. This picture is necessarily a personal one comprising facts available at the time of writing: it will change as new facts become available. Recent studies indicate that some recognized H-serotypes of *B.t.* contain more than one serologically and toxicologically differentiable crystal type (Krywienczyk *et al.*, 1978;

Fig. 1. Electron micrograph of a carbon replica of a crystal (courtesy Dr. C. Hannay).

Chapter 11). However, some literature identifies the isolates used only by H-serotype. To clarify and simplify references to *B.t.* types, I have chosen to use a three-letter code (Table I). Various aspects of the subject discussed here have been reviewed by Heimpel and Angus (1960), Heimpel (1967), Rogoff and Yousten (1969), Lecadet (1970), Cooksey (1971), Somerville (1973), Prasad and Shethna (1976), Somerville (1978).

II. Biogenesis

When *B.t.* is grown in or on artificial media, a period of rapid vegetative growth is followed by formation of an environmentally resistant endospore. It is during spore formation that the crystal is formed. Vegetative cells are atoxic and

TABLE I.

Identity of Bacillus thuringiensis *varieties referred to in this review*

H-antigen serotype	Variety	Code
1	*berliner, thuringiensis*	*thu*
2	*finitimus*	*fin*
3a	*alesti*	*ale*
3?	*anduze*	*and*
3a, 3b	*kurstaki* HD-1	*kur-1*
3a, 3b	*kurstaki* HD-73	*kur-73*
4a, 4b	*sotto*	*sot*
4a, 4c	*kenyae*	*ken*
5a, 5b	*galleriae*	*gal*
6	*entomocidus*	*ent*
7	*aizawai, pacificus*	*aiz*
9	*tolworthi*	*tol*

although a small amount of toxic protein antigenically related to the crystal can be detected in the particulate fraction of a vegetative cell homogenate (Luethy *et al.*, 1970; Luethy, 1975), there is good serological (Monro, 1961a) and biochemical evidence as given below that crystals are not formed by simple crystallization of a protein present in the vegetative cell or by assembly of a portion of the crystal formed in the vegetative phase and another part formed during sporulation.

The sequence of events during spore development (Fig. 2) are not very different from those in related bacilli. Crystal antigens can be detected and the start of crystal formation has been observed microscopically during the latter part of stage II (Somerville, 1971; Lecadet and Dedonder, 1971; Ribier and Lecadet, 1973), although Young and Fitz-James (1959b) and Bechtel and Bulla (1976) could not detect crystal formation until stage III when the forespore is being engulfed by the forespore membrane. In some asporogenous crystal-producing mutants spore formation is blocked at stage II but crystals are formed normally, supporting initiation of crystal formation during stage II (Somerville, 1971; Ribier and Lecadet, 1973). Once crystal formation has been initiated, crystal protein is synthesized until the end of stage IV (Lecadet and Dedonder, 1971) although crystals may continue to enlarge until stage VI (Ribier and Lecadet, 1973).

Somerville (1971; Somerville and James, 1970) found evidence that the crystal is formed in close association with the exosporium membrane, but Ribier and Lecadet (1973) and Bechtel and Bulla (1976) could not confirm this. They found nascent crystals an hour or more before exosporium membrane was detectable and in some cells the crystal was formed in the opposite pole of the cell to the prespore. The former authors show micrographs of nascent crystals surrounded by ribosomes and associated with plasma membrane and perhaps a mesosome but the last component was not confirmed by Bechtel and Bulla (1976). Somerville (1978) postulates that the exosporium membrane may serve as template for crystal assembly, but because

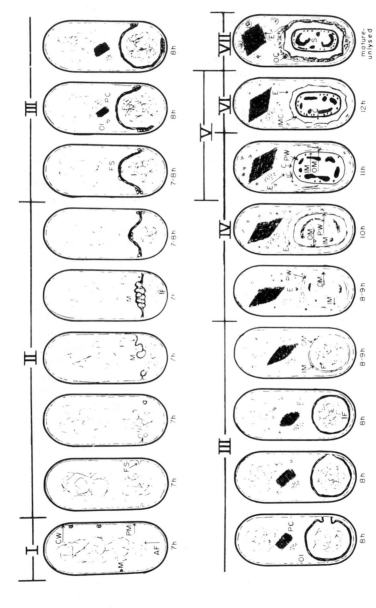

Fig. 2. Diagram of sporulation in *Bacillus thuringiensis*. M, mesosome; CW, cell wall; PM, plasma membrane; AF, axial filament; FS, forespore septum; IF, incipient forespore; OI, ovoid inclusion; PC, parasporal crystal; F, forespore; IM, inner membrane; OM, outer membrane; PW, primordial cell wall; E, exosporium; LC, lamella spore coat; OC, outer spore coat; C, cortex; IMC, incorporated mother cell cytoplasm; S, mature spore in an unlysed sporangium (Bechtel and Bulla, 1976).

crystal protein tends so strongly to aggregate such a template may not be needed.

Biochemical events leading to crystal production start at the end of logarithmic growth with a surge of nucleic acid synthesis (Young and Fitz-James, 1959a), when a stable messenger RNA (mRNA) coding for crystal protein is produced (Glatron and Rapoport, 1972). Stable mRNA is absent in vegetative cells but appears to be synthesized between 2 h and 3/4 h before stage I of Fig. 2, after which time addition of inhibitors of nucleic acid synthesis does not prevent sporulation or crystal production (Young and Fitz-James, 1959c). There is no net synthesis of DNA or RNA during sporulation but sporulation and crystal formation continue even when nucleic acid turnover is prevented. Stable mRNAs of sporulating cultures have been used to direct synthesis of crystal protein in the *E. coli* heterologous system (Gould *et al.*, 1973; Petit-Glatron and Rapoport, 1976). About 40% of the protein encoded by the stable mRNA was crystal protein.

Over 80% of the crystal protein produced during sporulation is synthesized *de novo* from amino acids supplied by breakdown of protein in the sporulation process (Monro, 1961b; Somerville, 1971). No more than 15% could be due to synthesis from free amino acids present in cells before sporulation. Indeed, only glutamine and alanine are present in significant quantities in this free amino acid pool (Young and Fitz-James, 1959a). Incorporation of radioactive amino acids into crystal protein is greatest during stages III and IV (Monro 1961b; Lecadet and Dedoner, 1971; Somerville 1971; Glatron and Rapoport, 1972; Herbert and Gould, 1973), and returns to near base-line levels by the time heat and octanol resistance are acquired (end of stage IV). All subunits of the crystal appear to be synthesized simultaneously (Lecadet and Dedonder, 1971; Herbert and Gould, 1973).

It is well established that a protein with solubility, antigenic and chromatographic characteristics as well as mol. wt. and amino acid composition similar to that of crystal protein is associated with the spore of *B.t.* and of closely related species (Somerville *et al.*, 1968, 1970; Delafield *et al.*, 1968; Lecadet *et al.*, 1972; Herbert and Gould, 1973). This protein has been localized to the inner side of the exosporium and to the spore coat by the labelled antibody technique (Short *et al.*, 1974). It is almost as toxic as crystal protein but whole spores are about 1000x less toxic than are crystals due to the small amounts of the protein extractable from spores (Somerville and Pockett, 1975; Schesser and Bulla, 1978). Since spores of *Bacillus cereus* contain a similar toxin of lower specific activity (Somerville and Pockett, 1975) and a related protein is present on the interior of the cell wall (Luethy, 1975), speculation that the crystal protein is a normal constitutent of cells and *Bacillus* spores that is overproduced to provide the organism with an ecological niche as an insect pathogen seems warranted (Somerville, 1978).

III. Chemistry

Separation of crystals from spores and debris has, until recently, been tedious involving repeated separations in biphasic systems that generally had to be designed for each H-serotype (review; Cooksey, 1971). Isopycnic centrifugation in sucrose (Monro, 1961a) or caesium chloride (Fast, 1972) gradients succeeded with several strains, but was still of limited application. Recently Renografin-76 (66.7% N-methylglucamine salt and 9.4% sodium salt of 3, 5-diacetamido-2,4,6-triiodobenzoic acid, supplied by E.R. Squibb and Sons) in linear (Sharpe et al., 1975; Milne et al., 1977) and discontinuous gradients (Milne et al., 1977) has permitted relatively large scale and rapid purification of crystals from a broad range of B.t. types and culture conditions. The buoyant densities were 1.32 for spores and 1.27 for crystals (Sharpe et al., 1975). The upper capacity limit of the discontinuous gradient has not been explored but exceeds 300 mg/tube. Two or three passes through the discontinuous gradient generally gives preparations pure enough for most purposes (total work time for harvesting, purification and washing − 1 day).

Quantitative analyses (Hannay and Fitz-James, 1955; Angus, 1956) showed the crystal to be composed largely of protein. Absence of nucleic acids was shown by lack of phosphorus or of an absorption peak at 260 nm. No lipids were detected by chloroform−methanol extraction (Bulla et al., 1977; Fast, unpublished observations).

The presence of carbohydrate is more problematic. Lecadet (1965) found none. Holmes and Monro (1965) reported 0.5% carbohydrate but suggested this could have been derived from contaminating spores. Bateson and Stainsby (1970) found 12% carbohydrate, mainly glucose with lesser quantities of mannose, xylose and arabinose, but no amino sugars, in thu crystals. With the phenol−sulphuric acid method Fast (unpublished observations) found that sot crystals contained $3.0 \pm 0.6\%$ and kur-1 crystals $1.2 \pm 0.2\%$ carbohydrate, but no amino sugars or sialic acids. Treatment of dissolved crystals with emulsin (a mixture of carbohydrases) or passage of dissolved toxin down a column of bound concanavilin A, which binds glucose-containing carbohydrates and glycoproteins, failed to significantly reduce toxicity. It was concluded that the carbohydrate content observed by Bateson and Stainsby was probably artifactual and, even if not, that the carbohydrate did not affect toxicity. Mettler (1975) detected no carbohydrate in carefully purified crystals of thu (cited in Luethy, 1975).

Recently Bulla et al. (1977) reported that kur-1 crystals contained 5.6% carbohydrate consisting of glucose and mannose in the ratio 2:1. Again no amino sugars were detected. In SDS gels stained with periodic acid−Schiff reagent, which stains glycoproteins, all major bands reacted. The bacterial isolate was the same as that used by Fast, and was grown in the same medium; only the purification procedures and the analytical method to detect carbohydrate differed. Covalent bonding of the carbohydrate to the protein chain has not been demonstrated, nor has it been shown that carbohydrate is involved in toxicity.

Data indicating the presence of carbohydrate, or of the metals discussed below, must be interpreted with considerable caution since according to protein physical chemists "crystals of globular proteins can be regarded as an ordered and open array of compact molecules that make minimal contact with an interstitial region that is about half of the crystal volume, continuous, filled with solvent that is similar to the bulk liquid, and comprised of channels that are spacious enough to accommodate compounds of molecular weight greater than 4000" (Rupley, 1969). Crystals should be extensively dialysed and, where appropriate, evidence for covalent bonding to the polypeptide chain should be presented before conclusions about the presence of non-protein components are drawn.

Detection of "significant concentrations" of Ca, Fe, Mg and Si by emission spectroscopy (Faust *et al.*, 1973) must be considered in the light of the foregoing. Colorimetric detection of Si had been reported earlier (Faust and Estes, 1966) and formed the basis for the hypothesis that Si is the template on which crystal protein is precipitated during formation and the basis for the crystals' relative insolubility (Faust and Estes, 1966; Heimpel, 1967). Heimpel (1967) states that sporulation of cells and formation of crystals was prevented in the absence of Si. No further evidence has been forthcoming. If the amount of Si reported is referred to the 230 000 dalton crystal subunit established by Holmes and Monro (1965), Si represents 26 atoms/subunit or 26 atoms/9295 carbon atoms.

In all published amino acid analyses (Table II), methionine is the amino acid present in smallest proportion. Assuming a single residue of methionine and proportional amounts, from the amino acid analyses, of the other amino acids a minimal mol. wt. can be calculated. The actual mol. wt. of the crystal subunit must be a multiple of this value. Using an average minimal mol. wt. of 13 000 from Table II the multiple that most closely corresponds to the mol. wt. calculated from the X-ray data is 18 for a mol. wt. of 234 000. Cys represents 1 or 2 residues in the min. mol. wt. and therefore 18 or 36 residues in the crystal subunit. Amide N comprised 13 μgN/mg of *thu* crystal (Glatron *et al.*, 1972), 14.00 ± 0.41 μgN/mg of *sot* crystal and 9.02 ± 0.16 μgN/mg *kur-1* crystal from

TABLE II
Amino acid analyses of crystals in terms of residue/minimal mol. wt.[k]

AA/type	thu (8)[a]	fin (2)[b]	sot (3)[c]	ale (3)[d]	tol (2)[e]	gal (2)[f]	ken (2)[g]	and (1)[h]	kur-1 (1)[j]
Asp	14	14	13	14	15	16	15	12	14
Thr	8	8	7	8	9	9	8	7	7
Ser	8	9	8	9	12	11	9	8	9
Glu	13	12	15	14	12	13	14	15	13
Pro	5	6	8	6	6	7	5	7	4
Gly	10	15	9	12	16	11	9	10	8
Ala	8	13	7	8	11	10	8	8	6
Val	8	9	6	9	8	8	8	7	8
Cys	1	2	1	2	1	2	2	2	2
Met	1	1	1	1	1	1	1	1	1
Ile	7	8	8	8	8	8	8	7	6
Leu	11	11	13	11	10	13	13	11	9
Tyr	5	4	5	5	3	5	5	4	5
Phe	5	4	6	5	4	5	6	5	4
Lys	4	6	4	4	3	4	5	4	3
His	2	3	2	2	2	3	2	3	2
Arg	6	3	8	7	5	7	7	6	9
Try	1	1	1	1	ND	1	ND	1	1
min. mol. wt.	12820	13231	13231	11895	12796	14291	13042	13282	12400

[a]Spencer, 1968; Glatron *et al.*, 1972; Lecadet, 1965; Lecadet *et al.*, 1972; Bateson and Stainsby, 1970; Holmes and Monro, 1965. Cooksey, 1968. Lilley *et al.*, 1980. [b]Lecadet *et al.*, 1972; Lilley *et al.*, 1978. [c]Angus, 1956; Lecadet, 1965; Spencer, 1968. [d]Somerville *et al.*, 1968; Spencer, 1968; Lilley *et al.*, 1980. [e]Herbert *et al.*, 1971; Lilley *et al.*, 1980. [f]Spencer, 1968; Chestukhina *et al.*, 1977. [g]Cooksey, 1971. [h]Lecadet, 1965. [j]Bulla *et al.*, 1977. [k]Analyses for each serotype were averaged and the averages compared to methionine calculated to the nearest integer. Minimal mol. wt. was calculated by summing integer × mol. wt. of the respective amino acid residues. Numbers in brackets refer to the number of analyses averaged.

Abbott Laboratory's Dipel (Fast, unpublished observations). Thus almost half the dicarboxylic amino acids are likely to be present as amides.

Bateson and Stainsby (1970) claim that proline is absent from their crystals and that the peak observed at 440 nm in their analyses was an artifact. If their claim is sound then perhaps the proline content reported by other workers is also invalid.

Although the amino acid analyses reported in Table II are quite similar, varietal differences can be discerned. Whether these apparent differences are real cannot be determined from the published data and must await a careful comparative study.

The relative insolubility of the crystal is surprising and is a major stumbling block to the study of crystal protein, which requires that the crystal must be dissolved without breaking covalent bonds, or at least only specified covalent bonds. Before summarizing work on crystals, therefore, it is worthwhile reviewing briefly how native conformation and solubility are determined.

Proteins are composed of one or more linear polymer chains each containing a number of chemically different amino acids covalently linked in a particular genetically determined sequence. In the native or functional state these chains are folded into a precise, compact but irregular structure governed by a combination of two types of bonds, (1) noncovalent interactions that are a sensitive function of residue composition, sequence and the solvent environment, and (2) covalent, principally disulphide, bonds. Since the force stabilizing the folded state is due to the net removal of non-polar atoms from unfavourable contact with the solvent (the so-called hydrophobic bond), the non-polar amino acids tend to be on the inside of the molecule, shielded from the solvent, while the polar and/or ionized side chains are nearly always on the outside. Polar groups in the interior of the molecule are paired in hydrogen bonds making them essentially non-polar in nature. However, many of the polar side chains, e.g. that of lysine, consist of more non-polar atoms than of polar atoms so it is inevitable that significant portions of the surface be non-polar. These non-polar domains on the surface of the protein molecule govern such phenomena as solubility, aggregation, ligand binding and conformation of active sites.

A protein is insoluble only in an environment where the non-polar domains on the surface of the protein cannot be accommodated in the structure of the solvent, i.e. where chain—chain interactions are favoured over chain—solvent interactions. The protein is solubilized when solvent structure is altered by addition of ions, or raising the temperature so that chain—solvent interactions predominate, or when the non-polar domains are made polar by addition of detergents. Further addition of ions or increases in temperature greater than those required to affect solubilization result in increasing unfolding of the surface chains until, in a rapid cooperative transition, the entire molecule unfolds and the chains approximate random coils — the denatured state.

The properties of water, more specifically the hydrogen bonded structure, are altered not only by the concentration of ions but also by the particular ion concerned. In the lyotropic or Hofmeister series (Fig. 3), ions to the right of acetate or tetramethylammonium ion are increasingly structure-breaking and thus increasingly denaturing and those to the left are increasingly structure-making or stabilizing. Their effects are linearly dependent on concentration and temperature and thus through judicious choice of ion, concentration and temperature it is possible to alter surface conformation in a subtle fashion.

The interested reader is referred to reviews by Anfinsen and Scheraga (1975) and to the excellent recent discussion of protein architecture by Creighton (1978) for more detailed discussion of protein folding and unfolding, and to von Hippel and Schleich (1969) for the effects of neutral salts on protein conformation.

The aggregation of dissolved protein molecules which has been such a problem in studies on crystal protein is due to the same forces and is remedied by similar alterations in the solvent environment as are required for solubilization. It must be emphasized that aggregation can destroy the validity of almost every analytical or preparative system unless it is prevented. I have observed (Fast, unpublished observations) that 30% or more of the dissolved crystal protein in enzymic dissolutions can precipitate in even short incubation times, that aggregation can occur during gel filtration with the aggregating protein remaining on the column, during storage of digests or purified fractions at room, refrigerator or freezer temperature, and that aggregated protein is no longer active in a cell assay although it may still be active in an insect assay due to redisaggregation in the gut of the insect. I have found that neutral salts disaggregate digested protein in accord with the lyotropic series shown in Fig. 3 and have somewhat empirically standardized on $1 M$ KSCN to prevent aggregation without loss of toxicity. It is routinely added to all solvents and buffers.

Cooksey (1971) and Heimpel (1967) argued that insolubility of the crystal cannot be due to the presence of disulphide bonds since the amount of cystine is considerably less than in wool. Such bonds do, however, play a significant role in stabilization of the crystal since the addition of disulphide reducing agents lowers the pH required for dissolution by some 2 pH units (Young and Fitz-James, 1959b; Lecadet, 1966, 1967, 1970). Since there are at least 18 half-cystines in each 230 000 dalton crystal subunit, or according to Bulla *et al.* (1977) 18/120 000 dalton principal peptide, there are 3.4×10^7 possible ways these can combine to form disulphide bonds (Anfinsen and Scheraga, 1975). Obviously both inter- and intra-chain and intersubunit disulphide bonds are possible. No studies on the role, number, or location of disulphide bonds in the

stabilizing ◄───► destabilizing
denaturing

$$SO_4^{2-} < CH_3COO^- < Cl^- < Br^- < No_3^- < ClO_4^- < I^- < CNS^-$$
$$(CH_3)^4N^+ < NH_4^+ < Rb^+, K^+, Na^+, Cs^+, < Li^+ < Mg^{2+} < Ca^{2+} < Ba^{2+}$$
$$(CH_3)_4N^+ < (C_2H_5)_4N^+ < (C_3H_7)_4N^+ < (C_4H_9)_4N^+$$

Fig. 3. Relative effectiveness of various ions in stabilizing or destabilizing the native form of collagen (adapted from von Hippel and Schleich, 1969).

crystal beyond demonstrating the need to reduce them for dissolution have been published. They are not required for toxicity since reduction sufficient for dissolution causes no loss in toxicity (Nishiitsutsuji-Uwo et al., 1977). However, carboxymethylation of reduced disulphides destroys toxicity (Bulla et al., 1977).

Complete denaturation of a protein using such agents as sodium dodecyl sulphate (SDS) or high concentrations of guanidine hydrochloride or urea can yield information on the mol. wt., peptide chain composition and interchain and intersubunit interactions within the protein. Crystals can be dissolved in $6\,M$ guanidine hydrochloride at pH 7.5 with disulphide reducing agents. Glatron et al. (1969) showed that thu crystals thus dissolved had single N-terminal and C-terminal amino acids, indicating a single peptide chain and determined the mol. wt. to be $80\,000 \pm 9000$ daltons. When the same protein was dissolved in pH 9.5 buffer with reducing agents, four N- and C-terminal amino acids were detected indicating breakage of peptide bonds at the higher pH. Fast and Martin (1970, an unpublished observation) followed this lead by dissolving thu and sot crystals in $6\,M$ guanidine hydrochloride at pH 7.5. Equilibrium centrifugation yielded a mol. wt. of $69\,000$ daltons for thu crystals, not significantly different from Glatron's results, and $77\,000$ daltons for sot crystals. The schlieren patterns did not reveal any gross inhomogeneity in either crystal solution but the resolving power in this solvent is not great. Crystals dissolved in guanidine hydrochloride solutions are sufficiently denatured so that little or no toxicity is retained (Truempi, 1976; Nishiitsutsuji-Uwo et al., 1977).

Crystals from the same batches of thu and sot as well as of kur-1 were also subjected to electrophoresis in SDS gels after dissolution by boiling in phosphate buffer pH 7.0 with 1% SDS and 1% mercaptoethanol. This separates the constituent peptide chains on the basis of their mol. wt. Densitometric scans of gels run under identical conditions can be compared (Fig. 4). This confirms that thu crystals are composed of a single peptide chain as found by Glatron et al. (1969, 1972) but that sot and kur-1 contain more than one peptide chain. Multiple bands in SDS gels have also been reported in crystals of tol (Herbert et al., 1971; Lilley et al., 1980), gal (Chestukhina et al., 1977), kur-1 (Bulla et al., 1977) and thu (Lilley et al., 1980; Truempi, 1976). These latter two references to thu crystals differ from the observations of Glatron et al. (1972), and of Fig. 4, indicating that there may be more than one crystal type within the thu variety. Serological and toxicity data (Krywienczyk et al., 1979; Chapter 11) strongly indicate at least two crystal types within thu but direct comparisons in SDS gels have not yet been made.

Common to all three crystals in Fig. 4 is a major band of mol. wt. $120\,000$ daltons in this system. In thu this is the only band and must be the component that had a mol. wt. of $80\,000$ daltons by sedimentation equilibrium centrifugation. Such a discrepancy in mol. wt. suggests that the crystal protein may not be completely denatured by the guanidine hydrochloride. A similar band is the principal component in kur-1 crystals (Bulla et al., 1977) and gal crystals (Chestukhina et al., 1977), and is a major band in tol crystals (Herbert et al., 1971; Lilley et al., 1980). This protein must contain the toxin in thu crystals

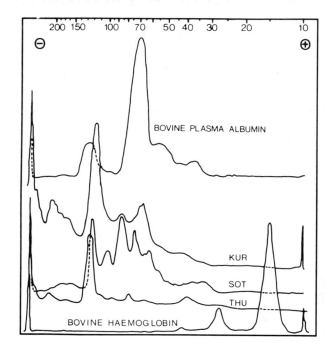

Fig. 4. Densitometric curves of SDS gels of *sot*, *kur-1* and *thu* crystals (Fast and W.G. Martin, unpublished observations).

and is toxic in *gal* crystals where it constitutes 65% of the protein. However, in *tol* crystals it is not toxic; all toxicity in these is in a peptide with a mol. wt. in the same SDS system of 55 000 daltons (Herbert *et al.*, 1971). The larger (120 000 dalton) protein may well be the antigen Krywienczyk and Angus (1967) found to be common to crystals of all types they tested.

Urea (8 M) can completely disperse *thu* crystals at pH 8.0 without, however, releasing any protein into the supernatant after centrifugation (Lecadet, 1967). If the urea is removed by dialysis or dilution the crystals are completely reformed. If disulphide reducing agents are added to the urea solution, irreversible dissolution occurs. Thus it appears that there are both interchain and intersubunit disulphide bonds that are stable at pH 8.0 and permit complete refolding of the crystal structure after dispersal by relaxing the forces holding the chains together. Chestukhina *et al.* (1977) found that urea extracted a 65 000 dalton fragment comprising 8–10% of the total protein from *gal* crystals without dissolving the rest of the crystal. Disulphide reducing agents at pH 9.0 dissolved the remainder of the crystal constituting the bulk of the protein. They did not, however, monitor whether this latter fraction was reversibly dispersed by the urea. The *gal* crystals were composed of at least four peptide chains whereas the *thu* crystals used by Lecadet contained only one chain. These two studies are not in conflict but emphasize the importance of the combination of disulphide reducing agents with urea for complete dissolution of crystals. Unfortunately much toxicity is lost during dissolution in urea-disulphide reducing solutions (Nishiitsutsuji-Uwo *et al.*, 1977; Truempi, 1976; Cooksey, 1971). Sayles *et al.* (1970) reported that *thu* crystals yielded only peptides of mol. wt. 1400 daltons when completely dissolved in $8\,M$ urea-0.5% dithiothreitol. Truempi (1976) found only high mol. wt. peptides in *thu* crystals dissolved under the same conditions. Perhaps this too is due to differing crystal types within *thu*.

The loss of toxicity caused by denaturing agents or high temperatures (Angus, 1956; Truempi, 1976) clearly indicates a requirement for non-covalent tertiary structure in the toxic moiety.

Dissolution of crystals with retention of toxicity has relied on highly alkaline solutions with or without reducing agents. Cooksey (1968), Faust *et al.* (1974b) and Truempi (1976), among others, used $0.01–0.1\,N$ NaOH, but the dissolved crystals lost half or more of the original toxicity (Angus, 1956; Cooksey, 1971; Nishiitsutsuji-Uwo *et al.*, 1977). Toxicity decreased with time at high pH. Such alkali-dissolved crystals are composed of highly aggregating proteins with ill-defined banding patterns in electrophoresis. Those from *thu* and *kur-1* are not toxic by injection (Angus, 1956; Lecadet and Martouret, 1967a; Truempi, 1976) or to tissue culture cells in an *in vitro* assay (Murphy, Sohi and Fast, unpubl.) and so are still protoxins. Furthermore, peptide bonds in *thu* crystals (Glatron *et al.*, 1972) and at least one of the disulphide bonds in *kur-1* (Bulla *et al.*, 1977) are broken even at pH values as low as 9.5.

Crystals of *thu* fragmented in more concentrated NaOH ($1\,M$), yielding peptides of mol. wt. around 10 000 daltons, close to that calculated as the minimal mol. wt. from amino acid analyses (Glatron *et al.*, 1972; Prasad and Shethna, 1974), but when *kur-1* crystals were used and the solubilized protein passed down a Sephacryl S-200 gel filtration column in media that prevent aggregation, only high mol. wt. material was found (Fast, unpublished observation).

Dissolution of crystal at pH 9.5–10 with reducing agents has been widely used but toxicity has been reported only qualitatively. Herbert *et al.* (1971) isolated a toxic fragment from *tol* crystals dissolved in "Ellis" universal buffer at pH 10 containing $0.1\,M$ mercaptoethanol. The fragment had a mol. wt. of 65 000 daltons. A 120 000 dalton fragment was atoxic. Nishiitsutsuji-Uwo *et al.* (1977) found this system best among those tried for dissolving *aiz* crystals, permitting complete recovery of toxicity. Indeed, the dissolved crystals were about 50% more toxic than the intact crystals.

Recently completed studies on the effect of various salts on crystal solubility (Fast, unpublished observations) permitted development of a system for dissolving crystals at near neutral pH with retention of toxicity. *Kur-1* crystals were completely dissolved in 1 h at $37°$ in $0.1\,M$ 3-(N-morpholino) propane-sulfonic acid (MOPS) buffer pH 7.8 that was $1\,M$ in KSCN and $0.5\,M$ dithiothreitol. The crystal solution was about 50% more toxic than intact crystals and toxicity was stable over a week and only mildly degraded over a month at room temperature (Fast and Milne, 1979).

IV. Proteolysis

Since the crystal itself is a protoxin, it has no effect if injected into the haemocoel of susceptible insects (Angus, 1956). Before discussing and attempting to evaluate efforts to isolate and characterize a toxic active fragment from the crystal the assay systems used to determine toxicity must be evaluated and their characteristics clearly kept in mind.

Because the crystal is "only" a protein with no distinctive non-protein characteristics on which to hang a chemical or physical assay system, reliance must be placed on bioassay or on techniques that distinguish one protein from another. The serological approach deserves consideration because of its great sensitivity and ease of application. Such an assay was used by Cooksey (1971) to follow digestion of the crystal and by Winkler *et al.* (1971) to determine toxicity of various preparations. Both used a precipitin reaction. For precipitation to occur, however, the antigen must be polyvalent and on a large protein such as the dissolved crystals used to elicit antibodies there may be many combining sites. If the toxicity determinant is only a short segment of a larger protein then precipitation could occur without blocking the toxic segment, or, alternatively, the toxic segment will be only one of several determinants participating in precipitation. In neither case is a precipitation reaction *necessarily* indicative of the presence of the toxic segment. While precipitation reactions may be crystal-protein-specific they cannot be relied upon to recognize the toxic segment in fragmentation studies, or peptide mapping, unless specificity for the toxic segment is demonstrated. Such an antiserum would be a great boon.

The feeding assay, assay by injection into the haemocoel of susceptible larvae, and assay on cultured cells are direct means of demonstrating the toxic segment. The feeding assay is the original and most commonly used assay. The test material may be applied to a leaf, incorporated into a diet, or force fed in a small aliquot of the test solution. Either mortality or feeding inhibition may be measured after an appropriate time. This assay has the advantage of generality, in that it detects the presence of the undenatured toxic segment, whether in protoxin or activated form. It cannot, however, establish whether the test material is active without further alteration because the test material is subjected to the enzymic and ionic environment of the gut which may hydrolyse, disaggregate, unfold and/or reduce the test material.

Several workers have circumvented the problems with the feeding assay by injecting toxin or fragments thereof. This system probably detects only active toxin, particularly since Lhoste and Martouret (1968) showed that injected toxin causes the same histological symptoms as does fed toxin. The implications of this will be discussed in the section on mode of action. The haemocoel is, however, still an unknown environment containing enzymes and other proteins as well as cells which may alter or inactivate the injected toxic fragments.

The ideal system is one in which the medium surrounding the target can be completely controlled and the target itself responds in the same manner as the insect gut epithelium of susceptible larvae. We (Murphy *et al.*, 1976) found certain cell cultures were susceptible to enzyme-digested crystal toxin showing symptoms resembling those of gut epithelium (Angus and Cooksey, 1971, p. 269), so a cell assay system for activated toxin that meets most of the criteria required was developed. Washed cells are suspended in buffered saline with the test preparation. After 30 min, boiling Tris buffer pH 7.5 is added and boiled for an additional 10 min to stop the reaction and inactivate any ATPases. Residual ATP from cells alive at the time of addition of boiling buffer is then measured by the luciferin—luciferase reaction (Strehler, 1968). Buffer controls and maximal response assays using a standardized dosage in excess of that required to give a maximal response are included in each assay. Loss of cellular ATP is dose-dependent and gives a straight line when probit percent loss is plotted against log dosage, permitting the calculation of an EC_{50} (concentration required to cause a 50% reduction in ATP). The assay is simple, sensitive, reasonably fast and has good accuracy. It was originally developed on the Cf-124 cell line with a maximal response around 45%, which was found to decrease with time in culture. In a survey of cell lines (Table III), Cf-230-1 (now designated Cf-1) gave a maximal response of 78% and is now used in all our assays. Whether the effect of toxin in these cells and in gut epithelium is rigidly parallel has not been established but crystal fragments that exhibit this cell effect have so far invariably also poisoned larvae in feeding assays.

Lecadet and Martouret (1962) showed that partially purified proteases from gut juice of *Pieris brassicae* solubilized crystals without reducing agents at 2 pH units lower than the pH at which such crystals dissolved non-enzymically in alkali. Such enzymically-hydrolysed toxin was lethal by injection into the haemocoel and so had been activated, whereas the alkali-dissolved toxin was not lethal. These proteases closely resembled trypsin and chymotrypsin except that their pH optimum was around 10 and they could dissolve intact crystals whereas the vertebrate enzymes could not (Lecadet and Dedonder, 1966a, b). *Bombyx mori* (Lecadet and Martouret, 1967b), *Trichoplusia ni, Heliothis zea* and *Spodoptera exigua* (Murphy, 1973) all have enzymes similar to those of *P. brassicae* with pH optima between 10 and 10.5 and the ability to dissolve and activate *thu* crystals. Murphy, Fast and Sohi (unpublished observations) showed that while alkali-dissolved crystals were atoxic to cultured cells, further treatment of these dissolved crystals with gut enzymes from *T. ni* resulted in toxicity to cells. Such preparations were considerably less toxic than crystals treated directly with gut enzymes, presumably due to the already-mentioned loss of toxicity in alkali. Enzymically dissolved crystal is toxic by injection or in cell

TABLE III

Assay of various cell cultures derived from Choristoneura fumiferana *against protease resistant protein obtained from kur-1 crystals by* Trichoplusia ni *gut proteases*

Cf cell line	230−5	230−3	230−1	124−0	230−8	230−10	230−11
Reduction in ATP (%)	39	0	78	15	7	54	16

culture whenever it is toxic by ingestion (Lecadet and Martouret, 1967a; Pendleton, 1968, 1973; Fast, unpublished observations), leading to the conclusion that enzymic hydrolysis is essential to the activation process at least in *thu* and *kur-1* crystals.

However, Lilley *et al.* (1980) reported that *tol* crystals dissolved in "Ellis" buffer, pH 10.5 with mercaptoethanol, were as toxic by injection as purified toxic fraction from an enzymic digest. Furthermore Herbert *et al.* (1971) reported that all toxicity in *tol* crystals resided in a fraction with a 55 000 dalton mol. wt., even though there is good evidence (Section III) that in *thu* and *kur-1* crystals the toxicity is in a larger fragment (120 000 daltons by SDS electrophoresis or 80 000 daltons by ultracentrifugation). Perhaps the broad host spectrum and enhanced toxicity reported for *tol* (Norris, personal communication) is due to a "naked" active toxin in the crystal.

Faust *et al.* (1967) and Heimpel (1967) have argued that with *aiz* (*pacificus*) the crystal must first dissolve in the alkaline and reducing gut juice before enzymic hydrolysis can take place. Lecadet (1970) has pointed out that many things could be happening in the gut juice but *in vitro* the dissolution is undoubtedly enzymic. Furthermore, Murphy (1973) noted that *thu* crystals were not dissolved in gut juice that had been heated at 85°C for 5 min to inactivate the enzymes.

Gut juice proteases may also contribute to the host toxicity spectrum exhibited by each taxonomic *B.t.* group. *And* crystals are much less toxic to *P. brassicae* than *thu* crystals and are dissolved much more slowly. When equal amounts of dissolved toxin are assayed, toxicity is the same indicating that the differences between the crystals are due to relative susceptibility to hydrolysis by the gut enzymes (Lecadet and Martouret, 1964). The amount of crystal dissolved by *T. ni* partially purified proteases at pH 10.5 differed for crystals of different types: *kur-1* > *thu* > *sot* > *gal* (Geiser, Fast and Sohi, unpublished observations), and *thu* crystal was much less susceptible to *P. brassicae* proteases than was *kur-1*. Further work is required to substantiate these observations.

The progress of proteolysis can be followed by SDS gel electrophoresis of samples removed from the digest at various times during the digestion. Figure 5 in which digestion was slowed by buffering to pH 9.5 instead of the enzyme optimum pH 10.5, is qualitatively similar to observations on *tol* crystals dissolved in "Ellis" buffer pH 10.5 and then treated with trypsin, chymotrypsin or *P. brassicae* gut juice (Lilley *et al.*, 1980). Always higher mol. wt. polypeptides rapidly disappeared while new bands in the range 70 000–200 000 daltons appeared. As digestion progressed, these new bands also disappeared except for a single band corresponding to the original band of lowest mol. wt. and some lightly staining material of still lower mol. wt., which presumably consisted of fragments formed from the larger peptide chains. At pH 10.5 dissolution was essentially complete within 6 h but hydrolysis of larger to smaller fragments continued between 6 and 20 h. Toxicity to cells increased from the 2 to the 20 h sample indicating that protoxin was being hydrolysed to active toxin during this period. Lecadet and Dedonder (1967) also reported continued hydrolysis of protein after dissolution of the crystals was complete.

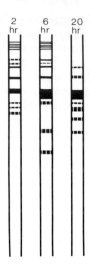

Fig. 5. Diagram of staining patterns of SDS gels of *kur-1* crystals exposed to 1 μg/mg *Trichoplusia ni* partially purified gut proteases at pH 9.5 for designated times (Geiser, Fast and Sohi, unpublished observations). Broken bands represent weakly staining bands.

With one exception, all studies in which enzymic digests have been fractionated by mol. wt. report that the bulk of the digested protein has a mol. wt. between 40 000 and 100 000 daltons (Lecadet and Dedonder, 1967; Cooksey, 1968; Pendleton, 1973; Faust *et al.*, 1973; Truempi, 1976; Huber, 1977; Fast, unpublished observations; Geiser, Fast and Sohi, unpublished observations; Lilley *et al.* 1980). Comparison of 20 h digests of *thu, gal, sot, kur-73, kur-1* in SDS gels run simultaneously and with marker proteins (Fig. 6) shows that the principal component in each case had a mol. wt. in the vicinity of 60 000 daltons

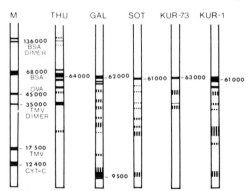

Fig. 6. Diagram of staining patterns of SDS gels of digests of crystals exposed to 1 μg/mg of *Trichoplusia ni* partially purified gut proteases pH 10.5 for 20 h. Broken bands as in Fig. 5.

when the electrophoresis buffer had a pH of 6.5. Only the 9500 dalton band in *gal* contained significant protein in addition to the 60 000 dalton protease resistant protein (PRP). Lilley *et al.* (1980) also found only a single PRP with a mol. wt. in the range 60 000–70 000 daltons in SDS gels of enzymic digests of *ale* and *tol* crystals but found two bands in this range in *thu* crystal digests. In *ale* and *tol* digests only one N-terminal amino acid, serine, was detected whereas in *thu* digests threonine was detected along with serine. Pendleton (1973) found a single PRP with a mol. wt. of 50 000 daltons by gel filtration and an isoelectric point of 6.4 by isoelectric focusing in digests of *ent* crystals treated with *Samia* (*Philosamia*) *ricini* gut juice. Huber (1977) found only a single PRP with a mol. wt. of 75 000 daltons by ultracentrifugation when *thu* crystals were treated with *P. brassicae* gut juice. *Aiz* crystals yielded a similar fraction when treated either with *P. brassicae* or *Spodoptera littoralis* gut juice. In Huber's hands no such PRP could be found in *kur-1* digests, presumably because it aggregated before elution from the column. We regularly produce such a PRP with a mol. wt. of 60 000–70 000 daltons from *kur-1* crystals by digestion with *T. ni* gut juice proteases, but aggregation must be prevented.

Reported mol. wts of these proteins must be accepted cautiously. We have observed that the relative position of this band in SDS gels is influenced by the pH of the gel system, and gel filtration is also subject to considerable uncertainty. From Fig. 6 it is reasonable to assume that every study was dealing with essentially the same PRP with possible minor differences in mol. wt. depending on source of crystal and enzyme. Some careful comparative studies using ultracentrifugation would be very helpful.

The one exception to a PRP in the range of 40 000–100 000 daltons is the report by Pendleton (1968) that *thu* crystals digested in *P. brassicae* gut juice yielded only 5000–10 000 dalton PRP and a very high mol. wt. protein excluded from Sephadex G-200. The excluded material could be either undigested protoxin or aggregated PRP since it was atoxic by injection. Resuspension of this fraction in fresh gut juice changed it to the 5000–10 000 dalton fraction.

The PRP contains the bulk of the toxicity in enzymic digests. In addition it has invariably been found to be toxic by injection as well as ingestion (Lecadet and Martouret, 1967a; Pendleton, 1973; Lilley *et al.*, 1980). All the digests in Fig. 6 were toxic to cells and so contained active toxin. Also PRP produced from *kur-1* crystals by *T. ni* proteases was used to develop the cell assay (Murphy *et al.*, 1976) and to demonstrate that activated toxin can act at the surface of susceptible cells (Fast *et al.*, 1978). In *tol* (Lilley *et al.*, 1980) and in *kur-1* digests (Fast and Sohi, unpublished observations) PRP accounts for all toxicity in the digests, and one can conclude that PRP is "an" and probably "the" active toxin both *in vitro* and *in vivo*.

That is not to say, however, that smaller toxin fragments cannot exist. The best evidence for such low mol. wt. toxic fragments was presented by Lecadet and Dedonder (1967) and Lecadet and Martouret (1967a). They found four peptides, representing about 10% of the digest, included in Sephadex G-25 and separable by ion exchange chromatography and paper electrophoresis, to be toxic by injection. The two principal peptides had aspartic acid as N-terminal amino acid and leucine as C-terminal acid. The mol. wt. by ultracentrifugation was 5000 daltons.

Much of the remaining evidence for low mol. wt. toxic fragments stems from gel filtration where toxicity eluting in low mol. wt. fractions could be due to precipitation of high mol. wt. peptides with resultant retardation (Pendleton, 1968; Cooksey, 1968; Faust et al., 1973; Rogoff and Yousten, 1969) or from membrane filtration of digests (Fast and Angus, 1970). When the particular batch of crystals used for the latter study was exhausted it became impossible to isolate any more low mol. wt. toxin and the results must be ascribed to a unique and unreproducible batch of crystals. We have never, despite considerable effort, detected any low mol. wt. toxin in digests of kur-1 treated either with T. ni proteases or silkworm gut juice. Several attempts to detect low mol. wt. peptides in digests of thu with P. brassicae gut juice or enzymes (Mettler, 1974; Truempi, 1976; Huber, 1977) have failed. Whether this is a question of scale (Lecadet and Dedonder treated as much as 350 mg of crystals to isolate their peptide fraction) or for other reasons, there is as yet no justification for discounting Lecadet and Dedonder's results.

The acid protease pepsin also attacked and dissolved crystals (Rogoff and Yousten, 1969) but the dissolved material was atoxic by ingestion to test larvae. Huber (1977) treated alkali-dissolved crystal protein with pepsin and subjected the digest to G-200 gel filtration. No PRP similar to that produced by trypsin or gut juice was obtained. However, toxicity of the digests was not reported.

V. Mode of Action

To present a coherent picture of the mode of action of crystal toxin as I at present visualize it, this discussion will more or less follow the chronological course of events within a treated larva as opposed to a historical perspective. In this way each major effect is seen to be a direct outcome of antecedent events.

Crystals ingested by a susceptible larva are rapidly hydrolysed, yielding (Section IV) an active toxin which must either act on the gut or pass through the gut into the haemocoel and there exert its effect. When radioactive crystals were fed to susceptible larvae of three different species no fragments large enough to contain the toxic moiety were detected in haemolymph only non-toxic dipeptides were radioactive (Fast and Videnova, 1974; Fast, 1975). The toxin therefore appears to be confined to the gut itself. Histological lesions are also confined to the gut. Swelling of microvilli of gut epithelium is detectable within 5 min and a general swelling of membranous organelles is apparent within 12 min after treatment in P. brassicae larvae (Ebersold et al., 1977). Apices of columnar cells of silkworm guts were swollen between 10 and 20 min after treatment (Angus, 1970). Eventually the cells burst spilling their cytoplasmic contents into the lumen and by 45 min after treatment there is extensive separation of cells from the basement membrane (Heimpel and Angus, 1959). Similar effects were observed in other susceptible insects (Hoopingarner and Materu, 1964; Martouret et al., 1965; Sutter and Raun, 1967). Iizuka (1974) suggests that goblet cells may be the site of the original lesions. Injection of active toxin into the haemocoel causes similar lesions in the gut epithelium (Lhoste and Martouret, 1968).

The earliest toxic symptom is a stimulation of glucose uptake by gut epithelial cells, detectable within one minute of dosing with crystal (Fast and Donaghue, 1971). This has been interpreted as due to a stimulation of respiration (Faust *et al.*, 1974b), observed in midgut homogenates as a result of the uncoupling of oxidative phosphorylation of midgut mitochondria by active toxin (Travers *et al.*, 1976). Such an interpretation is supported by a rapid (half-time 2 min) depletion of ATP caused by active toxin in cultured cells susceptible to the toxin (Fast and Geiser, unpublished observations). Such a depletion of ATP would stimulate respiration and subsequent uptake of glucose from the medium. The ATP loss is not due simply to a shortage of glucose in the medium since tripling the glucose concentration did not alter the rate of ATP loss in cultured cells (Fast, unpublished observations).

However, the obvious hypothesis that the toxin acts directly on mitochondria of gut epithelial cells is not supported by studies designed to determine whether toxin has to penetrate the cell before exerting its effect and by the relatively narrow specificity of active toxin. Fast *et al.* (1978) bound PRP to Sephadex beads too large to be ingested by target cells: the toxin was still active, indicating that toxin probably acts at the external surface of target cells. Furthermore, if toxin penetrated freely into cells and acted directly on mitochondria one would expect all cells, or at least all insect cells, to be susceptible. In fact, in our experience, susceptible cells are in a minority in most cultures (Table III; Murphy *et al.*, 1976; Fast and Sohi, unpublished observations) and Murphy (unpublished observations) has selected a totally unsusceptible cell line by treating three successive passages of a susceptible cell line with toxin. These data suggest that specific toxin receptors might be present, a suggestion strengthened by Huber's (1977) observation that *Spodoptera littoralis* larvae are not susceptible to whole *thu* crystals or to active PRP produced by *P. brassicae* gut juice even though *S. littoralis* gut juice dissolves crystals and produces a PRP of appropriate mol. wt. toxic to *P. brassicae* larvae. The observation that toxin injected into the haemocoel causes histological lesions equivalent to those produced by ingested toxin (Lhoste and Martouret, 1968) indicates that attack by toxin is not restricted to the lumenal surface of susceptible gut cells.

Whether we accept that activated toxin causes a reduction of ATP leading to cell destruction or not, and such a hypothesis is at present tentative at best, the mechanism by which this occurs is still open to question. Angus (1968) suggested, on the basis of the similarities between the gross symptoms produced by toxin and valinomycin, that the toxin may act as an ionophore — a substance that facilitates or causes transport of ions across cell membranes. Such activity would explain the observations discussed above since coupling of oxidation to respiration is critically dependent on ion concentration and many ionophores have uncoupling activity (Pressman, 1976). Such an effect could also be produced at the external surface of the cell, destroying the intracellular and subsequently the intramitochondrial ionic balance and leading to rapid depletion of ATP as the cell attempts to maintain homeostasis. Angus's hypothesis was supported by observations that within 10 min after treatment potassium ion levels are higher in haemolymph of treated insects than in the control (Ramakrishnan, 1968; Pendleton, 1970; Narayanan and Jayaraj, 1974). Levels of other ions increase more slowly and that of calcium ions decreases. Further support was provided by a preliminary report, never confirmed or published in detail, that enzymically hydrolysed toxin blocked nerve conduction in cockroach cercal nerve preparations (Cooksey et al., 1969). However, the increase in haemolymph K concentration begins after metabolism of gut epithelial cells has been effectively broken down and general leakage of ions from gut to haemolymph has begun (Louloudes and Heimpel, 1969 Fast and Donaghue, 1971). No accumulation or increased turnover of K ions, nor indeed, in the level of Ca, Mg, or Na ions, was detected in midgut epithelium within the first 10 min after treatment (Fast and Morrison, 1972). Ionophoric activity should be detectable within at most a minute or two of toxin administration. It seems that both toxin and valinomycin deplete cellular ATP but the mechanism by which this is achieved is different, and remains unknown for the toxin.

Luethy (1973, 1975) suggested that crystal toxin nullifies mechanisms by which gut cells are protected against autodigestion by contents of the gut lumen. His experiments do not differentiate between the effects of gut juice alone and gut juice containing toxin and so do not bear directly on the role of toxin. It is quite conceivable, however, that depletion of ATP might nullify such protective mechanisms.

The events within the gut cause larvae to stop feeding due to " extreme digestive discomfort" within as little as 2 min after ingestion of a droplet of toxin (Heimpel and Angus, 1959 and references therein; Luethy, 1975). This feeding inhibition is a universal response to the toxin among susceptible larvae and is a major factor in the successful application of B.t. for crop protection. Whether feeding is inhibited by paralysis of gut muscles observed by Heimpel and Angus (1959) or by discomfort engendered by early responses of epithelial cells to toxin has not been established.

Stimulation of glucose uptake seen in vivo reaches a maximum about 5 min after dosing, about the time that cellular ATP has been depleted in cultured cells. This stimulation ceases by 11 min after treatment and an inhibition of glucose uptake sets in. Leucine uptake into gut epithelial cells is also inhibited between 10 and 20 min after ingestion of toxin, which is also the period in which leucine transport into haemolymph is inhibited (Louloudes and Heimpel, 1969), and K ions and CO_3 ions begin to surpass normal levels in haemolymph (Fast and Angus, 1966; Louloudes and Heimpel, 1969; Pendleton, 1970; Fast and Morrison, 1972). These effects coincide with the first appearance of histological damage. Taken together these data indicate that the homeostatic mechanisms of gut epithelial cells cease functioning between 5–10 min after treatment with toxin.

When this occurs ions and other molecules can equilibrate across the gut wall. The resulting changes in composition and pH of haemolymph can cause a general paralysis, as observed in silkworms as early as 90 min after treatment, killing the larva; or, in those insects where a general paralysis is not evident, killing by a combination of starvation and ionic imbalance due to a nonfunctional gut. Often changes in gut and haemolymph permit vegetative propagation of normal gut microbial flora or of introduced organisms, resulting in a septicemia which can contribute to or cause death of the larva. Septicemia is, however, rarely seen in > 75% of dead larvae after a field application for crop protection.

VI. Summary

The crystal is a bipyramidal aggregate of protein molecules that are ellipsoidal or dumb-bell shaped with a length of ~ 15 nm, a diameter of ~ 5 nm, and a mol. wt. of 230 000 daltons. The crystal is held together by hydrophobic and inter-subunit disulphide bonds. It is produced after the bacterium is fully committed to sporulation from a stable messenger RNA formed about 3/4 h before stage I of sporulation. Crystal synthesis is first detected in stage II but the major growth occurs during stages III and IV of sporulation. A small amount of related protein is present on the inside of the cell wall during the vegetative phase but cannot kill insects until the cell is broken. A closely related protein is also present on the inner aspect of the spore coat, suggesting that crystal protein is a normal constituent of cell walls and spore coats overproduced in *B.t.* to provide the organism with an ecological niche as an insect pathogen.

The ellipsoidal subunit consists of a number of peptide chains, prominent among which is one having a mol. wt. of 120 000 daltons by SDS electrophoresis and 80 000 daltons by ultracentrifugation, also held together by hydrophobic and interchain disulphide bonds. Carbohydrate, if present, probably plays no role in the essential nature of the crystal or its toxin. Crystals from different serotypes differ in composition and quantity of various chains and in solubility and susceptibility to attack by gut enzymes of a host insect. The toxin is present in the crystal as a heat-labile protoxin that must be hydrolysed to be active just as trypsinogen must be hydrolysed to form active trypsin. There is suggestive evidence that in the *tol* crystal active toxin is present as a molecule with a mol. wt. of 55 000 daltons. When the protoxin is activated by hydrolysis a protease resistant peptide with a mol. wt. of ca. 60 000 daltons is produced in all tested serotypes. There is evidence as well that a much smaller, ca. 5000 dalton, toxic peptide is produced in small quantities during hydrolysis, but the large toxin can fulfil all the functions of the smaller one.

The toxin acts at the surface of gut epithelial cells to cause a rapid loss of ATP from the cells stimulating respiration and glucose uptake. Soon thereafter the microvilli swell and the cell apices begin to swell into the gut lumen. Similar effects occur in some cell cultures. Feeding is inhibited about the time that swelling of the cell apices begins. Metabolic breakdown of epithelial cells is virtually complete by 10–15 min after treatment and ions leak from gut lumen to haemolymph. Paralysis and/or death results from ionic imbalance in the haemolymph. Under field conditions septicemia is favoured but typically kills only 50–75% of the larvae.

VII. Rescarch Requirements

The foregoing is a personal and tentative interpretation of existing data. It is intended to stimulate testing of its inherent postulates and assumptions so that a complete and accurate picture of the mode of action and chemistry of the crystal can be built. Almost nothing beyond the physical structure of the crystal is firmly established; refinement and testing are required throughout. In particular, careful comparative studies using thoroughly tested techniques are essential to determine the limits within which each characteristic can vary. There is much work to be done on the chemistry of the crystal. Since means for dissolving the crystal under mild conditions with retention of toxicity are now at hand, isolation and characterization of the toxic peptides should proceed apace. Comparative crystal chemistry studies will shed light on differences in toxicity and toxicity spectra among the serotypes and should point the way to development of more effective strains. The role of carbohydrate, if any, in toxicity must be established. The strong hydrophobic interactions in the crystal and dissolved toxin will have to be considered in any structural studies. Precautions are needed to prevent aggregation as a result of chain—chain interactions. Careful account will have to be taken of the lyotropic action of buffers and salts if ion-exchange separations are attempted. The existence of monomeric peptides of mol. wt. 13 000 daltons should be confirmed and their relationship to the 80 000 or 120 000 dalton peptide or the toxic moiety established. Finally, nothing is known about the active site on the toxic peptide except that it has a tertiary structure. Such structures are possible in peptide chains as small as 5000 daltons. The isolation of such peptides must be repeated and when isolated they must be further characterized. There is therefore ample opportunity for sound research and, because of widespread and growing application of *B.t.* for insect control, an urgent need to know what we are dealing with so that commercial application does not outstrip our knowledge.

References

Anfinsen, C.B. and Scheraga, H.A. (1975). *Adv. Protein Chem.* **29**, 205–300.

Angus, T.A. (1956). *Can. J. Microbiol.* **2**, 416–426.

Angus, T.A. (1968). *J. Invertebr. Path.* **11**, 145–146.

Angus, T.A. (1970). Proc. IV Int. Colloq. Insect Path. College Park, Md. pp. 183–189.

Bateson, J.B. and Stainsby, G. (1970). *J. Fd Technol.* **5**, 403–415.

Bechtel, D.B. and Bulla, L.A. (1976). *J. Bact.* **127**, 1472–1481.

Bulla, L.A. Jr., Kramer, K.J. and Davidson, L.I. (1977). *J. Bact.* **130**, 375–383.

Chestukhina, G.G., Kostina, L.I., Zalunin, I.A., Kotova, T.S., Katrukha, S.P., Kuznetsov, Y.S. and Stepanov, V.M. (1977). *Biokhimiya* **42**, 1660–1667.

Cooksey, K.E. (1968). *Biochem. J.* **106**, 445–454.

Cooksey, K.E. (1971). *In* "Microbial Control of Insects and Mites" (H.D. Burges and N.W. Hussey, eds), pp. 247–274. Academic Press, London and New York.

Cooksey, K.E., Donninger, C., Norris, J.R. and Shankland, D. (1969). *J. Invertebr. Path.* **13**, 461–462.
Creighton, T.E. (1978). *Prog. Biophys. molec. Biol.* **33**, 231–298.
Delafield, F.P., Somerville, H.J. and Rittenberg, S.C. (1968), *J. Bact.* **96**, 713–720.
Ebersold, H.R., Luethy, P. and Mueller, M. (1977). *Mitt. schweiz. ent. Ges.* **50**, 269–276.
Fast, P.G. (1972). *J. Invertebr. Path.* **20**, 139–140.
Fast, P.G. (1975). *Bi-mon. Res. Notes, Can. For. Serv.* **31**, 1–2.
Fast, P.G. and Angus, T.A. (1966). *J. Invertebr. Path.* **7**, 29–32.
Fast, P.G. and Angus, T.A. (1970). *J. Invertebr. Path.* **16**, 465.
Fast, P.G. and Donaghue, T.P. (1971). *J. Invertbr. Path.* **18**, 135–138.
Fast, P.G. and Morrison, I.K. (1972). *J. Invertebr. Path.* **20**, 208–211.
Fast, P.G. and Videnova, E. (1974). *J. Invertebr. Path.* **23**, 280–284.
Fast, P.G., Murphy, D.W. and Sohi, S.S. (1978). *Experientia* **34**, 762–763.
Fast, P.G. and Milne, R. (1979). *J. Invertebr. Path* **34**, 319.
Faust, R.M. and Estes, Z.E. (1966). *J. Invertebr. Path.* **8**, 141–144.
Faust, R.M., Adams, J.R. and Heimpel, A.M. (1967). *J. Invertebr. Path.* **9**, 488–499.
Faust, R.M., Hallam, G.M. and Travers, R.S. (1973). *J. Invertebr. Path.* **22**, 478–480.
Faust, R.M., Travers, R.S. and Hallam, G.M. (1974a). *J. Invertebr. Path.* **23**, 259–261.
Faust, R.M., Halam, G.M. and Travers, R.S. (1974b). *J. Invertebr. Path.* **24**, 365–373.
Glatron, M.F., Lecadet, M.M. and Dedonder, R. (1969). *C. r. hebd. Séanc. Acad. Sci. Paris,* **269**, 1338–1341.
Glatron, M.F. and Rapoport, G. (1972). *Biochemie* **54**, 1291–1301.
Glatron, M.F., Lecadet M.M. and Dedonder, R. (1972). *Eur. J. Biochem.* **30**, 330–338.
Gould, H.J., Loviny, T.F.L., Vasu, S.S. and Herbert, B.N. (1973). *Eur. J. Biochem.* **37**, 449–458.
Hannay, C.L. and Fitz-James, P.C. (1955). *Can. J. Microbiol.* **1**, 694–710.
Heimpel, A.M. (1967). *A. Rev. Ent.* **12**, 287–322.
Heimpel, A.M. and Angus, T.A. (1959). *J. Insect Path.* **1**, 152–170.
Heimpel, A.M. and Angus, T.A. (1960). *Bact. Rev.* **24**, 266–288.
Herbert, B.N. and Gould, H.J. (1973). *Eur. J. Biochem.* **37**, 441–448.
Herbert, B.N., Gould, H.J. and Chain, E.B. (1971). *Eur. J. Biochem.* **24**, 366–375.
Hippel, P.H., von and Schleich, T. (1969). *In* " Structure and Stability of Macromolecules" (S.N. Timasheff and G.O. Nasman, eds), pp. 417–574. Marcel Dekker, New York.
Holmes, K.C. and Monro, R.E. (1965). *J. molec. Biol.* **14**, 572–581.
Hoopingarner, R. and Materu, M.E.A. (1964). *J. Insect Path.* **6**, 26–30.
Huber, H.E. (1977). Diplomarbeit abt. Naturw. ETH. Zurich. 55 pp.

Iizuka, T. (1974). *J. Fac. Agric. Hokkaido Univ.* **57**, 313–322.

Krywienczyk, J. and Angus, T.A. (1967), *J. Invertebr. Path.* **9**, 126–128.

Krywienczyk, J., Dulmage, H.T. and Fast, P.G. (1978) *J. Invertebr. Path.* **31**, 372–375.

Lecadet, M.M. (1965). Ph.D. Thesis: État Faculté de Science, Paris.

Lecadet, M.M. (1966). *C. r. hebd. Séanc. Acad. Sci. Paris* **262**, 195–198.

Lecadet, M.M. (1967). *C. r. hebd. Séanc. Acad. Sci. Paris* **264**, 2847–2850.

Lecadet, M.M. (1970). *In* "Microbial Toxins" (T.C. Montie, S. Kadis and S.J. Ajl, eds), Vol. III, pp. 437–471. Academic Press, New York and London.

Lecadet, M.M. and Dedonder, R. (1966a). *Bull. Soc. Chim. Biol.* **48**, 631–659.

Lecadet, M.M. and Dedonder, R. (1966b). *Bull. Soc. Chim. Biol.* **48**, 661–691.

Lecadet, M.M. and Dedonder, R. (1967). *J. Invertebr. Path.* **9**, 310–321.

Lecadet, M.M. and Dedonder, R. (1971). *Eur. J. Biochem.* **23**, 282–294.

Lecadet, M.M. and Martouret, D. (1962). *C. r. hebd. Séanc. Acad. Sci. Paris.* **254**, 2457–2459.

Lecadet, M.M. and Martouret, D. (1964). *Entomophaga Mem. Hors. Ser.* No. 2, 205–212.

Lecadet, M.M. and Martouret, D. (1967a). *J. Invertebr. Path.* **9**, 322–330.

Lecadet, M.M. and Martouret, D. (1967b). *C. r. hebd. Séanc. Acad. Sci. Paris.* **265**, 1543–1546.

Lecadet, M.M., Chevrier, G. and Dedonder, R. (1972). *Eur. J. Biochem.* **25**, 349–358.

Lilley, M., Ruffell, R.N. and Somerville, H.J. (1980). *J. gen. Microbiol.* **118**, 1–11.

Lhoste, J. and Martouret, D. (1968). *In* "Proc. 13th Int. Congr. Ent. Moscow" Vol. 2. pp. 80–81.

Louloudes, S.J. and Heimpel, A.M. (1969). *J. Invertebr. Path.* **14**, 375–380.

Luethy, P. (1973). *J. Invertebr. Path.* **22**, 139–140.

Luethy, P. (1975). *Vjschr. naturf. Ges. Zurich* **120**, 81–163.

Luethy, P., Hayashi, Y. and Angus, T.A. (1970). *Can. J. Microbiol.* **16**, 905–906.

Martouret, D., Lhoste, J. and Roche, A. (1965). *Entomophaga* **10**, 349–365.

Mettler, M. (1975). Diplomarbeit, abt. Naturw. ETH, Zurich.

Milne, R., Murphy, D.W. and Fast, P.G. (1977). *J. Invertebr. Path.* **29**, 230–231.

Monro, R.E. (1961a). *J. Biophys. Biochem. Cytol.* **11**, 321–331.

Monro, R.E. (1961b). *Biochem. J.* **81**, 225–232.

Murphy, D.W. (1973). Ph.D. Thesis, Univ. Calif., Riverside.

Murphy, D.W., Sohi, S.S. and Fast, P.G. (1976). *Science* **194**, 954–956.

Narayanan, K. and Jayaraj, S. (1974). *J. Invertebr. Path.* **23**, 125–126.

Nishiitsutsuji-Uwo, J., Ohsawa, A. and Nishimura, M.S. (1977). *J. Invertebr. Path.* **29**, 162–169.

Norris, J.R. (1971). *In* "Microbial Control of Insects and Mites" (H.D. Burges and N.W. Hussey, eds), pp. 229–246. Academic Press, London and New York.

Pendleton, I.R. (1968). *J. appl. Bact.* **31**, 208–214.

Pendleton, I.R. (1970). *J. Invertebr. Path.* **16**, 313–314.
Pendleton, I.R. (1973). *J. Invertebr. Path.* **21**, 46–52.
Petit-Glatron, M.F. and Rapoport, G. (1976). *Biochemie* **58**, 119–129.
Prasad, S.S.S.V. and Shethna, Y.I. (1974). *Biochim. Biophys. Acta* **363**, 558–566.
Prasad, S.S.S.V. and Shethna, Y.I. (1976). *Biochem. Rev.* **47**, 70–77.
Pressman, B.C. (1976). *A Rev. Biochem.* **45**, 501–530.
Ramkrishnan, N. (1968). *J. Invertebr. Path.* **10**, 449–450.
Ribier, J. and Lecadet, M.M. (1973). *Ann. Microbiol.* **124A**, 311–344.
Rogoff, M.H. and Yousten, A.A. (1969). *A Rev. Microbiol.* **23**, 357–386.
Rupley, J.A. (1969). *In* "Structure and Stability of Biological Macromolecules" (S.N. Timasheff and G.D. Fasman, eds), pp. 291–352. Marcel Dekker, New York.
Sayles, V.B. Jr., Aronson, J.B. and Rosenthal, A. (1970). *Biochem. Biophys. Res. Commun.* **41**, 1126–1133.
Schesser, J.H. and Bulla, L.A. Jr. (1978). *Appl. Environ. Microbiol.* **35**, 121–123.
Sharpe, E.S., Nickerson, K.W., Bulla, L.A. Jr. and Aronson, J.N. (1975). *Appl. Microbiol.* **30**, 1052–1053).
Short, J.A., Walker, P.D., Thompson, R.O. and Somerville, H.S. (1974). *J. Gen. Microbiol.* **84**, 261–276.
Somerville, H.J. (1971). *Eur. J. Biochem.* **18**, 226–237.
Somerville, H.J. (1973). *Ann. N.Y. Acad. Sci.* **217**, 93–105.
Somerville, H.J. (1978). *Trends Biochem. Sci.* **3**, 108–110.
Somerville, H.J. and James, C.R. (1970). *J. Bact.* **102**, 580–583.
Somerville, H.J. and Pockett, H.V. (1975). *J. Gen. Microbiol.* **87**, 359–369.
Somerville, H.J., Delafield, F.P. and Rittenberg, S.C. (1968). *J. Bact.* **96**, 721–726.
Somerville, H.J., Delafield, F.P. and Rittenberg, S.C. (1970). *J. Bact.* **101**, 551–560.
Spencer, E.Y. (1968). *J. Invertebr. Path.* **10**, 444–445.
Strehler, B.L. (1968). *In* "Methods of Biochemical Analysis" (D. Glick, ed), pp. 99–181. Interscience, New York.
Sutter, G.R. and Raun, E.S. (1967). *J. Invertebr. Path.* **9**, 90–103.
Travers, R.S., Faust, R.M. and Reichelderfer, C.F. (1976). *J. Invertebr. Path.* **28**, 351–356.
Truempi, B. (1976). *Zbl. Bakt. Abt. II* **131**, 305–360.
Winkler, V.W., Hansen, G.D. and Yoder, J. (1971). *J. Invertebr. Path.* **18**, 378–382.
Young, E. and Fitz-James, P.C. (1959a). *J. Biophys. Biochem. Cytol.* **6**, 467–482.
Young, E. and Fitz-James, P.C. (1959b). *J. biophys. Biochem. Cytol.* **6**, 483–498.
Young, E. and Fitz-James, P.C. (1959c). *J. biophys. Biochem. Cytol.* **6**, 499–506.

Note Added in Proof

Recent evidence shows that sporulation and crystal formation can be "de-linked" by thermal or antibiotic treatments. A spore⁻crystal⁺ mutant blocked at stage 0–1 of sporulation is reported (Meenakshi and Jayaraman, 1979). Dastidar and Nickerson (1979) failed to detect ester bonds or ϵ-NH$_2$-lysine crosslinks in *thu* crystals. No SH was detected in GuHCl but 12 disulphide bonds per 150 000 dalton subunit (18 per 235 000 subunit) were found. In independent unpublished studies in SDS I found 19 disulphides per 235 000 dalton subunit in *kur–1* crystals. Aronson and Tillinghast (1977) dissociated *thu* crystals in 20% SDS + 0.15% Dtt into subunits of < 10 000 daltons and detected nine N-terminal amino acids. Fast and Martin (1980) found that in 2-4M KSCN + 0.05 M Dtt at pH 7.8, *kur–1, thu* and *sot* crystals dissociated into peptides with molecular weights of ca. 1000 daltons with full retention of toxicity. Seki *et al.* (1978) reported that PRP attacked their sarcoma 180 cells releasing phospholipids and substances with UV characteristics similar to those of nucleic acids. Nishiitsutsuji-Uwo *et al.* (1979) showed that the effect of PRP on swelling of TN-368 cells (from *Trichoplusia ni*) depended on the presence of salts; both cations and anions at > 150 mM caused swelling of cells. Ultrastructural studies clearly differentiated between the effect of valinomycin and that of PRP. Electrophysiological measurements on midguts treated with alkali-solubilized *B.t.* (Harvey and Wolfersberger, 1979) or *in vivo* with *B.t.* followed by isolation of midgut (Griego *et al.*, 1979) demonstrated that the active transport of K⁺ is inhibited in one direction only: from gut lumen to blood. These data are of singular importance but do not entirely accord with other published data.

Aronson, J.N. and Tillinghast, J. (1977). *In* "Spore Research 1976" (A.N. Barker *et al.*, eds) Vol. I. pp. 351–357. Academic Press, London New York.
Dastidar, P.G. and Nickerson, K.W. (1979). *FEBS Lett.* **108**, 411–414.
Fast, P.G. and Martin, W.G. (1980). *Biochem. Biophys. Res. Comm.* **95**, 1314–1320.
Griego, V.M., Moffett, D., Spence, K.D. (1979). *J. insect Physiol.* **25**, 283–288.
Harvey, W.R. and Wolfersberger, M.G. (1979). *J. exp. Biol.* **83**, 293–304.
Meenakshi, K. and Jayaraman, K. (1979). *Arch. Microbiol.* **120**, 9–14.
Nishiitsutsuji-Uwo, J., Endo, Y., Himeno, M. (1970). *J. Invertebr. Path.* **34**, 267–275.
Seki, T., Nagamatsu, M., Nagamatsu, Y., Tsutsui, R., Ichimaru, T., Watanabe, T., Koga, K., Hayashi, K. (1978). *Sci. Bull. Fac. Agric.,* Kyushu Univ. **1**, 19–24.

CHAPTER 13

Thuringiensin, the Beta-exotoxin
of *Bacillus thuringiensis*

K. ŠEBESTA, J. FARKAŠ, K. HORSKÁ

Institute of Organic Chemistry and Biochemistry,
Czechoslovak Academy of Sciences, Prague, Czechoslovakia

and

J. VAŇKOVÁ

Institute of Entomology, Czechoslovak Academy of Sciences,
Prague, Czechoslovakia

I. Introduction

Almost two decades ago a low molecular weight, insecticidal compound was reported in culture media of *Bacillus thuringiensis*. In a review in the previous volume of this book (Bond *et al.*, 1971), the characteristics of this compound – named thermostable exotoxin or β-exotoxin – were defined and its structural formula proposed (Farkaš *et al.*, 1969). It lacked activity against microorganisms but was highly toxic to insects and, in certain circumstances, even against mammals (Šebesta *et al.*, 1968). Preliminary information was quoted on its inhibition of *Escherichia coli* DNA-dependent RNA polymerase, one of the enzymes essential for transfer of genetic information (Šebesta and Horská, 1968). These observations brought new ideas and encouraged further work; also they raised doubts as to whether the exotoxin would be used in practice as an insecticide (e.g. review by Faust, 1973).

The subsequent decade witnessed the solution of many problems. The proposed structural formula, however unusual, was confirmed by synthesis of the compound which also led to the preparation of analogues. Much information was obtained about the inhibitory action of exotoxin on bacterial DNA-dependent RNA polymerase *in vitro,* which is due to structural analogy with ATP. *In vivo* the exotoxin affects the terminal stages of RNA biosynthesis. All details of the interference of the exotoxin with the synthesis of the polyribonucleotide chain and its subsequent maturation are not yet elucidated, but enough is known to assess the implications and risks of using β-exotoxin as an insecticide.

Classification of this compound as a structural ATP analogue indicates that the term exotoxin does not correspond to its character and effect. American, French and Czechoslovak authors (Kim and Huang, 1970; Pais and de Barjac, 1974; Farkaš *et al.*, 1977) have therefore proposed that exotoxin is replaced by the term thuringiensin, which will be used throughout this review.

II. Production

Thuringiensin is usually prepared from high yielding isolates of H-serotype 1 *B. thuringiensis* (Šebesta *et al.*, 1967, 1968, 1969a, b). The names of the H-serotypes are given in Chapter 3. It is also produced by some isolates of serotypes 4a4c, 4a4b, 5, 9 and 10 (Bond *et al.*, 1971). Serotypes 11 and 12 produce 500-times less thuringiensin than serotype 1 isolates (de Barjac and Burgerjon, 1973). De Barjac and Bonnefoi (1968) did not obtain it from serotype 5 possibly because they used different isolates of this serotype. Of three serotype 1 isolates tested by Šebesta *et al.* (1973), all yielded the same amount, but the serotype 1 isolate called "gelechiae" yielded 25% less. Serotype 2 isolates yielded no thuringiensin. Thuringiensin is produced by some isolates of *B. thuringiensis* that do not form spores or crystals (de Barjac *et al.*, 1966), but not by isolates of *B. cereus* (Šebesta *et al.*, 1973). Reported yields vary from 50 mg (Bond *et al.*, 1969), 120 mg (Šebesta *et al.*, 1969a, 1973), to 250 to 300 mg thuringiensin/litre of cultivation medium (de Barjac and Lecadet, 1976).

Preferred media for isolation of thuringiensin are simple to avoid complicating the isolation procedure. They consist of a basal salts mixture (Chapter 15, III, B), sometimes supplemented with an essential amino acid, e.g. thiamine–HCl (Cantwell *et al.*, 1964), with casein digests as the simplest nitrogen source and Na-citrate or glucose as the carbon source (Carlberg, 1973).

A range of more complex ingredients has been added, e.g. peptone (de Barjac and Lecadet, 1976), or ill-defined substances such as corn steep liquor (Barbashova and Vladimirova, 1976) and beet molasses etc. (Kim and Huang, 1970) to make rich complex media when productivity rather than easy isolation is the objective. Development of production media is far from complete and genetic improvement by mutation and selection has barely started.

Thuringiensin is still being isolated by modifications of early successful methods. Kim and Huang (1970) modified the method of Šebesta et al. (1969a) by precipitating thuringiensin as the calcium salt. Thuringiensin and its lactone can thus be obtained as distinct fractions. Thuringiensin lactone is then converted into thuringiensin by an ion exchanger (Dowex 1 in OH⁻-form) or by NH_4OH (Pais and de Barjac, 1974, using H-serotype 4a4c). The effluent volume of thuringiensin is lower than the volume in which the lactone emerges (Šebesta et al., 1969a).

Methods of quantifying thuringiensin in culture media are far from satisfactory and a sensitive method for detecting small quantities is still lacking. Many bioassay methods are summarized by Bond et al. (1971). A currently employed assay uses the coccoid bacterium Sarcina flava as test organism (Rosenberg et al., 1971; Carlberg, 1973). A novel biochemical method is based on the inhibition of DNA-dependent RNA polymerase (de Barjac and Lecadet, 1976). An assay of great precision yet not suitable for routine work uses isotope dilution of pure ^{32}P-labelled thuringiensin (Šebesta et al., 1973). Labelling is achieved by growing the bacteria in a low-P medium with $Na_3{}^{32}PO_4$. This assay was used to measure the error of a simple method based on adsorption onto charcoal. The error is considerable, but because it is constant, the method is useful, e.g. to follow the process of production.

Šebesta et al. (1973) examined thuringiensin production during the bacterial life cycle and differences between individual serotypes. By the method of isotope dilution, they showed that in three serotypes production is greatest at maximal bacterial growth and is almost complete at the start of sporulation, although with a Soviet isolate of serotype 1 excretion into the medium continued into the sporulation phase (Fig. 1). In contrast, de Barjac and Lecadet (1976) assigned excretion in serotype 1 to the sporulation phase, but they used a less reliable assay.

The isotope dilution method was also used to detect thuringiensin in the cell of the producing bacterium (Horská et al., 1975), an important phenomenon when thuringiensin was found to inhibit the DNA-dependent RNA polymerase of its producer (Klier et al., 1973) as well as the enzyme from E. coli. Well washed cultures contained $0.1-0.4\,\mu g$ of thuringiensin/mg of bacteria (dry weight), i.e. ca. half the quantity of cell ATP (Fig. 2). A cell excretes 20–25% of its total thuringiensin content into the medium in 1 min. These contradictory facts can be reconciled by assuming strict compartmentalization of inhibitor synthesis in the cell or, alternatively, by assuming that thuringiensin is synthesized to its biologically active, i.e. phosphorylated, form during transport through the cell membrane and is not present in the cell. A decision between these two alternatives awaits further study.

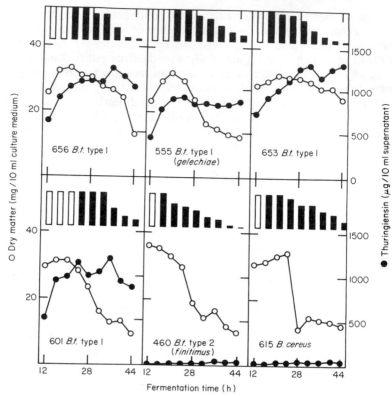

Fig. 1. Thuringiensin production by two H-serotypes of *Bacillus thuringiensis* and by *B. cereus*, numbered as in the culture collection of the Czechoslovak Academy of Sciences. ○ Dry matter. ● Thuringiensin. Empty columns, vegetative cells; full columns, spores inside cells.

III. Chemistry

A. CHEMICAL AND SPECTRAL PROPERTIES

The elemental composition of thuringiensin corresponds to the formula $C_{22}H_{32}N_5O_{19}P.3H_2O$ and dephosphorylated thuringiensin to $C_{22}H_{31}N_5O_{16}$ (Pais and de Barjac, 1974). Discrepancies in molecular weight estimates of thuringiensin (Bond *et al.*, 1971; Kim and Huang, 1970) were resolved by measurement of the mass spectra of thuringiensin derivatives. The mass spectrum of the permethyl derivative of dephosphorelated thuringiensin (10) shows a molecular peak at m/e 789, in view of which dephosphorylated thuringiensin was ascribed a molecular weight of 621 and thuringiensin 701 (Šorm, 1971; Farkaš *et al.*, 1977). These data agree with the mass spectrum of thuringiensin measured by the field desorption method (Pais and de Barjac, 1974). This

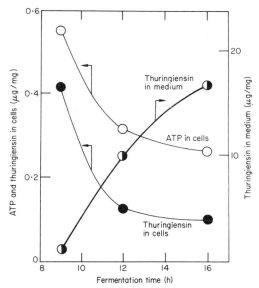

Fig. 2. Effect of time on thuringiensin and ATP in cells of *Bacillus thuringiensis* and of thuringiensin in the culture medium.

spectrum shows a peak at m/e 684 of the $(M + 1)^+$ ion, corresponding to thuringiensin lactone, and a peak at m/e 604 of the $(M + 1)^+$ ion, corresponding to the lactone of dephosphorelated thuriengsin.

The ^{1}NMR spectrum of thuringiensin (Fig. 3) and of dephosphorelated thuringiensin (Fig. 4) have been reported by Farkaš *et al.* (1977) and the ^{13}C NMR spectrum of thuringiensin by Pais and de Barjac (1974). The molar extinction coefficient of thuringiensin at λ_{max} 260 nm is $\epsilon = 13\,700$ (Pais and de Barjac, 1974). Kim and Huang (1970) established the infrared spectra of thuringiensin and thuringiensin lactone, and confirmed the observation of Šebesta *et al.* (1969a) that thuringiensin lactone **(36)** contains a γ-lactone ring.

B. STRUCTURE

Bond *et al.* (1971) felt that, in the tentative structure − formula 1 − of thuringiensin proposed by Farkaš *et al.* (1969), the structure of the "disaccharide" moiety and the link between glucose and allaric acid required additional proof. In 1970–1977, the original proposal was confirmed by chemical and spectral methods, also the stereochemistry on the anomeric center of glucose and on C_2''' carbon atom of the allaric acid residue was elucidated.

1. Sequence Ribose-Glucose

Convincing evidence of the structure of the "disaccharide" moiety of thuringiensin (Šorm, 1971) and later confirmatory results of degradation experiments (Farkaš *et al.*, 1977) are consistent with the data of Pais and de Barjac (1974).

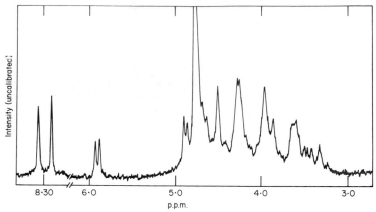

Fig. 3. 100 MHz-[1] H NMR spectrum of thuringiensin. Thuringiensin (22 mg) in 1 M-LiOD/D_2O (0.3 ml); tert-butyl alcohol as internal standard; scale related to tetramethylsilane using δ tert-butyl equal to 1.22 ppm.

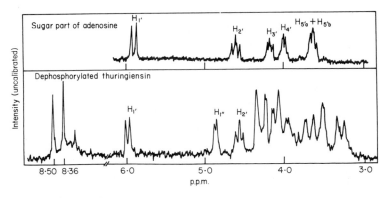

Fig. 4. 100 MHz-[1] H NMR spectrum of dephosphorylated thuringiensin. Solvent, hexadeuteriodimethyl sulfoxide plus trideuterioacetic acid; hexadimethyldisiloxane (HMDS) as internal standard; scale related to tetramethylsilane using δ HMDS equal to 0.06 ppm.

Dephosphorylated thuringiensin (2) was subjected to acid-catalysed methanolysis and the products isolated were as shown in Scheme I (Šorm, 1971; Farkaš et al., 1977). After the separation of adenine, the neutral saccharide fraction was submitted to alkaline hydrolysis. The resulting mixture was resolved into two components by ion-exchange chromatography on Dowex 1 (in acetate form). The acidic component was identified in accordance with earlier observations (Bond et al., 1971) as allaric acid. The neutral component contained a mixture of anomeric glycosides (3). Mixture (3) was converted into acetonides (4) and chromatographically separated into three homogeneous fractions. The mass spectra of these fractions showed the presence of the

1   R = P(OH)$_2$

2   R = H

3   R^1= OCH$_3$ R^2= OCH$_3$ R^3= H   R^4= H
4   R^1= OCH$_3$ R^2= OCH$_3$ R^3= H   R^4= C(CH$_3$)$_2$
5   R^1= α-OCH$_3$ R^2= ß-OCH$_3$ R^3= H   R^4= C(CH$_3$)$_2$
6   R^1= α-OCH$_3$ R^2= ß-OCH$_3$ R^3= Ac R^4= C(CH$_3$)$_2$
7   R^1= OCH$_3$ R^2= OH   R^3= H   R^4= H

8   R^1 = OCH$_3$ R^2 = H
9   R^1 = α-OCH$_3$, R^2 = Ac

11 R = H
12 R = CH$_3$

10
R = CH$_3$

13

14
R = CH$_3$

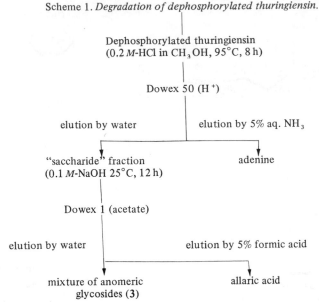

Scheme 1. *Degradation of dephosphorylated thuringiensin.*

Dephosphorylated thuringiensin
(0.2 M-HCl in CH_3OH, 95°C, 8 h)

Dowex 50 (H⁺)

elution by water   elution by 5% aq. NH_3

"saccharide" fraction   adenine
(0.1 M-NaOH 25°C, 12 h)

Dowex 1 (acetate)

elution by water   elution by 5% formic acid

mixture of anomeric   allaric acid
glycosides (3)

$(M-15)^+$ ion, characteristic of acetonides. Fraction (5), which was prevalent in mixture (4) was used as starting material for experiments designed to prove the "disaccharide" moiety. Acetonide (5) was converted into tri-O-acetyl derivative (6) and its structure determined by ^{1}H NMR spectroscopy (Fig. 5). The upfield region shows the presence of singlets of methyl groups indicating that compound (6) is an acetonide bearing two methoxyl groups. The rest of the spectrum represents signals of protons of the OCH-type. Frequency-swept decoupling experiments showed that this part of the spectrum consists of two isolated spin systems. A characteristic feature of the six-spin system is low values of $J_{1,2}$ and $J_{3,4}$ (≪1 Hz), which indicates the trans-configuration of H_1-H_2 protons and of H_3-H_4 protons. The presence of an isopropylidene group can be attributed to the cis-configuration of H_2-H_3 protons. The six-spin system agrees with the known spectra of 2, 3-isopropylidene ribofuranosides. The high values of interaction constants $J_{2',3'}$, $J_{3',4'}$ and $J_{4',5'}$ (9-10 Hz) of the seven-spin are indicative of a typical $^1C_4$ conformation of glucopyranosides. This part of the ^{1}H NMR spectrum agrees with the spectrum of methyl 2, 3, 4, 6-tetra-O-acetyl-α-D-glucopyranoside, the only difference being in the symmetry of the spin system of $H_{5'}$ and $H_{6'}$ protons and especially in the position of the $H_{4'}$ proton quartet. The signal of the H_4 proton in methyl 2, 3, 4, 6-tetra-O-acetyl-α-D-glucopyranoside occurs at 5.03 ppm and that of the $H_{4'}$ proton of compound (6) is localized at 3.46 ppm. This difference in chemical shifts indicates the presence of an ether bond on the $C_{4'}$ carbon atom. Hence, the ^{1}H NMR spectrum of compound (6) shows that the "disaccharide" moiety of the thuringiensin molecule consists of ribose and glucose, linked to one another by an anomalous ether bond between the C_5 atom of ribose and the $C_{4'}$ atom of glucose. The D-configuration of both sugars was determined earlier (Bond *et al.*, 1971).

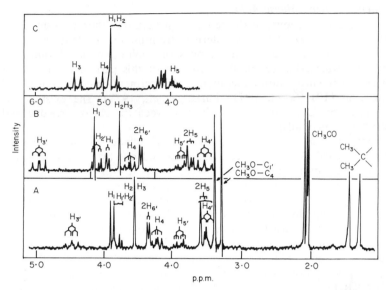

Fig. 5. 100 MHz-^{1}H NMR spectrum of compound (6). A. Measured in deuterio-chloroform (tetramethylsilane as internal standard). B. Sugar part of the bottom spectrum in hexadeuteriobenzene (100 MHz). C. Sugar part of methyl 2, 3, 4, 6-tetra-O-acetyl-α-D-glucopyranoside spectrum in deuteriochloroform (100 MHz).

Final proof of the structure of compound (6) and thus also of the "disaccharide" moiety of thuringiensin was provided by overlapping ^{1}H NMR spectra of a sample of compound (6), prepared from dephosphorylated thuringiensin (2), and of a sample obtained by unambiguous synthesis (Prystaš and Šorm, 1971a).

The presence of an ether bond on the $C_{5'}$ carbon atom of thuringiensin is also shown by the ^{13}C NMR spectrum (Pais and de Barjac, 1974). The signal of the $C_{5'}$ carbon atom of the ribose moiety in thuringiensin is shifted approximately 10 ppm to the lower field compared with the corresponding signal of adenosine.

Further evidence supporting the proposed structure of the "disaccharide" moiety was the ^{1}H NMR spectrum of compound (9) (Farkaš et al., 1977). The mixture of anomeric glycosides (3) was partially hydrolysed to a mixture of glycosides (7); the latter was submitted to mild oxidation and cyclized to γ-lactone (8). After acetylation of (8) and chromatography of the reaction mixture the individual α-glycoside (9) was isolated. The ^{1}H NMR spectrum of compound (9) suggests an isolated, five-spin system $ABMX_2$ resembling γ-ribose lactone. The remaining signals of the ^{1}H NMR spectrum are consistent with the presence of a glucopyranoside residue in $^1C_4$ conformation with an ether bond at carbon atom $C_{4'}$.

2. Sequence Glucose-Allaric Acid

This sequence as shown in the tentative structure of thuringiensin (Farkaš *et al.*, 1969; Bond *et al.*, 1971) was derived from indirect evidence based on the existence of two five-membered lactones of dephosphorylated thuringiensin. Direct supporting evidence for the structure of this part of the molecule was furnished by fragmentation experiments with permethylated dephosphorylated thuringiensin (**10**) (Šorm, 1971; Farkaš *et al.*, 1977). The existence of an ion at m/e 566 is a proof of the link between the glucose residue and the α-hydroxyl group of allaric acid.

A rigorous proof of the existence of glycosyl–allaric acid and the elucidation of stereochemical aspects yielded studies carried out by Prystaš *et al.* (1975). Dephosphorylated thuringiensin (**2**) was converted into its dimethyl ester by treatment with diazomethane. The hydroxyl groups were subsequently protected by trimethylsilyl groups. The trimethylsilyl derivative was reduced by lithium aluminium hydride, the trimethylsilyl groups were removed, and the resulting allitol derivative (**11**) converted into permethyl derivative (**12**). From a comparison of the ^{1}H NMR spectrum of compound (**12**) with that of permethyladenosine (**13**) and of the permethyl derivative of (2S)-2-O-(α-D-glucopyranosyl) allitol (**14**), prepared synthetically, the doublet at 5.25 ppm in the ^{1}H NMR spectrum of compound (**12**) was attributed to the anomeric center of the glucose component. The interaction constant ($J_{1',2'} = 3.3$ Hz) proves the α-configuration of the glucoside bond. The signal of the anomeric proton of glucose at 5.17 ppm ($J_{1'',2''} = 3$ Hz) can be seen in the ^{1}H NMR spectrum of thuringiensin (Bond *et al.*, 1971) yet its assignment was considered inconclusive in view of the complicated spectrum. Acid-catalysed methanolysis of compound (**12**) afforded (2S)-1, 3, 4, 5, 6-penta-O-methylallitol (**15**), whose structure was established as follows. The mass spectrum of compound (**15**) is identical with the mass spectrum of (2R)-1, 3, 4, 5, 6-penta-O-methylallitol (**16**), yet differs from the spectrum of (3S)-1, 2, 4, 5, 6-penta-O-methylallitol (**17**). (Compounds **16** and **17** were prepared by unambiguous syntheses.) The CD spectra of the 2-O-acetyl derivatives of compounds (**15**) and (**16**) show an opposite Cotton effect and are therefore optical antipodes. In view of these results the sequence glucose-allaric acid was ascribed the structure of (2R)-2-O-(α-D-glucopyranosyl) allaric acid. The stereochemistry of the allaric residue of thuringiensin was also demonstrated by another approach (Kalvoda *et al.*, 1973). Dephosphorylated thuringiensin (**2**), was oxidized in periodic acid under mild conditions and the resulting hydroxy aldehydes were reduced by $NaBH_4$. After mild acid hydrolysis, D-glyceric acid was isolated from the reaction mixture and subsequently identified as the crystalline 2-[(1R)-1, 2-dihydroxyethyl]-benzimidazole. This is another proof of the position of substitution and of the (2R) configuration of the allaric residue.

3. Sequence Allaric Acid-Phosphoric Acid

Oxidation of thuringiensin by sodium periodate at pH 7.5 and subsequent hydrolysis in a weakly acidic medium afforded 2(3)-phosphoallaric acid as a chromatographically pure compound. When the latter was digested with

15

16

17

5
4
3
2

20   $R^1 = COC_6H_5$ $R^2 = H$
21   $R^1 = R^2 = COC_6H_5$

18

19

$R = CH_2C_6H_5$

22   $R^1 = COC_6H_5$ $R^2 = CH_2C_6H_5$

23

$R^1 = COC_6H_5$
$R^2 = CH_2C_6H_5$

24 α $R^1 = Cl_3CCH_2O$ $R^2 = CH_2C_6H_5$ $R^3 = COC_6H_5$
25 α $R^1 = Cl_3CCH_2O$ $R^2 = H$ $R^3 = COC_6H_5$
26 α $R^1 = Cl_3CCH_2O$ $R^2 = COCH_3$ $R^3 = COC_6H_5$
27    $R^1 = OCOCH_3$ $R^2 = COCH_3$ $R^3 = COC_6H_5$
28    $R^1 = Br$ $R^2 = COCH_3$ $R^3 = COC_6H_5$
29    $R^1 = N^6$-BzAd $R^2 = COCH_3$ $R^3 = COC_6H_5$

24 β $R^1 = Cl_3CCH_2O$ $R^2 = CH_2C_6H_5$ $R^3 = COC_6H_5$
25 β $R^1 = Cl_3CCH_2O$ $R^2 = H$ $R^3 = COC_6H_5$
26 β $R^1 = Cl_3CCH_2O$ $R^2 = COCH_3$ $R^3 = COC_6H_5$

alkaline phosphatase, allaric acid and phosphoric acid were obtained (Farkaš, unpublished observations). This is direct evidence of the bond between phosphoric and allaric acid. The position of the phosphate bond was deduced indirectly from the uptake of periodic acid during oxidation of thuringiensin and dephosphorylated thuringiensin under comparable conditions (Šorm, 1971; Farkaš et al., 1977). Direct evidence of the position of the phosphate residue on the molecule of allaric acid was presented by Pais and de Barjac (1974). The signal of the $C_{4'''}$ atom in the ^{13}C NMR spectrum of thuringiensin showed splitting brought about by the nucleus of the phosphorus atom.

4. Complete structure

The structure of thuringiensin (1), deduced from structural studies on integral parts of its molecule, was also confirmed by spectroscopic examination of the intact thuringiensin molecule. The links between the individual structural components, with the exception of phosphoric acid, are evident from the fragmentation pattern of the permethyl derivative of dephosphorylated thuringiensin (10) (Farkaš et al., 1977). Participation of all these components in the structure of thuringiensin follows from the exact determination of its molecular weight by the measurement of its mass spectrum and from its ^{13}C NMR spectrum, which permitted 14 carbon atoms out of 22 to be identified unambiguously (Pais and de Barjac, 1974).

C. TOTAL SYNTHESIS

Recent experiments culminated in the completion of thuringiensin synthesis and paved the way for the preparation of thuringiensin analogues, interesting from the viewpoint of its mode of action. The key step in the synthesis of thuringiensin was the formation of the ether bond ($5' \rightarrow 4''$) between ribose and glucose and of the ($1''\alpha \rightarrow 2'''R$)-glycosidic bond between glucose and allaric acid. Both problems were approached by synthesis of model compounds containing the sequence ribose-glucose (Prystaš and Šorm, 1971a, b) and the sequence glucose-allaric acid (Prystaš et al., 1975). These studies led to two procedures for the complete synthesis of thuringiensin (Kalvoda et al., 1976a, b; Prystaš et al., 1976). The reaction schemes of both are essentially the same. The second is an improvement and shortening of the first by use of better protecting groups. The improved version only is presented here.

Strategy of the synthesis was dictated by the need to introduce individual structural components, with the exception of phosphoric acid, in order of the decreasing hydrolytic stability of the bonds by which they are linked to one another in the thuringiensin molecule (Prystaš et al., 1976). The order of the reactions is shown schematically as follows:

$$\{[(Glu + Ri) + Allaric \ a.] + Ad\} + Phosphoric \ a.$$

The disaccharide component (20) was afforded by the reaction of 2, 2, 2-trichloroethyl 2, 3-di-O-benzoyl-β-D-ribofuranoside (18) with 1, 6 : 3, 4-anhydro-2-O-benzyl-β-D-galactopyranose (19) in benzene, catalysed by $SnCl_4$.

The reaction is strictly stereospecific, since it proceeds via a trans-diaxial opening of the oxirane ring. The free 3′-hydroxyl group of compound (20) was benzoylated and product (21) was subjected to acetolysis in a mixture of acetic anhydride and catalytic quantity of H_2SO_4. The resulting anomeric mixture of acetyl derivatives (22) was reacted with the 1,4-lactone of (5S)-5-O-benzoyl-3-O-benzylallaric acid methyl ester (23) in benzene; the reaction was catalysed by BF_3 etherate. A mixture of anomeric glucosides (24) was isolated from the complicated reaction mixture by chromatography. After hydrogenolytic removal of benzyl groups from glucosides (24), a mixture of compounds (25) was obtained. From this, crystalline α-glycoside (25 α) was isolated chromatographically. Less of the other anomer (25 β) was obtained: it was characterized as its acetyl derivative (26 β). Preferential formation of the desired α-anomer (25 α) was permitted by the presence of a nonparticipating benzyloxy group near the anomeric center of compounds (22). Compound (25 α) was converted by acetylation in an acidic medium into its diacetyl derivative (26 α), a versatile intermediate for synthesis of thuringiensin and its analogues. For this reason the protecting 2, 2, 2-trichloroethyl group of compound (26 α) was removed by zinc and the resulting hydroxy derivative was acetylated *in situ* by a mixture of trifluoroacetic acid and acetic anhydride to compound (27). The latter was also obtained directly by treatment of anomer (25 α) with zinc and the acetylation mixture. Compound (27) was converted by treatment with HBr into bromide (28), which was reacted with the chloromercuric salt of 6-benzamidopurine to yield the completely protected nucleoside (29). To introduce the phosphoric acid residue, the lactone ring of compound (29) was opened under mild conditions in methanolic solution of Na acetate and yielded product (30). After the reaction of this with phosphorus oxychloride in benzene and subsequent mild hydrolysis of the derivative of the phosphorodichloridate group, the blocked sequence of thuringiensin (31) was obtained. Alkaline methanolysis of the protecting groups of product (31) afforded nucleotide (32) which was isolated chromatographically.

Nucleotide (32), prepared as described above, as well as the nucleotide obtained by the first version (Kalvoda *et al.*, 1976a, b), inhibited DNA-dependent RNA polymerase in concentrations identical to those of thuringiensin. Both nucleotides also showed the same chromatographic and electrophoretic mobility as thuringiensin. These facts, together with the unambiguity of the synthetic procedures used, support the conclusion that the synthetic nucleotides are identical to thuringiensin.

D. SYNTHESIS OF ANALOGUES

Certain thuringiensin analogues with modified base (compounds 33–35), or with modified allaric acid (compounds 36–40), were prepared as materials for studies on the relationship between structure and inhibitory activity. Some of the analogues were synthesized by simple transformations of thuringiensin prepared enzymatically. The inosine analogue (33) was prepared by deamination of thuringiensin with HNO_2. The reaction of the potassium salt of perphthalic acid afforded analogue (34) (Šebesta and Horská, 1970). Ammonolysis of the

30   R^1 = $COCH_3$ R^2 = COC_6H_5 R^3 = H
31   R^1 = $COCH_3$ R^2 = COC_6H_5 R^3 = $PO(OH)_2$
32   R^1 = H R^2 = H R^3 = $PO(OH)_2$

$$X = \overset{O}{\underset{}{P}}(OH)_2$$

R:

33   34   35

$$X = \overset{O}{\underset{}{P}}(OH)_2$$

R:

36   37   38   39   40

readily accessible lactone (36) yielded amide (37) (Horská *et al.*, 1976). Analogue (35) (Kalvoda, unpublished results) was synthesized from the saccharide intermediates used for the first procedure for total synthesis of thuringiensin (Kalvoda *et al.*, 1976a, b). Analogues (37) to (40) with modified allaric acid were prepared by a similar approach (Horská *et al.*, 1976).

IV. Pathogenicity in Invertebrates and Vertebrates

A. PATHOGENICITY IN INSECTS

Toxic effects ascribed to thuringiensin should be judged with utmost prudence since only a few studies (as indicated in the text) used pure or at least purified thuringiensin. Most results were obtained with autoclaved supernatants of *B. thuringiensis* cultures or with commercial preparations containing both spent culture medium, crystals and spores. Even though thuringiensin is present, possible effects of the other components should not be disregarded. Several authors have shown that pathological changes may be induced by the spent culture medium itself (e.g. Meretoja *et al.*, 1977) and Carlberg (1973) observed a mitostatic effect with autoclaved supernatant, but not with pure thuringiensin, even at high concentrations.

A comprehensive list of insects susceptible to thuringiensin (Burgerjon and Martouret, 1971, and Appendix 1) shows that it is toxic to a much broader range of insects than that of the *B. thuringiensis* crystal endotoxin. Thuringiensin kills species of the orders Lepidoptera, Diptera, Coleoptera, Hymenoptera (including bees), Isoptera and Orthoptera. Whether applied by injection or perorally, it is active mainly against insect larvae and certain adults. The effect varies greatly with dose, mode and time of application (Burgerjon and Biache, 1967a). Sublethal doses often produce anomalies, deformities, and teratologic changes. The parenteral LD_{50} of pure thuringiensin for *Galleria mellonella* larvae is $0.5\,\mu g/g$ body weight, i.e. ca. x30 more toxic than the parenteral LD_{50} for mice (Šebesta *et al.*, 1969b). As can be expected from the interference with RNA transcription by thuringiensin, the toxicity is most marked during physiologically critical developmental stages, e.g. moulting, pupation, or metamorphosis. The effect on larvae is either death just before the next moult or later atrophy of pupae or adults. Adults are infertile or have reduced fecundity and longevity (Burgerjon and Biache, 1967a, b; Greenwood, 1964; Ignoffo and Gregory, 1972).

Although the structures affected by thuringiensin vary a little in different insect orders, they are usually, in sequence, the mouthparts, antennae, thorax and wings in adults, and deformities are observed in pupae (reviewed: Burgerjon, 1974: recent studies; Burgerjon, 1972; Kim *et al.*, 1972a; Wolfenbarger *et al.*, 1972; Wasti *et al.*, 1973; Burges, 1975). Malformation of the mouthparts prevents feeding thus reducing longevity (Burgerjon and Biache, 1967a; Ignoffo and Gregory, 1972). Thuringiensin injected into the haemocoel of *Galleria mellonella* larvae produced pathological changes of haemocytes (Vaňková and Leskova, 1972) and was much less toxic perorally than parenterally (Schmid and Benz, 1969; Šebesta *et al*., 1969b). Vaňková *et al*. (1974) concluded that thuringiensin is rapidly degraded in the intestine, but only slightly in the haemolymph. Digestion by intestinal homogenates of *Musca domestica* and *G. mellonella* larvae, and *Apis mellifera* adults demonstrated that thuringiensin is degraded by intestinal phosphatases (Vaňková and Horská, 1975). In *Drosophila melanogaster* larval development is retarded and mortality is high especially in females. Adults formed from surviving larvae are small but not deformed, although fecundity and longevity are reduced (van Herrewege, 1970). Disturbance of ovogenesis can be the result of thuringiensin penetrating into the haemolymph and acting on the ovaries (David and Vago, 1967). Some adults have ovaries atrophied, with more or less abnormal follicles, while the ovaries of others are swollen with retained eggs. A marked necrosis of developing imaginal discs was also observed. A low dose of thuringiensin used with the nematode *Neoaplectana carpocapsae* against leatherjackets (*Tipula paludosa*) increased mortality obviously due to synergism (Lam and Webster, 1972). As yet, no Hemiptera have been found susceptible. *Perillus bioculatus* fed on Colorado beetle larvae poisoned with thuringiensin were not harmed (Burgerjon and Biache, 1966).

B. PATHOGENICITY IN OTHER INVERTEBRATES

Against the Acarina, autoclaved *B. thuringiensis* supernatant sprayed on bean leaves killed all carmine spider mites, *Tetranychus urticae* (as *telarius*), in 35 days (Krieg, 1968). Application of thuringiensin to orange tree leaves killed many adult citrus red mites, *Panonynchus citri*, also eggs and immature stages (Hall *et al*., 1971). With ixodid ticks, Grebelski *et al*. (1972) found that the effect of Bitoxibacillin, a thuringiensin-containing preparation, varied depending on use before or after bloodsucking. Thuringiensin was highly nematocidal against plant parasitic Nematodes, *Meloidogyne* spp. (Prasad *et al*., 1972) and both *M. incognita* and the free living myceliophagus *Aphelenchus avenae* (Ignoffo and Dropkin, 1977).

C. PATHOGENICITY IN VERTEBRATES

Šebesta *et al*. (1968) found a parenteral LD_{50} value of 18 µg/g body weight for mice and regarded inhibition of liver RNA biosynthesis as the underlying cause of the toxicity. Toxicity is lost completely after the phosphoric group has been split off (Šebesta *et al*., 1969b). De Barjac and Riou (1969) found very similar LD_{50} values (13.3 µg/g body weight for mice intraperitoneally and 16.6 µg/g subcutaneously). Pathological lesions occurred mainly in the liver (necrotizing degenerative hepatitis), kidneys and adrenal glands. An LD_{50} value *per os* could not be attained even when 200 µg/g body weight was fed consecutively for 8 days. Half this dose did not kill a single mouse. Orally applied ^{32}P-labelled thuringiensin did not penetrate the intestinal wall and was voided almost completely in faeces (de Barjac and Lecadet, 1975). Its parenteral metabolism is described in detail later. *Per os* toxicity of the calcium and sodium salts of thuringiensin in hens is considerably higher than that in mice

(Barker and Anderson, 1975). There were severe gizzard erosions then enteritis and proventriculitis, even with low consecutive daily doses (3 $\mu g/g$ body weight, hens; 1 $\mu g/g$, pullets) and some animals died in several weeks. These studies are significant because pure thuringiensin was used. Carlberg (1973) obtained similar results both with pure thuringiensin and autoclaved B. thuringiensis supernatant.

In cultures of human embryonic pulmonary cells and U-cells, pure thuringiensin was not cytotoxic up to 40 $\mu g/ml$ (Carlberg, 1973). In contrast, culture supernatant produced toxic effects in both types of cells, including inhibition of thymidine incorporation into DNA, but these effects cannot be attributed to thuringiensin. However, administration per os of spent culture medium to mice for 3 months caused no pathological changes.

Mutagenicity of thuringiensin in various mammalian systems was assayed by Meretoja et al. (1977) and confirmed by Meretoja and Carlberg (1977). Application of serotype 1 supernatant (containing thuringiensin) to human blood cultures increased chromosomal aberrations, serotype 3 supernatant (without thuringiensin) produced a weak clastogenic effect but no mitotic arrest. Pure thuringiensin had less effect, the number of cells showing chromosomal aberrations approximating the number in cultures treated with the control unfermented nutrient medium. In vivo, with rats, there were no clastogenic effects in bone marrow cells. The low concentrations of serotype 1 supernatant, containing thuringiensin, did not affect blood metaphases either. Clastogenic effects appeared when drinking water was substituted by 50 or 100% of the supernatant for 3 months or when rats were given lethal doses of serotype 1 supernatant. Weak mutagenic activity appeared in cells of the onion plant as chromatin condensation and pycnosis, cell rupture, and increase in the number of nucleoli (Linnainmaa et al., 1977). Application of serotype 1 supernatant to Drosophila melanogaster males significantly increased X-linked recessive lethals among the offspring. Tests with the yeast Saccharomyces cerevisiae, designed to reveal mitotic recombinations, were negative.

V. Mode of Action

In mammals, thuringiensin preferentially inhibits biosynthesis of RNA, which, together with its chemical structure, suggests that it is a specific inhibitor of DNA-dependent RNA polymerases. Studies on mode of action have followed basically two approaches: (a) detailed studies, especially in mammals, have contributed to a deeper understanding of RNA biosynthesis and, at the same time, helped to establish objective criteria of risks involved in practical insecticidal use; (b) thuringiensin was employed as a tool to investigate the character and conditions of inhibition of DNA-dependent RNA polymerase of various origins, especially bacterial, in studies designed to provide background data for studies on thuringiensin toxicity, to explain the complicated action of DNA-dependent RNA polymerases and to examine the properties of their active sites. As well as thuringiensin itself, a number of its derivatives and analogues were used.

A. MODE OF TOXIC ACTION *IN VIVO*

1. *Basic Data*

Thuringiensin is toxic both to insects and, especially parenterally, to higher animals (Bond *et al.*, 1971), due to inhibition of RNA biosynthesis (Šebesta *et al.*, 1969b; Kim *et al.*, 1972b; Čihák *et al.*, 1975). Most studies failed to prove an effect on biosynthesis of proteins and DNA, except in a few cases when high doses of pure thuringiensin (Kim *et al.*, 1972a) or impure preparations or medium concentrates were applied (Carlberg, 1973).

To observe the activity pathways in insects and mammals, ^{32}P-labelled thuringiensin was injected parenterally (Šebesta and Horská, 1973). Mice excreted intact thuringiensin in urine, reaching a maximal rate after 25 min and totalling about 20% of the amount applied in 30 min. Excretion was paralleled by rapid cleavage into a nontoxic product, mostly by enzymatic dephosphory-lation. The cleavage rate was very high since, 5 min after injection of 500 μg of thuringiensin/mouse, its concentration in the liver was only ca. 3 μg/g of tissue and, after 30 min, 1 μg/g of tissue at the most. Slightly more was detected only in the kidney. By contrast, cleavage of thuringiensin in larvae of *Galleria mellonella* is slower: 3 h after parenteral injection, 80% of intact thuringiensin was detected in the haemolymph (Vaňková *et al.*, 1974). When eaten, however, thuringiensin does not pass through the gut wall and is degraded by gut phosphatases (Vaňková and Horská, 1975). This explains why in insects thuringiensin is x10 less toxic perorally than parenterally (Šebesta *et al.*, 1969b).

2. *Effects on RNA Biosynthesis*

Present knowledge of the mode of action of thuringiensin is due to study of its effect on RNA biosynthesis. Data were obtained (a) by determining the level of the labelling of individual RNA species in relation to the dose of thuringiensin and (b) by examination of the inhibition of individual DNA-dependent RNA polymerases (especially in cell nuclei). Attention was focussed on this enzyme at the final stage of RNA synthesis because thuringiensin was its specific inhibitor and also because no other enzyme along the metabolic pathway prior to triphosphate condensation was affected. Experiments with cell nuclei may be divided into two groups: (1) nuclei isolated from normal animals were incubated with thuringiensin and with the nucleoside triphosphates as substrates of DNA-dependent RNA polymerases; (2) nuclei from animals treated with thuringiensin *in vivo* were reacted with substrates without thuringiensin. Thus the transport of thuringiensin, its crossing of the cell membrane, and its degradation rate in the animal were of primary importance in the latter experiments.

The data with untreated nuclei are not unambiguous (Table I; also Mackedonski and Hadjiolov, 1974). This is due to experimental differences, especially to variations of the composition of reaction mixtures and substrate concentrations relative to the number of cell nuclei used. Markedly different results were obtained with nuclei isolated from insect tissues (Beebee and Bond, 1973a, b). All results, however, share one feature in common, i.e. that

TABLE I.

Inhibition of DNA-dependent RNA polymerase in untreated nuclei by thuringiensin

Source of nuclei	Type of enzyme	Conditions of incubation	Conc. of thur. at 50% inhibition	Reference
Sarcophaga bullata	α-amanitin resistant	low ionic strength	10.0 nM	Beebee and Bond, 1973a
		low ionic strength + ecdysone (1.5 μg/ml)	50.0 nM	
	α-amanitin sensitive	low ionic strength	5.0 μM	
		low ionic strength + ecdysone (1.5 μg/ml)	80.0 μM	
Mouse liver	Mg^{2+}-activated	low ionic strength	12.5 μM	Smuckler and Hadjiolov, 1972
	Mn^{2+}-activated	high ionic strength	300.0 μM	
Rat liver	Mg^{2+}-activated	low ionic strength	10.0 μM	Čihák et al., 1975
	Mn^{2+}-activated	high ionic strength	10.0 μM	

α-amanitin resistant DNA-dependent RNA polymerase, activated by Mg^{2+}-ions, is inhibited by thuringiensin more strongly than α-amanitin sensitive polymerase, activated by Mn^{2+}-ions and a higher ionic strength of the medium. The former enzyme more or less corresponds to a nucleolar enzyme synthesizing ribosomal RNA, the latter to a nucleoplasmic enzyme synthesizing mostly messenger RNA. This finding agrees with the results of measurements of RNA labelling after treatment with thuringiensin *in vivo* (see below), which also involves preferential inhibition of ribosomal RNA synthesis.

Nuclei isolated from animals treated intraperitoneally with thuringiensin afforded consistent data (Table II). These did not show a difference in the sensitivity of the two nuclear RNA polymerases to the inhibitor such as that described above or that with isolated enzymes (see below).

It is difficult to correlate both types of experiment, i.e. with untreated and with treated nuclei. The essential difference is that with untreated nuclei the RNA polymerases are under simultaneous action of thuringiensin and the substrate, whereas in nuclei from the animals pretreated with thuringiensin the enzymes are apparently inhibited prior to their contact with the substrate. This is suggested by the time course of the inhibition, i.e. maximal inhibition at 1 h after injecting thuringiensin and return to normal enzyme activity after about 14 h (Čihák *et al.*, 1975; Fig. 6).

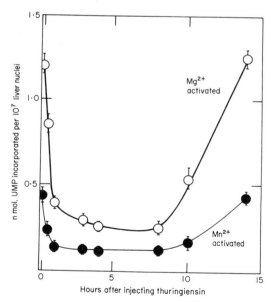

Fig. 6. Time course of inhibition of nuclear Mg^{2+}- and Mn^{2+}-activated RNA polymerases after a single dose of thuringiensin *in vivo*. Groups of 4–6 male rats (170 g) starved for 16 h were injected intraperitoneally with thuringiensin (0.5 μmol/100 g). RNA polymerases were assayed in isolated liver nuclei.

TABLE II.

Inhibition of DNA-dependent RNA polymerase in nuclei from animals pretreated by thuringiensin

Source of nuclei	$\mu g/g$ body weight	Time of action (min)	Enzyme activated by	Ionic strength	Inhibition (%)	Reference
Rat liver	0.75	180	Mn^{2+}		60.0	Čihák et al., 1975
	0.37	180	Mn^{2+}		37.5	
	0.75	180	Mg^{2+}		57.0	
	0.37	180	Mg^{2+}		40.0	
Mouse liver	1.00	120	Mn^{2+}	high	75.0	Smuckler and Hadjiolov, 1972
	1.00	120	Mn^{2+}	low	53.0	
	1.00	120	Mg^{2+}	high	63.0	
	1.00	120	Mg^{2+}	low	63.0	

Inhibition of RNA polymerases in nuclei by thuringiensin has also been exploited in studies on certain morphological changes of the nucleolus that could be related to the activity of these enzymes. Increasing thuringiensin concentrations transformed compact nucleoli to ring-shaped ones then to micronucleoli, indicating segregation of nucleolar components (Smetana *et al.*, 1974).

RNA labelling under the influence of thuringiensin (Table III) revealed preferential inhibition of rRNA biosynthesis. Labelling of $5S$ RNA extracted from ribosomes was inhibited to the same degree as labelling of rRNA. Labelling of DNA-like RNA in the nucleus and of tRNA extracted from whole cells was inhibited much less. These observations agree with the observed inhibition of the corresponding DNA-dependent RNA polymerases in the nucleus. A hitherto unknown effect was observed when nuclear RNAs were pulse labelled in the presence of low doses of thuringiensin (Mackedonski, 1975). At $25\,\mu$g thuringiensin/mouse, which otherwise has no effect on the labelling of rRNAs, a pronounced effect on the $45S$ pre-rRNA was revealed by electrophoresis in agar gel (Table III; Fig. 7). The author interprets this as a rapid breakdown of $45S$ pre-rRNA. From analysis of the degree of labelling of other RNA species he concludes that low thuringiensin doses also inhibit conversion of $38S$ pre-rRNA into $32S$ and $21S$ RNA. This effect on processing of RNA molecules during rRNA synthesis deserves additional examination, since it suggests another possible site where thuringiensin may interfere with RNA biosynthesis.

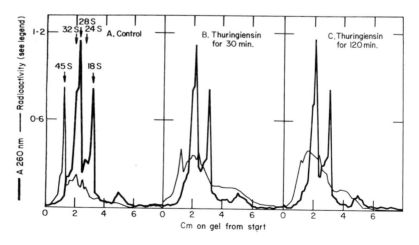

Fig. 7. Agar gel electrophoresis of mouse liver nuclear rRNA. A. Control mice; B. Pretreated with $25\,\mu$g, of thuringiensin/mouse for 30 and ca. 120 min. Then $25\,\mu$Ci per mouse of $[^{14}C]$-orotic acid was injected intraperitoneally. Labelling time was 15 min. Heavy line − absorbancy at 260 nm; thin line − radioactivity recorded from the blackening of radioautograms at 550 nm.; cm − mobility from start.

TABLE III

Labelling of RNA species after application of thuringiensin and $[^{14}C]$-orotic acid to mice[a]

µg/mouse	Labelling time (min)	Nuclear RNA			Cytoplasmic RNA			Reference
		rRNA	45 S preRNA	DNA-like RNA	5 S RNA	tRNA (4 S RNA)	pre-tRNA-4.6 S RNA	
500	60	7.4		24.8				Mackedonski et al., 1972a
150	120	12.0		52.6				
150	120	12.0		52.6	12.0	50.0		Mackedonski et al., 1972b
25	90	60.0						Mackedonski, 1975
200	90							
25	15	100.0	50.0					
25	15[b]	93.0	42.0					
25	30				50.0	90.0	100.0	
25	120				50.0	90.0	100.0	

[a] Results are given in % of control. $[^{14}C]$-orotic acid was applied 30 min after the intraperitoneal injection of thuringiensin. [b] Thuringiensin was applied 120 min before $[^{14}C]$-orotic acid.

B. INHIBITION OF ENZYMES *IN VITRO*

1. *Effect on Prokaryotic and Bacteriophage DNA-dependent RNA Polymerases*

Most data on the nature of inhibition of prokaryotic RNA polymerases were obtained using the enzyme from *E. coli* (Šebesta and Horská, 1968, 1970; Šebesta and Sternbach, 1970; Horská *et al.*, 1976). Inhibition could be decreased only if the concentration of ATP (yet not of the remaining three nucleoside triphosphates) in the reaction mixture was increased. Double reciprocal plots have shown that the inhibition is competitive with ATP (Fig. 8a), affecting the elongation stage of the polymerase action, i.e. the synthesis of the poly-nucleotide chain. It remains to be shown whether thuringiensin plays a role also in the initiation phase. The competitive nature of the inhibition indicates that thuringiensin and ATP occupy the identical substrate-binding site. Seen from this angle the adenosine moiety of thuringiensin mimics the adenosine component of ATP, whereas its acid moiety, i.e. phosphorylated allaric acid, mimics the triphosphate part. This notion is supported by loss of the inhibitory activity of thuringiensin on dephosphorylation.

The role of individual parts of the thuringiensin molecule in the inhibition of RNA polymerase activity has also been examined. Replacement of the base in the thuringiensin molecule and change of the template were chosen to elucidate the role of the nucleoside component. In experiments of the first type thuringiensin was deaminated by HNO_2 and an analogue was prepared containing inosine instead of adenosine (Šebesta and Horská, 1970). This analogue (formula 33) has the same inhibitory activity as thuringiensin; its effect, however, cannot be reversed by ATP but specifically by GTP in a competitive manner (Fig. 8b). Hence, the inosine analogue enters the GTP − binding site. Competition between the inosine part of the inhibitor and the guanosine part of the substrate complies with the rules of base pairing, according to which both inosine and guanosine are hydrogen-bonding with cytosine. They are therefore interchangeable at the binding site specified by cytosine of the template. Identical results were obtained with the synthetic analogue of thuringiensin in which adenine was replaced by uracil (Kalvoda, personal communication). Inhibitory effects of this analogue were the same as those of thuringiensin; however it competed exclusively with UTP (Horská, unpublished observations). Evidence supporting the effect of the base moiety of thuringiensin on the binding to the template was furnished by experiments with simple polydeoxyribonucleotides acting as templates in the RNA−polymerase catalysed reaction (Šebesta and Sternbach, 1970). There was practically no inhibition by thuringiensin with templates lacking the possibility of base pairing with ATP, and, *vice versa*, inosine-containing thuringiensin did not inhibit unless the template provided for base pairing with GTP or ITP. Strong inhibition was observed in opposite cases (Table IV). All these results fit the hypothesis that the base of the inhibitor functions as the triphosphate base, which forms hydrogen bonds with the corresponding base of the DNA template.

Results of the experiments to elucidate the role of the acidic moiety of thuringiensin were ambiguous (Horská *et al.*, 1976). As shown above, the

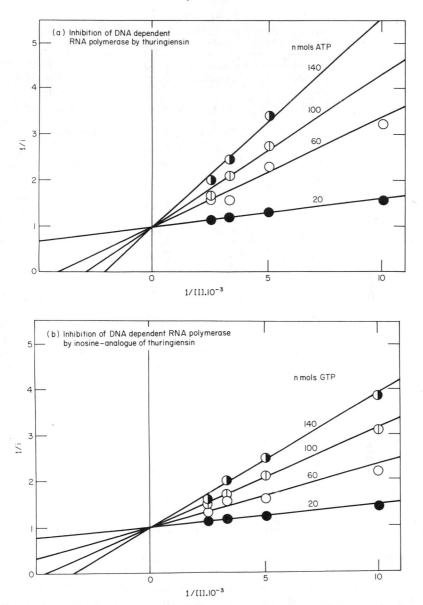

Fig. 8. Double reciprocal plot of *Escherichia coli* RNA polymerase activity at different ratios of thuringiensin:ATP (a) and inosine analogue of thuringiensin: GTP (b) (Šebesta and Horská, 1970). [I] denotes molar concentration of thuringiensin or its analogue; $i = (v_n - v_i)/v_n$ where v_n is the reaction rate of the control, v_i that of the reaction with inhibitor, both expressed in counts per min.

TABLE IV

Inhibition of substrate incorporation by thuringiensin (T) and inosine-analogue of thuringiensin (IT) using poly dT or poly dIdC as templates

Inhibitor	Concn. rel. to labelled substrate	Substrate [^{14}C]-ATP. Template poly dT.		Substrate [^{14}C]-ITP. Template poly dIdC^a.	
		Incorporation		Incorporation	
		nmol/ml	%	nmol/ml	%
–	–	88.0	100.0	326.0	100.0
T	0.10	59.0	12.1	265.0	81.0
T	0.02	165.0	34.8	376.0	115.0
T	0.01	207.0	42.5	372.0	114.0
IT	0.10	538.0	110.0	21.0	6.6
IT	0.02	572.0	117.0	106.0	32.4
IT	0.01	595.0	122.0	161.0	49.3

^{a} Wtih ITP as the only substrate merely the dC strand was transcribed.

necessary prerequisite of inhibition is the presence of a phosphoric acid residue. If this is split off, inhibition is lost. Equally when the carboxylic or carbonyl group at C_5 of allaric acid (i.e. the carboxylic group localized closer to the phosphoric acid residue) is replaced by an alcoholic group (formula 38), an analogue of thuringiensin is obtained which practically has no inhibitory activity against RNA polymerase. On the other hand, the closure of a five-membered lactone ring by a bond formed between this carboxylic group and the hydroxyl group on C_3 (formula 36), as well as the conversion of the carboxylic group into an amide group (formula 37), decreases the inhibition by about one half. A similar effect was produced by two alterations of the acidic moiety of thuringiensin, namely shortening of the distance between the phosphoric acid residue and the carboxyl at C_5 by one carbon atom (formula 40) or inversion of the configuration at C_3 of allaric acid (formula 39). This indicates that the presence of the phosphoric acid residue is insufficient by itself for the binding of thuringiensin to the active centre of the enzyme and that the participation of other functional groups of allaric acid is required. Apparently the three-dimensional structure of the inhibitor has considerable specificity to comply with the structural requirements of the ATP-binding site of RNA polymerase.

The DNA-dependent RNA polymerase of *B. thuringiensis* is inhibited by thuringiensin to the same degree as the enzyme of *E. coli*. The K_i value for thuringiensin of 6×10^{-5} M agrees with the order of the K_i value established for the *E. coli* enzyme (Horská, unpublished observation). Johnson *et al.* (1975) found that inhibition of the enzyme isolated from *B. thuringiensis* during sporulation is lower than the inhibition of the same enzyme isolated during exponential growth. Klier *et al.* (1973), however, found no such differences. Johnson (1978) recently found a decrease of inhibition of nearly 50% for the enzyme from sporulating *B. thuringiensis* cells. This difference is ascribed to lack of σ-factor in this enzyme. A similar decrease of sensitivity towards thuringiensin was found in *E. coli* polymerase from which σ-factor was removed by means of a phosphocellulose column.

Küpper *et al.* (1973) showed that thuringiensin does not inhibit DNA-dependent RNA polymerase from phage T7, which they regard as a proof of differences in the structure of the active centre between enzymes from phage and from *E. coli.* Accordingly, there are also other differences: the enzyme from phage T7 has a lower molecular weight, one polypeptide chain only, and its template requirements differ from those of the *E. coli* enzyme.

2. Effect on Eukaryotic DNA-dependent RNA Polymerases

Data with cell nuclei are considered when describing the toxic effect *in vivo* (Section VA) rather than here because the methods do not reliably distinguish between the individual enzymes. In work on isolated DNA-dependent RNA polymerases, the isolation was effected by $(NH_4)_2 SO_4$ precipitation according to Roeder and Rutter (1969, 1970) up to fraction 4, followed by chromatography on DEAE-Sephadex. Active fractions were collected in two peaks representing polymerase A and B and used in inhibition studies, either as such or after a brief dialysis against buffer. Beebee *et al.* (1972) found that enzyme A was more sensitive to thuringiensin than enzyme B and that the inhibition of both competed with ATP (Table V; Fig. 9). Smuckler and Hadjiolov (1972),

TABLE V

Inhibition of isolated mammalian RNA polymerases by thuringiensin

	RNA polymerase AI		RNA polymerase B		
	K_i (μM)	Concentration of thuringien-sin for 50% inhibition (μM)	K_i (μM)	Concentration of thuringien-sin for 50% inhibition (μM)	Reference
Rat liver	*a*	1.4	*a*	5.0	Beebee *et al.*, 1972
Mouse liver	2.78	117.0	0.204	0.18	Smuckler and Hadjiolov, 1972
Calf thymus	0.4	1.0	1.5	4.0	Horská unpub. observations.

a Not determined.

in contrast, reported that B was inhibited more strongly than A and that only the inhibition of A competes with ATP. Horská (unpublished observations) used further purified preparations A and B from calf thymus by the above procedures, with additional chromatography on phosphocellulose and hydroxyapatite columns as well as centrifugation using glycerol gradients (Gissinger and Chambon, 1972; Kedinger and Chambon, 1972). She confirmed (Table V) the finding of Beebee *et al.* (1972) that inhibition of both enzymes is competitive and that the nucleolar enzyme is inhibited preferentially. The higher

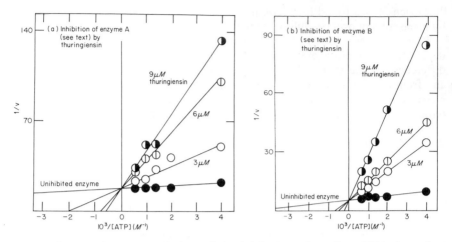

Fig. 9. Double reciprocal plots of the inhibition of rat liver DNA-dependent RNA polymerases by thuringiensin: v is expressed in counts per min incorporated into material precipitated by trichloracetic acid after 5 min of reaction.

sensitivity of the nucleolar enzyme agrees with the results of assays carried out with isolated nuclei.

3. Effect on Other Enzymes

Unpublished results from various laboratories showed that the following enzymes are not affected by thuringiensin: DNA polymerase, polynucleotide phosphorylase, tRNA pyrophosphorylase, orotate phosphoribosyl transferase, orotidine 5'-phosphate decarboxylase. The only enzyme besides DNA-dependent RNA polymerases reported as inhibited is adenyl cyclase from pigeon erythrocyte, rat adrenal cortex and rat brain (Graham-Smith et al., 1975). The inhibition is competitive and affects stimulation of this enzyme both by fluorides and enzymes.

C. CONCLUSIONS

Our present knowledge permits the following conclusions. (a) Thuringiensin does not affect directly the biosynthetic pathways of proteins and DNA. (b) It is a specific inhibitor of both prokaryotic and eukaryotic DNA-dependent RNA polymerases. (c) It inhibits by competing with ATP for the binding site of the enzyme because of a similar three-dimensional structure of both molecules decisive for the binding. (d) Eukaryotic DNA-dependent RNA polymerases are more sensitive to thuringiensin (K_i-values, $0.5-1.0\,\mu M$) than prokaryotic polymerases (K_i-values, $30-60\,\mu M$). (e) This inhibition is the main form of interference of thuringiensin with the metabolism of eukaryotes. It affects preferentially the nucleolar polymerase (polymerase A) which synthesizes ribosomal RNA. (f) In addition to the synthesis of the polyribonucleotide chain, thuringiensin apparently inhibits also the processing of synthesized RNA.

VI. Practical Use and Its Implications

Trials of thuringiensin as an insecticide have been limited. Burgerjon and Diache (1964) effectively used the culture filtrate from *B. thuringiensis* (i.e. without endotoxin) against larvae of the sawfly *Diprion pini* on pine.

Preparations containing both thuringiensin and endotoxin have been tested more often, e.g. the present product Bitoxibacillin in the USSR. In the field it may be effectively used at lower dose levels than preparations containing endotoxin only, against larvae of the moths *Agrotis segetum*, *Heliothis obsoleta* and *Spodoptera exigua*, also predictably against early instars of the beetle *Leptinotarsa decemlineata* (the endotoxin would have no effect on beetle larvae). It is much less toxic to the predators *Coccinella septempunctata*, *Chrysopa carnea*, *Harpalus pubescens* etc. (Stus *et al.*, 1976; Kandybin and Barbashova, 1978). Addition of thuringiensin to animal feeds controls flies in faeces (review: Laird, 1971), but pathological changes were observed in test animals (Harvey and Brethour, 1960; Burns *et al.*, 1961; Galichet, 1966; Ode and Matthysse, 1964) particularly in hens for which it is definitely unsuitable (Barker and Anderson, 1975).

Thuringiensin is persistent. Its sodium and calcium salts applied to the soil surface inhibit reproduction of *Spodoptera* (*Prodenia*) *eridania* and the activity remained unchanged for 4 months (Hitchings, 1960). It persisted on orange leaves for 45 days (Hall *et al.*, 1971) and on cotton-tree leaves for 6–12 days (Wolfenbarger *et al.*, 1972).

Criteria for the safe use of thuringiensin can be defined from knowledge of its toxicity, mode of action, biodegradation and physiological penetration into insects and vertebrates: (a) The gastrointestinal barrier affords sufficient protection against amounts of thuringiensin likely to be used in field and forest. These would not harm the natural environment. (b) Higher doses employed *per os* in animal feeds may result in toxic symptoms. These are not advisable. (c) Since thuringiensin affects the transfer of genetic information, factors that aid its penetration into the blood system should be avoided.

VII. Future Outlook

Many important theoretical and practical questions remain unanswered. Little is known about the biosynthetic pathway of thuringiensin which remains difficult to elucidate, but is extremely interesting because different parts of the structure of thuringiensin are very unusual and may be formed by new enzymatic reactions. Another essential problem is the uptake of this markedly acidic compound by the cell. Unlike the transport mechanism of other phosphoric esters, the involvement of dephosphorylation and rephosphorylation in the transport mechanism across the membrane seems to be excluded in this case. This is indicated by the finding that dephosphorylated thuringiensin is nontoxic since the cells are apparently unable to phosphorylate this compound. A problem deserving further study is the mode of penetration of thuringiensin into cells with different growth potentials: it apparently is taken up mainly by stationary, not proliferating, cells. *E. coli* and *B. subtilis* are normally completely resistant to thuringiensin, but its application to cells treated with EDTA and Tris, known to increase their permeability, leads to inhibition of bacterial RNA polymerase; its penetration into the cell could not, however, be demonstrated (Johnson, 1976). Its effect in plants needs more study since it has a mitostatic effect in the root meristem of the onion (Sharma *et al.*, 1976).

There remains the problem of its interference with RNA processing. Likewise, regarding the inhibition of polymerases, we are still lacking K_i-values for individual enzymes, especially those from insects. Lastly, little progress has been made in using thuringiensin as a specific inhibitor for studies on RNA biosynthesis in eukaryotes.

References

Barbashova, N.M. and Vladimirova, G.A. (1976). *In* "Voprosy Ekologii i Fiziologii Mikroorganizmov Izpolzuemykh v Selskom Khozyaistve", pp. 122—130. VNII Selskokhozyaistvenoi Mikrobiologii, Leningrad.

Barjac, H. de and Bonnefoi, A. (1968). *J. Invertebr. Path.* **11**, 335—347.

Barjac, H. de and Burgerjon, A. (1973). *J. Invertebr. Path.* **21**, 325—327.

Barjac, H. de and Lecadet, M.M. (1975). *C. r. hebd. Séanc. Acad. Sci. Paris* **280**, 677—679.

Barjac, H. de and Lecadet, M.M. (1976). *C. r. hebd Séanc. Acad. Sci. Paris* **282**, 2119—2122.

Barjac, H. de and Riou, J.Y. (1969). *Revue Path. comp. Méd. exp.* **6**, 367—374.

Barjac, H. de, Burgerjon, A. and Bonnefoi, A. (1966). *J. Invertebr. Path.* **8**, 537—538.

Barker, R.J. and Anderson, W.F. (1975). *J. med. Ent.* **12**, 103—110.

Beebee, T.J.C. and Bond, R.P.M. (1973a). *Biochem. J.* **136**, 1—7.

Beebee, T.J.C. and Bond, R.P.M. (1973b). *Biochem. J.* **136**, 9—13.

Beebee, T.J.C., Korner, A. and Bond, R.P.M. (1972). *Biochem. J.* **127**, 619—624.

Bond, R.P.M., Boyce, C.B.C. and French, S.K. (1969). *Biochem. J.* **114**, 477—488.

Bond, R.P.M., Boyce, C.B.C., Rogoff, M.H. and Shieh, T.R. (1971). *In* "Microbial Control of Insects and Mites" (H.D. Burges and N.W. Hussey, eds), pp. 275—303. Academic Press, London and New York.

Burgerjon, A. (1972). *Ent. Exp. Appl.* **15**, 112—127.

Burgerjon, A. (1974). *Annls Parasit, hum. comp.* **48**, 835—844.

Burgerjon, A. and Biache, G. (1964). *J. Insect Path.* **6**, 538—541.

Burgerjon, A. and Biache, G. (1966). *Entomophaga* **11**, 279—284.

Burgerjon, A. and Biache, G. (1967a). *Ann. Soc. ent. Fr. N.S.* **3**, 929—952.

Burgerjon, A. and Biache, G. (1967b). *C. r. hebd. Séanc. Acad. Sci. Paris* **264**, 2423–2425.

Burgerjon, A. and Martouret, D. (1971). *In* "Microbial Control of Insects and Mites" (H.D. Burges and N.W. Hussey, eds), pp. 305–325. Academic Press, London.

Burges, H.D. (1975). *J. Invertebr. Path.* 26, 419–420.

Burns, E.C., Wilson, B.H. and Tower, B.A. (1961). *J. econ. Ent.* **54**, 913–915.

Cantwell, G.E., Heimpel, A.M. and Thompson, M.J. (1964). *J. Insect Path.* **6**, 466–480.

Carlberg, G. (1973). Rep. Dept. Microbiol. Univ. Helsinki, 6 pp.

Čihák, A., Horská, K. and Šebesta, K. (1975). *Coll. Czech. Chem. Commun.* **40**, 2912–2922.

David, J. and Vago, C. (1967). *Entomophaga* 12, 153–159.

Farkaš, J., Šebesta, K., Horská, K., Samek, Z., Dolejš, L. and Šorm, F. (1977). *Coll. Czech. Chem. Commun.* **34**, 1118–1120.

Farkaš, J., Šebesta, K., Horská, K., Samek, Z., Dolejš, L. and Šorm, F. (1977). *Coll. Czech. Chem. Commun.* **42**, 909–929.

Faust, R.M. (1973). *Bull. ent. Soc. Am.* 19, 153–156.

Galichet, P.F. (1966). *Ann. Zootech. Paris* 15, 135–145.

Gissinger, F. and Chambon, P. (1972). *Eur. J. Biochem.* 28, 277–282.

Grahame-Smith, D.G., Isaac, P., Heal, D.J. and Bond, R.P.M. (1975). *Nature Lond.* 253, 58–60.

Grebelski, S.G., Kandybin, N.V., Sharafutdinov, Sh.A., Stus, A.A. and Simonova, L.A. (1972). *In* "Pathogennye Mikroorganizmy Vreditelei Rastenii", pp. 46–47. Zinatne, Riga.

Greenwood, E.S. (1964). *N.Z.J. Sci.* 7, 221–226.

Hall, I.M., Hunter, D.K. and Arakawa, K.Y. (1971). *J. Invertebr. Path.* 18, 359–362.

Harvey, T.L. and Brethour, J.R. (1960). *J. econ. Ent.* 53, 774–776.

Herrewege, J. van (1970). *Entomophaga* 15, 209–222.

Hitchings, D.L. (1960). *J. econ. Ent.* 60, 569–597.

Horská K., Kalvoda L. and Šebesta, K. (1976). *Coll. Czech. Chem. Commun.* **41**, 3837–3841.

Horská, K., Vaňková, J. and Šebesta, K. (1975). *Z. Naturforsch.* **30c**, 120–123.

Ignoffo, C.M. and Dropkin, V.H. (1977). *J. Kans. Ent. Soc.* **50**, 394–398.

Ignoffo, C.M. and Gregory, B. (1972). *Environ. Ent.* 1, 269–272.

Johnson, D.E. (1976). *Nature Lond.* 260, 333–335.

Johnson, D.E. (1978). *Can. J. Microbiol.* 24, 537–543.

Johnson, D.E., Bulla Jr, L.A. and Nickerson, K.W. (1975). *In* "Spores" (P. Gerhardt, R.N. Costilow and H.L. Sadoff, eds) Vol. VI, 248–254. Am. Soc. Microbiol.

Kalvoda, L., Prystaš, M. and Šorm, F. (1973). *Coll. Czech. Chem. Commun.* **38**, 2529–2532.

Kalvoda, L., Prystaš, M. and Šorm, F. (1976a). *Coll. Czech. Chem. Commun.* **41**, 788–799.

280   K. ŠEBESTA, J. FARKAŠ, K. HORSKÁ AND J. VAŇKOVÁ

Kalvoda, L., Prystaš, M. and Šorm, F. (1976b). *Coll. Czech. Chem. Commun.* **41**, 800–815.

Kandybin, N.V. and Barbashova, N.M. (1978). *In* "Trudy Soveshchaniya po Usloviyam i Perspektivam Biologicheskogo Metoda Borby protiv Nasekomykh, SEV, Liblice 1977" (J. Weiser, ed.) pp. 35–46. Inst. Ent., Prague.

Kedinger, C. and Chambon, P. (1972). *Eur. J. Biochem.* **28**, 283–290.

Kim, Y.T. and Huang, H.T. (1970). *J. Invertebr. Path.* **15**, 100–108.

Kim, Y.T., Gregory, B.G. and Ignoffo, C.M. (1972a). *J. Invertebr. Path.* **20**, 46–50.

Kim, Y.T., Gregory, B.G., Ignoffo, C.M. and Shapiro, M. (1972b). *J. Invertebr. Path.* **20**, 284–287.

Klier, A., Lecadet, M.M. and Barjac, H. de (1973). *C. r hebd. Séanc. Acad. Sci. Paris* **277**, 2805–2808.

Krieg, A. (1968). *J. Invertebr. Path.* **12**, 478.

Küpper, H.A., McAllister, W.T. and Bautz, E.K.F. (1973). *Eur. J. Biochem.* **38**, 581–586.

Laird, M. (1971). *In* "Microbial Control of Insects and Mites" (H.D. Burges and N.W. Hussey, eds.), pp. 387–406. Academic Press, London and New York.

Lam, A.B.Q. and Webster, J.M. (1972). *J. Invertebr. Path.* **20**, 141–149.

Linnainmaa, K., Sorsa, M., Carlberg, G., Grippenberg, U. and Meretoja, T. (1977). *Hereditas* **85**, 113–122.

Mackedonski, V.V. (1975). *Biochim. Biophys. Acta* **390**, 319–326.

Mackedonski, V.V. and Hadjiolov, A.A. (1974). *C. r. Bulg. Acad. Sci.* **27**, 1117–1119.

Mackedonski, V.V., Hadjiolov, A.A. and Šebesta, K. (1972a) *FEBS Lett.* **21**, 211–214.

Mackedonski, V.V., Nikolaev, N., Šebesta, K. and Hadjiolov, A.A. (1972b). *Biochim. Biophys. Acta* **272**, 56–66.

Meretoja, T. and Carlberg, G. (1977). *FEMS microbiol. Lett.* **2**, 109–111.

Meretoja, T., Carlberg, G., Grippenberg, U., Linnainmaa, K. and Sorsa, M. (1977). *Hereditas* **85**, 105–112.

Ode, P.E. and Matthysese, J.G. (1964). *J. econ. Ent.* **57**, 637–640.

Pais, M. and Barjac, H. de (1974). *J. Carbohyd. Nucleosides Nucleotides* **1**, 213–223.

Prasad, S.S.V., Tilak, K.V.B.R. and Gollakota, K.G. (1972). *J. Invertebr. Path.* **20**, 377–378.

Prystaš, M. and Šorm, F. (1971a). *Coll. Czech. Chem. Commun.* **36**, 1448–1471.

Prystaš, M. and Šorm, F. (1971b). *Coll. Czech. Chem. Commun.* **36**, 1472–1481.

Prystaš, M. Kalvoda, L. and Šorm, F. (1975). *Coll. Czech. Chem. Commun.* **40**, 1775–1785.

Prystaš, M., Kalvoda, L. and Šorm, F. (1976). *Coll. Czech. Chem. Commun.* **41**, 1426–1447.

Roeder, R.G. and Rutter, W.J. (1969). *Nature Lond.* **224**, 234–237.

Roeder, R.G. and Rutter, W.J. (1970). *Proc. natn. Acad. Sci. U.S.* **65**, 675–682.

Rosenberg, G., Carlberg, G. and Gyllenberg, H.G. (1971). *J. appl. Bact.* **34**, 417–423.

Schmid, E. and Benz, G. (1969). *Experientia* **25**, 96–98.

Šebesta, K. and Horská, K. (1968). *Biochim. Biophys. Acta* **169**, 281–282.

Šebesta, K. and Horská, K. (1970). *Biochim. Biophys. Acta* **209**, 357–367.

Šebesta, K. and Horská, K. (1973). *Coll. Czech. Chem. Commun.* **38**, 2533–2537.

Šebesta, K. and Sternbach, H. (1970). *FEBS Lett.* **8**, 233–235.

Šebesta, K., Horská, K. and Vaňková, J. (1967). *In* "Insect Pathology and Microbial Control", (P.A. van der Laan ed.), pp. 238–242. N. Holland Publ. Co., Amsterdam.

Šebesta, K. Horská, K. and Vaňková, J. (1968). *Abstr.*, 5th Meeting FEBS, Prague, p. 250. Czech. Biochem. Soc., Prague.

Šebesta, K., Horská, K. and Vaňková, J. (1969a). *Coll. Czech. Chem. Commun.* **34**, 891–900.

Šebesta K., Horská, K. and Vaňková, J. (1969b). *Coll. Czech. Chem. Commun.* **34**, 1786–1791.

Šebesta, K., Horská, K. and Vaňková, J. (1973). *Coll. Czech. Chem. Commun.* **38**, 298–303.

Sharma, C.B.S.R., Prasad, S.S.V., Pai, S.B. and Sharma, S. (1976). *Experientia* **32**, 1465–1466.

Smetana, K., Raška, I. and Šebesta, K. (1974). *Expl. Cell Res.* **87**, 351–358.

Smuckler, E.A. and Hadjiolov, A.A. (1972). *Biochem. J.* **129**, 153–166.

Šorm, F. (1971). *Pure appl. Chem.* **25**, 253–269.

Stus, A.A., Kandybin, N.V. and Zaytsev, O.F. (1976). *In* "Voprosy Ekologii i Fiziologii Mikroorganizmov Izpolzuemykh v Selskom Khozyaistve", pp. 131–137. VNII Selskokhozyaistvenoi Mikrobiologii, Leningrad.

Vaňková, J. and Horská, K. (1975). *Acta Ent. Bohemoslov.* **72**, 7–12.

Vaňková J. and Leskova, A. Ya. (1972). *Acta Ent. Bohemoslov.* **69**, 297–304.

Vaňková, J., Horská, K. and Šebesta, K. (1974). *J. Invertebr. Path.* **23**, 209–212.

Wasti, S.S., Mahadeo, C.R. and Knell, J.D. (1973). *Z. angew. Ent.* **74**, 157–160.

Wolfenbarger, D.A., Guerra, A.A., Dulmage, H.T. and Garcia, R.D. (1972). *J. econ. Ent.* **65**, 1245–1247.

CHAPTER 14

Potential of *Bacillus sphaericus* and Related Spore-forming Bacteria for Pest Control

S. SINGER

Department of Biological Sciences, Western Illinois University,
Macombe, Illinois, USA.

I. Introduction

Use of natural diseases of vectors to regulate vectors of human disease has always been intriguing because of its simplicity. What is more logical, and fruitful, than turning our attention to nature's parasite pool? To encourage the collection and survey of candidate organisms, since 1966, the World Health Organization (WHO) distributed pocket-size kits to collect dead insect vectors of medical importance (Cantwell and Laird, 1966). The material is sent for examination to a Collaborating Center for the Biological Control of Vectors of Human Disease (WHO/CCBC) at Ohio State University. If a bacterium is involved, the materials are sent to the Western Illinois University for detailed assessment.

We initially classify isolated bacilli by the original division of the genus *Bacillus* by Smith *et al.* (1952). Two groups of bacilli active in insects were isolated, a complex related to *Bacillus alvei–B. brevis–B. circulans* morphological group II and *Bacillus sphaericus* Group III (Singer, 1973, 1974). The sources of accessioned material and the presumptive identifications show that no one geographical area has a monopoly of morphological groups.

This once neglected area of pest control is the subject of this chapter. It is an area that appears to be developing with a rush, so that much of the information is either in press or not yet published. Although it mostly concerns human disease vectors, particularly mosquito larvae, the work on the Cigarette Beetle should illustrate the potential of this group of bacteria for control of important agricultural pests.

II. *Bacillus sphaericus*

A. BACTERIAL CHARACTERIZATION AND FERMENTATION

Sporeforming bacteria of the genus *Bacillus* have been popular for biological control of insect pests because: 1. Mass production by submerged fermentation is economical. 2. The endospore survives well in nature. 3. Members of this group (*B. thuringiensis*, *B. popilliae*) are specific, innocuous in the environment and mainly atoxic to mammals (Burges and Hussey, 1971). This is no less the case with *B. sphaericus.*

With little effort, thousands of isolates can be obtained from our accessions. Critical to success is knowing what to do afterwards, and judgment, born of experience, has evolved the following steps: 1. Initial bioassay to establish potency against several target vectors. 2. Fermentation studies to assess propagation capabilities. 3. Determination of the source of insecticidal activity — vegetative cell, an intermediate stage or product. This information feeds back to item 2. 4. Cultural and fermentation information about preservation of viability and activity. 5. Improvement of field potency by culture selection and fermentation manipulation (media, physical-chemical requirements, etc.) 6. Selection of several most suitable, easily propagated, active candidates for potential field use.

Our accessions usually come as three or four dried mosquito larvae, sometimes bacterial slant cultures, occasionally dead adult insects, or hide and hair scrapings from infested animals. Fortunately the spores survive well, and we still have material from all our original samples, which is important because of variation in composition of some of the bacterial populations. We also dry on filter paper some dead larvae from our bioassays to preserve bacterial inocula in the affected host specimens. We also maintain cultures in soil extract medium (Gordon *et al.*, 1973), or in sterile soil. Soon we shall have freeze-dried, vacuum dried, or acetone-precipitated dry powders, which maintain activity better, especially since our latest strains appear to be more amenable to this treatment.

Bacillus sphaericus is ubiquitous and cosmopolitan (Table I) in soil and in soil—aquatic systems (Buchanan and Gibbons, 1974). Strains with which we are working have widely spaced origins, e.g. India, California, El Salvador and the Philippines. Strain SSII-1 has received most attention. Both strains K and Q are slightly active against larvae of *Culex, Aedes* and *Anopheles*.

All strains except two (Table II) are *B. sphaericus* var *fusiformis* (Buchanan and Gibbons, 1974) and form urease. This is correlated with insecticidal activity, for strain 7054. No other characteristics of growth on the usual substrates or other taxonomic criteria are known to differentiate mosquito—pathogenic

TABLE I
Some bacilli from WHO accessions (Singer, 1977)

Origin	Host	Bacillus group	Accession number
Australia	*Culex annulirostris*	II	1443-2
Burma	*Culex quinquefasciatus (fatigans)*	II	1242
El Salvador	*Anopheles albimanus*	II	1246
		III	1691
		III	1881
India	*Aedes aegypti*	II, III	1321 (I–XV)
		Untyped	1434
		III	1436
India	*Culex quinquefasciatus*	Untyped	1430
		Untyped	1432
		Untyped	1435
		III	1438
		Untyped	1446
Indonesia	*Culex quinquefasciatus*	II	1482-2
Indonesia	*Culex quinquefasciatus*	III	1593-4
Korea	*Culex quinquefasciatus*	Untyped	1431-1
Nigeria	*Anopheles gambiae*	III	1595, 1537
The Philippines	*Culex quinquefasciatus*	II, III	1404-9
The Philippines	*Anopheles litoralis*	II	1545
Thailand	*Culex quinquefasciatus*	II	1427
		II	1428
Zambia	Cattle Hair,	Untyped	1407
	scrapings and *Hyalomma* ticks	Untyped	1408

TABLE II.
Insecticidal activity of Bacillus sphaericus *strains against larvae of*
Culex quinquefasciatus *and* Aedes aegypti *(Singer, 1977)*

Identification	Accession no.	Urease	Potency
Bacillus sphaericus *B. sphaericus*	ATCC 14577, 1537	—	Absent
var *fusiformis* *B. sphaericus*	ATCC 7054	+	Absent
var *fusiformis*	K(Kellen *et al.*, 1965), Q(sxK) SSII-1, 1404–9, 1593–4, 1693, 1881		Present

strains. Hopefully this handicap may be overcome by serology with flagella antigens which show differences between different strains (H. de Barjac, unpublished observations) or with DNA homology and specific bacteriophage (A. Youston and J. Hedrick, unpublished observations). For example, one phage that infects strain 1593 also infects 1881 and 1691, indicating some added similarity. However that phage also attacks several non-pathogens (Youston, personal communication). Identification is important not only to relate new to existing isolates but also for field testing and regulatory requirements.

The high insecticidal activity that may be envisaged from strain selection and manipulation may be different from that operating in the wild or for that matter present on initial isolation. It is not unusual to isolate from one dead larva several group II strains in addition to *B. sphaericus*. Thus one can visualize that wild populations of mosquito larvae that are under "natural" bacterial control (in epizootic or endozootic situations) are probably suffering an interaction of several entomogenous bacterial species.

Initial isolation of bacilli from accessions has been best with Brain Heart Infusion media. Initial insecticidal activity is determined by bioassay with our three laboratory-reared mosquito species, *Culex (pipiens) quinquefasciatus*, *Aedes aegypti*, and *Anopheles albimanus*. Collaborating workers have also tested our bacteria against these and related laboratory reared or field-collected vectors.

Our standard mosquito bioassay involves decimal dilutions of bacterial cultures (50 ml/assay cup or beaker) in commercially available spring water or in a synthetic spring water of our own design. The latter contained $EDTA \cdot Na_2$, 52.8; $CaCl_2$, 70.3; NaCl, 175.7; $NaHCO_3$, 105.6 μg/litre of either deionized or distilled water. Ten to 30 mosquito larvae are used per cup at 25°C. Any of several larval food materials, e.g. brewer's yeast, dried milk powder, ground dog or hog chow, liver powder, tropical fish food, is added daily. Three to five such cups without bacteria serve as a control. Death counts and other pertinent observations are made daily or at a minimum of days 2 and 7. LC_{50} and LC_{90} values are calculated by linear regression of mortality percentages, corrected using Abbott's formula (Finney, 1962), on dilution factors using a programmed Hewlett-Packard 25C hand calculator. Results are expressed as log dosage in terms of the dilution of the final whole culture (FWC), not in terms of numbers of viable bacterial cells since dead cells can be insecticidal (Singer, 1974; Davidson *et al.*, 1975; Myers and Youston, 1978). Total viable counts are made to insure consistency of the FWCs during an experimental series and to verify consistent populations under our standard fermentation conditions.

In our studies with the initial strain (SSII-1), media appeared to play a key role in potency. The best medium was Sphaericus Synthetic Medium (SSM):- $MnSO_4 \cdot H_2O$, 1 mg; $MgSO_4 \cdot 7H_2O$, 10 mg; $CaCl_2$, 10 mg; EDTA, disodium salt, 10 mg; biotin, 100 mg; thiamin HCl, 1 mg; nicotinic acid, 1 mg; DL calcium pantothenate 1 mg; DL lysine HCl, 100 mg; DL isoleucine, 100 mg; DL valine, 100 mg; DL methionine, 100 mg; L-glutamic acid, 600 mg; glycerol, 1 g; K_2HPO_4, 100 mg; TRIS, 1.21 g/100 ml; pH adjusted to 7.5 with HCl.

The best production procedure for all strains starts with a 2-day old SSM slant used to inoculate roller tubes containing 5 ml of SSM broth incubated for 16 h at 30°C, 26 RPM. These in turn are inoculated into seed flasks (25 ml SSM broth in 125 ml Erlenmeyer flasks), incubated at 30°C for 6–9 h, depending on growth, in a rotary shaker bath at 250 RPM, with 1–5% (V/V) inoculum level. The production flasks are identical in composition, volume of medium and conditions but are grown for up to 24 h to produce FWC.

TABLE III

Effects of media on potency in Culex quinquefasciatus *total viable and spore count of* Bacillus sphaericus, *SSII-1*

Media[a]	LC$_{50}$		Total viable count	Viable spore count
	Dilution	No. cells/ml		
SSM	3.2×10^{-7}	7.5×10^{1}	2.4×10^{9}	1.7×10^{5}
BHI	2.8×10^{-6}	5.7×10^{2}	1.6×10^{9}	5.4×10^{3}
RAW	4.2×10^{-4}	7.6×10^{3}	3.2×10^{8}	1.3×10^{8}
SSM + 1% milk	3.2×10^{-5}	5.6×10^{2}	1.8×10^{8}	1.6×10^{7}

[a]SSM = Sphaericus Synthetic Medium. BHI = Brain Heart Infusion Broth (Difco) + vitamins. RAW = 1% each fish meal, soy bean oil meal; vitamin mix; 0.01% $CaCl_2$. SSM + 1% milk is SSM with amino acids replaced by 1% dried milk.

TABLE IV

Potency of several strains of Bacillus sphaericus *against larvae of three laboratory-reared mosquito species (instar II).*

Strain	Expt.	LC$_{50}$ (log dilution of final whole culture)		
		Anopheles albimanus	*Aedes aegypti*	*Culex quinquefasciatus*
SSII-1	1	-6.75	-4.63	-9.19
	2	-3.51	-3.00	-4.15
	3	-7.02	-5.03	$-$
	4	-6.20	-5.76	-5.47
1404-9	1	-8.45	-5.49	-5.98
	2	-5.03	-4.78	-7.82
	3	-9.76	-5.44	$-$
	4	-7.66	-7.51	-4.90
1593-4	1	-9.00	-5.36	-5.72
	2	-6.05	-4.93	-8.00
	3	-9.00	-8.00	$-$
	4	-9.00	-6.51	-8.96

Growth and potency of strain SSII-1 are best in SSM medium (Table III). With the other strains, particularly 1593, RAW (Table III) is equally good, which augurs well for future industrial development of these new strains since RAW is an inexpensive industrial medium. Table IV illustrates some of our better results with three of the more active strains, the variation between experiments being typical (see also Singer, 1977).

Since their initial isolation in 1965 (Kellen *et al.*, 1965), the potency of cultures has declined during storage. FWC of SSII-1 in different media, treated or not with chloroform or some related solvent, and of different ages and spore levels lost 90% of potency every 2 to 3 months in refrigerated storage (Styrlund, 1974). Addition of urea, sodium acetate or mucin did not reduce this loss. Usually only spores survived after several weeks storage. Lyophilization met with mixed success. Fortunately strains 1593, 1691 and 1881 are more stable and amenable to lyophilization and other drying techniques, but less is known about strain 1404 (see also Section II, C, *4*). Pasteurization (60°C for 30 min) of untreated and chloroform-treated FWCs of all strains destroys potency, but germination of this pasteurized material in fresh media results in the usual high potency, which is thus heat labile.

B. NATURE OF *BACILLUS SPHAERICUS* INSECTICIDAL ACTIVITY

Studies on the nature of the insecticidal activity involve both biogenesis of the toxin and pathogenesis in the target insect. "Mode of action" studies on a biochemical level are perhaps premature at present. Work to date has mainly dealt wih strain SSII-1 and its effects on *Cx. quinquefasciatus* larvae.

1. Biogenesis of the Toxin(s)

Potency of *B. sphaericus* during growth of the culture peaked midway through the sporulation cycle and eventually decayed to $1/3-1/10$ of its maximum value (Singer, 1973, 1974). Bioassays of supernatant and cells of the FWCs showed that all insecticidal activity was associated with the cells, not with the supernatant.

Several solvents (chloroform, toluene, acetone and formaldehyde) were used to treat FWCs. One part chloroform to 10 parts FWC lowered the total viable count 10^3 to 10^6 fold down to about the spore count level, while the potency dropped 10-fold (Table V) (Styrlund, 1974). Toluene gave similar effects,

TABLE V

Effect of chloroform on potency in Culex quinquefasciatus *and cell counts of 16-h final whole cultures of* Bacillus sphaericus *in brain heart infusion broth*

Strain	LC$_{50}$		Total viable count	Viable spore count
	Dilution	No. cells/ml		
SSII-1	1.5×10^{-6}	6.8×10^3	2.2×10^9	1.0×10^3
7054	no activity	no activity	4.0×10^7	1.7×10^4
14577	no activity	no activity	5.5×10^6	1.2×10^3
Chloroform Treatment				
SSII-1	3.2×10^{-5}	< 1	1.5×10^3	7.8×10^2
7054	no activity	no activity	2.6×10^4	6.2×10^3
14577	no activity	no activity	1.4×10^3	1.2×10^3

acetone had very little effect on either the total viable count or the potency. Formaldehyde destroyed all viable cells and spores, as well as the potency. Therefore viable cell numbers may be irrelevant to the expression of *B. sphaericus* potency, since dead cells can still kill insects. The best expression of activity would be a chemical measurement of the toxin. Until this is available, expression in units of a standard material, either a dry stable powder (until recently unavailable) or a liquid prepared under standard conditions will have to suffice. If it really is stable, the powder would be better.

Myers and Youston (1978) have recently verified the effects of chloroform and the instability of the toxin, and have found similar effects with ultraviolet light. In contrast to our earlier findings, they present three lines of evidence indicating that toxic activity may not be related to sporulation: (1) cells grown in complex or defined media are equally toxic at all ages of the culture; (2) when supplemental Mn^{2+} was excluded from a complex medium, the culture yielded few spores but was of equal toxicity to a culture containing many spores; (3) several early-blocked oligosporogenous mutants had toxic activities comparable to the parent strain. Thus vegetative cells produce toxins. Our conclusion (Singer, 1973, 1974), that the vegetative cells were relatively inactive, with most of the activity being seen during the period of growth termed secondary metabolism (Weinberg, 1977), may have been caused by instability of the toxin under our conditions. Preliminary work by Youston, indicates that their effects may not be true of all toxic strains; strain 1593 increases slightly in activity during the early sporulation phases of growth. The above data indicate that the biogenesis of the *B. sphaericus* toxin(s) may be complex.

2. Pathogenesis

Culex pipiens larvae fed on strain SSII-1 first appear unwell as early as 12 h, with death usually occurring within 2 days (Singer, 1973, 1974; Davidson *et al.*, 1975). At lower bacterial concentrations, death may take as long as 5–7 days. To determine the earliest toxic symptoms, we examined several markers to trace passage of food down the midgut of *Cx. pipiens* and *Aedes aegypti* larvae. In feeding experiments with modified techniques of Dadd (1968, 1975) and of Wilton *et al.* (1972), Sudan black-dyed whole wheat flour revealed the food better than charcoal, crystal violet-stained nonpathogenic *B. sphaericus* strain 7054 or pigmented *Serratia marcescens*. Feeding of *C. pipiens* stops between 2–5 h, while partial inhibition may occur as early as 10 min (Table VI). Preliminary experients using *Ae. aegypti* larvae show less feeding inhibition.

The broad outlines of the pathogenesis of *B. sphaericus* strain SSII-1 have been revealed by histological studies using light and electron microscopy

TABLE VI

Uptake of Sudan black-dyed whole wheat by Culex pipiens *instar III larvae exposed to* Bacillus sphaericus SSII-1

Time after treatment	Mean no. of black segments (19 larvae)	
	No bacteria	Bacteria
0 min	5.9	6.0
10 min	5.9	3.2
1 h	5.8	2.6
2 h	5.8	2.9
3 h	5.8	0.2
5 h	6.0	0

(Davidson *et al.*, 1975; Davidson, 1977; Kellen *et al.*, 1965). The pathogen as well as the normal gut flora were contained within the peritrophic membrane. Bacteria invaded host tissue only long after death and in many larvae after autolysis of some body tissues. Posterior midgut cells swelled and deteriorated. This confinement of the bacteria entirely within the peritrophic membrane, even in some dead larvae, and the lack of insecticidal activity of supernatant from FWC led to the conclusion that one or more cell-associated toxins are involved. *B. sphaericus* cells are digested or modified within the peritrophic membrane of the gut, and the toxin released to pass through the peritrophic membrane and kill the larva. The ability of chloroform-killed SSII-1 bacteria to kill larvae and the drop in bacterial numbers in larvae after feeding (Davidson *et al.*, 1975) support these conclusions. Significantly, in all the above studies, as well as in studies of non-pathogenic strains (Holt and Leadbetter, 1969), no parasporal body equivalent to the *B. thuringiensis* delta endotoxin has been seen. The fate of populations of *B. sphaericus* insecticidal strains and the possible role of the normal gut flora will be discussed below in ecological terms.

C. DEVELOPMENT OF VECTOR CONTROL

The laboratory scale explorations so far discussed do not solve the essential practical problem namely, can these bacteria be effective insecticides in the field? Can we curb human tropical diseases by interrupting the vector life-cycle using these strains?

Two WHO activities bear on this: (1) a five-stage evaluation of candidate organisms (WHO, 1975) (Table VII); (2) a new Special Program for Research and Training in Tropical Diseases (Mahler, 1976) which stresses six infections, malaria, schistosomiasis, filariasis, trypanosomiasis, leishmaniasis and leprosy. In addition to these, one of several "trans-disease" areas emphasizes biological control of vectors by microbial pathogens and parasites, of which *B. sphaericus* has a high priority.

1. Efficacy

In general, based on laboratory results and preliminary field trials (including use of field-derived vectors), cultures of active strains can be diluted $10^6 - 10^9$ times for an LC_{50} value, i.e. $10-1000$ living or dead cells/ml. Some of the new experimental dry powders are active down to hundredths of a part per million — theoretically equivalent to fractions of a pound of FWC per acre-foot of water.

Information is now available for many mosquito species. Against *Anopheles gambiae*, strain SSII-1 had some effect although results from metre-square experimental plots were inconclusive (J. Briggs and D. Bown, 1975, unpublished observations). Against field-collected larvae LC_{50}s for strains SSII-1, 1404, and 1593 were $10^{-4} - 10^{-6}$ dilutions of FWCs (Singer, 1977). Against laboratory-reared *An. albimanus*, LC_{50} dilutions were ca. 10^{-4}, although there were no significant differences in susceptibility between the four instar stages (Singer, 1977).

Against *Aedes aegypti* and *Ae. albopictus* LC_{50} dilutions of SSII-1 were

TABLE VII

Preliminary scheme for screening and evaluating the efficacy and safety of biological agents for control of disease vectors (WHO, 1975)

STAGE I Laboratory	STAGE II Laboratory	STAGE III Preliminary field trials	STAGE IV Laboratory	STAGE V Large scale field trials
A.	A.		A.	
Identification, characterization[a]	Mammalian infectivity tests; safety to laboratory and field personnel[a]	Regulated pond tests under WHO supervision:[c] efficacy against disease vectors in natural conditions	More detailed tests on mammalian infectivity[b]	Under WHO auspices. Not at present defined, will vary accordingly to target vector habitat(s), mode of application, etc.
B.	B.		B.	
Assessment against target vectors	Preliminary assessment against non-target species		Laboratory and field trials	
C.			C.	
Preliminary evaluation of rearing in quantity			Detailed studies on non-target range, especially other fauna in habitat for Stage V trials	

REVIEW OF STAGES I AND II (between Stage II and Stage III)

REVIEW OF I to IV BY INFORMAL CONSULTATION GROUP (between Stage IV and Stage V)

[a]This study may vary from routine taxonomic determination (fish, nematodes, predatory insects) to the detailed serotyping necessary for microorganisms, especially viruses. [b]Not required on predator–prey situations, but more detailed tests on effects on non-target organisms could be substituted in trials of larvivorous fish, predatory insects, etc. [c]Especially if the biological control agent is not indigenous.

ca. 10^{-4}, with less activity against *Ae. polynesiensis*, obtained by Mouchet's group (Bondi, France) (Singer, 1977). At our laboratory samples of the same material gave LC_{50} dilutions of 10^{-6} against *Ae. aegypti*, the difference probably explainable by deterioration of the material in shipping. Against lab-reared *Ae. triseriatus* the LC_{50} of strain SSII-1 was ca. 10^{-4} (Malloy and Wraight, unpublished observations). Against field-collected and field tested *Ae. triseriatus*, activity was less. Older instars seemed to be more susceptible than younger. With field collected snow pool *Aedes* (primarily *Ae. canadensis*) LC_{50} dilutions ranged up to 10^{-5} of strain SSII-1 with greater susceptibility among the older instars. Against field collected and field tested *Culex pipiens* results were better than against *Aedes*, with mean LC_{50} dilutions of 10^{-5} to 10^{-6}.

In bioassays with strains 1593, 1404 and SSII-II against field-collected larvae, the LC_{50} dilutions were 2×10^{-4} for strain SSII-II against *Psorophora columbiae* and 2×10^{-7} for strain 1593 against *Culex nigripalpus*, both species being susceptible to the other strains also (Ramoska *et al.*, 1977, 1978). *Aedes taeniorhynchus* was less susceptible to all three strains but most susceptible to 1593 (1.5×10^{-2} dilution). The three strains applied to field populations of *Cx. nigripalpus* and strain 1593 to *Psorophora columbiae*, at ca. 10^{-4} dilution, reduced larvae by 89–100%. No live larvae were found 48 h after treatment with strain 1593 in one plot and after 30 h in another. Also, bite count data suggested a reduction in adult emergence. This was the first successful use of standard commercial spraying equipment in the application of *B. sphaericus*.

2. Safety

At this time several *B. sphaericus* strains are being tested for safety and appear to have no effect on test mammals via the usual routes of administration (J.A. Shadduck, personal communication). Further tests are in progress (Stage IVA, Table VII). In addition to the classical safety tests, the new concept of "high hazard testing" is being applied, which maximizes the opportunity of entomopathogens to produce mammalian lesions. This includes the administration of many viable organisms into organs and tissues not ordinarily exposed to the environment. As with all bacteria an effect can be obtained if enough material is injected, so the results of the high hazard testing are, of course viewed in the proper perspective.

3. Environmental and Ecological Considerations

This area is the least explored, with the least firm technical information, but the most publicized. It encompasses both the microecology of the larval gut and the ecology of the pathogen in water in relation to the water's flora and fauna.

The fate of a single dose of untreated and chloroform treated *B. sphaericus* strain SSII-1 and bacterial flora normally found in the larval gut of *Culex quinquefasciatus* was studied by monitoring larval mortality and population counts of bacteria (Davidson *et al.*, 1975). The non-pathogenic strain 7054 was observed as a control. At intervals following ingestion of bacteria, surface-sterilized larvae were ground and the resulting homogenate plated. The numbers of non-pathogenic 7054 fell to zero within 24 h after feeding: normal flora remained constant. Exposure to insecticidal SSII-1 cells saw the same drop in numbers to two cells/larva at 6 h. At the onset of mortality the numbers of SSII-1 began to rise ultimately to more than 10^5 cells/larva (72 h), accompanied by a rise in normal flora. When chloroform-treated SSII-1 culture was fed, no live *B. sphaericus* cells were isolated, but 98% of the larvae died, i.e. the cells must be rapidly eliminated from the larval gut either by digestion or defaecation. If a large dose was eaten, digestion and movement of gut contents stopped before all of the SSII-1 cells were eliminated leaving the normal flora to multiply in the dead larvae. L.J. Goldberg (personal communication) suggested that SSII-1 must build up to about 10^5/larva before larvae die, but Davidson's data does not support this. In addition, Myers and Youston (1978) found that $100\,\mu g/ml$ of bacitracin and $1\,\mu g/ml$ of rifampin inhibited bacterial replication but did not affect normal larval growth and development. The antibiotics did not alter mortality due to strain SSII-1 in bioassays, indicating that bacterial replication did not occur in the assay system prior to death of larvae. Periodic total viable counts of bioassay cups revealed no increase in bacteria after addition of the FWCs to the cups. There was over a 10-fold increase in deaths in one-litre bioassay water volumes compared to 50 ml volumes (Styrlund, 1974). The *Culex* larvae are apparently very efficient vacuum cleaners, at least in this laboratory situation. Thus larvae die if they eat enough bacteria and death is not dependent on bacterial multiplication in the gut. The role of normal gut flora of mosquito larvae needs to be explored more fully. There is little difference between bacterial numbers in live and moribund larvae (Davidson *et al.*, 1975; Davidson, 1977), and both *B. sphaericus* and other bacteria increase in dead larvae; therefore increases in background bacterial number are not important in the initial stages of intoxication. However, both Davidson (1977) and Goldberg (personal communication) suggest a role for the species composition of the normal gut flora. Goldberg (*in* Davidson, 1977) reported that the lethal concentration of SSII-1 varied as much as 10^4-fold depending on the microbial flora present or added to the bioassay. In preliminary studies, Davidson found that monoxenic larvae may vary markedly in response to SSII-1 depending on other microorganisms present. Unfortunately in neither instance is the identity of the "normal flora" known and, unless it is, one cannot reach legitimate conclusions. Davison does suggest however that, since microbes are a major part of the diet of newly-hatched mosquito larvae, floral differences could influence larval feeding and hence susceptibility to *B. sphaericus*. Also the more normal bacteria there are present, the less is the chance for ingestion of a lethal dose of *B. sphaericus*. One could also speculate that specific gut bacteria may affect the toxin generated within the peritrophic membrane.

Multiplication of *B. sphaericus* in larvae after death suggests that cadavers could spread the insecticidal bacteria in the environment, so we placed dead, SSII-1-infected larvae into our bioassay system. Fresh *Culex* larvae were infected with this material for up to two cycles. The importance of this event in the wild remains to be explored further.

Another facet of the ecology of *B. sphaericus* strains is their fate in water and there are very few studies on populations in aquatic systems. One difficulty is our inability to distinguish the added *B. sphaericus* culture from the indigenous one. Marked strains, as yet unavailable, would facilitate such studies and their development is included in the WHO's new Special Programme. However, preliminary studies with a selective SSM medium, formed by the addition of 2% urea + 3% NaCl, indicate that, at least with strain SSII-1, the bacterial count falls rapidly over 10 days in an aquatic system mimicking native systems. Thus cultures without protective additives do not survive well. One of the major aspects of the two WHO schemes concerns the effect of new microbial candidates on the environment. To date there appears to be no untoward effects of *B. sphaericus* in many non-target organisms. In a series of tests (J. Briggs, unpublished observations) strains SSII-1, 1404 and 1593 did not harm crayfish (*Orconectes rusticus*), larvivorus fish *Aphyosemion gardneri* and *Epiplatys bifasciatus*, Nepidae (Hemiptera), Libellulidae (Odonata) and Hydrophilidae (Coleoptera). We found no effects on the fruit fly, black biting fly *Simulium vittatum*, Indian meal moth *Plodia interpunctella* and cigarette beetle *Lasioderma serricorne*. Dosages of *B. sphaericus* lethal to mosquitoes were relatively safe for both invertebrates and vertebrates which normally occur in mosquito habitats, namely tadpoles, a minnow (*Gambusia affinis*), crayfish, damselfly naiads and corixids (H.C. Chapman, personal communication). It can be concluded that *B. sphaericus* should cause no disruption to fauna or flora of that environment.

In an insect of economic importance, the honey bee *Apis mellifera*, strain SSII-1 had no untoward effect over 3 months on the longevity of newly emerged bees, on brood production or on honey production (Davidson *et al.*, 1977).

4. Industrial Development

B. sphaericus can be produced by submerged fermentation which has a well developed technology, although much of this is confidential and not available for discussion. Small amounts of a stable, dried, commercial-style preparation have been made. Kilogram amounts can almost certainly be made. One batch was very potent (see above) and since acetone does not appreciably reduce potency the techniques of Dulmage *et al.* (1970) may be a good beginning. With careful temperature control, spray-drying and vacuum drying are suitable and we have been particularly successful with strain 1593. We do not anticipate undue difficulty with parameters of the fermentation cycle. *B. sphaericus* responds well, nutritionally and otherwise, to a whole series of inexpensive materials and methods with no expensive or exotic requirements. Even aeration is less demanding than in some other aerobic processes.

Formulation and delivery of *B. sphaericus* pose problems different to those of *Bacillus thuringiensis* for use in agriculture. Both must be ingested by larvae and in the gut a "protoxin" is transformed into a toxin and death ensues. The important difference lies in the eco-systems in which they operate: aquatic or terrestrial. The final whole culture of *B. sphaericus* can be added to a laboratory bioassay to give effective control under laboratory conditions. In the field we face a different situation. The organism and its toxin must be protected from inactivation and excessive dispersion in a water environment. *Anopheles albimanus* larvae feed on the surface. *B. sphaericus*/1593 needs to float on the surface of that eco-niche. A tree-hole *Aedes*, being photophobic, retreats to the depths of the hole. *B. sphaericus* must follow. It must not drop too deeply, however, or it will be lost in the crevices of the tree hole. *B. sphaericus* as a "food source" for *Culex* perhaps needs to be on a bait that stays in the right position to compete with the sewer "goodies" available to the larvae. And finally the bacterium needs to survive its trip to the flooded foot prints of Nigeria. These are essentially bioengineering problems which have already been faced with chemical insecticides, and which are now receiving attention for *B. sphaericus* with WHO support.

Standardization of material is a requirement that accompanies commercialization. As discussed in Section II,B,1, expression in units of a stable dry powder would be best. This would not only standardize commercial material but would also allow research groups and industrialists to measure and compare their efforts. Again, in the near future this will receive WHO support. Sufficient kilos of commercial dry powder will be set aside as a standard and its stability tested. A unit of specific biological activity can be established using a standardized *Culex* (and/or *Anopheles*) bioassay system.

III. Morphological Group II: *B. alvei—B. brevis—B. circulans* Complex

Over a hundred isolates belonging to *Bacillus* morphological group II formed a complex, matching or intermediate to, *B. alvei*, *B. brevis* and *B. circulans* (Singer, 1973, 1974). This complex, which should include *B. pulvifaciens* and *B. laterosporus,* is poorly studied, poorly understood and difficult to identify and classify (Gordon, 1977). Of the cultures studied, many were active against *Culex* larvae, at dilution of FWC of ca. $10^{-2}-10^{-4}$, and involved toxins, some heat labile, some heat stable. Many were unstable and may have lost their initial potency. Some attack *Simulium* larvae, others the cigarette beetle *Lasioderma serricorne*. The latter were found as contaminants among cultures of *B. cereus*, isolated from dead cigarette beetle larvae and sent by W. Kellen.

A. *SIMULIUM* STUDIES

Of the group II isolates active against *Simulium* larvae, some species of which carry Onchocerciasis, *B. alvei* had an LC_{50} dilution of ca. 10^{-4} to 10^{-5} (Table VIII: Jaska, 1977). Several of our WHO mosquito accessions were active, even at 5°C, in a new bioassay with field-collected *Simulium vittatum* larvae. The supernatant of FWC was not larvicidal. However, unlike the *B. sphaericus* mosquito larvicidal activity, which is found solely in the cells, the activity of *B. alvei* cells is inexplicably 8–15-fold less than that of *B. alvei* FWC. Also whereas pasteurization of *B. sphaericus* FWC (60°C for 30 min) destroys potency in mosquitoes (a 6–8 log drop), the same treatment lowers potency of *B. alvei* FWC to *Simulium* larvae only 8–15 fold. The nature of the postulated *B. alvei* toxin is unknown. Preliminary histopathology revealed no gross tissue damage prior to death of the *Simulium* larva, which is similar to the effect of

TABLE VIII

Potency of Bacillus alvei/II_3D_T-li2 *grown in plate count media plus vitamins in two vectors*

Test organism	Total viable count	Viable spore count	Dilution for	
			LC$_{50}$	LC$_{90}$
Simulium vittatum	3.0×10^7	3.0×10^5	$10^{-4.9}$	$10^{-3.0}$
Culex quinquefasciatus	3.0×10^7	3.0×10^5	$10^{-2.3}$	$10^{-0.41}$

B. sphaericus in mosquito larvae. Comparison of the two systems is most intriguing, insecticidal *B. sphaericus* strains having no effect against *S. vittatum* larvae, while *B. alvei* attacks some mosquito larvae. There has been no mammalian toxicity testing with *B. alvei,* nor testing against non-target arthropods, except that no ill effects were noted when honey bee colonies were challenged with *B. alvei* for 3 months.

B. CIGARETTE BEETLE, *LASIODERMA SERRICORNE*, STUDIES

Identified as an intermediate strain designated as *B. laterosporus/alvei (B. l/a)* by Bradshaw (unpublished observations), cultures were active only against neonate larvae. During the fermentation cycle, potency of the *B. l/a* cultures did not begin until about 18 h and lasted up to at least 48 h. The cell itself was potent, but not the supernatant. Potency was increased by insect passage (Table IX). Similar procedures have no effect with *B. sphaericus* or *B. alvei*, which suggests invasiveness as a factor in the insecticidal effects of the *B. l/a* cultures. A related morphological group II bacillus, *B. pulvifaciens*, also killed larvae of the cigarette beetle by a non-dialysable and heat labile factor extractable from the whole cells (Jackson and Long, 1965).

TABLE IX

Potency of Bacillus laterosporus/alvei *against neonate larvae of* Lasioderma serricorne *before and after passage in insects.*

Strain of *B. laterosporus/alvei*	LC$_{50}$ dilution of FWC	
	Before	After insect passage
01	1.9×10^{-2}	5.0×10^{-2}
02	1.8×10^{-2}	5.0×10^{-2}

IV. Potential and Research Requirements

Information about efficacy, safety, effects on non-target organisms, fermentation, formulation, field application, field testing — all the important aspects needed to develop the *Bacillus sphaericus* strains as effective insecticides — will expand in the immediate future. We anticipate standardization of *B. sphaericus* material with an international standard, particularly strain 1593, and also others, since unexpected requirements are often imposed during mass propagation and/or field use. Biogenesis, pathogenesis and eventually mode of action studies of the insecticidal component(s) of these cultures should continue, to improve our understanding of how this *Bacillus* works.

An area needing study and most often neglected is the ecology of using these strains in the wild. Our techniques and approaches are still too primitive to enable us to understand the relationships between the introduced populations and the competing indigenous populations. Unless we can probe the BIOLOGY of these relationships, *B. sphaericus* will be just another insecticide. Microbial control agents are not chemicals. They function in response to limnological (or terrestrial) conditions, population density of the target insect, and competition from micro- and macro-biota.

Still, the isolation and development of the entomogenous *B. sphaericus* strains remain a successful example of both international cooperation and the cooperation of microbiologists and entomologists—the interdisciplinary approach, largely sponsored by WHO. This is a model for the future development of other sporeformers without crystals, a still largely unexploited pathogen pool in nature, for use against important vectors of human disease. For vectors other than mosquitoes a determined search for potential candidates is required. The future use of sporeformers without crystals in agriculture can be expected if a concerted effort is made, but at present no research group is involved. The potency level of *B. laterosporus/alvei, B. alvei,* etc. is where *B. sphaericus* started before 1973.

References

Buchanan, E.E. and Gibbons, N.E. (1974). "Bergey's Manual of Determinative Bacteriology". 8th Edn. Williams and Wilkins, Baltimore 1268 pp.

Burges, H.D. and Hussey, N.W. eds (1971). "Microbial Control of Insects and Mites". 861 pp. Academic Press, London and New York.

Cantwell, G.E. and Laird, M. (1966). *J. Invertebr. Path.* 8, 442–451.

Dadd, R. H. (1968). *Mosquito News* 28, 226–230.

Dadd, R.H. (1975). *J. Insect Physiol.* 21, 1847–1853.

Davidson, E.W. (1977). *In* "Biological Regulation of Vectors". (J.D. Briggs, ed.) 19–30, Dept. Hlth, Education and Welf. Publ. No. (NIH) 77-1180.

Davidson, E.W., Singer, S. and Briggs, J.D. (1975). *J. Invertebr. Path.* 25, 179–184.

Davidson, E.W., Morton, A.L., Moffett and Singer, S. (1977). *J. Invertebr. Path.* 29, 344–346.

Dulmage, H.T., Correa, J.A. and Martinez, A.J. (1970). *J. Invertebr. Path.* 15, 15–20.

Finney, D.A. (1962). "Probit Analysis". Cambridge University Press, London. 318 pp.

Gordon, R.E. (1977). *In* "Biological Regulation of Vectors". (J.D. Briggs, ed.) 67–82, Dept. Hlth Education and Welf. Publ. No. (NIH) 77-1180.

Gordon, R.E., Haynes, W.C. and Hor-Nay, C. (1973). *In* "The Genus *Bacillus*". (R.E. Gordon, W.C. Haynes and C.H.N. Pang, eds.) 99–109. Agric. Handbk. No. 427. ARS. USDA.

Holt, S.C. and Leadbetter, R.R. (1969). *Bact. Rev.* **33**, 346–378.

Jackson, R.H. and Long, M.E. (1965). *Biochim. Biophy. Acta* **100**, 418–425.

Jaska, J.M. (1977). Unpubl. Masters thesis, Western Illinois University 74 pp.

Kellen, W.R., Clark, T.B., Lindegren, J.E., Mo, B.C., Rogoff, M.H. and Singer, S. (1965). *J. Invertebr. Path.* **7**, 442–448.

Mahler, H. (1976). WHO Chronicle 30 (June), 6.

Myers, P. and Youston, A. (1978). *Infect. Immun.* **19**, 1047–1053.

Ramoska, W.A., Singer, S. and Levy. R. (1977). *J. Invertebr. Path.* **30**, 151–154.

Ramoska, W.A., Burgess, J. and Singer, S. (1978). *Mosq. News* **38**, 57–60.

Singer, S. (1973). *Nature, Lond.* **244**, 110–111.

Singer, S. (1974). *In* "Developments in Industrial Microbiology". (E.D. Murray and A.W. Bourquim, eds.), S.I.M., A.I.B.S., Washington, D.C.

Singer, S. (1977). *In* "Biological Regulation of Vectors". (J.D. Briggs, ed.). 3–18, Dept. Hlth, Education and Welf. Publ. No. (NIH) 77-1180.

Smith, N.R., Gordon, R.E. and Clark, F.E. (1952). *In* "Aerobic Sporeforming Bacteria". Agric. Monogr. No. 16. USDA.

Styrlund, C.R. (1974) Unpubl. Masters thesis, Western Illinois University, 54 pp.

Weinberg, E.D. (1977). *In* "Biological Regulation of Vectors". (J.D. Briggs, ed.) 49–65, Dept. Hlth, Education and Welf. Publ. No. (NIH) 77-1180.

Wilton, E., Fetzger, L. and Fay, R. (1972). *Mosq. News* **32**, 23–27.

World Health Organization (1975). Twenty-first Report of the WHO Expert Committee on Insecticides. Tech. Rep. Ser. No. 561, WHO, Geneva, 35 pp.

Note Added in Proof

Recently it has been shown that some strains of *B. sphaericus* which are highly insecticidal in the spore stage, e.g. 1593, 1691, 2013 and MR-4, bear parasporal crystal-like polyhedral inclusions. It is not at present known whether these inclusions are toxic (E.W. Davidson, personal communication).

CHAPTER 15

Genetics and Genetic Manipulation of
Bacillus thuringiensis

PHYLLIS A.W. MARTIN and D.H. DEAN

Departments of Genetics, Microbiology and the Bacillus Genetic Stock Center,
The Ohio State University, Columbus, Ohio, USA

I. Introduction

The genetics of *Bacillus thuringiensis* has been dormant. There is now emerging evidence that genetic systems are available which will amplify the toxicity and broaden the host range of this important insect pathogen. Recent papers have demonstrated genetic mapping by generalized transduction in several serotypes and recombinant DNA technology may allow a totally new means of genetic exchange which would obviate the need for classical genetic exchange systems.

From a geneticist's view-point, *B. thuringiensis* offers exciting possibilities. One of the products of interest, δ endotoxin, is deposited as a protoxin into a distinct crystal, making its isolation and observation very simple. By most estimations δ endotoxin is a single polypeptide chain (80 000–120 000 daltons, see Chapter 12), i.e. the product of a single gene. The crystal may also consist of additional peptides. The impact of these observations is that one may bio-chemically or even visually determine if the crystal is absent, as an initial screen prior to biological assays. Antisera may be prepared to allow subtle analysis of mutant crystals which may not function as a toxin. Finally, being a single gene will facilitate the isolation and amplification of the δ endotoxin gene.

The task before geneticists is to develop a genetic system which will allow manipulation and amplification of pathogenicity. In industrial microbiology, genetics has gone hand in hand with fermentation to amplify productivity. In basic research, genetics is required to study the regulation of toxin synthesis and to determine the cellular (plasmid or chromosomal) location of genes coding for toxins such as δ endotoxin and β exotoxin.

This chapter examines the potential for developing a genetic system for *B. thuringiensis* by reviewing the attributes of its serotypes which are the foundation of a mutant collection. The present known genetic transmission systems will also be reviewed as a prelude to future prospects for genetic engineering with *B. thuringiensis*.

II. Prerequisites for Genetics

To establish a genetic system for *B. thuringiensis* two prerequisites must exist. The first is variety, either existing strain differences or induced mutations. The second is a transmission system so that the "variety" can be transferred between bacteria. Natural transmission systems of conjugation and transduction require the existence of plasmids and bacteriophages respectively. Plasmids also may contain a wealth of genetic information including antibiotic resistance, bacteriocin production and toxin production. Variety and transmission in *B. thuringiensis* are discussed below.

A. VARIETY

Natural variation is widely observed in *B. thuringiensis*. All the 22 described flagellar or H-serotypes named in Appendix 1, Table I produce a parasporal crystal, but their pathogenicity to various insects differs widely from very high activity to virtually none: some isolates that are highly active in one host species are inactive in another (Chapter 11).

Some *B. thuringiensis* types also form α exotoxin (Chapter 11, IB 2b and 3) and β exotoxin (an adenosine triphosphate (ATP) structural analogue). Other, less well known, toxins are produced only by a few types.

Natural antibiotic resistance of the H-serotypes have received little study. There is a great range of resistance to different drugs (Table I). Most types are resistant to acriflavin, ampicillin, neomycin, spectinomycin, sulfanilamide, tetracycline and trimethoprim at levels on which wild type *Bacillus subtilis* will not grow. Because of the widespread occurrence of resistance to these antibiotics, they may not be particularly useful in genetic studies. One might suspect that this resistance is due to membrane permeability barriers. However, the few strains carrying natural resistance to chloramphenicol, kanamycin and streptomycin do not show reduced insecticidal activity compared to sensitive strains of the same serotype. Selection of resistant mutants did not reduce

TABLE I

Natural antibiotic resistance of Bacillus thuringiensis

H-serotype (upper numbers) and strain (lower numbers)

Antibiotic	µg/ml	1	1	1	2	3a	3a	3a 3b	3a 3b	3a 3b	4a 4b	4a 4c	5a 5b	5a 5b	5a 5b	5a 5c	5c	6	7	7	7	9	9	10	10	11	12	13	14
		1	2	3	1	1	2	1	2	4	1	1	1	2	3	1	1	1	1	2	3	1	2	1	2	1	1	1	1
Acriflavin	24	+	—	—	+	+	+	+	+	+	+	+	+	+	+	+	+	+	+	+	+	+	+	+	+	+	+	±	+
Ampicillin	50	+	—	—	+	+	+	+	+	+	±	+	+	+	+	+	+	+	+	+	+	+	+	+	+	+	+	+	+
Neomycin	5	+	—	+	+	±	+	+	+	+	—	±	+	+	+	+	+	—	±	+	—	+	±	+	±	±	±		+
Spectinomycin	50	+	±	+	+	+	+	+	+	+	±	+	+	+	+	+	+	—	+	+	—	+	+	+	+	—	+	+	+
Sulfanilamide	20	+	+	+	+	+	+	+	+	+	+	+	+	±	+	+	+	—	+	±	+	+	+	+	+	±	+	+	+
Tetracycline	20	±	—	+	—	±	—	±	±	—	—	—	±	±	+	+	—	—	±	—	+	+	—	±	±	±	±	±	—
Azide	100	±	—	—	—	—	—	—	—	—	—	—	±	—	—	—	—	—	+	±	—	+	—	—	—	+	—	—	—
Bryamycin	50	—	—	—	—	—	—	—	—	—	+	+	—	—	—	—	—	+	+	—	+	+	±	—	—	+	—	—	—
Chloramphenicol	10	—	—	—	—	—	—	—	—	—	—	—	—	—	—	—	—	—	—	—	—	—	—	—	—	—	—	—	—
Erythromycin	50	—	—	—	—	—	+	—	—	—	—	—	+	—	—	—	—	—	—	—	—	—	—	—	—	+	—	—	—
Kanamycin	20	—	—	—	—	—	—	—	—	—	—	—	+	—	—	+	+	—	—	—	—	—	—	—	—	—	—	—	—
Nalidixic Acid	50	—	—	—	—	—	—	—	—	—	—	—	—	—	—	—	—	—	—	—	—	—	—	—	—	—	—	—	—
Rifampicin	50	—	—	—	—	+	—	—	—	—	—	±	—	—	—	±	—	—	—	—	—	—	—	—	—	—	—	—	—
Streptomycin	50	—	—	—	±	±	±	—	—	—	—	±	—	—	±	±	—	—	—	—	±	—	—	—	±	—	±	—	—
Lincomycin	50	—	+	—	—	—	—	—	—	±	—	—	+	—	—	—	—	—	—	—	—	—	—	—	—	—	—	—	—

+, colony growth > 0.5 cm; ±, colony growth visible—0.5 cm; —, no growth; all at 3 days 30°C on L agar.

activity either (H.D. Burges, per. comm., see below). Therefore these genetic markers may be useful. For example, 3a 3b–1 (HD–1, H.T. Dulmage) and 3a 3b–2 (from Dipel, Abbott Laboratories) originated from the same parent and despite different histories and growth conditions, their antibiotic resistance remained identical, illustrating the genetic stability of these characters. Few other isolates show identical resistance patterns.

The *B. thuringiensis* crystal is produced at sporulation. Three phenotypic types of sporulation-crystal mutants are known: (1) do not sporulate and produce no crystal (Spo$^-$Cry$^-$); (2) asporogenic, but retain the crystal (Spo$^-$ Cry$^+$) and (3) sporulate but have lost the ability to form crystals (Spo$^+$Cry$^-$). Historically Spo$^+$Cry$^-$ mutants were the first to be discovered in H-types 1 and 4a 4c (sotto) making them indistinguishable from *B. cereus* (Fitz-James and Young, 1959). Five Spo$^-$ Cry$^+$ mutants from H-types 3a 3b and 7 were obtained using nitrosoguanidine and selection for asporogenous mutants, which remained highly active against the silkworm (Nishiitsutsiji-Uwo *et al.*, 1975). They were least as active as the parental strains. Crystals in these mutants form within 48 h in culture and most of the cells autolyse. The few remaining vegetative cells could be killed by heat treatment at 55°C for 30 min. without harming spores or crystals (Burges and Bailey, 1968). This type of mutant would be best to use in the preparation of antibodies against the crystal although this involves the assumption that the crystal has undergone no subtle changes. There are several recent reports (Yousten, 1978, Azizbekyan *et al.*, 1978) of Spo$^-$ Cry$^-$ mutants, which do however form spores at a low level (< 5%). These types of mutant indicate a close relationship between crystal formation and sporulation but a cause-effect relationship is not yet known.

Few induced mutations, such as auxotrophs and antibiotic resistance, are available (Table II). These will nonetheless be very useful as selectable markers in developing a genetic system.

TABLE II
Auxotrophic and antibiotic mutants available in Bacillus thuringiensis

H-serotype	Mutant (−, requirement; R resistant)	Reference
1	thyamine$^-$, uracil$^-$	de Barjac, 1970
2	tryptophan$^-$, niacin$^-$	Thorne, 1978
3a	tryptophan$^-$, methionine$^-$ histidine$^-$ arginine$^-$, cysteine$^-$	Thorne, 1978
3a anduze	penicillin^R, streptomycin^R	Delafield *et al.*, 1968
3a 3b	streptomycin^R, sodium azide^R	Fettig, 1976
12	tryptophan$^-$, niacin$^-$, methionine$^-$ leucine$^-$	Thorne, 1978

$^-$requires nutrient; ^{R}resistant to antibiotic

The generation of a variety of auxotrophic markers has been hampered by the lack of a standard minimal medium on which all, or most, serotypes will grow, sporulate and germinate normally. Representative isolates of 12 serotypes have an auxotrophic requirement satisfied by glutamate, citrate or aspartate

(Nickerson and Bulla, 1974). These strains grew, sporulated and formed crystals on this supplemented glucose-salts medium, but a richer medium containing vitamin-free casamino acids was needed for prompt germination. A glucose salts medium supplemented with citrate, glutamate, glycine and thiamin hydrochloride was used for transduction in H-type 3a (Thorne, 1978). The mutants were isolated using UV irradiation and the different resistance of spores and cells to boiling water.

Other workers obtained antibiotic resistant mutants by plating large numbers of bacteria on an antibiotic medium and selecting the few survivors. Such a mutant of type 3a 3b, doubly resistant to sodium azide and streptomycin, was used to test for field persistence (Fettig, 1976). A mutant of type 3a-anduze, doubly resistant to penicillin and streptomycin, sporulated and formed crystals normally (Delafield et al., 1968). Selection on antibiotic-containing agar increased resistance of isolates by 1000-fold to streptomycin, kanamycin, oleandomycin, and rifampicin, with normal spores and crystals and no decrease in insecticidal activity (H.D. Burges, pers. comm.).

Ethylmethane sulfonate mutagenesis produced rifampicin resistant mutants of H-type 1 (Lecadet et al., 1974). Some of these mutants were also asporogenic, and in electron micrographs it appeared that, blocked early in sporulation, they did not form crystals. Availability of such resistant Spo⁻ Cry⁻ mutants should facilitate the development of a comprehensive genetic system.

B. PLASMIDS AND EXTRACHROMOSOMAL DNA

Plasmids are found in most serotypes of *B. thuringiensis* (Fig. 1) but until recently have not been extensively studied. All these extrachromosomal elements remain cryptic. Probably some may be found to carry antibiotic or heavy metal resistance determinants or produce bacteriocins. Possibly the large plasmids may be temperate or pseudotemperate forms of phage similar to phage P1 of *Escherichia coli* (Ikeda and Tomizawa, 1968).

There is growing evidence that plasmids are related to sporulation and/or crystal formation. Three methods of plasmid curing, heat (Yousten, 1978; Stahly et al., 1978a), ethidium bromide (Azizbekyan et al., 1978) and acriflavin (Martin and Dean, unpubl.) have been used to isolate Spo⁻ Cry⁻ mutants. There is loss of all plasmid DNA in these mutants (Azizbekyan et al., 1978; Stahly et al., 1978b). Plasmid loss and loss of crystal production are associated in that both events are frequent and that once lost the Cry⁻ phenotype does not revert (Stahly et al., 1978a). Both lines of evidence support the plasmid borne Cry gene hypothesis and are commonly used to identify products coded for by plasmid borne genes (Lacey, 1975). While these results are suggestive, two points should be considered: (1) the isolates used by Azizbekyan and by Stahly et al. carry several plasmids and all the plasmids were removed by the respective treatments; (2) ethidium bromide causes a specific class of Spo⁻ mutant blocked early in sporulation due to its binding to sporulation operators in *B. subtilis* (M.S. Rogolsky, pers. comm.).

The potential association of crystal δ endotoxin with a plasmid is a very exciting prospect but two pieces of evidence are currently missing: without them

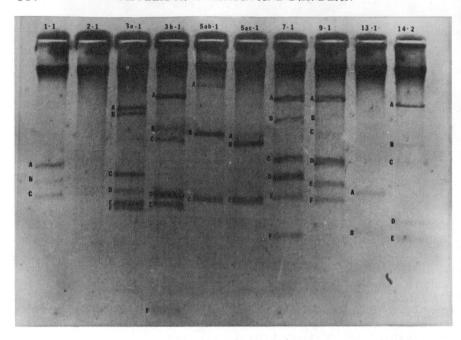

Fig. 1 Plasmid patterns of 10 serotypes of *Bacillus thuringiensis* in ethidium bromide-stained 0.7% agarose gel. Most show complex patterns. Top band in each lane is chromosomal DNA; lower bands are plasmids. At present it is unknown which bands are covalently closed circular forms and which represent open circular forms, but both are present in the patterns. In the lanes are representatives of the following H-serotypes; lane 1—serotype, 1, 2—2, 3—3a, 4—3a 3b, 5—5a 5b, 6—5a 5c, 7—7, 8—9, 9—13, 10—14. Extrachromosomal DNA has not yet been found only in serotype 2.

the toxin-plasmid hypothesis remains circumstantial. First, strains cured of only individual plasmids of several they harbour are needed in order to selectively study the altered phenotype and lessen the possibility of contaminants and other artifacts. Besides the treatments previously mentioned, rifampicin (Johnston and Richmond, 1970), SDS (Salisbury *et al.*, 1972), trimethoprim (Pinney and Smith, 1973), penicillin (Lacey *et al.*, 1973), mitomycin C (Chakrabarty, 1972) and glycine (Tomoeda *et al.*, 1976) also have been used to cure some types of plasmids. Second, a genetic mechanism of returning purified plasmids to the cured cell is needed to conclusively demonstrate the association of a phenotype, be it δ endotoxin or antibiotic resistance, with a particular plasmid. Without such a system, however, a character may be convincingly linked with an extrachromosomal plasmid in two ways. (1) A character on a single copy plasmid can be lost at a rate of 1 in $10^3 - 10^5$, an irreversible change as the DNA is lost. (2) If two or more characters are lost simultaneously this is also evidence as to their extrachromosomal origin (Lacey, 1975).

C. BACTERIOPHAGES

The role of phages in *B. thuringiensis* genetics is important. They provide a mechanism for specialized and generalized transduction which might provide a functional genetic transmission system. The discovery of lysogeny in *B. thuringiensis* has also opened the possibility that one of the toxins is coded for by a prophage as in *Corynebacterium diphtheriae* (Groman, 1953) and in *Clostridium botulinum* Type C (Eklund *et al.*, 1971).

Three types of phage lyse *B. thuringiensis*, virulent, pseudolysogenic and temperate. They respectively lyse cells immediately, or form temporary or permanent associations with the host bacterium. The H-serotypes differ in the phage associated with them. Most isolates are found in nature associated with phage (Table III). Only types 2, 9, 12 and 13 have lacked a phage.

TABLE III
Temperate and pseudotemperate phages of Bacillus thuringiensis

H-serotype	Associated phage	Inducible phage	Induction agent	Reference
1	+	−		de Barjac *et al.*, 1974
2	−	−		
3a	+	+	mitomycin C	de Barjac *et al.*, 1974
				Ackermann *et al.*, 1974
				Martin & Dean unpublished
3a 3b	+	+	mitomycin C	Martin & Dean, unpublished
4a 4b	+	+	mitomycin C	de Barjac *et al.*, 1974
				Martin & Dean, unpublished;
				Kochina *et al.*, 1977
4a 4c	+	+	mitomycin C	de Barjac *et al.*, 1974
			U.V. light	Martin & Dean, unpublished
5a 5b	+	+	mitomycin C	de Barjac *et al.*, 1974
			U.V. light	Martin & Dean, unpublished
			hydrogen	Kochina *et al.*, 1977
			peroxide	Colasito & Rogoff, 1969b
5a 5c	+	+	mitomycin C	Martin & Dean, unpublished
6	+	−		de Barjac *et al.*, 1974
7	+	+	mitomycin C	de Barjac *et al.*, 1974
				Martin & Dean, unpublished
8	+	−		de Barjac *et al.*, 1974
9	−	−		
10	+	−		de Barjac *et al.*, 1974
11	+	+	mitomycin C	Martin & Dean, unpublished
12	−	−		
13	−	−		
14	+	+	mitomycin C	Martin & Dean, unpublished

Lytic phages of *B. thuringiensis* have been isolated from the soil (C.L. Fort, unpubl.; Thorne, 1968; Thorne, 1978), or associated with bacteria from insects (de Barjac *et al.*, 1974; Colasito and Rogoff, 1969a; Ackermann *et al.*, 1974; Rautenshtein *et al.*, 1972). These phages have been characterized as to their morphology and some according to host range in *B. thuringiensis* and their serology as noted by the authorities above. Tobek *et al.* (1978) have done restriction analysis on one lytic phage. Restriction patterns (Dean *et al.*, 1978) should prove to be the best means to distinguish the different phages.

Temperate phages of *B. thuringiensis* have been found (de Barjac *et al.*, 1974; Colasito and Rogoff, 1969b; Rautenshtein *et al.*, 1972). They have usually been induced with mitomycin C or UV light, but none have formed a lysogenic relationship with the host. Ten of the phages isolated in our laboratory lysogenized isolate 3 of H-type 3a 3b, which harboured no phage. Phages induced from this one isolate were then studied without the added variable of isolate differences.

Phage host ranges were studied within the genus *Bacillus* and most *B. thuringiensis* phages also formed plaques on the closely related *B. cereus* (Table IV). Several were able to plate on *B. subtilis*, which has an established genetic

TABLE IV
Host range of τ phages of Bacillus thuringiensis

Host *Bacillus* species	τ phages of *B. thuringiensis*								
	3a–1	3a–2	3a 3b–1	4a 4c–1	5a 5b–2	5a 5b–4	5a 5c–1	7–1	11–1
amyloliquifaciens	–	–	–	–	–	–	–	–	–
cereus	+	+	+	–	+	+	+	+	+
globigii	–	–	–	–	–	–	–	+	+
licheniformis	–	–	–	–	–	–	–	–	–
megaterium	–	–	–	–	–	–	–	–	–
niger	–	–	–	–	–	–	–	–	–
pumilis	+	–	–	–	–	–	–	–	+
subtilis	–	+	–	+	+	+	–	– -	+
thuringiensis, type 3a 3b	+	+	+	+	+	+	+	+	+

transmission system. To try to establish relatedness among these phages, immunity studies were made. No clear pattern emerges and the study is continuing by looking at restriction patterns of their DNA.

These phages may prove useful for transduction as there seems to be some limitation to the transducing phages available at present. A system of transformation may be developed using a phage as a helper for the uptake of DNA as in *E. coli* (Kaiser and Hogness, 1960). The fact that some phages carry genes other than those needed for phage propagation prompts us to look for other markers, especially toxin production, that they may carry.

III. Transmission Systems

A. TRANSFORMATION

Genetic transformation was classically described in *Diplococcus pneumoniae* but one of the best described systems is that of *Bacillus subtilis* (Spizizen, 1958). This genetic transmission system depends upon the ability of the recipient cell to take up isolated DNA (become competent) by a mechanism not completely understood (Notani and Setlow, 1976; Dubnau, 1976). Spizizen discovered that, of the many strains available, one particular strain known as *B. subtilis* 168 had this attribute. The efficiency of transformation is ca. 1% or less. The single report of transformation in *B. thuringiensis* involved DNA from H-serotype 1 into serotype 2 (Reeves, 1966).

Two approaches used to force competence in other bacteria might prove useful for *B. thuringiensis*. *E. coli*, which normally will not take up DNA, can be transformed with phage DNA in the presence of 0.05 M $CaCl_2$ (Mandel and Higa, 1970). Polyethylene glycol (PEG) promotes the uptake of plasmid DNA into protoplasts in *Streptomyces* (Bibb *et al.*, 1978) and also plasmid DNA in *B. subtilis* protoplasts (Chang and Cohen, 1979). Exploiting the latter, we have efficiently (90–100%) protoplasted *B. thuringiensis*. The *Staphylococcus aureus* molecular cloning vehicle, pC194 (Ehrlich, 1978), which is a transposon (integrating plasmid) carrying chloramphenicol resistance we have introduced into *B. thuringiensis* at a high transforming frequency (10%) and shown to integrate into the *B. thuringiensis* genome. The pC194 resistance enzyme, chloramphenicol acetyl transferase, is easily measurable in these transformants. These experiments indicate that the era of genetic engineering has dawned upon *B. thuringiensis*.

B. CONJUGATION

Conjugation is gene transfer mediated by plasmids. It is well known in *E. coli* and related bacteria where self transmissible plasmids were initially studied, including F factors (Lederberg and Tatum, 1946) and R factors, which carry genes determining antibiotic resistance (Watanabe, 1963). A spore forming organism *Clostridium perfringens* also has plasmids capable of self transmission (Rood *et al.*, 1978). This system for gene transfer may be possible in *B. thuringiensis* as most serotypes carry plasmids which may bear transferable antibiotic resistance.

C. TRANSDUCTION

Transduction is bacteriophage-mediated gene transfer between bacteria. Two types of transduction are known, specialized and generalized. Specialized transduction, best known in coliphage λ (Morse *et al.*, 1965), is a property of certain temperate phages, i.e., phages which integrate into the chromosome of the host bacterium (lysogeny). When an integrated phage is induced it excises its DNA out of the chromosome and begins to replicate. Rarely a faulty excision occurs and a portion of the bacterial chromosome adjacent to the integrated phage is excised and, still linked to the phage DNA, is packaged in the phage head. Thus only selected genes are transferred. There have been no reports of specialized transduction in *B. thuringiensis* but the paucity of genetic mutants has prevented a means of detecting this genetic exchange mechanism.

Generalized transduction was first discovered in the Salmonella phage, P22 (Zinder and Lederberg, 1952). In this transmission system a bacteriophage erroneously packages bacterial DNA (without phage DNA) from any location on the chromosome into the phage head. There are three generalized transducing phages for *B. thuringiensis.*

de Barjac (1970) isolated phage Th1 from soil using *B. thuringiensis* H-type 1 as indicator. Th1 also plates on types 4a 4b-sotto, 4a 4c, 5a 5b, 7, 9 and 10. Transduction was demonstrated in serotype 1 by transferring prototrophy from wild type cells to thymine and uracil auxotrophic mutants. de Barjac also transduced the H_7 flagellar antigen from serotype 7 to type 1. Anti H_1 serum was then used to select against nontransductants. Important controls were included in this work, namely that genetic transfer was sensitive to antiphage serum but not to DNase. Transduction was also shown to be related to the concentration of phage per cell (multiplicity of infection). UV light was used to reduce the virulence of the phage. Also Yelton and Thorne (1970) mentioned, in a paper on *B. cereus* transduction, that phage CP51 could transduce undesignated genes from *B. thuringiensis* 1328 (serotype not given) into *B. cereus* 6464 and 569.

In a thorough study of generalized transduction in *B. thuringiensis*, broad-host-range phages CP51 and CP54, were analysed. All twelve serotypes serve as hosts for CP54 and all but serotypes 2, 3a, 5a 5b and 6 for CP51 (Thorne, 1978). Transduction of several genetic markers in type 3a demonstrated linkage between *trp*1, *cys*1 and *cys*2 and between *met*1, *arg*1 and *arg*2. These linkages represent the origins of a genetic map of *B. thuringiensis.*

Heterologous transduction between serotypes 2 and 3a as recipients was achieved by donor phage from 2, 3a, 5a 5b, 4a and 7-limassol. Type 12 could be transduced only by a phage prepared on 12 and no other serotype nor *B. cereus* 568 would serve as donors. We note here that although phages grown on one strain of *B. thuringiensis* (1328) will transduce *B. cereus* 6464 and 569 (Yelton and Thorne, 1970), phages grown on *B. cereus* 569 will not transduce any serotypes of *B. thuringiensis* (Thorne, 1978). Other attempts to use CP51 and CP54 transduction with other serotypes failed because of the very lytic nature of these phages. As with de Barjac's study, UV inactivation of the phage was necessary to reduce lysis of the recipient. Unfortunately, CP51 and CP54 are too lytic on the highly entomocidal commercial strain HD-1 (3a 3b) to promote transduction. Perhaps genetic alteration of the phages will allow their use in other serotypes.

IV. Future Work

This review appears during early development of the genetics of *B. thuringiensis.* We can easily predict that the next 5 years will bring exciting new discoveries. We have described the current background knowledge upon which future work will build. It would, of course, help our prediction if more researchers and grant money becomes available for genetic approaches to increase potency of *B. thuringiensis.* We have several suggestions that would greatly aid research on this organism. (1) Rapid communication is essential in any molecular genetic approach, to bring together biochemists, geneticists, fermentation microbiologists, and taxonomists who have a common interest in this organism. H.D. Burges has started a Newsletter for this purpose for which contributions are urgently requested. (2) Collaboration between scientists of different disciplines is imperative in the field of microbial control. Few geneticists can mount the effort to adequately bioassay mutants they construct. Insect pathologists should join with biochemists, to study the chemical nature of toxins. (3) Sharing of genetic mutants, and isolates of serotypes is a necessity for rapid progress in genetic research. The *"Bacillus* Genetic Stock Center" (BGSC) at The Ohio State University will stock a complete set of H-serotypes, genetic mutants, plasmid deficient stocks and bacteriophages of *B. thuringiensis.* Inquiries and requests should be made to its Director. Researchers should use this service to avoid duplication of effort and to standardize genetic loci and allele numbering. Likewise, researchers should freely contribute to the BGSC to promote progress. (4) A potential problem may arise in *B. thuringiensis* genetics. With 22 serotypes identified to-date, research should be concentrated on a few representative H-serotypes to avoid confusing genetic results. A wise choice is to concentrate on the commercially exploited isolate HD-1 of H-serotype 3a 3b. Whenever possible this particular culture (i.e. HD-1) should be used for genetic studies. This suggestion requires that the current generalized transducing phages be mutated to plate on this strain at lower virulence..

In summary, it is clear that genetic transmission systems are now available for some serotypes. Recombinant DNA techniques have also been applied. The prerequisite variety and mutants are present and extrachromosomal DNA is abundantly present, enhancing the variety and interest in *B. thuringiensis* as a genetic system.

References

Ackermann, H., Smirnoff, W.A. and Bilsky, A.Z. (1974). *Can. J. Microbiol.* **20**, 29–33.

Azizbekyan, R.R., Belykh, R.A. and Netyksa, Y.M. (1978). Abstr. Third Internatl. Symp. Genet. Ind. Microorganisms p. 14. American Society for Microbiology.

Barjac, H. de (1970) *C. r. hebd. Seanc. Acad. Sci., Paris* **270**, 2227–2229.

Barjac, H. de, Sisman, J. and Cosmao-Dumanoir, V. (1974). *C. r. hebd. Seance. Acad. Sci., Paris* **279**, 1939–1942.

Bibb, M.J., Ward, J.M. and Hopwood, D.A. (1978). *Nature Lond.* **274**, 398–400.

Burges, H.D. and Bailey, L. (1968). *J. Invertebr. Path.* **11**, 184–195.

Chakrabarty, A.M. (1972) *J. Bacteriol.* **112**, 815–823.

Chang, S. and Cohen, S.N. (1979). *Molec. Gen. Genet.* **168**, 111–115.

Colasito, D.J. and Rogoff, M.H. (1969a), *J. Gen. Virol.* **4**, 267–274.

Colasito, D.J. and Rogoff, M.H. (1969b). *J. Gen. Virol.* **4**, 275–281.

Dean, D.H., Perkins, J.B. and Zarley, C.D. (1978). *In* "Spores VII" (J.C. Vary and G. Chambliss, eds), pp. 144–149. American Society for Microbiology.

Delafield, F.P., Somerville, H.J. and Rittenberg, S.C. (1968). *J. Bacteriol.* **96**, 713–720.

Dubnau, D. (1976). In "Microbiology 1976" (D. Schlessinger, ed.), pp. 14–27. American Society for Microbiology.

Dulmage, H.T., Boening, O.P., Rehnborg, C.S. and Hansen, G.D. (1971). *J. Invertebr. Path.* **18**, 240–245.

Ehrlich, S.D. (1978). *Proc. Natl. Acad. Sci. USA* **75**, 1433–1436.

Eklund, M.W., Poysky, F.T., Reed, S.M. and Smith, C.A. (1971). *Science* **172**, 480–482.

Fettig, P. (1976). Ph.D. thesis, The Ohio State University, Columbus, Ohio.

Fitz-James, P.C. and Young, I.E. (1959). *J. Bacteriol.* **78**, 743–754.

Groman, N.B. (1953). *J. Bacteriol.* **66**, 184–191.

Ikeda, H. and Tomizawa, J.I. (1968). *Cold Spring Harbor Symp. Quant. Biol.* **33**, 791–798.

Johnston, J. and Richmond, M.H. (1970). *J. Gen. Microbiol.* **60**, 137–139.

Kaiser, A.D. and Hogness, D.S. (1960). *J. Molec. Biol.* **2**, 392–415.

Kochina, Z., Blokhina, T.P. and Rautenshtein, Y.I. (1977). *Mikrobiologiya* **46**, 599–604.

Lacey, R.W. (1975). *Bacteriol. Rev.* **39**, 1–32.

Lacey, R.W., Lewis, E., II and Grinsted, J. (1973). *J. Med. Microbiol.* **6**, 191–200.

Lecadet, M.M., Klier, A.F. and Ribier, J. (1974). *Biochimie.* **56**, 1471–1479.

Lederberg, J. and Tatum, E.L. (1946). *Cold Spring Harbor Symp. Quant. Biol.* **11**, 113–114.

Mandel, M. and Higa, A. (1970). *J. Molec. Biol.* **53**, 159–162.

Morse, M.L., Lederberg, E.M. and Lederberg, J. (1965). *Genetics* **41**, 142–156.

Nickerson, K.W. and Bulla, L.A. (1974). *App. Microbiol.* **28**, 124–128.

Nishiitsutsuji-Uwo, J., Wakiasaka, Y. and Eda, M. (1975). *J. Invertebr. Path.* **25**, 355–362.

Notani, N.K. and Setlow, J.K. (1976). *Prog. Nucleic Acid Res. Molec. Biol.* **14**, 39–100.

Pinney, R.J. and Smith, J.T. (1973). *Antimicrobial. Agents Chemother.* **3**, 670–675.

Rautenshtein, Y.I., Krukouskaya, G.E., Blokhina, T.P. and Soloveva, N.Y. (1972). *Mikrobiologiya* **41**, 150–151.

Reeves, E.L. (1966). Ph.D. Thesis, The Ohio State University. Columbus, Ohio.

Rood, J.I., Scott, V.N. and Duncan, C.L. (1978). *Plasmid* **1**, 563–570.

Salisbury, V., Hedges, R.W. and Datta, N. (1972). *J. Gen. Microbiol.* **70**, 443–452.

Spizizen, J. (1958). *Fed. Proc.* **18**, 957–965.

Stahly, D.P., Dingman, D.W., Bulla, L. and Aronson, A. (1978a). *Biochem. Biophys. Res. Comm.* **84**, 581–588.

Stahly, D.P., Dingman, D.W., Irgens, R.L., Field, C.C., Feis, M.G. and Smith, G.L. (1978b). *FEMS Microbiol. Lett.* **3**, 139–141.

Thorne, C.B. (1968). *J. Virol.* **2**, 657–662.

Thorne, C.B. (1978). *App. Environ. Microbiol.* **35**, 1109–1120.

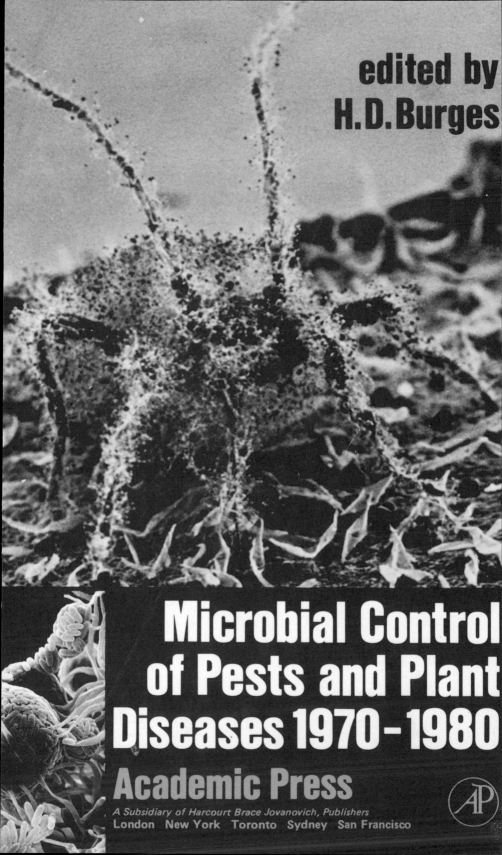

edited by
H.D.Burges

Microbial Control
of Pests and Plant
Diseases 1970-1980

Academic Press

A Subsidiary of Harcourt Brace Jovanovich, Publishers
London New York Toronto Sydney San Francisco

Tobek, I., Tichy, P. and Lipauska, H. (1978). *Folia Microbiol.* **23**, 137–139.
Tomoeda, M., Inuzuka, M. and Hayashi, C.M. (1976). *Jap. J. Microbiol.* **20**, 27–32.
Watanabe, T. (1963). *Bacteriol. Rev.* **27**, 87–115.
Yelton, D.B. and Thorne, C.B. (1970). *J. Bacteriol.* **102**, 573–579.
Yousten, A.A. (1978). *Can. J. Microbiol.* **24**, 492–494.
Zinder, N.D. and Lederberg, J. (1952). *J. Bacteriol.* **64**, 679–699.

CHAPTER 16

Production of Insect Viruses in Cell Culture

H. STOCKDALE and R.A.J. PRISTON

Shell Research Ltd., Sittingbourne Research Centre,
Sittingbourne, Kent, England

I. Introduction

Large-scale production of arthropod pathogens *in vitro* has several attractions when compared with methods using living insects or mites. Fungi and bacteria may be grown in artificial media, but at present a second group of pathogens, i.e. protozoa, rickettsia and viruses, demand either a living host or its cells for multiplication. Growth of pathogens in cells *in vitro* allows the process to be optimized by methods available in the fermentation industry. Goodwin *et al.* (1970) indicated the feasibility of using viruses produced *in vitro* to control insects. Even so, this chapter is of necessity highly speculative: at present neither a commercial nor a pilot plant for production of insect viruses *in vitro* exists. However, certain broad inferences concerning bulk production can be drawn from the extensive literature about insect cell culture and pathology in general. Most of the data in this chapter concern the production of nuclear polyhedrosis viruses (NPV) in lepidopteran cells. While this fairly reflects the value accorded NPVs as control agents, it in no way suggests that production of other pathogens by similar basic techniques is not equally feasible.

II. History

Culture of insect tissues *in vitro* has as venerable a lineage as that of vertebrate tissue culture. Goldschmidt's experiments with cecropia spermatocytes (1915) only slightly postdate the work of Carrel (1913) with vertebrate tissues. However, the rapid advances in the culture of vertebrate, especially mammalian, cells during the late 1940s and the 1950s were not complemented by similar progress with invertebrates. This doubtlessly reflects the immediate value of mammalian cell culture to medicine and virology. *In vitro* studies with invertebrates have largely been regarded as having little practical potential.

Early attempts to obtain continuously-dividing cultures of insect cells met with only limited success (Grace, 1958; Vago and Chastang, 1958). The cell populations obtained were often capable of subcultivation, but not indefinite survival. Establishment of four continuously-dividing lines from the silkmoth *Antheraea eucalypti* (Grace, 1962a) marked a significant advance in invertebrate cell culture. Lines from Diptera (Grace, 1966), Dictyoptera (Landureau, 1968) and Homoptera (Chiu and Black, 1967) followed. Hink's (1976) compilation lists 121 invertebrate lines, predominantly insect.

Viruses most frequently considered for control of insects (usually sawflies and Lepidoptera) are the occluded viruses, namely NPV, cytoplasmic polyhedrosis (CPV), granulosis (GV) and entomopoxviruses (EPV). Although the nonoccluded *Sericesthis* iridescent virus (SIV) was produced in *Anth. eucalypti* cells (Bellet and Mercer, 1964), several years elapsed before unequivocal infection of established cells with an occluded virus was demonstrated. In a review, Grace (1967a) described production of *Bombyx mori* NPV in *Anth. eucalypti* cells. Goodwin *et al.* (1970) established a line from the armyworm, *Spodoptera frugiperda*, and infected it with an homologous NPV. Previously *in vitro* growth of NPV (Trager, 1935) and CPV (Grace, 1962b) had been confined to primary cultures. At that time it was widely considered, without evidence, that established cells were inherently refractory to infection by occluded viruses. Availability of continuously dividing cells meant that scale-up procedures were possible. When Vaughn (1968) adapted Grace's *Anth. eucalypti* cells to growth in suspension culture the basic materials and methods were available for producing insect viruses in a bulk fermentation system. Subsequent work was directed to improving the scope and efficiency of the process by providing more cells and viruses, modifying media and increasing culture volumes.

III. Uses and Limitations of *in vitro* Methods

The reasons for working with cells, tissues and viruses *in vitro* are similar in the vertebrate and invertebrate fields. *In vitro* methods provide basic information on metabolism of animal cells free of complications from the specialized functions of the whole animal. In the screening of materials for metabolic activity the use of a homogenous population of cells eliminates the wide variability observed with whole animals.

At present viruses used in control programs are produced exclusively in whole insects. Infected insects may be collected from the field, and larvae are reared on foliage or diet then infected in the laboratory or in purpose-built breeding facilities. Projected *in vitro* methods would have several advantages over mass rearing techniques. Use of frozen seed-stocks prevents genetic drift in host material and guards against danger of an epizootic in breeding stock. Viral preparations from whole larvae can contain many bacteria as well as potentially-antigenic material such as hair and cuticle. As cell culture must be operated under sterile conditions a far cleaner product is obtained. In whole insects, viruses other than the one administered may appear during the course of infection. In cell culture there is a better opportunity for eliminating, or at least recognizing, such contaminants.

Mass rearing methods are far more labour-intensive than a proposed *in vitro* counterpart. Rearing, inoculation and harvesting of larvae involve several discrete operations, originally carried out manually. Although one or more of these operations can be automated, manpower requirements will probably remain high in comparison with those of a fermenter. A further factor in favour of *in vitro* production is its flexibility. Production of another virus by mass rearing requires the commissioning of a new breeding and/or inoculation unit. By contrast a fermenter can be diverted to production of another virus within a week.

Against these advantages must be balanced several drawbacks. The capital cost of fermentation plant could be high compared with that of breeding facilities. Sterile conditions are required throughout and, as growth of insect cells and production of virus are slow relative to the growth of many micro-organisms, if contamination occurs it will usually lead to the loss of a production run.

IV. Basic Methods

A. SELECTION OF CELLS

A prerequisite for large-scale production of insect viruses *in vitro* is the possession of cells that multiply fast enough to produce large batches regularly from seed stock. These cells can, with reservations, be considered as free-living microorganisms. In general they can be cultivated and handled with techniques similar, if not identical, to those reserved for the more slow-growing and fastidious organisms encountered in the environment.

When a section of animal tissue, a so-called "explant", with the potential for cell division is placed in a nutrient medium, migration of cells from the explant is often observed. Judicious choice of incubation conditions encourages cells to proliferate enough for subculturing. Cells capable of being subcultured repeatedly are said to be "established" and to constitute a cell line. It is generally, though not exclusively, true that lines can be established most often from material that has the potential for unrestricted growth, e.g. embryonic or tumour tissue.

Patterns of establishment vary widely between insect and vertebrate cells. Vertebrate cells, typified by the fibroblast, proliferate rapidly from the time of explanting and possess a diploid karyotype. Lines derived from these have a fixed lifespan, cells dying after ca. 50 doublings. Occasionally such lines may give rise to heteroploid cells which, in common with tumour-derived cells, multiply indefinitely. In contrast, explanted insect cells are inactive for a prolonged period lasting from a few months to over a year, after which they either proliferate rapidly or die. Most lepidopteran lines are heteroploid: within the Diptera diploid and heteroploid lines are found (Hink, 1976). To date no example of a diploid insect cell line of fixed life-span exists.

In attempts to establish insect cell lines workers have generally chosen material potentially able to multiply rapidly and easily handled under sterile conditions. Pupal ovaries have been popular (Grace, 1962a; Goodwin, 1975; Mitsuhashi, 1973); but larval (Vago and Chastang, 1958; Grace, 1967b) and adult (Hink, 1970; Hink and Ignoffo, 1970) ovaries have been used. Embryonic material is widely employed (Hink and Ellis, 1971; Peleg and Sahar, 1972; Echalier and Ohanessian, 1970; Landureau, 1968). Sohi (1973a) established a line of *Choristoneura fumiferana* from neonate larvae hatched from surface-sterilized eggs. Most mosquito lines are derived from minced larval tissues (Hink, 1976).

However, tissues not immediately considered as suitable material may have potential for producing lines. At least one line has been established from pupal fat body of *Trichoplusia ni* (Goodwin *et al.*, 1974). Larval and pupal haemocytes have given lines from *Chilo suppressalis* (Mitsuhashi, 1967), *Estigmene acrea* (Granados and Naughton, 1975), *Malacasoma disstria* (Sohi, 1973b) and *Samia cynthia* (Chao and Ball, 1971).

Why certain cell lines support production of occluded viruses and others do not is not known at present. Absence of polyhedral inclusion bodies (PIB) need not denote a complete failure in production. Replication may be partial. For instance, Ignoffo *et al.* (1971) could not demonstrate formation of polyhedral bodies (PIBs) of the singly-embedded NPV (SEV) of *Heliothis zea* in the *H. zea* cell line designated IMC-HZ-1 (Hink and Ignoffo, 1970) although they succeeded in raising the titre $> 10^8$ by serial passage. In the absence of observable virions, infectivity was ascribed to naked viral DNA. In contrast, Goodwin *et al.* (1974) produced *H. zea* SEV in another *H. zea* line (IPLB-1079). They made the curious observation that cell lines growing as attached monolayers produced PIBs; those growing in suspension did not. Volkman and Summers (1975) selected several clones from the *T. ni* line of Hink (1970) that gave varying yields of PIBs of the NPV of *Autographa californica* (ACNPV). They did not, however, report a clone producing no PIBs. Thus virus yield may, within limits, be influenced by cell type.

In vitro production of a particular virus need not depend on cells isolated from the primary host or target organism. *In vitro* a virus is rarely specific to a single species: existing cell lines may support production of a newly-isolated virus. The "broad spectrum" multiply-embedded NPV of *T. ni* (TNMEV) and *A. californica* replicates in several lepidopteran lines. More specific NPVs can also be produced in cells other than those of the original host (Sohi, 1973b). The same may be true of CPV, while iridescent viruses (IV) replicate in cells from different orders (Bellet and Mercer, 1964).

B. SELECTION OF VIRUSES

Those viruses normally considered for insect control are predominantly of the occluded category. Historically they (together with the IV) may have been chosen because of their characteristic appearance in diseased insects. Occluded viruses are considered, perhaps sometimes erroneously, to be more stable than nonoccluded viruses. Among occluded viruses the GV and NPV have been recommended for insect control as they were thought to be exclusively insect pathogens (WHO, 1973). However, they have since been found in other groups (Couch et al. 1975; Federici and Humber, 1977). Among the nonoccluded viruses, IV (Bellet and Mercer, 1964; Kelly, 1976a) 'Oryctes' (Kelly, 1976b) and cricket paralysis virus (Scotti, 1976) have been produced in cell culture, but evidently for research purposes only. Most occluded viruses handled in vitro are NPVs (details: Hink, 1976), although CPV (Granados et al. 1974) and entomopox viruses (Granados and Naughton, 1975) have been established in cell culture.

The usual mode of establishing occluded viruses in cell culture involves challenging cells with infectious fluid from the whole insect, either haemolymph (HL) or a tissue lysate. The infectious unit in vitro is presumed to be the complete, free virion (Volkman and Summers, 1977). It seems fair to suggest that virions occluded in a protein matrix are not infectious in vitro. Virions liberated from PIBs with alkali (Bergold, 1959) have been used to initiate in vitro infections with varying degrees of success (Granados, 1976). Once infection is instituted in vitro a mixture of free and occluded virions is produced and free virions transmit infection either to other cells or to cultures. Problems inherent in the use of cell culture-produced virus will be discussed later.

Viruses isolated from different localities can be expected to show different degrees of virulence. Shapiro and Ignoffo (1970) selected the most potent of 34 isolates of H. zea NPV on the basis of a per os bioassay. The number of infectious units in HL or cell culture supernates can be determined by either an end-point dilution assay (Brown and Faulkner, 1975) or a plaque-counting assay (Hink and Vail, 1973; Brown and Faulkner, 1977) (more detail in Section V; B2). Virulence of PIBs can be estimated with a larval bio-assay.

Storage of PIBs in the laboratory raises few if any problems. Cunningham (1970) noted gradual degradation of aqueous PIB suspensions at 4°C but dry PIB powders can be stored at −20°C for several years without detected loss of activity. Infectious HL and culture supernates require more careful handling. Phenoloxidase activity must be suppressed in HL. Free virus can be kept well at 4°C for several weeks but unexpectedly rapid degradation occurs at −20°C to −30°C (unpublished observation). Ultra-low temperatures (ca −90 or −196°C) are recommended for storage of free virions.

C. SELECTION OF MEDIA

Cell culture media are generally modelled on the body fluids of the donor animal. In the Insecta, HL composition varies markedly among the orders (Sutcliffe, 1963) and throughout development of a particular insect (Jeuniaux, 1971). Inorganic HL constituents of primitive insects (Apterygota) approximate to those of vertebrate sera. Higher orders, e.g. Coleoptera and Lepidoptera, diverge progressively with a decreasing Na^+ to K^+ ratio, a decreasing chloride content, an increasing proportion of divalent (Ca^{2+} and Mg^{2+}) cations and an increasing concentration of small organic molecules, notably amino acids (Sutcliffe, 1963). The practical result of this variation is that a single medium is unlikely to be suitable for the culture of cells from different orders.

Inorganic ion balance is probably one of the most important physicochemical factors determining whether or not an insect cell grows in a given medium. Animal cells *in vitro* require a favourable external ionic balance to maintain the internal milieu and allow transport across the cell membrane. Maizels *et al.* (1958) calculated that in mouse cells up to 20% of energy requirements are devoted to excluding Na^+ and maintaining internal K^+. This requirement presumably increases when the external balance is unfavourable. There are few published data on the effect of different ionic balance on growth of insect cells *in vitro*. McIntosh *et al.* (1973) adapted a line of the leafhopper, *Agallia constricta* to grow on a mammalian medium over six months. Lee (personal communication) adapted a line of *Antheraea pernyi* (moth) cells to a medium consisting mainly of Eagles' minimum essential medium. The only quantitative study of the effect of specific ions on established insect cells showed that while Na^+ concentration had no influence on *H. zea* cells independent of that of osmotic pressure (OsP), K^+ above a threshold concentration of 70 mM decreased the growth rate proportionally to its concentration (Kurtii *et al.* 1975). Insect cells adapt to relatively broad limits of pH and OsP *in vitro*, beyond which they are drastically inhibited. Buck (1953) recorded a wide range of pH values (from 5.9 to 7.7) for insect HL according to order and species. Generally the pH of media for lepidopteran cell cultures ranges from 6.2 to 6.5; for dipteran cells, 6.5 to 7.2; and for leafhopper cells (Chiu and Black, 1967) 7.0. Again little has been reported about the effects of pH on growth of insect cells. Kurtii and Brooks (1972) found little difference in final cell concentration of *H. zea* cells at pH 6.0 to 8.0, but an extended lag phase at high pH values. Cells inoculated into alkaline medium lowered the pH and *vice versa*. *H. zea* cells grew within an OsP range of 230 to 467 milliosmoles (mOsm) (Kurtii *et al.*, 1974). Growth rate was maximal between 300 and 310 mOsm and > 90% of this between 290 and 360 mOsm. *T. ni* cells grew to a similar final concentration and produced PIBs of ACNPV within certain broad limits of pH (5.5–6.5) and OsP (250–500 mOsm) (Gardiner *et al.*, 1976).

Within the restrictions imposed by these physicochemical parameters, insect cells adapt to a variety of nutritional regimes. Basic nutritional requirements of animal cells in culture are: a utilizable carbon and energy source, amino acids and vitamins. Mouse L cells grow in a medium containing salts, glucose, thirteen amino acids and eight vitamins (Merchant and Hellman, 1962). Use of chemically-defined media for large-scale production of animal cells and viruses would be advantageous in terms of ease of handling, cost and quality control. At present chemically-defined media able to support prolonged growth of insect cells have not been devised, with the possible exception of the medium of Landureau and Grellet (1972).

While animal cells may have fairly simple nutritional needs and may supply many constituents by biosynthesis, they are unable to concentrate small molecules to the degree typical of most free-living microorganisms. Consequently such molecules – synthesized by the cell, but necessary for growth – are lost by diffusion when excised tissue is placed in liquid medium. Growth is possible only when a balance between the extra- and intracellular concentrations of such molecules is achieved (Eagle and Piez, 1962). The problem is alleviated either by retaining a certain proportion of "conditioned" medium when medium in contact with the cells is renewed, or by using a "rich" medium containing a large proportion of nutrients of "biological" origin. Hink's (1970) medium, used in the establishment of lines from *H. zea, T. ni* and *Cydia pomonella*, contains chicken egg ultrafiltrate (CEU), bovine plasma albumin (BPA) and tissue culture grade yeastolate (TCY). That of Yunker *et al.* (1967) contains CEU and BPA. Most popular insect cell media contain varying amounts of foetal bovine serum (FBS). Chick embryo extract (CEE) has also been used (Peleg, 1965). Inclusion of homologous or heterologous HL has been a common practice in the past. Biological supplements can be costly and difficult to handle, limiting large-scale cell culture. A reasonable strategy may be to employ media rich in such supplements for the establishment of cell lines and small-scale growth, then to adapt cells to a cheaper (and probably) more meagre medium for large-scale growth.

Such adaptation can succeed if undertaken "stepwise" by either the sequential omission of a single expensive nutrient or its substitution by a cheaper or more readily-available product. Cell division may decrease or stop before resuming its former rate (Gardiner and Stockdale, 1975). Possibly only a few cells, of greater biosynthetic or accumulative capacity, reproduce after such a change, so a wide divergence between the original and final populations may result.

The realization that there was no absolute need for HL encouraged work on bulk production of insect cells. Cells have been adapted to grow without or with less HL (Vago and Chastang, 1962; Sohi, 1969) or have been established *de novo* in HL-free medium (Hink, 1970). FBS has usually been substituted for HL although other vertebrate sera support growth (Goodwin, 1975).

Growth-stimulants in HL and sera remain unidentified. Insect, but not vertebrate, cells require sterols. Insects are unable to synthesize sterols from squalene (Clark and Bloch, 1959) and must receive them in their food (Dadd, 1960). Sera and HL may transport sterols and essential fatty acids (Louloudes *et al.*, 1973). The recipe for a chemically-defined growth medium for insect cells may depend on the use of an artificial carrier of essential lipids.

At present, however, high cell concentrations are obtained with partially-defined media in which known quantities of amino acids and vitamins are supplemented with biological additives (Grace, 1962a; Mitsuhashi, 1973) or in which most nutrients are supplied as extracts or hydrolysates of biological origin (Mitsuhashi and Maramorosch, 1964; Schneider, 1964). It should be possible to improve cell and virus production in present media by identifying and modifying the limiting factors.

V. Optimization for Production: Some Problems

A. COMPARISON WITH *IN VIVO* PRODUCTION

Industrial fermentations are directed to the cheapest production compatible with quality. Factors affecting cost are: (i) Scale of operation: as the capacity of a reactor and ancilliary equipment increases, the proportion of the cost of a product ascribed to capital (plant) costs and labour decreases. (ii) Feedstock costs: profitability depends on the difference in value between a product and its precursors, and the conversion efficiency of the latter. (iii) Productivity: this depends on the concentration of product per unit volume and the production per unit time. Ideally, a process would give a rapid output of product at high concentration with maximal and efficient utilization of precursors. However, efforts towards optimization in a particular area may have to be tempered by consideration of other aspects of the process as well as of external factors. It may prove expedient to sacrifice maximal product concentration for an increase in rate of production.

These speculations are further complicated by the two-stage nature of virus production, dependent on the number and quality of cells per unit volume, factors themselves affected by culture conditions. Factual statements about maximum rates and concentrations are not possible for a process that as yet is undeveloped. However, the virus-infected insect provides a natural yardstick against which productivity can be measured. In addition the data of several workers can be used to indicate areas where improvements may be made and to predict factors that could decrease efficiency.

A *T. ni* larva can produce $6-13 \times 10^9$ PIB of TNMEV (Ignoffo, 1966), i.e. $20-43 \times 10^9$ PIB/g larval wet weight = ca. $12-26 \times 10^{10}$ PIB/g larval dry weight (dw). Given that TNMEV-infected *T. ni* (TN-368) cells produce a mean of 30 PIB/cell *in vitro* (MacKinnon *et al.*, 1974) and that the dw of 1×10^6 TN-368 cells is 448 μg (unpublished observation), then 1 g dw TN-368 cells can produce 66×10^9 PIB. Hink *et al.* (1977) however produced a mean of 100 PIB of ACNPV/TN-368 cell, or 22×10^{10} PIB/g dw: such data suggest that on a cell-for-cell basis PIBs are produced with similar efficiency *in vivo* and *in vitro*.

At present a direct comparison of *in vivo* and *in vitro* costs cannot be made, but an example may be illustrative. Ignoffo (1965) estimated a cost of 0.6 p (Stg) to rear one *T. ni* larva. Medium BML-TC/7A (Gardiner and Stockdale, 1975) with 2% (v/v) FBS and 0.2% (w/v) TCY costs 34p (Stg) per litre and produces $2-3 \times 10^9$ cells (0.9–1.3 g dw). Given this cell concentration, the equivalent number of PIBs from one larva could be produced from between 80 and 180 ml medium if one assumes a figure of 30 PIBs/cell. This suggests a price of 2.7–6.1 p (Stg)/larval equivalent. The data of Hink et al. (1977) indicating 100 PIBs/cell would lower this cost to 0.81–1.83 p (Stg). This would be a fair comparison if medium costs formed a high proportion of overall costs in the in vitro system and if efficiency of production could be maintained during scale-up, and at high cell concentrations.

However, this optimistic projection must be tempered by present realities and we shall, below, attempt to identify factors leading to decreased productivity and suggest possible remedies and topics for further investigation.

B. FACTORS AFFECTING PRODUCTIVITY

An operator in charge of either an industrial or a pilot-plant-scale virus production facility aims to obtain a large volume of culture, containing a high concentration of cells, capable of producing a large number of PIBs, in as short a time as possible. Problems encountered on the road to this desired goal are interlinked. An attempt to solve one problem frequently affects other aspects of the process, for good or ill. Some of these problems are connected with the process of in vitro virus production alone, others are caused or exacerbated by the scaling-up of the process. The following aspects of the process provide scope for improvement.

1. Choice of Cells

The population doubling times (PDT) of insect cells are comparable with those recorded for vertebrate lines. The fastest growth recorded is that of the T. ni (TN-368) cell with a PDT of 17 h (Hink, 1970); the PDT of the hamster kidney cell line BHK 21 is 12 h (Stoker and MacPherson, 1964). Possession of a fast-growing cell line is advantageous in that the period before inoculation with virus is reduced and virus may be produced at a faster rate. A higher yield of cells per unit of a given nutrient is produced because maintenance requirements are reduced. It is unlikely that fast-growing cells could be selected by cloning, but several cell lines from the same species may have very different PDTs. A further matter of concern is the mechanical strength of the cells. In any moving system cells are liable to disruption by shear forces at the wall–liquid and stirrer–liquid interfaces, or by bubble formation and bursting (Pollard and Khosrovi, 1978). NPV infection often results in cell lysis; and possibly mechanical damage of an already weakened cell could arrest further virus production. Assuming that the cell membrane is equally strong in all lines, it would appear likely that a population of largely spheroidal cells, e.g. that of S. frugiperda (Goodwin et al., 1970), is less susceptible to mechanical disruption than is a fusiform type with projections as typified by the T. ni cell of Hink (1970). To a certain extent the problem may be obviated by the cells themselves. Vaughn (1968) noted a rigorous initial selection during transfer of Anth. eucalypti cells from stationary to stirred culture, and it may prove possible to obtain more resilient cells by progressively increasing stirring speeds.

Some mechanical protection can be afforded by incorporating polymers, e.g. proteins, methylcellulose (MC) and polyvinylpyrrolidone (PVP), in the medium.

2. Selection and Manipulation of Viruses

The work of Goodwin et al. (1970) suggests that production of insect viruses in vitro is inherently feasible. Since then, however, several characteristics of viruses in vitro have been noted: these increase the complexity of the problem considerably.

Faulkner and Henderson (1972) observed changes in the quality of PIB of TNMEV during repeated passage in TN-368 cells. Results with ACNPV have been similar (Priston, unpublished observations). The number of PIBs/cell declines rapidly, also these PIBs contain fewer virions and matrix protein with a less ordered structure. In a cell culture plaque assay for ACNPV infectivity, Hink and Vail (1973) observed two plaque types, designated "many polyhedra" (MP) and "few polyhedra" (FP), according to the number of PIBs in the cells forming the plaque. Both MP and FP strains are found in vivo. However, maintenance in vivo of a predominantly MP population with many PIBs, many occluded and few free virions is favoured by the transmission in nature of infection by virions occluded in the PIBs. Free virions do not survive. The reverse is true in vitro. Occluded virions are not infectious, and infection is accomplished by non-occluded particles. Thus selection in vitro favours production of free virions at the expense of occluded virus. Potter et al. (1976) made an elegant demonstration of the relationship between MP and FP progeny of TNMEV. Individual MP and FP plaques were subcultured. While they obtained FP populations free from MP, they could not achieve the reverse. Growth studies on progeny from the two plaque types showed the in vitro decline in PIB production to be the result of selection of an FP population from a predominantly MP population. Potter et al. (1978) have shown FP PIBs to have a far lower per os infectivity than MP PIBs, a reflection of the lower virion content of FP PIBs (MacKinnon et al., 1974; Ramoska and Hink, 1974). If large-scale production of virus is contemplated, this problem must be solved, or at least minimized. Enough infectious HL is unlikely to be available for the inoculation of large fermenters and several cycles of virus replication will be necessary for maximal cell infection.

Another problem is the likelihood of restricted PIB production at high cell concentrations. Ideally an operator would prefer a high concentration of cells containing many PIBs. Several reports now suggest that this may not be achieved easily. Volkman and Summers (1975) found reduced plaque titres, indicated by PIB formation, in the lag- and stationary-, as opposed to logarithmic-phase of growth, when TN-368 cells were challenged with ACNPV. Vaughn (1976) noted a decrease in PIBs of ACNPV at high concentrations of S. frugiperda cells. Hink et al. (1977) and Stockdale and Gardiner (1977) examined this aspect of PIB production in more detail using ACNPV and TN-368 cells. Although their quantitative results differed, both groups observed a similar pattern, i.e. the number of PIBs/cell fell above a certain cell concentration. Stockdale and Gardiner (1977) found that medium in which cells had grown suppressed PIB formation. Stockdale (1977) also noted that PIB production was suppressed at high cell density in the presence of apparently adequate nutrients.

These problems, while difficult, may not prove insurmountable and there is great scope for improvement of *in vitro* techniques. More attention should be paid to metabolic aspects of cell and virus interaction in future. The advent of reliable plaquing systems suggests that a more wide ranging cloning and selection procedure could well yield viruses with superior performance *in vitro*. Deliberate modification of viruses, e.g. by mutagens, for field use should be contemplated with trepidation if at all.

3. Choice of Equipment

Little can be said about this topic owing to the paucity of published data. Vaughn (1976) produced ACNPV in *S. frugiperda* cells in roller-bottle culture. This, however, would probably be unacceptable for industrial production because of the labour involved. Many insect cell lines grow and produce virus in suspension culture. As there is no dependence on adhesion to a solid substrate for growth, we need not consider the various sophisticated devices to maximize surface area for the growth of diploid vertebrate cells. The most likely vessel of choice would appear to be a simple fermenter in which cells are suspended in liquid medium. In such a system the major problems are to maintain cells in suspension and ensure an adequate supply of nutrients and oxygen. Two general types of vessel can be envisaged (i) the stirred tank reactor, in which cells are kept in suspension with a rotating paddle and oxygen is supplied by sparging; (ii) the column fermenter in which cell suspension and aeration are effected by the ascent of gas bubbles from the perforated base of the vessel. Relative merits of these reactors for large-scale growth of insect cells have not been assessed experimentally. The column fermenter may present fewer problems with scale-up as stirred tank reactors show high shear forces near the stirrer, forces that increase with its size and speed. The column fermenter has a more even distribution of shear forces, but bubble formation, coalescence and bursting may in themselves prove highly disruptive of cells. Pollard and Khosrovi (1978) made a detailed theoretical study of the factors affecting growth of fragile cells in fermenters, with special reference to ACNPV production in TN-368 cells. Their results suggest that the best design is a tubular flow reactor in which cells are kept in suspension in a series of vertical repeating units by a gentle flow of pumped medium. Oxygen is supplied indirectly across tubular permeable membranes inside the fermenter.

4. Physicochemical Conditions

Physicochemical factors, i.e. ionic balance, OsP, pH and temperature, are unlikely to present grave difficulties during the process. Conditions optimal for cell growth and virus production differ but little, if at all. Growth of cells does not appear to alter the ionic balance and OsP of the medium beyond the limits suitable for virus production. Variation in pH caused by production of CO_2 and organic acids, and the possible deamination of amino acids is slight, owing to the relatively high buffering capacity imparted by the high amino acid content (ca. 60 mM) of lepidopteran media. Insect cells grow at a lower temperature range than mammalian cells with an optimum of 27–30°C. Production of virus (NPV) follows a similar pattern (Priston, unpublished observations).

5. Supply of Gases

This has been almost completely ignored by workers concerned with *in vitro* methods: yet it is one of great potential importance, which could produce some unexpected problems. The aerobic nature of insect tissue has long been recognized (Glaser, 1918), but little if anything has been published concerning the O_2 demand of cultured cells. Stockdale and Gardiner (1976a) obtained O_2 uptake values (QO_2) for *T. ni* cells ranging from 15 to 45 μl/mg dw/h. In the absence of an external O_2 source a culture containing 1×10^6 cells/ml could use all available dissolved O_2 within 17–50 min. This effect is not evident either in small-scale monolayer or suspension cultures where a large surface area to volume ratio allows ready diffusion of O_2 into the medium. However, when suspension cultures of 3–5 litre are required, cells rapidly become O_2-limited; O_2 must be supplied by sparging, a procedure which raises its own problems. For sparging to be effective air (or O_2) must be supplied as small bubbles with intimate mixing to promote rapid gas exchange. This can lead to mechanical damage of the cells by the stirrer and foam in which the cells are trapped and die in the head space of the vessel. Matters can be alleviated by use of a marine-type impeller, which produces turbulence with less shear force than the standard flat-bladed impeller, and by use of non-toxic antifoams. However, the problem remains a major one.

CO_2 production by insect cells is a similarly neglected topic. The low pH (6.1–6.3) of lepidopteran media means that only a small proportion of CO_2 is present as bicarbonate ion (Umbreit, 1972). CO_2 is a required nutrient for animal cells (Geyer and Chang, 1958) and its removal from the medium by a vigorous gas flow can kill them. However, the problem is easily circumvented either by delaying sparging until sufficient CO_2 is produced endogenously by the cells or by incorporating a little CO_2 (1–5%) into the gas supply. To date no work has been done on the effect of high dissolved CO_2 concentrations on cell growth or virus production.

6. Medium Modifications

Medium for large-scale cell and virus production should be designed for economy and ease of handling. Substitution of expensive nutrients, already discussed, enabled Hink *et al.* (1974) to reduce the cost of Hink's (1970) original medium by 50%. Some problems remain, notably that of animal sera, which are expensive, in short supply, difficult to store and handle, and may carry adventitious viruses. Stockdale and Gardiner (1976b) substituted the FBS of BML-TC/7A medium (Gardiner and Stockdale, 1975) with 2% (w/v) TC yeastolate. *T. ni* cells grew indefinitely in this medium, but did not produce PIBs when inoculated with ACNPV. Stockdale and Priston (unpublished observations) found that production of PIBs of ACNPV in *T. ni* cells was proportional to the concentration of FBS up to 0.5% (v/v). Thus, both FBS and TC yeastolate have cell growth-stimulating factors but the property of PIB stimulation is confined to FBS.

Another pertinent investigation is the identification of medium factors contributing to cellular and viral biomass. This can eventually lead to the formulation of balanced media in which high concentrations of expensive components do not remain at the end of growth or virus-production. Limiting nutrients can be selected at will, and medium can be regenerated for further use by replenishment with the previously-limiting factor. Re-use of medium is potentially attractive in the fermentation industry where disposal is a major problem. A comparison of lepidopteran and mammalian cell culture media suggests many amino acids are present in ca. x 10 excess of requirements for synthesis of the biomass normally produced ($2-3 \times 10^6$ cells/ml). Most of 21 amino acids were present in excess after growth of *Anth. eucalypti* cells in culture (Grace and Brzostowski, 1966). Of the normal sugars in lepidopteran media, fructose and glucose could be utilized for growth by *T. ni* cells while sucrose could not (Stockdale and Gardiner, 1976a). If this is so for other lepidopteran cells, the amounts of utilizable sugars in the media of Grace (1962a), Hink (1970) and Gardiner and Stockdale (1975) could well prove growth-limiting for cells. Stockdale (unpublished observations) found that 0.2% (w/v) glucose, the only sugar in medium BML-TC/7A, was depleted as the growth of *T. ni* cells ceased. Replenishment of glucose in the used medium allowed a second batch of cells to grow to the same concentration as previously. However, this "regenerated" medium did not support production of as many PIBs of ACNPV as did fresh medium. Presumably a PIB-production factor, as yet unidentified, was removed or modified during growth of the cells.

In summation although tentative steps are being taken towards rationalization of media, much remains to be done to ensure their effective and efficient use.

VI. Present Problems and Future Scope

The future of large scale *in vitro* production of insect viruses depends on three sets of factors, (i) commercial success (or guaranteed subsidization) of the viruses chosen, (ii) attitude of regulatory authorities to virus produced *in vivo*, (iii) as yet unsolved technical problems. Insect viruses are not expected to supplant chemical insecticides either entirely or rapidly. Rather one expects a gradual introduction of viruses in the company of many other alternative control methods as conventional insecticides become less effective and more expensive to develop. The way in which regulatory authorities react to insect viruses could well be crucial to the method of their production; if material produced *in vivo* is acceptable, producers will be understandably reluctant to sacrifice a known for an untried process. The main technical problems would appear to be: achieving high oxygen transfer rates, preventing PIB loss at high passage levels and cell concentrations, and balancing media for efficient nutrient utilization. There appears to be no *prima facie* case why these problems cannot be resolved.

Possibly *in vitro* production of insect viruses may first be achieved by an organization with an interest in fermentation and with spare fermenter capacity. Successful operation of such a process could stimulate research beyond the present predominant area of lepidopteran NPV production. It is highly probable that numerous CPVs and EPVs can be produced *in vitro*, although establishment of GV in cell culture is still intractable. With the notable exception of sawflies, little is known of viruses pathogenic for non-lepidopteran insects, and continuous cell cultures from such important groups as the Hymenoptera, Coleoptera and Orthoptera are lacking. While other methods may prove more effective, *in vitro* systems could contribute modestly to the understanding and possible application of the pathogens of these groups.

References

Bellet, A.J.D. and Mercer, E.H. (1964). *Virology* 24, 645–653.

Bergold, G.H. (1959). *In* "The Viruses" Vol I (F.M. Burnet and W.M. Stanley, eds), pp. 505–523. Academic Press, New York and London.

Brown, M. and Faulkner, P. (1975). *J. Invertebr. Path.* 26, 251–257.

Brown, M. and Faulkner, P. (1977). *J. gen. Virol.* 36, 361–364.

Buck, J.B. (1953). *In* "Insect Physiology" (K.D. Roeder, ed.), pp. 147–190. John Wiley, New York.

Carrel, A. (1913). *J. expl. Med.* 17, 14–19.

Chao, J. and Ball, G.H. (1971). *Curr. Top. Microbiol. Immun.* 55, 28–32.

Chiu, R.J. and Black, L.M. (1967). *Nature Lond.* 215, 1076–1078.

Clark, A.J. and Bloch, K. (1959). *J. Biol. Chem.* 234, 2578–2587.

Couch, J.A., Summers, M.D. and Courtney, L. (1975). *Ann. N.Y. Acad. Sci.* 266, 528–536.

Cunningham, J.C. (1970). *J. Invertebr. Path.* 16, 352–356.

Dadd, R.H. (1960). *J. Insect. Physiol.* 5, 161–168.

Eagle, H. and Piez, K.A. (1962). *J. exp. Med.* 116, 29–43.

Echalier, G. and Ohanessian, A. (1970). *In Vitro* 6, 162–172.

Faulkner, P. and Henderson, J.F. (1972). *Virology* 50, 920–924.

Federici, B.A. and Humber, R.A. (1977). *J. gen. Virol.* 35, 387–392.

Gardiner, G.R. and Stockdale, H. (1975). *J. Invertebr. Path.* 25, 363–370.

Gardiner, G.R., Priston, R.A.J. and Stockdale, H. (1976). Proc. 1st Int. Colloq. Invertebr. Path. Kingston, Canada, pp. 99–103.

Geyer, R.P. and Chang, R.S. (1958). *Arch. Biochem. Biophys.* 73, 500–506.

Glaser, R.W. (1918). *J. N.Y. ent. Soc.* 26, 1–3.

Goldschmidt, R. (1915). *Proc. Natl. Acad. Sci. U.S.,* 1, 220–222.

Goodwin, R.H. (1975). *In Vitro,* 11, 369–378.

Goodwin, R.H., Vaughn, J.L. Adams, J.R. and Louloudes, S.J. (1970). *J. Invertebr. Path.* 16, 284–288.

Goodwin, R.H., Vaughn, J.L., Adams, J.R. and Louloudes, S.J. (1974). *Misc. Publ. ent. Soc. Am.* 9, 66–72.

Grace, T.D.C. (1958). *J. gen. Physiol.* 41, 1027–1034.

Grace. T.D.C. (1962a). *Nature, Lond.* **195**, 788–789.

Grace, T.D.C. (1962b). *Virology* **18**, 33–42.

Grace, T.D.C. (1966). *Nature, Lond.* **211**, 366–367.

Grace, T.D.C. (1967a). *In Vitro* **3**, 104–117.

Grace T.D.C. (1967b). *Nature, Lond.* **216**, 613.

Grace, T.D.C. and Brzostowski, H.W. (1966). *J. Insect Physiol.* **12**, 625–633.

Granados, R.R. (1976). *Adv. Virus Res.* **30**, 189–236.

Granados, R.R., McCarthy, R.J. and Naughton, M. (1974). *Virology* **59**, 584–586.

Granados, R.R. and Naughton, M. (1975). *Intervirology* **5**, 62–68.

Hink, W.F. (1970). *Nature, Lond.* **226**, 466–467.

Hink, W.F. (1976). *In* "Invertebrate Tissue Culture: Research Applications" (K. Maramorosch, ed.), pp. 319–369. Academic Press, New York and London.

Hink, W.F. and Ellis, B.J. (1971). *Curr. Top. Microbiol. Immunol.* **55**, 19–28.

Hink, W.F. and Ignoffo, C.M. (1970). *Expl. Cell Res.* **60**, 307–309.

Hink, W.F. and Vail, P.V. (1973). *J. Invertebr. Path.* **22**, 168–174.

Hink, W.F., Strauss, E. and Mears, J.L. (1974). *In Vitro* **9**, 371.

Hink, W.F., Strauss, E.M. and Ramoska, W.A. (1977). *J. Invertebr. Path.* **30**, 185–191.

Ignoffo, C.M. (1965). *Entomophaga* **10**, 29–40.

Ignoffo, C.M. (1966). *In* "Insect Colonization and Mass Production" (C.N. Smith, ed.), pp. 501–530. Academic Press, New York and London.

Ignoffo, C.M. Shapiro, M. and Hink, W.F. (1971). *J. Invertebr. Path.* **18**, 131–134.

Jeuniaux, C. (1971). *In* "Chemical Zoology" Vol VI (M. Florkin and V.T. Scheer, eds.), pp. 64–118. Academic Press, New York and London.

Kelly, D.C. (1976a). *J. Invertebr. Path.* **27**, 415–418.

Kelly, D.C. (1976b). *Virology* **69**, 596–606.

Kurtii, T.J. and Brooks, M.A. (1972). *In* "Insect and Mite Nutrition" (J.G. Rodriguez, ed.), pp. 387–395. N. Holland Publishing Co.

Kurtii, T.J., Chaudhary, S.P.S. and Brooks, M.A. (1974). *In Vitro* **10**, 149–156.

Kurtii, T.J., Chaudhary, S.P.S. and Brooks, M.A. (1975). *In Vitro* **11**, 274–285.

Landureau, J.C. (1968). *Expl. Cell Res.* **50**, 323–387.

Landureau, J.C. and Grellet, P. (1972). *C. r. hebd. Séanc. Acad. Sci. Paris,* **274**, 1372–1375.

Louloudes, S.J., Vaughn, J.L. and Dougherty, K.A. (1973). *In Vitro* **8**, 473–479.

MacKinnon, E.A., Henderson, J.F., Stoltz, D.B. and Faulkner, P. (1974). *J. Ultrastruct, Res.* **49**, 419–435.

Maizels, M., Remington, M. and Truscoe, R. (1958). *J. Physiol.* **140**, 80–93.

McIntosh, A.H., Maramorosch, K. and Rechtoris, C. (1973). *In Vitro* **8**, 375–378.

Merchant, D.J. and Hellman, K.B. (1962). *Proc. Soc. expl. Biol. Med.* **110**, 194–198.

Mitsuhashi, J. (1967). *Nature, Lond.* **215**, 863–864.

Mitsuhashi, J. (1973). *Appl. Ent. Zool.* **8**, 64–72.
Mitsuhashi, J. and Maramorosch, K. (1964). *Contrib. Boyce. Thompson Inst.* **22**, 435–460.
Peleg, J. (1965). *Nature, Lond.* **206**, 427–428.
Peleg, J. and Sahar, A. (1972). *Tissue Cell* **4**, 55–62.
Pollard, R. and Khosrovi, B. (1978). *Process Biochem.* May, 31–37.
Potter, K.N., Faulkner, P. and McKinnon, A.E. (1976). *J. Virol.* **18**, 1040–1050.
Potter, K.N., Jaques, R.P. and Faulkner, P. (1978). *Intervirology* **9**, 76–85.
Ramoska, W.A. and Hink, W.F. (1974). *J. Invertebr. Path.* **23**, 197–201.
Schneider, I. (1964). *J. expl. Zool.* **156**, 91–104.
Scotti, P.D. (1976). *Intervirology* **6**, 333–342.
Shapiro, M. and Ignoffo, C.M. (1970). *J. Invertebr. Path.* **16**, 107–111.
Sohi, S.S. (1969). *Can. J. Microbiol.* **15**, 1197–1200.
Sohi, S.S. (1973a). Proc. Int. Colloq. Invertebr. Tissue Cult., 3rd, 1971, pp. 75–92
Sohi, S.S. (1973b). Proc. Int. Colloq. Invertebr. Tissue Cult. 3rd, 1971, pp. 27–39.
Stockdale, H. (1977). Abst. Xth Ann. Meet. Sci. Invertebr. Path. E. Lansing, Mich. p 29–30.
Stockdale, H. and Gardiner, G.R. (1976a). *In* "Invertebrate Tissue Culture: Applications in Medicine, Biology and Agriculture" (E. Kurstak and K. Maramorosch, eds.), pp. 267–274. Academic Press, New York and London.
Stockdale, H. and Gardiner, G.R. (1976b). Europ. Soc. An. Cell Technol. Proc. 1st Gen. Meet. Amsterdam, pp. 15–21.
Stockdale, H. and Gardiner, G.R. (1977). *J. Invertebr. Path.* **30**, 330–336.
Stoker, M.G.P. and MacPherson, I. (1964). *Nature, Lond.* **203**, 1335–1337.
Sutcliffe, D.W. (1963). *Comp. Biochem. Physiol.* **9**, 121–135.
Trager, W. (1935). *J. expl. Med.* **61**, 501–513.
Umbreit, W.W. (1972). *In* "Manometric and Biochemical Techniques" (W.W. Umbreit, R.H. Burris and J.F. Stauffer, eds.), pp. 30–47. Burgess Publishing Co., Minneapolis.
Vago, C. and Chastang, S. (1958). *Experientia* **14**, 110–111.
Vago, C. and Chastang, S. (1962). *C.r. hebd. Séanc. Acad. Sci. Paris* **255**, 3226–3228.
Vaughn, J.L. (1968). Proc. Int. Colloq. Invertebr. Tissue Cult. 2nd, 1967, pp. 119–125.
Vaughn, J.L. (1976). *J. Invertebr. Path.* **28**, 233–237.
Volkman, L.E. and Summers, M.D. (1975). *J. Virol.* **16**, 1630–1637.
Volkman, L.E. and Summers, M.D. (1977). *J. Invertebr. Path.* **30**, 102–103.
World Health Organization (1973). WHO Techn. Rep. No. 531, Geneva.
Yunker, C.E., Vaughn, J.L. and Cory, J. (1967). *Science* **155**, 1565–1566.

CHAPTER 17

The Nucleopolyhedrosis Virus of *Heliothis* Species as a Microbial Insecticide

C.M. IGNOFFO

Biological Control of Insects Research Laboratory,
U.S. Department of Agriculture, Columbia, Missouri, USA

and

T.L. COUCH

Abbott Laboratories, Chemical and Agricultural Products Division,
North Chicago, Illinois, USA

I. Introduction

Development of the nucleopolyhedrosis virus (NPV) of *Heliothis* spp. (*Baculovirus heliothis*)* began in 1961, progressed through various research and developmental phases, and attained technical realization as the first commercial viral pesticide (Ignoffo, 1973a). An exemption from the requirement of a tolerance was granted by the Environmental Protection Agency in May, 1973 (Anon., 1973) and a label approved in December, 1975. Currently, *B. heliothis* is marketed as safe and effective for use on cotton under the name Elcar[tm] (Sandoz Inc.); Nutrilite Products Inc. has an equivalent experimental product called Biotrol-VHZ.

There are several major reasons why a virus of *Heliothis* spp. was chosen as the first to be developed (Ignoffo 1973a): (1) species of *Heliothis* are major, worldwide pests which attack at least 30 different food and fibre crops (Fig. 1a); (2) considerable quantities of broadspectrum, toxic, chemical insecticides are used against *Heliothis;* (3) costs of control and loss from *Heliothis* spp. in the USA alone exceed 10^9 \$(U.S.); and (4) species of *Heliothis* are resistant to all major insecticides.

Research and development of *B. heliothis* from initial field isolation to a viral product were planned by using a program--evaluation--review technique (PERT) (Ignoffo, 1973a; 1975a). The PERT program included basic studies on *B. heliothis*, production of virus, evaluation of possible risks to man and the environment, and effectiveness against the target pest. Objectives, defined at each successive, developmental phase, were frequently evaluated to determine whether technical success could be attained and then translated into a commercial success. *B. heliothis* was considered a technical success if it could be continuously produced, was safe and would effectively suppress target pests.

Developmental costs were 2 to 5 X below those estimated for a chemical pesticide (Ignoffo, 1973a, 1975b). About 2×10^6 U.S. \$ and 33 research-team-years were needed to develop a proto-product. Most of the cost was borne by

---

* Editorial note: the binomial system has not been applied to viruses in the rest of this book, but has been used in this chapter on the personal preference of the authors.

industry (80%), the balance by federal and state agencies. About 6% of this cost was to obtain sufficient data to decide whether *B. heliothis* could be commercialized.

II. Identification, Description and Detection

A. NOMENCLATURE AND IDENTITY

Four viruses are reported from *Heliothis* species; three inclusion viruses and one non-inclusion virus. Two inclusion viruses (granulosis and NPV) are in the genus *Baculovirus* (Parsons, 1936; Smith and Rivers, 1956; Steinhaus, 1960; Ignoffo, 1965a; Falcon *et al.*, 1967; Gitay and Polson, 1971; Teakle, 1974). The other inclusion virus (a cytoplasmic polyhedrosis) is included within the Reoviridae (Smith and Rivers, 1956; Sikorowski *et al.*, 1971). The non-inclusion virus is in the genus *Iridovirus* (Stadelbacher *et al.*, 1978). Since the virion of the NPV of *Heliothis* spp. is rod-shaped, it is a *Baculovirus* and has been named *Baculovirus heliothis* (Ignoffo and Allen, 1972; Ignoffo *et al.*, 1974a, b; Ignoffo, 1975a; Vago *et al.*, 1974).

Parsons (1936) was the first to suggest that virus caused the wilt disease of *Heliothis* spp. (Fig. 1b) although wilt symptoms were observed earlier by Mally (1891, 1892), Lounsbury (1913), and Chapman and Glaser (1915). Identification as an NPV was later confirmed by electron micrographs of polyhedral inclusion bodies (PIBs) and virions (Smith and Rivers, 1956; Steinhaus, 1957, 1960; Bergold and Ripper, 1957). Current results (morphology, biochemistry, physico-chemistry, microbiology, pathology and host specificity) presented in this section suggest that *B. heliothis*, found in different parts of the world, is the same virus. For example, McCarthy *et al.* (1978) found no significant differences in molecular weight, melting temperature, guanine plus cytosine ratio and nucleotide dissociation value between DNA of *B. heliothis* isolated from four species of *Heliothis* (*H. zea*, USA; *H. virescens*, Colombia, South America; *H. armigera*, South Africa; *H. punctigera*, Australia). In addition, no significant differences in polypeptides were observed between purified virions or nucleocapsids of *H. zea* and *H. armigera* isolates of *B. heliothis* (polyacrylamide gel electrophoresis) nor when these virions were compared with virions from *H. zea*, passed once through *H. virescens* (Brown D.A., personal communication).

B. DESCRIPTION OF INCLUSION BODY

B. heliothis is a typical NPV which produces crystal-like, irregular, proteinaceous PIBs in nuclei of infected cells (Gregory *et al.*, 1969). Some faces of PIBs are rhomboid, and the edges are rounded. There are no apparent patterns or structures on the surfaces of PIBs, although regularly spaced projections are sometimes seen at the outer surface and purification may etch the surface to expose virions and virion sockets (Fig. 2). Although there is no true membrane covering a PIB (nor is there any change in the lattice pattern of the outer layer of a PIB), difficulties in staining PIB, the retention of their shape, and the

Fig. 1. *Heliothis zea*: (a) healthy larvae feeding on corn kernels, (ca. 2x), (b) larvae killed by *Baculovirus heliothis* (ca. 3x).

presence of a membrane-like coat following chemical or physical treatment indicate that the exterior portion of a PIB is different from the interior portion (Fig. 3).

Fig. 2. Carbon replica of purified polyhedral inclusion bodies (ca. 25,000 x); arrows point to virion sockets.

The proteinaceous lattice of the PIB-matrix appears as two sets of orthogonal, electron-dense lines (Scharnhorst *et al.*, 1977; Fig. 4). Matrix protein and protein at the exterior layer(s) of PIBs have major polypeptides of molecular weight 27 500 and 24 500 daltons, respectively (Ryel, 1973). Treatment with alkali yielded 20 bands of proteins.

The mean diameter of PIBs from *H. zea* is 916.4 ± 9.5 nm (S.E. = standard error) (Gregory *et al.*, 1969) and from *H. virescens* 1020 ± 33 nm (S.D. = standard deviation) (Tompkins *et al.*, 1969). The mean diameter of PIBs from *H. armigera* is 1100 nm (Bergold and Ripper, 1957), 1000–1900 nm with some as

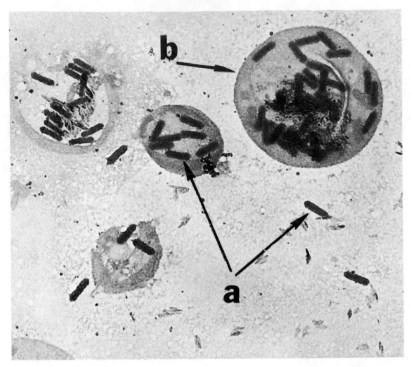

Fig. 3. Dissolved and partially dissolved inclusion bodies showing (a) released virions and (b) membrane-like covering of inclusion bodies (ca. 18,000x).

large as 3000 nm (Lipa, 1968) and 978 ± 11 nm (S.E.) (Teakle, 1973). Density of PIBs is 1.26–1.28 g/ml (Cline *et al.*, 1972; Ryel, 1973; Scharnhorst *et al.* 1977). Sedimentation coefficient for PIBs is 1.223×10^5 S, and 3.58, 12.12, 18.85 S for matrix protein (Ryel, 1973).

The mean CHN contents for PIBs are 50.7%, 7.2% and 13.1% respectively (Shapiro and Ignoffo, 1971a). Seventeen metals and at least four non-metals are present. Silicate, postulated as the skeletal lattice on which protein is deposited during PIB formation, was present at 0.12–0.16% (Estes and Faust, 1966; Shapiro and Ignoffo, 1971a). The amino acid pattern and content of PIBs is like that of other NPVs and is similar to the pattern for healthy larval tissue (Shapiro and Ignoffo, 1971b). Nucleic acid concentration of PIBs is: DNA 6.81 ± 0.14 (S.E.) μg/mg PIB; RNA 2.02 ± 0.03 μg/mg PIB (Estes and Ignoffo, 1965; Shapiro and Ignoffo, 1971b). The DNA is from occluded virions; RNA is probably a cellular contaminant trapped during PIB formation.

C. DESCRIPTION OF VIRION

Infectious, rod-shaped virions are randomly occluded and singly embedded in PIBs without any apparent disruption of the lattice (Fig. 4a); an 8-nm layer

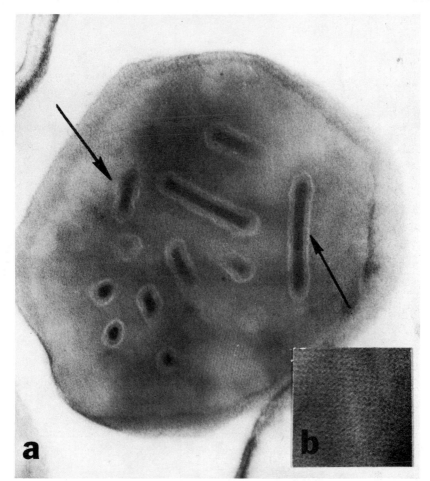

Fig. 4. Cross section through a polyhedral inclusion body showing (a) occluded virions and (b) lattice pattern (ca. 60 000x) (courtesy, J. Adams).

separates virion from protein matrix. Bundles of virions were never observed and the mean virions/PIB was 26.4 ± 5.8 (S.D.). There are no apparent differences in size or structure of virions of *B. heliothis* obtained from *H. zea, H. virescens* or *H. armigera*. Alkali-liberated virions (Fig. 5a) measured 336 ± 22 (S.E.) nm × 62 ± 4 nm (Gregory *et al.*, 1969); embedded virions measured 292 x 48 nm (Scharnhorst *et al.*, 1977). Alkaline-liberated virions readily lose their envelopes (Fig. 5b) to reveal nucleocapsids each made up of a capsid surrounding a DNA core. The capsid, in turn, consists of protein subunits arranged along its long axis. A distinct polypeptide of 18 000 daltons was detected from virions and virion envelope (Ryel, 1973). The angle (88%) and

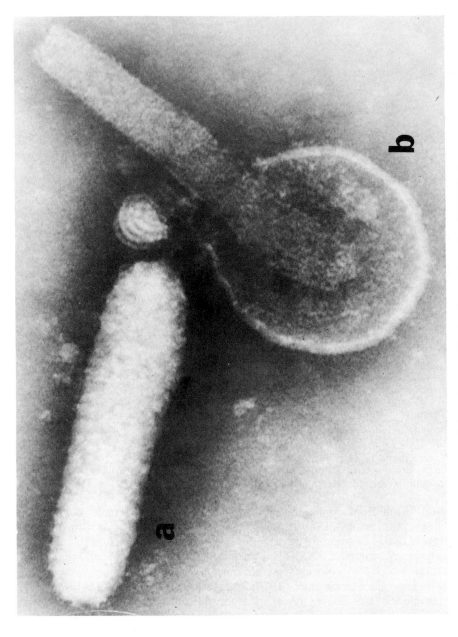

Fig. 5. (a) Alkali-liberated virion and (b) nucleocapsid losing its envelope (ca 450 000 x).

spacing (4.8 nm) are consistent with a helix pattern with a pitch of $2°$ (Scharnhorst *et al.*, 1977).

Virions from *H. armigera* with an envelope measured 320 ± 10 (S.E.) nm $\times$ 90 ± 10 nm (Bergold and Ripper, 1957). These early measurements were subsequently corroborated by Teakle (1973) i.e. virion with envelope 325 ± 4.9 mm $\times$ 76.2 ± 1.1 nm (S.E.), virion without envelope 368 ± 18.8 nm $\times$ 54.5 ± 1.5 nm. The density of virions is 1.16 g/ml on sucrose gradients (Scharnhorst *et al.*, 1977) and 1.296 g/ml on CsCl banding (Ryel, 1973). The sedimentation coefficient for virions is 1835 S (Ryel, 1973).

D. CHARACTERIZATION OF NUCLEIC ACID

Virions of *B. heliothis* contain double-stranded, circular DNA molecules (Estes and Ignoffo, 1965; Gregory *et al.*, 1969; Rubinstein *et al.*, 1976; Scharnhorst *et al.*, 1977). Microscopic observation of DNA extruding from virions (Fig. 6) and of supercoiled circular DNA indicated that the DNA was packed into the nucleocapsid (Fig. 7) and folded over on itself at least once (Scharnhorst *et al.*, 1977). Scharnhorst *et al.* (1977) and Burgess (1977) detected only two bands of DNA in Cs Cl gradients containing the intercalating dye ethidium bromide: a dense band (1.61 g/ml) of supercoiled DNA which was destroyed by shearing and a less dense band (1.57 g/ml) of nicked circles and linear DNA. Presence of supercoiled and open circular DNA also was confirmed by electron microscopy. McCarthy *et al.* (1978) obtained the same bands, plus a band of intermediate density containing undefined DNA molecules from *B. heliothis* isolated from four species of *Heliothis*. Thermal denaturation of DNA indicated a hyperchromicity of $> 25\%$ (McCarthy *et al.*, 1978). Treatment with DNase gave a hyperchromicity of 32% and RNase treatment did not indicate hyperchromicity (Rubinstein *et al.*, 1976). The bouyant density of DNA ($Cs_2 SO_4$) was 1.42 g/cm^3 (Rubinstein *et al.*, 1976).

DNA molecules measured 15 to 45 μm (real genome size) with most molecules at 20–25 μm (Scharnhorst *et al.*, 1977; Burgess, 1977; McCarthy *et al.*, 1978). Molecular weight of the DNA, based upon electron microscopy, reassociation kinetics and restriction endonuclease analysis (REN), ranged from 40 to 93 $\times 10^6$ daltons. Measurements of DNA molecules provided estimates of 80 $\times 10^6$ daltons (Ignoffo, unpublished observations), 77.2 ± 3.0 (S.E.) $\times 10^6$ (Burgess, 1977), 40–50 $\times 10^6$ (Scharnhorst *et al.*, 1977), and 87.9 ± 2.6 (S.D.) to $92.9 \pm 3.8 \times 10^6$ (McCarthy *et al.*, 1978). Reassociation kinetics and REN provided estimates of 50 $\times 10^6$ daltons, $62 \pm 7 \times 10^6$ and 72.6×10^6 (Scharnhorst *et al.*, 1977; Miller and Dawes, 1978; Smith and Summers, 1978).

E. DETECTION OF *BACULOVIRUS HELIOTHIS*

Agar gel diffusion, fluorescent antibody labelling, enzyme-linked immuno-sorbent assaying, haemagglutination and viral neutralization and REN have all been used to detect or identify *B. heliothis* (Shapiro and Ignoffo, 1970a, 1971c, 1975; Teakle, 1973; Davidson and Pinnock, 1973; Young *et al.*, 1975; Scott *et al.*, 1976; Kelly *et al.*, 1978; Miller and Dawes, 1978). Gel diffusion

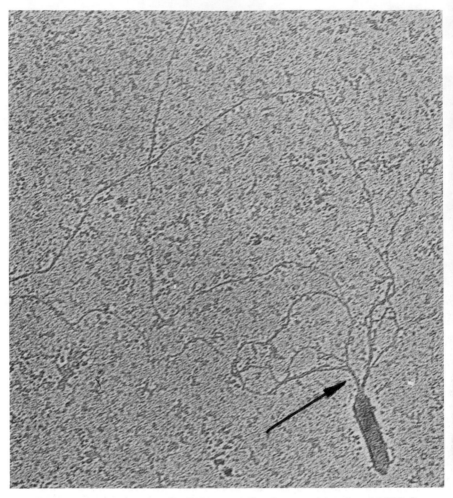

Fig. 6. Supercoiled, circular DNA extruding from a virion (ca. 95 000x).

techniques have not satisfactorily differentiated between PIB-protein or virion protein of isolates of *B. heliothis* (Teakle, 1973) but may be used to monitor *B. heliothis* in the host or in the environment (Shapiro and Ignoffo, 1971c; Davidson and Pinnock, 1973; Young *et al.*, 1975; Kelly *et al*, 1978). More recently, peptide mapping was used to identify the PIB protein of *B. heliothis* (Maruniak and Summers, 1978). An isolate of *B. heliothis* from *H. punctigera* was serologically different from four other NPVs but not from NPV of *H. armigera* (Teakle, 1973). Young *et al.* (1975) used gel diffusion to detect *B. heliothis* (48 h post-exposure) in instar II *H. zea* larvae but could not detect infection in instar I larvae or infections before 48 h post-exposure. Ultra-low volume applications of PIBs on cotton were monitored by a fluorescent

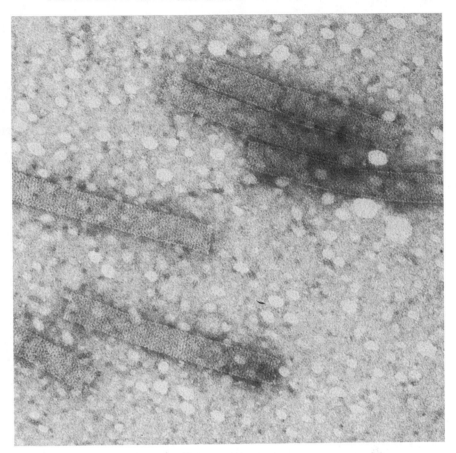

Fig. 7. Nucleocapsid sheaths after dissolution with alkali (ca. 250 000x).

antibody labelling technique, which detected PIBs on cotton plants 320 feet from the spray generator (Davidson and Pinnock, 1973). More recently, an enzyme-linked immunosorbent assay was suggested as a rapid, sensitive, simple technique to monitor baculoviruses (Kelly *et al.*, 1978). It was highly specific for virions of *B. heliothis* (reactions were not obtained with seven other baculoviruses) and as little as 1 ng/ml of virion antigen could be detected. Virion antigens also could be detected in larval extracts and in infected larvae 12 h post-exposure.

Since NPV agglutinated chicken red blood cells Shapiro and Ignoffo (1970a) reasoned that this phenomenon might be used to distinguish viral isolates. Haemagglutination occurred only when virions were present and it could be specifically inhibited by virion antiserum; however, differences between isolates were not detected. This technique was ca. 30X more sensitive than Ouchterlony but ca. 10^6 X less sensitive than bioassay. Haemagglutination, however, could be detected within 1 h whereas bioassay required at least 72 h.

Neutralization techniques also might be used to monitor activity of *B. heliothis* (Shapiro and Ignoffo, 1975). About 65% of the viral activity (solubilized PIBs) was neutralized after treatment with virion antisera while antiserum to intact PIBs, PIB-protein or *H. zea* haemolymph did not neutralize viral activity.

REN and gel electrophoresis resolved about 23 to 27 fragments of the viral DNA-genome (Miller and Dawes, 1978; Smith and Summers, 1978) 18 fragments were $> 10^6$ daltons and ca. 9 fragments were $< 10^6$ daltons (Miller and Dawes, 1978).

III. Activity and Host Susceptibility

A. MODE OF ACTION

The following mode of action of *B. heliothis* in *Heliothis* larvae was summarized from the studies of Harpaz and Zlotkin (1965), Lipa (1968), Ignoffo (1973a), Teakle (1973) and Whitlock (1974). Shortly after ingestion, an infectious entity, released from dissolved PIBs, passed from the midgut to the haemocoel and subsequently infected nuclei of susceptible cells. Within 1—2 days of ingestion, the usually clear haemolymph became cloudy and PIBs were detected in haemocytes and in a few cells of the midgut. Hyperaminoacidaemia and hyperproteinaemia were detected within 24 h and glycine increased steadily, coinciding with PIB formation and an increase in nucleic acid synthesis (Shapiro and Ignoffo, 1971d). The decrease in haemocytes (in larvae exposed to a high viral dose) also coincided with PIB formation (Shapiro *et al.*, 1969). Chromatin changes were the first indication of cellular infection (generally in haemocytes and fat body cells), followed by assembly of virions, formation of PIBs and hypertrophy of nuclei. Infected larvae at this stage appeared shiny, discoloured and sometimes pink-white. *B. heliothis* has been seen in fat body, hypodermis, tracheal matrix, epithelial and blood cells (leucocytes and lymphocytes but not spherule cells or oenocytoids). Cell lysis and disintegration of larval tissue began shortly after PIB formation. Young larvae died within 2 days but older larvae within 4 to 9 days after ingestion of virus. Shortly after death the larva becomes flaccid and the integument ruptures, releasing billions of inclusion bodies.

B. ACTIVITY AND SUSCEPTIBILITY

1. Activity of Viral Isolates

There was a 56-fold range in activity (LD_{50}) for 34 isolates of *B. heliothis*, among which two to eight activity classes were defined (Shapiro and Ignoffo, 1970b). Activity of one isolate was enhanced ca. 80% by *in vivo* serial passage, while no change was obtained with another isolate. Differences in isolates were attributed to physico-biochemical characteristics of PIBs or virions rather than to virulence of the viral genome.

2. Species Susceptibility

Seven species of *Heliothis* were susceptible to *B. heliothis* (cf. section VB). Although different bioassays were used there was no major difference in susceptibility between five of these species (Ignoffo, 1968a, b). Direct comparisons between two species indicated that *H. virescens* was ca. 30–50% more susceptible than *H. zea* regardless of whether *B. heliothis* was isolated from *H. zea* or *H. virescens* (Ignoffo, 1965a, c 1966c).

3. Effects of Age, Body Weight and Temperature

Decreased susceptibility (LD_{50} or LT_{50}) due to maturation was recorded for both *H. zea* and *H. virescens* larvae exposed to *B. heliothis* (Ignoffo 1966a). Of 1- to 6-day old *H. zea* (0.1 to 179 mg/larva) or *H. virescens* larvae (0.1 to 77.6 mg/larva), > 99% died after exposure to 2598 or 1247 PIB/mm^2 of diet respectively. Mortality thereafter steadily decreased as larval age and body weight increased. Two days were required for initial mortality regardless of dosage. Similar results were obtained with 1- to 13-day-old larvae of *H. armigera* (Daoust, 1974; Whitlock, 1977a). All of Daoust's larvae of *H. armigera* 1—7 days old (0.1 to 11 mg/larva) died after exposure to 1054 PIB/mm^2; mortality decreased with increase in age (98% for 5—8 day-old larvae, 28 mg/larva; 1.3% for 11-day-old larvae, 532 mg/larva). The LT_{50} increased from 3.6 days for 1-day-old larvae to 9.9 days for 8—9 day-old larvae (122 mg/larva). All of Whitlock's 1- to 5-day-old larvae of *H. armigera* (7 to 28 mg/larva) died after exposure to 2200 PIB/mm^2. Mortality decreased from 72% for 6-day-old larvae (49 mg/larva) to 0% for 12-day-old larvae (385 mg/larva). The LT_{50} increased from 2—3 days for 1-day-old larvae to 14 days for 8-day-old larvae.

No difference in mortality of 5-day-old *H. zea* larvae (at 5° intervals from 13° to 35°C) was detected when larvae were exposed to 1139 PIB/mm^2 (88—98% mortality) or 11 PIB/mm^2 (44—64% mortality) (Ignoffo 1966b). There was, however, some indication that higher temperatures inhibited viral development, especially at low dosages. Temperature did not substantially influence total mortality but did influence its rate.

4. Resistance and Inhibition

Selection pressures of LC_{50-70} maintained for 20 and 25 generations did not induce resistance of *H. zea* to *B. heliothis* (Ignoffo and Allen, 1972). Laboratory populations under selection pressure were as susceptible as a non-selected population or five wild populations of *H. zea*. Similar results were obtained with laboratory populations of *H. armigera* selected for resistance over 22 generations (Whitlock, 1977a, b).

Total mortality and rate decreased when *H. armigera* larvae were simultaneously infected with an NPV and a granulosis virus (GV) (Whitlock 1977c). Mortality curves for the simultaneous infections were similar to that for GV alone. The LT_{30} for NPV, NPV + GV, and GV was ca. 7, 21 and 23 days, respectively. Moulting hormone (β-ecdysone) also inhibited viral replication of *B. heliothis,* probably due to competition for available cellular nucleotides and amino acids (Keeley and Vinson, 1975).

IV. Stability, Sensitivity and Persistence

A. ULTRAVIOLET STABILITY

In the 1950's Watanabe (1951) and Aizawa (1955) demonstrated that short-wave UV (215–260 nm) inactivated PIBs; however, researchers generally assumed that PIB protein would shield virions from inactivation by sunlight-UV. We now know that natural sunlight-UV (> 290 nm) is the major environmental factor inactivating *B. heliothis* and probably most insect viruses. This has been conclusively demonstrated for *B. heliothis* exposed to natural sunlight (Bullock, 1967; Ignoffo and Batzer, 1971; Young and Yearian, 1974; Ignoffo *et al.*, 1967, 1977; Bull *et al.*, 1976) as well as to simulated sunlight and germicidal lamps (Ignoffo and Garcia, 1966a; Gudauskas and Canerday, 1968; Bullock *et al.*, 1970; Ignoffo and Batzer, 1971; Ignoffo *et al.*, 1976a, 1977; Bull *et al.*, 1976; McLeod *et al.*, 1977). The half-life of *B. heliothis* was < 1 day when PIBs were exposed to either simulated or natural sunlight. Total inactivation of PIBs occurred within minutes of exposure to germicidal lamps. Other environmental factors (i.e. temperature, humidity or pH) contributed only slightly to UV-inactivation of *B. heliothis* (McLeod *et al.*, 1977; Andrews and Sikorowski, 1973).

Since about 1965 there has been a concerted effort to minimize or prevent sunlight inactivation of *B. heliothis*. Commonly used sun-screens or sun-blocks generally did not provide protection, could not be adapted to aqueous, wettable-powder formulations, or were too expensive (Ignoffo, unpublished observations). Particulate sun-shields, e.g. carbon, carbon-based dyes, aluminium oxide, titanium dioxide, clays, flour (Ignoffo and Batzer, 1971; Bull *et al.*, 1976; Ignoffo *et al.* 1976a, b), flourescent materials, e.g. polyflavanoids, whitening agents (Ignoffo *et al.*, 1972; Ignoffo and Garcia, unpublished observations) and baits (McLaughlin *et al.*, 1971) have all increased the persistence of PIBs. Micro-encapsulation with sunlight protectants also increased the stability of PIBs; however, laboratory and field studies indicated that this increase was no better than that given by simple mixtures, or non-encapsulated or spray-dried, commercial formulations (Ignoffo and Batzer, 1971; Bull *et al.*, 1976; Ignoffo *et al.*, 1976 a, b,; Bull, 1978). Recently, Ignoffo *et al.* (1977) suggested that sunlight inactivation of *B. heliothis* probably is caused by peroxide or peroxide radicals produced by the UV irradiation of amino acid(s). Sunlight inactivation was inhibited but not completely stopped by addition of peroxidase (Ignoffo and Garcia, 1978).

B. TEMPERATURE STABILITY

Although field temperatures (15 to 45°C) had no effect on stability of PIBs (MacFarlane and Keeley, 1969; McLeod *et al.*, 1977), viral replication was inhibited at 40°C (Ignoffo, 1966c). In water, 10 min at 80°C completely inactivated PIBs whereas 10 min at 70°C reduced activity ca. 11% (Ignoffo, unpublished observations). Gudauskas and Canerday (1968) obtained similar results after 10 min, i.e. no loss of activity at 70–75°C, 40% loss at 75°C and

100% loss at 80°C. MacFarlane and Keeley (1969) detected no loss after 20 min at < 65°C, an initial loss at 70°C and total loss after 5 min at 85°C. In contrast, Stuermer and Bullock (1968) recorded 45% and 66% loss after 15 min at 71.1°C and 82.2°C respectively, and 66% and 96%, respectively after 2 h. Virions were relatively stable in water for 7 days at 5°, 37° or 50°C but lost significant infectivity thereafter, the half-life being 120, 60 and 50 days, respectively and the loss after 225 days about 16-, 100-, and 600-fold respectively (Shapiro and Ignoffo, 1969). Infectivity was still present after 4 weeks at 37°C in cultures of primate cells inoculated with virions (McIntosh and Maramorosch, 1973; McIntosh, 1975). Lyophilized preparations of PIBs at − 20°C and 5°C were very stable and lost no activity after ca. 15 years although an initial loss was detected after ca. 5 years for samples stored at 23–27°C (Ignoffo and Garcia, unpublished observations).

C. SENSITIVITY TO CHEMICALS

It has been known for some time that acids or alkalis disrupt PIBs and thus presumably destroy viral activity (Bergold, 1963). Ignoffo and Garcia (1966b) extended this generalization to *B. heliothis:* viral activity was not affected at pH 7.0, but was reduced slightly (< 15%) at pH 4.0 and 9.0 and reduced significantly (> 95%) at pH 1.2 and 12.4. These results were later corroborated by Gudauskas and Canerday (1968) using phosphate buffers: infectivity decreased as molarity increased; however, molarities < 0.01 at pH 2.0 did not reduce viral activity.

Early field and laboratory studies indicated that most insecticides or insecticidal adjuvants were compatible with *B. heliothis* (Tanada and Reiner, 1962; Wolfenbarger, 1964, 1965; Ignoffo *et al.* 1965; Ignoffo and Montoya, 1966). In a sensitive bioassay, only methyl parathion (of 12 insecticides or adjuvants) inactivated PIBs (Ignoffo and Montoya, 1966). Formalin and sodium hypochlorite destroy viruses and thus are commonly used as disinfectants for NPV (Vail *et al.*, 1968; Ignoffo, 1967; Ignoffo and Garcia, 1968).

D. PERSISTENCE

1. On Plants

Persistence of PIBs on foliage may be affected by the kind of leaf, the leaf texture, leaf exudates or dew. For example, PIBs persisted longer on tomato and soybean leaves than on cotton leaves: the percentage original activity of PIBs remaining (%OAR) after 48 h was 32.4%, 30.0% and 13.6% respectively (Young and Yearian, 1974). Greater stability on tomato was probably due to mechanical screening of PIBs by leaf-hairs and leaf-curling. Young and Yearian (1974) obtained similar results with hairy cotton leaves.

Control of *H. zea* larvae was increased in California (Falcon 1969) but not in Arkansas (Young and Yearian, 1976) when buffered *B. heliothis* was applied to cotton. Since dry periods increased pH of the leaf surface it was concluded that pH and ionic concentration of dew would be greater where minimal rainfall and flood-type irrigation prevailed (Young et al., 1977). Andrews and Sikorowski (1973) concluded that more than high pH was needed to inactivate PIBs since washes from cotton leaves (pH 9.6–10.1) caused only swelling of PIBs. Young et al. (1977) and McLeod et al. (1977) later reported no loss in activity when PIBs were placed in dew from cotton leaves (pH 7.4 to 8.8) or soybean leaves (pH 7.2 to 7.8). Although structural disruption of PIBs was not seen after 7 days on either cotton or soybean leaves, PIBs dissolved and lost activity after repeated drying and wetting in cotton dew.

The half-life of B. heliothis on soybean leaves was ca. 2 days and no activity was detected after 14 days exposure (Ignoffo et al., 1974a). It persisted better on sorghum than on cotton in tests in Africa (Roome and Daoust, 1976). The mean half-life on cotton was < 2 days, but on sorghum it was > 30 days and virus was still detected at harvest > 80 days after spraying. The long half-life on sorghum was probably the result of accumulation of virus from dying larvae and/ or shielding of PIBs from sunlight by the compact sorghum head. Ignoffo et al. (1965) estimated that death of one larva in a sorghum head could provide 1000X more virus than that originally applied.

Placement of B. heliothis on specific plant parts may also affect its stability. For example, in the USA PIBs on the calyx, bract or bloom of cotton and the underside of cotton leaves were more protected than on the upper surfaces of either mature or terminal leaves (Yearian and Young, 1974). Virus persistence was ca. 10X greater on the calyx and the inner surface of bracts than on mature leaves. The half-life at unprotected sites was < 24 h, at protected leaf sites 24–48 h and at protected floral sites > 96 h. In central Africa there was hardly any difference in persistence of PIBs placed on the upper or lower surface of cotton leaves; the half-life was ca. 6 days (McKinley, 1971). The half-life of PIBs on corn silks was similar to that on cotton (Ignoffo et al., 1972, 1973) and addition of carbon extended it from < 1 day to > 3 days. The half-life of carbon-protected PIBs on cotton was ca. 5 days (Ignoffo et al., 1972).

2. In Soil

Only one study evaluated residues of B. heliothis in soil (Roome and Daoust, 1976). Soil taken from unsprayed and sprayed sorghum plots (87 days post-spray) contained ca. 210×10^{12} and 1764×10^{12} infective PIB/ha respectively. After winter (252 days post-spray) the count was 468×10^{12} PIB/ha and 198×10^{12} PIB/ha for sprayed—cultivated and unsprayed—uncultivated plots respectively. This indicates that PIBs were surprisingly stable, with little or no loss in infectivity in unsprayed plots and only a 75% loss in infectivity in sprayed plots.

3. In Animals

Most of the infectivity was either destroyed or voided within 48 h of oral administration of PIB to white rats at doses equivalent to a man eating the number of PIBs used to treat ca. 35 ha of cotton (Ignoffo et al., 1971a). The half-life of PIB-infectivity (in the alimentary tract) was 30 to 40 min.; 99.9% of PIB-infectivity was lost within 4 h. These in vivo results are in agreement with an *in vitro* half-life of PIBs infectivity in undiluted human gastric juice of ca. 40 min with < 5% of the infectivity still present after 120 min (Chauthani *et al.*, 1968a). PIBs also persist on eggs laid by infected females of *H. zea* (Hamm and Young, 1974).

V. Specificity

A. BACTERIA, YEASTS AND HIGHER PLANTS

Wells and Heimpel (1970) and the patent of Wells (1970) claimed that *B. heliothis* replicated in yeasts and bacteria. This would be a significant advance (Ignoffo, 1967); unfortunately, early reports of replication in non-homologous hosts have not been confirmed (Rubinstein and Polson, 1976; unpublished results of: Ignoffo and Rhodes, Couch, Tinsley, and Faulkner). Higher plants, mostly economic species, were exposed to rates as high as 125 times the mean recommended rate of *B. heliothis* (2.4×10^{11} PIB/0.4 ha) (Ignoffo 1975b). No phytotoxicity or pathogenicity to cabbage, corn, cotton, radish, sorghum, soybean, tobacco or tomato were detected. Over 2 million hectare-treatments just on cotton in the USA produced no report of toxicity or pathogenicity.

B. INVERTEBRATES

Only seven species were susceptible to *B. heliothis:* all are species of *Heliothis*, namely, *armigera*, *paradoxa*, *peltigera*, *phloxiphaga*, *punctigera*, *virescens* and *zea* (Ignoffo, 1965b; Harpaz and Zlotkin, 1965; Ignoffo, 1968a; Teakle, 1973; Vandamme and Angelini, 1966; Patel *et al.*, 1968). In contrast, *B. heliothis* could not be transmitted to 37 other insects, spiders and mites (Table I), fed or topically treated at 10-100x the recommended field rate. Some insects (*Musca domestica, Galleria mellonella, Nomophila nearctica, Trichoplusia ni*) were even injected with virions (Ignoffo, 1968a, 1973a). Jafri and Khan (1969), however, reported that *G. mellonella*, exposed to gamma radiation, were susceptible to *B. heliothis*.

Although non-insect invertebrates have not been extensively tested (Ignoffo, 1968a, 1973a), grass shrimp, brown shrimp, oyster and carmine mite were not susceptible to $10^6 - 10^9$ PIB/animal (Heimpel, 1967; Ignoffo, 1968a). Also, cell lines derived from *Trichoplusia ni* (1 line), *Spodoptera frugiperda* (2 lines) and *Lymantria dispar* (3 lines) would not replicate *B. heliothis*; however, replication was obtained in lines derived from *H. zea* (Goodwin *et al.*, 1976). In another *in vitro* study an infectious entity (probably naked DNA) in *Heliothis* cells induced typical symptoms *in vivo* and could be serially passed *in vitro*; however, typical cytoplasmic effects (CPE), inclusion bodies or virions were never detected *in vitro* (Ignoffo *et al.*, 1971b). Granados *et al.* (1978) later detected a baculovirus from this same *H. zea* cell line.

TABLE I

Species of Insects Not Susceptible to Baculovirus heliothis[a]

LEPIDOPTERA

Agrotis segetum (as fucosa)	Papilio xuthus
Anthela varia	Pieris brassicae
Cactoblastis cactorum	Pieris rapae
Chrysodeixis chalcites	Plusia nigrisigna
Euproctis subflava	Plutella xylostella
Galleria mellonella	Spodoptera exigua
Lymantria dispar	Spodoptera frugiperda
Manduca quinquemaculata	Spodoptera litura
Manduca sexta	Spodoptera mauritia
Nomophila nearctica	Spodoptera ornithogalli
Orgyia anartoides	Trichoplusia ni

HYMENOPTERA

Apis mellifera	Chelonus blackburni
Brachmyeria intermedia	Meterous leviventris
Campoletis sonorensis	

DIPTERA

Musca domestica	Voria ruralis

COLEOPTERA

Hippodamia convergens	Phaedon brassicae

NEUROPTERA

Chrysopa carnea	

HETEROPTERA

Geocoris pallens	Nephotettix cinticeps
Laodelphax striatella	Nilaparvata lugens
Orius tristicolor	

[a] Data from: Chamberlin and Dutky, 1958; Falcon et al., 1965; Cantwell et al., 1966; Atkins and Anderson, 1967; Ignoffo, 1968a, b; Tompkins et al., 1969; Teakle, 1973; Irabagon and Brooks, 1974; Wilkinson, et al., 1975; Lewis, unpublished observations.

C. VERTEBRATES

In the first study of its kind on an insect virus, Ignoffo and Rafajko (1972) demonstrated that B. heliothis would not develop in four primate cell types (human primary embryonic kidney, human diploid embryonic lung, human carcinoma of the cervix, and African green monkey kidney) under conditions that would replicate mammalian viruses. All primate cells failed to: show CPE after three serial passages; agglutinate guinea pig erythrocytes; inhibit the replication of Echo-11 virus. McIntosh and Maramorosch (1973) obtained similar results using primary cultures of human embryo, amnion, foreskin, lung and leucocyte. They never observed evidence of viral replication, i.e. CPE, PIBs, reduced cell viability, presence of antigen (immunofluorescence), culture transformation, and differences in DNA synthesis (autoradiography). B. heliothis, however, persisted for 4 weeks in amnion, lung and leucocyte cultures and, in one of three tests, may have enhanced simian virus-40 transformation of amnion cells.

Results of studies against 18 vertebrates, 36 invertebrates and 24 plants, summarized by Ignoffo (1973a, b, 1975b), demonstrated that *B. heliothis* was safe for use on food crops. Later Ignoffo *et al.* (1975) showed that rhesus monkeys were not susceptible to 26 weekly doses (subcutaneous injection, inhalation, oral gavage) of *B. heliothis.* Clinical results, blood chemistry, and histopathology of treated and untreated monkeys were similar and infectious virus, viral antibodies, or viral antigens were not detected in blood from treated monkeys. Adverse clinical effects were not observed in six persons exposed during 26 months of virus production (Huang *et al.*, 1977). Chemical and serological examinations and bioassays of blood or urine did not reveal infective virus, viral antigens, or viral antibodies. In an unpublished skin-patch study of 48 persons (19 heavy exposure to virus; 19 moderate to occasional exposure; 10 no exposure) *B. heliothis* was not systematically or topically toxic or pathogenic to humans, nor was it a cutaneous or respiratory sensitizer (Ignoffo *et al.*, 1968).

VI. Production

A. FACILITIES AND PROCESS

B. heliothis can be produced only in *Heliothis* larvae (Ignoffo, 1965a, 1966e, 1973a; Ignoffo and Anderson, 1979) although several sophisticated processes have been suggested (Ignoffo, 1967, 1973b; Ignoffo and Hink, 1971; Wells, 1970; Wells and Heimpel, 1970). Mass production of *B. heliothis* began with production of a large quantity of the selected isolate; this provided inoculum for several years of production. This scheme minimized possible genetic changes of *B. heliothis* due to repeated passages through the host. Sub-samples of this inoculum stabilized, lyophilized and sealed under vacuum in individual glass vials were stored at $-20°C$ as reference standards. Production of *B. heliothis* was conducted in four isolated facilities: a diet–storage–preparation facility; a facility with a unit for rearing the stock culture of *Heliothis* and another unit for mass rearing larvae for virus production; a virus production facility; and a product recovery and formulation facility. Other production functions (quality control, bioassay, product characterization) were routinely done in an analytical laboratory.

Only the general process is described since details of diet, insect rearing and virus production were previously reported (Ignoffo, 1965b, 1966e; Ignoffo and Boening, 1970). Machinery and technology of the food industry were used to automate the process. Hot liquid diet was piped from a large mixing tank to a machine that formed trays from plastic sheets; a uniform volume of this diet was dispensed in individual cells of these trays, and the trays were sealed in a continuous automatic operation at up to 200 trays/h. Neonatal larvae were individually inserted through the plastic covering of each cell of each tray. If the process from tray-forming to larval infestation were fully automated, production of 10 000 larvae/h would be possible. Pilot-plant production was about 1 million larvae/month while maximum capacity for continuous operation of one production line could be 7 million larvae/month. Trays, infested with larvae, were stacked on mobile racks, incubated at 30°C for 5–7 days, inoculated with virus and then incubated at 26°C for another 5–7 days. Dead and dying larvae were sucked from each cell, triturated with water, and then concentrated with or without adjuvants and diluents by centrifugation, precipitation, filtration or spray drying. About 7–9X more active virus and 2–5X more PIB/larva were obtained from dead larvae than from living larvae (Ignoffo and Shapiro, 1978). Similar results were obtained when diet and frass were processed along with larvae. Although there were no differences in activity between aqueous, lyophilized or acetone-extracted virus (immediately after processing), the acetone-extracted preparations were not as stable as lyophilized preparations (especially after storage above 35°C).

After drying the virus was screened and processed into a technical product, formulated with various adjuvants and diluents and then packaged for sale. Commercial preparations were characterized as to identity, number of PIB/unit weight, activity/unit weight, number of secondary biotypes/unit weight, particle size and range, % lipids and % moisture. Safety of each production batch of virus was confirmed by serology, injections into mice, and a determination of type and level of microbial contaminants (Ignoffo, 1973c; Duggan, 1970). Bacterial counts for preparations processed from dead larvae, although 50X higher than counts for preparations from living larvae, were still considerably lower than standards for non-fat dry milk (Ignoffo and Shapiro, 1978).

Bioassays were used to determine the activity of each batch of virus (Ignoffo, 1965b). Several modes of administration were tried: surface treatment, diet incorporation and direct feeding using different larval instars (Ignoffo, 1965c; Montoya and Ignoffo, 1966; Chauthani et al., 1968b; Allen and Ignoffo, 1969; Ignoffo and Batzer, 1971; Daoust and Roome, 1974). The easiest large-scale, analytical method, however, was to incorporate virus into a diet and feed it to neonatal or 24 h-old larvae. Insecticidal activity, measured as an LD_{50} or LC_{50} with 95% fiducial limits, was related to a standard and recorded as number of insecticidal units per unit weight of product (Ignoffo, 1966d).

B. PURIFICATION

The production process was designed to exclude rather than remove secondary biotypes. Presence of larval extracts and contaminants thus was not considered a problem if shown to be harmless (Ignoffo, 1973c). Nevertheless, mass purification was studied to ensure that a method was available if needed, and thus it was demonstrated that B. heliothis could be purified isopycnically (Martignoni et al., 1968; Anderson et al., 1969; Breillatt et al., 1969, 1972; Cline et al., 1972). Initially, batch quantities of lyophilized virus and larval homogenates were purified in B-XXIII and B-XXIX rotors and then larger scale purification (100 g PIB/day; a bacteria:PIB ratio of 1:10 000) was attained in a continuous-flow system and isopycnic banding in a K-X rotor. Highly purified PIBs also could be obtained from homogenates of living diseased larvae, living diseased—fermented larvae, dead virus—killed larvae and dead virus—killed—fermented larvae by a single-step isopycnic banding (discontinuous sucrose gradient) in a B-XIV zonal centrifuge rotor.

C. CAPACITY AND COSTS

Current larval production technology easily meets the estimated product demand and costs to make it competitive with chemical insecticides. Assuming an average of three applications of virus/0.4 ha/season, and about 5% of the cotton being treated, then about 0.8×10^6 ha-treatments of virus would be required annually (Anon., 1975; Ignoffo, 1979). Since the average rate of use on cotton is 240×10^9 PIB/0.4 ha (equivalent to ca. 25 larvae), then ca. 50×10^6 larvae/year would satisfy the demand for only cotton in the USA. The current cost for commercial Elcar^{tm} is \$3.12/0.4-ha-treatment. Additional process improvements could easily lower production costs about 50 to 90%.

VII. Efficacy

Between 150 and 200 field tests were conducted with *B. heliothis* during the last two decades. (Since there are many reports, unpublished and published, on the efficacy of *B. heliothis* they are listed, by commodity, in an addendum at the end of the chapter.) About 60% of the tests were on cotton, 30% on corn and the rest on either soybean, sorghum, tobacco or tomato. Field tests before 1967 were done with triturated diseased larvae, or with unformulated freeze-dried or acetone preparations of diseased larvae. Field experiments from ca. 1967 to 1970 were made mainly to evaluate diluents, adjuvants, sunlight protectants and gustatory stimulants in an attempt to increase effectiveness and consistency. Development toward more active, stable formulations began in about 1969 and culminated with registration of Elcar^{tm} in 1973.

A. TESTS ON COTTON

1. Rate—Yield Response

Most of the rate-response tests on cotton were conducted from 1963 to 1967 using the same unformulated viral isolate. A summary of these tests indicated that an increase in rate above 60×10^{10} PIB/0.4 ha did not materially increase yields (Table II). The optimum rate for a light to moderate infestation of *Heliothis* was ca. 10×10^{10} PIB/0.4 ha or for a heavy infestation 45×10^{10} PIB/0.4 ha.

2. Comparisons with Untreated Plots and Insecticide Standards

The mean ratios of cotton yield (*B. heliothis* to untreated check) for 137 comparisons of heavy, moderate or light infestations of *Heliothis* larvae were 2.22, 1.48 and 1.11, respectively (Table III). Generally, rates $> 60 \times 10^{10}$ PIB/0.4 ha gave better control than rates $< 60 \times 10^{10}$ PIB/0.4 ha. The equivalent mean ratios of cotton yields of *B. heliothis* compared to the insecticide standard

TABLE II

Efficacy ratios for yields of cotton from virus treatment to the untreated check for 22 heavy and 16 light to moderate infestations of Heliothis *spp. from 1963–1967*

Dose/0.4 ha	Infestation Level[a]	
(× 10^{10} PIB)	Light to moderate	Heavy
6	1.06	1.77
60	1.24	2.17
600	1.34	2.26

[a] Criteria: untreated checks for heavy levels averaged < 453.6 kg of seed-cotton/0.4 ha; and for moderate to light > 453.6 kg.

TABLE III

Efficacy ratio of virus treatment for yields of cotton at different levels of infestation of Heliothis *spp., 1960 to 1978*

Dose/0.4 ha × 10^{10} PIB	Ratio virus: untreated check[a]			Ratio virus: insecticide[a]		
	Heavy	Moderate	Light	Heavy	Moderate	Light
6	1.77(3)	–	–	0.89(4)	1.08(4)	–
12	–	–	1.07(5)	–	–	0.97(15)
24	2.08(35)	1.68(6)	1.03(15)	0.90(15)	1.12(24)	0.91(17)
30	–	–	1.17(4)	–	1.16(4)	1.10(8)
48	–	–	1.12(8)	0.98(3)	0.85(4)	0.95(8)
60	2.37(38)	1.27(9)	1.18(9)	0.96(30)	0.83(4)	0.94(10)
600	2.26(5)	–	–	1.14(2)	–	–
Mean	2.22	1.48	1.11	0.97	1.01	0.97

[a] Mean yield for untreated checks for heavy infestations < 453.6 kg seed-cotton/0.4 ha; for light, > 680.4 kg. Figures in parentheses are the numbers of comparisons.

(152 comparisons) were 0.97, 1.01 and 0.97, respectively (Table III). Thus, control with *B. heliothis* generally was as effective as with the standard insecticide. Rates below 48×10^9 PIB/0.4 ha, however, were not as effective as the standard insecticides for controlling heavy infestations of *Heliothis* spp.

3. Use of Stimulants and Protectants

Laboratory tests indicated that gustatory stimulants should increase the efficacy of *B. heliothis*. The mean ratio (yield, % reduction in damage, number of larvae) of virus + stimulant to virus in laboratory tests (11 comparisons) was 2.18 (range 1.20 to 2.28). Field results, however, did not reflect this anticipated increase. The equivalent mean ratio (12 comparisons) was 0.92 (range 0.43 to 1.69). A better relationship between laboratory and field studies was obtained with sunlight protectants; mean ratios for virus + protectant to virus in the laboratory (17 comparisons) and field (20 comparisons) were 2.43 and 1.28, respectively. One direct comparison between laboratory and field results (Ignoffo *et al.*, 1976a) indicated that, although field results did corroborate laboratory results, the degree of protection could not be correlated with yield. Mean ratios of virus + protectant to virus in the laboratory and in the field were 2.14 (range 1.52–4.03), and 1.51 (range 1.14–2.14), respectively.

B. TESTS ON CORN

Field tests on corn against *H. zea* used both conventional over-the-plant methods as well as hand application of *B. heliothis* to ears, silks or silk-channels. Degree of control was generally related to application rate (Tanada and Reiner, 1962; Klostermeyer, 1968; Young and Hamm, 1966) and dusts generally performed better than aqueous formulations (Tanada and Reiner, 1962), although paraffinic oils did increase effectiveness (Wolfenbarger, 1964). Control, as measured by either larval counts, % worm-free ears, % damaged ears or % marketable ears was generally most effective with hand application (Table IV). Although a remarkable increase in control over untreated corn (26.97X) was obtained with virus at high levels of *H. zea* infestation, control was still less (0.62) than that with standard insecticides. Results were better than with the

TABLE IV

Efficacy ratio for virus treatments in field tests on sweet corn at various infestation levels of Heliothis zea[a]

Infestation level (% worm-free ears)	HAND APPLIED (5–200 × 10⁶ PIB/ear)		CONVENTIONALLY APPLIED (10–500 × 10¹⁰ PIB/0.4 ha)	
	Virus: check	Virus: insecticide	Virus: check	Virus: insecticide
< 15%	26.97 (15)	0.62 (4)	9.29 (7)	0.85 (7)
> 15% < 60%	3.77 (8)	0.94 (6)	1.48 (8)	0.62 (8)
> 60%	–	–	1.08 (19)	1.56 (4)

[a] Values in parenthesis are numbers of comparisons since 1960.

insecticide standard for moderate and light infestations of *Heliothis*. Oatman *et al.* (1970) concluded that the use of virus on sweet corn might be commercially acceptable, especially when parasites and predators are present.

C. TESTS ON SOYBEANS

Seven studies were conducted against *Heliothis* spp. on legumes. In one laboratory study, > 90% of *H. armigera* larvae feeding on potted gram were killed before reaching instar III (Patel *et al.*, 1968). In another study a gustatory stimulant increased feeding so that an instar III *H. zea* larva consumed 2.1 × an LC$_{90}$ dose (870 PIB) rather than 645 PIB/larva (Allen and Ignoffo, 1969; Ignoffo *et al.*, 1976b).

Two simulated field experiments established performance specifications for *B. heliothis* and evaluated several formulations and nozzle-pressure combinations (Smith *et al.*, 1977, 1978). A combination of small droplets, high density and high PIB concentration was best for control of *H. zea*. Smith *et al.* (1977) concluded that, at 12.7×10^9 PIB/litre, $270\,\mu$ droplets should not be used; coverage of > 23.26 droplets/cm² and > 34.88 droplets/cm² could be achieved using $180\,\mu$ and $90\text{-}\mu$ droplets respectively. Nozzles (TX-4) at 54 psi were better than TX-1 nozzles at 80 psi for aqueous suspensions of PIB. A polyvinyl alcohol (PVA) + Shade® formulation was better than a mineral oil emulsion, PVA, or two aqueous suspensions and also increased stability in sunlight.

Only three field tests have been made on soybeans, all in the USA. Two studies (1975, 1976) on caged soybeans showed that *B. heliothis* effectively controlled *H. zea* (Ignoffo *et al.*, 1978). In 1975 the reduction in number of larvae, seed-damaged pods/plant, mean % damaged pods and mean % damaged seeds increased with rates from 6, 30 300 to 600×10^9 PIB/0.4 ha. Larval populations were reduced by 92% (6×10^9 PIB/0.4 ha) to 100% (600×10^9 PIB/0.4 ha) and seed-damaged pods by ca. 75% (6×10^9 PIB/0.4 ha) to 90% (600×10^9 PIB/0.4/ha). Typical dose—mortality responses were also obtained in 1976. Larval populations were reduced by 84% (30×10^9 PIB/0.4 ha) to 96% (240×10^9 PIB/0.4ha): reduction in adult populations ranged from 70% (30×10^9 PIB/0.4 ha) to 90% (240×10^9 PIB/0.4 ha). Four applications of 30×10^9 PIB were better than one; however, one application of 120×10^9 PIB was better than two at 60×10^9 PIB or four at 30×10^9 PIB/0.4 ha. Although all spray treatments were very effective (90 to 96% control) they were poorer than the release of virus-infected larvae (96—98% control). In the third field test on soybeans the percent plants with damaged pods at 0, 15, 30 and 60×10^{10} PIB/0.4 ha was 6.3, 7.0, 4.7, and 1.0%, respectively (Ditman, unpublished observations).

D. TESTS ON SORGHUM

Six field trials were made on sorghum, one in the USA against *H. zea* (Ignoffo *et al.*, 1965) and five in Africa against *H. armigera* (Roome, 1975). A single application of *B. heliothis* (9.96×10^{11} PIB/0.4 ha) in the USA reduced larvae by 88% and infested sorghum heads by 86% (Ignoffo *et al.*, 1965). Sorghum was proposed as an ideal crop for application of *B. heliothis* since one or two applications easily covered sorghum heads; new growth did not dilute virus; larval feeding was confined to the heads; and death of one larva in a head provided ca. 1000X more virus than that originally applied. Although rates below 6×10^{11} PIB/ha in Africa did not control *H. armigera*, higher rates were consistently good and control was comparable to that of carbaryl and endosulfan.

E. TESTS ON TOBACCO AND TOMATO

Four field tests were conducted on tobacco (*H. virescens*) and one on tomato (*H. zea*). Desired levels of control were not obtained in any of these tests. For example, *B. heliothis* at 2.4×10^{13} PIB/0.4 ha caused 100% mortality; however, control was slow and tobacco plants were severely damaged (Chamberlin and Dutky, 1958). Mistric (1965, 1966 unpublished) made two tests on tobacco; in the first, the numbers of larvae/plant treated with *B. heliothis* (6.4×10^{11} PIB/0.4),

Rhothane® (0.7 kg/0.4 ha) and the check were 0.63, 0.17 and 0.97, respectively; in the 2nd test, the numbers of larvae/plant, as before, for *B. heliothis* (30 × 10^{11} PIB/0.4 ha), Rhothane® (0.7 kg 0.4 ha) and the check were 0.38, 0.35 and 1.04, respectively. In the 4th test on tobacco six applications of a dust (30 × 10^{11} PIB/0.4 ha) at weekly intervals was better than the check but not as good as methyl parathion (1%) + DDT (10%) (Gentry, 1966 unpublished observations). Little or no control of *H. zea* on tomatoes was obtained at 6, 30 and 60 × 10^{11} PIB/0.4 ha while carbaryl (0.9 kg/0.4) provided complete protection (Lange, 1966, unpublished observations).

VIII. Future Developments

A. PRODUCTION

Several procedures for producing baculoviruses in either homologous or non-homologous hosts have been proposed (cf. sections V, A and VI). Although virus produced in cell lines was as effective as that from larvae (Ignoffo *et al.*, 1974b) and *B. heliothis* will replicate in cell lines (Hink and Ignoffo, 1970), this technology will not be commercially practical for a long time (Chapter 16; Ignoffo and Hink, 1971). Significant development in production of *B. heliothis* will probably come from some radically new, exotic technique, such as production in non-homologous hosts coupled with recombinant-DNA techniques.

B. SPECIFICITY AND SAFETY

Although all current information demonstrates that *B. heliothis* is very specific, absolute safety cannot be guaranteed and, therefore, we must continually re-evaluate probabilities of risks. Thus, future research should clarify the nature of this specificity and the way *B. heliothis* infects and replicates. Efforts also should be made to improve or develop methods to characterize and to identify isolates, and to improve techniques to detect low levels of *B. heliothis* in organisms and in the environment.

C. EFFICACY

Future development on efficacy will undoubtedly concentrate on improving and preserving the activity of *B. heliothis* and on improving effectiveness and consistency through better dispersal and coverage. One obvious future development is extension of the current label from effective use on cotton to use on soybeans, sorghum and corn. The selection of soybeans as the first commodity is based on larval behaviour and the way the plant is expected to respond to damage. For example, on corn *Heliothis* lays eggs on the silk and larvae penetrate the ear soon after hatching. Thus, ca. 90% of the larval life is in the protective confines of the ear. Also, corn does not readily rebound from damage by developing either extra ears, larger ears or larger kernels. On cotton, eggs are laid on or near terminal buds. Larvae thus feed on terminal leaves and buds exposed to virus ca. 40% of their larval life and cotton compensates more than corn for feeding damage. Soybeans are examples at the other extreme. Larvae are exposed ca. 90% of the time, move freely over the entire plant and the plant rebounds remarkably from insect damage by producing either more pods or larger seeds.

IX. Conclusions

One obvious benefit of the *B. heliothis* program was the development of a safe viral product, the first ever registered for use to control an insect pest. Besides being a unique achievement the program demonstrated that industrial, federal, and state scientists working together could develop a naturally occurring entomopathogen into a safe product.

Another benefit was the creation of industrial facilities, technology, and trained personnel not in existence before 1965. Technology specifically developed for *B. heliothis* is being applied to other promising microbial insecticides. Pilot plants are operational and larger plants have been designed. Semi-automated systems and inexpensive diets developed to produce *B. heliothis* are being used to produce other insect viruses. Major biological and pathological problems associated with the continuous rearing of insects as well as major technical problems of formulation, stability, and application are defined and many have been resolved.

Another significant benefit of this program was the establishment of protocols and guidelines for evaluating viral insecticides. Prior to 1968, when petitions were first submitted to the FDA by industry, there were no guidelines or protocols for registering microbial insecticides. We now have functional guidelines and a mechanism for review of microbial insecticides.

References

Aizawa, K. (1955). *J. Sericult. Sci.* **24**, 398–399.

Allen, G.E. and Ignoffo, C.M. (1969). *J. Invertebr. Path.* **13**, 378–381.

Anderson, N.G., Waters, D.A., Nunley, C.E., Gibson, R.F., Schilling, R.M., Denny, E.C., Cline, G.B., Babelay, E.F. and Perardi, T.D. (1969). *Analytical Biochem.* **32**, 460–494.

Andrews, G.L. and Sikorowski, P.P. (1973). *J. Invertebr. Path.* **22**, 290–291.

Anonymous. (1973). Code of Federal Regulations 38: 10643.

Anonymous. (1975). Agricultural Statistics, 1975. U.S. Dept. Agric., U.S. Govt. Printing Office.

Atkins, E.L. and Anderson, L.D. (1967). *Calif. Agric. Ext. Serv.* AXS-M-16.

Bergold, G.H. (1963). *In* "Insect Pathology, An Advanced Treatise" (E.A. Steinhaus, ed.), 1, pp. 413–456 Academic Press, New York and London.

Bergold, G.H. and Ripper, W.E. (1957). *Nature Lond.* **180**, 764–765.

Breillatt, J.P., Martignoni, M. and Anderson, N.G. (1969). *Biophys. J.* **9**, A-262.

Breillatt, J.P., Brantley, J.N., Mazzone, H.M., Martignoni, M.E., Franklin, J.E. and Anderson, H.G. (1972). *Appl. Microbiol.* **23**, 923–930.

Bull, D.L. (1978). *In*: "Microbial Control of Insect Pests: Future Strategies in Pest Management." (G.E. Allen, C.M. Ignoffo, R.P. Jaques, eds.), pp. 118–121, USDA, Univ. Florida, Gainesville.

Bull, D.L., Ridgway, R.L., House, V.S. and Pryor, N.W. (1976). *J. econ. Ent.* **69**, 731–736.

Bullock, H.R. (1967). *J. Invertebr. Path.* **9**, 434–436.

Bullock, H.R., Hollingsworth, J.P. and Hartstack, Jr., A.W. (1970). *J. Invertebr. Path,* **16**, 419–422.

Burgess, S. (1977). *J. gen. Virol.* **37**, 509–510.

Cantwell, G.E., Knox, D.A., Lehnert, T. and Michael, A.S. (1966). *J. Invertebr. Path.* **8**, 228–233.

Chamberlin, F.S. and Dutky, S.R. (1958). *J. econ. Ent.* **51**, 560.

Chapman, J.W. and Glaser, R.W. (1915). *J. econ. Ent.* **8**, 140–150.

Chauthani, A.R., Claussen, D. and Rehnborg, C.S. (1968a). *J. Invertebr. Path.* **12**, 335–338.

Chauthani, A.R., Murphy, A., Claussen, D. and Rehnborg, C.S. (1968b), *J. Invertebr. Path.* **12**, 145–147.

Cline, G.B., Ryel, E., Ignoffo, C.M., Shapiro, M. and Straehle, W. (1972). Proc. 4th Int. Colloq. Insect Pathol., College Park, MD. pp. 363–370.

Daoust, R.A. (1974). *J. Invertebr. Path.* **23**, 400–401.

Daoust, R.A. and Roome, R.E. (1974). *J. Invertebr. Path.* **23**, 318–324.

Davidson, A. and Pinnock, D.E. (1973). *J. econ. Ent.* **66**, 586–587.

Duggan, R.E. (1970). Federal Register 35, 18690.

Estes, Z.E. and Faust, R.M. (1966). *J. Invertebr. Path.* **8**, 145–149.

Estes, Z.E. and Ignoffo, C.M. (1965). *J. Invertebr. Path.* **7**, 258–259.

Falcon, L.A. (1969). *In* "Biological Control" (C.B. Huffaker, ed.), pp. 346–364.

Falcon, L.A., Kane, W.R., Etzel, L.K. and Leutenegger, R. (1967). *J. Invertebr. Path.* **9**, 134–136.

Falcon, L.A., Leigh, T.F., van den Bosch, R., Black, J.H. and Burton, V.E. (1965). *Calif. Agric.* **19**, 13–14.

Gitay, H. and Polson, A. (1971). *J. Invertebr. Path.* **17**, 288–290.

Goodwin, R.H., Adams, J.R., Vaughn, J.L. and Louloudes, S.J. (1976). Proc. 1st Int. Colloq. Invertebr. Path., Kingston, Canada, pp. 94–98.

Granados, R.R., Nguyen, T. and Cato, B. (1978). *Intervirology* **10**: 309–317.

Gregory, B.G., Ignoffo, C.M. and Shapiro, M. (1969). *J. Invertebr. Path.* **14**, 186–193.

Gudauskas, R.T. and Canerday, D. (1968). *J. Invertebr. Path.* **12**, 405–411.

Hamm, J.J. and Young, J.R. (1974). *J. Invertebr. Path.* **24**, 70–81.

Harpaz, I. and Zlotkin, E. (1965). *Annls Soc. ent. Fr.* **1**, 963–972.

Heimpel, A.M. (1967). Proc. Joint U.S. – Japan, Semin. Microbial Control Insect Pests, Fukuoka, Japan, pp. 44–48.

Hink, W.F. and Ignoffo, C.M. (1970). *Exptl Cell Res.* **60**, 307–309.

Huang, H., Ignoffo, C.M. and Shapiro, M. (1977). *J. Kans. ent. Soc.* **50**, 200–202.

Ignoffo, C.M. (1965a). *J. Invertebr. Path.* **7**, 209–216.
Ignoffo, C.M. (1965b). *J. Invertebr. Path.* **7**, 217–226.
Ignoffo, C.M. (1965c). *J. Invertebr. Path.* **7**, 315–319.
Ignoffo, C.M. (1966a). *J. Invertebr. Path.* **8**, 279–282.
Ignoffo, C.M. (1966b). *J. Invertebr. Path.* **8**, 290–292.
Ignoffo, C.M. (1966c). *J. Invertebr. Path.* **8**, 531–536.
Ignoffo, C.M. (1966d). *J. Invertebr. Path.* **8**, 547–548.
Ignoffo, C.M. (1966e). *In* "Insect Colonization and Mass Production." (C.N. Smith, ed.), pp. 501–530. Academic Press, New York and London.
Ignoffo, C.M. (1967). *In* "Insect Pathology and Microbial Control" (P.A. van der Laan, ed.), pp. 91–117, Wageningen, The Netherlands.
Ignoffo, C.M. (1968a). *Bull. ent. Soc. Am.* **14**, 265–276.
Ignoffo, C.M. (1968b). Petition for Exemption from the Requirement of a Tolerance for Usage of the Nuclear-Polyhedrosis Virus of the Genus *Heliothis.* Petition No. 8F0697. International Minerals & Chemical Corporation, Skokie, Ill., USA.
Ignoffo, C.M. (1973a). *Expl Parasit.* **33**, 380–406.
Ignoffo, C.M. (1973b). *Ann. N.Y. Acad. Sci.* **217**, 141–164.
Ignoffo, C.M. (1973c). *Misc. Publ. ent. Soc. Am.* **9**, 57–61.
Ignoffo, C.M. (1975a). *Environ. Lett.* **8**, 23–40.
Ignoffo, C.M. (1975b). *In* "Baculoviruses for Insect Pest Control: Safety Considerations," (M. Summers, R. Engler, L.A. Falcon, P.V. Vail, eds), Am. Soc. Microbiol., Washington. 186 p.
Ignoffo, C.M. (1979). *In* "Developments in Industrial Microbiology". pp. 105–115, Society for Industrial Microbiology. Lubrecht and Cramer, Montichello.
Ignoffo, C.M. and Allen, G.E. (1972). *J. Invertebr. Path.* **20**, 187–192.
Ignoffo, C.M. and Anderson, R. (1979). *In* "Microbial Technology" 2nd Ed. (H.J. Peppler and D. Perlman, eds.), Vol. 1, pp. 1–28 Academic Press, New York and London.
Ignoffo, C.M., Anderson, R.F. and Rostenberg, A. (1968). *In* "Petition Viron/H" 8F0697. International Minerals and Chemical Corporation, Skokie, Ill. USA.
Ignoffo, C.M. and Batzer, O.F. (1971). *J. econ. Ent.* **64**, 850–853.
Ignoffo, C.M., Batzer, O.F., Barker, W.M. and Ebert, A.G. (1971a). *Proc. 4th Int. Colloq. Insect Path.*, College Park, MD, pp. 357–362.
Ignoffo, C.M. and Boening, O.P. (1970). *J. econ. Ent.* **63**, 1696–1697.
Ignoffo, C.M., Bradley, J.R., Jr., Gilliland, F.R., Jr., Harris, F.A., Falcon, L.A., Larson, L.V., McGarr, R.L., Sikorowski, P.P., Watson, T.F. and Yearian, W.C. (1972). *Environ. Ent.* **1**, 388–390.
Ignoffo, C.M., Chapman, A. and Martin, D. (1965). *J. Invertebr. Path.* **7**, 227–235.
Ignoffo, C.M. and Garcia, C. (1966a). Q. Rep. II. International Minerals and Chemical Corporation, pp. 3–6.
Ignoffo, C.M. and Garcia, C. (1966b). *J. Invertebr. Path.* **8**, 426–427.
Ignoffo, C.M. and Garcia, C. (1968). *J. Invertebr. Path.* **10**, 430–432.

Ignoffo, C.M. and Garcia, C. (1978). *Environ. Ent.* **7**, 270–272.

Ignoffo, C.M. and Hink, W.F. (1971). *In* "Microbial Control of Insects and Mites." (H.D. Burges and N.W. Hussey, eds.), pp. 541–580. Academic Press, London and New York.

Ignoffo, C.M., Hostetter, D.L., Biever, K.D., Garcia, C., Thomas, G.D., Dickerson, W. and Pinnell, R. (1978). *J. econ. Ent.* **71**, 165–168.

Ignoffo, C.M., Hostetter, D.L. and Pinnell, R.E. (1974a). *Environ. Ent.* **3**, 117–119.

Ignoffo, C.M., Hostetter, D.L. and Shapiro, M. (1974b). *J. Invertebr. Path.* **24**, 184–187.

Ignoffo, C.M., Hostetter, D.L., Sikorowski, P.P., Sutter, G. and Brooks, W.M. (1977). *Environ. Ent.* **6**, 411–415.

Ignoffo, C.M., Hostetter, D.L. and Smith, D.B. (1976b). *J. econ. Ent.* **69**, 207–210.

Ignoffo, C.M., Huang, H.T., Shapiro, M. and Woodard, G. (1975). *Environ. Ent.* **4**, 569–573.

Ignoffo, C.M., Knapp, J. and Shirar, C.S. (1967). Q. Rep. II. International Minerals and Chemical Corporation, 58–63.

Ignoffo, C.M. and Montoya, E.L. (1966). *J. Invertebr. Path.* **8**, 409–412.

Ignoffo, C.M., Parker, F.D., Boening, O.P., Pinnell, R.E. and Hostetter, D.L. (1973). *Environ. Ent.* **2**, 302–303.

Ignoffo, C.M. and Rafajko, R.R. (1972). *J. Invertebr. Path.* **20**, 321–325.

Ignoffo, C.M. and Shapiro, M. (1978). *J. econ. Ent.* **71**, 186–188.

Ignoffo, C.M., Shapiro, M. and Hink, W.F. (1971b). *J. Invertebr. Path.* **18**, 131–134.

Ignoffo, C.M., Yearian, W.C., Young, S.Y., Hostetter, D.L. and Bulla, D.L. (1976a). *J. econ. Ent.* **69**, 233–236.

Irabagon, T.C. and Brooks, W.M. (1974). *J. econ. Ent.* **67**, 229–231.

Jafri, R.H. and Khan, A.A. (1969). *J. Invertebr. Path.* **14**, 104.

Keeley, L.L. and Vinson, S.B. (1975), *J. Invertebr. Path.* **26**, 121–123.

Kelly, D.C., Edwards, M.L., Evans, H.F. and Robertson, J.S. (1978) *J. gen. Virol.* **40**, 465–469.

Klostermeyer, E. (1968). *J. econ. Ent.* **61**, 1020–1023.

Lipa, J.J. (1968). *Bull. Entomologique Pologne* **38**, 611–616.

Lounsbury, C.P. (1913). *J. Agric. S. Afr.* **5**, 448–452.

McCarthy, W.J., Mercer, W.E. and Murphy, T.F. (1978). *Virology*, **90**, 374–378.

MacFarlane, J.J., Jr. and Keeley, L.L. (1969). *J. econ. Ent.* **62**, 925–929.

McIntosh, A.H. (1975). *In* "Baculoviruses for Insect Pest Control: Safety Considerations." (M. Summers, R. Engler, L.A. Falcon, P.V. Vail, eds.), *Am. Soc. Microbiol.* Washington, D.C. 186 pp.

McIntosh, A.H. and Maramorosch, K. (1973). *J. N.Y. ent. Soc.* **81**, 175–182.

McKinley, D.J. (1971). *Cotton Growing Rev.* **48**, 297–303.

McLaughlin, R., Andrews, G. and Bell, M.R. (1971). *J. Invertebr. Path.* **18**, 304–305.

McLeod, P.J., Yearian, W.C. and Young, S.Y., III. (1977). *J. Invertebr. Path.* **30**, 237–241.

Mally, F.W. (1891). *U.S. Bur. Ent. Bull.* **24**, 48–50.

Mally, F.W. (1892). *U.S. Bur. Ent. Bull.* **26**, 54–56.

Martignoni, M.E., Breillatt, J.P. and Anderson, N.G. (1968). *J. Invertebr. Path.* **11**, 507–510.

Maruniak, J.E., and Summers, M.D. (1978). *J. Invertebr. Path.* **32**, 196–201.

Miller, L.K. and Dawes, K.P. (1978). *Appl. environ. Microbiol.* **35**, 411–421.

Montoya, E.L. and Ignoffo, C.M. (1966). *J. Invertebr. Path.* **8**, 251–254.

Oatman, E.R., Hall, I.M., Arakawa, K.Y., Platner, G.R., Bascom, L.A., and Beegle, C.C. (1970). *J. econ. Ent.* **63**, 415–421.

Parsons, F.S. (1936). *Prog. Rep. Expl. Stn Empire Cotton Growing Corp.,* Barberton, S. Afr., 1934–1935, 24–31.

Patel, R., Singh, R. and Patel, P. (1968). *J. econ. Ent.* **61**, 191–193.

Roome, R.E. (1975). *Bull. ent. Res.* **65**, 507–514.

Roome, R.E. and Daoust, R.A. (1976). *J. Invertebr. Path.* **27**, 7–12.

Rubinstein, R., Harley, E.H., Losman, M. and Lutton, D. (1976). *Virology* **69**, 323–326.

Rubinstein, R. and Polson, A. (1976). *J. Invertebr. Path.* **28**, 157–160.

Ryel, E.M. (1973). *Diss. Abst. Int.* **33**, 4414.

Scharnhorst, D.W., Saving, K.S., Vuturo, S.B., Cooke, P.H. and Weaver, R.F. (1977). *J. Virol.* **21**, 292–300.

Scott, H.A., Yearian, W.C. and Young S.Y. (1976). *J. Invertebr. Path.* **28**, 229–232.

Shapiro, M. and Ignoffo, C.M. (1969). *J. Invertebr. Path.* **14**, 130–134.

Shapiro, M. and Ignoffo, C.M. (1970a). *Virology* **41**, 577–579.

Shapiro, M. and Ignoffo, C.M. (1970b). *J. Invertebr. Path.* **16**, 107–111.

Shapiro, M. and Ignoffo, C.M. (1971a). *J. Invertebr. Path.* **17**, 449–450.

Shapiro, M. and Ignoffo, C.M. (1971b). *J. Invertebr. Path.* **18**, 154–155.

Shapiro, M. and Ignoffo, C.M. (1971c), *Proc. 4th Int. Colloq. Insect Path.* pp. 147–151.

Shapiro, M. and Ignoffo, C.M. (1971d). *J. Invertebr. Path.* **17**, 327–332.

Shapiro, M. and Ignoffo, C.M. (1975). *J. Kans. ent. Soc.* **48**, 362–365.

Shapiro, M., Stock, R.D. and Ignoffo, C.M. (1969). *J. Invertebr. Path.* **14**, 28–30.

Sikorowski, P.P., Broome, J.R. and Andrews, G.L. (1971). *J. Invertebr. Path.* **17**, 451–452.

Smith, D.B., Hostetter, D.L., and Ignoffo, C.M., (1977). *J. econ. Ent.* **70**, 437–441.

Smith, D.B., Hostetter, D.L., and Ignoffo, C.M. (1978). *J. econ. Ent.* **71**, 814–817.

Smith, G.F. and Summers, M.D. (1978). *Virology* **89**, 517–527.

Smith, K.M. and Rivers, C.F. (1956). *Parasitology* **46**, 235–242.

Stadelbacher, E.A., Adams, J.R., Faust, R.M. and Tompkins, G.J. (1978). *J. Invertebr. Path.* **32**, 71–76.

Steinhaus, E.A. (1957). *Hilgardia* **26**, 417–430.

Steinhaus, E.A. (1960). *J. Insect Path.* **2**, 327–333.

Stuermer, C.W., Jr. and Bullock, H.R. (1968). *J. Invertebr. Pathol.* **12**, 473–474.

Tanada, Y. and Reiner, C. (1962). *J. Insect Path.* **4**, 139–154.
Teakle, R.E. (1973). *Queensland J. Agric. Anim. Sci.,* Bull. **648**, pp. 161–177.
Teakle, R.E. (1974). *J. Invertebr. Path.* **23**, 127–129.
Tompkins, G.J., Adams, J.R. and Heimpel, A.M. (1969). *J. Invertebr. Path.* **14**, 343–357.
Vago, C., Aizawa, K., Ignoffo, C.M., Martignoni, M.E., Tarasevitch, L. and Tinsley, T.W. (1974). *J. Invertebr. Path.* **23**, 133–134.
Vail, P.V., Henneberry, T.J., Kishaba, A.N. and Arakawa, K.Y. (1968). *J. Invertebr. Path.* **10**, 84–93.
Vandamme, P. and Angelini, A. (1966). *Coton Fibr. trop.* **21**, 333–338.
Watanabe, S. (1951). *Jap. J. exp. Med.* **21**, 299–313.
Wells, F.E. (1970). Pat. No. 807. 967, Repub. S. Africa.
Wells, F.E. and Heimpel, A.M. (1970). *J. Invertebr. Path.* **16**, 301–304.
Whitlock, V.H. (1974). *J. Invertebr. Path.* **23**, 70–75.
Whitlock, V.H. (1977a). *J. Invertebr. Path.* **30**, 80–86.
Whitlock, V.H. (1977b). *J. ent. Soc. S. Afr.* **40**, 251–253.
Whitlock, V.H. (1977c). *J. Invertebr. Path.* **29**, 297–303.
Wilkinson, J.D., Biever, K.D. and Ignoffo, C.M. (1975). *Entomophaga* **20**, 113–120.
Wolfenbarger, D.A. (1964). *J. econ. Ent.* **5**, 732–735.
Wolfenbarger, D.A. (1965). *J. Invertebr. Path.* **7**, 33–38.
Yearian, W.C. and Young, S.Y. (1974). *Environ. Ent.* **3**, 1035–1036.
Young, J.R. and Hamm, J.J. (1966). *J. econ. Ent.* **59**, 382–384.
Young, S.Y. and Yearian, W.C. (1974). *Environ. Ent.* **3**, 253–255.
Young, S.Y. and Yearian, W.C. (1976). *J. Georgia ent. Soc.* **11**, 277–282.
Young, S.Y., Yearian, W.C. and Kim, K.S. (1977). *J. Invertebr. Path.* **29**, 105–111.
Young, S.Y., Yearian, W.C. and Scott, H.A. (1975). *J. Invertebr. Path.* **26**, 309–312.

Addendum: Citations for Section on Efficacy

(Citations appearing in the main reference list are given by author and date only.)

A. REFERENCES FOR COTTON

Allen, G.E., and Pate, T.L. (1966). *J. Invertebr. Path.* **8**, 129–131.
Allen, G.E., Gregory, B. and Brazzel, J. (1966). *J. econ. Ent.* **59**, 1333–1336.
Allen, G.E., Gregory, B.G. and Pate, T.L. (1967). *J. Invertebr. Path.* **9**, 40–42.
Alvaro, C.M., Guillermo, A.A., Rafael, E.M. and Hernan, A.V. (1972). Federacion Nacional Algodoneros, Departamento Tecnico — Agricola, pp. 7–41.
Andrews, G.L., Harris, F.A., Sikorowski, P.P. and McLaughlin, R.E. (1975). *J. econ. Ent.* **68**, 87–90.

Anonymous. (1976). *Acta ent. sin.* **19**, 167–1972.

Arant, F.S. and Ivey, W.D. (1967). *Ann. Rep.* Jan. 1 to Dec. 31, 1966. Project: Hatch (Ala. 512).

Atger, P. (1969). Experimentation sur les Viroses D'Insectes en vue de la lutte contre les Avageurs du Cotonnier au Tchad. pp. 6–8.

Bell, M.R. and Kanavel, R.F. (1978). *J. econ. Ent.* **71**, 350–352.

Brazzel, J.R., Adair, M., Pate, T.L., Allen, G.E. and Worley, G. (1966). Cotton Insect Res. Control Conf., Dallas, Texas, Jan. 10–11, 1967.

Bull, D.L., Ridgway, R.L., House, V.S. and Pryor, N.W. (1976).

Chapman, A.J. and Bell, R.A. (1964). *J. econ. Ent.* **60**, 655–656.

Chapman, A.J. and Ignoffo, C.M. (1972). *J. Invertebr. Path.* **20**, 183–186.

Chapman, A.J., Bell, R.A. and Ignoffo, C.M. (1967). *J. econ. Ent.* **60**, 655–656.

Cleveland, T. (1967). A summary prepared for the Conference at Dallas, Texas, Jan. 10–11, 1967.

Coppedge, J.R., Kinzer, R.E. and Ridgway, R.L. (1972). *Texas Agric. Stn Prog. Rep.* 3082–3091. pp. 43–49.

Falcon, L.A. (1965). Summary of cotton research program.

Falcon, L.A. (1966). Summary of results of two field tests with *Heliothis* virus for the control of cotton bollworm in 1966.

Falcon, L.A. (1967). Summary of results of research conducted with insect pathogens for the control of lepidopterous pests on cotton in the San Joaquin Valley in 1964, 1965, and 1966.

Falcon, L.A., Leigh, T.F., van den Bosch, R., Black, J.H. and Burton, V.E. (1965). *California Agriculture* **19**, 13–14.

Falcon, L.A., Leigh, T.F., van den Bosch, R., Black, J.H. and Burton, V.E. (1966). *Agri. Pest Control,* January 1966.

Falcon, L.A., Leigh, T.F., van den Bosch, R., Black, J.H. and Burton, V.E. (1966). *Pest Control Operator News,* pp. 23–25.

Falcon, L., van den Bosch, R., Leigh, T., Etzel, L. and Stinner, R. (1966). *In* "Cotton Insect Control" Rep. Agric. Exp. Stn Agric. Ext. Serv. Univ. Calif. pp 17–22.

Fernandez, A.T., Graham, H.M., Jukefahr, M.J., Bullock, H.R. and Hernandez, Jr. N.S., (1969). *J. econ. Ent.* **62**, 173–177.

Harris, F.A. (1973). Cotton Yield Results from Annual Viron-H Field Tests, 1969–1972, in Mississippi.

House, V.S., Bull, D.L., Coppedge, J.R. and Ridgway, R.L. (1976). *S West. Ent.* **1**, 81–84.

Ignoffo, C.M. (1966). *J. Invertebr. Path.* **8**, 531–536.

Ignoffo, C.M. (1968b).

Ignoffo, C.M. (1972). *J. Invertebr. Path.* **19**, i–ii.

Ignoffo, C.M. and Batzer, O.F. (1971).

Ignoffo, C.M., Bradley, J., Jr., Gilliland, F. A., Jr., Harris, F.A., Falcon, L.A., Larson, L.V., McGarr, R.L., Sikorowski, P.P., Watson, T.F. and Yearian, W.C. (1972). *Environ. Ent.* **1**, 388–390.

Ignoffo, C.M., Chapman, A.J. and Martin, D.F. (1965). *J. Invertebr. Path.* **7**, 227–235.

Ignoffo, C.M. and Garcia, C. (1966a).
Ignoffo, C.M., Hostetter, D.L., and Smith, D.B. (1976).
Ignoffo, C.M., Hostetter, D.L., Garcia, C., Biever, K.D. and Thomas, G.D. (1979). *In* "Microbial Control of Insect Pests: Future Strategies in Pest Management Systems" (Allen, G.E., Ignoffo, C.M., Jaques, R.P. eds), pp. 69–71. Proc. N.S.F.-U.S.D.A.-Univ. Fla. Workshop.
Ignoffo, C.M., Yearian, W.C., Young, S.Y., Hostetter, D.L. and Bull, D.L. (1976). *J. econ. Ent.* **69**, 233–236.
Kinzer, R.E., Bariola, L.A., Ridgway, R.L. and Jones, S.L. (1976). *J. econ. Ent.* **69**, 697–701.
Knapp, J.L. (1967). Summary of Viron-H Tests on Cotton. Q. Rep. – Jan. to March, 1967, International Minerals & Chemical Corp.
Leigh, T.F., Black, J.H. and Jackson, C. (1966). Bollworm experiment – Palla Rosedale Ranch.
McGarr, R.L. and Ignoffo, C.M. (1966). *J. econ. Ent.* **59**, 1284–1285.
McKinley, D.J. (1971). *Cotton Growing Rev.* **48**, 297–303.
McLaughlin, R.E., Andrews, G. and Bell, M.R. (1971).
Montoya, E.L., Ignoffo, C.M. and McGarr, R.L. (1966). *J. Invertebr. Path.* **8**, 320–324.
Roome, R.E. (1975). *Bull. ent. Res.* **65**, 507–514.
Stacey, A.L., Yearian, W.C., and Young, III, S.Y. (1977). *Arkan. Farm Res.* **26**, 3.
Stacey, A.L., Yearian, W.C. and Young, III, S.Y. (1977). *J. econ. Ent.* **70**, 779–784.
Stacey, A.L., Young, III, S.Y., and Yearian, W.C. (1977). *J. econ. Ent.* **70**, 383–386.
Stacey, A.L., Young, III, S.Y. and Yearian, W.C. (1977). *J. ent. Soc.* **12**, 167–173.
Taft, H.M. and Hopkins, A.R. (1969). Results of tests with experimental compounds furnished by International Minerals and Chemical Corp.
van den Bosch, R., Gonzalez, D., Falcon, L.A., Leigh, T.F., Hagen, K.S. Stinner, R.E. and Etzel, L.K. (1966). *In* "Cotton Insect Control". Rep. Agric. Exp. Stn Agric. Ext. Serv. Univ. Calif. pp. 9–15.
Yearian, W.C. (1966) Cotton Insects in Arkansas. Cotton Insect Res. Control Conf., Dallas, Texas, Jan. 10–11, 1967.

B. REFERENCES FOR CORN

Anderson, L.D. and Nakakihara, H. (1966). Earworm control studies on sweet corn at the University of California, Riverside.
Anderson, L.D. and Nakakihara, H. (1966). IMC Viron-H corn tests – field results from California.
Anonymous. (1964). Valley Packers (East St. Louis, IL). Report on Field Testing of Earworm Virus – reported on 9/28/64.
Brazzel, J.R. (1965). Polyhedrosis virus test at Crystal Springs in 1965 for control of *Heliothis* larvae on sweet corn.

Hall, I.M. (1965). Results of hand applications of dusts and liquid suspensions of a commercial preparation of nuclear polyhedrosis virus to silk of sweet corn for control of *Heliothis zea*. Orange County, CA.

Hall, I.M. (1966). Results of hand spray applications of commercial formulations of nuclear polyhedrosis virus on tassel and silk for control of *Heliothis zea*. Orange County, CA.

Hantsbarger, W.M. (1966). Field results from Colorado.

Hantsbarger, W.M. and Swartz, J. (1966). Field tests with *Heliothis* polyhedral virus on control of corn earworm — Colorado 1966.

Ignoffo, C. (1968b).

Ignoffo, C.M., Chapman, A.J. and Martin, D.F. (1965).

Randell, R. (1965). Corn earworm control plots — Watson Rouch Farm, Collinsville, IL.

Semel, M. (1966). Field Results from New York.

Semel, M. (1966). Long Island Vegetable Research Farm — Biological Control of the Corn Earworm.

White C. and Gard, I. (1964). Virus Field Study Results. Petition No. 8F0697.

C. REFERENCES FOR SORGHUM, SOYBEANS, TOBACCO AND TOMATO

Chamberlin, F.S. and Dutky, S.R. (1958).

Ditman, L.P. (1966). Control of the corn earworm on beans by polyhedrosis virus. Expt No. 1, 1966.

Gentry, C.R. (1966). Small plot replicated test for polyhedral virus in *Trichoplusia ni* and *Heliothis* control.

Ignoffo, C.M. (1968b).

Ignoffo, C.M., Chapman, A.J. and Martin, D.F. (1965).

Lange, H. (1969). Field test results — Yolo, California.

Mistric, W.J. (1965). Summary of research on control of tobacco insects.

Mistric, W.J. (1966). Summary of research on control of tobacco insects.

CHAPTER 18

Control of the Gypsy Moth by a Baculovirus

F.B. LEWIS

Forest Insect and Disease Laboratory, Hamden, Connecticut, USA

I. Introduction

The gypsy moth, *Lymantria dispar*, is a significant forest defoliator that causes severe problems particularly in central Europe and the USA. It also causes damage in Western Europe, the Mediterranean region, and Japan. Throughout its geographical range, it is limited by a variety of natural control factors including a nucleopolyhedrosis virus (NPV; baculovirus) (Lewis and Daviault, 1967). The gypsy moth is susceptible to several diseases (Lewis and Etter, 1978) and disease appears to be an important regulating mechanism of the insect (Campbell, 1963; Doane, 1970; Podgwaite and Campbell, 1971, 1972). However, Bess (1961) concluded that the NPV was only a minor factor in regulating the insect populations.

The virus disease was reported as the wilt disease in the USA in the early 1900's by Reiff (1911) and Glaser (1915). Not until Bergold's 1947 report was the true nature of the wilt disease elucidated. Reiff (1911) and Glaser and Chapman (1913) conducted some field tests that led them to conclude that the disease had potential for biological control. In the latter half of the 1900's, the establishment of insect pathology by Steinhaus (1949, 1963) and the problems generated by the use of synthetic pesticides caused an upsurge of interest in biological control, particularly by baculoviruses. Worldwide research on the NPV of the gypsy moth intensified (Orlovskaja, 1961; Vasiljevic, 1961; Rollinson *et al.*, 1965; Magnoler, 1967, 1968a, 1974; Doane, 1970; Yendol *et al.* 1977; Wollam *et al.*, 1978; Lewis *et al.*, 1979). In the USA, a widespread outbreak of the gypsy moth in the early 1970's led to intensified gypsy moth research, followed in 1974 by the U.S. Department of Agriculture's accelerated gypsy moth Research, Development, and Application Program. Studies on the baculo-virus were part of this program, ultimately leading to the registration of the NPV product, Gypchek®.

II. Laboratory Studies

Production and registration of baculovirus in the USA requires the establish-ment of a standard bioassay technique and the calculation of potency of the product. Many methods have been used to bioassay baculoviruses against lepidopterous larvae. These generally fall in the categories: diet incorporation (Ignoffo, 1972; Lewis *et al.* 1979), diet surface contamination (Vail *et al.*, 1971; Burgerjon *et al.*, 1975; Harper, 1976; Lewis *et al.*, 1979), and *per os* force feeding (Paschke *et al.*, 1968; Martignoni and Iwai, 1977). Reviews by Dulmage (1973), Vail (1975), and Dulmage and Burgerjon (1977) discuss concepts of bioassay, international standards, and the merits and disadvantages of techniques.

Development of a standard bioassay for gypsy moth larvae presented diffi-culties that required the selection of a potent virus strain and a standard insect strain. Wide variations in virus potency and insect response have been reported for different geographical sources (Doane, 1967; Magnoler, 1968b; Rollinson and Lewis, 1973; Vasiljevic and Injac, 1973). One to three log differences occurred in the LC_{50} of a given NPV isolate tested against different gypsy moth stocks. There were similar variations for several NPV sources tested against a single gypsy moth population. These virus potency differences must be differentiated from the increase in dosage required as the instar increases. In addition, host susceptibility is confounded by food, population history and age, physiologic condition, and the genetic constitution of local populations.

A. BACULOVIRUS STRAIN SELECTION

Since there were many gypsy month NPV sources available, it was paramount to select the most active one for development and registration for use, to ensure that it could be traced back to the original isolate and that it had main-tained its viral identity. The isolate selected generally exhibited the strongest potency of the isolates tested against the U.S. populations. It was isolated from a 1967 Connecticut population with a natural NPV epizootic. It was used in all our field and laboratory studies and called the Hamden strain. Purified and concentrated by the procedures described by Breillatt *et al.* (1972), it is the primary U.S. gypsy moth NPV standard and has been designated the "Hamden standard (K-rotor)". Secondary laboratory standards, calibrated to the primary standard, are used in all bioassay and potency determinations in our laboratory. This NPV strain is the active ingredient in Gypchek.

B. GYPSY MOTH LARVAL STOCK SELECTION

Equally important is the establishment of a laboratory stock of larvae for assay. It must respond within reasonable limits. Such a stock, originating from a New Jersey (NJ) population, has been established and is currently in the sixteenth generation (Bell and Shapiro, 1979). It is used in all laboratory bioassays. A secondary Pennsylvania (PA) stock is also used.

C. BIOASSAY PROCEDURES

After selection of the NPV and the laboratory insect, two laboratory bioassay protocols were developed: a diet incorporation technique to determine LC_{50} dosages for use in formulation studies and comparative product evaluations, and a diet plug method to determine LD_{50}s needed for more precise dose evaluations. Both methods use free feeding and were chosen because of their relevance to field situations, normal method of larval acquisition of the NPV, operational ease, and lack of restriction of larval numbers practicable to test.

The NJ strain of the gypsy moth is reared at $24 \pm 2°C$. Usually 4–5 days after eclosion, larvae (instar II) are large enough for bioassay. They must be vigorous and 6 ± 2 mg in weight before transfer to NPV-treated artificial diet (ODell and Rollinson, 1966). They are held for 14 days at $24-25°C$ in a photoperiod of 16 h daily then mortality is recorded and evaluated by Berkson's logit Chi square method (Paschke et al., 1968).

A 10-mg sample of NPV powder is placed in a sterile tissue homogenizer with 10 ml sterile Tris buffer (25 ml of $0.2 M$ 2-amino-2-hydroxymethyl-1-3 propanediol mixed with 47.0 ml of $0.1 N$ HCl and diluted to 100 ml). The mixture is blended for 3 min and diluted 1:100 to contain 0.01 mg NPV powder /ml. Polyhedral inclusion body (PIB) counts of the diluted samples are made with an improved Levy–Neubauer haemocytometer. The charged chamber is allowed to stand for 10 min before counting at 440X. Twenty charges are averaged to estimate the PIB/ml. Five suspensions of decreasing concentration are then prepared. These calibrated suspensions are incorporated into diet or applied to the top of diet plugs for bioassay.

1. Diet Incorporation Method

The essential parameters for this procedure are:

Exposure to the virus	Virus incorporated into diet at 52°C
Inoculum	1 ml test suspension NPV to 99 ml diet. Two 1.25 cm³ diet cubes/sterile plastic petri dish (100 × 15 mm)
Dosage Unit	ng/ml of diet and PIB/ml of diet

Number of test insects	10 larvae/petri dish (5 dishes/test concentration)
Number of dosages tested	5 per test and controls, 10× dilution for screening, 2× for precise dosage determination
Control	One untreated group of 50 larvae
Period on virus diet	48 h, then untreated diet given

2. Diet Plug Method

Larvae are placed individually on diet plugs each in a 1-oz (30-ml) capped plastic creamer. The diet plugs are prepared by cutting into a 1-mm thick sheet of diet with a number 1 cork borer. The virus dose is administered, 1 μl per plug, by a microapplicator with magnetic stirring. At least 50 larvae are used at each concentration. Control plugs are dosed with 1 μl sterile water without virus. As the diet plugs are consumed they are replaced with fresh untreated diet. Concentrations of NPV are expressed as PIB/μl.

For the diet plug method with the NJ stock, the LD_{50} for the Hamden standard was 473 ± 110 PIB. Hedlund (1974), who used PA stock larvae and a leaf disc technique, reported an LD_{50} of 220 PIB. Magnoler (1974), using 23.9 mg-larvae on diet plugs, reported an LD_{50} of 1729 PIB. With instar II larvae, Doane (1967) reported LT_{50} values (time required to kill 50% of the test larvae) of 9.0, 7.6, and 5.1 days with concentrations of 698, 6987, and 69 879 PIB respectively. Hedlund (1974) had LT_{50} values of 12.5 days with 798 PIBs on leaf discs and 24 000 PIB/ml of diet by the diet incorporation method.

D. GYPSY MOTH POTENCY UNIT

The potency of a baculovirus usually has been expressed as the concentration of PIBs per surface or volume. These figures have been used to construct dosage–response curves and also in comparative potency determinations. These techniques have serious inherent errors since the number and viability of the infectious unit (the occluded virion) are unknown (Burgerjon and Dulmage, 1977). In addition, the U.S. Environmental Protection Agency (EPA) requires that the labels of registered NPV products display potency as well as PIB concentration. Consequently, a gypsy moth potency unit (GMPU) has been established. One GMPU is defined as the nanograms of NPV product per ml of diet that produce an LC_{50} in newly moulted instar II gypsy moth larvae. The standard diet incorporation bioassay is used with NJ F_{16} larvae as test insects. The potency per gram, kilogram, or ounce is calculated from the nanogram LC_{50} data. Since the NJ strain has been established as the standard test strain and is more resistant than the previously used PA strain, a conversion factor of 2.989× has been calculated to convert potency determined on the NJ larvae to the potency for PA larvae. Other stocks of gypsy moth can be similarly calibrated.

E. STORAGE OF GYPSY MOTH BACULOVIRUS

Suspensions, air-dried powders, or lyophilized powders lost activity rapidly at high temperatures, and slowly at ambient temperatures. Acceptable activity was maintained at 4°C as a suspension and as a lyophilized powder. Air-dried powder should be kept frozen (Lewis *et al.*; 1979).

F. EARLY PRODUCTION OF GYPSY MOTH BACULOVIRUS

Gypsy moth NPV must be propagated in the living host or in an established cell line. At present the *in vivo* system is the only economical system available. Early NPV producers focused on maximizing the production of PIBs rather than producing a product that would meet rigid safety standards.

Pioneering production techniques (Rollinson and Lewis, 1962) required field collections of gypsy moth larvae from populations with NPV epizootics. In heavily diseased areas, 1 pint (0.55 litres) of cadavers could be collected/person/h. This pint yielded about 10 g of dried powder containing 6×10^{11} PIB when processed by differential centrifugation and screening. This method had several drawbacks. NPV epizootics were often difficult to locate and the diseased insects were almost always dead for several days before collection; as a result, they were usually contaminated with opportunistic bacteria and fungi. Thus, there was no practical way to ensure that some of the insects collected were not killed by another pathogen, and early NPV production batches were often mixtures of many gypsy moth pathogens. These methods were too time consuming and crude for the production of large quantities of highly purified NPV. Micro-biological quality control on early NPV productions was limited to standard plate counts, which indicated the levels of bacterial contamination in the finished product. However, without guidelines, these counts were useful only in indicating the relative purity of the product.

Concurrent reporting of an artificial gypsy moth diet by ODell and Rollinson (1966) and Leonard and Doane (1966) had a major impact on NPV production methods. The diet enabled year-round moth rearing and the subsequent selection of strains suitable for maximizing NPV production. Further, it led to the use of the diet incorporation method of infecting larvae, so that the minimum NPV dosage could be used to yield the maximum number of PIBs for various larval stages.

Mass propagation of NPV from a laboratory-reared strain of gypsy moth and mass rearing techniques were developed by the Agricultural Research Service's (ARS) Methods Improvement Laboratory, Otis, Massachusetts (Lewis, 1971). Larvae hatched from surface-sterilized eggs were reared at 25°C on artificial diet in petri dishes. Larvae IV were placed in 0.55-litre containers and fed diet containing 1×10^5 PIB/ml. Infected larvae were collected before death to yield 2×10^9 PIB/larva. Two to three people could infect and rear the larvae necessary to produce $10^{14}-10^{15}$ PIB/year, at a cost of 2–3 ¢ (US)/larval equivalent (2×10^9 PIB) of final product. Since early experimental NPV applications were made at 500 larval equivalents/0.4 ha (acre), the cost for 0.4 ha (excluding formulation and application costs) was (US) $10–$15.

Concurrent with efforts to improve and economise production came an increasing awareness of the potential hazards associated with the product – in particular, its high bacterial load and the presence of allergens, primarily from gypsy moth hair and scales. In anticipation of stringent EPA safety guidelines, isopycnic centrifugation in the zonal "k'" centrifuge (Breillatt et al., 1972) was introduced to remove allergens and undesirable biotypes. This was quite effective, but it increased the cost of one larval equivalent from 2 ¢–3 ¢ (US) to 8 ¢–10 ¢ (US), an alarming (US) $50–$60 when translated to per 0.4-ha costs.

By 1972, substantial progress had been made. A standard virus (Rollinson and Lewis, 1973) had been established and studies were well underway to determine the best combinations of dosage, larval stage, and method of infection. Several large-scale productions, essentially scale-ups of laboratory systems, were made by The Pennsylvania State University, Boyce Thompson Institute, and Syracuse University. Larvae were reared to instar IV on artifical diet then fed diet containing $0.3–1.0 \times 10^7$ PIB/ml. Moribund larvae were harvested and suspensions prepared by blending, screening, and differential centrifugation. These suspensions were further purified and concentrated by zonal centrifugation at the Atomic Energy Commission Laboratory, Oak Ridge Tennessee. Although cost–benefit improvement by large-scale rearing and processing was encouraging, the high fixed cost of zonal centrifugation made the final product noncompetitive per hectare with conventional, chemical insecticides.

In 1975 the U.S. Forest Service contracted for the commercial production of 5×10^{15} PIB at a cost of 6 ¢ (US)/larval equivalent, exclusive of zonal centrifugation. The technique was essentially that of a commercial operation – large capacity rearing chambers, refrigerated vacuum system for collecting cadavers, and large capacity swinging-bucket centrifuges for concentrating PIBs. It produced a PIB slurry ready for zonal centrifugation.

Simultaneously the EPA released guidelines for safety testing and registering insect viruses (Summers et al., 1975). After consulation with the EPA, it was concluded that zonal centrifugation would not be required as long as the final NPV product met the prescribed safety requirements for registration so this step was eliminated in future NPV mass productions. This led to modifications to produce a dry NPV powder with a bacterial load and other properties commensurate with EPA specification. This was the product finally registered (U.S. EPA registration #27586-2).

Elimination of zonal centrifugation resulted in a more intense focus on the microbial quality of the less refined product. Quality control tests were developed (Podgwaite and Bruen, 1978) to monitor batch preparations of NPV. These tests included examinations for potentially dangerous bacteria and other undesirable biotypes by aerobic and anaerobic bacterial count/g, bacterial spore count/g, coliform bacteria count/g, faecal coliform detection, and detection of primary pathogenic bacteria (*Salmonella, Shigella, Vibrio, Streptococcus, Staphylococcus, Clostridium*). Also, a mouse safety test, required by EPA, was developed to determine the presence of toxic or infectious substances in the NPV product. The presence of faecal coliforms or primary pathogens, or the failure of the mouse test resulted in rejection of the batch. The estimated cost for one larval equivalent of this product was about 2.5 ¢ (US) and virus to treat 1 acre (0.4 ha), at a dosage of $2 \times 10^{11}/0.4$ ha, cost < (US) $5.

In late 1976, the Forest Service, Science and Education Administration,

Animal and Plant Health Inspection Service initiated a joint research and development program on mass NPV propagation. The major objective was to design and test a mass NPV production method (using, where possible, cleanroom technology) to yield a product that met all the EPA requirements. This cooperative effort developed the present "state of the art" of gypsy moth NPV production.

G. PRESENT GYPSY MOTH BACULOVIRUS PRODUCTION

The colonized New Jersey F_{16} strain was chosen as the standard host insect, the modified hornworm diet as the standard diet, and the Hamden standard of *L. dispar* NPV as the virus inoculum. The sequence of operations in NPV production is summarized below (Shapiro 1979). The scheme includes the single transfer of 14-day-old larvae (instar IV) to new diet. For best cost-efficiency, it is desirable to maintain larvae in the same containers throughout the production period. Preliminary data indicate that this is feasible.

1. Infested newly hatched larvae (10 per 1.75-oz cup).
2. Larvae held from day 1–14 at 26°C.
3. At day 14 (instar IV), larvae transferred to 6-oz cup (10 larvae/cup) with 90 ml diet; surface of diet previously coated with 1 ml of aqueous suspension of virus inoculum (5×10^6 PIB/ml).
4. Incubated at 29°C for 10 days.
5. At day 24 containers with larvae placed in freezer at $-30°C$.
6. At day 25, frozen larvae harvested into plastic bags and returned to freezer until processing.

Prior work had established that applying the virus inoculum to the surface of the insects' food required least virus. A dosage of 5×10^6 PIB/ml/cup gave the best yield/larva and inoculation of instar IV gave a minimum yield of 1×10^9 PIB/larva. For routine rearing, larvae are kept at 25°–26°C. NPV yields and activity from larvae at 23°, 26° and 29°C were similar, however the LT_{50} decreased as the temperature increased up to 29°C. At 32°C both yields and activity fell, so 29°C was used for virus production. Both living infected and virus-killed insects were frozen at day 10 post-inoculation, at the time of ca. 30% mortality, to minimize the loss of recoverable virus by rupture of fragile integuments, to minimize the bacteria/larva, and to facilitate removing wilted larvae next day from containers.. Ten persons can harvest 10 000 larvae/h.

For processing, larvae were blended in distilled water to free the PIB's from host tissues, then filtered through cheesecloth. Since about 25% of the PIBs were retained by the cheesecloth in the tissue "mat" this was resuspended in distilled water, reblended, and refiltered. Combined filtrates were centrifuged at 7000 g for 25 min. Pellets were air-dried overnight in a vertical laminar flow hood under ambient conditions, then milled into a fine powder. The pertinent data (Shapiro, 1979) are:

Total cadavers produced	502 500
Total PIBs recovered	1.07×10^{15}
Mean yield (PIB/larva)	2.04×10^9 (range; $1.29–2.90 \times 10^9$)

Mean yield/male larva 1.57×10^9 (range; $0.813 - 2.80 \times 10^9$)
Mean yield/female larva 3.99×10^9 (range; $2.74 - 6.55 \times 10^9$)

Research continues to increase efficiency of NPV production. However, the pilot run demonstrated that gypsy moth baculovirus can be produced for (US) $1/10^{11}$ PIB (25×10^6 GMPU) at most, and processed for (US) $0.75/10^{11}$, totalling (US) $1.75/application of 0.4 ha.

III. Formulation and Application

The use of baculovirus is far more complex in forestry than in agriculture. The requirements of maximum coverage, maximum life of the virus, and minimum inhibition of feeding by formulation ingredients are similar for all baculoviruses, but use in forests is compounded by terrain inaccessible to ground equipment, with feeding arenas that may be 30 m thick, yet need thorough coverage at all levels. Also repeated applications are uneconomical, so the fewer applications required, the more effective the formulation needs to be. The egg hatching period may be 2–4 weeks in a given area, so the baculovirus formulation should be effective for this period. Formulation and application of baculoviruses have been studied by Boving et al. (1971); Smith et al. (1977); Ignoffo and Falcon (1978); Lewis (1978); and Smith et al. (1978); see also Chapters 34 and 35.

Certain features must be considered when formulating the NPV of the gypsy moth. NPV must be eaten to take effect, so the feeding activity of larvae must not be impaired, the feeding area — including both leaf surfaces — must be thoroughly covered, and leaf expansion should be well advanced to minimize untreated areas. NPV is quickly degraded by ultraviolet light (UV), so the formulations must provide protection from it for several days to ensure contact between NPV and host insect. Susceptibility of gypsy moth larvae decreases rapidly as they grow larger. The formulation currently in use is: feed-grade molasses, 1.9 litres; Shade®, International Mineral and Chemical Corp., Chicago, Illinois, 454 g; Chevron Spray Sticker®, 180 ml; H_2O, 5.8 litres; Gypchek®, 25 million GMPU.

Sprays of 7.7 litres/acre are applied twice, 7–10 days apart, by small fixed-wing aircraft with 8006 flat-fan nozzles, or preferably Beecomist® model/275 motorized nozzles with 100 μm perforated sleeves. Flight speed is 90–95 mph (145–153 km/h) at a height of 50 ft (15.2 m), to cover a width of 75 ft.

IV. Field Trials with Gypsy Moth Baculovirus

Rollinson et al. (1965) and Magnoler (1968b) found that a direct broadcast spray of NPV gave limited control of the gypsy moth. Vasiljevic and Injac (1973) and Rollinson and Lewis (1973) obtained wide variations in insect susceptibility to the virus and in insect response to different NPV isolates, depending on the insects' geographic source. This, coupled with formulation problems (Smith *et al.*, 1977 and others), probably contributes to the erratic or unacceptable results with baculovirus up to 1972.

In 1973, Yendol *et al.* reported the use of gypsy moth NPV at 10^{13} PIB/acre (0.4 ha), applied twice by a truck-mounted mist blower. Post-spray egg-masses were 52/0.1 ha in the treated and 823/0.1 ha in the untreated plot. Defoliation (ca. 15%) was about the same in both plots.

Based on these ground-spray results and laboratory evaluations, aerial treatment was planned for 1974. Earlier experience with timing and application equipment for *Bacillus thuringiensis* (*B.t.*) was used to plan the test (Lewis *et al.*, 1974). An attempt was made to keep the range of population densities low to medium in the test plots; this avoided both naturally collapsing dense populations, and sparse populations where differences between treatment and control are difficult to assess. K-rotor cleaned Hamden standard virus was applied by fixed-wing aircraft. Pre-spray egg-mass density ranged from 1479 to 1976 masses/0.4 ha. The formulation materials per 0.4 ha were: commercial sunscreen Shade®, 454 g; Chevron Spray Sticker®, 177 ml; CIB® (Cargill Insecticide Base, a stabilized molasses material), 1892 ml; NPV, 10^{12} PIB; and water, 5.7 litres. Each 0.4 ha received 7.6 litres of spray. In June, there were about twice as many larvae in untreated as in virus-treated plots. Adverse weather delayed the first spray and prevented the second so defoliation levels reached 50–75% in treated compared with 95–100% in control plots. Fall egg-mass densities were significantly less in treated (495/ha) than in control plots (13 000/ha), whereas spring egg-mass densities (SEM) had all been about the same. It was decided from these encouraging results that the next virus trials should have two virus applications at a lower concentration (Lewis *et al.*, 1979).

A larger aerial test in 1975 (Wollam *et al.*, 1978) involved one compared with two sprays of the same virus material, ca. 7–10 days apart at 18.7 litres/ha. Of two formulations one was 1.12 kg Shade UV screen, 0.88 litres Chevron Spray Sticker, 2.38 litres CIB, and 15.4 litres water/ha: the other was 9.34 litres SVA® (Sandoz Virus Adjuvant) and 9.34 litres water/ha. Each of the four treatments had three replicate (14.2 ha) blocks, and there were three control blocks. In ten 0.01-ha subplots in each block, the following data were collected: egg-mass numbers (pre- and post-treatment), number of larvae and pupae under burlap bands, defoliation estimates, spray deposit, and bioassay of leaves collected after spraying.

Treated blocks had significant increases in virus incidence compared with control blocks and there were significant differences in defoliation (Table I). All treatments provided acceptable foliage protection except double SVA. The SVA formulation resulted in 3X as much coverage as the CIB formulation (33.0 cf 10.2 drops/cm^2). The mmd (mass median diameter) was 320 μ for the SVA and 310 μ for the CIB. Bioassay of sprayed and unsprayed foliage showed a high incidence (63–76%) of virus in the treated plots immediately after spray as determined by microscopic examination of dead larvae. This fell to about 25% after 3 days and continued at this level for 2 weeks or more. The second spray caused an immediate rise in the incidence of NPV to about 75%, with a 2-day fallback to the 25% level.

TABLE I

Effect of formulation and number of aerial sprays with gypsy moth NPV on defoliation in 1975

Formulation	Sprays	Mean % defoliation ± standard deviation
Sandoz Virus Adjuvant ®	1	17 ± 13
	2	49 ± 31
Cargill Insecticide Base ®	1	23 ± 31
	2	36 ± 36
Control		69 ± 26

Population reduction was achieved with the single and double application of SVA but not with the CIB, mainly because of the poorer coverage with CIB (12.5% of total formulation). These low-density plots treated in 1975 were examined in 1976 for carry-over effect of the treatment. Virus treatment did not cause an increase in eggs/egg mass as with *B.t.* (Kaya *et al.*, 1974) or with chemical insecticides (Doane, 1968). Second-year defoliation in all treated plots (20%) was considerably less than in untreated plots (40%).

In the fall of 1975, a new NPV product (Gypchek) was developed, not K-rotor cleaned but air-dried. This change in product necessitated further efficacy evaluation. The dried slurry contained more protein and free virions, and was expected to be more effective because of the protective properties of the proteins.

The 1976 aerial trial was designed to test lower dosages of NPV, further test SVA, and, because of the relatively poor performance of the molasses formulation in 1974, to test a return to 25% molasses from the 12.5% used in 1974. Three 14.2 ha-replicates were treated with two dosages; 10^{11} PIB (25 million GMPU)/0.4 ha and 5X this level. Single and double applications were made, all in the evening. Two formulations were tested: feed-grade molasses, 1.9 litres; Chevron Spray Sticker, 177 ml; Shade 454 gm; water, 5.7 litres: and SVA, 3.8 litres; water, 3.8 litres. A double application at low dosage achieved a foliage protection and population reduction as acceptable as the double high dosage, regardless of formulation (Table II). The single application of the high dosage gave significantly less foliage protection and population reduction than the double application.

Another aerial test was made in 1977 with Gypchek to evaluate use of conventional flat-fan nozzles in place of spinning-sleeve nozzles; reduction of number and volume of sprays; morning versus evening spraying. Due to an unanticipated collapse of population in the forest, no conclusions could be drawn on reduced applications or morning versus evening spraying. Larvae grew rapidly due to unseasonably hot weather, soon stripping untreated areas around the sprayed 14.2-ha plots, and degradation of the spray from UV allowed larvae from the outside to invade treated plots, obliterating the effect of treatments (Lewis *et al.*, 1979). A successful part of this test was carried out in central Pennsylvania on 14.2-ha plots replicated three times. One dosage (10^{11} PIB/0.4

TABLE II

Effect of dosage of gypsy moth NPV and formulation on defoliation and egg masses in 1976

Dosage (PIB/0.4 ha)	Formulation	% final defoliation	Egg masses/0.4 ha		
			Spring	Fall	% change
1×10^{11}	Molasses twice	41.7	2060	380	-82
5×10^{11}	Molasses once	59.0	2073	897	-57
5×10^{11}	Molasses twice	46.7	1667	331	-80
5×10^{11}	SVA twice	42.3	2236	293	-87
0	Control	82.7	2606	1369	-47

ha; 25 million GMPU) was applied to all treated plots, using a single formulation (Feed-grade molasses, 1.9 litres; Shade, 454 gm; Chevron Spray Sticker, 180 ml; water, 5.7 litres), at 7.6 litres/0.4 ha.

The results (Table III) indicate no significant (0.05-level) effect of nozzle or formulation efficacy. Both foliage protection and population reduction were

TABLE III

Effect of dosage of gypsy moth NPV and nozzle on defoliation and egg-masses in 1977

Dosage (PIB/0.4 ha)	Nozzle	% final defoliation	Egg masses/0.4 ha		
			Spring	Fall	% change
10^{11} twice	Beecomist®	49	1337	303	-77
10^{11} twice	Flat-Fan®	55	1184	410	-65
Control	0	80	1154	2293	$+37$

acceptable. Droplet sizes were 300–320 for both nozzles. Evening application of the NPV twice gave comparable results (Fig. 1) to those in 1976.

V. Benefits of Use of Gypsy Moth Baculovirus

The cost–benefit relationship for gypsy moth NPV is not simply that between the costs of materials plus application and expected benefits. Other considerations, not easily quantified, play a role in the decision of whether to use the NPV. Environmental concerns, needs and desires of the land manager, and political and socio-economic factors are important.

On the positive side, the NPV has no demonstrable effects on beneficial forms of life. It is a natural component of the gypsy moth ecosystem and does not adversely affect other natural mortality factors. It protects foliage and is a sound pest management tool because of its safety and compatibility with other forms of insect control, and because the gypsy moth has not developed resistance to its

Fig. 1. Aerial infra-red photograph of block treated with Gypchek 1977.

effect. This is in marked contrast to most synthetic insecticides and is an important consideration when contemplating treatment of the same populations over time.

On the negative side, the NPV acts relatively slowly and does not kill immediately. Also, its application requires considerable care. At present, it is somewhat more expensive than conventional insecticides, but improvements in production and use patterns will significantly reduce the cost.

One main attribute of the NPV can be considered either a positive or negative benefit, i.e. the NPV is selective against gypsy moth larvae. This is a positive attribute if a user has only a gypsy moth problem; however, if there is a serious infestation of mixed pest insects, only the gypsy moth will be killed, which is a negative aspect. Another such attribute is the rapid loss of virulence in broadcast application. For a user who desires or needs short residual activity this is a positive aspect; however, if development of the insect is extended or longer residual activity is needed, it is a negative attribute.

Thus, a cost—benefit analysis falls on the positive side if environmental considerations are strong, if minimal damage to other natural factors is desired, if the treated area is mainly infested with gypsy moth, and if careful timing and application are not constraints. The reverse occurs if environmental considerations are low, infestations are mixes of pests each capable of causing significant damage, immediate kill is required, or long residual activity is necessary.

Although the data are not overwhelming, reported information indicates that the NPV can exert effects for more than one consecutive year. Rollinson et al. (1965) found only a few virus-killed larvae in the treated plot and no apparent spread of the gypsy moth NPV in the year following ground application of a heavy dosage (4×10^{12} PIB per 0.4 ha). However, Yendol et al. (1977) evaluated a series of plots in Pennsylvania which were treated by truck-mounted mist blower with the gypsy moth NPV (1×10^{13} PIB/ha). The year following application twice as many larvae (7/tree) and twice as many pupae were found in the untreated check plots as in the treated blocks (3.7 larvae/tree). Magnoler (1974) described NPV carry-over in plots treated with NPV, and the spread of the virus into untreated areas..

It is tempting to compare the use of B.t. versus baculovirus for control of the gypsy moth since both are microbial entities. However, they differ in their effects on gypsy moth in the following ways:- 1. Baculovirus is more specific. 2. B.t. acts quicker and can achieve foliage protection without large population reduction. 3. Baculovirus can exhibit carryover effects, B.t. cannot. 4. B.t. can affect alternate parasite hosts whereas baculovirus will not. 5. B.t. is less affected by UV radiation than baculovirus. 6. B.t. is more nearly analogous to a synthetic pesticide than is baculovirus.

VI. Safety of Gypsy Moth Baculovirus

Details about safety are given by Lewis and Podgwaite (1979) and Chapter 40. Essentially the baculovirus of the gypsy moth exhibited no harm to mammals, fish, birds, beneficial insects, parasitic insects; no relationship to arboviruses, and no effects on aquatic invertebrates. A slight eye irritation was caused when Gypchek was tested at extremely high dosages, an effect not due to microorganisms, but to insect parts in the product.

VII. Future Work with Forest Insect Baculoviruses

Gypchek is the third baculovirus product to be registered in the USA, and the first to be registered by the EPA for use on deciduous hardwood trees in the heavily populated eastern USA. An important implication of the approval of the gypsy moth NPV is that much remains to be done to optimize its use. Formulation improvements to extend its field activity are critically needed. More attention should be placed on alternative uses such as baiting, trapping, and vector enhancement. More work is needed in developing and evaluating its role in pest management systems. Further research should be conducted on the means of transmission of this insect virus, the mode of replication, and further elucidation of the basic nature of the baculovirus. Refinements and improvements in the production of the baculovirus are needed to reduce cost, improve effectiveness, and meet anticipated needs for the product.

References

Bell, R. and Shapiro, M. (1979). *In* "Gypsy Moth: Research Toward Integrated Pest Management" U.S. Dept. Agric. Tech. Bull. 1584. (In press.)

Bergold, G.H. (1947). *Z. Naturf.* **2B**, 122–143.

Bess, H.A. (1961). Conn. Agric. Exp. Stn Bull. **646**, 1–43.

Boving, P.A., Maksymuik, B., Winterfield, R.G. and Orchard, R.D. (1971). *Trans. Am. Soc. agric. Engrs* **14**, 48–51.

Breillatt, J.P., Brantley, J.N., Mazzone, H.M., Martignoni, M.E., Franklin, J.E. and Anderson, N.G. (1972). *Appl. Microbiol.* **23**, 923–930.

Burgerjon, A. and Dulmage, H.T. (1977). *Entomophaga* **22**, 121–129.

Burgerjon, A., Biache, G. and Chaufaux, J. (1975). *Entomophaga* **20**, 153–160.

Campbell, R.W. (1963). *Can. Ent.* **95**, 426–434.

Doane, C.C. (1967). *J. Invertebr. Path.* **9**, 376–386.

Doane, C.C. (1968). *J. econ. Ent.* **61**, 1288–1291.

Doane, C.C. (1970). *J. Invertebr. Path.* **15**, 21–33.

Dulmage, H.T. (1973). *Ann. Acad. Sci.* **217**, 187–199.

Dulmage, H.T. and Burgerjon, A. (1977). *Entomophaga* **22**, 131–139.

Glaser, R.W. (1915). *J. agric. Res.* **4**, 101–128.

Glaser, R.W. and Chapman, J.W. (1913). *J. econ. Ent.* **6**, 479–488.

Harper, J.P. (1976). *J. Invertebr. Path.* **27**, 275–277.

Hedlund, R.C. (1974). Ph.D. Thesis Pennsylvania St. Univ., University Park PA.

Ignoffo, C.M. (1972). *J. Invertebr. Path.* **6**, 318–326.

Ignoffo, C.M. and Falcon, L.A. (1978). *Misc. Publ. ent. Soc. Am.* **10**, 1–2.

Kaya, H., Dunbar, D., Doane, C., Weseloh, R. and Anderson, J. (1974). *Conn. Agric. Exp. Stn Bull.* **744**, 1–22.

Leonard, D.E. and Doane, C.C. (1966). *Ann. ent. Soc. Am.* **59**, 462–464.

Lewis, F.B. (1971). IV Int. Colloq. Microbial Control Insect Path. 320–326.

Lewis, F.B. (1978). Proc. Aerial Spray Conf., Columbus, Ohio, 1977.

Lewis, F.B. and Daviault, L. (1967). Can. Dep. For. Rur. Dev., 153–156.

Lewis, F.B. and Etter, D.O., Jr. (1978). *In* "Microbial Control of Insect Pests: Future Strategies in Pest Management Systems" (G.E. Allen, C.M. Ignoffo and R.P. Jaques, eds), pp 261–272. NSF-USDA-Univ. Florida, Gainesville.

Lewis, F.B. and Podgwaite, J.D. (1979). *In* "Gypsy Moth: Research Toward Integrated Pest Management" U.S. Dept. Agric. Tech. Bull. 1584. (In press)

Lewis, F.B., Dubois, N.R., Grimble, D., Metterhouse, W. and Quimby, J. (1974). *J. econ. Ent.* **67**, 351–354.

Lewis, F.B., Rollinson, W.D. and Yendol, W.B. (1979). *In* "Gypsy Moth: Research Toward Integrated Pest Management" U.S. Dep. Agric. Tech. Bull. 1584. pp. 503–512.

Magnoler, A. (1967). *Entomophaga* **12**, 199–207.

Magnoler, A. (1968a). *Entomophaga* 13, 335–344.

Magnoler, A. (1968b). *Annls. ent. Soc. France,* New Series 4, 227–232.

Magnoler, A. (1974). *Z. PfKrankh. PflSchutz* 81, 497–511.

Martignoni, M.E. and Iwai, P.J. (1977). U.S. Dep. Agric. Forest Serv. Res. Pap. PNW-222.

ODell, T.M. and Rollinson, W.D. (1966). *J. econ. Ent.* 59, 741–742.

Orlovskaja, E.V. (1961). *Bull. Vses inst. Zasc Rast.* (3–4), 54–57.

Paschke, J.D., Lowe, R.E. and Giese, R.L. (1968). *J. Invertebr. Path.* 10, 327–334.

Podgwaite, J.D. and Bruen, R.B. (1978). U.S. Dept. Agric. N East Forest Serv. Exp. Stn Gen. Tech. Rep. NE-38.

Podgwaite, J.D. and Campbell, R.W. (1971). Proc. 4th Int. Colloq. Insect Path. 279–284. College Park, Maryland.

Podgwaite, J.D. and Campbell, R.W. (1972). *J. Invertebr. Path.* 20, 303–308.

Reiff, W. (1911). "The Wilt Disease Flacherie of the Gypsy Moth". Wright and Potter, Boston.

Rollinson, W.D. and Lewis, F.B. (1962). U.S. Dept. Agric. N East Forest Serv. Exp. Stn Res. Note 130.

Rollinson, W.D. and Lewis, F.B. (1973). *Pl. Prot.* 24, 163–168.

Rollinson, W.D., Lewis, F.B. and Waters, W.E. (1965) *J. Invertebr. Path.* 7, 515–517.

Shapiro, M. (1979). *In* "Gypsy Moth: Research Toward Integrated Pest Management" U.S. Dept. Agric. Tech. Bull. 1584. (In press.)

Smith, D.B., Hostetter, D.L. and Ignoffo, C.M. (1977). *J. econ. Ent.* 70, 437–441.

Smith, D.B., Hostetter, D.L. and Ignoffo, C.M. (1978). *Misc. Publ. ent. Soc. Am.* 10, 44–66.

Steinhaus, E.A. (1949). "Principles of Insect Pathology". McGraw Hill, New York.

Steinhaus, E.A. (1963). "Insect Pathology, An advanced Treatise". 2 vols. Academic Press, New York and London.

Summers, M.D., Engler, R., Falcon, L.A. and Vail, P.V., eds. (1975). "Baculoviruses for Insect Pest Control: Safety Considerations". Am. Soc. Microbiol., Washington, D.C.

Vail, P.V. (1975). *In* "Baculovirus for Insect Pest Control: Safety Considerations" (M.D. Summers, R. Engler, L.A. Falcon and P.V. Vail, eds), pp. 44–46. Am. Soc. Microbiol., Washington, D.C.

Vail, P.V., Jay, D.L. and Hunter, D.K. (1971). Proc. 4th Int. Colloq. Insect Path., pp. 297–303. College Park, Maryland.

Vasiljevic, L. (1961). *Arh. poljopr. Nauke* 14, 103–118.

Vasiljevic, L. and Injac, M. (1973). *Pl. Prot.* 24, 124–125.

Wollam, J.D., Yendol, W.G. and Lewis, F.B. (1978). U.S. Dep. Agric. N East Forest Serv. Exp. Stn Res. Pap. NE-396.

Yendol, W.G., Hamlen, R.A. and Lewis, F.B. (1973). *J. econ. Ent.* 66, 183–186.

Yendol, W.G., Hedlund, R.G. and Lewis, F.B. (1977). *J. econ. Ent.* 70, 598–602.

CHAPTER 19

Control of Sawflies by Baculovirus

J.C. CUNNINGHAM and P.F. ENTWISTLE

Forest Pest Management Institute, Sault Ste. Marie, Ontario, Canada and Institute of Virology, Natural Environment Research Council, Oxford, England.

I. Sawflies and their Baculoviruses

A. SAWFLIES AS PESTS

Most sawfly (Hymenoptera:Symphyta) genera attack plant families which are "predominantly arborescent rather than herbaceous" (Benson, 1950). They are also largely a northern group, with few in the tropics and, in Australia, represented mainly by one family, the Pergidae, on eucalyptus. This is clearly reflected in their occurrence as pests, most sawflies attacking trees and shrubs in the Holarctic. The pest species are dominated by the family Diprionidae, the

91 members of which exclusively attack conifers, mainly *Pinus* and *Picea* (Smith, 1974). At least a quarter of diprionid species rank as serious pests. Conifers are also attacked by sawflies in other families, notably species of *Pristiphora* (Tentheredinidae) and of *Cephalcia* (Pamphiliidae).

Broad-leaved trees suffer less sawfly damage than conifers. However, birch is often heavily defoliated and large areas killed in northern Europe by species of *Fenusa* (Blennocampinae), *Heterarthus* (Heterarthinae), *Dineura* and *Croesus* (Nematinae). The hypermetamorphic larvae of *Hoplocampa testudinea* (Nematinae) attack apple and a complex of three *Hoplocampa* species attacks plum. Gooseberries and currants (*Ribes* spp.) are often severely infested with various species of *Nematus* and with *Pristiphora pallipes*.

Attacking non-woody plants, members of the tribe Cephini of the stem boring family Cephiidae are important pests of cereals and grasses. For instance, *Cephus cinctus* and *C. pigmaeus* are the infamous Western and European wheat-stem sawflies, respectively, and two species of *Trachelus* attack rye. Members of the genus *Athalia* (Blennocampinae) are frequent pests of Cruciferae, notably turnips, rape and mustard.

Some of the most economically important sawfly species not known to have baculoviruses are listed in Table I and all those with viruses are listed in Appendix 2. It is not yet clear if the difference between these two tables is more significantly related to sawfly taxonomy or to host plants. Most species with virus diseases are Diprionidae associated with conifers. On the other hand most pest species without viruses are Cephiidae and Tentheredinidae, few of which are important on conifers.

TABLE I.

Pest species of sawflies for which no virus diseases are known

CEPHIIDAE		TENTHEREDINIDAE (continued)	
Cephus cinctus	Western wheat stem sawfly	*Croesus septentrionalis*	Birch
		Dineura virididorsata	Birch
C. pigmaeus	European wheat stem sawfly	*Fenusa pusilla*	Birch
		Heterarthus nemoratus	leaf-mining sawflies
Syrista parreysii	Rose stem sawfly		
Trachelus tabidus	Rye	*Hoplocampa brevis*	Pear "slug"
T. troglodyta	Barley and Rye	*H. flava*	Plum "slug"
CIMBICIDAE		*H. minuta*	
		H. testudinea	Apple "slug"
Cimbex americana	Elm		
Trichiosoma lucorum	Birch	*Messa hortulana*	Poplar leaf miner
TENTHEREDINIDAE			
Athalia rosae	Cabbage and turnips	*Nematus leucotrochus*	Gooseberry, currant
		N. ribesii	
A. lugens proxima	Mustard, rape and radishes	*Pristiphora pallipes*	
		Psylloides chrysocephalus	Rape
Ardis brunniventris	Rose tip sawfly		
Caliroa cerasi	Cherry sawfly		

B. SAWFLY BACULOVIRUSES

Sawfly viruses have been listed by Hughes (1957), Martignoni and Langston (1960), Martignoni and Iwai (1977) and in Appendix 2. Nuclear polyhedrosis viruses (NPVs) are now known from 25 sawfly species. A granulosis virus (GV) has been reported from *Cephalcia fascipennis* (Smirnoff and Juneau, 1973). Some cytoplasmic polyhedrosis viruses (CPVs) are also known, notably from *Anoplonyx destructor* (Tentheredinidae), *Neodiprion merkeli* (Diprionidae) and five species of wood wasp (Siricidae). In some early records NPV and CPV were not distinguished and so are recorded as "polyhedroses" in Appendix II.

Taxonomic confusion in the Diprionidae sheds doubt on some host records. Some information on this topic was supplied by Dr. D.R. Wallace. *N. virginiana* was described in Virginia in 1918, but there is now no species bearing this name. In Ontario, there are two species called the "*N. virginiana* complex", *N. dubiosus* and *N. rugifrons*. The "*N. pratti* complex" consists of three sub-species or geographical races, *N. pratti banksianae* (north western Canada and the USA through Manitoba and Saskatchewan), *N. p. periodoxica* (central Canada, the north-central USA and along the east coast) and *N. p. pratti* (the Virginias, Carolinas and Kentucky). Geographical overlap of these sub-species is common and intermediate forms occur. The "*N. abietis* complex" is large, confusing and little studied. *N. abietis* occurs throughout North America from Mexico to Alaska in the West, across the centre of the continent and down the east coast. Many host species are involved with temporal differences in life histories. In Ontario an early race and late race feed on *Abies balsamea*; elsewhere *Pinus* spp., *Picea* spp., *Tsuga* spp., and *Pseudotsuga* spp. are the host trees. *N. abietis* can be divided into several sub-species if not distinct species.

Although sawfly NPVs are considered highly host specific, there are few reports on cross-infectivity tests. *Cephalcia lariciphila* was not susceptible to six diprionid and one tentheredinid NPV (Entwistle, unpublished observations) whilst no alternative host could be found for *Gilpinia hercyniae* NPV either in Canada (Bird, 1949) or the UK (Beck, unpublished observations). However, *N. p. banksianae* NPV infected *N. nanulus* (Bird, 1955). The NPVs of *N. sertifer* and *N. lecontei* have mild reciprocal cross-infectivity but are much less pathogenic in non-homologous than in homologous hosts (Cunningham, unpublished observations). Both these NPVs infected *N. abietis* larvae but high concentrations (10^7 PIB/ml) gave little mortality (Olofsson, unpublished observations). Polyhedra were observed in the midgut epithelia of *Pristiphora erichsonii* larvae fed *P. geniculata* NPV (Smirnoff, 1968).

Histopathological examination of larvae of several sawfly species has revealed replication of NPV only in midgut epithelial cells. Larval feeding declines and ceases long before death. Possibly, however, the NPV of *Cephalcia abietis* may replicate mainly in the fat body as do most NPVs of Lepidoptera (Jahn, 1962).

C. SAWFLY VIRUS ECOLOGY IN RELATION TO CONTROL

Features of virus ecology especially relevant to the role of viruses as control agents are the rates of growth of infection in relation to host population density and the means of horizontal spread and of vertical passage of disease. Vertical passage encompasses both spread by adults to progeny and persistence between generations in the extra-host environment, notably on plant surfaces and in soil and secondary hosts, if any. Understanding such factors reveals the extent to which virus persists and hence the durability of its controlling effect (Franz, 1956; Entwistle, 1976).

1. Infection Growth and Horizontal Passage

The rate at which natural infection grows is one of the main determinants of epizootic progress. The typical growth curve (proportion of hosts infected $[x]$ vs time) is sigmoid but can be straightened by use of $\text{Log}_{10} \dfrac{x}{1-x}$ instead of x, allowing comparison of infection growth curves. The slope of this straightened line indicates the rate of infection. Alternatively the infection rate "r" can be directly calculated from:

$$r = \frac{1}{t} \log_e \frac{x_2 (1-x_1)}{x_1 (1-x_2)}$$

where t is the time from the first to the peak level of infection and x_1 and x_2 are first and final levels of disease (van der Plank, 1963; Evans and Entwistle, 1976). Similar growth rates in different years or areas do not necessarily imply growth of infection to the same level, for this is influenced by the time of onset of infection. The inception of infection growth is a function of the quantity of available inoculum and is discussed below as a component of vertical passage since it concerns inoculum persistence from one generation to the next.

Mechanisms of horizontal spread of virus involve both abiotic and biotic agents. The main abiotic agent appears to be rain splash which is very localized. The question of possible dispersal by wind, in dust when dry and especially in aerosols when wet, has not been resolved either in laboratory or field. Biotic agents may be classified as specific (adult sawflies) or casual (parasites, predators and saprophages). Specific dispersal by adults is dealt with below under Vertical Passage. Spread of sawfly viruses by parasites has not been specifically identified as it has with lepidopterous hosts. A Welsh epizootic of NPV disease in *Gilpinia hercyniae* spread substantially without parasites and despite the introduction of parasites, the same was essentially true in the well known N. American outbreak. Clearly parasites are not always essential for adequate virus spread. Dispersal by predators is greatly aided by the capacity of NPVs to survive passage through the gut of birds, mammals and a wide variety of predatory invertebrates. Bird (1955) reported active *N. sertifer* NPV in the guts of two birds in Canada, *Dunatella carolinensis* and *Bombycilla cedrorum.* Franz *et al.* (1955) noted that in Germany the robin, *Erithaecus rubecula,* passed infective droppings after eating NPV-diseased *N. sertifer* larvae. In Wales spread of *G. hercyniae* NPV by birds has received detailed study (Entwistle *et al.*, 1977a, b; 1978a). In an epizootic area, sixteen species of birds in six families comprising 80% of bird numbers had infective virus in their faeces during September—October. Of droppings collected from trees, 90% were infective. Though most bird species were arboreal, so that live sawfly adults and larvae were accessible to them only between July and October, droppings collected from trees outside this period and from birds netted in January were infective, indicating birds dispersed virus all year. Outside July–October birds probably ingested virus in dead, infected, larvae adhering to the trees. Though NPV may first be defecated within < 1 h of ingestion, defecation of infective virus may continue for three or more days (Entwistle et al., 1978a). Depending on the habits of the bird species involved, this prolonged retention could clearly effect wide dispersal of the virus. In terms of the contribution of birds to horizontal spread, the late larval period of G. hercyniae has particular significance because the breeding territoriality of most birds has ceased and flocks of mixed species range widely. The relevance of birds to the fate of the total inoculum pool is further considered below in relation to virus persistence on trees.

2. Vertical Passage of NPVs: Environmental Persistence

NPVs have a capacity for prolonged persistence in the extra-host environment, mainly on plants and in soil. In Canada, Bird (1961), referring to G. hercyniae NPV on Picea glauca stated, "Virus that accumulated on foliage during a virus epizootic is inactivated by the following summer, except possibly that virus which is contained within desiccated cadavers", and he supported this by adding, "The infection of a population on a tree one year did not ensure that the population on the same tree would be infected in the following year". If the significance of the corpses is considered, these views are not necessarily at variance with survival of an NPV of the same insect host on Picea abies and P. sitchensis in Wales. Here infective virus was demonstrated on foliage throughout the winter to the beginning of the new larval generation (Entwistle and Adams, 1977; Evans and Entwistle, 1976, and in press). Virus applied to Picea foliage decayed at rates not entirely incompatible with this finding. For instance, the half lives for the physical presence of polyhedra, applied as pure or impure suspensions to P. abies foliage, were 38.3 ± 3.0 and 55.1 ± 19.4 days, indicating that deposits with a nominal initial value of 100 polyhedra would decline to, respectively, 1.5 and 6.2 during the 240 days between the end of one larval generation and the beginning of the next. However, decay measured biologically seems more rapid than this, probably mainly due to the viricidal effects of solar ultra-violet light. The long term presence of infective virus appears therefore to need a source of supply to replace attritional losses occasioned by both physical depletion of polyhedra and biological decay of the virus in those polyhedra remaining. Such a source probably lies in the corpses of virus-killed larvae adhering to the trees, the gradual decay of which liberates polyhedra. Such corpses are present up to at least the commencement of the new generation. This could explain the observed movement of virus over the foliage of trees during winter (Evans and Entwistle, in press). The presence of infective NPV in the faeces of birds in winter suggests they help to liberate virus from larval corpses and mobilize this inoculum pool. Evidence indicates mid-winter dispersal over 7 km by birds (Entwistle et al., 1977b).

The overwinter persistence of *N. sertifer* NPV on young *Pinus contorta* trees in Scotland has been experimentally demonstrated. The virus can be found on needles, in needle sheaths and on bark (Kaupp and Entwistle, unpublished observations).

The extent of overwinter persistence of NPVs in soil is considerable and though there is no direct information on the durability of sawfly NPVs in this environment there is also no reason to suppose they differ from those of Lepidoptera. Infective NPV of *G. hercyniae* in Wales was found to a depth of 15 cm but was largely restricted to the top 7 cm. The relative roles of arboreal and soil inocula in the epizootiology of NPV diseases of sawflies on conifers must differ strongly. Arboreal inoculum is bound up in the day to day processes of infection growth in larval populations whilst its overwinter survival has a big effect on the rise and fall of infection in successive generations. The long persistence, but limited availability, of inoculum in the soil probably confers an important role in perpetuating virus especially while host population density is too low to maintain actively spreading, albeit enzootic, disease.

3. Vertical Passage: Persistence in the Host Developmental Sequence

In sawflies, transmission of NPV diseases by adults to their progeny has been studied only in certain diprionids. In *G. hercyniae* a comparatively large dose of virus, $10^5 - 10^6$ PIB, is required to induce infection late in the last feeding instar (V). These larvae do not die but moult to instar VI, a non-feeding stage, which soon cocoon in forest-floor litter where they may diapause for one or more winters (Prebble, 1941a, b). At this moult the midgut lining of secretory cells is replaced by one composed of small undifferentiated cells. These are not susceptible to infection. In the relatively brief pupal period just before adult emergence the secretory epithelium of the midgut reforms and can support replication of the virus, the inoculum initiating this infection perhaps originating in virus persisting in the gut lumen of the diapausing instar VI or eonymph (Bird, 1953a). Replication of NPV continues in adult midgut (Entwistle and Adams, unpublished observations). Such infection does not impair adult size, weight or reproductive biology (James, 1974). All available evidence suggests that transmission of virus from adults to progeny is neither within nor on the egg surface: virus has not been detected in eggs either by electron microscopy or bioassay. Sawfly eggs hatch by bursting and larvae of *G. hercyniae*, in particular, do not feed on those spruce needles in which eggs have been laid. However, during oviposition adults contaminate foliage with virus from which larvae may become infected (Bird, 1961; Neilson and Elgee, 1968). The conditions leading to adult infection, i.e. a high dose of virus ingested by the larva within a very critical time period, raise the question of what epizootic conditions can give rise to an appreciable proportion of infected survivors. Optimal conditions for production of not only the highest proportion, but also the greatest number of infected adults occur when larval infection is high but before the population is reduced to a low density. Then the total inoculum available is considerable and most likely to provide the high infecting doses necessary. This mechanism could contribute strongly to the explosive nature of NPV disease epizootics observed in *G. hercyniae* populations. Geographically its influence will be related to the scale of adult dispersal flight: this occurs over at least 2.5 km and probably much further (Entwistle, 1976).

Passage of NPV in *N. sertifer* is similar, virus being found in the hindgut contents of adults, but not in embryos (Gulii, 1971). In *N. swainei* exposure of late larval instars to weak concentrations of NPV was largely non-fatal but allowed survivors to transmit virus disease to their progeny. By introducing cocoons which produced infective adults, Smirnoff (1962, 1972) started epizootics in larval populations of *N. swainei*.

4. Dispersal of Sawfly Virus Diseases

Epizootic capacity of a virus disease is shown by the rate at which infection grows and the speed with which it disperses spatially. Where, as is usual, the expression of full epizootic effect requires several host generations, the extent of persistence of inoculum between generations, discussed above, is of great importance. Studies in the dispersal of NPV disease from initially small epicentres into surrounding healthy larval populations of *G. hercyniae* suggest an initial phase similar in form to the development of epiphytotics (Gregory, 1952; Bird and Burk, 1961; Entwistle, 1972, 1974, 1976). Here the relationship of disease incidence (y) to distance (x) from the epicentre follows a gradient in the form of a concave curve. Such curves may be readily compared by conversion to straight lines through logging both units, i.e.

$$\log y = a + b \log x$$

where the regression coefficient, b, expresses the gradient of dispersal and is always negative. Gradients of primary dispersal for *G. hercyniae* NPV disease were remarkably similar in Canada and Wales and were maintained throughout the development of this phase of dispersal, suggesting that for this particular disease—host—environment association there was a consistent dispersal capacity.

The primary phase of dispersal matured into a wave form, presumably expressing the delayed density dependent relationship of disease and host. Again form and scale were very consistent in Canada and Wales. A third phase, less definite in form, may be due to the interaction of the initial epicentre with the expanding influence of secondary epicentres. The overall rate of epizootic growth in Wales exceeded the rate of expansion of an individual disease centre — presumably a partial consequence of discontinuous dispersal causing fresh epicentres. The agencies effecting both the continuous growth of epicentres and the discontinuous generation of new epicentres are presumably those described for horizontal spread and vertical passage of NPV. One dispersal agent, the adult sawfly, is also one of the means of persistence whereby epizootics may develop from season to season. The natural dispersal of sawfly NPVs has been applied in practice, e.g. in 1943—44 and 1948 the NPV of *G. hercyniae* was introduced to Newfoundland where it spread rapidly (Clark and Clarke, 1973). Smirnoff (1972) advocated spot spraying *N. swainei* larvae with weak doses to induce infection in adults which would start epizootics. He suggested that after 4 years the sawfly population in treated areas would be reduced and the disease would spread and coalesce with other treated areas. In Canada, the spread of *N. lecontei* NPV from tree to tree was more rapid than *N. sertifer* NPV, and this was attributed to enhanced spread by a larger complex of natural enemies in the native *N. lecontei* (Bird, 1961).

D. BIOLOGY OF SAWFLIES IN RELATION TO CONTROL BY VIRUSES

The length of time from hatching of sawfly eggs until cocoon spinning is an important factor in considering the use of viruses to regulate populations. Viruses have their greatest impact when directed against early instars with low dosages giving high mortality and foliage protection. When sawflies have a short life cycle, timing of sprays becomes critical. Developmental time depends on several factors, sawfly species, latitude of origin, host plant, temperature and sex of larvae. Wallace and Sullivan (unpublished observations) have extensively studied the biology of sawflies in Ontario. A few examples are given at constant conditions in Table II; the shortest larval period was found in *Neodiprion nigroscutum* and the longest in *N. lecontei*. Development of *N. sertifer* larvae at $21°C$ was fastest on *P. sylvestris* foliage, about 1 day slower on *P. banksiana* and 2 days slower on *P. resinosa*. At $10°C$ on *P. resinosa*, male *N. sertifer* larvae took 55 days to develop and females 61 days, while at $28°C$ males took 15.5 days and females 17.5 days. On *P. sylvestris*, the median length of the larval period was 45.7 days at $12°C$, 29.8 days at $18°C$ and 22.1 days at $24°C$ (Tvermyr, 1969).

TABLE II

Larval development times for some common diprionid sawflies in Ontario. Times are in days from hatch to cocoon spinning at $22 ± 1.5°C$ in a 16-h daylength (Wallace and Sullivan, unpubl.)

Species of *Neodiprion*	Latitude of origin	Host tree, *Pinus*	Males mean ± S.D.	Females mean ± S.D.
N. lecontei	46° 15'N	*P. resinosa*	29.9 ± 4.01	31.7 ± 4.86
N. nigroscutum	50°N	*P. banksiana*	15.2 ± 1.86	17.1 ± 1.23
N. sertifer	44° 30'N	*P. resinosa*	18.0 ± 1.43	21.6 ± 1.41
N. swainei	47°N	*P. banksiana*	24.6 ± 2.35	30.3 ± 2.81

Following virus sprays on sawfly populations, several workers observed that many more males emerge from surviving pupae than females e.g. *N. taedae linearis* (Yearian and Young, 1971), *N. p. pratti* (McIntyre and Dutky, 1961), *N. p. banksianae* (Bird, 1961) and *N. abietis* (Olofsson, 1973). This is because males develop faster than females (Table II) and are thus exposed to virus for a shorter period (Bird, 1961).

Most sawflies have only one generation per year although there are exceptions. In N. America *G. hercyniae* has from 1–3 generations whilst in southern USA *N. lecontei* can have > 3 generations (Benjamin, 1955). At high altitudes and in northern regions *N. sertifer* may have a 2-year life cycle passing the first winter as the egg and the second in the cocoon: arctic–alpine forms differ morphologically (Pschorn-Walcher, 1970). Prolonged diapause is known in *N. sertifer* (Lyons, 1964), and in *G. hercyniae* may extend up to 7 years (Prebble, 1941b). This is probably a common phenomenon in diprionid sawflies, perhaps to ensure population survival in adverse years. It also permits partial escape from applied baculoviruses.

Sawflies overwinter either as eggs on foliage or as eonymphs in cells or cocoons, often in the soil. More synchronous hatch is found in species with overwintering eggs, e.g. in Scotland over 90% of *N. sertifer* eggs hatch during one week, than in species with an eonymphal diapause, e.g., *N. lecontei*. Synchronous hatch is desirable when virus sprays are planned. Most coniferous-feeding sawflies devour only old foliage, eating the current year's growth only when old needles are exhausted. One exception is *N. swainei* which eats the current year's foliage. Neither feeding habit confers any protection from infection since virus released from larval corpses contaminates both old and new foliage within a year, e.g., NPV of *G. hercyniae* (Evans and Entwistle, in press) and of *N. sertifer* (Kaupp and Entwistle, unpublished observations).

II. Laboratory Assessment of Baculoviruses

A. DIAGNOSTIC METHODS

Most epizootiological studies and control operations involving virus–host interactions require diagnosis of numerous sawfly larvae. In the past, microscopic examination of dissected guts or smear preparations has been used, but recently the enzyme-linked immunosorbent assay (ELISA) has proved rapid and reliable (Chapter 5), for example for screening *N. sertifer* larvae for the presence of NPV (Kaupp, personal communication).

For microscopic diagnosis two methods have been used, phase contrast at low magnification for fresh, dissected gut material and either phase contrast for unstained smears or bright field for stained smears at very high magnification. In dissected midgut squash preparations of live lightly infected larvae, infected nuclei are seen as white spots at $\times 100$. When the infection is more advanced the tissue breaks down and this method becomes difficult. Buffalo black and Giemsa's stain are commonly used for tissue smears containing polyhedra. The former stains polyhedra black and the latter acts as a counterstain to polyhedra by staining host proteinaceous material blue/mauve. The generally small polyhedra of sawfly NPVs can be diagnosed only at $\times 1500$ or more.

The ELISA technique is preferable to microscopic diagnosis when only NPV is sought and other pathogens are to be excluded. "Human error" is removed from routine screening. Using the indirect antibody method and antiserum prepared against polyhedra protein, 10 ng of polyhedra protein can be detected. A larva containing 5×10^5 polyhedra or more gives a positive reaction above background and only one serum conjugate is required. Results can be read either optically or in a spectrophotometer (Kaupp, unpublished observations).

B. PATHOGENICITY TESTS

Little information has been published on the LD_{50}s (median lethal doses) of sawfly viruses, probably because many economically important species of sawflies are gregarious and do not take kindly to individual dosing and rearing nor to the synthetic diets so useful in LD_{50} tests on Lepidoptera. Hence most data are on LC_{50}s (median lethal concentrations) established by spraying foliage carrying colonies of larvae. These, being relative, are of little value in inter-species comparisons, especially with sawflies on different plant hosts.

Dubois (1976) described a technique to compare the efficacy of crude and sodium omadine decontaminated *N. sertifer* NPV. Bouquets of red-pine foliage were sprayed with 2 ml of four different dilutions of polyhedra and infested with 20 instar II larvae. Based on all the foliage being consumed by the 20 larvae (or 0.1 ml/larva), LC_{50}s with 95% fiducial limits were 86 (47–149) PIB for the decontaminated preparation and 155 (60–311) for the crude. Bird and Whalen (1953) fed 0.5 μl droplets containing 5, 50, 500 and 5000 PIB to individual larvae and established an LD_{50} between 100 and 500 PIB. The larval instar was not stated but was probably late. Bird compared several NPVs from colonial species of sawflies in the 1960s (Table III). Larvae were reared in

TABLE III

Potency of five sawfly baculoviruses: days at 22°C to 100% mortality,
+ P indicates some pupated, P indicates all (Bird, unpublished observations).

Species	Instar	Polyhedral inclusion bodies per ml.					
		10^7	10^6	10^5	10^4	10^3	10^2
	II	9	10	12	17	17 + P	16 + P
Neodiprion	III	9	10	14	14	22 + P	22 + P
swainei	IV	8 + P	11 + P	11 + P	11 + P	10 + P	9 + P
	V	7 + P	8 + P	P	P	P	P
	II	8	8	9	9	9	13
Neodiprion	III	8	8	8	11	14	14
lecontei	IV	8	9	11	10	9	11
	V	10	10	14 + P	14 + P	14 + P	14 + P
	I	4	5	7	7	11	12
Neodiprion	II	5	7	9	11	12	13
sertifer	III	5	6	7	11 + P	11 + P	14 + P
	IV	6	6	7	10 + P	12 + P	10 + P
	I	5	8	9	10	12 + P	12 + P
Neodiprion	II	6	8	10	12 + P	12 + P	12 + P
abietis[a]	III	10	9	10 + P	10 + P	10 + P	10 + P
	IV	6 + P	6 + P	P	NT[b]	NT	NT
	I	11	13	13	P	P	P
Neodiprion	II	11	10	17 + P	P	P	P
virginiana	III	13 + P	13 + P	P	P	P	P
complex	IV	8 + P	P	P	P	P	P
	V	P	P	P	P	P	P

[a] Data from Olofsson, 1973, using the same technique.   [b] NT = not tested.

lantern globes on sprigs of foliage sprayed with serial dilutions of virus. The NPVs of *N. lecontei* and *N. sertifer* were the most potent, then *N. abietis* and *N. swainei*, and finally *N. virginiana*, which in comparison is a mildly pathogenic virus.

European spruce sawfly, *Gilpinia hercyniae*, larvae are solitary and hence easier to handle in bioassays. Larvae can be dosed individually with drops of virus suspension ($0.5 \mu l$) dried on the tip of Norway spruce, *Picea abies*, needles, each held individually in plasticine or microtitration plates in a hole drilled at the base of each well. Larvae are placed on each needle and covered with a gelatine capsule. Those that eat the virus-contaminated tip within 24 h are transferred to individual containers with a sprig of foliage and incubated at $21-22°C$. In instars II to V 40% of larvae ingest the dose, but $< 30\%$ in instar I. Preliminary LD_{50} values ranged between 20 PIBs for instar I and 800 PIBs for early V (Entwistle, unpublished observations), values which substantiate the results of Bird (1961) with an undescribed technique.

The relationship between LD_{50}s and larval instar illustrates the susceptibility of a particular insect species to a particular virus and allows inter-species comparisons. However, both within and between species the relationship of virus dose to host mortality reveals more if examined in terms of LD_{50}/mg larval body weight. Values of LD_{50} *per se* and LD_{50}/mg are compared instar by instar for NPVs of *G. hercyniae* (Entwistle, unpublished observations) and *Operophtera brumata* (Lepidoptera:Geometridae) (Wigley, 1976) (Fig. 1). On a weight basis the susceptibilities of both species are consistent over instars I to IV. Then the situation changes, *O. brumata* showing strong maturation resistance and being virtually non-susceptible, whilst *G. hercyniae* has a high but not extreme resistance in the last instar.

The incubation period (LT_{50}) in larvae of *N. sertifer* fed on *P. sylvestris* foliage dipped in an NPV suspension containing 10^6 PIB/ml varied inversely with temperature but was relatively more prolonged at lower temperatures (Tvermyr, 1969).

III. Production of Sawfly Baculoviruses

A. PRODUCTION IN LABORATORY AND IN PLANTATIONS

There are no synthetic diets or established cell cultures for sawflies. Hence, for virus production, larvae must be reared on their host food plant. Basically two approaches have been used: (a) to collect insects in the field, infect them in the laboratory, feed until death and harvest the cadavers; or (b) to locate a suitable infested plantation, spray it with virus, harvest larvae as they die and process them in the laboratory.

Bird (1950) mass produced sawfly viruses by spraying virus on several thousand *N. sertifer* larvae on large open trays. He later changed to spraying heavy natural infestations of both *N. sertifer* and *N. lecontei*. In 10 man-days 45 000 diseased *N. lecontei* larvae were culled, yielding virus to produce 48 000 litres of virus suspension of unstated concentration (Bird, 1971) but probably ca. 1.1×10^8 PIB/litre.

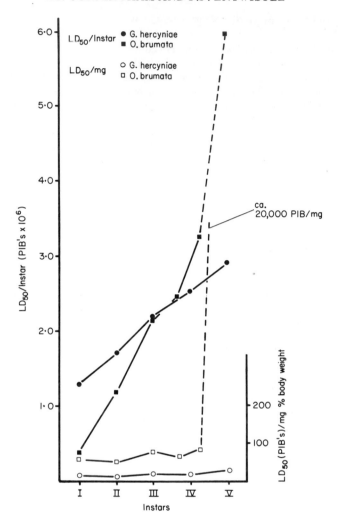

Fig. 1. LD_{50}/mg larval body weight of nuclear polyhedrosis viruses in different instars of *Gilpinia hercyniae* (Entwistle, unpublished observations) and *Operophtera brumata* (Wigley, 1976).

An operation to produce *N. sertifer* virus is described by Rollinson *et al.* (1970). Larvae, 1/2 to 2/3 grown, were collected in plantations and placed in cardboard rearing boxes with *Pinus resinosa* foliage at ca. 100 colonies /box. A suspension containing 10^7 PIB/ml was applied to "a point of near-dripping". Deaths began in 7 days and peaked in 8. For larger larvae 10^8 PIB/ml were sprayed, killing in about 6 days. Hand gathering cadavers was the lengthiest task. Two annual mass rearings produced a total of 1.39×10^{14} PIBs from 1.35×10^6

cadavers — a return of 10^8 PIB/larva at ca. 1 cent (US)/larva, or 75 cents/ha with a dosage of 7.5×10^9 PIB/ha.

N. sertifer virus production was studied in detail in 1973 (Cunningham, unpublished observations). Larvae IV and V were sprayed with a mistblower at 2×10^6 PIB/ml in 37.6 litre/ha and harvested 7 days later. From 100 420 larvae, 4×10^{12} PIBs cost $1000 (US), including 350 man-hours in wages, meals and lodging at 1973 prices. It treated 800 ha at 5×10^9 PIB/ha. Inflation would double this cost by 1979.

Smirnoff (1964) produced *N. swainei* NPV both indoors and by an outdoor "carousel" rearing method. Indoors, wooden racks with wire mesh bottoms holding 1200 sprigs with 75 000 larvae were used. Each screen was sprayed daily with 50 ml of suspension at 10^6 PIB/ml. Defoliated twigs were replaced by fresh sprayed foliage. At $25°-30°C$ death occurred in 10 to 12 days and larvae were harvested before they liquefied. The outdoor "carousel" had 8 arms pinned with sprigs carrying up to 5×10^5 larvae. It was rotated frequently to prevent larvae gathering on the sunny side. Screened trays below caught fallen larvae but passed frass. Between 0.65 and 1.3×10^{12} PIBs were harvested from 10^5 larvae. Indoor rearing was more reliable as the temperature was controlled. Low temperatures outdoors retarded development, but the "carousel" method was easier to use, required less daily attention and was cheaper. Production and processing 10^6 larvae, enough to treat 1300 to 2000 ha, cost ca. $500 to $600 (US) at 1962/3 prices.

Production of *N. taedae linearis* NPV was costed by Yearian and Young (1971). They sprayed infested trees at 5×10^7 PIB/ml and collected diseased colonies after 5 days. Larvae were picked off when dead, mascerated and filtered through organdie. Ca. 3.6×10^{12} polyhedra incurred a labour cost of $262.34. At 18.8 litre/ha with 5×10^7 PIB/ml treatment cost $6.75 (US)/ha and with 10^7 it cost $1.38/ha.

B. COMMERCIAL PRODUCTION

In the 1960s, an Indiana farmers' co-operative produced and marketed *N. sertifer* NPV, but stopped several years ago. From 1972 a Finnish state-owned chemical company, Kemira Oy, produced and marketed a partially purified suspension of *N. sertifer* polyhedra packaged in amounts to treat from 0.1 to 10 ha. They recommend treating instars I and II at 8×10^9 PIB/ha. In April 1978 this cost ca. $11.00 (US)/ha (A. Aapola, personal communication).

C. PURIFICATION METHODS

In most early studies virus-killed larvae were left to decompose in water and release polyhedra which eventually settled to the bottom of the vessel. Then, the supernatant was poured off and the polyhedra resuspended in water and either resettled or semi-purified by differential centrifugation.

In the 1970s freeze-drying became accepted for preserving baculoviruses, often as whole larvae. Just before spraying finely ground larvae are suspended in water. Due to the presence of host plant resin in sawflies, the bacterial count is much lower than in many lepidopterous insects (F. B. Lewis, personal communication). However, a bacteriocide, sodium omadine, has been used to chemically decontaminate *N. sertifer* NPV preparations but, as it also kills larvae, it must be washed from the suspension (Dubois, 1976).

Very pure suspensions of *N. sertifer* virus have been obtained with zonal rotors (Mazzone *et al.*, 1970; Breillatt *et al.*, 1972). The model K zonal centrifuge at the Oak Ridge National Laboratory took 6 h to process 20 litres of crude *N. sertifer* polyhedra. A 4-litre input contained 7.1×10^{12} PIBs and the cleanout 5.2×10^{12} PIBs, a 73% recovery. The ratio of bacteria to PIBs fell from 1:2.5 to 1:400 000.

Very pure virus has been used recently in field trials in the UK. Cadavers were triturated in very dilute sodium dodecyl sulphate (SDS), filtered through muslin and centrifuged at low speed to remove debris. The supernatant was decanted, PIBs pelleted at 10 000 g for 20 min and resuspended. The first purification used an "A" zonal rotor loaded with a 5–40% w/v sucrose gradient and 100 ml of concentrated, crude suspension and was centrifuged for up to 15 min. The gradient was fractionated, the PIBs pelleted from sucrose, resuspended and subjected to quasi-equilibrium centrifugation in a 50%–60% w/w sucrose gradient. Finally purified PIBs were harvested and pelleted; if SDS or other chemical aids were used, the pellets were extensively washed in deionized water or buffer (D. Brown, personal communication).

IV. Characterization and Safety Testing

A. BIOCHEMICAL CHARACTERIZATION

Biochemical studies on sawfly NPVs have lagged behind those of lepidopterous NPVs partly due to initial supply and purification problems. The yield of PIBs from a sawfly varies from ca. 0.05 to 0.02 of that from a lepidopterous larva and much material is required, especially for DNA studies. It is planned to register the NPVs of *N. sertifer* and *N. lecontei* for general use in the USA and Canada and *N. sertifer* in the UK. This requires detailed studies of viral proteins and DNA, so special efforts are being made to amass sufficient virus.

In serological studies using a complement fixation test virions from *N. sertifer* NPV and *G. hercyniae* NPV were identical when compared using *N. sertifer* antiserum, but were distinct from those from lepidopterous NPVs (Krywienczyk and Bergold, 1960a). PIB proteins from *N. sertifer* NPV and *G. hercyniae* NPV cross reacted strongly, but not at all with those from lepidopterous NPVs (Krywienczyk and Bergold, 1960b). In an ELISA test, *N. sertifer* NPV PIB-protein antiserum reacted strongly with PIB protein from both *N. lecontei* NPV and *G. hercyniae* NPV but not with that from lepidopterous NPVs (W.J. Kaupp, personal communication).

Preliminary biochemical studies on purified components of *N. sertifer* NPV using polyacrylamide gel electrophoresis (PAGE) established the molecular weight of PIB protein as 28 000 daltons. Virions had 9 structural proteins ranging from 7600 to 96 700, two major components being 12 000 and 33 000.

Sedimentation analysis gave a $S_{20}w$ value of 1100 S for virions, and 985 S for nucleocapsids prepared by treating virions with the detergent NP40. The buoyant density of these nucleocapsids was 1.267 g/cm (D. Brown *in* Entwistle *et al.*, 1978b). With PAGE on slab gels many similarities and some differences were detected between virion proteins of *N. lecontei* NPV (16 structural proteins) and those of *N. sertifer* NPV (11) (B.M. Arif, personal communication).

B. SAFETY TESTING

Extensive safety testing of *N. lecontei* NPV and *N. sertifer* NPV has been undertaken by contract research sponsored by the Forest Pest Management Institute, Sault Ste. Marie, Ontario, Canada and the USDA Forest Service, Hamden, Connecticut, USA, respectively. Generally, protocols outlined in a joint WHO/FAO meeting (Anon, 1973) were adopted.

In the following series of tests *N. lecontei* NPV had no detectable, harmful effects on mammals: acute oral tests on rats and rabbits, multiple small dose 90-day exposure of rats, acute dermal tests on rabbits, acute eye irritation on rabbits and acute inhalation tests in hamsters. In an oral lethal dosage trial mice were killed by the physical effects of feeding material equivalent to a 2400 ha dosage for a 70 kg man (Forsberg *et al.*, 1978). No ill effects were observed in acute oral toxicity tests in two species of bird, chickens and turkeys (Valli and Claxton, 1976). Rainbow trout were exposed in acute oral toxicity tests and by contaminating the water in fish tanks. Again no ill effects were observed (Geraci and Hicks, 1979). Two mammalian, two avian and two fish cell cultures were challenged with virions released from polyhedra. Using radio-active precursors, no replication of viral protein, DNA or RNA was detected at either 28°C or 35°C (Arif and Dobos, unpublished observations).

N. lecontei NPV was tested on an aquatic invertebrate, *Daphnia pulex,* by contaminating the water and no histological lesions or reduced reproductive capacity were observed (Geraci and Hicks, 1979). In the course of field trials, bee hives were studied and pre-spray and post-spray counts made of birds and aquatic fauna. No ill effects were noted (Kingsbury *et al.,* 1978).

Safety testing of *N. sertifer* NPV has also been extensive (F.B. Lewis, personal communication). It included acute oral tests on rats; primary skin irritation, eye irritation, and acute dermal tests on rabbits; skin sensitization tests on guinea pigs; acute inhalation tests on rats, carcinogenicity tests on new born hamsters and a two-year carcinogenicity test on rats. Acute oral tests on mallard ducks and bobwhite quail were undertaken and bluegill sunfish and rainbow trout were exposed to virus in aquaria. There was no evidence of the viral preparation having any harmful effects.

In a small trial, *N. swainei* NPV was fed to mice for 9 days and 4 to 6 months later there was no difference between groups of treated and untreated animals (Smirnoff and MacLeod, 1964).

V. Control — Methods and Examples

A. INTRODUCTION

Sawfly pest species are found mainly in forests or tree plantations. Often the profit margin on this resource is small and it is of prime importance in forest pest regulation to understand the likely outcome, in economic terms, of a sawfly outbreak and the pest density threshold above which financial loss occurs.

Where an NPV of considerable epizootic potential is enzootic in the sawfly population it may be economically feasible to make no positive control moves: such a passive policy could be based only on a thorough understanding of the situation and this seldom applies. *G. hercyniae* and its NPV is probably the only example known at present. There are three possible strategies for the use of NPV to control sawflies. Firstly, NPV may be dispersed as a classical biological control agent from relatively few points of introduction. There is always a lag phase between introduction and control, during which some economic loss may occur. It has been used with *G. hercyniae* NPV in Newfoundland and *N. swainei* in Quebec. Secondly, the NPV may be seeded at points on a lattice at a distance calculated from prior knowledge of the rate of dispersal of the virus to give an estimated level of control within an estimated period of time. Here potential savings in spray inoculum must be measured against economic losses between NPV introduction and control establishment. There is enough information to implement such a method only with *G. hercyniae*. Finally, the general practice has been overall spraying as with chemical pesticides. Where a quick kill is required, e.g. with heavy infestations of *N. lecontei* and *N. sertifer* on small trees, this is the obvious method.

B. EXAMPLES

1. *European Pine Sawfly,* Neodiprion sertifer

The NPV of *N. sertifer* is the most widely tested and operationally used sawfly baculovirus. At least 10 000 ha have been treated during the last 25 years. Trials have been conducted in Canada, USA, Germany, UK, Sweden, Finland, Norway, USSR, Austria, Poland, Yugoslavia and Italy (Table IV). Outbreaks of virus disease in *N. sertifer* were reported in Germany as early as 1913 (Escherich) and in Sweden in 1945 (Forrslund). The virus is endemic in Europe and the USSR and natural epizootics have caused severe mortality, e.g. 85% mortality in larvae on *Pinus sibirica* in the Tomsk region of the USSR (Gulii, 1971). *N. sertifer* was accidentally introduced into the Nearctic, first recorded in New Jersey in 1925 and is now widely distributed in Eastern North America. NPV obtained from Sweden in 1949 was tested in Ontario in 1950 (Bird, 1953b). Bird distributed the virus to collaborators in the USA (Dowden and Girth, 1953; Benjamin *et al.*, 1955; Drooz, 1959), in the UK (Rivers and Crooke, 1960; Rivers, 1962) and in Italy (Cavalcaselle, 1974). Completely independent trials successfully employing endemic NPV were conducted in Germany (Franz, 1954; Franz and Niklas, 1954).

TABLE IV.
Spray trials using nuclear polyhedrosis virus to control Neodiprion sertifer

Authors	Country	Spray method	PIB/ml	Litre/ha	PIB/ha
Austara, 1978	Norway	air	1.6×10^5	50.0	8×10^9
Benjamin et al., 1955	USA	ground	4.94×10^2	16.82	8.3×10^6
Bird, 1953b	Canada	ground	2×10^6	4.49	8.98×10^9
			2×10^7	4.81	9.63×10^{10}
		aerial	2×10^5	5.62	1.2×10^9
			1×10^6	5.62	5.6×10^9
			5×10^6	5.62	2.8×10^{10}
Cavalcaselle, 1974	Italy	ground	7×10^6, 3.5×10^6		
			7×10^5		
Cunningham et al., 1975	Canada	aerial	5.3×10^6	9.4	5.1×10^{10}
Donaubauer and Schönherr, 1972	Austria	ground	1×10^6, 1×10^5		
Donaubauer, 1973	Austria	ground	1×10^6, 2×10^5	6 ml/colony	
Dowden and Girth, 1953	USA	ground	6.25×10^4	1200	7.5×10^{10}
			1.11×10^5		
		aerial	1.38×10^5	28.0	3.86×10^9
Eidmann, 1970	Sweden	ground	1×10^6, 2×10^6		
		aerial	0.5 to 2.0×10^5		
Entwistle et al., 1978b	UK	ground	5×10^7		3×10^{11}
			5×10^5	6.24	3×10^9
			5×10^3		3×10^7
Franz, 1954; Franz and Niklas, 1954	Germany	ground	1×10^5, 1×10^6		
Glowacka-Pilot, 1972	Poland	ground	5×10^6	10 and 20	5×10^{10} and 10^{11}
Gulii and Zimerikin, 1971	USSR	aerial			4×10^9
					2×10^9
					5×10^8
Kemira Oy, commercial company, 1974	Finland	ground	1.6×10^4	500	8×10^9
		ground	4.0×10^4	200	8×10^9
		aerial	8.0×10^4	100	8×10^9
Nuorteva, 1972	Finland	ground	1.3×10^4 to		
			6.5×10^4		
			1.5×10^5		
			1.0×10^5		
			1.0×10^6		
Rivers, 1962	UK	ground	6.0×10^5	2.0	1.2×10^9
Rivers and Crooke, 1960	UK	ground	2.0×10^5	230 ml/3.3 m tree	
			4.0×10^5	45	1.8×10^{10}
			5.0×10^5	14.1	7.0×10^9
Sidor et al., 1975	Yugoslavia	ground	1.0 to 2.0×10^6		
Zarin et al., 1974	USSR		2.0×10^5 to	60	1.2×10^{10} to
			4×10^6		2.4×10^{11}

In North America the NPV was extensively used by Forest Service personnel, Christmas tree growers and private individuals until about 1970 when authorities began considering the registration of viruses. At present, Canadian and US scientists are collaborating on the registration of this virus and the NPV of N. lecontei so that they will again be generally available. In the UK early use of impure N. sertifer NPV preparations ceased about 1970, but in 1977 a joint Institute of Virology/Forestry Commission program of redevelopment aimed at official registration was begun.

Extensive literature on spray trials with N. sertifer NPV is summarized in Table IV. Both ground and aerial sprays gave generally satisfactory results. Best results were obtained when larvae were treated in instars I and II or even as early as 90% egg hatch. Then defoliation was minimal. Dosages ranged from 8.3×10^6 to 2.4×10^{11} PIB/ha. Current experience suggests that

$5-9 \times 10^9$ PIB/ha is adequate for moderate infestations. For severe infestations more may be needed to prevent appreciable defoliation in the year of application. Wide variation exists in the volumes sprayed — 2 litre to 1200 litre/ha. Probable sufficient volumes are: using aerial application, 10 litre/ha; using ground equipment such as mistblowers, 50—100 litre/ha; or by ultra low volume with anti-evaporant oil:water 1:4 using controlled size micro-droplets, 1.0 to 1.5 litre/ha (Entwistle, Evans, Harrap and Robertson, unpublished observations).

Formulations have normally been water alone although some workers have added wetting and sticking agents. A sunlight protectant Sandoz Shade® (= IMC 90-001) at 30 g/litre with 1.25 ml/litre Chevron® sticker, used in an aerial spray, was of doubtful benefit other than marking spray cards. It is clear that additives are not essential, excellent control being achieved with highly purified PIBs suspended in water (Entwistle et al., 1978b).

The effectiveness of N. sertifer NPV may possibly be enhanced by the ingestion of certain biologically inert chemicals. The incubation period of the virus was reduced by the addition of sodium fluoride, hydroxylamine, thioglycolic acid, hydrazine and potassium nitrite (Kreig, 1956) and by 0.05%—0.1% copper sulphate (Wellenstein, 1973). The incidence of NPV in N. sertifer larvae was increased by spraying finely divided quartz (Nuorteva, 1972), and the addition of ferrous sulphate or copper sulphate to NPV sprays markedly increased mortality (Luhl, 1974).

In almost all trials total spray coverage of the infestation has been used. However, 300 ha in an infested area of 10 000 ha was treated from the air in Sweden at 0.5 to 2×10^6 PIB/ml in a "zebra stripe" application (Benz, 1976). Action of the virus was relatively fast (Eidmann, 1970). In Canada, horizontal spread of N. sertifer NPV is much slower than that of N. lecontei NPV, which was attributed to N. sertifer being an introduced species with a poorer predator and parasite complex than N. lecontei. There was little vertical passage of this virus in Canada on small trees and annual sprays may be required: passage was better on larger trees (Bird, 1961). However, in Scotland, needles, needle bases and bark were infective at the start of a new sawfly season after a spray trial in the previous year (Kaupp and Entwistle, unpublished observations).

In Russia, mature trees have been identified as a source both of the sawfly and its NPV in young plantings (Victorov and Borodin, 1975). Heavy N. sertifer populations commonly begin to build up on pines which are between 4 and 6 years old. Possibly a single NPV application, or at most two, will give adequate protection until the age of 10—12 years, when susceptibility to attack declines.

2. Red-headed Pine Sawfly, Neodiprion lecontei

A wilt disease of this sawfly was observed in Wisconsin by S.A. Rohwer in 1912 and it was considered to be a bacterial infection (Middleton, 1921). From the limited description of the disease symptoms, the agent is almost certainly NPV. More recently, this NPV was reported in the USA by Steinhaus (1951) and in Canada by Bird (1970). Several ground sprays were applied between 1950 and 1970 (Bird, 1961, 1970, 1971). In the first trials on the north shore of Lake Huron, 2m-tall *Pinus resinosa* were sprayed with 50 ml of suspensions of 10^4, 10^5, 10^6, 10^7 and 5×10^7 PIB/ml. Ca. 98% of the larvae died in all treatments and the mean time until death ranged from 8.6 to 13.1 days (Bird, 1971). In Quebec in 1970, purified virus at 1.3×10^6 PIB/litre and lyophilized cadavers of unknown PIB count at 0.26, 0.13, 0.06 and 0.03 g/litre were tested against instar III larvae. The first deaths were seen 14 days after treatment. By 48 days all larvae were dead in areas treated with lyophilized material and 94.9% to 97.4% in areas treated with purified virus (De Boo and McPhee, 1970). Virus from Canada was used in Florida, the first trial of a microbial agent to control a forest pest in this state. With the high mean temperature of 30°C, larvae died in 5 days (Wilkinson, 1969).

In 1975, it was decided to promote use of this virus and gather data required to register it officially in Canada. Aerial sprays were applied in Ontario to 117 ha in the next 3 years (Kaupp and Cunningham, 1977; Kaupp et al., 1978; deGroot et al., 1979) and in 1978 to 700 ha in Quebec (Desaulniers and Cunningham, unpublished observations). A fixed-wing aircraft with boom and nozzle, another with Micronair and a helicopter with Beecomist equipment were used. At 1.25, 3.75, 5.0, 5.5 and 6.25×10^9 PIB/ha in 9.4 litre/ha results were excellent when the virus was applied before instar IV and satisfactory foliage protection obtained. The first larvae died 10 to 11 days post-spray and by 22 to 26 days very few uninfected sawfly colonies remained. With sprays at instars II and III, most died in instar IV and only a few colonies reached instar V. In all treated areas examined the following year, not a single sawfly colony was found, whereas populations were still high in untreated areas. The currently recommended spray rate is 5.0×10^9 polyhedra/ha at a cost of ca. $2.50 (US)/ha for virus.

In ground trials between 1975 and 1978, mainly to propagate more virus, ca. 18 ha were treated with mistblowers. In 1977, a Leco ULV cold fogger with the virus in an oil/water emulsion gave very impressive coverage over 0.25 ha and was considered a feasible application method (Johnson et al., 1978).

Rapidity of spread from colony to colony and from treated to untreated areas is a notable ecological feature of this virus. Thus introduction, as opposed to spraying an entire infested area, is a feasible strategy. In 1977, virus spread from a treated 13-ha plantation to an adjoining plantation of equal size: in 1978, no sawfly colonies were found in either. Introduction strategy was practiced in Quebec in 1978. Of the 700 ha treated, introductions were made in ca. 400 ha where the sawfly population was lightest. Here, instead of flying at the normal 30-m intervals used for complete coverage, swaths spaced at 100-m intervals gave strip coverage. Results were highly satisfactory.

All formulations for aerial application have been aqueous sometimes with a sunlight protectant, Sandoz Shade®, at 60g/litre and molasses ranging from 100 ml to 250 ml/litre. Comparisons in Quebec in 1978 between formulated and pure aqueous suspensions on instars II and III revealed no detectable differences (Desaulniers and Cunningham, unpublished observations). However, simulated aerial sprays on single trees at 5.5×10^9 PIB/ha on large instar IV larvae, a stage considered too old for good control with NPV, gave the following mortalities: purified virus, 34%; crude suspension of freeze-dried cadavers, 56%; crude virus + molasses, 87%; crude virus + molasses + Shade, 87%. These results, though preliminary, indicate that crude virus is better than purified, that molasses improves the formulation and that the UV screen gives no extra advantage (Hopewell and Cunningham, unpublished observations).

3. Swaine's Jack Pine Sawfly, Neodiprion swainei

An NPV was found in *N. swainei* larvae in N.E. Ontario in 1953 and another in Quebec in 1956 (Smirnoff, 1962). These isolates were of low virulence but this was considerably enhanced by selecting larvae with the heaviest midgut infections and those which died first. To develop the operational use of the virus, Smirnoff first tested it in small trials on isolated single colonies, on entire tree crowns and on small plots applying 5×10^5 and 10^6 PIB/ml at 37.6 litre/ha. Effective virus concentration ranged from 5×10^5 PIB/ml to 3×10^6 PIB/ml and higher dosages did not increase the rate of disease development or the eventual mortality. Mortality increased with virus concentration and decreased with larval age. Results were best from sprays applied before instar III. Spread of the virus in the year of application was only several hundred metres, perhaps resulting from larval migration. Trans-ovum transmission was presumed when the virus appeared the next year in progeny of surviving larvae (Smirnoff, 1961).

The first aerial spray trial in 1960 compared an aqueous and an oil based formulation each with 2×10^6 PIB/ml at 4.7 and 37.6 litre/ha respectively (= 5.4×10^9 and 7.5×10^{10} PIB/ha) at instar II (Smirnoff *et al.*, 1962). The aqueous formulation contained Geon 652 latex sticker (B.F. Goodrich Chemical Co.) and dried blood. The oil formulation was an invert emulsion with Span 80 in 80% no. 2 fuel oil. A Stearman aircraft was used with boom and Spraying Systems Swirl-jet 1/8 B2 nozzles calibrated to deliver 37.6 litre/ha on an 8.5-m swath. Deaths were first seen after 7 days with peaks at 17–18 and 22–25 days. Both sprays gave excellent results with cumulative mortality reaching 97%. Data from areas affected only by spray drift indicated that NPV in the emulsion may be more effective at low dosage levels. In the next year the plots and a surrounding 0.4 km zone were completely protected from defoliation and the disease spread 3.2 km from the plots.

In 1964, a larger aerial trial involved 27 000 litres of virus suspension at 10^6 to 1.5×10^6 PIB/ml with latex and blood. Two Stearman aircraft sprayed 1600 ha at 18.8 litre/ha (= $1.75 - 3.5 \times 10^{10}$ PIB/ha). Unseasonably cold and wet weather afterwards delayed mortality which was quite variable within the sprayed areas, at no point exceeding 76%. Microscopic examination of larval guts showed 20% infection at 10 days, 50% at 18 and 85% at 24 days post spray. In many samples brought to the laboratory 25% to 50% of larvae spun cocoons. Failure to decimate the insect population was attributed to the cold weather and possibly to too low a concentration of virus or volume of spray. Follow-up studies in 1965 and 1966 revealed less disease than expected. In 1966, disease was still present and the insect population had declined (McLeod and Smirnoff, 1971; Smirnoff and McLeod, 1975).

When 5×10^5 PIB/ml was sprayed on instar III, IV and V larvae, some spun cocoons (Smirnoff, 1962). About 60% developed into apparently normal adults but, unlike several other sawfly species, the sex ratio remained unchanged. Among their progeny, 33% to 95% died of NPV disease, deaths starting ca. 13 days after hatching. Smirnoff recommended spot spraying with weak dosages of virus to start epizootics, suggesting that in 4 years the sawfly population would be reduced and that disease would spread and coalesce with other treated areas. The virus can also be introduced by disseminating infected cocoons. The use of a flagellate *Herpetomonas swainei*, which co-infects with NPV in this insect, has been suggested as a marker to distinguish progeny of introduced infected adults from virus-infected insects already present (Smirnoff, 1972).

4. *Loblolly Pine Sawfly,* Neodiprion taedae linearis

An NPV of this sawfly was found in South Arkansas in 1962 (Young *et al.*, 1972). In 1969 trees bearing instar II and III larvae were sprayed to run off from the ground (Yearian and Young, 1971). Miller Nu Film B.T.® sticker at 0.25% was added to virus suspensions of 10^5, 5×10^5, 10^6, 10^7, 5×10^7 and 10^8 PIB/ml. After 14 days the mortalities were 39.2, 67.1, 66.1, 71.7, 84.7 and 91.1%, respectively. Due to their shorter larval feeding time more males than females survived.

Helicopter spray trials on 8-ha plots followed in 1972 (Yearian *et al.*, 1973). Aqueous sprays of 5×10^5, 10^6, 5×10^6, 10^7 and 5×10^7 PIB/ml were applied at 18.8 litre/ha (= 9.4×10^9 to 9×10^{11} PIB/ha) with the same sticker at 0.25%. Larvae were mainly instar I with some II. Colony size, cocoon numbers and defoliation were evaluated. No deaths were seen after 2 weeks but there were some due to virus after 3 weeks. At 4 weeks 5×10^6, 10^7 and 5×10^7 PIB/ml gave 97.9, 98.8 and 99.9% mortality compared with the control. The adult male to female ratio increased with increasing dosage. The mean reduction in cocoons ranged from 75.7% at 5×10^5 PIB/ml to 98.9% at 5×10^7. Although these dosages considerably exceed those for viruses of other *Neodiprion* species, estimated virus production costs (see Section III, A) were inexpensive.

5. *Virginia Pine Sawfly,* Neodiprion pratti pratti

An NPV found in Maryland in 1954 (Anon, 1956) gave 77% mortality 11 days after single trees were sprayed with 5×10^8 PIB/tree. Aerial sprays in 1958 covered 40 ha in three plots with 5×10^{10}, 2.5×10^{10} and 1.25×10^{10} PIB/ha applied by helicopter with boom and nozzle delivering 9.4 litre/ha. Larvae were treated 19 days after hatch. Disease appeared 8 to 10 days after spraying and mortality was almost complete by 16 days with no differences between treatments. Infection was seen in 93% of the colonies and 88% of larvae in sample collections died. Later 4.9% of cocooned larvae died giving a total mortality of 92.9%. The male to female ratio of surviving adults was 9:1. Defoliation was heavy in the treated area, attributed to insufficient spray coverage. The next year no diseased larvae or evidence of a virus epizootic were found in the treated area, and none has been found since in the large area of infestation under observation (McIntyre and Dutky, 1961).

6. *Jack Pine Sawfly,* Neodiprion pratti banksianae

Disease almost wiped out a population of *N. pratti banksianae* (= *N.*

banksianae) in Minnesota (Graham, 1925) and Steinhaus (1949) considered it to be caused by a virus. After small field trials in 1953 and 1954 in Ontario, Bird (1955) was pessimistic about use of this virus for control and concluded that it was impractical to aim for mortality > 70%. Moreover, sprays would have to be repeated annually, as epizootics or vertical passage did not result from small virus introductions. Later, suspensions of 10^3, 10^4, 3×10^4, 3×10^5 and 3×10^6 PIB/ml were sprayed on larval instar I and II causing 9.7%, 55.9%, 83.8%, 94.6% and 99.1% mortality, respectively (Bird, 1961). These figures appear quite encouraging. More males than females survived (Bird, 1955; 1961). Stairs (1971) suggested further tests before discarding the virus as unusable and he also suggested that its apparent failure may have resulted from spraying a sawfly population of mixed species.

7. Balsam Fir Sawfly, Neodiprion abietis

First recorded by Steinhaus (1949), this NPV was collected from the Prairie Provinces and Ontario in the 1950s and 60s and stored at Sault Ste. Marie until 1972. Olofsson (1973) conducted laboratory tests and single tree experiments. In the forest, larval development in *N. abietis* is completed in ca. 3 weeks making timing of sprays rather critical. Applications of 360 ml of virus suspension were applied with a mistblower to 2.5-m *Abies balsamea* at two larval stages, late instar I and early III, 8 days apart. At 10^4, 10^5 and 10^6 PIB/ml on stage I mortalities were 72, 96 and 100% and at 10^5, 10^6 and 10^7 on stage III 60, 88 and 100% respectively. All instar I larvae were infected after 4 days with 10^7 PIB/ml and after 8 days with 10^6 PIB/ml, also all instar III after 6 days with 10^7 PIB/ml.

Olofsson obtained 10^8 PIBs/diseased larva. He recommended spraying as soon as all eggs have hatched with little leeway in timing. This virus has considerable control potential.

8. Red-headed Jack Pine Sawfly, Neodiprion virginiana, *complex*

In a small trial in Ontario in 1966 on *Pinus banksiana* four concentrations ranging from 10^4 PIB/ml to 10^7 PIB/ml were each sprayed on 20 trees with larvae in instars II and III (Bird, personal communication). None died which is not surprising in view of the pathogenicity of this NPV in laboratory tests (Section II, B). However, the original source of the virus is unknown and it is uncertain if the target insect was *N. dubiosus* or *N. rugifrons* (Section I, B), so the virus may have been tested on a different sawfly species.

9. European Spruce Sawfly, Gilpinia hercyniae

A severe outbreak of *G. hercyniae* in eastern Canada and north-eastern USA was controlled by an NPV and by introduced parasites (Dowden, 1940; Balch and Bird, 1944; Bird and Elgee, 1957). Probably virus was fortuitously introduced with parasites or sawflies from Europe. Early in 1936 the disease occurred in laboratory cultures and in 1939 diseased larvae became numerous in parts of New Brunswick and New Hampshire. By 1952, the virus had spread through most infested areas and, except in the more northerly regions, the population was drastically reduced.

There were only a few planned transfers of the virus between 1939 and 1950, three in Quebec, three in Ontario and several in Newfoundland. In Quebec and Newfoundland the virus established successfully but not in Southern Ontario from releases in 1941 and 1945. As the disease became established at widely separated points by natural dispersal, the benefit of planned introductions is questionable (McGugan and Coppel, 1962). The last virus release in a healthy sawfly population at Kirkwood, near Sault Ste. Marie, Ontario, was carefully monitored. Levels of NPV disease and parasitism were recorded each year from 1950 to 1959 (Bird and Burk, 1961). Seven trees were sprayed with 285 ml of virus suspension (10^6 PIB/ml) per tree which gave only partial foliage cover. From these small introductions the disease became established and spread rapidly through the infested area. Virus epizootics recurred each year, effectively suppressing the sawflies below the threshold for serious defoliation and economic damage. For > 25 years the European spruce sawfly has remained endemic in eastern North America, held in check by the NPV and by parasites.

On a smaller scale, an outbreak of G. hercyniae in Wales defoliated 20 ha of sitka spruce (Picea sitchensis) in 1970 and spread to 1800 ha by 1972. A natural NPV was observed in the population in 1970 and by 1974 the outbreak was controlled by natural spread of the virus (Entwistle, 1972, 1974, 1976).

If further outbreaks of this pest occur in North America, Europe or elsewhere this potent virus is available to control them. Probably it has never been aerially sprayed but could be easily disseminated by air or by spot introductions. Larvae of G. hercyniae are solitary feeders, so mass collections for virus production take longer than for gregarious species of sawfly. However, because females are parthenogenetic and diapause can be avoided by a long day (18h) laboratory culture is simple.

10. Web-spinning Sawflies, Cephalcia spp.

Baculoviruses have been found in three members of this genus, the European Cephalcia abietis (= Lyda hypotrophica), the Japanese C. issiki (both NPVs) and C. fascipennis (GV) in N. America, whilst a "polyhedrosis" has been noted in C. lariciphila (= alpina) in Europe.

Estimates of infection levels in C. lariciphila in Schleswig-Holstein ranged up to 50% but were too low to terminate the sawfly outbreak (Röhrig, 1954). In Austria, outbreaks of C. abietis occur approximately on a 3-year cycle as a result of a prolonged diapause pattern. Incidence of disease appears to depend on larval density so that high rates of infection were found only in areas of heavier infestation where they seemed effective both in curbing the numbers of larvae maturing and reducing defoliation (Jahn, 1962, 1977). Despite the long period between host outbreaks, disease in C. abietis is more successful than in C. lariciphila, possibly because the gregarious behaviour of its larvae favours disease spread.

C. DISCUSSION OF APPLICATION METHODS

As noted in Section V, B, a very wide range of spray equipment, from hand-held pressure sprayers to aircraft fitted with sophisticated spraying systems, has been used to disseminate sawfly baculoviruses. In North America mainly plantations have been sprayed. Here, aerial operations resemble those in agriculture, the aircraft flying about 6 m from the ground. Aerial spraying is expensive [$12 to $25 (US)/ha] but uniform coverage is obtained and an area of several 100 ha can be treated in one morning. From aircraft the volume of carrier must be minimized and recently 9.4 litre/ha has been routine in Canada for virus applications. Possibly this can be reduced to 5 litre/ha and still give excellent results. Using ground spray equipment parallel passes are made every fifth to tenth row of trees giving a graded coverage across swaths, which in turn leads to variation in the incubation period of disease and so to the degree of foliage protection. Also, it is difficult to calibrate ground spray equipment to deliver an exact volume/ha and, though this is unimportant in operational applications, it is troublesome in experiments.

Against gregarious sawfly species such as *N. sertifer* and *N. lecontei,* good coverage is not very important since only one larva per colony need be infected to destroy that colony. When one infected *N. sertifer* larva was placed in a healthy colony, infection was detected in some larvae after 4 days and in the entire colony after 6 days (Kaupp, unpublished observations). When single colonies were infected, the disease spread rapidly to all other colonies on the tree (Bird, 1961).

A suspension of purified virus alone in water appears generally adequate to obtain satisfactory control. Oil formulations have been used with the NPVs of *N. swainei* (Smirnoff *et al.,* 1962), *N. lecontei* (Johnson *et al.,* 1978) and *N. sertifer* (Entwistle, Evans, Harrap and Robertson, unpublished observations), without harming the virus. Oil is preferable in aerial applications where prolonged periods of low humidity are encountered, and anti-evaporant oils are probably essential in ultra low volume controlled droplet applications.

Timing of application is the most important factor when using baculoviruses for sawfly control. Applications should be made as soon as the eggs hatch, 90% hatch suffices, and should not be delayed beyond instar III. At instar IV, dosages must be increased and even then severe defoliation will result from a heavy infestation.

VI. Assessment of the Value of Baculoviruses for Sawfly Control

Compared with baculovirus dosages required to control lepidopterous pests on conifers, dosages for coniferous sawflies are extremely low. Recommended rates for *N. sertifer* NPV and *N. lecontei* NPV are 8×10^9 and 5×10^9 PIB/ha, respectively, in comparison to 2.5×10^{11} for Douglas-fir tussock moth, *Orgyia pseudotsugata* (Stelzer *et al.,* 1977), two sprays of 2.5×10^{11} for gypsy moth, *Lymantria dispar* (F.B. Lewis, personal communication) and 7.5×10^{11} for eastern spruce budworm, *Choristoneura fumiferana* (Cunningham *et al.,* 1978). However, yield of PIB/larva from sawflies, about 10^8, is much lower than yields from lepidopterous species which often range from 10^9 to 10^{10}. Despite this low yield, sawfly viruses produced in the field, or in field collected larvae brought to the laboratory, cost only ca. $2.50 (US)/ha for unpurified NPV of *N. sertifer* and *N. lecontei*. Commercially produced, purified *N. sertifer* NPV from Kemira Oy, costs ca. $11.00 (US)/ha and highly purified NPV for field trials in the UK ca. $5.00. A double application of unpurified *L. dispar* NPV costs ca. $40/ha (F.B. Lewis, personal communication), a single application of unpurified *O. pseudotsugata* NPV ca. $10 (M.E. Martignoni, personal communication) and, with present production methods, unpurified *C. fumiferana* NPV > $125 (Cunningham *et al.*, 1978).

Persistence of control by virus from year to year has been clearly demonstrated with *G. hercyniae*, *N. sertifer*, *N. swainei* and *N. lecontei* NPVs but not with *N. pratti pratti* and *N. pratti banksianae* NPVs. With sawfly species notably associated with younger trees, probably one application will give control for the whole life of the crop as, with age, trees rapidly become less susceptible to attack. Thus sawfly control in the relatively stable environment of coniferous forest contrasts agreeably to control of insect pests in the seasonally turbulent agricultural environment, where the control effect persists much less from year to year.

An added bonus in the use of some sawfly NPVs is their horizontal spread in the first and subsequent years. The extent of this capacity depends on dispersal mechanisms in specific geographical localities so generalizations cannot be made. Virus introductions, as opposed to total coverage, were effective with *N. sertifer* NPV in Sweden (Eidmann, 1970) and with *N. lecontei* NPV in Canada (Desaulnier and Cunningham, unpublished observations), and doubtlessly if spot introductions of *G. hercyniae* NPV had been made in the UK or Canada, results would have been similar (Entwistle, 1976).

The effective vertical passage system of sawfly NPVs through the adults contrasts markedly to NPVs in Lepidoptera and helps to explain the apparently greater epizootic potential of sawfly NPV diseases. The biological control potential is high for viruses which can passage vertically and spread horizontally. These include the baculovirus of the rhinoceros beetle which was spread by infected adults quite spectacularly in the South Pacific (Chapter 20).

However, sawfly baculoviruses have some disadvantages, i.e. the timing of application and the technicalities of virus registration. These problems are shared with lepidopterous NPVs and some other pathogens. To generalize, sawfly NPVs should be applied before instar IV to obtain maximum impact and foliage protection. Often infestations are unseen until damage to trees is clearly visible by which time larvae are usually in their final instar. In this event either chemicals should be used or a virus spray delayed until the next generation when egg hatching can be noted.

Some barriers exist to the laboratory production of sawfly viruses. No artificial diet has been found for sawflies and all sawfly eggs are laid within plant tissues from which they obtain moisture necessary for their development. The problem of a synthetic substrate for oviposition has not been examined so laboratory rearing of sawflies at present requires foliage. It is thus time-consuming and inefficient compared with rearing Lepidoptera on semi-synthetic diets and this has hampered laboratory studies of sawfly NPVs. Biochemical studies on NPVs would be greatly facilitated by sawfly cell cultures, but none has yet been developed.

The NPVs of *N. sertifer*, *N. lecontei*, *N. swainei*, *N. pratti pratti*, *N. taedae linearis* and *G. hercyniae* are at present available as safe, non-polluting, environmentally acceptable and very efficient alternatives to chemical insecticides for control of these pests. More data must be accumulated for their registration as commercial viral insecticides.

References

Anon. (1956). *Ent. Soc. Am. Bull.* **2**, 10.

Anon. (1973). *Wld. Hlth. techn. Rep. Ser.* No. 531.

Austara, O. (1978). *Norw. J. Ent.* **19**, 91–92.

Balch, R.E. and Bird, F.T. (1944). *Sci. Agric.* **25**, 65–80.

Benjamin, D.M. (1955). *U.S.D.A. Tech. Bull.* No. 1118.

Benjamin, D.M., Larson, J.D. and Drooz, A.T. (1955). *J. For.* **53**, 359–362.

Benson, R.B. (1950). *Trans. Soc. Brit. Ent.* **10**, 45–142.

Benz, G. (1976). Proc. Ist. Int. Colloq. Invertebr. Path., Kingston, Canada, 52–58.

Bird, F.T. (1949). Ph.D. Thesis McGill Univ., Quebec.

Bird, F.T. (1950). *Bi-mon. Prog. Rep. Can. Dep. Agric.* **6**, 2–3.

Bird, F.T. (1953a). *Can. J. Zool.* **31**, 300–303.

Bird, F.T. (1953b). *Can. Ent.* **85**, 437–446.

Bird, F.T. (1955). Can. Ent. **87**, 124–127.

Bird, F.T. (1961). *J. Insect Path.* **3**, 352–380.

Bird, F.T. (1970). Can. For. Serv. Inf. Rep. DPC-X-1, 8–10.

Bird, F.T. (1971). *In* "Biological Control Programmes against Insects and Weeds in Canada" Commonw. Inst. Biol. Control, Tech. Commun. No. 4, 148–150.

Bird, F.T. and Burk, J.M. (1961). *Can. Ent.* **92**, 228–238.

Bird, F.T. and Elgee, D.E. (1957). *Can. Ent.* **89**, 371–378.

Bird, F.T. and Whalen, M.M. (1953). *Can. Ent.* **85**, 433–437.

Breillatt, J.P., Brantley, J.N., Mazzone, H.M., Martignoni, M.E., Franklin, J.E. and Anderson, N.G. (1972). *Appl. Microbiol.* **23**, 923–930.

Cavalcaselle, B. (1974). *Cellulosa e Carta* **25**, 27–32.

Clark, R.C. and Clarke, L.J. (1973). *Can. For. Serv. Bi-mon. Res. Notes* **29**, 2–3.

Cunningham, J.C., Kaupp, W.J., McPhee, J.R., Sippell, W.L. and Barnes, C.A. (1975). Can. For. Serv. Inf. Rep. IP-X-7.

Cunningham, J.C., Kaupp, W.J., Howse, G.M., McPhee, J.R. and deGroot, P. (1978). Can. For. Serv. Inf. Rep. FPM-X-3.

DeBoo, R.F. and McPhee, J.R. (1970). Can. For. Serv. Inf. Rep. DPC-X-1, 11–26.

deGroot, P., Cunningham, J.C. and McPhee, J.R. (1979). Can. For. Serv. Inf. Rep. FPM-X-20.

Donaubauer, E. (1973). *Eur. Pl. Prot. Bull.* **3**, 105–110.

Donaubauer, E. and Schonherr, J. (1972). *Zbl. ges. Forstw.* **89**, 26–33.

Dowden, P.B. (1940). *J. For.* **38**, 970–972.

Dowden, P.B. and Girth, H.B. (1953). *J. econ. Ent.* **46**, 525–526.

Drooz, A.T. (1959). *Am. Christmas Tree Growers' J.* **3**, 6–24.

Dubois, N. (1976). *J. econ. Ent.* **69**, 93–95.

Eidmann, H.H. (1970). *Skogsägaren nr.* **12**, 7–9.

Entwistle, P.F. (1972). Abstr. 14th Int. Congr. Ent., 1972, 221.

Entwistle, P.F. (1974). *Land,* Oxford, U.K. 1, 84–88.

Entwistle, P.F. (1976). Proc. Ist. Internat. Colloq. Invertebr. Path. Kingston, Canada, 184–188.

Entwistle, P.F. and Adams, P.H.W. (1977). *J. Invertebr. Path.* **29**, 392–394.

Entwistle, P.F., Adams, P.H.W. and Evans, H.F. (1977a). *J. Invertebr. Path.* **29**, 354–360.

Entwistle, P.F., Adams, P.H.W. and Evans, H.F. (1977b). *J. Invertebr. Path.* **30**, 15–99.

Entwistle, P.F., Adams, P.H.W. and Evans, H.F. (1978a). *J. Invertebr. Path.* **31**, 307–312.

Entwistle, P.F., Evans, H.F., Harrap, K.A. and Robertson, J.S. (1978b). Unit Invertebr. Virol., Oxford, Tech. Rep. No. 1.

Escherich, K. (1913). *Z. Land-Forst.* **11**, 86pp.

Evans, H.F. and Entwistle, P.F. (1976). Proc. Ist Int. Colloq. Invertebr. Path., Kingston, Canada, 350–351.

Evans, H.F. and Entwistle, P.F. (in press). *In* "Microbial Pesticides" (E. Kurstak, ed.), pp. xxx–xxx, Marcel Dekker, Basel.

Forsberg, C.M., Valli, V.E.O. and Dwyer, P. (1978). Can. For. Serv. Contract Rep. OSU76–00226.

Forslund, K.H. (1945). *Meddn. Skogsfor.* **34**, 365–390.

Franz, J. (1954). *Gesunde Pfl.* **6**, 173–175.

Franz, J. (1956). *Verh. dt. Zool. Ges. Erlangen.* 1955, 407–412.

Franz, J. and Niklas, O.F. (1954). *NachrBl. dt. Pflschutzdienstes.* **6**, 131–134.

Franz, J., Krieg, A. and Langenbuch, R. (1955). *Z. Pflanzenkrankh.* **62**, 721 726.

Geraci, J.R. and Hicks, B. (1979). Can. For. Serv. Contract Rep. OSU78–00041.

Glowacka-Pilot, B. (1972). *Pr. Badaw. Inst. badaw. Lesn.* **425**, 66–77.

Graham, S.A. (1925). *J. econ. Ent.* **18**, 337–345.

Gregory, P.H. (1952). *Ann. Rev. Phytopath.* **6**, 189–212.

Gulli, V.V. (1971). *Ser. Izvestiya Nauk.* **5**, 72–78.

Gulli, V.V. (1972). *In* "Zaschita Cesa ot ureditelet i boleznei", Moscow, USSR, Kolos 96–102.

Gulli, V.V. and Zimerikin, U.N. (1971) *Lesovedenie* **3**, 87–89.

Hughes, K.M. (1957). *Hilgardia* **26**, 597–629.

Jahn, E. (1962). *Anz. Schadlingsk.* **35**, 99–102.

Gulli, V.V. and Zimerikin, V.N. (1971). *Lesovedenie* 3, 87–89.

Jahn, E. (1977). *Anz. Schadlingsk. Pflanzenschutz, Umweltschutz.* 49, 145–149.

James, R.A. (1974). M.Sc. Thesis, Imperial College Science, London.

Johnson, W.T., Cunningham, J.C., Kaupp, W.J. and Edwards, J.C. (1978). *Can. For. Serv. Bi-mon. Res. Notes* 34, 25–26.

Kal'vish, T.K. and Zhimerikin, V.I. (1974). *Ser. Biol. Logicheskikn. Nauk.* 15, 66–70.

Kaupp, W.J. and Cunningham, J.C. (1977). Can. For. Serv. Inf. Rep. IP-X-14.

Kaupp, W.J., Cunningham, J.C. and deGroot, P. (1978). Can. For. Serv. Inf. Rep. FPM-X-1.

Kemira Oy. (1974). *Eur. Pl. Prot. Publ. series B.* 81, 29.

Kingsbury, P., McLeod, B. and Mortensen, K. (1978). Can. For. Serv. Inf. Rep. FPM-X-11.

Krieg, A. (1956). *Arch. ges. Virusforch.* 6, 472–481.

Krywienczyk, J. and Bergold, G.H. (1960a). *Virology* 10, 308–315.

Krywienczyk, J. and Bergold, G.H. (1960b). *J. Immunol.* 84, 404–408.

Luhl, R. (1974). *Z. angew. Ent.* 76, 49–65.

Lyons, L.A. (1964). *Proc. ent. Soc. Ontario* 94, 5–37.

McGugan, B.M. and Coppel, H.C. (1962). Commonw. Inst. Biol. Control Trinidad Tech. Comm. No. 2, 90–109.

McIntyre, T. and Dutky, S.R. (1961). *J. econ. Ent.* 54, 809–810.

McLeod, J.M. and Smirnoff, W.A. (1971). *In* "Biological Control Programmes against Insects and Weeds in Canada 1959–1968" Commw. Inst. Biol. Control Tech. Commun. No. 4, 162–167.

Martignoni, M.E. and Iwai, P.J. (1977). USDA For. Serv. Gen. Tech. Rep. PNW-40, 2nd ed.

Martignoni, M.E. and Langston, R.L. (1960). *Hilgardia* 30, 1–40.

Mazzone, H.M., Breillatt, J.P. and Anderson, N.G. (1970). Proc. IVth Int. Coll. Insect Path. College Park, Md., USA, 371–379.

Middleton, W. 1921. *J. Agric. Res.* 20, 741–760.

Neilson, M.M. and Elgee, D.E. (1968). *J. Invertebr. Path.* 12, 132–139.

Nuorteva, M. (1972). *Silva Fennica* 6, 172–178.

Olofsson, E. (1973). Can. For. Serv. Inf. Rep. IP-X-2.

Prebble, M.L. (1941a). *Can. J. Res.* 19, 295–322.

Prebble, M.L. (1941b). *Can. J. Res.* 19, 323–346.

Pschorn-Walcher, H. (1970). *Z. angew. Ent.* 66, 64–83.

Rivers, C.F. (1962). *Entomophaga Mém. hors. Sér.* No. 2, 476–480.

Rivers, C.F. and Crooke, M. (1960). Proc. Vth Wld. Forest Cong. Forest Prot. 951–952.

Röhrig, E. (1954). *Z. angew. Ent.* 35, 207–245.

Rollinson, W.D., Hubbard, H.B. and Lewis, F.B. (1970). *J. econ. Ent.* 63, 343–344.

Sidor, C., Zivojinovic, D., Dusanic, L., Stajkovic, B., Sekulic, D. and Vujm, M. (1975). *Zbornik Radova "Delibatski Pesak"* 3, 57–75.

Smirnoff, W.A. (1961). *J. Insect Path.* 3, 29–46.

Smirnoff, W.A. (1962). *J. Insect Path.* 4, 192–200.

Smirnoff, W.A. (1964). *Forest Chron.* **40**, 187–194.

Smirnoff, W.A. (1968). *J. Invertebr. Path.* **10**, 436–437.

Smirnoff, W.A. (1972). *Bio Science* **22**, 662–663.

Smirnoff, W.A. and Juneau, A. (1973). *Ann. Soc. ent. Quebec* **18**, 147–181.

Smirnoff, W.A. and MacLeod, C.F. (1964). *J. Insect Path.* **6**, 537–538.

Smirnoff, W.A. and McLeod, J.M. (1975). *In* "Aerial Control of Forest Insects in Canada" (M.L. Prebble, ed.). Environ. Can. 235–240.

Smirnoff, W.A., Fettes, J.J. and Haliburton, W. (1962). *Can. Ent.* **94**, 477–486.

Smith, D.R. (1974). *Proc. ent. Soc. Wash.* **76**, 409–418.

Stairs, G.R. (1971). *In* "Microbial Control of Insects and Mites" (H.D. Burges and N.W. Hussey, eds.), 97–124, Academic Press, New York and London.

Steinhaus, E.A. (1949). Principles of Insect Pathology. McGraw-Hill, New York.

Steinhaus, E.A. (1951). *Hilgardia* **20**, 629–678.

Stelzer, M., Neisess, J., Cunningham, J.C. and McPhee, J.R. (1977). *J. econ. Ent.* **70**, 243–246.

Tvermyr, S. (1969). *Entomophaga* **14**, 245–250.

Valli, V.E.O. and Claxton, M.J. (1976). Can. For. Serv. Contract Rep. OSS5-0194.

van der Plank, J.E. (1963). "Plant Diseases: Epidemics and Control" Academic Press, New York and London.

Victorov, G.A. and Borodin, A.L. (1975). *Zoologicheskii Zhurnal* **54**, 1092–1095.

Wellenstein, G. (1973). *Eur. Pl. Prot. Bull.* **9**, 43–52.

Wigley, P.J. (1976). D. Phil. Thesis, Oxford Univ., Oxford.

Wilkinson, R.C. (1969). Sunshine State Agric. Res. Rep. Nov. 13–15.

Yearian, W.C. and Young, S.Y. (1971). *Arkansas Farm Res.* **20**, 3.

Yearian, W.C., Young, S.Y. and Livingston, J.M. (1973). *J. Invertebr. Path.* **22**, 34–37.

Young, S.Y., Livingston, J.M., McMasters, J.A. and Yearian, W.C. (1972). *J. Invertebr. Path.* **20**, 220–221.

Zarin', I., Rituma, I. and Vitola, R. (1974). *Lesnoe Khozyaistvo* **10**, 79–83.

CHAPTER 20

Control of the Rhinoceros Beetle by Baculovirus

G.O. BEDFORD

School of Biological Sciences, Sydney Technical College,
Broadway, New South Wales 2007, Australia

I. Introduction

The palm rhinoceros beetle, *Oryctes rhinoceros*, occurs throughout southeast Asia and has been accidentally introduced into a number of South Pacific countries and into Mauritius. Other species of *Oryctes* attack palms in Africa, the Comores Archipelago, Madagascar, Seychelles, Malaysia and Papua-New Guinea. Adult *O. rhinoceros* fly to the central crown of the palm, crawl down the axil of a young frond and then bore through the heart of the palm into the unopened fronds. The fronds so damaged unfold later to reveal tattering and V-shaped cutting of the leaflets. (a) Palms may be killed by repeated or heavy beetle attacks which destroy the apical meristem (Fig. 1), or (b) in some countries the beetle holes may provide entry points for lethal infestations of palm weevils (*Rhynchophorus* spp.). Coconut palms (*Cocos nucifera*) of all ages and young oil palms (*Elais guineensis*) are the main economic palms attacked.

Fig. 1. Palms killed by *Oryctes rhinoceros* at Drauniivi on Viti Levu, Fiji (August 1973). Courtesy of PANS.

The eggs are laid, and the larvae develop and pupate, in the tops of dead standing palms, decaying trunks of palms and other wood, and in heaps of compost, sawdust, manure or other decomposing vegetable matter. Searches conducted over many years for biocontrol agents for *Oryctes* have produced a baculovirus capable of significantly reducing *Oryctes* populations, and a fungus, *Metarhizium anisopliae*, of some use in special situations.

II. Discovery, Structure and Properties of the Baculovirus

A. DISCOVERY

The virus was first discovered in *O. rhinoceros* larvae in Malaysia (Hüger, 1966b). It has since been found throughout the Philippines, and in Indonesia on the islands of Sumatra and west Kalimantan (Zelazny, 1977b). It did not exist in any of the South Pacific countries.

B. STRUCTURE AND TISSUES INFECTED

The virus was first observed in nuclei of cells of larval fat body (Hüger, 1966a); it was later observed in nuclei of midgut epithelium of larvae and adults (Hüger, 1973; Payne, 1974) (Fig. 2), and also in the wall of the ovarian sheath and the

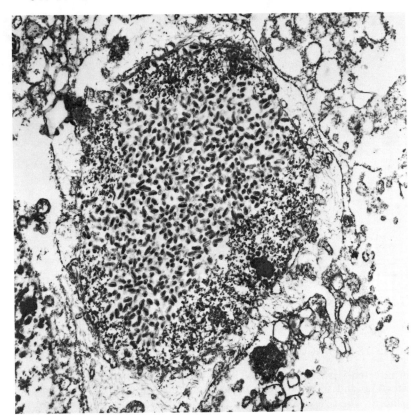

Fig. 2. An infected midgut epithelial cell nucleus of *Oryctes rhinoceros* filled with baculovirus particles. Courtesy of Payne (1974).

inner wall of the spermatheca (Monsarrat *et al.*, 1973a). It multiplies in the nuclei of cultured larval *O. rhinoceros* heart and blood cells (Quiot *et al.*, 1973), in the nuclei of cultured moth (*Spodoptera frugiperda*) cells and in the cytoplasm of cultured mosquito (*Aedes albopictus*) cells (Kelly, 1976).

The virus particle consists of a nucleocapsid surrounded by an envelope (235 × 110 nm, Fig. 3). The genome is a double-stranded supercoiled DNA molecule with a molecular weight of 60×10^6 to 92×10^6 daltons (Monsarrat *et al.*, 1973a, b; Revet and Monsarrat, 1974; Payne, 1974; Payne *et al.*, 1977). These characteristics strengthened the view that the virus should be included in the Baculoviridae, and use of the original name *Rhabdionvirus oryctes* (Hüger, 1966a) has now been discontinued (David, 1975).

The virus is normally non-occluded; however, inclusion bodies containing virions have been seen in the cytoplasm of cells of fat body, ovarian sheaths, inner spermathecal walls and midgut epithelium (Monsarrat *et al.*, 1973a), but these were not like polyhedra.

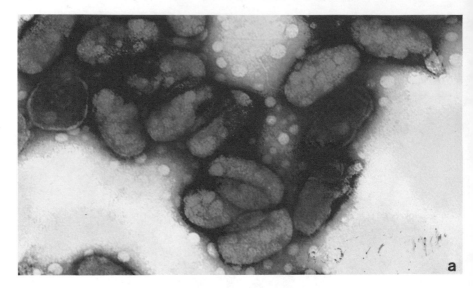

Fig. 3. Purified baculovirus particles from *Oryctes rhinoceros*. Courtesy of
Payne *et al.* (1977).

C. EFFECTS OF BACULOVIRUS ON LARVAE

In infected larvae the abdomen becomes turgid and glassy, while the fat body
disintegrates and the amount of haemolymph increases so that the larvae appear
translucent when viewed against light. Internal turgor may increase, extroverting
the rectum (Fig. 4). In the terminal phase chalky white bodies may appear under
the abdominal integument (Hüger, 1966a). Much virus multiplication occurs in
the midgut epithelium (Payne, 1974).

On virus-contaminated food medium the lethal infection time depends on
the instar, larva I dying after 9 days, II after 13 days and III after about 23 days.
Given similar doses by force-feeding, young larva III died after 18 days and old
III after 25 days (Zelazny, 1972). High temperatures speeded death (32° cf 25° or
27°C). The virus also killed larvae (II and early III) of the dynastid *Scapanes
australis grossepunctatus*, a pest of palms in New Britain. These larvae died
within 13–15 days of infection, but some older larva III appeared to be resistant
(Bedford, 1973a). Larvae of the European *O. nasicornis* could be infected with
virus (Hüger, 1966a). In the Ivory Coast, larvae of *O. boas* showed a sensitivity to
the virus close to that of *O. rhinoceros* (Julia and Mariau, 1976b). In artificial
conditions most larvae died between 1 and 2 months after feeding on virus-
treated medium. In contrast, *O. monoceros* was much less susceptible, and even
at high doses the larvae did not show typical disease symptoms although some
died. The progeny of adults developing from surviving larvae were normal. In
view of these results, and because *O. monoceros* is much more important than
O. boas, field trials were not undertaken in the Ivory Coast.

Fig. 4. Larva of *Oryctes rhinoceros* freshly dead from baculovirus. Note prolapsed rectum. Courtesy of PANS.

In the Philippines several strains of the baculovirus were found in *O. rhinoceros* which differed in their pathogenic effects, some seeming to infect larvae more easily than did the strain present in the South Pacific which originated from Malaysia. Differences were also noted in the survival times of larvae following inoculation with different Philippine strains (FAO, 1978).

D. EFFECTS OF BACULOVIRUS ON ADULTS

In adults the virus multiplies in nuclei of midgut epithelial cells. New cells are produced in regenerative crypts, become infected, their nuclei hypertrophy and the gut lumen eventually fills with disintegrating cells and virus particles (Hüger, 1973; Monty, 1974). Infected adults defaecate virus into the surrounding medium (Zelazny, 1973a) thus adults are virus reservoirs, spreading infective virus into the insect's natural habitats (Hüger, 1973). It is estimated that up to 0.3 mg virus/day may be produced in the faeces of an infected adult (Monsarrat and Veyrunes, 1976).

Diseased beetles generally show no external symptoms. In virus-treated breeding sites, Monty (1974) found a total of nine beetles with malformed elytra, wings or abdominal wall and one of these contained virus. However, Zelazny (1976) considers that virus is not normally carried over in development from an infected larva to the adult stage. In the laboratory Zelazny (1973b) found that virus-infected adults died sooner and laid fewer eggs than healthy controls (25 cf 70 days; 1.25 ± 0.13 cf 14.7 ± 1.5 eggs/female).

III. Production and Inactivation of Baculovirus

A. PRODUCTION AND STORAGE

Various authors have produced virus by basically similar methods (Hammes, 1971; Young, 1974), the following being suitable for bulk virus production (Bedford, 1976b). A plastic box 61 × 27 × 41 cm is two-thirds filled with steam-sterilized rotted sawdust mixed with eight virus-killed ground *O. rhinoceros* larvae. Fungicides may be added to inhibit the pathogenic fungus *M. anisopliae*. Some 100–150 healthy larvae are added to the box and allowed to feed for 5–7 days. They are then transferred to a similar holding box containing medium without virus, in which they gradually die, many coming to the surface just before death. The virus-packed cadavers are removed daily for indefinite deep-frozen storage. Virus can also be produced by infecting adults.

B. INACTIVATION

Storage of virus as ground cadavers mixed in sawdust at 26°C reduced viral activity to 0.091% of its initial value in one week, to 0.027% in 2 weeks, and no activity was detectable after one month. Drying the mixture or raising the temperature increased the rate of inactivation. Virus suspensions were also inactivated at 70°C for 10 min, or by a 1% solution of formaldehyde or Dettol, a commercial germicide (active ingredient 4.8% chloroxylenol) (Zelazny, 1972).

IV. Transmission and Application of Virus

A. TRANSMISSION

In laboratory studies (Zelazny, 1976) *O. rhinoceros* adults became infected *per os* in a mixture of sawdust and ground virus-killed larvae, or when kept together with other virus-infected adults. Adults developing from larvae that had survived exposure to various dosages of virus were not infected, nor were larvae hatching from eggs surface-contaminated with virus. Larvae hatching from eggs laid by infected females were very rarely infected. Nevertheless, in adults virions have been found in the nucleus and cytoplasm of spermatids, in cells and lumens of accessory glands and in the ejaculatory canal, as well as in chorionated oocytes and follicle cells (Monsarrat *et al.*, 1974).

In *O. rhinoceros* populations the virus is transmitted most frequently during mating, possibly when the healthy partner contacts by mouth virus defaecated by the infected partner. Virus can be transmitted similarly when infected and healthy beetles feed together in palms. Beetles visiting larval breeding sites containing freshly virus-killed larvae become infected and such beetles pass the infection to healthy larvae when visiting a breeding site (Zelazny, 1976).

B. METHODS OF VIRUS APPLICATION

Three methods of applying virus have been used, each superseded by the subsequent one as the mode of transmission became more clearly understood.

1. Artificial Compost Heap

A heap of decaying leaves and compost or sawdust, $3-4\,m^2$, was mixed with 10–50 ground virus-killed larvae in water, and walled by coconut logs. Such breeding sites attracted beetles and infected them as they crawled through the compost. Later they moved away and spread the virus elsewhere. There were many disadvantages: a large labour force was needed to construct the heaps; the virus quickly broke down; the vegetable matter decayed into soil unattractive to beetles unless more compost was added; the heaps required weeding. Another serious hazard was the breeding of many beetles in the coconut log walls if not adequately supervised.

2. Split Coconut Log Heap

Another sort of artificial breeding site was mainly used in Fiji (Bedford, 1976b, 1977). A half sack of sawdust mixed with 10–50 cadavers was tipped on the ground and 6 to 10 lengths of split coconut log each about 1.2 m long were laid side by side on top. To prolong the presence of virus in the heaps, recently infected live grubs could also be added. While easier to construct than the compost heaps they still required a large labour input, frequent replenishment with virus, renewal of rotted logs, and weeding. Their main disadvantage was dependence on the chance of beetles visiting them before the virus broke down, and having to compete with more attractive natural breeding sites such as dead standing palms. Usually any dead standing palms in the vicinity were felled to reduce competition.

3. Release of Virus-Infected Adults

Zelazny (1973a, b) and Hüger (1973) showed that virus multiplied in the midgut cells of adults and was defaecated so it became clear that adults were the natural vectors of the disease, responsible for its spread and transmission. Thus, the simplest, most economical and direct method of virus dissemination was to release laboratory-infected beetles, a method introduced early in 1972. Beetles were obtained by field collection, attractant trapping, rearing from field collected larvae, or mass rearing from eggs as discussed by Bedford (1976a). Bedford (1976b) immersed live beetles for 2–3 min in a suspension of two ground cadavers/litre of water, then allowed them to crawl for 24 h through about 1 kg of sterilized sawdust mixed with half a virus grub in $500\,cm^3$ of water. Zelazny (1977a) obtained 90% infection by swimming beetles for 10 min in a 10% suspension of ground freshly virus-killed larvae. For release, the beetles were allowed to crawl under logs or into vegetation (Fig. 5). They took wing at night.

Thus all the disadvantages of the artificial breeding sites were eliminated. The infected beetles dispersed widely before death, spreading the disease directly

Fig. 5. Baculovirus-infected adults of *Oryctes rhinoceros* being released. Courtesy PANS.

into the wild population, contaminating breeding sites which may have contained larval broods and other beetles, and also the palm crowns.

V. Methods for Sample Collection and Virus Detection

A. COLLECTION OF SAMPLES

Beetles for testing and observation may be caught in traps (Bedford, 1973b; Julia and Mariau, 1976c) with the attractant ethyl chrysanthemumate. Breeding sites and convenient feeding sites may be searched for live or freshly dead *Oryctes* material which *must* be kept singly in containers from the moment of collection to avoid cross infection.

B. METHODS FOR DETECTING VIRUS

Selection of a particular method or combination of methods will depend on available time and facilities, and the objective and level of accuracy required.

1. Detection of Virus in Adult Faeces

Monsarrat and Veyrunes (1976) demonstrated virions in faeces by electron microscopy of centrifuged faeces.

2 Histological Examination

For light microscopy, tissues, particularly adult midgut, have been fixed in Bouin's fluid, sectioned and stained in haematoxylin (Hüger, 1973; Marschall, 1973; Monty, 1974). Methods for electron microscopy have been described by Hüger (1966a, 1973), Monsarrat *et al.* (1973a), Payne (1974) and Kelly (1976).

3. Immunological Technique

Croizier and Monsarrat (1974) injected virus (antigen) from larval midguts into rabbits and used the resulting serum containing antibody to reveal virus in infected tissues by immunofluorescence.

4. Bioassay Method

Zelazny (1972) cut the test larva or adult into small pieces which were mixed with about 50 g of sterilized sawdust and fed to two healthy larvae in one tin, while two control larvae were kept on sawdust alone in a second tin. The test insect was regarded as virus-free if both larvae survived for 5 weeks, or if one died but the other survived for 10 weeks. The test insect was considered infected if both larvae died within 5 weeks while both controls survived. Although very useful, this technique would not be specific in countries where other diseases with similar symptoms might exist, precautions must be taken against death by *Metarhizium* infection and sufficient healthy larvae are difficult to obtain without a rearing program. Also, the test needs up to 10 weeks for completion.

5. Smear Method

To make smears of the adult midgut content, Zelazny (1973c) removed the midgut cleanly and placed it on tissue paper under a dissecting microscope. A fine pipette with a smooth tip 0.6—0.8 mm wide was used to transfer a little material from the midgut to a clean slide. This is smeared out, quickly air-dried, fixed for 5—7 min in acetone-free methanol, flooded with freshly prepared 3% Giemsa solution in distilled water for about 40 min, rinsed briefly in distilled water, dried and mounted. The gut content of virus-infected adults shows cells with usually very large nuclei surrounded by a thin layer of cytoplasm, both staining mauve—pink. In healthy adults the nuclei stain mauve—pink, each nucleus may be surrounded by abundant light blue cytoplasm, and there may be many free mauve—pink nuclei of small uniform size but without cytoplasm (FAO, 1978).

Using the same beetles for smear tests and bioassay, Zelazny (1974a) detected virus by the smear test earliest in adults which had been injected with the highest dose. Adults swum in a virus suspension received a wide range of dosages, and their infections developed at different rates. Whether infected by injection or swimming, over 90% of infections were detected by the smear method within 18 days. Accordingly, experimental or field-collected beetles should be kept for at least 2 weeks before smear testing. The results of bioassasy and smear test always agreed.

6. Macroscopic Examination of the Adult Midgut

The thorax is opened with strong scissors and the midgut exposed with forceps. In advanced virus infections the midgut is characteristically swollen several-fold and bright white. Healthy adults have a thin midgut, the contents varying from light to dark brown depending on how recently the adults have fed. This method is quick and simple but detects only about 50–75% of all virus infections. It is adequate where only relative values are needed, e.g. in practical control programs (Zelazny, 1974a).

VI. Effect of Virus on Beetle Populations

A. METHODS FOR ASSESSING THE EFFECT OF VIRUS

Traps must be checked frequently and the catch may be affected by factors other than population size (Bedford, 1973b, 1975; Zelazny, 1977a). Convenient breeding sites for regular sampling by the methods used in Wallis Island by Hammes (1974) and in Tonga by Young (1974) become progressively harder to find. Often the most convenient method for following *O. rhinoceros* population changes is to monitor damage in the palm crowns. Two survey methods have been used in Tonga and Western Samoa (Young, 1974) and in Fiji (Bedford, 1976b). Detailed surveys involved periodic counts of fronds above horizontal level in the crown of each of 20–30 paint-numbered palms, and also of fronds cut by beetles (Young, 1974). Results were expressed as percentage of damaged fronds at one site, or a group of sites combined.

Rapid surveys considered only the central, uppermost three to four most recently opened fronds of the crown, scoring up to 100 palms at an observation point as either damaged or undamaged (Young, 1974). The results of two or three observers standing at the same point and viewing the same general area of palms, should be averaged then expressed as the percentage of palms recently damaged. Monsarrat (1974) has described in detail a method for comparing crown damage before and after virus release in Wallis Island.

B. RESULTS OF VIRUS APPLICATION

The results can best be summarized country by country.

1. Western Samoa

Introduced into artificial log heaps in 1967, in one year the virus became established in the *O. rhinoceros* population and spread to other parts of the islands (Marschall, 1970). It apparently reduced the beetle population considerably. However, lack of quantitative damage studies caused certain doubts and questions, which stimulated work elsewhere. In 1969, 73% of larvae collected from natural breeding sites died from virus, but it is now known that the larvae were bulk-collected with cross contamination, so that the results did not represent the mortality in the wild population. These early results mainly sufficed to show that the virus was established.

Careful studies in 1970—71 showed that in fact only about 3% of larvae and 35% of adults in the wild were infected, while a mean of 7.3% of the breeding sites contained infected insects (Zelazny, 1973b). Larval breeding sites were most likely to contain infected insects if the sites bore two broods instead of one, and if the breeding was extensive in the area. Sites were less frequently visited by infected than by healthy females. Mated females collected from palms were more often infected than egg-laying females from breeding sites. Infections in male adults increased with age, but young females were more often infected than very young or old females (Zelazny, 1973b). Later, Zelazny (1977b) showed that from 1971 to 1974, breeding sites with infected larvae fluctuated between 5.6 and 11.2% (mean 8.2%). Detailed surveys showed that the pattern of 12.5—18% palm fronds damaged seemed to follow these fluctuations most of the time. Between 1973 and 1975 the percentage of infected beetles in traps fell from 63% to 35% (mean 51%), corresponding to a significant decline in the number of beetles trapped. Significantly more males than females were found infected.

2. Fiji

Virus application began in 1970 using the three methods in sequence as they evolved (Bedford, 1972, 1976b, 1977). Sampling in 1971—72 showed that on the islands of Beqa and Vatulele 2—3% of breeding sites contained infected *O. rhinoceros* specimens. On the main island of Viti Levu, of 68 to 88 beetles trapped in different areas, the proportions infected were 68% around Suva in January—March 1974, 66% between Lautoka and Nadi from September 1973 to March 1974, and 57% at Caboni from June 1973 to February 1974. Trappings on numerous other islands showed virus established in the populations. Surveys before release and at intervals later showed that in many localities palm damage fell very significantly indeed 12 to 18 months after virus establishment, and some examples are presented in Fig. 6 and illustrated in Fig. 7.

3. Wallis Island

Virus was applied in artificial log heaps from September 1970 to June 1971. In < 2 months after the start, virus had spread over the whole island (Hammes, 1971) and in one year the adult beetle population fell by 60—80% (Hammes, 1974). The percentage of fronds damaged fell by a mean of about 82%, ranging from 90% in densely planted groves to 76% in more open groves (Monsarrat, 1974).

4. Tongatapu

Virus was released into artificial sawdust and log heaps on the western tip of the island in November 1970—January 1971 (Young, 1974). The epizootic spread at a rate of about 3 km/month. Breeding sites with infected *O. rhinoceros* ranged from 2 to 40%, varying with the time elapsed since release and distance from the release point. Behind the zone of spread the virus incidence fell again.

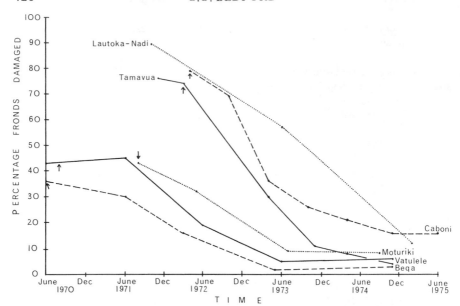

Fig. 6. Effect of baculovirus on palm damage by *Oryctes rhinoceros* at localities in the Fiji Islands. Arrows indicate date of virus introduction. Virus had spread naturally into the Lautoka–Nadi area by mid 1973. Damage survey is based on 170 palms at 11 sites for Vatulele, 176 palms at 9 sites for Beqa, 53 palms for Tamavua, 157 palms for Caboni, 188 palms for Lautoka–Nadi, and 20 palms for Moturiki.

A declining beetle population was indicated by fewer occupied sites and fewer larvae in later surveys. In the release zone rapid damage surveys showed that the palms had noticeably improved in appearance at 350 days post-introduction, and from the known rate of frond replacement, a significant change in beetle numbers must have occurred at about 200 days. In the release zone, palms with central crown damage fell from 28% at 150 days to 5% at 455 days, and in the next zone of spread from 27% to 10%.

5. American Samoa

Virus-infected beetles were released at one site in January–April 1972 (Swan, 1972), and here the upper damaged fronds fell from 89% to 47% in 33 months (Zelazny, 1974b). The virus spread at 0.8–1.6 km/month; within 3 km of the release area damage fell from 13% to 2% in 33 months; at 3–8 km away it fell from 15% to 5%; at 8–13 km away, from 16% to 7% – the further the locations were away from the release point, the later did the damage decline.

6. Tokelau Islands

The virus was introduced in artificial log heaps in 1967 (Marschall, 1968). An experiment to observe the effect of additional virus application was commenced

Fig. 7. Group of palms at Waisomo, Beqa Island, Fiji, photographed (above) in March 1971, and (below) in April 1972. Note replacement of damaged fronds by healthy ones. Courtesy of PANS.

in January 1973 (Zelazny, 1977a). The islets of Nukunonu atoll, which lie in an ellipse 8 × 13 km in size, were divided into two treatment groups and one control group. No treatment was applied to the east and west control islets which separated the treated north (90 ha) and south (40 ha) islets by 10.5 km to the east and 4.5 km to the west. On both treated islets beetles were caught with attractant traps and larvae collected every 2–3 months. Female beetles and larvae were killed, but male beetles were saved as a means of releasing virus on the south islets only. The males were released in groups of 20 each week after injection with haemolymph from infected larvae, the program continuing for 20 months. Damage to 1000 marked palms was recorded initially and after 23 months. On the control islets damage increased slightly. On those with beetle traps and larval collections, beetles trapped fell from 240 to 149/month, and upper fronds damaged fell slightly from 3.7 to 2.4%. On the islets with extra virus as well as trapping and larval collection, beetles trapped fell significantly more (82 to 14 beetles/month), the decline starting after 10 months, significantly fewer females being trapped, while damage fell significantly (6.5 to 1.9% damaged upper fronds).

7. Mauritius

Virus applied in manure heaps in 1970–1972 (Monty, 1972, 1978) reduced the mean number of larvae per heap from 24.6 in 1970 (before virus introduction) to 4.6 in 1976–1977. Infected larvae were found in dead standing palms indicating transmission by adults. No more virus has been released since 1972. Damage to palms has been reduced by 60 to 95% (Hammes, 1978). With the beetle population now regulated by the virus, the only way damage could be further reduced would be to decrease further the ratio of beetles to palms, by increasing the overall number of coconut palms through replanting.

8. Change in Level of Natural Enemies of O. rhinoceros in all countries

There is little information on this and no evidence so far that other insects co-habiting breeding sites are affected by the virus.

VII. Safety Testing

Safety was tested in France in 1975 using purified virus (FAO, 1978). No pathogenicity was observed in eight tissue cultures: two human and two pig cell cultures, and one each from mouse, hamster, fish and calf. No pathogenicity was observed in living inoculated mice or in brain, liver, spleen, lung, heart, stomach, bladder, intestine, muscle, kidney, gonads or blood of mice up to 60 days after inoculation. (See also Chapter 40, Section IV, D.)

VIII. Role of Virus in Integrated Control Programs

Virus, released as above, is easily integrated into control programs against O. rhinoceros, and can assume the primary role. It is easy to produce in larvae and adults. Another aspect of integrated programs would be destruction of breeding sites, which would greatly reduce the Oryctes problem, even without virus. However this is laborious and so is usually not done efficiently, although Monsarrat (1974) pointed out that where virus is used, some natural breeding sites should always be left. As virus is soon inactivated, it is clear that it can persist in an area only if there is an adequate insect population to propagate and transmit it.

Cover crops to conceal breeding sites may be very useful in some situations. In replanted areas on oil palm plantations in Malaysia, Wood (1969, 1976) found almost ten times as many O. rhinoceros larvae in exposed decaying oil palm logs on bare ground plots as on plots with breeding sites covered by natural vegetation or planted legumes. The cover also curbed damage to the replanted young oil palms. In the Ivory Coast, Julia and Mariau (1976a) have also shown the benefit of *Pueraria javanica* cover in reducing *O. monoceros* breeding in newly felled forest timber prior to planting coconuts. Coconut logs may be removed to make charcoal (Little, 1973; FAO, 1978). In breeding sites such as large sawdust heaps around sawmills, application of *M. anisopliae* spores may be a useful supplement to initial virus release (Latch, 1976). It does not seem that virus would interfere with control methods used for other types of palm pests.

IX. Economics of Virus Release

The economics of using baculovirus must be related to the overall costs caused by the presence of *O. rhinoceros*. Catley (1969) estimated that in 1968 this pest cost six South Pacific countries US$1 179 800 at that time in loss of copra, costs of control, quarantine, and additional plantation establishment costs for palm replacement. In Fiji alone, despite a cumulative quarantine expenditure of at least US$2.7 million from 1953 to the end of 1971, the beetle spread and established itself in much of the Fiji group (Bedford, 1976b).

Distribution of damage to palms may be very patchy and its severity varies locally depending on the availability of breeding sites. Where the virus has been established for several years, e.g. Western Samoa, light damage may cause only small crop losses (Zelazny, 1975). However where beetles multiply unchecked, severe damage may cause stunting, little or no nut production, and eventual death of the palms (Fig. 1).

Release of virus is difficult to cost because the work was done in experimental research projects by international staff, or government staff as only part of their duties. The only hard information available concerns the mass rearing of adult beetles for release in Fiji, where each beetle cost about US$0.92 in 1974, the main cost component being labour (Bedford, 1976a). In general, costs would vary from one country to another, and would be absorbed in the general agriculture budgets. Since, once introduced, the virus spreads rapidly to give a high degree of control, its use in integrated control programs is very attractive economically.

X. Suggestions for Future Work

Since the original discovery of the virus in Malaysia, little work has been done on it there (Bedford, 1971; Marschall, 1971). Where extensive breeding sites occur, virus alone is insufficient to reduce beetle populations to a subeconomic level, hence the work by Wood (1969, 1976) on cover crops. As the virus has presumably been long established there, it would be of interest to map its distribution and incidence across the country, and its correlation with beetle populations and palm damage levels.

Virus transmission from larva through pupa to adult, where it might sometimes cause deformity, is suggested by the results of Monty (1974) but not by those of Zelazny (1973a, 1976). Likewise the problem of virus transmission through the egg is not yet fully resolved. Although the work of Zelazny (1976) suggested that this does not normally occur, Monsarrat *et al.* (1973a, 1974) reported virions in the female genital organs.

The virus could be tested on *Oryctes* species of lesser economic importance, such as *O. gnu* in Malaysia and *O. centaurus* on the Papua–New Guinea mainland, and on related genera. Although *O. monoceros* in the Ivory Coast does not appear to be very susceptible to virus (Julia and Mariau, 1976b), the pathogen has been released in the Seychelles against this beetle (Windsor, 1975), and an examination of the eventual results would be interesting. The lesser *Oryctes* pests may have different strains of virus which could be investigated.

If an unusually large number of breeding sites is created locally, e.g. by felling palms, the *O. rhinoceros* population can resurge in these "outbreak" areas, increasing damage to neighbouring palms despite the virus. Long term investigations should examine the value of repeated releases of virus-infected beetles in the outbreak areas, following the lead of Zelazny (1977a) in the Tokelau Islands, where damage was further reduced although damage levels were already initially quite low. However the islets of Nukunonu are small and isolated and, on large islands or land masses, released beetles could disperse from the outbreak areas, so diluting their effect.

There is as yet no evidence that *O. rhinoceros* populations in the South Pacific are developing resistance to the virus, but this should be monitored in future. New virus strains may need to be introduced.

References

Bedford, G.O. (1971). "UN/SPC Rhinoceros Beetle Project, Ann. Rep.", pp. 225–230. S. Pacif. Commn, Noumea.

Bedford, G.O. (1972). "UN/SPC Rhinoceros Beetle Project, Ann. Rep.", pp. 7–18. S. Pacif. Commn, Noumea.

Bedford, G.O. (1973a). *J. Invertebr. Path.* **22**, 70–74.

Bedford, G.O. (1973b). *J. econ. Ent.* **66**, 1216–1217.

Bedford, G.O. (1975). *Bull. ent. Res.* **65**, 443–451.

Bedford, G.O. (1976a). *PANS, London* **22**, 5–10.

Bedford, G.O. (1976b). *PANS, London* **22**, 11–25.

Bedford, G.O. (1977). *S. Pac. Bull.* **27**, 27–34.

Catley, A. (1969). *PANS, London* **15**, 18–30.

Croizier, G. and Monsarrat, P. (1974). *Entomophaga* **19**, 115–116.

David, W.A.L. (1975). *Ann. Rev. Ent.* **20**, 97–117.

FAO. (1978). Research on the Control of the Coconut Palm Rhinoceros Beetle, Phase II, Fiji, Tonga, Western Samoa. Tech. Rep. 94 pp. Rome.

Hammes, C. (1971). *C. r. Hebd. Seanc. Acad. Sci. Paris* D **273**, 1048–1050.

Hammes, C. (1974). *Cah. ORSTOM Ser. Biol.* **22**, 52–91.

Hammes, C. (1978). *Rev. agric. sucr. Maurice* **57**, 4–18.

Hüger, A.M. (1966a). *J. Invertebr. Path.* **8**, 38–51.

Hüger, A.M. (1966b). Z. Angew. Ent. **58**, 89–95.
Hüger, A.M. (1973). Z. Angew. Ent. **72**, 303–319.
Julia, J.F. and Mariau, D. (1976a). Oleagineux **31**, 63–68.
Julia, J.F. and Mariau, D. (1976b). Oleagineux **31**, 113–117.
Julia, J.F. and Mariau, D. (1976c). Oleagineux **31**, 263–272.
Kelly, D.C. (1976). Virology **69**, 595–606.
Latch, G.C.M. (1976). Entomophaga **21**, 31–38.
Little, E.C.S. (1973). "UNDP/FAO Rhinoceros Beetle Project Ann. Rep.", pp. 25–53. Apia.
Marschall, K.J. (1968). "UN/SPC Rhinoceros Beetle Project, Semiann. Rep., Nov. 1967 to May 1968", pp. 21–26. S. Pacif. Commn, Noumea.
Marschall, K.J. (1970). Nature **225**, 288–289.
Marschall, K.J. (1971). "UN/SPC Rhinoceros Beetle Project, Ann. Rep.", pp. 231–233. S. Pacif. Commn, Noumea.
Marschall, K.J. (1973). "UNDP/FAO Rhinoceros Beetle Project, Ann. Rep.", pp. 60–61. Apia.
Monsarrat, P. (1974). Cah. ORSTOM Ser. Biol. **22**, 92–111.
Monsarrat, P. and Veyrunes, J.C. (1976). J. Invertebr. Path. **27**, 387–389.
Monsarrat, P., Meynadier, G., Croizier, G. and Vago, C. (1973a). C.r. hebd. Seanc. Acad. Sci., Paris D **276**, 2077–2080.
Monsarrat, P., Veyrunes, J.C., Meynadier, G., Croizier, G. and Vago, C. (1973b). C. r. hebd. Seanc. Acad. Sci., Paris D **277**, 1413–1415.
Monsarrat, P., Duthoit, J.L. and Vago, C. (1974). C. r. hebd. Seanc. Acad. Sci., Paris D **278**, 3259–3261.
Monty, J. (1972). Abst. 14th Int. Congr. Ent., Canberra. p. 220.
Monty, J. (1974). Bull. ent. Res. **64**, 633–636.
Monty, J. (1978). Rev. agric. sucr. Maurice **57**, 60–76.
Payne, C.C. (1974). J. gen. Virol. **25**, 105–116.
Payne, C.C., Compson, D. and de Looze, S.M. (1977). Virology **77**, 269–280.
Quiot, J.M., Monsarrat, P., Meynadier, G., Croizier, G. and Vago, C. (1973). C. r. hebd. Seanc. Acad. Sci., Paris D **276**, 3229–3231.
Revet, B. and Monsarrat, P. (1974). C. r. hebd. Seanc. Acad. Sci., Paris D **278**, 331–334.
Swan, D.I. (1972). "UN/SPC Rhinoceros Beetle Project, Ann. Rep.", pp. 166–169. S. Pacif. Commn, Noumea.
Windsor, K. (1975). Introduction of Rhabdionvirus oryctes into the Oryctes monoceros population of the Seychelles. 4th sess., FAO Techn. Working Party on Coconut Prod., Pro. and Processing, Kingston, Jamaica, 14–25 Sept. 1975, 2 pp.
Wood, B.J. (1969). Bull. ent. Res. **59**, 85–96.
Wood, B.J. (1976). In "Oil Palm Research" (R.H.V. Corley, J.J. Hardon, B.J. Wood, eds), pp. 347–367. Elsevier, Amsterdam.
Young, E.C. (1974). J. Invertebr. Path. **24**, 82–92.
Zelazny, B. (1972). J. Invertebr. Path. **20**, 235–241.
Zelazny, B. (1973a). J. Invertebr. Path. **22**, 122–126.
Zelazny, B. (1973b). J. Invertebr. Path. **22**, 359–363.

Zelazny, B. (1973c). "UNDP/FAO Rhinoceros Beetle Project, Ann. Rep.", pp. 62–63. Apia.

Zelazny, B. (1974a). UNDP/FAO/SPC Rhinoceros Beetle Project, Tech. Docum. 104, 19 pp.

Zelazny, B. (1974b). UNDP/FAO/SPC Rhinoceros Beetle Project, Tech. Docum. 108, 4pp.

Zelazny, B. (1975). UNDP/FAO/SPC Rhinoceros Beetle Project, Tech. Docum. 102, 9 pp.

Zelazny, B. (1976). *J. Invertebr. Path.* **27**, 221–227.

Zelazny, B. (1977a). *FAO Plant. Prot. Bull.* 73–77.

Zelazny, B. (1977b). *J. Invertebr. Path.* **29**, 210–215.

NOTE ADDED IN PROOF

For a recent review of the biology, ecology and control of palm rhinoceros beetles see Bedford, G.O. (1980). *Ann. Rev. Ent.* **25**, 309–339.

CHAPTER 21

Control of Mites by Non-occluded Viruses

D.K. REED

USDA, SEA, Agricultural Research, Fruit and Vegetable Insects Research Laboratory, Vincennes, Indiana, USA

I. Introduction

Although mites are one of the most important groups of pests in agriculture, very few pathogens are known which may be utilized in their control. This is particularly true of viruses. Since the review by Lipa (1971) which listed three viruses, no new viruses have been reported. One would expect there to be important viruses in mites besides that of the citrus red mite (*Panonychus citri*), which is the only mite virus with a useful part in pest—management practices. However, as such, it forms the subject of this chapter.

II. Advances in Basic Knowledge

A. ETIOLOGY

The etiological agent of the citrus red mite virus disease, first characterized by Smith *et al.* (1959) as a spherical particle, 35 nm in diameter, has more recently been identified by Reed and Hall (1972) as rod-shaped, 58 × 194 nm, and enclosed in envelopes, 111 × 266 nm. Such rods are non-occluded and are formed within the nuclei of midgut epithelial cells. The rods have consistently been associated with this disease, whereas the spherical particles, although present within the mites' bodies, are just as prevalent in apparently healthy as in obviously infected mites. In fact, Reed and Desjardins (1978) report that three sizes of spherical particles are present — 18 nm spheres in crystalline array and 30 and 37 nm particles — and the high 260:280 nm ultra violet absorbance ratios of both the large [1.73] and small [1.48] spheres indicate the presence of nucleic acid. Spherical particles occur only within laboratory-reared mites, and host—pathogen relationships of the bodies are unknown.

Rods isolated by density—gradient centrifugation (Reed, 1971) are elongate— oval in shape, 81 × 206 nm, usually with a rounded projection or knob at one end and remnants of the envelope at the other end. Morphology is very similar to that of rods extracted from virus infected *Oryctes rhinoceros* by Monsarrat *et al.* in 1973 and Payne in 1974. No chemical determinations have been made of citrus red mite virus due to the extremely low yield; however, from the striking similarity to *Oryctes* virus, it will probably be placed in the Baculoviridae when such work is accomplished.

B. TRANSMISSION

Knowledge of how the pest acquires the disease is basic to any microbial control technique. Reed *et al.* (1975) showed that the citrus red mite does not acquire virus when probing cells previously fed on by infected mites, as generally postulated by early workers, but encounters the virus in faeces and debris on the plant surface. The "suction cup" feeding mechanism of these mites (Jeppson *et al.*, 1975) tends to substantiate this observation. Also, the abundance of virus rods within hindgut cells that are continually sloughed into the gut lumen to be defaecated indicates a highly infectious faecal pellet, making such diseased mites in effect "virus factories". Envelopment of the virions in a matrix of cellular material may explain why natural infective material on contaminated surfaces lasts so much longer than aqueous sprays (Reed *et al.*, 1975). Gilmore and Munger (1963) reported 8 days while Tashiro *et al.* (1970) reported 28 days residual activity from contaminated surfaces of lemons. Aqueous sprays generally survive only 2—4 h (Gilmore and Munger, 1963).

C. EFFECTS OF ENVIRONMENT ON VIRUS STABILITY

As with most viruses, inactivation occurs very rapidly in the orchard where ultraviolet rays rapidly destroy the infectivity of aqueous sprays (Gilmore and Munger, 1963). Tashiro *et al.* (1970) reported that virus in intact mites retained viability for > 28 days in the laboratory and postulated that such intact dead bodies could adhere to plant surfaces after epizootics as reservoirs of infection. Shaw *et al.* (1972) found complete viability of virus in infected mites after 6.5 years storage at −5°C. Acaricides in the glasshouse had no effect on initiation of subsequent epizootics (Shaw and Pierce, 1972). Shaw *et al.* (1969)

obtained excellent control of red mites in glasshouses near San Diego with aqueous sprays of virus. However, poor control in a glasshouse in Riverside using the same techniques was attributed to high temperatures. In the laboratory, virus within dead intact mites was relatively unaffected by temperatures found in the orchard micro-climate, and full activity was retained at 40.5°C but was lost after 6 h exposure to 46°C and 1 h exposure to 60°C (Reed, 1974). Thus, applications of diseased mobile mites were better than other application techniques because of greater persistence and because virus was continually being produced and dispersed (Gilmore and Munger, 1963). Reed (1974) also noted that the much greater longevity of infected mites in cooler temperatures might explain the failures of virus epizootics to control mite populations in early spring and late fall.

III. Advances in Applied Concepts

A. IDENTIFICATION

Identification of virus-diseased mites is relatively easy, since birefringent crystals are usually present within the bodies of infected mites (Smith and Cressman, 1962). Reed et al. (1972a) reported that these crystals formed within the midgut are apparently related to normal production of faecal pellets, also birefringent, since both form more rapidly in starved than in fed mites, and both begin formation at about the same time. High humidity has been shown to inhibit production of these crystals so that care must be exerted in disease diagnosis during periods of fog or sustained high moisture levels (Reed et al., 1974). A portable apparatus (Reed et al., 1972b) can be used by growers and orchard personnel to detect the birefringent crystal formation in the orchard, without the expense of a polarizing microscope.

B. EFFECT ON OTHER ORGANISMS

Shaw et al. (1967) fed virus-infected red mites to three species of predatory mites common in Californian citrus, Amblyseius hibisci, A. limonicus and Typhlodromus occidentalis. There was no difference in consumption of diseased or healthy mites by the predators, and no apparent reduction of longevity or activity of the predators. Experiments with seven other tetranychid mites by Beavers and Reed (1972) indicated that only the Carmen spider mite, Tetranychus cinnabarinus, exhibited signs of disease after feeding on virus-contaminated substrates. These signs consisted of presence of birefringent crystals in a few mites but no apparent mortality. Although no adverse effects were noted, these infected mites contaminated the surfaces of lemons while feeding so that subsequent populations of citrus red mites on these surfaces acquired the disease.

C. COLLECTION AND STORAGE

Since the virus can be cultured only in living citrus red mites, laboratory production depends on the harvest, storage and utilization of green lemons for rearing sufficient healthy mites for inoculation, incubation and harvest of diseased mites. This is adequate for laboratory studies but materials, space and labour make it too costly for large-scale orchard applications. A cheaper alternative is to collect diseased mites in the orchard with vacuum-suction machines, a saving of 95–98% over laboratory cultured mites (Shaw *et al.*, 1971). These collected mites may then be held on green lemons for 6–7 days to increase infection to levels obtained in the laboratory (Reed *et al.*, 1972c).

D. EXTENDERS OF VIRUS VIABILITY

Since aqueous suspensions of red mite virus are rapidly inactivated, Reed *et al.* (1973) tested various spray additivies. Carbohydrate-rich materials such as honey, molasses and sucrose prolonged activity for several more hours, but extracted body fluids from *Trichoplusia ni* larvae and pupae extended activity for 144 h in the laboratory. Results of orchard trials were inconclusive so more research is required to decide whether aqueous sprays will ever be a practical control technique.

IV. Practical Considerations for Use of the Virus

A. CURRENT USE IN CONTROL

The virus occurs naturally in mite populations throughout the citrus-growing areas of California and Arizona and exerts considerable natural control of large populations of mites. It is not effective against small populations, nor during periods of wet, cool weather. The only practical application technique at this time is the spread of infected mites taken from epizootic areas, a practice not in general use due to labour requirements, the danger of moving scale or other insects, and ignorance of benefits. However, the highly-trained and motivated insect ecologists who advise Californian citrus growers on pest management well know the benefits of natural epizootics in high mite densities and recommend delaying acaricidal applications to allow such epizootics to develop.

B. FUTURE NEEDS

As the situation now exists, the best manner of using the citrus red mite virus is the avoidance of premature acaricidal applications. The key to such an approach is education. Research has been sufficient to prove the benefits of the virus to pest management, information which should be better disseminated to citrus growers and managers to show them this alternative in pest control.

More basic research is needed on the infectious particles. This would include isolation in sufficient quantities for complete and physical characterization of the virions, which would in turn, expedite further identification and classification of the virus. Too little is known about the epizootiological principles in the orchard. The work on temperature relationships has been a start, but much remains to be learned in order to best utilize the virus disease as a control agent. Also, to use the virus as a microbial insecticide, better extenders and protectants should be developed, along with better application techniques. Cell culture techniques may be useful for mass production of the virus. Safety criteria should be investigated at the earliest practical time.

The demonstrated potential of the citrus red mite virus should stimulate research on other known viruses such as that of the European red mite, *Panonychus ulmi* (Bird, 1967), to determine its role against this serious mite pest. It should also stimulate a search for more viruses.

V. Conclusions

The virus disease of citrus red mites has been intensively studied since its discovery in 1958. Basic and applied research has indicated its great potential in the practical control of mites. Reports from orchard personnel and pest-management advisors indicate that, in most years, the disease will decimate large populations of mites, particularly in the San Joaquin Valley of California, and that it has a definite place in the citrus pest-management program.

References

Beavers, J.B. and Reed, D.K. (1972). *J. Invertebr. Path.* **20**, 279–283.

Bird, F.T. (1967). *J. Insect Path.* **1**, 270–280.

Gilmore, J.E. and Munger, F. (1963). *J. Insect Path.* **5**, 141–151.

Jeppson, L.R., Kiefer, H.H. and Baker, E.W. (1975). "Mites Injurious to Economic Plants." Univ. Calif. Press, Berkeley and Los Angeles.

Lipa, J.J. (1971). *In* "Microbial Control of Insects and Mites." (H.D. Burges and N.W. Hussey, eds), pp. 357–373. Academic Press, London and New York.

Monsarrat, P., Veyrures, I.C. Meynadier, G., Croizier, G. and Vago, C. (1973). *C. r. hebd. Séanc. Acad. Sci. Paris* **277**, 1414–1415.

Payne, C.C. (1974). *J. gen. Virol.* **25**, 105–116.

Reed, D.K. (1971). Ph.D. Thesis. Univ. Calif., Riverside. 88 p.

Reed, D.K. (1974). *J. Invertebr. Path.* **24**, 218–223.

Reed, D.K. and Desjardins, P.R. (1978). *J. Invertebr. Pathol.* **31**, 188–193.

Reed, D.K. and Hall, I.M. (1972). *J. Invertebr. Path.* **20**, 272–278.

Reed, D.K., Hall, I.M., Rich, J.E. and Shaw, J.G. (1972a). *J. Invertebr. Path.* **20**, 170–175.

Reed, D.K., Rich, J.E. and Shaw, J.G. (1972b). *J. econ. Ent.* **64**, 890–891.

Reed, D.K., Shaw, J.G. and Rich, J.E. (1972c). *J. econ. Ent.* **65**, 1507.

Reed, D.K., Hendrickson, R.M., Jr., Rich, J.E. and Shaw, J.G. (1973). *J. Invertebr. Path.* **22**, 182–185.

Reed, D.K., Rich, J.E. and Shaw, J.G. (1974). *J. Invertebr. Path.* **23**, 285–288.

Reed, D.K., Tashiro, H. and Beavers, J.B. (1975). *J. Invertebr. Path.* **26**, 239–246.

Shaw, J.G. and Pierce, H.D. (1972). *J. econ. Ent.* **65**, 619–620.

Shaw, J.G., Moffitt, C. and Scriven, G.T. (1967). *J. econ. Ent.* **60**, 1751–1752.

Shaw, J.G., Beavers, J.B., Pappas, J.L. and Hampton, R.B. (1969). *J. econ. Ent.* **62**, 1154–1156.

Shaw, J.G., Reed, D.K., Stewart, J.R., Gorden, J.M. and Rich, J.E. (1971). *J. econ. Ent.* **64**, 1223–1224.

Shaw, J.G., Rich, J.E. and Reed, D.K. (1972). *J. econ. Ent.* **65**, 1512.

Smith, K.M. and Cressman, A.W. (1962). *J. Insect Path.* **4**, 229–236.

Smith, K.M., Hills, G.J., Munger, F. and Gilmore, J.E. (1959). *Nature Lond.* **184**, 70.

Tashiro, H., Beavers, J.B., Groza, M. and Moffitt, C. (1970). *J. Invertebr. Path.* **16**, 63–68.

CHAPTER 22

Pest Control by Cytoplasmic Polyhedrosis Viruses

K. KATAGIRI

Forestry and Forest Products Research Institute, Ushiku, Ibaraki, Japan

I. Introduction

Since the first electron microscopic demonstration of a cytoplasmic polyhedrosis virus (CPV) (Smith and Wickoff, 1950), CPVs have been found in ca. 200 species of insects, some 85% of which are Lepidoptera, 9% Diptera, 4% Hymenoptera and the rest Coleoptera (Martignoni and Iwai, 1977; Appendix 2). Of these CPVs, < 10% were suggested for use as control agents and only a few have been used in practice. A CPV of the pine processionary caterpillar *Thaumetopoea pityocampa* was the first application of CPV in biological control (Grison *et al.*, 1959). A CPV of the pine caterpillar *Dendrolimus spectabilis* causing a flacherie disease is the only one to be explored and registered as a microbial insecticide. Several authors have reviewed the CPVs of insects (Aruga, 1973; Aruga and Tanada, 1971; Smith, 1976). Here, therefore, they will be discussed from the viewpoint of their practical use in the field.

433

II. Unique Characteristics of CPVs as Control Agents

A. RELATION TO HOST POPULATIONS

Most CPVs do not reduce host populations as drastically as nuclear poly-hedrosis viruses (NPVs) in the field. Rather, CPVs are chronic and persist at low rates, with some exceptions such as the CPV of *Thaumetopoea pityocampa* which infected field populations at a high rate in Corsica, France, where CPV incidence was density dependent, i.e. more disease occurred in the more densely populated parts of the insect population (Dusaussoy and Geri, 1968).

CPV sprays may produce lethal infections directly, or infection in a larva may build up to a lethal level by contagion. In *D. spectabilis* the effect depended largely on pest population density and the best dosage was 10^{11} polyhedra/ha for intermediate population densities during the increasing phase. At the lowest ebb of the population, more virus was required to obtain an equal effect. In an overcrowded population, mortality from flacherie was high compared with natural deaths in an unsprayed population but not enough to obtain control. Thus, Katagiri (1969, 1975) concluded that population cycles should be taken into consideration and that the *D. spectabilis* CPV is most effective when sprayed before peak population density.

Mortality increases with heavier dosage. Larvae surviving small dosages are less active than healthy ones and feed less (Slizynski and Lipa, 1975). Diseased larvae often pupate and form adults which are abnormal in morphology, physiology, and ecology (Bell and Kanavel, 1976; Magnoler, 1974; Maleki-Milani, 1970; Simmons and Sikorowski, 1973). They lay few or no eggs and so their reproductive potential is greatly reduced. Thus immediate mortality is a poor measure of the effect of CPV: its real controlling power must be evaluated over several generations, including its effect on fecundity and the ability of the insects to resist other natural mortality factors. CPVs should be used on crops where some insects can be tolerated so that these characteristics can be exploited.

B. HOST RANGE

In general, CPVs have a wide host range, wider than NPVs or granulosis viruses. For instance, CPV from *D. spectabilis* attacks *D. superans*, *D. punctatus*, *Lymantria dispar*, *L. fumida*, *Malacosoma neustria testacea*, *Pygaera* (*Clostera*) *anastomosis*, *Hyphantria cunea*, *Bombyx mori*, etc. Aruga (1973) cited 17 lepidopterous species susceptible to the CPV of *Bombyx mori*. Host range, however, is confused by the possibility of induction of a CPV by inoculation with another CPV (Aruga *et al.*, 1961).

C. INFECTIVITY, SUSCEPTIBILITY AND TRANSMISSION

Infectivity of a CPV with a wide host range varies according to the host species. Low infectivity in a species different from the initial host is increased by passage through the new species (Koyama and Iwata, 1963; Koyama and Kushida, 1963). Different isolates of a CPV from a single host species differ in virulence (Magnoler, 1970). Also ecological and geographical strains of a host vary in susceptibility, e.g. the LC_{50} of a CPV in *D. spectabilis* from Iwate Pref., northern part of Japan, was 10 times that in larvae from other parts, suggesting that the Iwate strain of the moth was less susceptible. Thus it may be possible to select CPVs more suitable for use in biological control programs (Aizawa, 1971).

Larvae of both *Choristoneura fumiferana* and *Malacosoma disstria,* are more resistant to CPV and NPV in later instars, more so with NPVs than CPVs (Bird, 1969). The NPVs were more lethal than the CPVs. Most young larvae infected with CPV died or became stunted and retarded in development.

CPVs infect the next host generation in nature by contamination of food and habitat, and by transmission from ovary to egg and on the outside of the egg. CPVs persisted in soil (Hukuhara, 1972), on plants or in alternate hosts (Tanada and Omi, 1974). The CPV of *Lymantria fumida* persisted also on tree branches for at least one year in the forest (Katagiri, 1977). Many eggs of *Trichoplusia ni* moths treated with CPV produced infected larvae, but when eggs were surface-sterilized, little or no infection was found in larvae (Vail and Gough, 1970). Adults of *Heliothis virescens* infected with CPV produced healthy offspring when their eggs were surface-sterilized (Sikorowski *et al.*, 1973). Faeces from diseased larvae can transmit the disease (Watanabe, 1968; Iwata and Katagiri, 1976). Polyhedra are voided together with destroyed cells of the epithelium of *Mamestra brassicae* (Maleki-Milani, 1970), and of *Hyphantria cunea* (Yamaguchi, 1976). In field studies the percent of CPV infection in larvae from virus-infected parents was density dependent (Sikorowski *et al.*, 1973).

III. Application of CPVs

A CPV of *Thaumetopoea pityocampa,* mixed with bentonite dust, was applied by helicopter over ca. 320 ha in France, with satisfactory results (Grison *et al.*, 1959). In Japan, a CPV also controlled *Dendrolimus spectabilis* in repeated forest tests since 1961 which led to its development as a registered commercial microbial insecticide (see below). A CPV from *Dendrolimus pini* showed promise against both *D. pini* and *Malacosoma neustria* (Slizynski and Lipa, 1975). After describing laboratory studies on the CPVs of the spruce budworm *Choristoneura fumiferana* and the forest tent caterpillar *Malacosoma disstria,* Bird (1969) also suggested their possible use in biological control.

In general, CPVs do not kill insects until some weeks after field application, but they are one of the most active mortality factors over the entire generation. In *D. punctatus* (Ying, 1970) and more recently in *D. spectabilis,* a combination of CPV and *Bacillus thuringiensis* killed more insects more quickly than CPV or *B. thuringiensis* alone (Katagiri and Iwata, 1976; Katagiri *et al.*, 1977), as also with NPV and *B. thuringiensis* (Stelzer, 1965; Oatman *et al.*, 1970; Lipa *et al.*, 1975a, b). Spraying a mixture of CPV and *B. thuringiensis* was synergistic. They killed many larvae quickly and continued killing throughout the pupal stage, reducing the pest population to a very low level. Moreover, the CPV concentration in the mixture was effective even at 1/10 that in the virus suspension alone. Reduction of the CPV in the mixture is considered a distinct advantage, in addition to the better control.

CPV interfered with and retarded the development of NPV in *C. fumiferana* and *M. disstria*, although the greatest mortality of the pest would be expected from using both viruses together (Bird, 1969). In long term studies on *C. fumiferana* populations sprayed with NPV contaminated with CPV, initially the CPV infected many insects and then declined while the NPV persisted longer (Cunningham *et al.*, 1975).

CPV and NPV inoculated simultaneously into *Lymantria fumida* interfered with each other. After laboratory and field investigations, Katagiri (1977) considered that the interaction plays a dominant role in terminating an epizootic in the field population, an important fact when applying CPV in the field where NPV already exists and often prevails epizootically. For example, a 1:1 mixture of NPV and CPV was sprayed on young *L. fumida* larvae to suppress an outbreak. NPV produced an initial infection then a secondary one. Mortality by CPV began to appear 2 weeks after spraying and its rate did not vary widely with time. The incidence of the CPV was slower than that of NPV and chronic. The CPV prevailed in an area where the NPV was infrequent, with some double infections in all areas. In these areas the incidence of virus diseases disappeared faster than in other areas. Since natural epizootics of CPV were not observed, its frequent incidence appeared to be caused by CPV spraying. Total mortality was 99.98% of the initial population, NPV killing about 59% and CPV 24% of the larvae, which destroyed the pest outbreak. Incidence of the virus diseases continued at a low rate for several generations (one generation/year) and as a whole disappeared after 6 years (Katagiri, 1977). In *Cryptophlebia leucotreta*, a mixture of CPV and a granulosis virus (GV) killed more insects than CPV or GV alone (Angelini and Labonne, 1970).

A mixture of boric acid or waterglass and NPV induced a CPV of *Lymantria monacha*, which played an important role in controlling the pest before the occurrence of NPV (Yadava, 1970).

No detrimental effects of CPVs on natural enemies have been reported. In many field experiments in Japan against *D. spectabilis*, no differences were seen between the levels of parasitism in treated and untreated areas, and it was concluded that the CPV did not affect parasitism by Hymenoptera and Diptera (Katagiri, 1975). Judging from mortalities due to native natural enemies in trials with *B. thuringiensis*, CPV, and a mixture of the two on the pine caterpillar, Katagiri *et al.* (1977) concluded that these microbes did not harm these natural enemies. Moreover, in a *L. fumida* population, the viruses (CPV and NPV), disseminated merely to reduce the density of pests, in fact curbed the pest population without harming beneficial insects, which resulted in an increase in the percentage of parasitism by egg parasites (Koyama and Katagiri, 1967).

IV. Commercial Preparation of Pine Caterpillar Virus

The CPV of *Dendrolimus spectabilis* was registered in 1974 as a microbial insecticide in Japan.

A. MASS PROPAGATION

In the laboratory, the polyhedra yield per *D. spectabilis* larva was maximal in 10 to 14 days after feeding on virus-sprayed pine needles. In the forest the most economical method for mass propagation was to rear instar VII or older larvae on pine branches sprayed with CPV (10^6 polyhedra/ml) and covered with cloth bags. After 2 to 3 weeks the larvae were macerated with water and filtered with cheese-cloth. The average polyhedra per larva was 5×10^8 or more (Iwata and Katagiri, 1974). Other CPVs are propagated in a similar way and yield 10^8 to 10^{10} polyhedra/larva.

B. FORMULATION

Silica powder intensified the virulence of CPV (Katagiri and Takamura, 1966) and so it was used as the main formulation component. The suspension of polyhedra was mixed with silica powder, then both a sticker and a wetting agent was added.

C. BIOASSAY OF POTENCY

A working standard CPV of *D. spectabilis* was prepared following a protocol of the Ministry of Agriculture, Forestry and Fisheries, Tokyo (Okada *et al.*, 1974), enclosed in an ampoule with $1/15$ M phosphate buffer (pH 7.0) and stored in the dark at 0 to 5°C. Each ampoule contains 1,000 units of CPV. Production batches of *D. spectabilis* CPV are compared with the working standard.

According to Okada *et al.* (1974), activities (P DCV-units; 1 DCV-unit denotes the infectivity of 10^8 virions of standard CPV to the *D. spectabilis* larvae of instar VII) are related to infectivities ($I\%$ infection rates) by

$$I = a \log P + b,$$

where a and b are constants, and $a = 28.0$ for the standard preparation. Infectivities (% infection rates on day 13 from inoculation) of 0.1% suspension of a test preparation (Iu), 0.1% (1 DCV-unit) and 1% (10 DCV-units) suspensions of working standard are obtained by bioassay using *D. spectabilis* 2-day-old larvae of instar VII. From the two infectivities of working standard two constants (a' and b' stand for a and b for the standard) are calculated. If a'/a is < 0.5 or > 1.5, i.e. if a' is < 14 or a' is > 42, the test must be repeated. If it is between 0.5 and 1.5, i.e. if $14 < a' < 42$, the formula is rewritten as $Iu = a' \log Pu + b'$ for the test preparation. As Iu is obtained experimentally from the above test, the activity of the test preparation as % suspension (Pu) can be calculated by this formula. Then the activity of the test preparation is $Pu \times 10^3$ DCV-units/g.

V. Safety of CPV

A. VIRUS INJECTED INTO TEST VERTEBRATES

The CPV of *D. spectabilis* was applied to white mice, hamsters, and rabbits as polyhedral inclusion bodies and/or virus particles by intravenous, intracerebral, and intraperitoneal injections. No pathological change in the organs of the animals was observed. Chick embryos were also exposed by intrayolksac and intraallantoic cavity injections. Some embryos died, probably due to the growth of contaminating bacteria. In serial passage tests through mice and chick embryos, all the mice and embryos survived and the organs were normal. Tanaka *et al*. (1967) concluded that the CPV of *D. spectabilis* was not pathogenic to the test vertebrates.

B. VIRUS APPLIED ORALLY TO TEST MAMMALS

Tests were made by Matsu-ura and Akabori (1969).

In acute toxicity tests, rats and mice of both sexes were fed with up to 3.3×10^{10} polyhedra/kg body weight, and reared for 7 days. None died, so the LD_{50} was $> 3.3 \times 10^{10}$ polyhedra/kg body weight. There were no signs of paralysis, convulsions, loss of appetite, low vitality, weakening, reduction in body weight, etc, or pathological changes in any internal organs.

In subacute toxicity tests, rats and mice of both sexes were inoculated with CPV by stomach probe daily for 90 days at 1.8×10^{10}, 0.9×10^{10}, 0.45×10^{10}, 0.23×10^{10} and 0 polyhedra/kg body weight. None died or showed toxic symptoms. Growth was normal except for a little suppression in male mice with 0.45×10^{10} polyhedra/kg or more. Food and water consumption was normal. There were no abnormalities in blood tests except a smaller than usual alkaliphosphatase activity, suggesting an effect of the CPV. There were no abnormal weight increases of internal organs, e.g. heart, brain, liver, lungs, spleen, stomach, adrenal bodies, kidney, uterus, ovary, testis and bladder, but the spleen and ovary of female rats were slightly enlarged. There were no histopathological abnormalities attributable to the CPV in internal organs, though a weak blood congestion occurred in the intestines and kidney of treated mice and a weak leucocyte infiltration in the stomach of treated rats.

C. TOXICITY OF CPV TO FISH

Carp and killifish were kept in water suspensions of polyhedra at 10^3, 10^4, 10^5, and 10^6 polyhedra/ml for 72 h. None died. It was concluded from these data that the CPV of *D. spectabilis* is safe for forest application against pine caterpillars.

VI. Conclusions and Research Requirements

Most CPVs are less potent than NPVs in natural conditions, but CPVs are very infective and kill when they are sprayed. They are transmitted easily *per os*.

Larvae usually take > 10 days to die in the field and some may die as pupae and young adults. Thus CPVs are active mortality factors throughout the insect life cycle, surviving adults are less fecund and their offspring are likely to be infected. CPVs are suitable for long term population regulation, e.g. in the forest, where some insects can be tolerated and pests need to be maintained at a tolerable level with the least expense and effort.

Research should continue. In agriculture where rapid insect death is desirable, more CPVs should be developed as microbial insecticides for use with other biological agents and chemicals in integrated programs. In contrast, in forests, where long term control is desirable, the ecology of CPVs should be studied to assess their role and interactions in the dynamics of host and related insect populations. This knowledge is important in deciding when to apply CPVs at different ecological sites.

References

Aizawa, K. (1971). In "Microbial Control of Insects and Mites" (H.D. Burges and N.W. Hussey, eds), pp. 655–672, Academic Press, London.

Angelini, A. and Labonne, V. (1970). Coton Fibr. Trop. 25, 497–500.

Aruga, H. (1973). In "An Introduction to Insect Pathology". Yokendo, Tokyo. 614 pp.

Aruga, H. and Tanada, Y. (eds) (1971). In "The Cytoplasmic Polyhedrosis Virus of the Silkworm". Univ. Tokyo Press, Tokyo, 234 pp.

Aruga, H., Hukuhara, T., Yoshitake, N. and Israngkul Na Ayudhya, A. (1961). J. Insect. Path. 3, 81–92.

Bell, M.R. and Kanavel, R.F. (1976). J. Invertebr. Path. 28, 121–126.

Bird, F.T. (1969). Can. Ent. 101. 1269–1285.

Cunningham, J.C., Howse, G.M., Kaupp, W.J., McPhee, J.R., Harnden, A.A. and White, M.B.E. (1975). Information Report Ip-X-10, Canad. For. Ser., Ontario, 1–32.

Dusaussoy, G. and Geri, C. (1968). Res. Rep. Stn. Rech. Lutte Biologique Biocoenotique, Inst. natn. Res. Agronomique, La Minière 18 pp.

Grison, P., Maury, R. and Vago, C. (1959). Rev. Forestière Fr. No 5, 353–370.

Hukuhara, T. (1972). J. Invertebr. Path. 20, 375–376.

Iwata, Z. and Katagiri, K. (1974). Trans. 85th Meet. Jap. For. Soc. 208–209.

Iwata, Z. and Katagiri, K. (1976). Trans. 87th Meet. Jap. For. Soc. 291–292.

Katagiri, K. (1969). Rev. Pl. Prot. Res. 2, 31–41.

Katagiri, K. (1975). Proc. 1st Intersectional Congr., Int. Ass. Microbiol. Socs, Tokyo 1974, 2, 613–620.

Katagiri, K. (1977). Bull. Govt For. Expl Stn No. 294, 85–135.

Katagiri, K. and Iwata, Z. (1976). Appl. Ent. Zool. 11, 363–364.

Katagiri, K. and Takamura, N. (1966). J. Jap. For. Soc. 48, 209–212.

Katagiri, K., Iwata, Z., Kushida, T., Fukuizumi, Yasu and Ishizuka, H. (1977). J. Jap. For. Soc. 59, 442–448.

Koyama, R. and Iwata, Z. (1963). Trans. 74th Meet. Jap. For. Soc. 350–352.

Koyama, R. and Katagiri, K. (1967). Proc. Joint US-Japan Seminar on Microbial Control of Insect Pests, Fukuoka, 63–69.

Koyama, R. and Kushida, T. (1963). Trans. 74th Meet. Jap. For. Soc. 352–354.

Lipa, J.J., Slizynski, K., Ziemnicka, J. and Bartkowski, J. (1975a). *In* "Environmental Quality and Safety. Supplement Volume III" (F. Coulston and F. Korte, eds), 668–671. Georg Thieme, Stuttgart.

Lipa, J.J., Slizynski, K., Ziemnicka, J. and Bartkowski, J. (1975b). *Proc. VIII Int. Plant. Prot. Congr.,* Moscow, 129–137.

Magnoler, A. (1970). *Entomophaga* 15, 407–412.

Magnoler, A. (1974). *J. Invertebr. Path.* 23, 263–274.

Maleki-Milani, H. (1970). *Entomophaga* 15, 315–325.

Martignoni, M.E. and Iwai, P.J. (1977). USDA For. Serv. Gen. Tech. Rep. PNW-40, 28 pp.

Matsu-ura, K. and Akabori, F. (1969). Unpubl. document submitted to Min. Agric. For., Japan, for registration of CPV-insecticide.

Oatman, E.R., Hall, I.M., Arakawa, K.Y., Planner, G.R., Bascom, L.A. and Beagle, C.C. (1970). *J. econ. Ent.* 63, 415–421.

Okada, T., Matsutani, S., Imamura, K., Yoshida, K. and Tanaka, N. (1974). *Bull. Agr. Chem. Insp. Stn* No. 14, 97–102.

Sikorowski, P.P., Andrews, G.L. and Broome, J.R. (1973). *J. Invertebr. Path.* 21, 41–45.

Simmons, C.L. and Sikorowski, P.P. (1973). *J. Invertebr. Path.* 22, 369–371.

Slizynski, K. and Lipa, J.J. (1975). *Proc. VIII Int. Pl. Prot. Congr.,* V. Moscow, 167–168.

Smith, K.M. (1976). *In* "Virus-Insect Relationships". Longman, New York, 291 pp.

Smith, K.M. and Wickoff, R.W.G. (1950). *Nature Lond.* 166, 861.

Stelzer, M.J. (1965). *J. Invertebr. Path.* 7, 122–125.

Tanada, Y. and Omi, E.M. (1974). *J. Invertebr. Path.* 23, 360–365.

Tanaka, N., Naiki, M. and Mitsui, T. (1967). *Ann. Meet. Jap. Appl. Zool. Ent. Soc.* (Abstr.), 36–37.

Vail, P.V. and Gough, D. (1970). *J. Invertebr. Path.* 15, 397–400.

Watanabe, H. (1968). *J. Sericult. Sci. Japan* 37, 385–389.

Yadava, R.L. (1970). *Z. Angew. Ent.* 65, 175–183.

Yamaguchi, K. (1976). *J. Sericult. Sci. Japan* 45, 60–65.

Ying, S.-L. (1970). *Quart. J. Chin. For.* 4, 51–68.

CHAPTER 23

Toxins of Entomopathogenic Fungi

D.W. ROBERTS

Boyce Thompson Institute for Plant Research, Tower Road, Cornell University, Ithaca, New York, USA

I. Introduction

Many entomopathogenic fungi overcome their hosts after only limited growth in the haemocoel, so toxins are presumed to cause host death. The importance of toxins to the virulence of an entomopathogenic fungus is difficult to evaluate, however, because toxin production in the host must be preceded by several activities of the fungus. With most fungi disease development can be divided into nine steps: (1) Attachment of the infective unit (e.g. conidium or zoospore) to the insect epicuticle. (2) Germination of the infective unit on the cuticle. (3) Penetration of the cuticle, either directly by germ tubes or by infection pegs from appressoria. (4) Multiplication in the yeast phase (hyphal bodies) in the haemocoel. (5) Production of toxic metabolites. (6) Death of the host. (7) Growth in the mycelial phase with invasion of virtually all host organs. (8) Penetration of hyphae from the interior through the cuticle to the exterior of the insect. (9) Production of infective units on the exterior of the insect. Obviously fungi which fail to complete steps 1–4 will have low virulence regardless of high toxin biosynthetic capability.

There are four principal objectives of studies on fungal metabolites toxic to insects: (a) to elucidate the mode of action of fungi pathogenic to insects, (b) to search for new chemicals for insect control, (c) to evaluate the safety of specific fungi proposed for use in pest control, or (d) to conduct basic natural-products chemistry studies. In addition to other uses, the information gained in these studies can be helpful in selecting strains of entomopathogenic fungi for insect control. Thus, given equal capacity to penetrate the host's integument, highly toxigenic fungal isolates kill their hosts more quickly, and so protect plants sooner than isolates without toxin. Programs to identify toxins and to select isolates with high toxin production eventually will help provide superior fungal strains for use in insect control. At present, there are no fungal metabolites under commercial development as insecticides (Ciegler, 1977).

Insects may be exposed to toxic fungal metabolites in three ways: (a) *per os* (ingestion), (b) cuticular contact, or (c) development within or injection into the haemocoel. *Per os* and contact exposure occurs in nature when insects eat food, such as grains, in which toxins have been produced by saprobic growth of fungi. Haemocoelic exposure is by elaboration of toxins in the haemocoel by fungi after infection of the insect. In laboratory studies, the mode of exposure must be selected which coincides with the objectives of the investigation. If the purpose is to discover insecticides, then *per os* and contact exposure should be utilized. However, if the purpose is to elucidate the mode of action of insect-invading fungi, then the only valid tests are those conducted by intra-haemocoelic injection.

The terminology associated with toxic compounds is complicated and, on the surface, illogical. The initial scientific definition of "toxin", proposed about 1888 by Ludwig Brieger, designated it for poisonous substances produced by pathogenic organisms (Skinner, 1949; Wain, 1958). Mammalian bacteriology has restricted usage of "toxin" to antigenic compounds, based on that field's early experience with bacterial exo- and endotoxins. Plant pathologists, however, have tended to use "toxin" for non-enzymatic, low-molecular-weight (non-antigenic) products of microorganisms, or microorganism–host inter-actions, harmful to plants in low concentrations (Rudolph, 1976). This definition, with "arthropod" in place of "plant", is used in this chapter.

The prefixes "myco-", "phyto-", "patho-", and "vivo"- are used with "toxin" by various groups, but the definitions are based more on historical events than logical semantics. For example, "mycotoxins" usually is used for saprophytically produced fungal compounds toxic to vertebrates by ingestion; whereas "phytotoxins" are microbial compounds toxic to plants but not of primary importance during pathogenesis. "Pathotoxin" is a plant pathology term referring to a host-specific toxin which, in reasonable concentration, induces all the typical disease symptoms and its production is correlated with pathogenicity. Only ten pathotoxins are known, all associated with widespread introductions of new, susceptible plant genotypes (Scheffer, 1976). "Vivotoxin", also a plant pathology term, may be useful in insect pathology. It is defined as "a substance produced in the infected host by the pathogen and/or its host, which functions in the production of disease, but is not itself the initial inciting agent of disease" (Dimond and Waggoner, 1953). As pointed out by Rudolf (1976), the term should be applied only to compounds produced during the disease development stages up to and including death, and not to compounds produced afterwards. One advantage of the vivotoxin concept is that it essentially requires the application of Koch's postulates to toxins. Unfortunately, isolation of toxins from diseased plants has proved difficult to impossible with some diseases because of toxin instability, irreversible binding to the host, or because they affect the host at very low concentrations. As will be discussed later, virtually no searches have been conducted for toxins in fungus-infected insects. At present, only destruxin B and desmethyldestruxin B produced by *Metarhizium anisopliae* in silkworm larvae (Suzuki *et al.*, 1971) can be referred to as vivotoxins in insect pathology.

In addition to low-molecular-weight compounds, proteases and other enzymes are produced by entomopathogenic fungi. Injection of filtrates of *Entomophthora* spp. cultures into *Galleria mellonella* larvae caused blackening similar to that noted in infected larvae (Yendol *et al.*, 1968; Prasertphon and Tanada, 1969). Preliminary chemical studies indicated that the active components were proteases. Two proteases from *Metarhizium anisopliae* cultures also blackened *Galleria* larvae (Kucera, 1979). A combination of fungal protease(s), lipase(s), and chitinase(s) was necessary for *in vitro* digestion of excised insect cuticle (Samsinakova and Misikova, 1973). The existence of a positive correlation between high enzyme production *in vitro* and high virulence for insects has been suggested, with emphasis on chitinases and proteases (e.g. Samsinakova and Misikova, 1973; Samsinakova *et al.*, 1979; Pavljushin and Evlakhova, 1978); but there are fungal isolates for which correlation is impossible (Paris and Segretain, 1975). Since the most important function of such enzymes may be to aid penetration of insect cuticle, probably virulence is more likely to be correlated *in vitro* with enzyme production by germinating conidia than by mycelium. Enzymes, particularly proteases, presumably are important also for rapid fungal growth after death of the host. Enzymes of entomopathogenic fungi are sometimes referred to as toxins (e.g., Kucera and Samsinakova, 1968; Yendol *et al.*, 1968; Prasertphon and Tanada, 1969; Lysenko and Kucera, 1971; Kucera, 1971; Leopold *et al.*, 1973; Samsinakova *et al.*, 1976). Even so, as mentioned above, enzymes will not be defined as toxins in this chapter and will not be discussed further.

The following account will emphasize toxins produced by fungi able to invade the haemocoel of insects. In general, the lower entomopathogenic fungi (e.g. *Coelomycidium, Coelomomyces, Lagenidium,* and *Entomophthora*) appear to overcome susceptible hosts primarily by utilization of the available nutrients in the haemocoel, rather than by low-molecular-weight toxins. However, *E. virulenta* produces *in vitro* an azoxybenzenoid compound toxic to insects (Claydon and Grove, 1978). Since most entomopathogenic higher fungi are Deuteromycetes (Roberts and Yendol, 1971), that group will be emphasized.

II. Fungal Genera

A. ENTOMOPATHOGENIC FUNGI

Certain fungal genera occur in nature primarily as pathogens of insects. Most can be cultured on artificial media, and for some the medium, or extracts of medium and/or mycelium, have been examined for toxicity to insects. For example, in *per os* tests of 6th day culture filtrates of a wide range of Deuteromycete pathogens of silkworms, only *Aspergillus ochraceus* and *Sterigmatocystis japonica* produced toxic substances *in vitro* (Kodaira, 1961). Intrahaemocoelic injection of extracts from fungus-killed silkworm larvae into healthy larvae indicated that three entomopathogenic fungi, *Beauveria bassiana, Metarhizium anisopliae* and *Aspergillus ochraceus,* produced significant amounts of toxic compounds within their hosts. Although the principal active compound(s) is known for some such fungi, this is not so for most. Some compounds of unknown biological activity during disease development have been characterized. Very few of the many strains of *Aspergillus* infect insects, and toxins produced by these will be discussed here, but the non-pathogenic isolates will be discussed in Section II, B on facultative parasites and saprophytes.

1. *Beauveria*

Species of the genus *Beauveria* are often isolated from diseased insects and frequently used in microbial control tests. Toxin production has been partially examined for the two commonest species, *B. bassiana* and *B. brongniartii* (= *tenella*). From descriptions of symptoms, some strains obviously produce toxic compounds which rapidly debilitate the host after invasion of the haemolymph. *In vitro* studies have indicated many unidentified toxic compounds (e.g., West and Briggs, 1968; Velitskaya, 1973; Wojciechowska, 1973; Pavliushin, 1976). Since there are many isolates of these fungi in culture and only a few of the toxin studies have reported purified and chemically characterized active compounds, probably new toxins will be found in fungi of this genus.

a. *Beauvericin.* The compound from *B. bassiana* which has had the most research attention is the depsipeptide beauvericin (Fig. 1). Related chemically to the enniatins, it comprises a cyclic repeating sequence of three molecules of N-methyl phenylalanine alternating with three molecules of 2-hydroxyisovaleric acid (Hamill *et al.*, 1969). LD_{50} doses (doses required to kill 50% of test insects) have not been published, but it is somewhat toxic to mosquito larvae, brine shrimp (*Artemia salina*), bacteria (Hamill *et al.*, 1969), housefly adults (Roberts, unpublished observations), and cockroach cardiac cells *in vitro* (Vey *et al.*, 1973). Beauvericin did not affect silkworm larvae at 1000 ppm in their artificial diet nor at $100 \, \mu g/1.2$-g larva by injection (Kanaoka *et al.*, 1978). The mode of action presumably involves the ionophoric characteristic of the molecule which permits cation transport through cell membranes, with the cation specificity being altered with changes in the anions present (Dorschner and Lardy, 1968; Estrada *et al.*, 1972; Bystrov *et al.*, 1972; Estrada, 1974; Prince *et al.*, 1974;

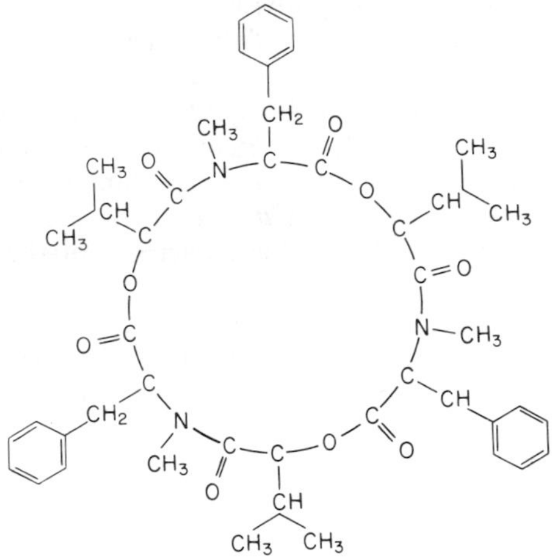

Fig. 1. Beauvericin.

Roeske *et al.*, 1974; Hamilton *et al.*, 1975; Geddes and Akrigg, 1976). Not all isolates of *B. bassiana* produce beauvericin *in vitro* (Frappier *et al.*, 1975). It has been isolated also from mycelium of *Paecilomyces fumosoroseus* (Bernardini *et al.*, 1975).

b. *Beauverolides*. Two very similar cyclotetradepsipeptides, beauverolides H and I, were isolated from mycelium of a *B. bassiana* from South Africa (Elsworth and Grove, 1974, 1977). Mixtures of H and I containing primarily H were isolated in France from *B. bassiana* and *B. brongniartii* and named beauvellide (Frappier *et al.*, 1975). This name has been discarded (Frappier *et al.*, 1978). The 1975 study established that a mixture of the compounds was toxic to cockroach cardiac cells *in vitro*, but the 1974 study found the mixture to be innocuous to mosquito (*Aedes*) and fly (*Calliphora*) larvae. The mode of administration to insects was not mentioned, but presumably it was by feeding.

c. *Bassianolide*. A cyclodepsipeptide, bassianolide, composed of four moles each of L-N-methyl leucine and D-α-hydroxyisovaleric acid has been isolated from mycelium of *B. bassiana* and *Verticillium lecanii* (Suzuki *et al.*, 1977). Both fungal isolates were obtained from dead silkworm (*Bombyx mori*) larvae. Bassianolide killed instar V silkworm at 13 ppm by feeding in an artificial diet, and at 5 μg/1.2-g larva by injection.

d. *Isarolides*. Isarolides A, B, and C, also cyclodepsipeptides, were found (Briggs *et al.*, 1966) in a *B. brongniartii* from New Zealand misidentified as *Isaria* sp. (R.A. Samson, personal communication). These, or very similar compounds were found in the mycelium of both *B. bassiana* and *B. brongniartii* in France (Frappier *et al.*, 1975). Unfortunately, toxicity tests were not reported.

e. *Pigments*. Two very similar yellow pigments, tenellin and bassianin, produced *in vitro* by some strains of both *B. bassiana* and *B. brongniartii* (= *tenella*), are concentrated in the mycelium rather than released into the medium. Their structure and biosynthesis have been studied extensively (El Basyouni *et al.*, 1968; McInnes *et al.*, 1974a, b; Leete *et al.*, 1975; Wright *et al.*, 1977; Wat *et al.*, 1977). Tenellin is 3-(4, 6-dimethyl-*E, E*-octa-2, 4-dienoyl)-1, 4-dihydroxy-5-(p-hydroxyphenyl)-2(1*H*)-pyridone and bassianin is 3-(6, 8-dimethyl-*E, E, E*-deca-2, 4, 6,-trienoyl)-1, 4-dihydroxy-5-(p-hydroxy phenyl)-2(1*H*)-pyridone. They represent a new family of fungal biochromes. Ilicicolin, a related compound from *Cylindrocladium ilicicola* (Wat *et al.*, 1977), is an antifungal antibiotic, but the biological activity of the *Beauveria* pigments is unknown.

The dibenzoquinone pigment oosporein, produced by many isolates of *Beauveria* (El Basyouni *et al.*, 1968; Vining *et al.*, 1962; El Basyouni and Vining, 1966; Zajic, 1963), probably accounts for the reddish colour of infected caterpillars. Production of this pigment can be lost after repeated transfer of isolates on artificial media. It is released into the medium, colouring the reverse of agar plates a deep red. Oosporein from *Chaetomium trilaterale* isolated from mouldy peanuts had an LD_{50} of 6.12 mg/kg in day-old cockerels (Cole *et al.*, 1974), but its toxicity to insects is unknown.

f. *Oxalic Acid.* Approximately 20% of the original solids in a peptone medium were found to be converted to oxalic acid (primarily ammonium oxalate) by an isolate of *B. brongniartii* (Cordon and Schwartz, 1962). The same isolate did not produce oxalic acid on glucose medium. This acid is a general poison and its role in the mode of action in certain plant pathogenic fungi is well documented (Kritzman *et al.*, 1977). Oxalate crystals have been noted on the surface of insects killed by *B. bassiana* (Kodaira, 1961; Müller-Kögler, 1965). Possibly oxalic acid is an important toxin in the haemolymph of *Beauveria*-infected insects.

The yield in peptide-rich medium can be high, viz. ca. 1 mg/ml with two isolates after 9 days in submerged culture (Roberts, W. McCarthy, and P.R. Hughes, unpublished observations). It was estimated by $KMnO_4$ titration of CaCl precipitates from culture filtrates previously extracted with CCl_4 and ethyl acetate. The identity of the precipitate was verified by gas chromatography. Production was much reduced when the carbohydrate (glucose) was reduced by half in media containing Bacto-Peptone (Difco) in place of Neopeptone (Difco). Similar effects of nitrogen sources on oxalate production were noted by Kodaira (1961). Although insect haemolymph presumably is sufficiently rich to support a high yield of oxalate, oxalic acid was not detected in blood from silkworm larvae within one hour of death from *B. bassiana* infection (Kodaira, 1961). The infections were initiated by injection of conidia into the haemolymph. The detection was by paper chromatography of ether extracts of acidified haemolymph. It is not known if the method was sensitive enough to detect the minimum amount of oxalate needed to cause serious physiological derangement. The test was positive, however, with samples taken from larvae held at room temperature for 24 h after death.

g. *Other Compounds.* A gas chromatographic "aroma analysis" of the volatiles over a *B. bassiana* culture yielded acetone, methylacetate, methanol, ethylacetate, ethanol, 6-methyl-5-hepten-2-on, and 6-methyl-5-hepten-2-ol

(Binder, 1966). The existence, and amounts if present, of these compounds in diseased insects is unknown. Houseflies were paralysed in 30 min by methylheptenon in petri dish tests (Schaerffenberg and Winkler, 1969). An unidentified, methanol-extractable contact poison(s) was obtained from mycelium, but not culture filtrates. Dresner (1949, 1950) claimed houseflies and other insects died within hours of exposure to *B. bassiana* conidia formulated in wheat flour. The toxicity was lost at < 100% relative humidity. Steinhaus and Bell (1953), however, could not reproduce Dresner's results. Crystals of oxalic acid or ammonium oxalate on moist filter paper, although killing flies in 1–2 days at high doses (up to 100 mg/petri dish) did not cause rapid knockdown (Roberts, unpublished observations). Beauvericin (up to 2 mg/petri dish) tested in the same way also did not duplicate Dresner's results, nor did combinations of oxalic acid and beauvericin.

2. *Metarhizium*

Thin-section electron microscopic studies of elaterid larvae infected by *Metarhizium anisopliae* revealed many ultrastructural alterations in cells not invaded by the fungus (Zacharuk, 1971, 1973, 1974). The changes presumably were induced by fungal toxins, but these were not identified. In general, membranous sites of energy production (e.g. mitochondria) and synthesis (e.g. rough endoplasmic reticulum) were disrupted. Filtrates of *M. anisopliae* cultures were toxic to coleopterous (*Oryctes rhinoceros*) haemocytes *in vitro*, producing organelle changes similar to those described by Zacharuk *in vivo* (Vey and Quiot, 1975). Culture filtrates were toxic to wax moth larvae (*Galleria mellonella*) by injection (Roberts, 1966a) and mycelium extracts toxic by contact to housefly adults (Schaerffenberg and Winkler, 1969). Heat-treated blood from infected moribund silkworm (*Bombyx mori*) larvae paralysed *Galleria* larvae on intrahaemocoelic injection, indicating toxins were present in moribund silkworms (Roberts, 1966b). Also, solvent extracts of *Metarhizium*-killed silkworms were toxic to healthy silkworms by intrahaemocoelic injection (Kodaira, 1961).

a. *Destruxins.* Six cyclodepsipeptides with the same five-member amino acid skeleton of β-alanine, alanine, valine, isoleucine, and proline have been isolated from filtrates of *M. anisopliae* cultures (Fig. 2) (Suzuki *et al.*, 1970; Suzuki and Tamura, 1972). The apparent biosynthetic precursor is protodestruxin (PD), a compound without N-methylation of any amino acids and harmless to silkworms (Lee *et al.*, 1975). (Suzuki and Tamura, 1972, incorrectly reported silkworm toxicity in their report on the discovery of PD.) N-methylation of alanine converts protodestruxin into desmethyldestruxin B (DMDB) and fixes the molecule in a specific configuration (Naganawa *et al.*, 1976). This, and the following four destruxins, apparently due to the fixed configuration, are toxic to silkworm larvae on intrahaemocoelic injection. N-methylation of valine in addition to alanine, affords destruxin B (DB). Destruxins C (DC) and D (DD) differ from destruxin B in that the residue on the ester is a hydroxy acid in C and a carboxyl in D. DA differs from DB by the insertion of a double bond in the ester residue with the loss of a methyl group. The major destruxin present in young (1-week-old) cultures was DB, but DA and DB predominated in old (3-week-old) cultures (Suzuki and Tamura, 1971; Tamura and Suzuki, 1976). Both DA and DB were synthesized in a medium with inorganic nitrogen only (Roberts, unpublished observations). Isolates from Europe, Asia, and America produced destruxins (Roberts, 1969 and unpublished observations).

	R_1		R_2	R_3
PD	$-CH_2-CH$ $\begin{smallmatrix}CH_3\\CH_3\end{smallmatrix}$		$-H$	$-H$
DMDB	$-CH_2-CH$ $\begin{smallmatrix}CH_3\\CH_3\end{smallmatrix}$		$-H$	$-CH_3$
DB	$-CH_2-CH$ $\begin{smallmatrix}CH_3\\CH_3\end{smallmatrix}$		$-CH_3$	$-CH_3$
DC	$-CH_2-CH$ $\begin{smallmatrix}CH_3\\CH_2OH\end{smallmatrix}$		$-CH_3$	$-CH_3$
DD	$-CH_2-CH$ $\begin{smallmatrix}CH_3\\COOH\end{smallmatrix}$		$-CH_3$	$-CH_3$
DA	$-CH_2-CH=CH_2$		$-CH_3$	$-CH_3$

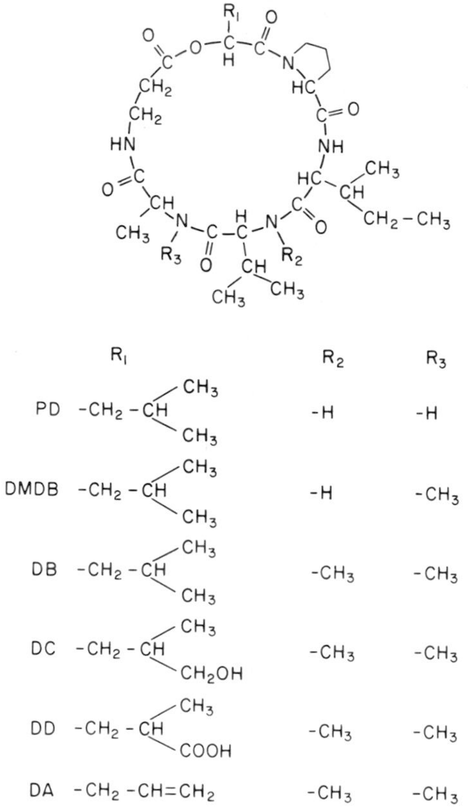

Fig. 2. The destruxins. Destruxin A = DA, destruxin B = DB, destruxin C = DC, destruxin D = DD, desmethyldestruxin B = DMDB, and proto-destruxin = PD.

Insect species vary considerably in susceptibility to destruxins. The LD_{50} in silkworm larvae for either DA or DB administered by intrahaemocoelic injection was 0.15–0.30 µg/g at 24 h (Kodaira, 1961; Tamura and Takahashi, 1970; Suzuki and Tamura, 1971). *Galleria* larvae were ca. 300 and 500 times less susceptible to DA and DB, respectively (Roberts, unpublished observations). The LD_{50} of DA for phasmid (*Carausius morosus*) immatures was about the same as that for *Galleria* (Roberts, unpublished observations). The ED_{50}, based on immediate tetanic paralysis, was $0.015-0.030$ $\mu g/g$ in silkworm larvae for both DA and DB (Kodaira, 1961), whereas with *Galleria* larvae DA was 10–30 times less active and DB 5–10 times less active (Roberts, unpublished observations). The LT_{50} values [time to 50% mortality with constant exposure to 50 ppm of DA + DB (70% DA)] were 6.75, 8.5, 3.5, and 2 days for larvae of the mosquito species *Anopheles stephensi, Aedes aegypti, Ac. epactius* (= *atropalpus*), and *Culex pipiens,* respectively (Roberts, 1970). The LC_{50} values (in 5 days) were 10–100 ppm. Mixtures of DA and DB in various ratios were not synergistic in *Galleria* (Roberts, unpublished observations).

Insect species also vary in symptoms caused by injected destruxins. The most striking symptom is an immediate tetanus induced in most lepidopterous larvae. With low doses the tetanus developed over a period of up to 3 min. With high doses tetanic paralysis was very brief or absent and a flaccid paralysis occurred. In *Galleria* larvae tetanus seemed to result from direct action on muscles without intervention of nerves. Muscles with neuromuscular junctions blocked with *Bracon hebetor* venom or with 5-hydroxytryptamine, or with the appropriate central-nervous-system ganglion removed 7 days before testing still responded to injected destruxins with tetanus. In *Galleria* tetanus could be very brief (10 min) with low doses. After recovery larvae responded to further injections with susceptibility remaining unchanged with six consecutive injections at 30-min intervals (Roberts, unpublished observations). Orthoptera and Coleoptera, although susceptible, did not respond to destruxins (DA + DB) with overall tetanus.

Destruxin solutions on leaves were phagodepressants for larvae of the potato lady beetle (*Epilachna sparsa*) (Kodaira, 1961) and immature *C. morosus* (Roberts, unpublished observations). It is not clear whether death was due to starvation or to destruxins ingested *per os*.

The LD_{50} in mice, one hour after intraperitoneal injection of DA containing benzene from crystalization was $1.0-1.35$ mg/kg, and of DB was $13.2-16.9$ mg/kg (Kodaira, 1961). The LD_{50} of DA without benzene was similar to that of DB (N. Takahashi, personal communication). DA and DB have no antibacterial or antifungal activity (Tamura and Takahashi, 1970; S. Naef-Roth, personal communication).

DB was detected, but not quantified, by mass spectrometry in extracts of fungus-exposed but still living mosquito larvae (Roberts, 1970). Levels of destruxin production *in vitro* correlated directly with virulence for mosquito larvae with several mutants of *M. anisopliae* (Al-Aidroos and Roberts, 1978). In the only study of its kind, Suzuki *et al.* (1971) isolated DB and DMDB from *M. anisopliae*-infected silkworm larvae and estimated the total amount of destruxins in the larvae by β-alanine analysis of positive fractions. The larvae were infected by topical application of conidia. Cuticular spots and hyphal bodies in the haemolymph were noted on day 4 post-infection and the majority died on day 5. Living larvae contained 0.0004 μmole destruxin/larva on day 4 and 0.0006 on day 5, approaching the amount of destruxin that causes paralysis and death of silkworm larvae on injection of pure DB

(Kodaira, 1961). Analysis of larva which died and were frozen "immediately" indicated 0.02 μ mole destruxins/larvae. This is more than enough to cause death, and indicates that considerable saprobic fungal growth may have occurred in a few of these specimens before freezing. Also, tetanic paralysis, similar to that induced by injection of destruxins, occurred in silkworms immediately before death from *M. anisopliae* infection (Roberts, 1966b). These experiments establish destruxins as the probable cause of death of silkworms infected by at least two strains of *M. anisopliae*, and therefore DB and DMDB can be designated as vivotoxins.

b. *Cytochalasins.* Cytochalasins are fungal metabolites with shared structural features (perhydro-indoles having a macrocyclic ring). Their biosynthetic origin probably is from phenylalanine or tryptophan linked to a C_{16} or C_{18} polyketide chain. Their biological activities include inhibition of cytoplasmic cleavage in cultured mammalian cells producing polynucleate cells, inhibition of cell movement, and nuclear extrusion (Steyn, 1977). The major interest in these compounds at present is in their use as tools in cell biology and medicine. There are 10 known cytochalasins produced by a wide variety of fungi (Natori, 1977). Two, cytochalasins C and D, were isolated from cultures of *Metarhizium anisopliae* (Aldridge and Turner, 1969). Yields, especially of D, were low. However, zygosporin A, a major metabolite of *Zygosporium masonii*, identical to cytochalasin D, afforded sufficient material to determine its mammalian toxicity. Acute toxicities (LD_{50}) in mice were 1.85 mg/kg by subcutaneous injection, 36 mg/kg *per os* (Hayakawa *et al.*, 1968), and 10 mg/kg by intraperitoneal injection (Minato *et al.*, 1973). Injection of cytochalasin B into *Galleria* larvae at 150 μg/g had no discernable immediate or long term effects (Roberts, unpublished observations). These compounds may affect other insect species, but tests have not been conducted. If produced in sufficient quantities in *M. anisopliae*-infected insects, the cytochalasins might assist in disease development, e.g. inhibition of cell movement would reduce phagocytic activity and granuloma formation.

3. *Nomuraea*

Gypsy moth larvae (*Lymantria dispar*) succumbed very rapidly after exposure to large numbers of *Nomuraea rileyi* conidia: 50% and 100% mortality in 24 and 48 h, respectively (Wasti and Hartman, 1978). Conidial germ tubes penetrated the cuticle between 17 and 28 h after exposure. Unidentified compounds were extracted from mycelium produced in submerged culture. The extract on injection into *L. dispar* larvae caused no mortality in 24 h, 47% in 3 days, 60% in 7, and 63% in 20 days. Topical application also caused no mortality in 24 h, but 37% in 3 days, 40% in 7, 80% in 11, and 83% in 20 days. Topical application to *Galleria* killed 50% in 20 days. Toxic action from topical application is unusual with the known toxins from entomopathogenic fungi. It is possible that the active compound(s) in these extracts was not the same as that which rapidly killed the fungus-infected *L. dispar*. Only the mycelium, and not the culture filtrate, was examined; and the mycelial extract was partially purified by silica gel chromatography before toxicity was evaluated.

Injection of filtrates of haemolymph from *N. rileyi* (= *Spicaria prasina*)-infected silkworm larvae into healthy silkworm larvae caused precocious pupation (Mikuni and Kawakami, 1975). Injections of culture filtrates had the opposite effect, viz. pupation of instar V larvae was inhibited, as was emergence of adults from pupae. The active compounds were not identified. Culture filtrates were detoxified by heat (50°C, 10 min) or by fairly nonpolar solvents but not by polar solvents or pH 1.7 and 9.7. It is not known if the active compound in infected haemolymph was produced by the fungus, the host, or by the combined action of both.

4. *Aspergillus*

Aspergillus spp. are among the most thoroughly studied organisms for mycotoxins produced during saprobic growth. A small minority of strains from a few *Aspergillus* spp. can infect insects. Culture filtrates of an *A. flavus* isolated from a moribund mosquito larva contained two chloroform-extractable compounds toxic to *Culex* larvae by addition to their water (Toscano and Reeves, 1973). The compounds were heat stable (121°C, 15 min). Although not identified, it was established that they were not aflatoxins B_1, B_2, G_1 or G_2. Unidentified heat-stable compounds toxic to the beetle *Epilachna vigintioctopunctata* were present in culture filtrates of an *A. flavus* pathogenic to this insect (Krishnamoorthy *et al.*, 1972). The filtrates were strongly phagodepressant when sprayed on leaves. An isolate of *A. flavus* pathogenic to honey bees (*Apis mellifera*) produced unidentified substances *in vitro* which were toxic to honey bee adults *per os* (Burnside, 1930).

a. *Aflatoxins.* Aflatoxins are the most extensively studied mycotoxins (Detroy *et al.*, 1971; Rodricks *et al.*, 1977). They are produced on human and animal food products by saprobic growth of fungi, particularly fungi of the *A. flavus* group. Fifteen silkworm-pathogenic strains of this fungus produced aflatoxins *in vitro* and in fungus-killed larvae (Murakoshi *et al.*, 1971, 1977; Ohtomo *et al.*, 1975). Aflatoxins B_1, B_2, G_1, and G_2 were sought in living, infected silkworm larvae 3 days after exposure to conidia, which was one day before death (day 4) (Murakoshi *et al.*, 1977). Two strains were used. All four aflatoxins were found in larvae infected with one strain, and none in larvae infected with the other strain. It is not known if the levels detected (B_1 = 0.027 µg/g, and G_1 = 0.005 µg/g) were sufficient to cause mortality.

b. *Aspochracin.* *Aspergillus ochraceus* produced substances toxic to silkworm larvae both *in vivo* and *in vitro* (Kodaira, 1961). The active compounds initially were incorrectly identified as amino acid anhydrides. More recently, a novel cyclotripeptide, aspochracin, composed of N-methyl-L-alanine, N-methyl-L-valine, L-ornithine and an octratrienoic acid side chain was isolated from culture filtrates (Myokei *et al.*, 1969). Aspochracin was less toxic to insects than destruxins A and B. The minimum intrahaemocoelic dose to cause immediate paralysis followed by death was 17 µg/g for last instar fall webworm larvae (*Hyphantria cunea*). Aspochracin was not phago-depressant. Silkworm larvae readily ate mulberry leaves coated with 50–200 µg of the compound, resulting in immediate knockdown and occasional death. Weak contact toxicity to silkworm eggs and instar I larvae was also noted. Mammalian toxicity was low. Intravenous injection into mice had no effect at 165 mg/kg.

c. *Asperentin*. A strain of *A. flavus* isolated from a pupa of *Galleria mellonella* did not produce aflatoxins *in vitro* (Grove, 1972), but culture filtrates were insecticidal. Analysis of these filtrates revealed a number of phenolic compounds. Asperentin, 3,4-dihydro-6,8-dihydroxy-3-(6-methyltetra-hydropyran-2-ylmethyl) isocoumarin, was a minor metabolic product, as were asperentin 6-O-methylether and three closely related phenolic compounds, including 5′-hydroxyasperentin. The major product was asperentin 8-O-methylether. Extending the culture period from 3–4 weeks to 5–6 weeks increased the amount of 5′-hydroxy-asperentin (Grove, 1973). Sussman (1952) suggested that phenolic compounds were involved in *A. flavus* disease development in *Cecropia* pupae. Asperentins are likely candidates.

5. *Verticillium*

Recent studies revealed that the mycelium of *Verticillium lecanii*, a common pathogen of homopterous insects, can contain a cyclodepsipeptide originally isolated from *B. bassiana:* bassianolide (Suzuki *et al.*, 1977; Kanaoka *et al.*, 1978; see section II, A, 1, c).

6. *Paecilomyces*

Pyridine-2,6-dicarboxylic acid was isolated from *Paecilomyces* (= *Isaria*) *farinosus* and *P.* (= *Isaria*) *fumosoroseus* (Shima, 1955). The cyclodepsipeptide beauvericin was isolated from the mycelium of *P. fumosoroseus* (Bernardini *et al.*, 1975; see section II, A, 1, a).

7. *Isaria*

Isaria felina (as *cretacea*) is frequently found associated with insects. A cyclopentadepsipeptide, isariin, has been isolated from mycelium of two strains of *I. felina* (Vining and Taber, 1962), one of which also produced partially characterized isariin-like compounds differing only in amino acid composition (Taber and Vining, 1963). The biological activities of these compounds are unknown.

8. *Fusarium*

Two strains of *Fusarium solani,* one pathogenic to bark beetles (*Scolytus scolytus*) and the other to lobsters (*Homarus americanus*), were produced in liquid medium substances toxic to adult blowflies (*Calliphora erythrocephala*) by intrahaemocoelic injection (Claydon *et al.*, 1977b). The insecticidal activity was fully accounted for by the naphthazarin pigments fusarubin (= hydroxy-javanicin) and anhydrofusarubin from both strains plus javanicin and fusaric acid (5-*n*-butylpyridine-2-carboxylic acid) (Claydon *et al.*, 1977a) from the lobster strain.

9. *Cordyceps*

A weak antibiotic, cordycepin, was isolated from *Cordyceps militaris* cultures after it was noted that infected insects were bacteria-free (Cunningham *et al.*, 1951); and the compound was identified as 3'-deoxyadenosine by Hansessian *et al.* (1966). Cordycepin is used in molecular biology to block RNA synthesis. The LD_{50} for last instar *Galleria* larvae at 21 days by intrahaemocoelic injection was ca. $30 \mu g/g$ (Roberts, unpublished observations). Sublethal doses did not affect egg production, but higher doses caused formation of pupae with tangled legs which did not become adults (Fig. 3). Culture filtrates of this fungus were toxic to mosquito larvae (*Culex pipiens, Aedes epactius* (= *atropalpus*), and *Ae. aegypti*) and to mosquito tissue culture cells (*Ae. albopictus*) (Belloncik and Parent, 1976; Belloncik and Gharbi-Said, 1977). It is not known whether cordycepin or other compounds caused the toxicity.

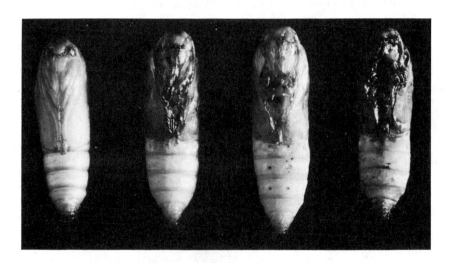

Fig. 3. *Galleria mellonella* pupae formed after intrahaemocoelic injection of cordycepin into last instar larvae. Dosages, from left to right, were 0, 37.5, 75, and $150 \mu g/g$.

10. *Entomophthora*

As mentioned previously, *Entomophthora* spp. culture filtrates or concentrates killed Lepidoptera and Diptera on intrahaemocoelic injection (Prasertphon, 1967; Yendol *et al.*, 1968; Prasertphon and Tanada, 1969). The potency, however, was low and partial characterization indicated the activity was due to enzyme(s). More recently, solvent extracts of *E. virulenta* cultures have yielded two compounds, azoxybenzene-4, 4'-dicarboxylic acid and 4'-hydroxymethylazoxybenzene-4-carboxylic acid (Claydon and Grove, 1978). The hydroxyacid was rather toxic to blowfly adults (*Calliphora erythrocephala*) by intrahaemocoelic injection, whereas the diacid was non-toxic (Claydon,

1978). The *E. virulenta* isolate used killed blowfly adults in 4 days or less with very little fungal development in the host. The presence of the hydroxyacid may have been the cause of death, but it has not been sought in moribund hosts as yet. This compound has not been previously identified from a natural source. It is particularly interesting because of its structural similarity to insecticides in the DDT group. (See also Chapter 28, Section III, B).

B. FACULTATIVE PARASITES AND SAPROBIC FUNGI

A very extensive literature exists on toxic fungal metabolites. These are primarily compounds produced on contaminated foodstuffs, particularly nuts and grains. They were isolated, identified and tested for toxicity to vertebrates as part of the very active field of mycotoxin research which is some 20 years old. Only those cases where metabolites from non-entomopathogenic fungal isolates were tested for toxicity to insects, usually by *per os* administration, will be discussed here. This topic was reviewed by Ciegler (1977).

1. *Aspergillus*

a. *Aflatoxins.* Thirteen aflatoxins are now known (Stoloff, 1977), but only four, B_1, B_2, G_1 and G_2, have been tested against insects. The older literature on the affects of aflatoxins on insects has been summarized by Lalor *et al.* (1976). *Drosophila melanogaster* has been used frequently and wild types were not equally susceptible to aflatoxin B_1 (AFB_1) incorporated into the larval medium (Llewellyn and Chinnici, 1978). Effects included reduced size (Lalor *et al.*, 1976; Chinnici *et al.*, 1976), mortality (Matsumura and Knight, 1967; Reiss, 1975; Lalor *et al.*, 1976), reduced egg viability (Matsumura and Knight, 1967; Chinnici *et al.*, 1976), extended developmental periods (Lalor *et al.*, 1976; Llewellyn and Chinnici, 1978), induction of recessive lethals (Lamb and Lilly, 1971), and suppression of cerebral neurosecretory cell activity (Kirk *et al.*, 1971). The amount of genetic crossing over was not increased by rearing female larvae on AFB_1-containing medium (Chinnici *et al.*, 1976). Other Diptera, house flies (*Musca domestica*) and yellow-fever mosquitoes (*Aedes aegypti*), also were susceptible to aflatoxins (Matsumura and Knight, 1967; Al-Adil *et al.*, 1972, 1973). AFG_1 was more toxic than AFB_1 (Al-Adil *et al.*, 1972), in contrast to mammalian systems where AFB_1 is the more toxic.

A mixture of AFB_1, AFB_2, AFG_1 and AFG_2 at 0.6 ppm in food of adult boll weevils, *Anthonomus grandis*, was lethal and at 0.06 ppm it was a chemosterilant (Moore *et al.*, 1978). A moth, *Heliothis virescens*, was susceptible to AFB_1 alone or to a mixture of the four major aflatoxins at 10 ppm in diet (Gudauskas *et al.*, 1967). Sensitivity decreased with larval age. Aflatoxins have been proposed but never verified as the cause of "autumn collapse" of honey bees (*Apis mellifera*) (Foote, 1966). In laboratory studies , AFB_1 and AFG_1 were produced on several substrates found in bee hives (pollen, brood comb, and dead larvae and adults) (Hilldrup *et al.*, 1977). Cockroach (*Periplaneta americana*) adults were not susceptible to AFB_1 at $12 \mu g/ml$ of diet (Llewellyn *et al.*, 1976). They did not accumulate AFB_1 in their bodies, and so *P. americana* was proposed as a model for investigating detoxification of AFB_1.

b. *Kojic Acid*. Kojic acid, 5-hydroxy-2-(hydroxymethyl)-4*H*-pyran-4-one, a well known toxic metabolite formed by several microorganisms, is best known from *Aspergillus* spp. (Wilson, 1971a). It inhibited development of *Drosophila* (Dobias and Nemec, 1975; Dobias *et al.*, 1977); had low acute toxicity to and retarded development of house flies (*Musca domestica*) and milkweed bugs (*Oncopeltus fasciatus*) (Beard and Walton, 1969); and it enhanced the effectiveness of nicotine bentonite (Mayer *et al.*, 1946). The compound, however, has not shown sufficient promise for development as an insecticide.

c. *Other Compounds*. The dihydroisocoumarin, ochratoxin A, was lethal to moth (*Plodia interpunctella*) and beetle larvae (*Tribolium confusum*) at 10 ppm (Sargent, 1975). It is produced by several *Aspergillus* spp. (Steyn, 1971). Nigragillin, isolated from *A. niger* culture filtrates, killed silkworm larvae at 80 ppm in artificial diet, and 5 µg/g by topical application caused immediate knockdown followed by occasional deaths (Isogai *et al.*, 1975). An unidentified brownish-red pigment from *A. flavipes* culture filtrates killed rice moth (*Corcyra cephalonica*) at 500 µg/ml in artificial diet (Jayaraman and Shanmugasundaram, 1972). A new, unnamed indole from *A. ruber* mycelium killed silkworm larvae at 1000 ppm in artificial diet (Nagasawa *et al.*, 1975). Water soluble compounds, extracted from *A. flavus* var *columnaris* culture medium, killed house fly larvae and milkweed bug, retarded growth of the beetle, *Tribolium confusum*, and were not toxic to earwigs, cockroaches, Lepidoptera and termites (Beard and Walton, 1971). The toxic compounds were not identified, but it was established that there were probably three present: two high molecular weight heat-labile substances and one heat-stable substance. Aflatoxins and kojic acid were not detected in the toxic extracts. A compound with an LD_{50} of 50 µg/ml of insect medium against the fly, *Lucilia sericata*, was produced *in vitro* by an isolate of *A. ochraceous* (Cole and Rolinson, 1972). Its identity is unknown, but it was not aspochracin. A metabolite of *A. versicolor*, versimide [methyl-α-(methylsuccinimido)acrylate], was slightly toxic to a wide range of adult insects by contact and vapour action (Cole and Rolinson, 1972), and all *D. melanogaster* adults were knocked down in 3 to 4 h at 5 µg/cm². It was inactive in a larval test against *L. sericata*. The mildly toxic compound cordycepin was isolated from the medium of *A. nidulans* cultures (Kaczka *et al.*, 1964; see Section II, A, 9).

Aspergillus tamarii, a wound parasite of pupae of the trepetid fly *Ceratitis capitata*, developed in integumentary wounds and produced unidentified toxins which diffused into the haemolymph, killing pupae before fungal invasion of the haemocoel (Vey *et al.*, 1967).

2. *Penicillium*

Many mycotoxins have been isolated from *Penicillium* spp. cultures (Ciegler *et al.*, 1971), but very few have been tested for toxicity to insects. Patulin and rubratoxins A and B were toxic to *Lucilia sericata* larvae and/or *Drosophila* adults (Cole and Rolinson, 1972; Reiss, 1975). Patulin had no activity in a *L. sericata* larval test, but did have weak toxicity to *Drosophila* in a 24-h knockdown test. A metabolite of *Penicillium* spp., griseofulvin, which inhibits growth of chitinous fungi and is used for chemotherapy of subcutaneous vertebrate fungal infections (Wilson, 1971b), affected mosquito larvae, *Aedes epactius* (= *atropalpus*) and *Ae. vexans* (Anderson, 1966). Rearing larvae in medium with 18.5 ppm or more griseofulvin caused gross anatomical changes in the cuticle, detachment of somatic muscles from the integument, and prolongation of the moulting cycle. The first two changes were apparent only after a moult following exposure to the chemical.

3. Fusarium and Myrothecium

Fusarium- and *Myrothecium*-contaminated grains affected growth of yellow mealworm (*Tenebrio molitor*) larvae (Davis *et al.*, 1975; Davis and Smith, 1977). Many fungal strains were used in this study. Insect growth usually was unaffected or retarded, but with a few strains it was promoted. Toxin production varied with culture conditions as well as fungal strain. The active compounds were not identified, but they probably were trichothecenes and zearalenone or related compounds.

a. Trichothecenes. The trichothecenes (12, 13-epoxytrichothecene family) are a very important group of mycotoxins from several genera of fungi, but primarily *Fusarium* and *Myrothecium* (Rodricks *et al.*, 1977). Kishaba *et al.* (1962) noted that two compounds, 379X and 379Y (roridin A and verrucarin A, respectively) produced by *M. roridum,* were phago-depressants and contact poisons for Mexican bean beetle (*Epilachna varivestis*) larvae and adults. Roridin A was less toxic than verrucarin A. These findings were confirmed by Grove and Hosken (1975) who tested 17 naturally occurring trichothecenes primarily from *Fusarium* spp. and 19 of their derivatives and transformation products for toxicity to mosquito (*A. aegypti*) larvae. Toxicity varied considerably, and structure:toxicity relationships were discussed. Of particular interest was trichodermin, which had moderately high toxicity to insects but low cytotoxicity to vertebrates and vertebrate cells as well as low dermatitic activity. Diacetoxyscirpenol, trichothecin, and trichothecolone were toxic to larval *L. sericata* in diet (Cole and Rolinson, 1972). Diacetoxyscirpenol was weakly toxic to larval and adult *Drosophila* (Reiss 1975). T-2 toxin was weakly toxic to the beetle *Tribolium confusum* at 100 ppm in flour, causing slow larval development, moderate larval mortality, increased fecundity, but reduced egg hatch (Wright *et al.*, 1976).

b. Zearalenone. Zearalenone (also known as F-2) from *Fusarium* spp. is the only phytoestrogen produced by a fungus (Mirocha *et al.*, 1977). After feeding to larvae of *T. confusum* and the lesser mealworm (*Alphitobius diaperinus*) it was detected during metamorphosis, and after starvation and death of the adults (Eugenio *et al.*, 1970). It did not harm *T. confusum*. The beetles, however, had a prolonged reproductive life, i.e. increased egg production with age (Wright *et al.*, 1976), in contrast to mammalian systems where zearalenone ingestion reduces fecundity.

4. *Alternaria*

Alternaria tenuis produced tenuazonic acid (α-acetyl-γ-sec-butyltetramic acid) *in vitro,* toxic to *Lucilia sericata* larvae but not to *Drosophila* adults (Cole and Rolinson, 1972). The LC_{50} at 48 h was 120 μg/ml of larval diet.

5. *Mucor*

Filtrates of 36-h-old cultures of the wound parasite *Mucor hiemalis* contained unidentified substances strongly toxic to primary cockroach (*Leucophaea maderae*) dorsal vessel cell cultures (Vey and Quiot, 1976).

6. *Tricholoma* and *Amanita*

The Basidiomycetes *Tricholoma muscarium, Amanita strobiliformis,* and *A. muscaria* have been used in Japan as toxicants to control flies (summarized by Huang and Shapiro, 1971). The active compounds were novel amino acids, in *T. muscarium* mushrooms tricholomic acid (α-amino-3-oxo-5-isoxazolidine-5-acetic acid) and in *Amanita* mushrooms, ibotenic acid (α-amino-3-oxo-5-isoxazoline-5-acetic acid). Both were highly toxic to adult house flies.

III Research Needs

It is apparent from the foregoing that information on insect-affecting fungal toxins, particularly those produced by entomopathogenic fungi, is far from complete. As mentioned previously, there are four major reasons for conducting research on insecticidal fungal metabolites. Some research areas which need investigation, some research approaches, and some caveats on each of the four topics are given below.

A. MODE OF ACTION

1. *Entomopathogenic fungi*

(a) The mode of administration of compounds to estimate their potential role in fungal disease must be by injection into the insect haemocoel – the site of elaboration during fungal infections. Contact and *per os* modes are appropriate only in searches for insecticidal compounds or to evaluate the effects of mouldy foodstuffs on their insect pests.

(b) Surveys are needed to detect, isolate and characterize toxins from entomopathogenic fungi. Very little has been done with most such fungi. Some fungal metabolites have been fully characterized, but not tested for toxicity. The toxicity of others has been tested, but with a very few insect species, sometimes not including the host from which the fungus was first isolated. On the other hand, toxicity has been detected in culture filtrates, extracts of fungus-killed insects, etc. with some fungi; but the active compounds have not been isolated and characterized.

(c) Test insect species for survey and evaluation of toxins should be selected with care. The effects of toxins, particularly the depsipeptides, on different insect species can be dissimilar, indicating that the modes of action or host detoxification mechanisms may differ. Species susceptible to the fungal strain under investigation are the most appropriate indicator animals.

(d) The modes of action of most toxic fungal metabolites are unknown. The mode of action (e.g., uncoupling of oxydative phosphorylation) of some toxins can be conjectured from studies in other systems, but almost none have been studied in insect systems. Actual verification in insects is needed.

(e) It is presumed that the various low-molecular-weight toxins produced by entomopathogenic fungi *in vitro* are also produced *in vivo*, but usually the presence of these compounds in moribund, fungus-infected insects in sufficient quantities to harm them has not been verified. Unequivocal classification of compounds which meet the criteria of vivotoxins is needed.

2. *Non-entomopathogenic fungi*

The suggestions in Section III, A, 1, except (e), pertain also to the toxins from non-entomopathogenic fungi.

B. INSECT CONTROL

(1) Entomopathogenic fungal strains able to rapidly produce biologically significant amounts of toxins should be sought in programs to find better fungi for insect control. Early production should be stressed, since rapid host mortality enhances plant protection. Genetic recombination should be used to transfer the genes for toxin production into strains which infect the host quickly.

(2) Very few metabolites of fungi, entomopathogenic or otherwise, have been tested for contact or *per os* toxicity to insects. An energetic project to search for metabolites with such toxicity, particularly if conducted with pest insect species as the test animals, may provide compounds which can be developed for insect control such as has been done with *Bacillus thuringiensis* endotoxin worldwide and exotoxin in the USSR.

C. SAFETY

Mammalian toxicity of most compounds produced by entomopathogenic fungi has been studied only superficially to date. Further investigations on this topic are needed, particularly for fungi considered for field use, as part of their overall safety evaluation. Vertebrate toxicology studies are needed before registration for each toxin proposed for use alone (without the living fungus) for insect control.

D. CHEMISTRY

Several of the above points clearly rely on competent chemical studies. More collaboration between chemists interested in identifying fungal metabolites and insect pathologists with skills in evaluating toxicity would expedite our understanding of fungal toxins.

IV. Conclusions

Toxins are presumed to be important in disease development of entomopathogenic fungi; and, if they have the capacity to invade the host, strains with significant early production of rapidly acting toxins probably will afford superior crop protection in microbial control projects. Numerous fungal metabolites produced by saprophytic growth of non-entomopathogenic fungi also are insecticidal. Unfortunately, their potency usually is low and many were known to be toxic to vertebrates before being tested against insects. Fungal metabolites probably would be readily biodegradable if used in field applications for pest control. There are, however, no fungal toxins under development at present for pest control. An extended search for such compounds with high potency against pest species, low toxicity to vertebrates, and low cost appears warranted.

References

Al-Adil, K.M., Kilgore, W.W. and Painter, R.R. (1972). *J. econ. Ent.* **65**, 375–378.

Al-Adil, K.M., Kilgore, W.W. and Painter, R.R. (1973). *Toxicol. Appl. Pharmacol.* **26**, 130–136.

Al-Aidroos, K. and Roberts, D.W. (1978). *Can. J. Genet. Cytol.* **20**, 211–219.

Aldridge, D.C. and Turner, W.B. (1969). *J. Chem. Soc.,* C. 923–928.

Anderson, J.F. (1966). *J. econ. Ent.* **59**, 1476–1482.

Beard, R.L. and Walton, G.S. (1969). *J. Invertebr. Path.* **14**, 53–59.

Beard, R.L. and Walton, G.S. (1971). Bull Conn. Agric. Exp. Stn., New Haven, No. 725. 26 pp.

Belloncik, S. and Gharbi-Said, R. (1977). *Entomophaga* **22**, 243–246.

Belloncik, S. and Parent, N. (1976). *Entomophaga* **21**, 343–347.

Bernardini, M., Carilli, A., Pacioni, G. and Santurbano, B. (1975). *Phytochemistry* **14**, 1865.

Binder, H. (1966). *J. Chromatog.* **25**, 189–197.

Briggs, L.H., Fergus, B.J. and Shannon, J.S. (1966). *Tetrahedron, Suppl.* **8**, Part 1, 269–278.

Burnside, C.E. (1930). U.S. Dep. Agric. Tech. Bull. No. 149, 1–42.

Bystrov, V.F., Ivanov, V.T., Kos'min, S.A., Mikhaleva, I.I., Khalilulina, K. Kh. and Ovchinnikov, Yu. A. (1972). *FEBS Lett.* **21**, 34–38.

Chinnici, J.P., Booker, M.A. and Llewellyn, G.C. (1976). *J. Invertebr. Path.* **27**, 255–258.

Ciegler, A. (1977). *In* "Biological Regulation of Vectors" (J.D. Briggs, ed.), pp. 135–150. U.S. Dep. Hlth. Education and Welf. Publ. No. (NIH) 77–1180, Washington, D.C.

Ciegler, A., Kadis, S. and Ajl, S.J. (eds.) (1971). "Microbial Toxins. Vol. VI. Fungal Toxins". Academic Press, New York and London.

Claydon, N. (1978). *J. Invertebr. Path.* **32**, 319–324.

Claydon, N. and Grove, J.F. (1978). *J. Chem. Soc. (Perkin Trans. I)* 1978, 171–173.

Claydon, N., Grove, J.F. and Pople, M. (1977a). *Phytochemistry* 16, 603.
Claydon, N., Grove, J.F. and Pople, M. (1977b). *J. Invertebr. Path.* 30, 216–223.
Cole, M. and Rolinson, G.N. (1972). *Appl. Microbiol.* 24, 660–662.
Cole, R.J., Kirksey, J.W., Cutler, H.G. and Davis, E.E. (1974). *J. Agric. Fd Chem.* 22, 517–520.
Cordon, T.C. and Schwartz, J.H. (1962). *Science* 138, 1265–1266.
Cunningham, K.G., Hutchinson, S.A., Manson, W. and Spring, F.S. (1951). *J. Chem. Soc.* 1951, 2299–2300.
Davis, G.R.F. and Smith, J.D. (1977). *J. Invertebr. Path.* 30, 325–329.
Davis, G.R.F., Smith, J.D., Schiefer, B. and Loew, F.M. (1975). *J. Invert. Path.* 26, 299–303.
Detroy, R.W., Lillehoj, E.B. and Ciegler, A. (1971). *In* "Microbial Toxins. Vol. VI. Fungal Toxins" (A. Ciegler, S. Kadis and S.J. Ajl, eds), pp. 3–178. Academic Press, New York and London.
Dimond, A.E. and Waggoner, P.E. (1953). *Phytopathology* 43, 229–235.
Dobias, J. and Nemec, P. (1975). *Biologia Bratis.* 30, 727–732.
Dobias, J., Nenec, P. and Brtko, J. (1977). *Biologia Bratis.* 32, 417–421.
Dorschner, E. and Lardy, H. (1968). *Antimicrob. Agents Chemother.* 1968, 11–14.
Dresner, E. (1949). *Contr. Boyce Thompson Inst.* 15, 319–335.
Dresner, E. (1950). *J. N.Y. ent. Soc.* 58, 269–278.
El Basyouni, S.H. and Vining, L.C. (1966). *Can. J. Biochem.* 44, 557–565.
El Basyouni, S.H., Brewer, D. and Vining, L.C. (1968). *Can. J. Bot.* 46, 441–448.
Elsworth, J.F. and Grove, J.F. (1974). *S. Afr. J. Sci.* 70, 379.
Elsworth, J.F. and Grove, J.F (1977). *J. Chem. Soc.* (Perkin Trans. I) 1977, 270–273.
Estrada-O.,S. (1974). *Fed. Proc.* 33, 1257.
Estrada-O.,S., Gomez-Lojero, C. and Montal, M. (1972). *J. Bioenerg.* 3, 417–428.
Eugenio, C., Casas, E. de Las, Harein, P.K. and Mirocha, C.J. (1970). *J. econ. Ent.* 63, 412–415.
Foote, H.L. (1966). *Am. Bee J.* 106, 126–127.
Frappier, F., Ferron, P. and Pais, M. (1975). *Phytochemistry* 14, 2703–2705.
Frappier, F., Pais, M., Elsworth, J.F. and Grove J.F. (1978). *Phytochemistry* 17, 545–546.
Geddes, A.J. and Akrigg, D. (1976). *Acta Crystallogr.*, Sect. B, 32, 3164–3171.
Grove, J.F. (1972). *J. Chem. Soc.* (Perkin Trans. I) 1972, 2400–2406.
Grove, J.F. (1973). *J. Chem. Soc.* (Perkin Trans. I) 1973, 2704–2706.
Grove, J.F. and Hosken, M. (1975). *Biochem. Pharmacol.* 24, 959–962.
Gudauskas, R.T., Davis, N.D. and Diener, U.L. (1967). *J. Invertebr. Path.* 9, 132–133.
Hamill, R.L., Higgens, C.E., Boaz, H.E. and Gorman, M. (1969). *Tetrahedron Lett.* 49, 4255–4258.

Hamilton, J.A., Steinrauf, L.K. and Braden, B. (1975). *Biochem. Biophys. Res. Commun.* **64**, 151–156.

Hanessian, S., DeJongh, D.C. and McCloskey, J.A. (1966). *Biochim. Biophys. Acta* **117**, 480–482.

Hayakawa, S., Matsushima, T., Kimura, T., Minato, H. and Katagiri, K. (1968). *J. Antibiot.* **21**, 523–524.

Hildrup, J.A.L., Eadie, T. and Llewellyn, G.C. (1977). *J. Ass. Off. Anal. Chem.* **60**, 96–99.

Huang, H.T. and Shapiro, M. (1971). *Prog. Ind. Microbiol.* **9**, 80–112.

Isogai, A., Horii, T., Suzuki, A., Murakoshi, S., Ikeda, K., Sato, S. and Tamura, S. (1975). *Agric. Biol. Chem.* **39**, 739–740.

Jayaraman, S. and Shanmugasundaram, E.R.B. (1972). *Indian J. Exp. Biol.* **10**, 399–400.

Kaczka, E.A., Dulaney, E.L., Gitterman, C.O., Woodruff, H.B. and Folkers, K. (1964). *Biochem. Biophys. Res. Commun.* **14**, 452–455.

Kanaoka, M., Isogai, A., Murakoshi, S., Ichinoe, M., Suzuki, A. and Tamura, S. (1978). *Agric. Biol. Chem.* **42**, 629–635.

Kirk, H.D., Ewen, A.B., Emson, H.E. and Blair, D.G.R. (1971). *J. Invertebr. Path.* **18**, 313–315.

Kishaba, A.N., Shankland, D.L., Curtis, R.W. and Wilson, M.C. (1962). *J. econ. Ent.* **55**, 211–214.

Kodaira, Y. (1961). *J. Fac. Text. Sci. Technol.*, Shinshu Univ., No. 29, Ser. E. (5), 1–68.

Krishnamoorthy, C., Sankar Naidu, M., Siva Rao, D.V. and Thippeswamy, M. (1972). *Proc. Indian Sci. Cong. Ass.* **59** (Part 3), 570–571.

Kritzman, G., Chet, I. and Henis, Y. (1977). *Expl. Mycol.* **1**, 280–285.

Kucera, M. (1971). *J. Invertebr. Path.* **17**, 211–215.

Kucera, M. (1979). Proc. Int. Colloq. Invertebr. Path., Prague, Sept. 1978. pp. 101–102.

Kucera, M. and Samsinakova, A. (1968). *J. Invertebr. Path.* **12**, 316–320.

Lalor, J.H., Chinnici, J.P. and Llewellyn, G.C. (1976). *Dev. Ind. Microbiol.* **17**, 443–449.

Lamb, M.J. and Lilly, L.J. (1971). *Mutat. Res.* **11**, 430–433.

Lee, S., Izumiya, N., Suzuki, A. and Tamura, S. (1975). *Tetrahedron Lett.* **11**, 883–886.

Leete, E., Kowanko, N., Newmark, R.A., Vining, L.C., McInnes, A.G. and Wright, J.L.C. (1975). *Tetrahedron Lett.* **47**, 4103–4106.

Leopold, J., Samsinakova, A. and Misikova, S. (1973). *Zentbl. Bakt. Parasitenk. Infektionskr. Hyg. Abt. II*, **128**, 31–41.

Llewellyn, G.C. and Chinnici, J.P. (1978). *J. Invertebr. Path.* **31**, 37–40.

Llewellyn, G.C., Sherertz, P.C. and Mills, R.R. (1976). *Bull. Environ. Contam. Toxicol.* **15**, 391–397.

Lysenko, O. and Kucera, M. (1971). *In* "Microbial Control of Insects and Mites" (H.D. Burges and N.W. Hussey, eds), pp. 205–227. Academic Press, London and New York.

McInnes, A.G., Smith, D.G. and Walter, J.A. (1974a). *J. Chem. Soc. Chem. Commun.* 1974 (8), 284–284.

McInnes, A.G., Smith, D.G., Wat, C.-K., Vining, L.C. and Wright, J.L.C. (1974b). *J. Chem. Soc. Chem. Commun.* 1974 (8), 281–282.

Matsumura, F. and Knight, S.G. (1967). *J. econ. Ent.* **60**, 871–872.

Mayer, E.L., Talley, F.B. and Woodward, C.F. (1946). "Nicotine Insecticides. Part II. Search for Activators". U.S. Dep. Agric., Bur. Ent. Plant Quar., F709. 16 pp.

Mikuni, T. and Kawakami, K. (1975). *Jap. J. appl. Ent. Zool.* **19**, 203–207.

Minato, H., Katayama, T., Matsumoto, M., Katagiri, K., Matsuura, S., Sunagawa, N., Hori, K., Harada, M., and Takeuchi, M. (1973). *Chem. Pharm. Bull.* (Tokyo) **21**, 2268. (Cited in Natori, 1977).

Mirocha, C.J., Pathre, S.V. and Christensen, C.M. (1977). *In* "Mycotoxins in Human and Animal Health" (J.V. Rodricks, C.W. Hesseltine, and M.A. Mehlman, eds), pp. 345–364. Pathotox Publishers, Park Forest South, Illinois.

Moore, J.H., Hammond, A.M. and Llewellyn, G.C. (1978). *J. Invertebr. Path.* **31**, 365–367.

Müller-Kögler, E. (1965). "Pilzkrankheiten bei Insekten". Paul Parey, Berlin.

Murakoshi, S., Sugiyama, J. and Ohtomo, T. (1971). *J. seric. Soc. Japan* **40**, 157–175.

Murakoshi, S., Ichinoe, M., Kumata, H. and Kurata, H. (1977). *Appl. Ent. Zool.* **12**, 255–259.

Myokei, R., Sakurai, A., Chang, C.-F., Kodaira, Y., Takahashi, N. and Tamura, S. (1969). *Agric. Biol. Chem.* **33**, 1491–1500.

Naganawa, H., Takita, T., Suzuki, A., Tamura, S., Lee, S. and Izumiya, N. (1976). *Agric. Biol. Chem.* **40**, 2223–2229.

Naganawa, H., Isogai, A., Ikeda, K., Sato, S., Murakoshi, S., Suzuki, A. and Tamura, S. (1975). *Agric. Biol. Chem.* **39**, 1901–1902.

Natori, S. (1977). *In* "Mycotoxins in Human and Animal Health" (J.V. Rodricks, C.W. Hesseltine and M.A. Mehlman, eds), pp. 559–581. Pathotox Publishers, Park Forest South, Illinois.

Ohtomo, T., Murakoshi, S., Sugiyama, J. and Kurata, H. (1975). *Appl. Microbiol.* **30**, 1034–1035.

Paris, S. and Segretain, G. (1975). *Entomophaga* **20**, 135–138.

Pavlijushin, V.A. (1976). *Mikol. Fitopathol.* **10**, 225–227.

Pavljushin, W.A. and Evlakhova, A.A. (1978). *Abst. Int. Colloq. Invertebr. Path.*, Prague, Sept. 1978, 84.

Prasertphon, S. (1967). *J. Invertebr. Path.* **9**, 281–282.

Prasertphon, S. and Tanada, Y. (1969). *Hilgardia* **39**, 581–600.

Prince, R.C., Crofts, A.R. and Steinrauf, L.K. (1974). *Biochem. Biophys. Res. Commun.* **59**, 697–703.

Reiss, J. (1975). *Chem.-Biol. Interact.* **10**, 339–342.

Roberts, D.W. (1966a). *J. Invertebr. Path.* **8**, 212–221.

Roberts, D.W. (1966b). *J. Invertebr. Path.* **8**, 222–227.

Roberts, D.W. (1969). *J. Invertebr. Path.* **14**, 82–88.

Roberts, D.W. (1970). *Misc. Publ. ent. Soc. Am.* **7** (1), 140–155.

Roberts, D.W. and Yendol, W.G. (1971). *In* "Microbial Control of Insects and Mites" (H.D. Burges and N.W. Hussey, eds), pp. 125–149. Academic Press, London and New York

Rodricks, J.V., Hesseltine, C.W. and Mehlman, M.A., eds. (1977). "Mycotoxins in Human and Animal Health" Pathotox Publishers, Park Forest South, Illinois.

Roeske, R.W., Isaac, S., King, T.E. and Steinrauf, L.K. (1974). *Biochem. Biophys. Res. Commun.* **57**, 554–561.

Rudolph, K. (1976). *In* "Encyclopedia of Plant Pathology. Vol. 4. Physiological Plant Pathology" (R. Heitefuss and P.H. Williams, eds), pp. 270–315. Springer-Verlag, New York.

Samsinakova, A. and Misikova, S. (1973). *Ceska Mykol.* **27**, 55–60.

Samsinakova, A., Leopold, H. and Kalalova, S. (1976). *Zentbl. Bakt. Parasit. Infektionskr. Hyg. Abt.* II, **131**, 60–65.

Samsinakova, A., Bajan, C., Kalalova, S., Kmitowa, K. and Wojciechowska, M. (1979). *Proc. Int. Colloq. Invertebr. Path.*, Prague, Sept. 1978. (In press).

Sargent, J.E. (1975). *Proc. N. Cent. Brch. ent. Soc. Am.* **29**, 155.

Schaerffenberg, B. and Winkler, R. (1969). *Nova Hedwigia* **17**, 203–218.

Scheffer, R.P. (1976). *In* "Encyclopedia of Plant Pathology. Vol. 4. Physiological Plant Pathology" (R. Heitefuss and P.H. Williams, eds), pp. 247–269. Springer-Verlag, New York.

Shima, M. (1955). *Sanshi Skikingo Hokoku* **14**, 427. (Cited in Briggs *et al.*, 1966).

Skinner, H.A. (1949). "The Origin of Medical Terms". Williams and Wilkins, Baltimore, Maryland.

Somerville, H.J. (1973). *Ann. N.Y. Acad. Sci.* **217**, 93–108.

Steinhaus, E.A. and Bell, C.R. (1953). *J. econ. Ent.* **46**, 582–598.

Steyn, P.S. (1971). *In* "Microbial Toxins. Vol. VI. Fungal Toxins" (A. Ciegler, S. Kadis and S.J. Ajl, eds), pp. 179–205. Academic Press, New York and London.

Steyn, P.S. (1977). *In* "Mycotoxins in Human and Animal Health" (J.V. Rodricks, C.W. Hesseltine and M.A. Melhman, eds), pp. 419–467. Pathotox Publishers, Park Forest South, Illinois.

Stoloff, L. (1977). *In* "Mycotoxins in Human and Animal Health" (J.V. Rodricks, C.W. Hesseltine and M.A. Mehlman, eds), pp. 7–28. Pathotox Publishers, Park Forest South, Illinois.

Sussman, A.S. (1952). *Mycologia* **44**, 493–505.

Suzuki, A. and Tamura, S. (1971). Proc. 2nd Internat. IUPAC Congr. Pesticide Chem., Tel-Aviv, 1971, 1, 163–177.

Suzuki, A. and Tamura, S. (1972). *Agric. Biol. Chem.* **36**, 896–898.

Suzuki, A., Taguchi, H. and Tamura, S. (1970). *Agric. Biol. Chem.* **34**, 813–816.

Suzuki, A., Kawakami, K. and Tamura, S. (1971). *Agric. Biol. Chem.* **35**, 1641–1643.

Suzuki, A., Kanaoka, M., Isogai, A., Murakoshi, A., Ichinoe, M. and Tamura, S. (1977). *Tetrahedron Lett.* **25**, 2167–2170.

Taber, W.A. and Vining, L.C. (1963). *Can. J. Microbiol.* **9**, 136–139.

Tamura, S. and Suzuki, A. (1976). *Tampakushitsu Kakusan Koso, Bessatsu* **76**(5), 225–240.

Tamura, S. and Takahashi, N. (1970). *In* "Naturally Occurring Insecticides" (M. Jacobson and D.G. Crosby, eds), pp. 499–539. Marcel Dekker, New York.

Toscano, N.C. and Reeves, E.L. (1973). *J. Invertebr. Path.* **22**, 55–59.

Velitskaya, I.S. (1973). *Mikol Fitopathol.* **7**, 229–232.

Vey, A. and Quiot, J.M. (1975). *C. r. hebd. Seanc. Acad. Sci., Paris*, Ser. D, **280**, 931–934.

Vey, A. and Quiot, J.M. (1976). *Entomophaga* **21**, 275–279.

Vey, A., Vago, C. and Delanoue, P. (1967). *Rev. Mycol.* **32**, 300–305.

Vey, A., Quiot, J.M. and Vago, C. (1973). *C. r. hebd. Seanc. Acad. Sci., Paris*, Ser. D, **276**, 2489–2492.

Vining, L.C. and Taber, W.A. (1962). *Can. J. Chem.* **40**, 1579–1584.

Vining, L.C., Kelleher, W.J. and Schwarting, A.E. (1962). *Can. J. Microbiol.* **8**, 931–933.

Wain, H. (1958). "The Story Behind the Word". Thomas, Springfield, Illinois.

Wasti, S.S. and Hartman, G.C. (1978). *Appl. Ent. Zool.* **13**, 23–28.

Wat, C.-K., McInnes, A.G., Smith, D.G., Wright, J.L.C. and Vining, L.C. (1977). *Can. J. Chem.* **55**, 4090–4098.

West, E.J. and Briggs, J.D. (1968). *J. econ. Ent.* **61**, 684–687.

Wilson, B.J. (1971a). *In* "Microbial Toxins. Vol VI. Fungal Toxins" (A. Ciegler, S. Kadis and S.J. Ajl, eds), pp. 207–295. Academic Press, New York and London.

Wilson, B.J. (1971b). *In* "Microbial Toxins. Vol. VI. Fungal Toxins" (A. Ciegler, S. Kadis and S.J. Ajl, eds), pp. 459–521. Academic Press, New York and London.

Wojciechowska, M. (1973). *Ekol. Pol.* **21**, 699–704.

Wright, J.L.C., Vining, L.C., McInnes, A.G., Smith, D.G. and Walter, J.A. (1977). *Can. J. Biochem.* **55**, 678–685.

Wright, V.F., Casas, E. de las and Harein, P.K. (1976). *Environ. Ent.* **5**, 371–374.

Yendol, W.G., Miller, E.M. and Behnke, C.N. (1968). *J. Invertebr. Path.* **10**, 313–319.

Zacharuk, R.Y. (1971). *Can. J. Microbiol.* **17**, 281–289.

Zacharuk, R.Y. (1973). *In* "Some Recent Advances in Insect Pathology" (D.W. Roberts and W.G. Yendol, eds), pp. 112–119. *Misc. Publ. ent. Soc. Am.*, Vol. 9, No. 2.

Zacharuk, R.Y. (1974). *J. Invertebr. Path.* **23**, 13–21.

Zajic, J.E. (1963). *J. Insect Path.* **5**, 16–27.

CHAPTER 24

Pest Control by the Fungi Beauveria and Metarhizium

P. FERRON

INRA, Station de Recherches de Lutte Biologique, La Minière
78280 Guyancourt, France

I. Introduction

Although the mycoses caused by the entomopathogenic Fungi Imperfecti (Moniliales), *Beauveria bassiana*, *Beauveria brongniartii (= B. tenella)*, *Metarhizium anisopliae* var *anisopliae* and var *major* — the white or green muscardines depending on the colour of the spores — have been studied for about a century, it is principally during the last 15 years that special attention has been focused on them to develop new methods of biological control of insects. For many years they were regarded as biological agents of secondary interest, due to pessimistic conclusions from the first field trials in several countries at the end of the last century. However, in the 1950s East European countries started investigations, particularly with *B. bassiana*, as part of a general strategy to control the Colorado beetle, *Leptinotarsa decemlineata*. (Review: Roberts and Yendol, 1971).

These investigations resulted in original developments, often unknown to specialists in western countries (Ferron, 1970a), in the prognosis of insect infectious diseases, conditions of muscardine development, mode of application in the field and technology of mass production of the fungi.

While this practical approach is part of a current program with *M. anisopliae* against some *Cercopidae (Homoptera)*, pests of sugar cane and fodder crops in Brazil, most western research has taken an analytical approach, first trying to gain a better knowledge of the mode of action of the pathogen, the host defence reactions and the epidemiology of muscardines. Together these results now offer real prospects for practical application of these pathogens as part of an integrated control strategy. There has been parallel interest in other genera of entomopathogenic fungi reflecting a general development in the principles of their use, a trend also revealed by the increasing number of publications on this subject (Ferron, 1978a). These researches have not led to commercial application in the west since no commercial preparations of these fungi are available yet, even though one of them has been tested on areas of ca. 50 000 ha.

II. Etiology of Muscardine Diseases

A. MODE OF INFECTION

In contrast to bacteria and viruses that pass through the gut wall from contaminated food, fungi have an unique mode of infection. They reach the haemocoel through the cuticle or possibly through the mouth parts. Ingested fungal spores do not germinate in the gut and are voided in the faeces. Infection therefore results from contact between a virulent infectious inoculum and a susceptible insect cuticle, germination, penetration of the germ tubes through the integument and finally spread of the pathogen through the host tissues.

1. Sites of Infection

In the laboratory, the most sensitive sites on larvae of the cockchafer *Melolontha melolontha* to the fungus *B. brongniartii* are the mouth and anus (Delmas, 1973). On larvae of the same species collected in nature where white muscardine is endemic, the most frequent sites of infection revealed by melanization following hyphal penetration, occurred on the membranes between head capsule and thorax or between segments of the appendages (Ferron, 1978b). This apparent contradiction is possibly due to the larval behaviour of burrowing in soil so that soil particles continuously scrape infectious inoculum off the exposed cuticle whereas the membrane is protected from this mechanical action. Both *Hylobius pales* adults (Schabel, 1976b) and *Schistocerca gregaria* (Veen, 1968) are very sensitive to oral infection by *M. anisopliae*, which suggests practical use of the fungi mixed into food baits against such pests as the Noctuidae.

2. Germination of Infectious Inoculum

Spore germination on the cuticle surface has often been regarded as closely subject to macroclimatic factors, especially temperature and humidity, by reference to *in vitro* data. The optimal growth temperature is near 23–25°C for *Beauveria* and near 27–28°C for *Metarhizium*. That is why muscardine outbreaks appear in nature chiefly during summer and earlier for aerial insects than for subterranean insects, because the soil warms slowly. Thus in temperate zones it is better to select strains with below average thermal optima (Kalvish, 1974; Latch, 1976).

Although conidia require relative humidity of 92% to germinate, adult bean weevils *Acanthoscelides obtectus* are infected by *B. bassiana* conidia ireespective of relative humidity (Ferron, 1977). Though this cannot be regarded yet as a general rule it suggests that there may be a microclimate in the boundary layers of the host cuticle.

3. Penetration of Hyphae Through the Integument

Penetration of hyphae through the cuticle involves both mechanical and enzymatic factors. McInnis (1975) discussed the biochemical bases of pathogenesis by mycoses. Only ca. 10% of the total proteins of the integument are not linked, therefore the proteins must be hydrolysed before the action of chitinases can begin (Samšiňáková *et al.*, 1971, 1976; Leopold *et al.*, 1973; Samšiňáková and Misiková, 1973). According to Sannasi (1969) the PAS-positive nature of the fungus-infected endocuticle is probably due to the action of fungal chitinase which seems to break down chitin into many monoacetylated chitin residues, but McCauley *et al.* (1968) have not observed this in infected wire-worms. Electron microscopy of the histopathology of *M. anisopliae* infection of wireworms is given in detail by Zacharuk (1970a,b, 1971a,b, 1973a,b, 1974).

A particular sign of mycosis is secondary melanization when hyphae attack epidermal cells, but it is not known whether accompanying tyrosinase comes from integument or hyphae. Since melanization is also controlled by ecdysone, an endocrine disfunction may be involved (Sannasi and Oliver, 1971). However, the possibility that the fungus itself secretes ecdysone and juvenile hormone cannot be eliminated (Schneiderman *et al.*, 1960). The observations of Mikuni and Kawakami (1975) on the interactions between infection and moulting or metamorphosis can be perhaps correlated with this hypothesis.

4. Relationship Between Infection and Moulting

Because an arthropod moults, an infection penetrates the haemocoel only if the hyphae reach the hypodermis. When *B. bassiana* infects instar III larvae of *Leptinotarsa decemlineata* at ecdysis, hyphae and blastospores invade the exuvial fluid and infect the newly formed integument (Fargues and Vey, 1974; Vey and Fargues, 1977). If infection is more advanced at ecdysis, the new cuticle is wounded during ecdysis because of adhesion of the two cuticles by mycelium. These wounds are an open door both to mycelial elements in the exuvial fluid and to bacteria that can cause lethal septicemia. However, if penetration into the old integument is only superficial, the insect can escape infection by casting the infectious inoculum. Thus the application of fungi in the field must be synchronized with the life cycle of a pest and its ecdysis frequency.

B. DEVELOPMENT OF MUSCARDINE DISEASE

After crossing the integument (Kawakami and Mikuni, 1965), the muscardine develops in the haemocoel in the presence of cellular defensive reactions of the host (Seryczynska and Bajan, 1975). Plasmatocytes surround the mycelium as a pseudotissue or granuloma as described by Vey et al. (1975) in invertebrate cell culture.

True pathogens such as *Beauveria* and *Metarhizium* produce toxins which erode these granuloma and allow blastospores to invade the haemocoel. Phagocytosis is not always observed (Vey and Quiot, 1975). Hyphal bodies proliferate only just before death of the host. Thus the role of entomogenous toxins is particularly important (Roberts, 1966; Evlakhova and Rakitin, 1968; Samšiňáková and Samšiňák, 1970; Evlakhova, 1974; Pavlyushin, 1974). Several toxic cyclodepsipeptides, such as destruxins A, B, C and D and desmethyl-destruxin, have been isolated from *M. anisopliae* cultures (Suzuki et al., 1970, 1971; Suzuki and Tamura, 1972; Naganawa et al., 1976); also beauvericin has been isolated from *B. bassiana* (Hamill et al., 1969; Elsworth ahd Grove, 1974, 1977) and beauvellide from *B. brongniartii* (Frappier et al, 1975). The cytological effect of beauvericin was studied in vitro in insect cell cultures (Vey et al., 1973). Toxin production in different strains of *B. bassiana* is correlated with their virulence to *Galleria mellonella* (Sikura and Bevzenko, 1972). Histopathological studies of elaterid tissues infected by *M. anisopliae* suggest that toxins kill the host by inciting progressive degeneration of the host tissues due to loss of the structural integrity of membranes and then dehydration of cells by fluid loss (Zacharuk, 1971a). Disturbances of the electrical activity in nerves (Evlakhova and Rakitin, 1968) were observed as an increase of the oxygen consumption (Pristavko and Yanishevskaya, 1971). (See also Chapter 23).

C. SAPROPHYTIC DEVELOPMENT OF THE FUNGUS

Host death marks the end of the parasitic phase of fungal development. Then the mycelium grows saprophytically through all tissues in competition with the intestinal bacterial flora. White muscardine produces oosporein, a red antibiotic pigment which colours the cadaver and curbs bacteria. Two yellow pigments bassianin and tenellin were also identified (McInnes et al., 1974) (see also Chapter 23). The fungus grows outside the integument and develops conidiophores only when the atmosphere is saturated with water (Ferron, 1977). Tinline (1971) and Zacharuk (1973a) found chlamydospores inside a cadaver killed by *M. anisopliae*.

III. Favourable Conditions for Muscardine Development

It is well established, now, that entomopathogenic fungi have a certain specificity (Fargues and Remaudière, 1977). In the same species of fungus, different strains can have very different activity spectra (Ferron and Diomandé, 1969; Ferron et al., 1972). Study of this specificity by Fargues et al. (1976) showed the importance of mechanisms acting at the integument and in the body cavity. Therefore inoculation of insects by injection into the haemocoel is not recommended for experimental research, nor, together with direct contamination of the cuticle, for bioassay of the infectious inoculum.

A. QUANTITY OF SPORES

Positive correlation between the number of infective spores and mortality by mycosis has been established by many authors (Ferron, 1978a), but the influence of sublethal doses has been insufficiently studied.

In practice, Soviet authors recommend 2–4 kg Boverin/ha, a biological preparation containing 6×10^9 conidia of *B. bassiana*/g (= 1.2–2.4×10^{13} spores/ha), against *Leptinotarsa decemlineata*. For soil insects, e.g. *Melolontha melolontha*, Ferron (1978b) used about 5×10^{14} conidia of *B. brongniartii*/ha. For a small, short-lived, soil insect, more inoculum of *B. bassiana* or *M. anisopliae* must be used: 10^{16} to 10^{17} conidia/ha (Müller-Kögler and Stein, 1970, 1976). With fewer spores, muscardine disease develops slowly and affects only the older larvae or adults: disturbance in fecundity and diapause of surviving adults can occur (Primak, 1967; Müller-Kögler and Stein, 1970; Bajan and Kmitowa, 1972; Sikura and Gritsaenko, 1973; Litvinenko, 1974; N'Doye, 1976; Faizy, 1978).

In the laboratory the disease normally develops after contamination of insects either directly by spore suspensions (10^6 to 10^8 spores/ml), or by mixing the soil with 10^5 to 10^8 spores/g or cm^3 for soil insects.

B. HEALTH OF THE INSECT-HOST

To reduce the number of spores needed to kill the host, Telenga et al. (1967) recommended the combination of *B. bassiana* with low doses of chemical insecticides. In these conditions haemocytes were modified and the microflora of the digestive tract penetrated into the haemolymph (Pristavko, 1966). When larvae of *Leptinotarsa decemlineata* were treated simultaneously with *B. bassiana* and a low dose of DDT, the larvae least resistant to the chemical were killed by chemical poisoning, the more resistant died from bacterial septicemia and the survivors were killed by muscardine. Using the same material Fargues (1973) concluded that there was a very weak synergistic action.

With a long lived soil insect, *Melolontha melolontha*, synergism occurred between *B. brongniartii* and low doses of organochlorides or organophosphates (Ferron, 1970b, 1971). For example, mortality by mycosis reached ca. 55% after 3 months at 20°C in turf simultaneously treated with 5×10^5 conidia of *B. brongniartii* and a reduced dosage of HCH, whereas only 15% of instar III larvae were killed by the fungus alone in the same conditions. Similar phenomena resulted from combination of the fungus with the *Entomopoxvirus* of *Melolontha* or with *Bacillus popilliae* (Ferron et al., 1969; Ferron and Hurpin,

1974). Of particular interest is the observation that the fungus developed along-side both the virus and the bacterium.

IV. Epizootiology of Muscardine Diseases

This fundamental ecological approach has had little attention: research workers were concerned either with an immediate effect in the field or with the basic studies in controlled laboratory conditions. There are few data on the persistence of an infectious inoculum artificially applied to the soil (Bell, 1975; Milner and Lutton, 1976; Joussier, 1977; Roberts and Campbell, 1977).

Young (1974) studied whether *M. anisopliae* would be spread by adults of the coconut palm rhinoceros beetle, *Oryctes rhinoceros*, in Western Samoa. Conidia were added to heaps of rotted sawdust containing larvae: many larvae were killed but the disease was not dispersed. However Ferron (1978b) suspects that adults spread white muscardine disease in populations of *Melolontha melolontha*. Benz (1976) reports preliminary experimental application of this type of treatment in nature by Keller in Switzerland, but it was without effect on the pathological state of the larval population. A natural grass-land in eastern France was inoculated with laboratory produced conidia of *B. brongniartii* (20 to 90 x 10^9 spores/m²). Rapid infection within two months remained chronic for > 12 months until the end of the larval cycle, when it became acute and killed many larvae. Most of the collected surviving adults proved to be vectors of infection and consequently capable of spreading mycosis. In the initial contaminated area, the disease reappeared in the next generation of the pest. Also the amount of fungus in the soil was monitored throughout the year using Joussier and Catroux's (1976) isolation technique on selective agar medium. The infectious inoculum fluctuated from minimal in spring to maximal in autumn.

Thus natural epizootics in this pest develop more rapidly because the contamination of young larvae is both early and abundant. Because generations overlap one another in the same biotope, such an epizootic can result from the fungus multiplying on cadavers or from infected females which die in the soil after egg-laying.

V. Development of Fungal Preparations

It is necessary to mass produce a virulent specific inoculum capable of long storage, and also to define the conditions of utilization in the field. The compatibility of the fungi in modern integrated pest control programs is described in Chapter 38, Section II, C and their safety in Chapter 40.

A. CHARACTERIZATION AND SELECTION OF STRAINS

Conidiogenesis studies (Hammill, 1972; Reisinger and Olah, 1974) are inadequate to identify strains. Fargues *et al.* (1974, 1975) used electrophoretic and immunoelectrophoretic techniques combined with enzyme activity.

Alioshina *et al.* (1975) used serological techniques and Tumarkin *et al.* (1974) gave biochemical peculiarities of some strains. The virulence of different strains of *B. brongniartii* was correlated with an arc of precipitation obtained by immunoelectrophoresis (Segretain *et al.*, 1971; Paris *et al.*, 1975a; Paris and Segretain, 1975). The same authors obtained a connection between virulence and lipase activity, and this result is discussed by Samšiňáková and Misiková (1973).

Especially in the USSR there has been much research on artificially modifying virulence by mutagenesis, hybridization or production of heterokaryons (Andreev *et al.*, 1972; Usenko *et al.*, 1973; Kirsanova and Usenko, 1974; Nikolaev and Akimkina, 1974; Yurchenko *et al.*, 1974), but the new strains do not seem to be better than the original strains of *B. bassiana*. Tinline and Noviello (1971), and Paris (1977) described heterokaryosis by both *M. anisopliae* and *B. brongniartii*.

B. TECHNOLOGY OF MASS PRODUCTION

Production of blastospores in submerged culture was developed 10 years ago but has now been abandoned because of the difficulties of storing this type of spore (*B. bassiana*, Samšiňáková, 1964, 1966; *B. brongniartii*, Catroux *et al.*, 1970; Blachère *et al.*, 1973, medium given in Table I). Approximately 10^9 blastospores/ml of nutritive medium are produced after 72 h. These are harvested

TABLE I

Liquid medium for the mass production of Beauveria brongniartii *blastospores in a fermentor (Blachère* et al., *1973. French patent no. 72 35 452)*

Corn steep liquor	20 g	$FeSO_4 \cdot 7H_2O$	0.023 g
Sucrose	30 g	$ZnSO_4$	0.020 g
KH_2PO_4	2.26 g	K_2SO_4	0.174 g
$Na_2HPO_4 \cdot 12H_2O$	3.8 g	$CaCl_2 \cdot 2H_2O$	0.147 g
$MgSO_4 \cdot 7H_2O$	0.123 g	Water to make up volume to 1 000 ml	

by centrifugation and the paste is dried at low temperature in a ventilated drying closet after mixing with silica powder, osmotically active materials (e.g., sucrose and sodium glutamate), anti-oxidizing agents (e.g., sodium ascorbate) and a mixture of liquid paraffin—polyoxyethylene glyceryl oleate. This preparation is stored under vacuum or in nitrogen gas. Blastospores in such preparations dried at $4°C$ were viable after 8 months storage (Blachère *et al.*, 1973).

A two stage technique for mass-production of *B. bassiana* conidiospores is used in the USSR (Zakharchenko *et al.*, 1963; Telenga and Goral, 1966; Zakharchenko, 1967). First the biomass is produced as mycelium in a fermentor. This is then surface-cultured in trays of nutrient medium for sporulation. A pilot-factory in Krasnodar produces annually 22 tons Boverin (*B. bassiana* conidia plus an inert carrier, standardized at 6×10^9 conidia/g). A similar technique is used for the mass production of *M. anisopliae* (Goral and Lappa, 1973).

Technical difficulties have been partly solved in Brazil by replacing the trays with autoclavable polypropylene bags containing rice grains as nutritive substrate (Aquino *et al.*, 1977). A production unit in laboratories at Recife (State of Pernambuco) provides daily ca. 100 kg Metaquino. The estimated content is 1 or 2×10^9 conidia/g crushed rice grain carrier (Aquino, 1974; Guagliumi *et al.*, 1974; Aquino *et al.*, 1975). Other techniques of *M. anisopliae* mass-production are also used in Brazil; either in autoclavable plastic bags on standard nutritive medium, e.g. dextrose agar (Moura-Costa *et al.*, 1974), or a mixture of rice flour (5%) and agar-agar (1%) (Moura-Costa and Magalhaes, 1974), or in milk or plasma bottles on rice grains. These different methods are not standardized yet, but several production units with a daily capacity of 5 000 bottles, corresponding to ca. 250 kg biological preparation daily, are being built by the Instituto de Açucar e do Alcool. Other cereal grains have been suggested for the mass production of several entomopathogenic Fungi Imperfecti (Bajan *et al.*, 1975b; Latch and Falloon, 1976; Villacorta, 1976) but these have not been used apparently on so large a scale.

In the USSR, chiefly in the VNIBACPREPARAT in Moscow and in the Ukrainian Institute for Plant Protection in Kiev, an industrializable fermentor technique is being investigated for *B. bassiana* mass production in the conidial form. Two patents (Goral, 1971; Kondryatiev *et al.*, 1971) differ essentially in the composition of culture media (Table II). Under optimal conditions, yields vary from 5×10^8 to 2×10^9 conidia/ml but the preparation has a limited viability of ca. 2 or 3 months, a serious failure that considerably limits its industrial potential.

TABLE II

Composition of aqueous fermentation media for the mass production of
Beauveria bassiana *conidia*

Goral (1973a)		Alioshina *et al.* (1975)	
NaNO$_3$	0.2 to 0.5%	NaNO$_3$	0.9%
KH$_2$PO$_4$	0.2 to 0.5%	KH$_2$PO$_4$	0.075%
MgSO$_4$	0.05 to 0.2%	MgSO$_4$	0.075%
Sucrose	2%	CaCl$_2$	1.25%
Aminonitrate in beer wort			
or corn extract	0 to 0.080%	Sucrose	1.0%

Soviet publications describe strain selection (Alioshina *et al.*, 1972; Kazakova *et al.*, 1974), choice of culture media (Babayan *et al.*, 1974; Goral, 1972c) pH (Goral and Lappa, 1972), temperature (Goral, 1973b; Isaenko *et al.*, 1974; Il'Icheva *et al.*, 1976), aeration (Goral, 1972a), culture duration (Goral, 1972b), evolution of fungus morphology (Goral, 1975), viability of conidia (Goral and Lappa, 1975; Mayorova *et al.*, 1976), the technique for drying the preparation and addition of chemical additives which could improve its conservation (Belova *et al.*, 1974; Bulatova *et al.*, 1974).

Laboratory studies on defined media give further information about nutrients that secure the optimal growth and sporulation of *B. bassiana* and *M. anisopliae* (Aoki and Chigusa, 1968; Aoki and Yanase, 1970; Kucera, 1971; Barnes *et al.*, 1975; Tjulpanov *et al.*, 1977).

The spores are formulated either in an inert carrier e.g. kaolin, or mixed in the crushed nutritive substrate when this comprises cereal grain. Doubtlessly further additives would improve the quality of these preparations, by protecting them either from ultra violet radiations of the sun (Ignoffo *et al.*, 1977) or from biodegradation by soil microorganisms (Reisinger *et al.*, 1977).

C. QUALITY TESTING

Quality control of fungal preparations is a too often neglected necessity (Gruner and Abud-Antun, 1976; Müller-Kögler, 1960, 1966). The three basic criteria are spore count, viability and virulence. It would be opportune to standardize testing techniques in the various laboratories concerned. Estimation of virulence is complicated by difficulties such as strain specificity and lack of a standard reference preparation (Evlakhova and Kirsanova, 1968; Shamraj *et al.*, 1974; Ferron and Robert, 1975; Kirsanova *et al.*, 1975).

VI. Practical Use against Insect Pests

Due to the present technical difficulties in the mass-production of these two fungi, only two preparations of conidia can be regarded as in practical use: Boverin (*B. bassiana*), and Metaquino and other Brazilian experimental preparations (*M. anisopliae*). According to oral sources of information, enough material has been made to treat ca. 10 000 ha with *Beauveria* and ca. 50 000 ha with *Metarhizium*. These numbers, though not high, are the concrete expression of recent progress. Considerable use in China is described in Chapter 42.

A. USE OF BOVERIN IN THE USSR

1. Biological Control of the Colorado Beetle

Soviet research with this beetle started 20 years ago (Ferron, 1970b). Boverin is recommended for use with reduced dosage of chemical insecticides; reproducible results can thus be obtained, irrespective of climatic conditions in the year concerned. Mean mortality during four consecutive years reached 92% (85.7% to 97.6%). Infection also appeared in larvae hiding in the ground, which caused a noticeable reduction of the number of surviving adults (Sikura, 1974).

Generally two treatments are made at 15 day intervals, according to the age of the pest population, young larvae being more susceptible to muscardine than older ones. Now 1,5 kg/ha concentrated Boverin (30×10^9 conidia/g) is used in each treatment combined with a reduced dosage of chlorophos (= trichlorphon). (General discussion by Bajan *et al.*, 1975a; Fargues, 1975).

2. Biological Control of the Codling Moth

Recently Soviet research was extended to *Cydia* (*Carpocapsa*) *pomonella*. Boverin with reduced dosage of Dipterex (= trichlorphon), is used to limit the annual second generation outbreaks. Fungicides for the control of *Venturia inaequalis* in spring prevent simultaneous use of an entomopathogenic fungus to control the first generation (Sikura, 1974).

According to Drozda and Lappa (1974), the climate in the Ukraine justifies the combined use of Boverin and reduced dosage of chemical insecticides, e.g. chlorophos or carbophos (= malathion), at 1/5 of the usual dosage. In 1971, Boverin + chlorophos gave 95.3% undamaged fruit, cf. 96.2% with chlorophos alone at the usual dosage. With carbophos, the results were 92.4% and 94.7% respectively, 91.8% with Boverin alone and 78.2% for the untreated trees. In 1972, however, as the summer was particularly hot and dry, the results were poorer: chlorophos, usual dosage, 88.7%; carbophos, usual dosage, 86.2%; Boverin + chlorophos reduced dosage 80.3%; Boverin + carbophos, reduced dosage 84.5%; untreated, 75.5%.

These orchard experiments, aimed at defining the optimal conditions for fruit protection are complemented by studies on the susceptibility to muscardine of several species of orchard pest Lepidoptera, e.g. *Cydia* (*Carpocapsa*) *pyrivora*, *Cydia* (*Laspeyresia*) *funebrana*, *Zeuzera pyrina* (Sikura *et al.*, 1974), on the physiological alterations in adults surviving infection (Sikura and Gritsaenko, 1973; Lappa, 1975) and on the influence of climatic conditions on treatment efficiency (Lappa and Goral, 1975).

B. USE OF METAQUINO IN BRAZIL

Because of the fear of insect strains resistant to HCH, the chemical insecticide of choice, *M. anisopliae* was produced to control spittlebugs, such as *Mahanarva posticata* on sugar cane (Guagliumi, 1973; Guagliumi *et al.*, 1974). From 1972 to 1978, the area treated with the product Metaquino was increased from 500 to ca. 50 000 ha in the State of Pernambuco alone. Aerial sprays at 50 litres/ha (containing 6×10^{11} to 1.2×10^{12} conidia) were followed by mortality tests at 30, 60 and 90 days. In 1977, in 65 places (total area 4000 ha) green muscardine had killed means of 40%, 55%, 65% respectively in the three successive tests (oral communication from authorities in Recife). A similar program has recently been adopted by the Institute of Alcohol and Sugar which has a production unit in Maceio (State of Alagoas) able to provide material for daily treatment of 100 ha of sugar cane. Notably, in both programs the cost of fungal treatment is appreciably lower than that of the usual chemical treatments.

Moreover, a national program to control the landscape *Cercopidae* is being prepared with the prospect of treating several million ha with *Metarhizium* (Viega *et al.*, 1972; Araujo *et al.*, 1975).

C. POTENTIAL USES AGAINST INSECT PESTS

Besides the three examples described above, many field experiments reveal the potential uses of the muscardine fungi. Provided that the present technical difficulties limiting their mass-production are overcome, a noticeable increase can be expected in their use in integrated pest control programs. Indeed dosages recommended by different workers agree remarkably: 10^{13} to 10^{14} spores/ha for aerial insects, and about 10^{15} spores/ha for soil insects. The efficacy of muscardine fungi is thus comparable to that of chemical insecticides, with the additional advantage of noticeable long-term insect limitation.

A concrete example is control of *Melolontha melolontha* by *B. brongniartii* (Ferron, 1978b). Soil treatment with 20×10^9 conidia/m² $(= 2 \times 10^{14}$ spores/ ha) caused an epizootic only one year later, at the end of the larval stage; but during the next generation, some 4 years after the treatment, muscardine appeared again in the same area and thus caused a noticeable decrease of the pest population (Table III).

TABLE III

Number of Melolontha melolontha$/m^2$ *in two areas of 500 m², one treated with* Beauveria brongniartii *conidia in April 1974*

Generation n			Generation $n + 1$		
Dates	Untreated	Treated	Dates	Untreated	Treated
April, 1974	77 larva II	77 larva II	Sept., 1976	33 larva II	25 larva II
July, 1974	53 larva III	48 larva II	May, 1977	36 larva II	22 larva II
Sept., 1974	38 larva III	37 larva III	Sept., 1977	25 larva III	7 larva III
Aug., 1975	16 pupae + adults	3 pupae + adults	July, 1978	10 larva III	1 larva III

The essential features of this example are the perennial, undisturbed nature of the meadow agrosystem and the behaviour of the insect which, generation after generation, colonizes the same biotope. These features lead to a durable residual efficacy, because the fungus multiplies and persists on the insects that it kills. Of the adult survivors of generation n (Table III), 38% died of muscardine when in laboratory quarantine compared with $< 2\%$ from the untreated area, so these adults are muscardine vectors able to spread the disease. Thus the cost of the considerable quantities of spores needed to treat the soil, ca. 1×10^{15}/ha, is ameliorated because such a treatment is effective for several years.

Attention must therefore be paid to records of natural disease that reveal the susceptibility to muscardine of diverse soil pests, e.g. larvae of Scarabaeidae (Diomandé, 1969; Gruner, 1973; Young, 1974; Martynenko, 1975; Latch and Falloon, 1976; Fujishita and Kushida, 1976), of Curculionidae (Müller-Kögler and Stein, 1970, 1976; Beavers *et al.*, 1972; Tedders *et al.*, 1973; Marchal, 1977), of Elateridae (Zacharuk and Tinline, 1968; Dolin and Krasyukova, 1974) or caterpillars pupating in the soil (Halperin and Shanuni, 1974). Infections also occur in insects that live in habitats with microclimatic similarities to soil, e.g. boring Lepidoptera, bark beetles and spittelbugs of sugar cane (Berrios and Hidalgo-Salvatierra, 1971a,b; Taylor and Franklin, 1973; Moore, 1973; Schabel, 1976a; Barson, 1977). Finally the prospects of using muscardines for biological control of vector insects, especially mosquitoes, cannot be neglected (Pinnock *et al.*, 1973; Roberts, 1974, 1975).

VII. Future Research Requirements

It can be stated that the publications devoted to *Beauveria* and *Metarhizium* represent the majority of the scientific literature about the entomopathogenic fungi. This subject, however, is not worked out: the number of publications devoted to it has been greatly increasing for the last ten years, so extending the investigation field, due to more proven methods and new techniques. Although the problems encountered with these two genera are the same as those with such other genera as *Nomuraea, Paecilomyces* and *Verticillium*, the greater present knowledge about the muscardines enables their future research trends to be defined more precisely.

If we consider first the applied aspects aimed at defining conditions in which muscardine fungi could be used in integrated control programs against agricultural insect pests, mass production technology must be improved and the safety of the biopreparations towards vertebrates and beneficial insects studied further. This implies more thorough studies on the physiology of these fungi as well as on their host specificity and their behaviour in the ecosystems.

Ecological studies, hitherto particularly neglected, should provide much information on the epizootic processes as well as on the population dynamics of these microorganisms artificially put into competition with the indigenous flora in natural ecosystems. Such studies, by revealing the ecotype–specificity properties may contribute to the comprehension of the genetic processes determining strain aggressiveness and specificity. Some immuno-electrophoretic techniques have already been proposed for pathotype characterization, but the applied approaches to genetic manipulations are still very little investigated. These manipulations involve a more precise analysis of the germ aggressiveness: its possible polygenic nature should be examined by sequential studies at successive infection stages. In recent years, a growing attention has been focused on the enzymatic system as well as toxin production.

Such efforts must, however, be justified by demonstrating the practical potential of these entomopathogenic fungi. Field experiments in different agronomical systems should be performed under various ecological conditions, in order to establish the nature and extent of the applications likely to interest the phytosanitary industry.

References

Alioshina, O.A., Il'Itcheva, S.N., Kononova, E.V. and Koliada, N.A. (1972). *Mikol. Fitopatol.* 6, 341–344.

Alioshina, O.A., Boyarskij, B.G., Il'Itcheva, S.N., Kononova, E.V. and Ozerova, L.V. (1975). *Dokl. TSKhA* 209, 121–125.

Andreev, S.V., Evlakhova, A.A., Kirsanova, R.V. and Levitin, M.M. (1972). *Tsitol. Genet.* 6, 395–399.

Aoki, J. and Chigusa, K. (1968). *J. seric. Sci. Jap.* 37, 288–294.

Aoki, J. and Yanase, K. (1970). *J. seric. Sci. Jap.* 39, 285–292.

Aquino, de M. (1974). *Boll. tec. Inst. Pesq. Agron., Recife,* 72, 1–26.

Aquino, de M., Cavalcanti, V.A., Sena, R.C. and Queiroz, G.F. (1975). *Boll. Tec. Commissao Executiva de Defesa Fitossanitaria de Lavoura Canavieira de Pernambuco* **4**, 1–31.

Aquino, de M., Vital, F., Cavalcanti, B. and Nascimento, G. (1977). *Boll. Tec. Commissão Executiva de Defesa Fitossanitaria de Lavoura Conavieria de Pernambuco* **5**, 7–11.

Araujo, D.O.B. de, Bezerra, D.O. and d'Aguiar, Z.M.F. (1975). *Boll. Inst. Biol. Bahia, Salvador* **14**, 1–5.

Babayan, E. Yu, Mal'Kova, A.I., Bulatova, A.A. and Konovalova, G.N. (1974). *In* "Méthode Biologique dans la Protection des Plantes. Thèses et Conférences des Jeunes Chercheurs". Vses. Ord. Lenina Akad Sel'Skhokhoz. Nauk. Im. V.I. Lenina. Vses. nauch.-issled Inst. Biol. Meth. Zash. Rast., Kichinev, pp. 9–10.

Bajan, C. and Kmitowa, K. (1972). *Ekol. Polsk.* **20**, 423–432.

Bajan, C., Bilewicz-Pawinska, T., Fedorko, A. and Kmitova, K. (1975a). Rep. 8th Int. Plant Prot. Congr., Moscow, Sect. V, pp. 33–40.

Bajan, C., Kmitowa, K. and Wojciechowska, M. (1975b). *Bull. Acad. Pol. Sci., Ser. Sci. biol.* **23**, 45–47.

Barnes, G.L., Boethel, D.J., Eikenbary, R.D., Criswell, J.T. and Gentry, C.R. (1975). *J. Invertebr. Path.* **25**, 301–305.

Barson, G. (1977). *J. Invertebr. Path.* **29**, 361–366.

Beavers, J.B., McCoy, C.W., Kanavel, R.F., Sutton, R.A. and Selhime, A.G. (1972). *Fla. Ent.* **55**, 117–120.

Bell, J.V. (1975). *J. Invertebr. Path.* **26**, 129–130.

Belova, R.N., Shnurkina, L.A., Shamraj., L.G., Zhukov, V.N. and Kuzhleva, L.M. (1974). *In* "Méthode Biologique dans la Protection des Plantes. Thèses et Conférences des Jeunes Chercheurs". Vses. Ord. Lenina Akad. Sel' Skhokhoz. Nauk. Im. V.I. Lenina. Vses. nauch.-issled Inst. Biol. Meth. Zash. Rast., Kichinev, pp. 15–16.

Benz, G. (1976). Proc. 1st Int. Colloq. Invertebr. Path., and IXth Ann. Meet. Soc. Invertebr. Path. Kingston, pp. 52–58.

Berrios, F. and Hidalgo-Salvatierra, O. (1971a). *Turrialba* **21**, 214–219.

Berrios, F. and Hidalgo-Salvatierra, O. (1971b). *Turrialba* **21**, 451–454.

Blachère, H., Calvez, J., Ferron, P., Corrieu, G. and Peringer, P. (1973). *Ann. Zool. Ecol. anim.* **5**, 69–79.

Bulatova, A.A., Dmitrakova, N.S. and Savina, V.E. (1974). *In* "Méthode Biologique dans la Protection des Plantes. Thèses et Conférences des Jeunes Chercheurs". Vses. Ord. Lenina Akad. Sel'Skhokhoz. Nauk Im. V.I. Lenina. Vses. nauch.-issled Inst. Biol. Meth. Zash. Rast., Kichinev, pp. 13–14.

Catroux, G., Calvez, J., Ferron, P. and Blachère, H. (1970). *Ann. Zool. Ecol. anim.* **2**, 281–294.

Delmas, J.C. (1973). *C.R. hebd. Séanc. Acad. Sci. Paris.*, Ser. D **277**, 433–435.

Diomandé, T. (1969). *Bull. Inst. Fondam. Afr. Noire Ser. A* **21**, 1381–1405.

Dolin, V.G.D. and Krasyukova, Y.F. (1974), *In* "Pathologie des Arthropodes et Moyens Biologiques de Lutte", Kiev, pp. 62–63.

Drozda, V.F. and Lappa, N.V. (1974). *In* "Pathologie des Arthropodes et Moyens Biologiques de Lutte", Kiev, pp. 64–67.

Elsworth, J.F. and Grove, J.F. (1974). *S. Afr. J. Sci.* 70, 379.

Elsworth, J.F. and Grove, J.F. (1977). *J. Chem. Soc. Perkin Trans I* 3, 270–273.

Evlakhova, A.A. (1974). "The Entomopathogenic Fungi: Systematics, Biology and Practical Importance". Nauka ed. M.K. Khokhriakov. Leningrad

Evlakhova, A.A. and Kirsanova, R.V. (1968). *Plant Protection*, Moscow 3, 53–54.

Evlakhova, A.A. and Rakitin, A.A. (1968). *Dokl. Akad. Nauk. SSSR* 178, 485–488.

Faizy, R. (1978). Thèse 3e Cycle, Univ. Paris-Sud, Centre d'Orsay, 74 pp. polycop.

Fargues, J. (1973). *Ann. Zool. Ecol. anim.* 5, 231–246.

Fargues, J. (1975). *Ann. Zool. Ecol. anim.* 7, 242–264.

Fargues, J., Duriez, T., Andrieu, S. and Popeye, R. (1974). *C.R. hebd. Séanc. Acad. Sci. Paris, Ser. D,* 278, 2245–2247.

Fargues, J. and Remaudière, G. (1977). *Mycopathologia* 62, 31–37.

Fargues, J. and Vey, A. (1974). *Entomophaga* 19, 311–323.

Fargues, J., Duriez, T., Andrieu, S., Popeye, R. and Robert, P.H. (1975). *C.R. hebd. Seanc. Acad. Sci. Paris, Ser. D* 281, 1781–1784.

Fargues, J., Robert, P.H. and Vey, A. (1976). *C.R. hebd. Séanc. Acad. Sci. Paris, Ser. D* 181, 2223–2226.

Ferron, P. (1970a). *Ann. Zool. Ecol. anim., numéro hors sér.* 3, 117–134.

Ferron, P. (1970b). Proc. 5th Int. Colloq. Insect Path., College Park, pp. 66–79.

Ferron, P. (1971). *Ent. Exp. Appl.* 14, 57–66.

Ferron, P. (1977). *Entomophaga* 22, 393–396.

Ferron, P. (1978a). *Ann. Rev. Ent.* 23, 409–442.

Ferron, P. (1978b). Thèse Doctorat d'Etat, Univ. P. et M. Curie, Paris 6, 296 pp.

Ferron, P. and Diomandé, T. (1969). *C.R. hebd. Séanc. Acad. Sci. Paris, Ser. D* 268, 331–332.

Ferron, P. and Hurpin, B. (1974). *Ann. Soc. Ent. Fr.* 10, 771–779.

Ferron, P. and Robert, P.H. (1975). *J. Invertebr. Path.* 25, 379–388.

Ferron, P., Hurpin, B. and Robert, P.H. (1969). *Entomophaga* 14, 429–437.

Ferron, P., Hurpin, B. and Robert, P.H. (1972). *Entomophaga* 17, 165–178.

Frappier, F., Ferron, P. and Païs, M. (1975) *Phytochemistry;* 14, 2703–2705.

Fujishita, A. and Kushida, T. (1976). *Bull. Shizucka Pref. Forest Exp. Stn.* 8, 1–14.

Goral, V.M. (1971). *USSR Patent No.* 301 142.

Goral, V.M. (1972a). *Mikrobiol. Zh. Kiev* 24, 90–91.

Goral, V.M. (1972b). *Zakist Roslin Kiev* 16, 34–40.

Goral, V.M. (1972c). *Zakist Roslin Kiev* 15, 35–43.

Goral, V.M. (1973a). *Proc. 5th Int. Colloq. Insect Path. Microbiol Control,* Oxford. p. 73.

Goral, V.M. (1973b). *Zakist Roslin Kiev* 17, 48–51.

Goral, V.M. (1975). *Mikol. Fitopatol.* 9, 98–103.

Goral, V.M. and Lappa, N.V. (1972). *Mikrobiol. Zh. Kiev* 34, 454–457.

Goral, V.M. and Lappa, N.V. (1973). *Zash. Rast. Kiev* **1**, 19.

Goral, V.M. and Lappa, N.V. (1975). *Zakist Roslin Kiev* **22**, 36–40.

Gruner, L. (1973). *Ann. Zool. Ecol. anim.* **5**, 335–349.

Gruner, L. and Abud-Antun, A. (1976). *Turrialba* **26**, 241–246.

Guagliumi, P. (1973). *Bull. FAO* No. AT 3216, 26 pp.

Guagliumi, P., Marques, E.J. and Vilas Boas, A.M. (1974). *Boll. tecn. Commissão Executiva de Defesa Fitossanitaria de Lavoura Canavieira de Pernambuco, Recife* **3**, 54 pp.

Halperin, J. and Shanuni, Y. (1974). *In* "Institute of Plant Protection – Scientific Activities 1971–1974". Bet Dagan, Israël, p. 33.

Hamill, R.L., Higgens, C.E., Boaz, H.E. and Gorman, M. (1969). *Tetrahedron Lett.* **49**, 4255–4258.

Hammill, T.M. (1972). *Am. J. Bot.* **59**, 317–326.

Ignoffo, C.M., Hostetter, D.L., Sikorowski, P.P., Sutter, G. and Brooks, W.M. (1977). *Environ. Ent.* **6**, 411–415.

Il'Icheva, S.N., Alioshina, O.A., Kononova, E.V. and Yurshenene, Ya. E. (1976). *Mikrobiol. Fitopatol.* **10**, 87–92.

Isaenko, N.I., Kozachenko, V.I., Kononova, E.V. and Sarkisova, N.B. (1974). *In* "Méthode Biologique dans la Protection des Plantes. Thèses et Conférences des Jeunes Chercheurs". Vses. Ord. Lenina Akad. Sel'Skhokhoz. Nauk. Im. V.I. Lenina. Vses. nauch.-issled Inst. Biol. Meth. Zash. Rast., Kichinev, pp. 12–13.

Joussier, D. (1977). Thèse de 3e cycle Faculté des Sciences de la Vie et de l'Environement, Univ. Dijon, 125 pp. polycop.

Joussier, D. and Catroux, G. (1976). *Entomophaga* **21**, 223–225.

Kalvish, T.K. (1974). *Izv. Sib. Otd. Akad. Nauk SSSR* **5**, 67–77.

Kawakami, K. and Mikuni, T. (1965). *Acta Sericol.* **56**, 35–41.

Kazakova, T.A., Il'Icheva, S.N., Kononova, E.V. and Savina, V.E. (1974). In "Méthode Biologique dans la Protection des Plantes. Thèses et Conférences des Jeunes Chercheurs" Vses. Ord. Lenina Akad. Sel'Skhokhoz. Nauk. Im. V.I. Lenina. Vses. nauch.-issled Inst. Biol. Meth. Zash. Rast., Kichinev, pp. 15–16.

Kirsanova, R.V. and Usenko, L.I. (1974). *Genetika* **10**, 97–102.

Kirsanova, R.V., Levitin, M.M., Lekarkina, L.P., Usenko, L.I. and Sharygin, V.I. (1975). *Zh. Obshch. Biol.* **36**, 251–258.

Kondryatiev, N.N., Alioshina, O.A., Il'Itcheva, S.N., Perikhanova, A.G., Sinitsina, L.P., Oupenskaia, A.A. and Chagov, E.M. (1971). USSR Patent No. 313 531.

Kucera, M. (1971). *J. Invertebr. Path.* **17**, 211–215.

Lappa, N.A. (1975). *Zakhist Roslin Kiev* **21**, 61–68.

Lappa, N.A. and Goral, V.M. (1975). *Zakhist Roslin Kiev* **21**, 54–61.

Latch, G.C.M. (1976). *Entomophaga* **21**, 31–38.

Latch, G.C.M. and Falloon, R.E. (1976). *Entomophaga* **21**, 39–48.

Leopold, J., Samšiňáková, A. and Misiková, S. (1973). *Zentbl. Bakt. Parasitkde. Infectionskr. Hyg. Abt.* 2 **128**, 31–41.

Litvinenko, A.I. (1974). *In* "Biological Methods for Plant Protection". pp. 20–21, Kichinev, Inst. Biol. Met. Zash. Rast.

480　　　　　　　　　　　　　P. FERRON

McCauley, V.J.E., Zacharuk, R.Y. and Tinline, R.D. (1968). *J. Invertebr. Path.* 12, 444–459.
McInnes, A.G., Smith, D.G., Wat, C-K., Vining, L.C. and Wright, J.L.C. (1974). *J. Chem. Soc. Chem. Commun.* 8, 281–282.
McInnis, Th. Jr. (1975). *In* "Biological Regulation of Vectors. The Saprophytic and Aerobic Bacteria and Fungi. A Conference Report". (J.D. Briggs, ed.) US Dep. Hlth, Education and Welf Publ. no. (NIH) 77-1180, 95-110.
Marchal, M. (1977). *Rev. Zool. Agric. Path. Vég.* 76, 101–108.
Martynenko, V.V. (1975). *Mikol. Fitopatol.* 9, 333–337.
Mayorova, I.P., Solomatina, P.S., Yarovenko, M.L. and Kononova, E.V (1976). *Mikol. Fitopatol.* 10, 353–356.
Mikuni, T. and Kawakami, K. (1975). *Jap. J. Appl. Ent. Zool.* 19, 203–207.
Milner, R.J. and Lutton, G.G. (1976). Proc. 1st Int. Colloq. Invertebr. Path. and IXth Ann. Meet. Soc. Invertebr. Path. Kingston, 428–429.
Moore, G.E. (1973). *Environ. Ent.* 2, 54–57.
Moura Costa, M.D. de and Magalhaes, C.D. (1974). *Boll. Inst. Biol. Bahia Salvador* 13, 57–60.
Moura Costa, M.D. de, Matta, E.A.F. da, Magalhaes, C.D. and Matos, D.P. de. (1974). *Boll. Inst. Biol. Bahia Salvador* 13, 85–89.
Müller-Kögler, E. (1960). *Z. PflKrankh. PflPath. PflSchutz* 67, 663–668.
Müller-Kögler, E. (1966). *In* "Insect Pathology and Microbial Control". Proc. Int. Colloq. Insect Path. Microbiol. Control, Wageningen, pp. 339–353.
Müller-Kögler, E. and Stein, W. (1970). *Z. Angew. Ent.* 65, 59–76.
Müller-Kögler, E. and Stein, W. (1976). *Z. PflKrankh. PflPath. PflSchutz* 83, 96–108.
Naganawa, H., Takita, T., Suzuki, A., Tamura, S., Lee, S. and Izumiya, N. (1976). *Agric. Biol. Chem.* 40, 2223–2229.
N'Doye, M. (1976). *Entomophaga* 21, 371–376.
Nikolaev, A.N. and Akimkina, L.I. (1974). *In* "Biological Methods for Plant Protection", pp. 18–19. Kichinev: Inst. Biol. Met. Zash. Rast.
Paris, S. (1977). *Mycopathologia* 61, 67–75.
Paris, S., Bizzini, B. and Segretain, G. (1975) *Ann. Inst. Pasteur, Paris* 126 (A), 193–201.
Paris, S. and Segretain, G. (1975). *Entomophaga* 20, 135–138.
Pavlyushin, V.A. (1974). *In* "Biological Methods for Plant Protection", pp. 16–18, Kichinev: Inst. Biol. Met. Zash. Rast.
Pinnock, D.E., Garcia, R. and Cubbin, C.M. (1973). *J. Invertebr. Path.* 22, 143–147.
Primak, T.A. (1967). *Zashch. Rast, Kiev,* 5, 25–28.
Pristavko, V.P. (1966). *Entomophaga* 11, 311–324.
Pristavko, V.P. and Yanishevskaya, L.V. (1971). *Zool. zh.* 50, 1255–1256.
Reisinger, O. and Olah, G.M. (1974). *Can J. Microbiol.* 20, 1387–1392.
Reisinger, O., Fargues, J., Robert, P. and Arnould, M.F. (1977). *Ann. Microbiol. (Inst. Pasteur)* 128 B, 271–287.
Roberts, D.W. (1966) *J. Invertebr. Path.* 8, 222–227.

Roberts, D.W. (1974). *In* "Le Contrôle des Moustiques" (A. Aubin, J.P. Bourassa, S. Belloncik, M. Pellissier and E. Lacoursière, eds), Univ. Quebec, pp. 143–193.

Roberts, D.W. (1975). *In* "Biological Regulation of Vectors. The Saprophytic and Aerobic Bacteria and Fungi. A Conference Report". Publ. No. (NIH) 77-1180, 85–93.

Roberts, D.W. and Campbell, A.S. (1977). *Misc. Publs. Ent. Soc. Am.* 10, 1–80.

Roberts, D.W. and Yendol, W.G. (1971). *In* "Microbial Control of Insects and Mites" (H.D. Burges and N.W. Hussey, eds), pp. 125–149. Academic Press, London.

Samšiňáková, A. (1964). *Naturwissenschaften* 51, 121–122.

Samšiňáková, A. (1966). *J. Invertebr. Path.* 8, 395–400.

Samšiňáková, A. and Misiková, S. (1973). *Ceska Mykol.* 27, 55–60.

Samšiňáková, A. and Samsinak, K. (1970). *Z. ParasitKde* 34, 351–355.

Samšiňáková, A. Misiková, S. and Leopold, J. (1971). *J. Invertebr. Path.* 18, 322–330.

Samšiňáková, A., Leopold, H. and Kâlalová, S. (1976). *Zentbl. Bakt. ParazitKde. Infektionskr. Hyg. Abt 2* 131, 60–65.

Sannasi, A. (1969). *J. Invertebr. Path.* 13, 11–14.

Sannasi, A. and Oliver, J.H. Jr. (1971). *J. Invertebr. Path.* 17, 354–365.

Schabel, H.G. (1976a). *Z. Angew Ent.* 81, 413–421.

Schabel, H.G. (1976b). *J. Invertebr. Path.* 27, 377–383.

Schneiderman, H.A., Gilbert, L.I. and Weinstein, M.J. (1960). *Nature Lond.* 188, 1041–1042.

Segretain, G., Paris, S., Ferron, P. and Arcouteil, A. (1971). *C.R. hebd. Séanc. Acad. Sci., Paris* Ser. D 173, 140–142.

Seryczynska, H. and Bajan, C. (1975). *Bull. Acad. Sci. Pol. Ser. Sci. Biol.* 23, 267–271.

Shamraj, L.G., Sinitsina, L.P. and Vrublevskaya, L.S. (1974). *In* "Méthode Biologique dans la Protection des Plantes. Thèses et Conférences des Jeunes Chercheurs". Vses. Ord. Lenina Akad. Sel'Skhokhoz. Nauk Im. V.I. Lenina. Vses. Nauk.-issled Inst. Biol. Meth. Zash. Rast., Kichinev, pp. 10–12.

Sikura, A.I. (1974). *In* "Moyens Biologiques de Protection des Végétaux" (E.M. Chumakova, E.V. Gusev and N.S. Fedorinchik, eds), Moscow, Kolos, pp. 68–74.

Sikura, A.I. and Bevzenko, T.M. (1972). *Vopr. Biol. Zacht. Rast., Kichinev*, pp. 68–74.

Sikura, A.I. and Gritsaenko, N.N. (1973). *Vopr. Zatch. Rast., Kichinev*, 2, 90–94.

Sikura, A.I., Cherebedova, M.A. and Korina, A.V. (1974). *In* "Entomophages, phytophages et micoorganismes dans la protection des plantes". Vses. Ord. Lenina Akad. Sel'Skhokhoz. Nauk. Im. V.I. Lenina. Vses. Nauchn.-issled. Inst. Biol. Metod. Zash. Rast., Kichinev, p. 56–99.

Suzuki, A., Taguchi, H. and Tamura, S. (1970). *Agric. Biol. Chem.* 34, 813–816.

Suzuki, A., Kawakami, K. and Tamura, S. (1971). *Agric. Biol. Chem.* **35**, 1641–1643;

Suzuki, A. and Tamura, S. (1972). *Agric. Biol. Chem.* **36**, 896–898.

Taylor, J.W. and Franklin, R.T. (1973). *Can. Ent.* **105**, 123–125.

Tedders, W.L., Weaver, D.J. and Wehunt, E.J. (1973). *J. Econ. Ent.* **66**, 723–725.

Telenga, N.A. and Goral, V.M. (1966). *In* "Conf. Samarcande". Iz. Fan. R.S.S.Uz., Tachkent, 171–173.

Telenga, N.A., Sikura, A.I. and Smetnik, A.I. (1967). *Zashch. Rast. Kiev*, **4**, 3–23.

Tinline, R.D. (1971). *Mycologia* **63**, 713–721.

Tinline, R.D. and Noviello, C. (1971). *Mycologia* **63**, 701–712.

Tjulpanov, V.G., Kovrov, B.K. and Kovalev, V.S. (1977). *Prikl. Biokhim. Mikrobiol.* **13**, 255–259.

Tumarkin, R.I., Kononova, E.I. and Markova, N.G. (1974). *Prikl. Biokhim. Mikrobiol.* **10**, 74–79.

Usenko, L.I., Kirsanova, R.V. and Levitin, M.M. (1973). *Genetika* **9**, 1421–1425.

Veen, K.H. (1968). *Meded. LandbHoogesch. Wageningen* **68**, 1–117.

Vey, A. and Fargues, J. (1977). *J. Invertebr. Path.* **30**, 207–215.

Vey, A. and Quiot, J.-M. (1975). *C.R. hebd. Séanc. Acad. Sci. Paris, Ser. D* **280**, 931–934.

Vey, A., Quiot, J.-M. and Vago, C. (1973). *C.R. hebd. Séanc. Acad. Sci. Paris, Ser. D* **271**, 1489–2492.

Vey, A., Bouletreau, M., Quiot, J.-M. and Vago, C. (1975). *Entomophaga* **20**, 337–351.

Viega, A.S.L., Aquino, M.L.A. and Pereira, G. (1972). *Pesq. Agrop. Nord, Recife* **4**, 71.

Villacorta, A. (1976). *Ann. Soc. Ent. Bras.* **5**, 102–104.

Young, E.C. (1974). *J. Invertebr. Path.* **24**, 82–92.

Yurchenko, L.V., Zakharov, I.A. and Levitin, M.M. (1974). *Genetika* **10**, 95–101.

Zacharuk, R.Y. (1970a). *J. Invertebr. Path.* **15**, 81–91.

Zacharuk, R.Y. (1970b). *J. Invertebr. Path.* **15**, 372–396.

Zacharuk, R.Y. (1971a). *Can. J. Microbiol.* **17**, 281–289.

Zacharuk, R.Y. (1971b). *Can. J. Microbiol.* **17**, 525–529.

Zacharuk, R.Y. (1973a). *Misc. Publ. Ent. Soc. Am.* **9**, 112–119.

Zacharuk, R.Y. (1973b). *J. Invertebr. Path.* **21**, 101–106.

Zacharuk, R.Y. (1974). *J. Invertebr. Path.* **23**, 13–21.

Zacharuk, R.Y. and Tinline, R.D. (1968). *J. Invertebr. Path.* **12**, 294–309.

Zakharchenko, N.L. (1967). *Zash. Rast., Kiev* **4**, 134–138.

Zakharchenko, N.L., Primak, T.A. and Goral, V.M. (1963). *In* "Kolonadskii jouk ta novy metodi borot'bi s nim". Naukovi fratsi, 12: Institut Ukrainien de la Protection des Plantes. Ministère de l'Agriculture de la République d'Ukraine. Editions de litterature agricole de la Rèpublique d'Ukraine, Kiev, pp. 102–106.

CHAPTER 25

The Fungus *Verticillium lecanii* as a Microbial Insecticide against Aphids and Scales

R.A. HALL

Glasshouse Crops Research Institute, Littlehampton, England

I. Introduction

Verticillium lecanii is a well-documented, extremely widespread entomopathogen. However, spectacular epizootics are observed only amongst its most common hosts, aphids and scales, in tropical and sub-tropical regions. Thirteen varied successes in controlling aphids and scales by *V. lecanii* outdoors from 1910 to 1941 are summarized by Baird (1958). By contrast, in temperate climates, it has never been implicated in epizoosis, although it is frequently isolated from individual insects (Petch, 1948; Leatherdale, 1970; Barson, 1976). Indoors, in the somewhat tropical environment of greenhouses in N. Europe and the USA, it frequently decimates populations of scales and aphids, e.g. 100% mortality of several target insects was reported in limited trials (Neuzilová, 1957; Samsiñáková and Kálalová, 1976).

In view of this potential, it is surprising that *V. lecanii* has attracted so few and limited studies. The efficiency with which aphid and scale populations are sometimes controlled in greenhouses in England, prompted research on the ability of *V. lecanii* to reliably control greenhouse pests. The object of this chapter is to present up-to-date information on *V. lecanii* and an appraisal of its feasibility as a microbial insecticide of protected crops. Most of the author's own work used a single-spore isolate of *V. lecanii* designated as strain C-3.

II. Classification and Description

Corda in 1839 (Gams, 1971) placed all hyphomycetes bearing heads of single-celled spores at the tips of undifferentiated conidiophores in the genus *Cephalosporium*. However, this broad categorization caused some confusion in the interpretation of this genus.

In a more recent and detailed study, Gams (1971) re-examined the cephalosporia, placing a group of species containing many entomogenous fungi in a new section, *Verticillium* sect. *Prostrata*. The main characteristic of fungi in this new section is the deep velvet- or cotton-like structure of the aerial mycelium which may occasionally contain mesotonous to acrotonous whorls of phialides. One of these, *Verticillium lecanii*, contains numerous hitherto separate species, many of which are insect pathogens described by Petch (1925, 1931, 1948). Gams (1971) concluded that Petch's taxonomic criteria did not justify separation into species. In contrast, Balazy (1973) believed that Gams' work generalized too much, even questioning his inclusion of certain entomogenous cephalosporia in *Verticillium* sect. *Prostrata*. Balazy separated some strains from Gams' *V. lecanii* complex into separate species. However, in my experience, such is the continuous nature of the variability between strains of the *V. lecanii* complex, that I prefer Gams' somewhat broader concept of the fungus. The following is taken from Gams' (1971) description of *V. lecanii*: Colonies on agar, after 10 days, 18–22 mm in diameter, white or pale yellow, cottony–velvety. Phialides awl-shaped, very variable in size, single or in small groups of verticillate whorls on aerial mycelium. Conidia, in terminal heads of slime on phialides, are cylindrical to elipsoidal with symmetrically rounded ends. Chlamydospores absent.

III. Natural Occurrence

The most frequently recorded hosts are scale insects (McClelland and Tucker, 1929; Viégas, 1939; Ganhão, 1956) and aphids (Petch, 1948; Wilding, 1972;

Nagaich, 1973). Much less frequent are reports of hosts in other orders of insects, e.g. Coleoptera (Leatherdale, 1970; Lipa, 1975; Barson, 1976), Collembola, Diptera and in one Ichneumonid from spiders (Leatherdale, 1970; Petch, 1948; Gams, 1971) and eriophyid mites (Massee *in* Taylor, 1909). Also, *V. lecanii* sometimes hyperparasitizes phytopathogenic fungi, mostly rusts (Heterobasidiomycetaceae — Kotthoff, 1937; Hassebrauk, 1936; Silveira and Rodriguez, 1971; Garcia Acha *et al.*, 1965) and powdery mildews (Euascomycetidae — Leeming, 1976). Finally, it has been found on many other substrates, e.g. soil, wood, *Helvella lacunosa* (Gams, 1971) and, perhaps most interestingly, growing on and etching plastic contact lenses (M.P. English, personal communication). Obviously, strains of *V. lecanii* are widespread.

IV. Laboratory Culture

V. lecanii is non-fastidious and will grow on all conventional mycological media so far tested, e.g. Czapeck-Dox, Malt extract, Sabouraud and Potato dextrose agars, including a medium containing chitin as the sole source of carbon and nitrogen (Fig. 1a). On solid media, conidia are produced (Figs 1b and c). In contrast, *V. lecanii* assumes a semi-yeast morphology in liquid media probably in response to accumulation of CO_2 (Hall and Latgé, unpublished observations), forming budding elements known as blastospores (Fig. 1d).

V. Effect of Humidity and Temperature

A. HUMIDITY

Virtually all fungi require high humidity for spore germination, growth and sporulation. *V. lecanii* conidia require a high humidity to germinate and possibly do so only in a water film (Hall, unpublished observations). Thus, to ensure maximum germination of spores and hence highest possible levels of infection of insects, spore sprays should be synchronized with optimal humidity, which for most crops should occur in the evening as ambient temperature falls.

B. TEMPERATURE

Spore germination and colony growth of the C-3 isolate of *V. lecanii* both had similar temperature optima. Conidia germinated most rapidly between 20° and 25°C (Fig. 2). Following overnight incubation, germ tubes did not differ significantly in length between 20° and 25°C but at 15° and 27°C were only 1/2–1/3 as long, while at 11.5°C they were only just visible (Table I). Colonial growth rate was optimal at 23°–24°C (Fig. 3). Both germination and growth declined steeply above 25°C and ceased above 30°C (Figs 2 and 3). Sporulation responded over a slightly narrower temperature range than growth or germination, ceasing at 30°C. Narrower temperature limits for reproduction than for growth are almost universal in the fungi (Hawker, 1950; Cochrane, 1958).

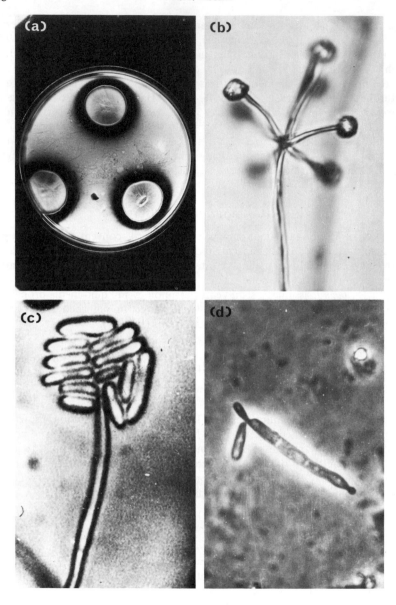

Fig. 1 (a) *Verticillium lecanii* colonies on colloidal chitin agar with clearing zones where chitin has been digested. (b) Slime heads of *V. lecanii* in a verticillate whorl (× 700). (c) Slime head of *V. lecanii* showing conidia inside (× 3000). (d) Dipolar blastospore formation (× 2000).

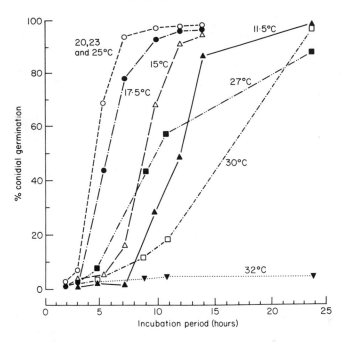

Fig. 2. Effect of temperature on germination of *Verticillium lecanii* conidia.

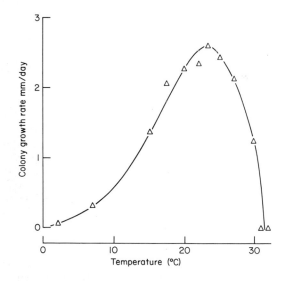

Fig. 3. Effect of temperature on growth of *Verticillium lecanii* colonies on Sabouraud dextrose agar.

TABLE I

Effect of temperature on germ tube lengths of Verticillium lecanii *conidiospores after 14 h incubation*

Temperature, °C	No. of observations	Mean ± standard error
11.5		Germ tubes just visible
13	54	12.1 ± 0.1
15 expt. 1	60	13.9 ± 0.6
expt. 2	65	18.2 ± 0.9
20 expt. 1	30	48.4 ± 3.8
expt. 2	42	51.5 ± 3.7
23	31	55.2 ± 5.1
25	31	56.4 ± 6.3
27	33	25.6 ± 2.4

It is concluded that the overall optimum temperature band for the C-3 isolate is 20°–25°C. The optimal temperature for infection by the C-3 isolate coincided with this band (Table II). However, the effect of temperature on the development of the target insect must be considered as well. For example, a sub-optimal

TABLE II

Effect of temperature on LT_{50} *s of apterous* Macrosiphoniella sanborni *adults treated with aqueous suspensions of* Verticillium lecanii *conidiospores*

Concentration (spores/ml)	11°C	15°C	20°C	25°C
10^6	13	6.87	4.52	4.25
10^7	–	4.69	3.30	3.29

temperature for a pathogen may be equally sub-optimal for the pest. Thus, on chrysanthemums, temperature affected *V. lecanii*-induced aphid mortality to a degree similar to its effect on the rate of increase of *Macrosiphoniella sanborni* populations (Hall, unpublished observations). Alternatively, a sub-optimal temperature for a pathogen may have more effect on the insect pest. Thus, with *Myzus persicae*, the effect of temperature on *V. lecanii*-induced lethal times was less relative to the effect on aphid population increase (Hall, unpublished observations). These results further suggested that in crop conditions where high humidity prevails, control by *V. lecanii* should be similar at temperatures between 11° and 25°C for *M. sanborni* and perhaps better for *My. persicae* at 11°C and 15°C than at 20°C. However, where infection of insects occurs mostly at high humidity at night, and not during the daytime at low humidity, an overnight temperature of 11°C may impair control since germ tubes and presumably invasive appressoria do not form within 14 h, whereas at higher temperatures they grow by this time (Table I). This was demonstrated in the laboratory by a high mortality of spore-treated aphids after 14 h at 100% relative humidity (r.h.) and 15° or 20°C followed by the rest of a 6-day period at 63–72% r.h. and 20°C, compared with a very low mortality where the overnight moist period was at 11°C (Hall, unpublished observations). Therefore, in theory, minimum greenhouse temperatures of 15°C used for crops like chrysanthemums should not impair aphid control by *V. lecanii*. Indeed, in greenhouses, *V. lecanii* controls *My. persicae* on chrysanthemums equally well at minimum temperatures of 15° or 20°C (Hall, 1975; Hall and Burges, 1979). Thus, given similar temperature relationships between host and pathogen on crops with temperature minima in the range 15°–20°C, prospects for microbial control should be good, whereas crops that are grown with minima < 15°C may be less amenable. At the other extreme, daytime temperature in greenhouses in hot weather can rise above 30°C particularly for some greenhouse crops, e.g. peppers, which are not ventilated until the temperature reaches 35°C. In theory, this may damage C-3 (Figs 2 and 3) but previous experience with chrysanthemums in greenhouses in exceptionally hot weather has shown that sustained daytime air temperatures around 35°C did not harm *V. lecanii* (Hall, unpublished observations).

The upper temperature limits for different strains of *V. lecanii* vary greatly (Hall, unpublished observations; Table III). Several strains grew well at 31°C, four grew at 34°C and one grew very slightly at 36°C. Easwaramoorthy and

TABLE III

Upper temperature limits for growth[a] of Verticillium lecanii *strains*

Host and source	Temperature, °C			Author [b]
	31	34	36	
Uromyces appendiculatus, UK	+++	++ (+)	(+)	1
Erysiphe graminis, UK	+++	++	−	1
Brachycaudus helichrysi, UK	++	(+)	−	1
Puccinia graminis, Netherlands	+	(+)	−	1
Rust on chrysanthemums, ?	+++	−	−	1
Brachycaudus helichrysi, UK	+++	−	−	1
Pulvinaria floccifera, Turkey	++	−	−	1
Saissetia oleae, Israel	++	−	−	1
Puccinia graminis, ?	+	−	−	1
Contact lens, UK	(+)	−	−	1
Hemileia vastatrix, New Caledonia	(+)	−	−	1
Hemileia vastatrix, India	−	−	−	1
Ceroplastes floridensis, Israel	−	−	−	1
Myzus persicae, India	−	−	−	1
Macrosiphoniella sanborni, UK	−	−	−	1
Scolytus scolytus, UK	−	−	−	2
	−	−	−	3

[a] Growth +++ luxuriant, ++ moderate, + poor, (+) slight, − no growth.

[b] 1. Hall, unpublished observations. 2. Barson, 1976. 3. Ganhão, 1956 (extrapolated data).

Jayaraj (1977) reported that their strain of *V. lecanii* still grew appreciably at 35°C and so may have an upper temperature limit higher than 36°C. Of the strains in Table II, unexpectedly those from topical and arid countries were less tolerant of high temperatures than those from temperate zones. Strains most tolerant of high temperatures originated from non-insect material, although some strains from rust fungi did not grow at 31°C. The optima and minima of most of these strains has not yet been determined but they are probably related to the maxima. It is possible that in extreme conditions of temperature, some strains may be better than others for different crops.

VI. Virulence and Strain Specificity

Virulence of *V. lecanii* spores can be measured by bioassay (Hall, 1976a). The LC_{50} for a 100% viable conidia suspension in which adult, apterous *M. sanborni* are immersed briefly is ca. 10^5 spores/ml (Hall, 1976a) and that for blastospores is slightly lower (Hall, 1979). However, when untreated aphids are placed on still-wet spore-treated chrysanthemum leaves, the LC_{50} is increased by more than X 100 (Hall, 1979). Furthermore, after drying, spores are presumably not readily dislodged from the leaves by the aphids. This suggests that a prophylactic spore-spray applied to an aphid-free crop may be wasted and ineffective when aphids invade. Infection of aphids by entomogenous fungal spores on leaf surfaces may be generally inefficient; Thoizon (1970) failed to infect aphids placed on leaves or glass slides covered with a layer of *Entomophthora* conidia.

Virulence of single- and multi-spore isolates of the strain C-3 remained remarkably stable on most artificial media, even after as many as 98 subculturings (Hall, unpublished observations). Passaging of *V. lecanii* through an aphid host did not increase virulence (Hall, unpublished observations). There are many reports in the literature which suggest that the virulence of other entomogenous fungi sometimes attenuates during only a few subculturings on media or enhances with passaging through a host (Fox and Jacques, 1958; Kawakami, 1960; Schaerffenberg, 1964; Latch, 1965; Hartmann and Wasti, 1974; Wasti and Hartmann, 1975).

Of ten insect-pathogenic strains of *V. lecanii*, isolated from various insects including aphids, scales and a beetle (*Scolytus scolytus*), all − except one − were similarly pathogenic for the aphid *M. sanborni* (Hall, unpublished observations) Of eight strains isolated from rust and mildew fungi, three were pathogenic for *M. sanborni* (Hall, unpublished observations). Interestingly, a strain isolated from a contact lens was also pathogenic for aphids (Hall, unpublished observations). Of the aphid pathogens, most were as pathogenic per spore as strain C-3 and none significantly more so. Thus, prospects of increasing virulence by selection from existing strains are remote. However, some strains produce smaller spores than C-3 but, taking account of the relative spore sizes, are more prolific. Therefore, small-spore strains of similar pathogenicity to C-3 may be preferred.

VII. Longevity of Spores

So far, a stable storage formulation for *V. lecanii* spores has not been developed. Indeed, present knowledge is sparse. The half-life of conidia in distilled water varies, both at 2°C (110–160 days) and −17°C (60–120 days) (Hall, unpublished observations); blastospores on the whole are even shorter-lived and more variable (10–150 days at 2°C). When conidia were equilibrated at a range of humidities at 20°C, only high humidity permitted good survival. In contrast, dried conidia (whether in slime-heads separated from their parent mycelium or washed) at 58% r.h. died in < 24 h. However, conidia in slime-heads still attached to the parent mycelium on aphids or on culture mycelium (without agar) survived for up to 13 days at 58% r.h. but this may have been due to a microclimate effect under the conditions of study (Hall, unpublished observations). A favourable microclimate humidity was probably responsible for good survival of spores on aphid bodies killed by *V. lecanii* in greenhouses; in one experiment, 80–90% of conidia survived to the end of the experiment for at least 30 days after death of aphids despite daytime air temperatures of well over the upper temperature limit for growth (Hall, unpublished observations).

VIII. Spore Dispersal and Spread of Infection

Fungal spores in slime-heads adhere firmly to the mycelium when dry (Gregory, 1952). *V. lecanii* conidia did not become airborne from dried cultures or from *V. lecanii*-killed aphids (Hall, 1977). However, possibly following initial dispersal of spores in slime by wetting, spores may be further dispersed in currents of air after drying (Gregory, 1952) but, since desiccation rapidly kills detached *V. lecanii* spores (see above), airborne spores in dry air would probably sooner or later die. Probably, therefore, infection of insects from naked airborne *V. lecanii* spores is virtually impossible. Presumably, infection of a new aphid population on a new crop originates from soil. *V. lecanii* has occasionally been isolated from soil (Gams, 1971), and Barron (1968) stated that *Verticillium* spp. are extremely common in this environment. Itinerant aphids or aphids dislodged from plants could perhaps contact *V. lecanii*-bearing material or insect remains on soil. Alternatively, *V. lecanii* may be splash-dispersed on to aphid-infested plants by rain or irrigation. Other creatures moving from soil to plants, e.g. mites, ants, sciarids, may also carry spores to aphids or scales – particularly those insects, such as ants (Hussey *et al.*, 1969) and sciarids (Tiensuu, 1936), attracted to honeydew. Although the probability of an individual host insect becoming infected by any of these routes may be low, the likelihood of populations being infected increases as host numbers increase.

Once disease is introduced either naturally, or artificially by spore sprays, it probably spreads by a combination of contagion and dispersal of spores as described above. The relative importance of these modes of spread will depend upon the behaviour and ecology of the insect species.

IX. Control of Aphids and Scales in Greenhouses

A. APHIDS ON CHRYSANTHEMUMS

In extensive greenhouse experiments to control three aphid pests, infested chrysanthemums (cultivar "Tuneful", which is very susceptible to aphids) were sprayed to "run-off" with *V. lecanii* conidia or blastospores suspended in phosphate buffer containing 0.02% Triton X-100 as wetting agent. After evening sprays (17.30–18.30 h) plants were covered with polythene blackout sheets, as used in commerce, to restrict daylength for flower initiation. These blackouts also boosted overnight humidity. In small greenhouses control of *M. sanborni*, an occasional pest of the stems of chrysanthemums, was variable; in some experiments aphids were virtually eliminated whereas in others control was only partial and thus commercially unsatisfactory, but good when compared with untreated aphid populations (Hall, 1975; Hall and Burges, 1979). Also in small greenhouses, another occasional aphid pest, *Brachycaudus helichrysi*, was entirely controlled inside vegetative plant tips, which the species preferentially infested on immature plants, but later control was only partial when aphids were forced to feed on the outside of tightly-closed flower buds, where doubtlessly the microclimate humidity was lower (Hall, 1975; Hall and Burges, 1979). However, in trials in both small and large greenhouses, single sprays of *V. lecanii* spores have successfully and consistently controlled the major aphid pest of chrysanthemums, *Myzus persicae*, even sparse populations — whether introduced or natural — being virtually wiped out (Hall and Burges, 1979). A single spray eliminates low-density aphid populations scattered over leaves in any position on the plant and maintains control for the duration of the crop (3 months) (Fig. 4).

The differences observed in aphid susceptibility in trials in small greenhouses are not apparently inherent. In laboratory assays (Hall and Burges, 1979) the

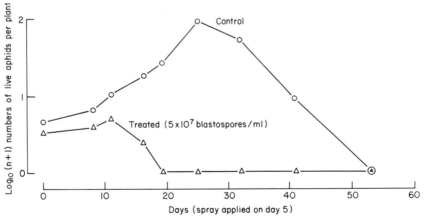

Fig. 4. Control of *Myzus persicae* on chrysanthemums by *Verticillium lecanii*. Fungus spread naturally from treated to control plants.

LC$_{50}$ s of these three species were similar, *My. persicae* if anything being the most resistant. Therefore, ecological and behavioural factors must account for the differences observed in the greenhouse environment. These are probably a combination of differences such as aphid mobility and possibly a different microclimate humidity in the different feeding positions.

V. lecanii is expected to give better control of *M. sanborni* and *B. helichrysi* in large commercial greenhouses, since humidities should be higher than in small houses owing to greater foliage density, fewer draughts and greater ratio of volume to glass area. This is also suggested by good control in commercial greenhouses of fortuitous infestations of the black bean aphid, *Aphis fabae* (Hall, unpublished observations), which occupies a similar habitat to *M. sanborni* on the chrysanthemum stem. Another aphid pest, *Aphis gossypii*, unexpectedly invaded experiments in a large greenhouse intended to assess control of *My. persicae*, completely swamping this aphid; it was controlled within 3 weeks and maintained at low levels (< 1 aphid/plant) for the crop duration, despite heavy waves of natural re-infestations (Hall, unpublished observations).

For aphids, ultimate control of populations depends on contagion. A spray only introduces the disease into the population. Infection appears 5–10 days after spraying. During this time the aphids increase in numbers, since even heavily diseased aphids produce healthy offspring at near-normal rates (Table IV; Hall, 1976b). That contagion can be extremely efficient is evinced by the death

TABLE IV

Production of Macrosiphoniella sanborni *progeny by diseased, adult aphids in a 24-h period one day before death due to* Verticillium lecanii

Expt. no.	Spore concentration/ml			
	0 (control)[a]	5 × 10^5	10^6	2 × 10^6
1	1.5	1.1	1.2	1.2
2	1.7	1.6	1.4	1.2
3	1.7	1.1	1.7	—
4	2.7	2.5	2.2	1.9

[a] Mean daily progeny/control aphid during period over which *V. lecanii* killed treated aphids.

of sparse *My. persicae* populations on the undersides of lower leaves in the middle of dense chrysanthemum beds where the spray could not possibly have reached (Hall and Burges, 1979).

Because of the role of contagion, the lowest spore concentration which will elicit reliable control must be assessed from trials in greenhouses. In laboratory assays (Hall, 1977), aphids briefly but wholly immersed in high blastospore concentrations (10^7–10^8 spores/ml) died 2–3 times faster (LT$_{50}$ s, 1.9–2.3 days) than those treated with a lower concentration, 10^5 spores/ml (LT$_{50}$ s, 4.8–6.2 days). Since in greenhouses a spore-spray would miss some aphids on plants, practical LT$_{50}$ s should be considerably longer and a high spray inoculum should be preferable. Nevertheless, in greenhouses, sprays of 10^5 and 10^7

spores/ml produced very similar levels of *My. persicae* control (Hall, unpublished observations), which emphasizes the importance of contagion.

B. APHIDS ON CUCUMBERS

In two trials with *Aphis gossypii*, a spore spray of *V. lecanii* controlled sparse aphid populations within 14 days (Hall, unpublished observations). However, 4–6 weeks later, aphids increased again, but the fungus did not spread efficiently, especially to aphids on new foliage – in contrast to the above results on chrysanthemums. Possibly, *A. gossypii* is less mobile on cucumbers than on chrysanthemums. Thus, more than one spray will probably be necessary to control this pest on cucumbers.

C. WHITEFLY ON CUCUMBERS

In trials with whitefly, *Trialeurodes vaporariorum*, results were similar to those for *A. gossypii* (P. Kanagaratnam, Hall and Burges, unpublished observations). All stages except eggs in an initial infestation wetted by a spore-spray are controlled but, since immature stages are immobile, contagion is not efficient within a whitefly population and healthy survivors missed by the spray lay eggs on new foliage, causing a recovery in infestation. So, again, more than one spray is necessary to control this pest. However, since adults tend to lay eggs on young leaves only, just the new foliage need be sprayed.

X. Appraisal of *Verticillium lecanii* as a Microbial Insecticide

A. PRODUCTION AND STORAGE

The choice of infectious material is between conidia and blastospores. Production of conidia on agar is too expensive and it is also difficult to ensure culture purity. Alternatively, conidia can be produced on a cheap granular solid substrate such as grain for 5–6 days in aerated vessels.

Submerged culture of blastospores would be cheapest, producing between 10^9 and 10^{11} blastospores/g of substrate in liquid media, depending on the strain. *V. lecanii* blastospores are less stable than conidia as with some other entomogenous fungi, e.g. *Beauveria bassiana* (Samsiňáková, 1966; Müller-Kögler, 1967) and *B. brongniartii* (*tenella*) (Ferron, 1967; Blachère *et al.*, 1973). Undoubtedly, longevity of stored *V. lecanii* blastospores could be improved using methods similar to those of Blachère *et al.* (1973) for *B. brongniartii* blastospores. Be that as it may, existing knowledge does not permit production of a commercially feasible long-term storage formulation for either conidia or blastospores of *V. lecanii*. However, the fungus could be cultured and distributed for immediate use by commercial growers as at present with other biological agents, such as insect predators or parasites. *V. lecanii* would have the advantage of longer storability than these.

B. STANDARDIZATION

The unwelcome risk of variation in virulence can be minimized by storing in deep-freeze or liquid nitrogen a large stock of replicate inocula grown from a single-spore isolate of a standard strain. This is a wise precaution even though the variation in *V. lecanii* pathogenicity is not great (Hall, unpublished observation).

Despite theoretical strain stability, a commercial product must be shown to have constant potency. This should be measured by viable spore count and bioassay. Viability of *V. lecanii* conidia and blastospores can easily be assessed by an agar—slide technique (Hall, 1976a). Pathogenicity of production batches of spores should be measured by bioassay in comparison with a standard since variation between consecutive assays is often significant (Hall, 1976a).

C. RELIABILITY OF PRACTICAL CONTROL

In small greenhouses, results of trials were variable with the minor aphid pests, *M. sanborni* and (on budded plants only) *B. helichrysi*, but results may be better in large commercial greenhouses. However, in extensive trials — large and small — by the author, at all seasons, *V. lecanii* always established and maintained excellent control even of sparse populations of *My. persicae*, the major aphid pest on chrysanthemums.

D. COMPATIBILITY WITH EXISTING CONTROL MEASURES

The pathogenicity of *V. lecanii* for parasitic and predatory arthropods, already widely-used for biological control of greenhouse pests, has not yet been tested in depth. However, in trials against aphids and whiteflies, *V. lecanii* has not been observed to attack the greenhouse red spider mite, *Tetranychus urticae*, its predator, *Phytoseiulus persimilis* or *Encarsia formosa,* the whitefly parasite, when the fungus occurred naturally in crops where these agents were active.

Thus, parasites and fungus can probably be integrated. For example where a whitefly infestation is too large for control by its parasite, pest numbers could be reduced by a spray of *V. lecanii* spores to a level at which the parasite could achieve a sub-economic equilibrium with its host.

Laboratory results *in vitro* on the effect of 33 chemical insecticides and fungicides on spore germination and growth on agar suggest that a judicious selection of insecticides — and even some broad-spectrum fungicides — can be effectively integrated with *V. lecanii* (Hall, unpublished observations). Moreover, results in greenhouses suggest that some chemicals, toxic to *V. lecanii in vitro*, may not upset pest control by the fungus, providing they are wisely applied. For example, dioxathion used against leaf miner on chrysanthemums, totally inhibits *V. lecanii* spore germination on agar at the recommended dosage. However, even when applied on plants at normal strength 24 h before *V. lecanii* spores, it caused virtually no diminution in pest control by the fungus (Hall, unpublished observations).

E. SAFETY TO MAN AND OTHER ANIMALS

The absence of records of *V. lecanii* in man and other vertebrates is impressive evidence for its innocuity. All *V. lecanii* strains so far examined by the author cannot grow at 37°C (Table III) and so the likelihood of infecting warm-blooded vertebrates internally is exceedingly remote. Before the start of greenhouse experiments, test mice each received 10^6 conidia of C-3, injected intravenously. No adverse symptoms were observed and, 28 days later, no gross pathological changes were apparent in the internal organs and no signs of the fungus could be found either in sectioned organs or in agar cultures from these (P.K.C. Austwick, personal communication). The author and others, who have worked closely with the fungus for several years, have never suffered any ill-effects, e.g. allergic reactions, whatsoever. *V. lecanii* was, however, amongst several fungi found in grain dust that caused respiratory symptoms among combine harvester drivers (Darke *et al.*, 1976). Safety tests are being carried out preparatory to commercial exploitation.

Most useful insects, including pollinators, are probably not infected by *V. lecanii* since it has been recorded from only one hymenopteron (Leatherdale, 1970). The few reports in the literature of it infecting arachnids have not to the author's knowledge been confirmed experimentally. Infections were never encountered in spiders which were always prevalent in crops treated with *V. lecanii* or in other denizens of the greenhouse, e.g. ladybirds, ants, woodlice and earwigs (Hall, unpublished observations). A sense of perspective should prevail when one realizes that *V. lecanii* is ubiquitous and already causes epizootics amongst insects on food crops, thereby liberating more spores into the environment than any spray program is likely to achieve.

XI. Conclusions, Future Prospects and Research

V. lecanii is clearly a promising biological control agent against aphids on chrysanthemums in greenhouses. Most encouraging is that success can be very spectacular and exceed expectations, as with *My. persicae* on chrysanthemums. Further trials in commercial greenhouses are in progress and there are prospects for commercial development.

Probably *V. lecanii* could be used in greenhouses against other insect pests and on other crops. It must be emphasized here that the standards of aphid control on ornamentals like chrysanthemums are very high, i.e. < 1 aphid/plant and, indeed, this was consistently achieved for *My. persicae*, but not always for the occasional pests, *M. sanborni* and *B. helichrysi*. However, the degree of control of *M. sanborni* was very good relative to untreated aphids and this control level would be satisfactory on other crops, e.g. food crops in which the infested part is not marketed.

Outdoors, critical parameters like temperature and humidity cannot be readily manipulated. Temperate weather is often unpredictable and often unfavourable to fungi, except perhaps where the pest occupies a moist micro-climate. However, great potential may exist in tropical and subtropical zones where high humidity is normal for prolonged periods. Indeed, van Brussel (1975) working with *Hirsutella thompsonii* successfully controlled the eriophyid mite, *Phyllocoptruta oleivora*, even in the dry season, in Surinam.

The most pressing research need at present is to develop a long-term viable storage formulation of *V. lecanii* with a shelf life comparable with those of conventional insecticides. Secondly, the development of a spray formulation containing nutrients and water-retaining agents (e.g. alginates) should be investigated. This may obviate the need for multiple aqueous sprays of spores necessary at present to control certain predominantly sedentary pests, by permitting limited growth and sporulation of *V. lecanii* on leaf surfaces. This would give the fungus a measure of persistence and increase the chances of insects (and their offspring), which have been missed by a spray, becoming subsequently infected.

References

Baird, R.B. (1958). "The Artificial Control of Insects by Means of Entomo-genous Fungi: a Compilation of References with Abstracts". Ent. Lab., Belleville, Ontario.

Balazy, S. (1973). *Bull. Soc. Amis. Sci. Lett., Poznan,* D **14**, 101–137.

Barron, L. (1968). "The Genera of Hyphomycetes from Soil". Williams and Wilkins, Baltimore, Md.

Barson, G. (1976). *Ann. appl. Biol.* **93**, 207–214.

Blachère, H., Calvez, J., Ferron, P., Corrieu, G. and Peringer, P. (1973). *Annls. Zool. Ecol. anim.* **5**, 69–79.

Brussel, E.W. van (1975). Ph.D. thesis, Agric. Univ. Wageningen, Netherlands.

Cochrane, V.W. (1958). "Physiology of Fungi". Wiley, New York. Chapman and Hall, London.

Darke, C.S., Knowelden, J., Lacey, J. and Milford Ward, A. (1976). *Thorax* **31**, 294–302.

Easwaramoorthy, S. and Jayaraj, S. (1977). *J. Invertebr. Path.* **29**, 399–400.

Ferron, P. (1967). *Entomophaga* **12**, 257–293.

Fox, C.J.S. and Jacques, R.P. (1958). *Can. Ent.* **90**, 314–315.

Gams, W. (1971). "*Cephalosporium* — artige Schimmelpilze (Hyphomycetes)" Gustav Fischer Verlag, Stuttgart.

Ganhão, J.F.P. (1956). *Brotéria* **25**, 71–135.

Garcia Acha, I., Leal, J.A. and Villanueva, J.R. (1965). *Phytopathology* **55**, 40–42.

Gregory, P.H. (1952). *Trans. Br. mycol. Soc.* **35**, 1–18.

Hall, R.A. (1975). Proc. 8th Br. Insecticide Fungicide Conf. 93–99.

Hall, R.A. (1976a). *J. Invertebr. Path.* **27**, 41–48.

Hall, R.A. (1976b). *J. Invertebr. Path.* **28**, 389–391.

Hall, R.A. (1977). Ph.D. thesis. Univ. Southampton.

Hall, R.A. (1979). *Entomophaga*, **24**, 191–198.

Hall, R.A. and Burges, H.D. (1979). *Ann. appl. Biol.* **93**, 235–246.

Hartmann, G.C. and Wasti, S.S. (1974). *Entomophaga* **19**, 353–360.

Hassebrauk, K. (1936). *Phytopath. Zeits.* 9, 513–516.

Hawker, L.E. (1950). "Physiology of Fungi". University of London Press. London.

Hussey, N.W., Read, W.H. and Hesling, J.J. (1969). "The Pests of Protected Cultivation". Arnold, London.

Kawakami, K. (1960). *Bull. seric. Exp. Stn., Japan* 16, 83–99.

Kotthoff, P. (1937). *Ang. Bot.* 19, 127–130.

Latch, G.C.M. (1965). *N.Z.J. Agric. Res.* 8, 384–396.

Leatherdale, D. (1970). *Entomophaga* 15, 419–435.

Leeming, A.R. (1976). D. Phil. Oxford.

Lipa, J. (1975). "An Outline of Insect Pathology". U.S. Dept. Commerce, Natn. Tech. Inform. Serv., Springfield, Virginia.

McClelland, T.B. and Tucker, C.M. (1929). *Agric. notes. Puerto Rico agric. Exp. Stn.* 48, 1–2.

Müller-Kögler, E. (1967). *In* "Insect Pathology and Microbial Control" (P.A. van der Laan, ed.), pp. 339–353. North Holland Publ. Co., Amsterdam.

Nagaich, B.B. (1973). *Indian Phytopath.* 26, 163–165.

Neuzilová, A. (1957). *Univ. Carol.* 3, 7–29.

Petch, T. (1925). *Trans. Br. mycol. Soc.* 10, 152–182.

Petch, T. (1931). *Trans. Br. mycol. Soc.* 16, 55–75.

Petch, T. (1948). *Trans. Br. mycol. Soc.* 31, 286–304.

Samšiňáková, A. (1966). *J. Invertebr. Path.* 8, 395–408.

Samšiňáková, A. and Kálalová, S. (1976). *Entomophaga* 20, 361–364.

Schaerffenberg, B. (1964). *J. Insect Path.* 6, 8–20.

Silveira, H.L. and Rodriguez, C.J. (1971). *Agronomia lusit., Lisboa* 33, 391–396.

Taylor, A.M. (1909). *J. econ. Biol.* 4, 1–8.

Thoizon, G. (1970). *Annls. Soc. ent. Fr. (N.S.).* 6, 517–562.

Tiensuu, L. (1936). *Suom. Hyont. Aikak.* 2, 161–169.

Viégas, A.B. (1939). *Rev. Inst. Café Estado S. Paulo* 14, 754–772.

Wasti, S.S. and Hartmann, G.C. (1975). *Parasitology* 70, 341–346.

Wilding, N. (1972). *Pl. Path.* 21, 137–139.

CHAPTER 26

Pest Control by the Fungus *Hirsutella thompsonii*

C.W. MCCOY

*University of Florida, IFAS, Agricultural Research and Education Center
Lake Alfred, Florida, USA*

I. Introduction

A fungal pathogen of the citrus rust mite, *Phyllocoptruta oleivora*, was first reported by Speare and Yothers (1924), who wrote: "about 2 weeks after mite populations reached a peak, in mid-September at the height of the rainy season, mites disappeared as if by magic." An hour of diligent search was sometimes required to find a single specimen. Yothers and Mason (1930) refer to "masses of mites stuck to one another like numerous angleworms." Fisher (1950) described the fungus as *Hirsutella thompsonii*. Both Fisher *et al.* (1949) and Muma (1955, 1958) reported dead mites containing internal hyphae suspected to be *Hirsutella* from field samples. Finally, McCoy and Kanavel (1969) isolated the fungus on an artificial medium and confirmed pathogenicity against the mite.

This chapter reviews research on the development of *H. thompsonii* as a potential mycoacaricide and the role it plays in the natural control of the citrus rust mite.

II. Taxonomic Status

Speare (1920) and Mains (1951) concluded that the genus *Hirsutella* belonged in the Fungi Imperfecti and had a relationship to some *Cordyceps* spp. (Ascomycetes, Hypocreales). An ascogenous association of *H. thompsonii* is not known (Mains, 1951). Most species of *Hirsutella* have been identified from the tropics and are pathogens of invertebrates. Recently, several strains of *H. thompsonii* have been isolated from different species of mites (McCoy, unpublished observations; R.A. Hall, personal communication). Four strains from the citrus bud mite, *Eriophyes sheldoni*, coconut flower mite, *Eriophyes guerreronis* and from *Colomerus novahebridensis* produce synnemata vertically and/or laterally on solid media but not on mites. All strains produce conidia singly on denticulate projections. At present, ultrastructure is being studied to elucidate conidium ontogeny and to confirm taxonomic status.

III. Gross Morphology

Gross morphology of *H. thompsonii* on various solid media varies considerably. On potato dextrose agar, corn meal agar and other media containing a major carbon source, colonies appear as a grey, loose, fluffy aerial growth slightly elevated above the medium (McCoy and Kanavel, 1969). On media high in nitrogen (brain–heart infusion and 1% peptone agar) colonies are characteristically flat with little aerial growth. Substratum colouration varies from tan to brown to greyish green. At all stages of growth, the mycelium is fine and hyaline. Mature hyphae are septate; mean diameter 2.4 μm, range 1.7 to 3.3 μm (Fisher, 1950). Hyphae generally develop singly, except for the above strains that produce synnemata, a feature lost after repeated culture.

H. thompsonii sporulates only moderately. Conidiogenous cells (phialide-like) are olivaceous to greyish, ca. 10–15 μm long, most arising laterally from hyphae (Fig. 1B, C). Basally phialides are narrowly conoid or ellipsoid, abruptly narrowing at the apices into one or more necks (Fisher, 1950). Conidia are spherical, ca. 2–4 μm in diameter, with a verrucose surface (Fig. 1D). Fisher (1950) identified a gelatinous matrix surrounding the conidia.

IV. Host-Parasite Relationship

A. HOST SPECIFICITY AND GEOGRAPHICAL DISTRIBUTION

H. thompsonii is a specific fungal pathogen of Acarina, particularly eriophyid and tetranychid mites, inhabiting citrus and other plants worldwide (Table I).

B. SYMPTOMS OF DISEASE

Early symptoms of disease are difficult to see particularly in small mites. The first evidence of infection of a citrus rust mite is sluggish movement. Later mites may change from a normal lemon-yellow to a deep yellow to brown (Muma,

1958; Burditt *et al.*, 1963). The whitish integument of the blueberry bud mite darkens following infection (Baker and Neunzig, 1968). Discolouration alone is not reliable for disease diagnosis, however, because citrus rust mites darken with age.

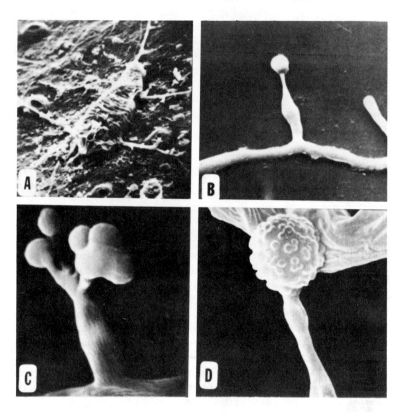

Fig. 1. Asexual reproductive stages of *Hirsutella thompsonii*. (A) external mycelia radiating laterally and posteriorly from citrus rust mite body onto fruit surface (scanning electron microscope, SEM, × 550); (B) single conidium on lateral phialide arising from hyphae (SEM, × 1,500); (C) phialide with multiple conidia (SEM, × 6,250); (D) spherical conidium with verrucose surface (SEM × 15,000).

C. DEVELOPMENT OF INFECTION

There is little information about how *H. thompsonii* invades a host. Conidia usually germinate on contact with any part of the mite cuticle. Invasion of the haemocoel is solely through the cuticle. Mites lack spiracles and entrance through ectodermal openings is therefore highly improbable. Within the host the hyphae ramify the haemocoel, showing no tissue selectivity. In eriophyids,

TABLE I.

Acarina susceptible to Hirsutella thompsonii *and their food plants*

Mite species	Food Plant	Location	References
Phyllocoptruta oleivora	citrus	USA, China	Fisher (1950), Yen (1974)
		Surinam, Cuba	van Brussel (1975), Cabrera (1977)
		Texas, USA	Villalon and Dean (1974)
Acalitus vaccinii	blueberry	USA	Baker and Neunzig (1968)
Eriophyes sheldoni	citrus	Rhodesia	McCoy and Selhime (1974)
Eriophyes sp.	poison ivy	USA	McCoy and Selhime (1974)
Eutetranychus banksi	citrus	USA	McCoy and Selhime (1974)
Eutetranychus sexmaculatus	citrus	USA	McCoy and Selhime (1974)
Panonychus citri	citrus	USA	McCoy and Selhime (1974)
Tetranychus cinnabarinus[a]	beans	Israel	Gerson *et al.* (1979)
Eutetranychus orientalis[a]	beans	Israel	Gerson *et al.* (1979)
Typhlodromalus peregrinus	citrus	USA	McCoy and Selhime (1974)
Tydeus gloveri	citrus	USA	McCoy and Selhime (1974)

[a]Occurrence in laboratory, others natural.

hyphae that develop initially in the centre of the haemocoel, generally grow both forwards through the legs and mouth and backwards along the inner body wall through the anus. In tetranychids, hyphae tend to remain localized and fill the body cavity. Hyphae emerge from mite cadavers through the mouth, anus, appendages, genital opening and at times laterally through the body wall about 3–4 days after infection at 27°C (Figs 1A, 2A). Typical of all entomopathogenic hyphomycetes, *H. thompsonii* produces infective conidia on conidiophores arising from external mycelia growing away from the dead host on a leaf (Fig. 1A).

The extent of external mycelial growth and asexual reproduction depends on the size of the host, tetranychid mites producing considerably more than eriophyids (Fig. 2). Also, fungal growth and conidia per phialide are affected by temperature. *H. thompsonii* is a mesothermophile with an optimum temperature of 27°C; range, 25–30°C (Kenneth *et al.*, 1979). *In vitro,* it grew between 5° and 37°C. Maximum infection of the citrus rust mite occurs in the presence of free water and also at 90–100% r.h. (McCoy, 1978). Gerson *et al.* (1979) obtained maximum conidial germination, penetration and infection of *Tetranychus cinnabarinus* at 27°C and very little at 13° and 35°C; conidial germ tubes survived temporary exposure to relative humidities from 5–100%. Once the fungus has penetrated into the mite body, most deaths occur at an optimum range of 25–30°C.

At 26–27°C, 4h are needed for a conidium to penetrate the integument of a citrus rust mite and about 72h for total fungal invasion and sporulation to commence (McCoy *et al.*, 1972). Under similar conditions, it takes about 60–72h from conidium contact to sporulation in the carmine mite (Gerson *et al.*, 1979). At least 98% r.h. is necessary for sporulation from cadavers.

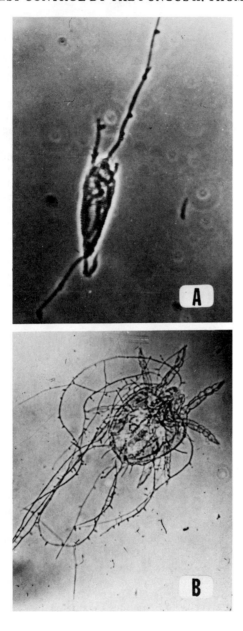

Fig. 2. Phase contrast micrographs of phytophagous mites infected with *Hirsutella thompsonii*. (A) Citrus rust mite (× 160). (B) Texas citrus mite (× 100).

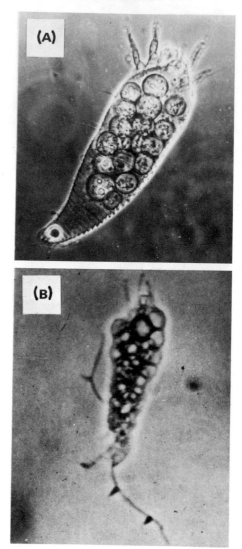

Fig. 3. Phase contrast micrographs of *Hirsutella thompsonii*. (A) Multi-nucleate spherical chlamydospores in haemocoel (× 200) and (B) germinated chlamydospores with mycelia and phialides outside citrus rust mite (× 400).

Approximately 9 h are needed for the fungus to develop and sporulate outside an infected mite at 26°C and 100% r.h. (Gerson *et al.*, 1979).

In citrus groves, the internal hyphae may break up and form multinucleate spherical chlamydospores, diameter ca. 13–22 μm (Fig. 3A). These can be found in diseased mites throughout the year. They germinate and produce mycelia that grow outside the host and reproduce asexually (Fig. 3B).

V. Natural Control

Lipa (1971) reviewed literature before 1960 on the importance of *H. thompsonii* in the natural control of *P. oleivora* in Florida citrus groves and the reduction of this control by pesticides; so my review covers later research.

Abundance of the blueberry bud mite in North Carolina was inversely correlated with seasonal deaths caused by *H. thompsonii* (Baker and Neunzig, 1968). The bud mite usually increased in October, reached a maximum in February and sharply declined in June to August while by late June through July infected mites reached a maximum.

Villalon and Dean (1974) found infected citrus rust mites in all citrus groves surveyed in the Rio Grande Valley of Texas, but epizootics there have not been documented. However, in Surinam, van Brussel (1975) found a distinct relationship between the increase of *H. thompsonii* in this mite and population decline, although he feels another regulatory factor causes the decline in citrus groves where the incidence of *Hirsutella* is < 15%. In Florida epizootics caused by interaction of weather, mite and fungus occur regularly in summer (Fig. 4).

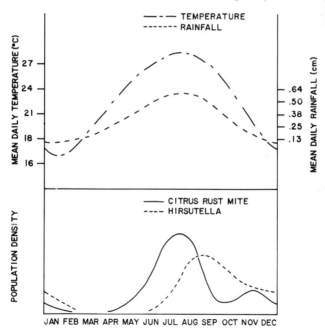

Fig. 4. Generalized scheme of the seasonal relationship between citrus rust mite population density, incidence of *Hirsutella thompsonii* and weather parameters (25 year means) in Florida citrus groves.

Diseased mites can be found on fruit and foliage throughout the year. Mites move from the old to the new foliage and to fruit in April and increase during May. By mid-June, there are a few diseased mites on the new growth. As mites reach peak densities after mid-June, epizootics lasting 2—3 weeks develop regularly. Elimination of the mites results in a high fungal residue that usually prevents further mite build-up during the fall and winter.

VI. Microbial Control

A. MICROBIAL INSECTICIDE

In recent years, *H. thompsonii* has been tested as a potential mycoacaricide against *P. oleivora* and other acarine pests in the USA, China, Surinam and Israel (McCoy et al., 1978). In the USA, a commercial formulation of conidia has been produced by Abbott Laboratories (North Chicago, Illinois) and an Experimental Use Permit for expanded field trials has been granted by the U. S. Environmental Protection Agency.

1. Culture, Mass Production and Storage

H. thompsonii has been cultured on agar media. These include potato—dextrose, modified-soil fungus, Sabouraud-dextrose (McCoy and Kanavel, 1969); potato carrot (Kenneth et al., 1979); Sabouraud maltose—peptone, grapefruit and egg yolk agar (van Brussel, 1975). Growth and sporulation on these media are rather slow, maximum vegetative growth occurring on media high in carbon, and maximum conidiation on nutrient-deficient agar (McCoy and Kanavel, 1969). In nutritional studies with liquid media, dextrose at 5 mg/ml and sucrose at 10 mg/ml were the best carbon sources (McCoy et al., 1972). Organic, rather than inorganic nitrogen, as yeast extract and peptone at 5 mg/ml and 0.5 mg/ml, was better for growth. A simplified aerated medium for producing mycelium contained molasses and soybean meal at an optimum pH of 7.5 (McCoy et al., 1978). Conidia were not formed.

Several large-scale mass production methods for conidia and mycelia have been developed using semi-solid and submerged fermentation (McCoy et al., 1971; Yen, 1974; McCoy et al., 1975; van Brussel, 1975; McCoy, 1978; Kenneth et al., 1979).

Kenneth et al. (1979) added a spore suspension inoculum to a semi-solid medium of sterile wheat bran (60 g), distilled water (60 ml) and chloramphenicol (250 ppm) per litre-conical flask. This was incubated at 25°C under ordinary light and shaken every other day. After 2 weeks, the bran—fungal mat was air-dried and blended to small particles. After holding in moisture chambers for further sporulation, the fungus was stored at 18—20°C until field use. Villalon and Dean (1974) and van Brussel (1975) scraped mycelium and conidia from an agar medium into distilled water or rainwater, then blended to fragment the aerial mycelium and applied it directly into the field without storage.

By-passing the sporulation phase of laboratory production, McCoy et al. (1975) mass-produced by submerged fermentation and applied in the field ca. 230 kg (wet wt.) of mycelia during a 2-year period. Seed cultures of mycelium were grown in a standard broth in Erlenmeyer flasks at 26–27°C for 48 h on a rotary shaker. Inoculum (75 ml), 600 mg of antibiotic, 12 litres of a dextrose–yeast extract–peptone broth and 25 ppm antifoam were contained in 19-litre sterilizable vessels stopped with aeration heads. Broth in vessels attached in series was aerated and agitated by sterile compressed air (10 psi) at 22–26°C. A mean of 400 g (wet wt.) mycelia/vessel was produced in 96 h, harvested by filtration and stored in stainless steel pans at 10°C as mat-mycelia with no loss of viability for 64 days.

In 1972, Abbott Laboratories developed commercial methodology for mycelium production; however, hyphae lysed within 2 weeks (McCoy, 1978). This loss of viability required cold storage and costly shipment which stopped commercial development. At present, development of a commercial conidial preparation by Abbott Laboratories (McCoy, 1978) appears more feasible for storage and formulation. After 2 years of storage, air-dried conidia at − 20°C had 100 × more viable conidia than those at 27°C or freeze-dried conidia at − 20°C and 27°C (McCoy, unpublished observations).

2. Formulation and Field Application

Adequate methodology for formulation, standardization and field application has yet to be developed for *H. thompsonii*. Both mycelial and conidial formulations have been produced in the laboratory and by industry. However, the pathogenicity of the various fungal batches against the citrus rust mite cannot be assayed in the laboratory until the mite can be reared there. At present, *Tetranychus* species are being tested as alternative hosts (McCoy, unpublished observations). Both mycelial and conidial germination have been monitored before and after formulation and application (McCoy *et al.*, 1975; McCoy, 1978) and infectivity tested immediately after mass-production (Gerson *et al.*, 1979). No report of a loss in viability has been found for formulations used in field studies.

Sprays of fragmented mycelium, with or without conidia, have been formulated and applied experimentally in the field in various ways. Adjuvants (Table II) included molasses plus Dacagin which appeared to actually stimulate conidiogenesis (McCoy *et al.*, 1975). Formulations including a commerical wettable powder were applied with a high pressure sprayer at 400–600 psi for thorough coverage of citrus foliage and fruit (McCoy, 1978). The pH of the tank mix was always maintained above 6. Van Brussel (1975) blended a mycelia–conidial suspension in rainwater and applied it with an "atomizer" hand sprayer. Gerson *et al.* (1979) applied a conidial–bran–distilled water spray with a hand sprayer.

3. Efficacy

Since 1969, many laboratory and field tests have been conducted using laboratory and commercial preparations of *H. thompsonii* to reduce or inhibit the build-up of mite populations. Results were variable. McCoy *et al.* (1971) sprayed laboratory-produced mycelia in rainwater (30 litres/tree; 1.0–10.0%,

TABLE II
Adjuvants used in the formulation of Hirsutella thompsonii *as a mycoacaricide*
(Muttath, 1974; McCoy and Selhime, 1974)

Material	Rate	Characteristic
Dacagin, Diamond Shamrock Corp.	0.25%	sticker—protectant
Molasses	1—4%	sticker—protectant—nutrient
Nutrient agar	0.1%	sticker—protectant—nutrient
Tween 20	— — —	spreader

w/v, at 500 psi) on mature citrus trees harboring moderate to high populations of citrus rust mites on leaves. After 2 weeks, mite populations were reduced to low levels and remained low for 10—14 weeks, while control populations increased for 2—4 weeks before declining. Conidiation of the mycelial inoculum on foliage appeared in 48 h as small patches of fuzzy grey fungus together with dead, fungus covered mites. Further trials in Florida from 1970—1973 with different laboratory and commercial formulations and concentrations of mycelia involved various weather and grove conditions (McCoy and Selhime, 1974). Although weather conditions influenced conidiation and survival resulting in inconsistent results, 0.5—2.5% fragmented mycelia in rainwater with different adjuvants (molasses, Dacagin, nutrient agar) at 20—30 litres/tree often reduced high citrus rust mite populations on fruit and foliage to low levels in 2 weeks and kept them low for 6 months to a year.

In China (Chekiang Province), laboratory-produced mycelium was applied at dosages of 0.5—1.0 g/litre to citrus trees (Yen, 1974; Chiang and Huffaker, 1976). One application caused 90% deaths in 3 days.

Direct application of *H. thompsonii* will prevent build-up or suppress moderate populations of the citrus rust mite in Surinam (van Brussel, 1975). In the laboratory, an 0.05% spore-fragmented mycelial suspension reduced mites on leaves to low levels in 2 days but lost effectiveness after 8 days. In similar studies, mites were reduced from 54.6 to 7.6/fruit in 5 days: meanwhile control populations increased from 46.0 to 61.2. In field studies, sprays of 0.025%, 0.05% and 0.1% spore-mycelial suspension on fruit prevented an increase in mite populations during the dry season. Also, a spore—mycelia rainwater formulation at 0.05—0.1% controlled mites on fruit for at least 3 weeks in dry weather.

In greenhouses, Gerson *et al.* (1979) sprayed a conidial—bran wettable powder (2.5×10^6 conidia/ml with Tween 20) on ground nut plants infested with low, medium and high populations of carmine spider mite. Extreme temperatures of $35°C$ and relative humidity seldom reaching 100% in the greenhouse prevented infection. However, the fungus survived in the bran residue. Subsequently, mites exposed to the same leaves under optimum environmental conditions were infected.

In 1976, Abbott Laboratories produced a wettable powder containing mostly conidia. Initial small-scale field tests were made at Plymouth, Florida in May (McCoy, 1978). Three rates (2.5, 5.0 and 10.0 g/500 ml water at 1.9×10^5 viable spores/g) were sprayed at ca. 30 litres/tree. At the highest rate, the mean number of citrus rust mites after 4 weeks (9.6/leaf) was significantly lower than the control (26.3). After 1 week, actively sporulating mycelium had radiated out from fungal residue on the leaf surface. Thus, the carrier in the formulation functioned as a substrate for further fungal growth and sporulation in the field. In 1977, the wettable powder was sprayed for the first time by high pressure sprayer at 0.1, 0.2 and 0.4 kg/litre to mature citrus trees (30 litres/tree) in three locations. Where environmental conditions were optimal for infection, citrus rust mites were reduced after 4 weeks (McCoy, 1978).

4. Safety to Vertebrates

Both direct and indirect evidence suggest that *H. thompsonii* is safe to mammals. Rats were fed a large dose of mycelia and whole culture filtrate equivalent to that consumed by a 70 kg man eating a dose that would be applied to an acre. They were in excellent condition for 21 days and exhibited no toxicity, pathogenicity or gross pathology (Ignoffo *et al.*, 1973). Eye irritation, acute inhalation, primary skin irritation, acute dermal and acute oral tests with massive doses of mycelia and conidia, were negative including hematograms and histology (McCoy and Heimpel, unpublished observations). For 3 years, six researchers handling conidia, mycelia and whole culture filtrate experienced no ill effects (McCoy *et al.*, 1975).

B. USE IN INTEGRATED PEST MANAGEMENT

1. Compatibility with Pesticides

Some chemicals have, on occasion, caused secondary mite outbreaks by reducing the natural incidence of *H. thompsonii* in citrus groves (Fisher *et al.*, 1949; McCoy *et al.*, 1976a; McCoy, 1977a; 1977b). Various *in vitro* techniques using treated agar discs and plates or assay discs have quantified the effect of various chemicals on the germination of conidia (Table III). Seven fungicides were scored highly active and were strongly fungistatic at rates below the recommended level. Various copper compounds, PCNB, binapacryl, carboxin, dinocap and thiabendazole had less effect. Light and medium oils were not fungistatic (McCoy *et al.*, 1976a) at recommended dosages. All insecticides at recommended rates, excluding sulphur, caused less inhibition than fungicides (Table III). MnO_2, a plant nutrient commonly applied to citrus foliage in Florida, moderately inhibited *H. thompsonii* (McCoy *et al.*, 1976a). Although results from field tests are meager, McCoy *et al.* (1976b) found significantly more infection in plots without fungicidal treatments. These same plots, however, received fewer pesticide treatments and so had a higher mite population. Field effects of chemicals on other fungi have often been less severe than effects in the laboratory (Chapter 25 and 27 to 29).

2. Compatibility with Natural Enemies

Although the host range of *H. thompsonii* is primarily limited to the Acarina, only *Typhlodromalus peregrinus* among the predatory mites has been found

TABLE III

Effect of agricultural chemicals at recommended field rates on germination of conidia of Hirsutella thompsonii *in the laboratory (Muttah, 1974; McCoy et al. 1976a)*

Group	Common name	Sensitivity		
		Low	Moderate	High
Fungicides	benomyl, captafol, captan, Daconil			+
	ferbam, maneb, zineb			+
	binapacryl, carboxin, basic $CuSO_4$		+	
	pentachloronitrobenzene (PCNB)		+	
	thiabendazole		+	
	$Cu(OH)_2$, copper oxychloride, dinocap	+		
Insecticides	dichlorvos, dicofol, Omite, sulphur		+	
	phenisobromolate		+	
	Plictran	+		
Nutritional	MnO_2			+
elements	$MnSO_4$		+	
	borax, $ZnSO_4$	+		

infected with it. Other invertebrate predators and parasites have never been found adversely affected in artificial applications of *H. thompsonii*.

3. Manipulation of the Citrus Ecosystem

In summer, *H. thompsonii* causes dramatic natural epizootics in unsprayed (with chemicals) and some sprayed citrus groves of Florida when citrus rust mites abound. To better utilize this natural control, McCoy *et al.* (1976a, b) tested an integrated pest management (IPM) strategy designed to conserve *H. thompsonii* in groves where external fruit quality was not critical since the fruit was marketed for processing. For 4 years, use of oil to control phytopathogenic fungi and maintenance of more citrus rust mites in groves in summer significantly increased control of rust mite by *H. thompsonii*. The overall effect of the IPM strategy reduced the amount and cost of pest control on citrus varieties for canning. This strategy along with others is being applied to an overall management model for citrus (McCoy, 1978).

VII. Appraisal

Where citrus rust mite is an economic pest, *H. thompsonii* is a key factor in its natural control, but the fungus is sometimes suppressed by pesticides. As IPM strategies become more sophisticated through the development of better spray thresholds by monitoring, development of more selective pesticides and predictive modelling, unnecessary pesticide treatments will be reduced. In turn, this will result in more disease in local mite populations. Thus, *H. thompsonii* can be used in practical IPM strategies for citrus particularly that for processing.

As a mycoacaricide, *H. thompsonii* is potentially useful against citrus rust mite and other acarines. Its advantages are: (1) production and formulation commercially in an economic manner; (2) activity against major economic pests; (3) safe to mammals and beneficial invertebrate parasites; (4) application with conventional spray equipment; (5) kills quickly and suppresses mite populations in the grove and (6) survives in the environment and prevents pest resurgence. Its disadvantages are: (1) requires specific environmental conditions for maximum survival and infection; (2) may require cold storage to insure extended viability and virulence; (3) usage may be limited to certain times of the year and (4) not compatible with some conventional pesticides.

Numerous research and development studies are needed to maximize the reliability of *H. thompsonii* in the field: (1) clarification of its taxonomic status; (2) development of a sound bioassay technique, probably using an alternative host to the citrus rust mite to standardize the commercial product as well as to identify and select more virulent strains for different acarines; (3) develop and evaluate new formulations, protected somewhat from environmental stress, but more importantly, persistent in the host's microclimate in a more natural disease inducing state when applied artificially; (4) design quantitative assays for environmental levels of *H. thompsonii;* (5) evaluate spray application methodology, e.g. optimum pH, spray coverage, necessity and compatibility of adjuvants; and (6) further study the pathogen's effect at various mite population densities in different climates on different acarines in different ecosystems.

References

Baker, J.R. and Neunzig, H.H. (1968). *J. econ. Ent.* **61**, 1117–1118.

Brussel, E.W. van (1975). *Landbouwproefstation Surinam/Agric. Exp. Stn. Surinam Bull.* **98**, 30–40.

Burditt, A.K. Jr., Reed, D.K. and Crittenden, C.R. (1963). *Fla Ent.* **46**, 1–5.

Cabrera, R.I.C. (1977). *Agrotecnia de Cuba* **9**, 3–11.

Chiang, H.C. and Huffaker, C.B. (1976). Proc. 1st. Int. Colloq. Invertebr. Path., Kingston, Canada. pp. 42–47.

Fisher, F.E. (1950). *Mycologia* **42**, 290–297.

Fisher, F.E., Griffiths, J.T. and Thompson, W.L. (1949). *Phytopathology* **39**, 510–512.

Gerson, U., Kenneth, R. and Muttath, T.I. (1979). *Ann. Appl. Biol.* **91**, 29–40.

Ignoffo, C.M., Barker, W.M. and McCoy, C.W. (1973). *Entomophaga* **18**, 333–335.

Kenneth, R., Muttath, T.I. and Gerson, U. (1979). *Ann. Appl. Biol.* **91**, 21–28.

Lipa, J.J. (1971). *In* "Microbial Control of Insects and Mites" (H.D. Burges and N.W. Hussey, eds), pp. 357–373. Academic Press, London.

Mains, E.B. (1951). *Mycologia* **43**, 691–718.

McCoy, C.W. (1977a). *J. econ. Ent.* **70**, 748–752.

McCoy, C.W. (1977b). *Proc. Int. Soc. Citriculture* **2**, 459–462.

McCoy, C.W. (1978). *In* "Microbial Control Insect Pests: Future Strategies in Pest Management Systems" (G.E. Allen, C.M. Ignoffo and R.P. Jaques, eds), pp. 211–219, NSF–USDA–Univ. Florida, Gainesville.

McCoy, C.W. and Kanavel, R.F. (1969). *J. Invertebr. Path.* **14**, 386–390.

McCoy, C.W. and Selhime, A.G. (1974). Proc. Int. Citrus Congr., Murcia, (1973) Vol. 2, 521–527.

McCoy, C.W., Selhime, A.G., Kanavel, R.F. and Hill, A.J. (1971). *J. Invertebr. Path.* **17**, 270–276.

McCoy, C.W., Hill, A.J. and Kanavel, R.F. (1972). *J. Invertebr. Path.* **19**, 370–374.

McCoy, C.W., Hill, A.J. and Kanavel, R.F. (1975). *Entomophaga* **20**, 229–240.

McCoy, C.W., Brooks, R.F., Allen, J.C. and Selhime, A.G. (1976a). *Proc. Tall Timbers Conf. Ecol. Anim. Control Habitat Man.* **6**, 1–17.

McCoy, C.W., Brooks, R.F., Allen, J.C., Selhime, A.G. and Wardowski, W.F. (1976b). *Proc. Fla State Hort. Soc.* **89**, 74–77.

McCoy, C.W., Couch, T.L. and Weatherwax, R. (1978). *J. Invertebr. Path.* **31**, 137–139.

Muma, M.H. (1955). *J. econ. Ent.* **48**, 432–438.

Muma, M.H. (1958). Proc. 10th Int. Congr. Ent. 4, 633–647.

Muttath, M.T. (1974). M. S. Thesis, Hebrew Univ. Jerusalem, Rehovot.

Speare, A.T. (1920). *Mycologia* **12**, 62–76.

Speare, A.T. and Yothers, W.W. (1924). *Science* **60**, 41–42.

Villalon, B. and Dean, H.A. (1974). *Entomophaga* **19**, 431–436.

Yen, H. (1974). *Acta Ent. Sin.* **17**, 225–226.

Yothers, W.W. and Mason, A.C. (1930). U.S.D.A. tech. Bull. No. 176, 54 pp.

NOTE ADDED IN PROOF

Cultural features and morphology of 11 isolates of *H. thompsonii* have been described, distinguishing three varieties *thompsonii, vinacea* and *synnematosa* (Samson, R.A., McCoy, C.W. and O'Donnell, K.L., 1980, *Mycologia* **72**, 359–377).

CHAPTER 27

The Fungus *Nomuraea rileyi* as a Microbial Insecticide

C.M. IGNOFFO

Biological Control of Insects Research Laboratory,
U.S. Department of Agriculture, Columbia, Missouri, USA

I. Introduction

Although the entomopathogenic fungus *Nomuraea rileyi* was first described ca. 100 years ago (1883), no attempt was made to use it experimentally for biological control until 1955 (Chamberlin and Dutky, 1958). This is surprising since fungi as a group were being extensively studied and used in the field prior to the advent of synthetic organic insecticides. The germ theory of disease was based on a study of an entomopathogenic fungus (Bassi, 1835); the first effort to mass-produce and artificially apply an entomopathogenic agent was with a fungus (Metchnikoff, 1879; Krassilstschik, 1888); and large-scale field experiments were being conducted in Europe and North America with fungi (Forbes, 1898; Snow, 1891; Giard, 1892). Indeed, it was also commonly known that *N. rileyi* induced extensive epizootics in caterpillar pests on cabbage, clover, soybeans, and velvetbeans and thus was a potential candidate for use as a microbial insecticide. The objective of this chapter is to review and summarize data on *N. rileyi* and to evaluate the feasibility of its use to control insect pests.

II. Taxonomic Status And Description

The name of *N. rileyi* has been recently changed from *Spicaria rileyi* (Farlow) Charles to *Nomuraea rileyi* (Farlow) Samson (Kish *et al.*, 1974). In fact, the taxonomy has not been very stable since *N. rileyi* was first described by Farlow (1883) as *Botrytis rileyi*. It was transferred to *Spicaria* (= *Beauveria rileyi*) by Charles (1936). Later, Hughes (1951) and then Brown and Smith (1957) proposed that many species of *Spicaria* be transferred to *Paecilomyces* and that the name *Spicaria* be placed in synonymy with other taxa. However, *S. rileyi* has an atypical phialidic stage and a green colony colour, so it was not transferred to *Paecilomyces*, but placed near *Penicillium* (section *Divaricata*). As late as 1966, *N. rileyi* still did not have a valid name (Behnke and Paschke, 1966); however, most scientists still continued to use the old generic name *Spicaria*. In 1974, Kish *et al.* resurrected the genus *Nomuraea* (originally the monotypic genus for *N. prasina*; Maulblanc, 1903; Petch, 1925), changed the name from *Spicaria rileyi* to *Nomuraea rileyi* and placed *S. prasina* (Sawada, 1919) into synonymy with *N. rileyi*.

The following description of *N. rileyi* is modified from Samson (1974) and Kish *et al.* (1974). On malt agar at 25°C, colonies grow very slowly and in a month reach a diameter of 0.7–1.2 cm. Initial growth is by yeast-like budding from the germ-tube of a conidium. After a few days these yeast-like hyphal bodies produce a cream coloured, sticky growth and a sweet, musty odour. Sporulation is initially localized and then spreads throughout the colony. The colour of the colony progresses from white to pale green to malachite green. Vegetative hyphae, 2–3 µm in mean diameter, are smooth, septate, and hyaline to slightly pigmented. Conidiophores, which grow from submerged hyphae, are erect, septate, up to 160 µm in length and 2–5 µm in diameter. Branches, formed near a septum, develop in whorls each giving rise to 2–4 phialides. The branches, 5–8 by 2–4 µm, are usually cylindrical, occasionally with a swollen base. Conidia, in dry divergent chains, are smooth, ellipsoidal, pale green and 3.4–4.5 by 2–3.1 µm.

Sporulation is more rapid (ca. 10 days) at 25°C on Sabouraud maltose agar fortified with 1% yeast extract (SMAY) and light has no effect on either growth or sporulation (Bell, 1975a). The optimum temperature for rapid mycelial growth and sporulation on SMAY is 25°C (Ignoffo *et al.*, 1976a). The average time to initial sporulation at 15°, 20°, and 25°C is 21.0 ± 0.4 (S.E.), 10.4 ± 0.4, and 8.8 ± 0.7 days, respectively. Conidia did not germinate at 5°, 35°, 37°, or 40°C.

III. Mode of Action

A. INFECTION CYCLE

The following description of the infection cycle of *N. rileyi* is based upon laboratory and field observations of disease in several species of caterpillar pests (Kawakami 1962a, 1973; Kawakami and Mikuni 1965; Kish and Allen, 1978; Mohamed *et al.* 1978a; Hostetter *et al.*, unpublished observations; Ignoffo and Garcia, unpublished observations).

A conidium initiates invasion of susceptible larvae and the entire life cycle takes ca. 8–12 days (Table I). Conidia once lodged on the integument swell and then produce a slender, invasion hypha (Fig. Ia, b, c). Nutrients (maltose, neopeptone, insect extracts) are required for germination (Kish and Allen,

TABLE I

A speculative developmental cycle of a Nomuraea rileyi *infection in larvae of* Trichoplusia ni *at 25°C*

Developmental stage	Days after instar I larvae exposed to 1 000 conidia/mm² of leaf surface
Germination of conidia	1/6 to 1/2
Invasion of haemocoel	1/2 to 1
Yeast-like growth phase	1 to 3
Mycelial-fusiform cell growth phase	3 to 5
Death of larvae	5 to 8
Development of conidiophores	7 to 9
Conidiogenesis	8 to 12

1978; Getzin, 1961; Ignoffo and Garcia, unpublished observations). Getzin (1961) reported that conidia germinated as a linear function of time when exposed to a 1:3 larval–water extract of macerated *Trichoplusia ni* larvae, but not at 35°C. Ignoffo *et al.* (1976a) observed some germination at 35°C; however, the invasion hypha never developed more than a few millimeters. Germination was optimal between 15–25°C, inhibited at 30°C, and it took 73.5 ± 15.5 h to attain 50% germination at 8°C. The percent germination/h at 8°, 15°, 20°, 25° and 30°C was 1.1, 6.2, 6.5, 5.8, and 7.6%, respectively (Getzin, 1961).

A portion of the invasion hypha sometimes enlarges into what superficially resembles the appressoria described by Berisford and Tsao (1975). Most investigators believe invasion of the host is primarily through the integument (Fig. 1d); however, some penetration via the alimentary tract is also possible. For example, of ten *Anticarsia gemmatalis* larvae fed a microdrop of water

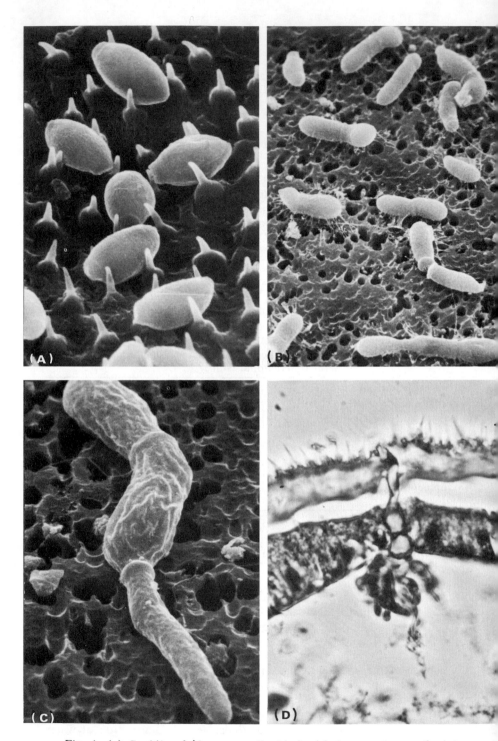

Fig. 1. (a) Conidia of *Nomuraea rileyi* lodged between setae on the integument of *Trichoplusia ni* (SEM, 7315 ×); (b) Germinating conidium (SEM, 2830 ×); (c) Invasion hypha of conidium (SEM, 6090 ×); (d) Invasion of haemocoel (section, 900 ×).

containing conidia three died in 8 days, the identical number that died among those given a microdrop externally (Kish and Allen, 1978). The actual method of penetration (physical and/or enzymatic) of the integument by the invasion hypha is unknown. The presence of an acid–Schiff reaction at the site of penetration of the endocuticle of *Malacosoma alpicola* by *Spicaria* spp. (Benz, 1963) and the more recent data of Mohamed *et al.* (1978a), however, indicate that *N. rileyi* produces chitinase, protease and lipase and thus invades enzymatically. Studies on enzymatic penetration of the integument have generally been conducted on the mycelium or mycelial extracts. This is surprising since penetration is accomplished only by germinating conidia. In view of this, studies of this type should focus on the conidial-germ tube and not the mycelium. The invasion hypha penetrates the integument within 24 h of initial exposure to conidia.

Small yellow to brown spots on the integument are the first (1–2 days post-exposure) overt sign of infection of a larva. Reduced feeding, lethargy, paling of the body colour, and a slight swelling of the posterior abdominal segments are symptoms that may be encountered 2–5 days post-exposure. Growth of the fungus after it reaches the haemocoel is by budding, which produces discrete yeast-like hyphal bodies (blastospores, Fig. 2a), the vegetative assimilative stage of *N. rileyi*. These are eventually transported throughout the haemocoel and give rise to localized concentrations of mycelia. Toxin(s) may also be produced by hyphal bodies or early mycelial growth (Mikuni and Kawakami, 1975; Ignoffo and Garcia, unpubl.). Toxins may speed death, delay metamorphosis and are active topically as well as by injection (Chapter 23, II A 3). These toxins might kill cells, thus enabling *N. rileyi* to grow saprophytically. A heavy growth of intertwining mycelia, with fusiform cells, develops in the haemocoel ca. 5 days post-exposure (Fig. 2b; Kish, 1975). Death generally occurs about 1–2 days later depending upon dose, temperature, and larval stage. After death, the larval body is completely mummified and covered by a dense white mycelial mat, from which conidiophores arise close together (Fig. 2c) and, 1 to 2 days later, produce a pile of pale green conidia (Fig. 2d). Conidia are easily dislodged and distributed by wind (Garcia and Ignoffo, 1977).

B. EFFECTS OF TEMPERATURE AND HUMIDITY ON PATHOGENICITY

The optimum temperature for invasion and infection of neonatal *T. ni* larvae sprayed with 10^7 conidia/ml was 25°C and the LT_{50} at 15°, 20°, 25° and 30°C was 13.4, 7.5, 6.9 and 8.8 h, respectively (Getzin, 1961). Similar results were obtained with *T. ni*, *Heliothis zea* and *A. gemmatalis* (Yen and Chu, 1963; Mohamed *et al.*, 1977; Kish and Allen, 1978). Against *H. zea* in all instars between I and V, *N. rileyi* generally was most effective at 20° and 25°C, while both the mortality rate and total were less at 15° and 30°C. Larvae of *A. gemmatalis* dusted with conidia and incubated at 90% relative humidity (r.h.) were not infected after 10 days exposure at 10°C; in contrast, 56% of the larvae at 25°C and 90% r.h. died in the same time (Kish and Allen, 1978).

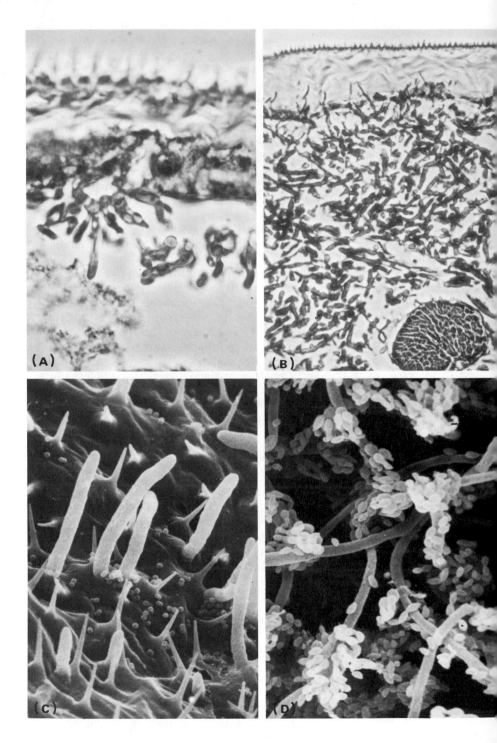

Fig. 2. (a) Budding from invasion hypha of *Nomuraea rileyi* to produce yeast-like blastospores (section, 900×); (b) Extensive growth in haemocoel of intertwining mycelium (section, 345×); (c) Conidiophore penetration of external surface of integument (SEM, 2 790×); (d) Conidiogenesis on external surface of integument (SEM, 1 100×).

The lower humidity limit for infection of larvae is probably between 40 and 60% r.h. Getzin (1961) reported that infection of *T. ni* was "severely affected by relative humidity," probably as a result of failure of conidia to germinate. Percent mortality at 42, 80, 90, and 100% r.h. was 4, 3, 49 and 100% respectively. Thus, Getzin (1961) obtained some deaths at 42% r.h., but Kish and Allen (1978) observed no dead *A. gemmatalis* at 50% r.h. This difference probably results from the differences in susceptibility since larvae of *T. ni* are about 15 × more susceptible to *N. rileyi* than *A. gemmatalis* (Puttler *et al.*, 1976).

IV. Specificity

A. INVERTEBRATE HOSTS

1. Host Spectrum

The spectrum of *N. rileyi* is primarily limited to Lepidoptera, although at least two species of Coleoptera (*Hypera punctata* and *Leptinotarsa decemlineata*) are susceptible (Table II). Natural epizootics have been observed in all but three of the listed species, *L. decemlineata, Bombyx mori* and *Peridroma saucia*. *B. mori* was infected by injection of culture filtrate and the other three species by surface contact with conidia. Epizootics in caterpillars have occurred in pastures (clover and alfalfa), many row and grain crops and even on weeds (Table II).

Eighteen species, mostly pests, in six orders of insects (Coleoptera, Lepidoptera, Diptera, Hymenoptera, Neuroptera, Hemiptera) were not susceptible to high concentrations of conidia (Table III). Also three predators (*Hippodamia convergens, Chrysopa carnea, Podisus maculiventris*), three parasites (*Voria ruralis, Apanteles marginiventris, Campoletis sonorensis*) and one egg parasite, *Telenomus proditor*, were not susceptible when exposed at rates ca. 25 × higher than that used in field experiments to induce epizootics (Ignoffo *et al.*, 1976b; Phadke and Rao, 1978). Thus, most beneficial insects should not be adversely affected by applications of *N. rileyi*. On the other hand, in the field, Burleigh (1975) reported less parasitization by *Microplitis croceipes* because of a high incidence of *N. rileyi*. The data of King and Bell (1980), however, indicated that reduced parasitization would occur only if larvae were simultaneously parasitized by *M. croceipes* and *N. rileyi*.

2. Instar Susceptibility

Reports about comparative instar susceptibility vary. In the laboratory the susceptibility of *T. ni* larvae decreased with age (Getzin, 1961), whereas instars I and II of *H. zea* were less susceptible than III and V (Mohamed *et al.* 1977). *T. ni* larvae exposed on sprayed field soybeans for 24 h and then reared in the laboratory as well as field studies indicated smaller differences in susceptibility with instars II and III being most susceptible (Ignoffo, 1976b). Perhaps more important, an increase in the age when larvae are exposed results in a

TABLE II

Species of insects reported as being susceptible to Nomuraea rileyi

Insect Species	Occurrence	Food Plant	Location	Reference
Coleoptera				
Hypera punctata	natural	clover	USA	Charles, 1941
Leptinotarsa decemlineata	laboratory	—	France	Fargues, 1976
Lepidoptera				
Achoea janata	laboratory	caster bean	India	Phadke and Rao, 1978
Agrotis ipsilon	natural	soybean	USA	Garcia and Ignoffo, Unpublished
	laboratory			
Amathes badinodis	natural	—	USA	Crumb, 1929
Anticarsia gemmatalis	natural,	velvetbean,	Brazil,	Watson, 1916; Hinds and
	laboratory	soybean	Puerto Rico,	Osterberger, 1931; Wolcott and
			USA	Martorell, 1940; Charles, 1941;
				Allen *et al.*, 1971; Johnson *et al.*,
				1976; Kish *et al.*, 1976; Ignoffo
				et al., 1976c
Bombyx mori	laboratory	—	Japan	Mikuni and Kawakami, 1975;
				Kawakami, 1960
Chrysodeixis eriosama	natural	cabbage	Ceylon	Steinhaus and Marsh, 1962
Cosmia nr. *exigua*	natural	—	Fiji	Steinhaus and Marsh, 1962
Feltia ducens, F. gladiaria	laboratory	—	USA	Crumb, 1929
Glyphodes phyloalis	natural	mulberry	Japan	Kawakami *et al.*, 1969
Heliothis armigera	natural	cotton	Africa	Vandamme and Angelini, 1966
Heliothis zea	natural	alfalfa, corn,	USA	Charles, 1941; Burleigh, 1975;
		cotton, soybean		Smith *et al.*, 1976;
				Mohamed *et al.*, 1977
Heliothis virescens	natural	cotton, soybean	USA	Sprenkel and Brooks, 1975;
				Smith *et al.*, 1976
	laboratory	soybean	USA	Puttler *et al.*, 1976
	insectary	tobacco	USA	Chamberlin and Dutky, 1958

TABLE II (*continued*)

Species of insects reported as being susceptible to Nomuraea rileyi

Insect Species	Occurrence	Food Plant	Location	Reference
Heliothis zea – virescens	natural	Dolichos, hairy vetch, morning glory, smart-weed, spring amaranth	USA	Roach, 1975
Hyphantria cunea	natural	mulberry	Japan	Kawakami *et al.*, 1969
Lymantria dispar	laboratory	–	USA	Wasti and Hartman, 1978
Mythimna (Pseudaletia) unipuncta	natural	rice	USA, Japan	Charles, 1941; Steinhaus and Marsh, 1962
Ostrinia nubilalis	natural	corn	USA	Steinhaus and Marsh, 1962
Peridroma saucia	laboratory	soybean	USA	Puttler *et al.*, 1976
Pieris rapae	natural	cabbage	USA	Biever and Hostetter, unpublished
Pionea forficalis	natural			Maublanc, 1903
Plathypena scabra	natural, laboratory	soybean	USA	Charles, 1941; Sprenkel and Brooks, 1975; Puttler *et al.*, 1976, Ignoffo *et al.*, 1976b
Pseudaplusia includens	natural, laboratory	cotton, soybean	USA	Gudauskas and Canerday, 1966; Harper and Carner, 1973; Newman and Carner, 1975; Sprenkel and Brooks, 1975
Spodoptera exigua	laboratory	soybean	USA	Puttler *et al.*, 1976
Spodoptera frugiperda	natural	gram, millet	India	Phadke *et al.*, 1978
	natural	–	Colombia, USA	Charles, 1941; Steinhaus and Marsh, 1962
Spodoptera litoralis	natural	cotton	Israel, Madagascar	Kenneth and Olmert, 1975

TABLE II (continued)

Species of insects reported as being susceptible to Nomuraea rileyi

Insect Species	Occurrence	Food Plant	Location	Reference
Spodoptera litura	natural	tobacco	India	Rao and Phadke, 1977
Spodoptera ornithogalli	natural	–	USA	Charles, 1941
Stenachroia elongella	laboratory	jowar	India	Phadke and Rao, 1978
Trichoplusia ni	natural, laboratory	cabbage, soybean, cotton	Dominican Republic, USA, Taiwan	Charles, 1941; Yen, 1960; Getzin, 1961; Steinhaus and Marsh, 1962; Behnke and Paschke, 1966; Harper and Carner, 1973; Puttler et al., 1976; Mohamed et al., 1977

TABLE III

Species of insects reported as being not susceptible to Nomuraea rileyi

Insect Species	Administration			Reference
	Route	Conidia	Stage[a]	
Coleoptera				
Centonia aurata	sprayed	10^8/ml	L-III	Fargues, 1976
Chalcodermus aeneus	soil	9.9×10^{11}/g	L-IV	Bell and Hamalle, 1970
Hippodamia convergens	filter paper	3.5×10^4/mm^2	A, L-III	Garcia and Ignoffo (unpublished observations)
Hypera postica	filter paper	3.5×10^4/mm^2	A, L	Garcia and Ignoffo (unpublished observations)
Melolontha melolontha	sprayed	10^8/ml	L-III	Fargues, 1976
Oryctes rhinoceros	sprayed	10^8/ml	L-III	Fargues, 1976
Tenebrio molitor	dipped	10^7/ml	L	Aizawa et al., 1976
Tribolium confusum	dipped	10^7/ml	L	Aizawa et al., 1976
Lepidoptera				
Manduca sexta	dusted	high	L-II, III	Chamberlin and Dutky, 1958
Manduca quinquemaculata	dipped	high	L-II, III	Chamberlin and Dutky, 1958
Pieris rapae	foliage	3×10^3/mm^2	L-I	Puttler et al., 1976
Plodia interpunctella	dipped	10^7/ml	L	Aizawa et al., 1976
Diptera				
Glossina morsitans	dusted	high	A	Poinar et al., 1977
Voria ruralis	filter paper	3.5×10^4/mm^2	A	Garcia and Ignoffo (unpublished observations)
Hymenoptera				
Apanteles marginiventris	filter paper	3.5×10^4/mm^2	A	Garcia and Ignoffo (unpublished observations)

TABLE III (*continued*)

Species of insects reported as being not susceptible to Nomuraea rileyi

| Insect Species | Administration | | Stage[a] | Reference |
	Route	Conidia		
Campoletis sonorensis	filter paper	$3.5 \times 10^4/\text{mm}^2$	A	Garcia and Ignoffo (unpublished observations)
Neuroptera				
Chrysopa carnea	filter paper	$3.5 \times 10^4/\text{mm}^2$	A, L	Garcia and Ignoffo (unpublished observations)
Hemiptera				
Podisus maculiventris	filter paper	$3.5 \times 10^4/\text{mm}^2$	A, N	Garcia and Ignoffo (unpublished observations)

[a] A = adult, L = Larva (number is instar), N = nymph

corresponding increase in the percentage of larvae reaching the pupal stage. *T. ni* larvae, like the majority of caterpillar pests, do > 75% of their feeding during the last instar. Thus, infection must occur during instar I and II if damage to the crop is to be minimized. Later infections permit most of the larvae (> 90%) to reach damaging size and thus seriously limit the utilization of *N. rileyi* as a bioinsecticide.

Temperature profoundly affects larval susceptibility (Mohamed *et al.*, 1977). Except in instars I and II, mortality of *H. zea* larvae increased with both temperature (15°, 20°, 25°, 30°C) and dose (1.4×10^9 to 7.7×10^9). Similar trends in instar susceptibility, although not as defined, were detected when daily temperatures fluctuated between 20° and 30°, 25° and 30°, and 20° and 35°C.

3. Activity of Fungal Isolates

When conidia of a single isolate of *N. rileyi* were tested against eight caterpillar pests, there was ca. a 23-fold difference in activity between the least and most susceptible species. The bioassay procedure was uniform: surface-treatment of foliage, 3.0 conidia/mm^2 and instar II test larvae (Puttler *et al.*, 1976). The relative susceptibility, equating *T. ni* to 1.00, was: *Sp. exigua*, 0.64; *Plathypena scabra*, 1.08; *H. zea*, 1.15; *Heliothis virescens*, 1.15; *Per. saucia*, 2.01; *Pseudoplusia includens*, 7.00; and *A. gemmatalis*, 14.50. *A. gemmatalis* and *P. includens*, which are frequently regulated by *N. rileyi* epizootics in Southern USA, were ca. 13 and 6 × less susceptible, respectively, than *Pl. scabra*, the most frequent defoliating pest of soybeans in Central USA. These relative values support field observation of a low incidence of *N. rileyi* in *A. gemmatalis* and a high incidence in *Pl. scabra* from Midwestern USA soybeans. Although *A. gemmatalis* was the least susceptible species to one isolate of *N. rileyi*, it often has annual epizootics in southern soybeans (Hinds and Osterberger, 1931; Allen *et al.*, 1971). Ignoffo *et al.* (1976c) demonstrated that there were differences in geographical isolates of *N. rileyi*. *Trichoplusia ni* was equally susceptible to isolates from Missouri, Mississippi, and Brazil, but an isolate from Florida was 7 to 17 times less active than isolates from other locations. Larvae of *A. gemmatalis* were either only slightly or not susceptible to isolates from Missouri, Florida and Mississippi. In contrast, larvae from Missouri, Florida and Brazil were all equally susceptible to a Brazilian isolate.

4. Loss in Activity

In an early study Rockwood (1950) reported that the entomogenous fungi *Metarhizium*, *Beauveria* and *Spicaria rileyi* lost their virulence after 1 year of growth on media. Kawakami (1960) found that the infectivity of *N. rileyi* (as *S. prasina*) decreased with each successive transfer on media, but could be recovered after passage through an insect. The extent and rapidity of the decrease were apparently dependent upon the culture medium. In addition to infectivity, various developmental stages (mycelial-mat growth, conidiophore formation, conidiogenesis) were reduced by each successive transfer on media. However, in preliminary experiments virulence has not decreased after 15 successive transfers of *N. rileyi* on SMAY (Ignoffo and Garcia unpublished observations).

B. VERTEBRATE HOSTS

1. Temperature Profile

The temperature profile for germination, growth and sporulation of *N. rileyi* indicates that the fungus probably will not infect homoiothermic vertebrates. Mammalian and avian rectal temperatures generally range from 34.4°C (opossum) to 38.8°C (Rhesus monkey) and 39.0°C (penguin) to 44.0°C (titmouse) (Spector, 1956). In contrast, *N. rileyi* grows poorly at 30°C, and it will not germinate or grow at or above 35°C (Getzin, 1961; Ignoffo *et al.*, 1976a). Although some mycelial growth was observed at 30°C, sporulation, which normally takes 8–9 days at 25°C, did not occur after 35 days at 30°C.

2. Mammalian Studies

Comprehensive, single-exposure, short-term tests in mammalian systems have recently been completed (Table IV). In these tests, conidia were rapidly inactivated ($\frac{1}{2}$ life ca. 20 min.) when exposed to human gastric juices at a rate equivalent to 10^9 conidia/person. Mice were administered (gastric intubation) 24.8×10^8 conidia/mouse, the equivalent of a 70-kg man receiving a 0.4-ha application of conidia. No deleterious effects were observed and > 99.99% of the infectivity of conidia, assayed against *T. ni* larvae (Ignoffo *et al.*, 1976c), was lost during passage through the alimentary tract. Also, > 99% of the total infectivity that was recovered from faeces was recovered 1 day after feeding conidia. No apparent clinical, pathological, or histological abnormalities were observed when white rats were exposed to 1.1×10^7 viable conidia/litre of air for 1 h, the equivalent of a 70-kg man inhaling conidia applied to ca. 0.02 ha. Eyes and the abraded skin of rabbits treated with a dose of 1.2×10^8 conidia/ eye or 3×10^9 conidia/cm^2 of skin, respectively, showed no apparent clinical, pathological, or histological abnormalities.

V. Stability and Sensitivity

A. ENVIRONMENTAL STABILITY

Conidia can withstand extended periods at low temperatures. Once dried, they have been stored ca. 3–4 years at 5° to − 20°C without any apparent loss in infectivity (Table V). However, outdoors most infectivity is lost within 1 year. On soil surfaces about 10% of the original activity was lost after 10–14 days, half after ca. 40–65 days, and > 99% after 250–350 days of exposure (Table V). Buried conidia survived better and had a half-life of 80 days.

Natural sunlight is the major environmental factor affecting survival of conidia reducing the half-life on foliage to only 2–3 days, although some infectivity may remain for 2–3 weeks following one application of 2 300

TABLE IV
Summary of results which evaluated possible risks in the use of Nomuraea rileyi

Test System	No. Tested	Route	Conidia × 10^6 / Animal	Evaluation Criteria	Results
Human gastric juice[a]	3	*in vitro*	10/ml	bioassay	$\frac{1}{2}$-life, 23 min
Albino mice	20	stomach intubation	2 440	clinical signs, daily body weight and food consumption, rectal temperature, necropsy, organ weight, bioassay of faeces	negative
Albino mice	20	stomach intubation	91	as above	negative
White rabbits	12	eye instillation	122	Draize's (1959) grading system, clinical signs	negative
White rabbits	12	dermal application	3 000/ cm^2	as above, histology, bioassay of skin	negative
Albino rats	30	inhalation	10	clinical signs, daily body weight and food consumption, rectal temperature, lung histology and bioassay	negative

[a] Ignoffo and Garcia (1977); all other data from Ignoffo and Garcia (1978; unpublished observations).
[b] Whole culture (substrate, mycelium, conidia) used; all other tests with conidia only.

conidia/mm^2 of soybean leaf surface (Table V). The half-life under a simulated-sunlight-source (Ignoffo and Batzer, 1971) on glass plates was ca. 2.4 h; 1 h under this source being biologically equivalent to ca. 1 day exposure under natural sunlight (Ignoffo et al., 1977a).

B. SENSITIVITY TO CHEMICALS AND THERAPEUTICS

Ignoffo et al. (1975b) used an *in vitro* paper-disc technique on culture media to determine the sensitivity of conidia to 44 chemical pesticides. Seven of eight fungicides (at 1/10 the recommended rate) inhibited growth of *N. rileyi* (Table VI). Chlorothalonil and ferbam were the most active fungicides. Pyroxychlor did not inhibit *N. rileyi*. Thirteen of 25 insecticides caused some inhibition, but none except methyl parathion were as active as the fungicides (Table VI). The three most inhibiting insecticides were monocrotophos, phenthoate and methyl

TABLE V

Stability of conidia and sclerotia (mummified cadavers) of Nomuraea rileyi under various environmental conditions

Stage	Substrate	Environmental Conditions	Stability[a]	Reference
Conidia	agar culture	storage − 20°C	infectivity > 4 years	Kawakami, 1970
Conidia	silica gel	storage − 20°C	infectivity > 3 years	Bell and Hamalle, 1974
Conidia	glass vial	storage 5°C	consistent LC_{50} > 2 yr	Ignoffo and Garcia, unpublished observations
Conidia	culture tube	outdoor shade[b]	½-life ca. 6−9 months	Bell, 1975b
Conidia	glass plate	simulated sunlight	½-life of 2.4 h	Ignoffo et al., 1977a
Conidia	glass vial	open field[c]	infective to 209 days	Sprenkel and Brooks, 1977
Conidia	soil surface	open field[d]	infective to 138 days	Sprenkel and Brooks, 1977
Sclerotia	soil surface	open field[c]	infective > 281 days	Sprenkel and Brooks, 1977
Conidia	soil surface	soybean field[e]	½-life of 40−65 days	Ignoffo and Garcia, unpublished observations
Sclerotia	soil-buried	open field[c]	infective to 194 days	Sprenkel and Brooks, 1977
Conidia	soil-buried	soybean field[e]	½-life of 80 days	Ignoffo and et al., 1978c
Conidia	soybean leaf	natural sunlight[a]	½-life of 2−3 days	Ignoffo et al., 1976b
Conidia	soybean leaf	natural sunlight[f]	½-life of 5−10 days	Gardner et al., 1977

[a] Determined by bioassay. [b] Stoneville, MS. [c] Raleigh, NC. [d] Clayton, NC. [e] Columbia, MO. [f] Blackville, SC.

TABLE VI

Agricultural chemicals, antibiotics, and chemotherapeutic substances that have inhibited growth and development of Nomuraea rileyi (Ignoffo et al. 1975b; Garcia and Ignoffo, 1979)

Fungicides	Insecticides
benomyl	carbophenothion
chlorothalonil	chlorpyrifos
fentin hydroxide	DBCP
ferbam	ethion
formalin	heptachlor
maneb	leptophos
methylparabenzoate	methyl parathion
potassium sorbate	monocrotophos
sulphur + zineb	phenthoate
zinc iron-maneb	toxaphene

Herbicides	Antibiotics
2,4-DB	methenamine
dinoseb	mandelate
linuron	nystatin
oxadiazon	

parathion. N. rileyi was sensitive to 4 of 11 herbicides (Table VI). The herbicide dinoseb, at concentrations equivalent to the fungicides, was about 30 to 50% as active as the fungicides. Some compatible pesticides are listed in Chapter 38, Section II, C).

In addition to the in vitro tests, bioassays were conducted with the fungicide benomyl, and the pre-postemergence herbicide dinoseb (Ignoffo et al., 1975b). Both are recommended on crops where caterpillar pests suffer regular natural seasonal epizootics of *N. rileyi* (Ignoffo *et al.*, 1976b). Benomyl, at 1/6 the lowest recommended rate and with 30.7 conidia/mm^2 of leaf surface, significantly inhibited infection of *T. ni* larvae. On average conidia alone killed 79.6 ± 7.3 (S.E.)% of larvae, conidia + benomyl 21.8 ± 11.1% and benomyl alone 0.6 ± 0.3%. Results with dinoseb, at a rate equivalent to one early-season soil application, were similar: with soil containing 5.4 × 10^6 conidia/g + dinoseb, a mean of 5.7 ± 3.3% larvae died, with conidia alone 31.8 ± 5.2% and dinoseb alone 0%.

Results from field tests were similar to laboratory results (Johnson *et al.*, 1976). Natural epizootics in populations of *A. gemmatalis* were delayed at least 23 days after soybean fields were treated with benomyl, benomyl + methyl parathion, or benomyl + carbaryl. Carbaryl and methyl parathion killed so many caterpillars that survivors could not provide sufficient conidia to sustain an epizootic. Johnson *et al.* (1976) concluded: "There is little doubt that, under field conditions in Florida, effective natural control by *N. rileyi* of *A. gemmatalis* on soybeans is adversely affected by some agrochemicals currently in use."

The sensitivity of *N. rileyi* to 47 antibiotics and chemotherapeutics was determined by the paper disc technique (Garcia and Ignoffo, 1979). Only methenamine mandelate and nystatin inhibited growth (Table VI) and in another test, they were as inhibitive (at equivalent concentrations) as the fungicides methylparabenzoate and potassium sorbate, but not as inhibitive as formalin.

VI. Epizootiology

A. PROBABLE CYCLIC EVENTS

A probable sequence resulting in seasonal epizootics of *N. rileyi* in caterpillars was proposed by Ignoffo *et al.* (1977b): (1) transmission of soil-borne conidia in early spring to seedlings or new growth of perennial plants; (2) infection of susceptible larvae feeding on the contaminated plants; (3) dispersal of infected larvae over a plant; (4) death and conidia from cadavers provide the first increase in inoculum and foci of infection; (5) repetitive phases of infection and inoculum increase throughout late spring and summer, with wind dispersal of conidia to initiate the late summer epizootic; (6) elimination of the local population of susceptible larvae in late summer and early fall; (7) contamination of soil in the fall by conidia from dead larvae; (8) overwintering and survival of infectious conidia in the soil. Thus, the load of overwintering inoculum, early availability of hosts, proper environmental conditions, and extent of dispersal of conidia early in the season largely determine when the epizootic peak is reached each season.

A model and equation to predict the incidence of *N. rileyi* on *A. gemmatalis* in soybeans were developed by Kish and Allen (1978). In 67 trials to test the model, the Chi square indicated verifications in 77% of the trials significant at the 20% level, 53% at the 5% level and 36% at the 1% level. Also, correlation was high (0.72 to 0.82) between the predicted and observed infection levels for all trials both in 1975 and 1976.

B. INOCULUM RESERVOIR

Ignoffo et al. (1977b) proposed that soil is the natural reservoir of conidia that start annual epizootics. Inoculum sufficient to contaminate soil each fall could be produced by as few as 1 diseased caterpillar/m^2; natural epizootics produce $> 10 \times$ that number. The inoculum could overwinter on cadavers (Sprenkel and Brooks, 1976) or as free conidia (Ignoffo et al., 1977b; Ignoffo et al., 1978c). There is little vertical movement of conidia through agricultural soil, so most conidia in undisturbed soil (90%) would remain in the upper 2 cm (Ignoffo et al., 1977c).

C. TRANSMISSION OF CONIDIA

1. Soil to Plant

Seedling plants or new growth from perennial plants probably are contaminated by soil-borne conidia. Leaflets from commercial fields of soybeans were contaminated with N. rileyi only 25 days after soybeans were planted when they were 10.2 cm tall with 2 trifoliates/plant (Ignoffo et al., 1975a). In the laboratory, larvae of T. ni were infected when fed on soybean seedlings germinated in soil sprayed with conidia (Ignoffo et al., 1977b; Montross and Carner unpublished observations). About 5% of the larvae became infected when they fed on cotyledons or unifoliates of seedlings germinated on soil surface-treated with 9×10^3 conidia/mm^2 (yield from ca. 0.00001 larva), and as much as 30–46% mortality was obtained when the conidial rate was increased 100-fold.

2. Throughout the Plant

As the plants grow infected larvae disperse and die. This is probably the major means of spreading N. rileyi over an entire plant. Diseased larvae released on the 1st trifoliate of soybean plants fed and dispersed to all trifoliates of plants 32–70 cm high with 6–22 trifoliates/plant (Ignoffo et al., 1977b). Over 90% of the larvae died within 7 days (ca. 65% in instar III) and sporulation occurred on these cadavers 2–3 days later. Subsequent bioassays demonstrated that trifoliates of all soybean plants were contaminated with N. rileyi. The mortality of T. ni larvae feeding on trifoliates 48–60 cm from the release point was 23.9 ± 5.3% (S.E.).

3. Throughout the Field

Larvae are initially infected by conidia on plants contaminated from soil-borne conidia. Spread of conidia from cadavers is probably the first of two to three sequential increases in inoculum that eventually provide sufficient conidia to start a late summer epizootic of N. rileyi in caterpillars.

(a) *Conidiogenesis*. The production of conidia per insect was computed from the equation

$$y = -0.07544 + 0.00586x + 0.0000259x^2$$

where y = conidia x 10^9 and x = mm^2 of host surface area (Kish and Allen, 1976). Production of conidia from cadavers occurs at most for 3 days; "no further production [occurs thereafter] because of depletion of nutrients" (Kish and Allen, 1976).

As few as 1 sporulating larva/100 plants would provide sufficient conidia to induce a rapidly developing epizootic similar to that reported by Ignoffo *et al.* (1976b). Probably less than 1 diseased larva/10 000 plants (= ca. 10 diseased larvae/0.4 ha) would produce the increase of *N. rileyi* often observed in natural epizootics.

Of all the meteorological variables that influence an epizootic, none is more critical for sporulation, germination and invasion of the host than high humidity ($> 90°$ RH) or water ($2-6$ h of dew) (Getzin, 1961; Allen *et al.*, 1971; Ignoffo *et al.*, 1977b; Kish and Allen, 1978). Although free water is important, an excess as heavy rains or long-standing dew can be adverse. Excesses can reduce the density of air-borne conidia, wash them from cadavers, and prevent their dissemination by wind (Garcia and Ignoffo, 1977; Kish and Allen, 1978). Cadavers under overhead irrigation for 2 h (= 0.72 inches of rain) lost 91% of their conidia while 49% of the conidia of cadavers not thus exposed was dislodged by air (Kish and Allen, 1978).

Field temperatures between $20°$ and $30°$C should not limit growth and sporulation. On the other hand, since *N. rileyi* will not grow at $< 15°$C and will not grow or sporulate at $> 30°$C, long periods at these extremes might limit both the initiation and development of epizootics (Ignoffo *et al.*, 1976c; Mohammed *et al.*, 1977; Kish and Allen, 1978).

(b) *Dispersal of Conidia*. The density of air-borne conidia is significantly less after rain; high levels of conidia are correlated with periods of dry, gusty winds and dry foliage (Kish and Allen, 1978). For example, the passage of a warm, dry front resulted in a count of $> 150 000$ conidia/slide for a 2 h period (1200–1400 h).

When the inoculum increases rapidly, most plants are undoubtedly contaminated by wind-borne conidia. The minimum wind velocity to dislodge conidia from the cadaver of an instar V, *T. ni* larva was 2.7 km/h (Garcia and Ignoffo, 1977). The percentage of the total dislodged conidia after 1, 2, 3, 5, 15, and 30 min exposure to a wind of 5.9 km/h was 66.2, 16.3, 6.0, 4.5, 4.6, and 2.4%, respectively. Thus, ca. 90% of the conidia that could be dislodged were released within the first 3 min of exposure. After the first sighting of *Nomuraea*-killed *A. gemmatalis* larvae on soybeans (Kish and Allen, 1978), the level of air-borne conidia rose slowly for 10 days, rapidly reached a peak in 33 days, then dropped precipitously (Table VII).

(c) *Survival of Conidia*. The disease progresses until by the fall larvae of susceptible species have either pupated or have been eliminated by *N. rileyi*. Kish (1975) proposed that *N. rileyi* might survive the winter in southern Florida on

TABLE VII

Relative number of conidia of Nomuraea rileyi *collected from air over soybeans in Florida* (*Kish and Allen, 1978*)

| Calendar date | Day after first diseased larva found | Number of | |
		Conidia/slide/h[a]	Infected larvae/ 30 cm row
August 26	0	< 100	0.2
Sept. 5	10	1 000	0.5
Sept. 16	22	16 000	2.9
Sept. 20	26	20 000	1.3
Sept. 27	33	70 000	0.2
Sept. 30	36	5 000	0.2
Oct. 3	39	1 000	0.2

[a] Highest number recorded in spore trap of Hirst (1952)

alternative hosts and then spread northward in spring via contaminated migrating adults. Although this is possible, it is more likely that mummified caterpillars or free conidia in soil provide the overwintering reservoir of conidia.

High loss in winter might indicate too few active conidia to start an epizootic the next field season. However, even without additional sporulation from cadavers (Sprenkel and Brooks, 1976), a reduction of 99.9% translates to ca. 6×10^6 active conidia on June 1st for each larva that died in the previous October. This equals 5×10^4 conidia/mm^2 of soil surface, which should be more than sufficient to start an epizootic. Active conidia were found on field soybeans only 25 days after planting (Ignoffo et al., 1975a) and, in the laboratory, about 12% of larvae died when fed on soybean seedlings germinated in soil surface-treated with ca. 9×10^4 conidia/mm^2 (Ignoffo et al., 1977b). These results further suggest that sufficient conidia to initiate an epizootic can come from larvae killed the previous year.

VII. Production

Production of limited quantities of conidia (< 5 kg) has been reported by Bell (1975a), St. Julian et al. (unpublished observations), Couch (unpublished observations), Ignoffo and Garcia (unpublished observations). Bell (1975a) routinely cultured *N. rileyi* on laboratory SMAY slants in culture tubes (40 × 5 cm). Sporulation began after 10 days at 24° ± 1°C, and conidia were harvested by suction after 21 days (Hamalle and Bell, 1976). The yield was 5.1 mg conidia/cm^2 of surface (= 6.3×10^8 conidia/cm^2). Production costs for materials were ca. 60.4 cents U.S./g of conidia (33.8¢ of this for the medium). Ignoffo and Garcia (unpublished observations) had similar results with disposable, plastic petri dishes (diam. 8.5 cm). Yields were 4.1 mg (= ca 5.1×10^8) conidia/cm^2. Material costs were 45.0 cents U.S./g of conidia (27.2¢ of this for dishes).

Blastospores or mycelium were grown in a "pupal–decoction" liquid fortified with 2% peptone (Kawakami, 1962b). Optimum production (2×10^3 blastospores/mm^3) took ca. 160 h at 25–29°C. Yen and Chu (1963) investigated nutritional requirements both on solid and liquid media. Growth was best on Sabouraud, potato-dextrose and yeast extract agars. A comparison between surface and submerged fermentation was made by Bell (1975a). Yields (dry weight) after 21 days at 24 ± 1°C averaged 386 mg of conidia/100 ml of agar and 575 mg of blastospores/100 ml of broth. However, blastospores were not infective against 6 day-old *H. zea* larvae (10^9 conidia/larva), whereas conidia killed 98% of the larvae within 14 days.

St. Julian *et al.* (unpublished observations) produced conidia in metal covered glass trays (42.5×15.5 cm; inside depth 1.9 cm) designed by Lindenfelser and Ciegler (1969). The trays, presterilized for 14 h in polyethylene bags saturated with ethylene oxide, were filled with 200 ml of sterile medium (agar 1.5%; maltose 4.0%; peptone 1.0%; yeast extract 0.05%; water 93.45%) and aseptically inoculated with 5 ml of suspension (10^8 conidia/ml; 0.01% Triton$^®$ X − 100). Conidia were harvested after 9 to 14 days at 25 ± 1°C by suction. Each tray produced ca. 3.3 g or 4×10^{11} conidia/tray. The medium cost 20.4 cents U.S./g.

Production batches of *N. rileyi* have been standardized by percent germination, bioassay, conidial counts and number of bacterial and fungal contaminants (Ignoffo and Garcia, unpublished observations).

Since blastospores are easily produced in submerged culture, they might be used as field-applied inoculum if a stable formulation of blastospores plus nutrients could be developed to permit sporulation of conidia in the field. The possible use of the toxin(s) as a microbial insecticide (Mikuni and Kawakami, 1975; Ignoffo and Garcia, unpublished) should be further explored.

VIII. Control of Insect Pests

A. INSECTICIDAL APPROACH

Of seven reported field tests, that of Chamberlin and Dutky (1958) showed that *N. rileyi* (used at ca. 7.5×10^{15} conidia/0.4-ha) would infect *H. virescens* larvae feeding on tobacco but it "failed to give adequate control under a considerable range of climatic conditions". On artificial infestations of *T. ni* on cabbage, Getzin (1961) sprayed 1.2×10^{10}, 1.2×10^{11}, and 1.2×10^{12} conidia/ 0.4 ha with 0.005% Triton$^®$ X−100 and 0.005% Du Pont spreader sticker$^®$. Although the highest application rate killed 67% of the larvae, he concluded that "spore distribution would be ineffective" because of the specific environmental requirements for infection, growth, and sporulation and the long time required to kill larvae. It could, however, reduce "the reproductive capacity of the cabbage-looper population . . . thus facilitating chemical control measures". Better results on spring and fall cabbage were reported by Bell (*in* Hostetter and Ignoffo, 1978). Seven weekly applications of dust formulation of conidia (5.6×10^{13} conidia/9.1 kg pyrophyllite/0.4 ha) gave significant control of *T. ni*, as measured by larval mortality and damage at harvest; however, damage on upper leaves excluded all heads from USDA No. 1 grade cabbage.

In replicated field cage studies on an artificial infestation of instar I, II and III, *H. zea* on soybeans (Ignoffo *et al.*, 1978), 10^{10}, 10^{11}, 10^{12}, 10^{13} conidia/ 0.4 ha were applied 3, 7, 10, and 14 days after the release of larvae. All rates except the lowest significantly reduced the population below the check. Reduction in seed-damaged pods ranged from 20% (10^{10} conidia/0.4 ha) to 47% (10^{12} conidia/0.4 ha). The authors concluded that *N. rileyi* "probably has little value as a microbial insecticide" for control of mid-instar to mature *Heliothis* larvae on soybeans, although it might be effective against instars I and II or as a prophylactic agent if used before the occurrence of damaging populations of *H. zea* (Ignoffo *et al.*, 1976b).

Mohamed *et al.* (1978b) conducted a replicated field cage study on instars I to V of *H. zea* on sweet corn. Conidia were applied 1 to 3 times (2–3 day intervals) at 1.6×10^{13} conidia/0.4 ha. Larval deaths (instars II to V) for early mid and late summer averaged 90.2%, 83.2% and 61.8%, respectively. The reduction in ear damage averaged 24.8%, 26.2% and 36.0% for 1, 2, and 3 applications, respectively. Thus, *N. rileyi* did infect many *H. zea* larvae but it did not significantly reduce economic damage to sweet corn.

B. INDUCED EPIZOOTIC

The above studies indicate that direct application of *N. rileyi* will not immediately control heavy populations of caterpillars unless it is directed against young instars and, even then, high concentrations would probably be needed. Ignoffo *et al.* (1975a, 1978), however, suggested that an early heavy prophylactic application of conidia or release of infected larvae might induce an earlier than normal epizootic and thus suppress caterpillar pests when plants are most sensitive to insect feeding. In the earliest such attempt, Watson (1916) could not induce an epizootic in *A. gemmatalis* on soybeans, but stated that "the *cholera* [*N. rileyi*] was by far the most efficient check of this insect. . . Although there may be a partial recovery. . . the caterpillars never again. . . became sufficiently numerous to be troublesome. The cholera often arrives too late to save the crop; it is nevertheless a great help as it reduces to a few weeks the time during which the farmer needs to apply arsenicals [insecticides]."

Later, Sprenkel and Brooks (1975) distributed *H. virescens* cadavers to control *H. zea*, *P. includens*, and *Pl. scabra* on soybeans. Cadavers were cut into 3 mm sections, mixed with dry vermiculite and distributed by a manually operated grass-seed spreader at 3 360 cadavers/ha (ca. 2–3 pieces of cadaver/ 1.5 m of row). Results from all experiments demonstrated that an epizootic could be produced ca. 14 days before the natural epizootic.

Ignoffo *et al.* (1976b) sprayed one heavy application of conidia (1.1×10^{13} conidia/0.4 ha) onto soybeans when half of the plants had at least one flower. The dose and volume/ha of a second experiment (when pods were 6 mm long at one of the four uppermost nodes) were doubled. Both experimental applications of conidia significantly altered the epizootic pattern. Initial detection and peak incidence of infected *Pl. scabra* in treated plots were advanced at least 14 days compared with untreated plots. The peak in treated plots therefore occurred prior to and during the stages of soybean growth that were most sensitive to defoliation. The percent infected *Pl. scabra* at the critical stage of soybean growth after the first experiment averaged 82.5% for treated and 7.4% for untreated plots. The equivalent percentages after the second spray were 90.0 and 18.5%, respectively. Although Bucher (1964) could not differentiate artificial from natural dispersal, Ignoffo *et al.* (1976b) did not observe any lateral spread of *N. rileyi* from treated to untreated plots. They concluded that "the greatest natural increase in incidence of *N. rileyi* in a field results from limited progressive outward spread of conidia from dead larvae," and that *N. rileyi* might be an effective microbial insecticide if directed against instars I and II; however, it offers more potential if used as a prophylactic system in insect pest management programs.

C. MANIPULATION OF ECOSYSTEM

It might be possible to start an early epizootic by early planting of a row of soybeans alternated with ca. 100 late-planted rows. For example, Sprenkel *et al.* (1973) reported a significantly higher incidence of *N. rileyi* in early-planted soybeans (June 5) than in late ones (July 12) − 2.02 to 8.70 infected larvae cf. 0.25 to 0.67 per 4×10^{-4} ha, respectively. No differences could be related to row spacing (0.75 vs 1.2 m rows) and seedling rate (6 vs 12 seeds/0.3 m of row). The incidence of *N. rileyi* in *Heliothis* spp. in a closed canopy of cotton ("Delta Pine") was ca. 1.3 × higher than in an open canopy (Okra-leaf variety) (Burleigh, 1975).

IX. Conclusion

The entomopathogenic fungus *Nomuraea rileyi* is a likely candidate for further consideration and development as a microbial insecticide. Factors in its favour include: (1) occurs naturally in several different kinds of agroecosystems (pastures, row crops, grain); (2) induces epizootics; (3) attacks many insect pests; (4) is not virulent against beneficial insects; (5) has temperature requirements below the body temperatures of homoiothermic vertebrates; (6) is not toxic or pathogenic when given by any route to mammals; (7) as an insecticide has been successfully used to suppress insect pests; (8) can be produced on artificial media at a cost that should encourage industrial participation.

Adverse factors include: (1) kills slowly so that caterpillars, especially older ones, may cause considerable damage before dying; (2) requires free water for germination, growth and sporulation; (3) its temperature range is 15° to 30°C, so extreme field temperatures outside this range may impair effectiveness; (4) may not be suited to the insecticidal-control approach unless large doses of spores are directed at young insects. Research and development efforts to further evaluate the feasibility of use of *N. rileyi* as a microbial insecticide are continuing in the USA.

References

Aizawa, K., Shimazu, T. and Shimizu, S. (1976). *In* "Proceedings of the Joint United States-Japan Seminar on Stored Product Insects" (Manhattan, Kansas, Jan. 5–8, 1976), pp. 59–67.

Allen, G.E., Greene, G.L. and Whitcomb, W.H. (1971). *Fla. Ent.* **54**, 189–191.

Bassi, A. (1835). *Orcesi Lodi.*

Behnke, C.N. and Paschke, J.D. (1966). *J. Invertebr. Path.* **8**, 103–108.

Bell, J.V. (1975a). *J. Invertebr. Path.* **26**, 129–130.

Bell, J.V. (1975b). *J. Ga. ent. Soc.* **10**, 357–358.

Bell, J.V. and Hamalle, R.J. (1970). *J. Invertebr. Path.* **15**, 447–450.

Bell, J.V. and Hamalle, R.J. (1974). *Can. J. Microbiol.* **20**, 639–642.

Benz, G. (1963). *In* "Insect Pathology, An Advanced Treatise" (E. A Steinhaus, ed.) Vol. 1. pp. 299–331. Academic Press, New York and London.

Berisford, Y.C. and Tsao, C.H. (1975). *Ann. ent. Soc. Am.* **68**, 111–112.

Brown, A. and Smith, G. (1957). *Trans. Br. Mycol. Soc.* **40**, 17–89.

Bucher, G.E. (1964). *Ann. ent. Soc. Queb.* **9**, 30–42.

Burleigh, J.G. (1975). *Environ. Ent.* **4**, 574–576.

Chamberlin, F.S. and Dutky, S.R. (1958). *J. econ. Ent.* **51**, 506.

Charles, V.K. (1936). *Mycologia* **28**, 397–398.

Charles, V.K. (1941). *US Dep. Agric. Insect Pest Survey Bull.* **21**, 707–785.

Crumb, S.E. (1929). US Dep. Agric. Tech. Bull. No. 88, 1–179.

Draize, J.H. (1959). *In* "Appraisal of the Safety of Chemicals in Foods, Drugs, and Cosmetics," pp. 46–59. Ass. Food Drug Officials US, Austin, TX.

Fargues, J. (1976). *Entomophaga* **21**, 313–323.

Farlow, W.G. (1883). Rep. US Comm. Agric., 121.

Forbes, S.A., II (1898). Sixteenth Rep. State Ent. on Noxious and Beneficial Insects of the State of Illinois.

Garcia, C. and Ignoffo, C.M. (1977). *J. Invertebr. Path.* **30**, 114–116.

Garcia, C. and Ignoffo, C.M. (1979). *J. Invertebr. Path.* **33**, 124–125.

Gardner, W.A., Sutton, R.M. and Noblet, R. (1977). *Environ. Ent.* **6**, 616–618.

Getzin, L.W. (1961). *J. Insect Path.* **3**, 2–10.

Giard, A. (1892). *Bull. Sci. Fr. Belg.* **24**, 1.

Gudauskas, R.T. and Canerday, T.D. (1966). *J. Invertebr. Path.* **8**, 277.

Hamalle, R.J. and Bell, J.V. (1976). *J.Ga. ent. Soc.* **11**, 221–223.

Harper, J.D. and Carner, G.R. (1973). *J. Invertebr. Path.* **22**, 80–85.

Hinds, W.E. and Osterberger, S.A. (1931). *J. econ. Ent.* **24**, 1168–1173.

Hirst, J.M. (1952). *Ann. appl. Biol.* **39**, 257–265.

Hostetter, D.L. and Ignoffo, C.M. (1978). *In* "Monograph on *Trichoplusia ni* (Hübner)". (G.L. Greene, ed.). *Fla. Agric. Exp. Stn.* Monogr. Ser. (In press).

Hughes, S.J. (1951). *Mycol. Paper.* No. 45, Commonwealth Mycological Institute, Kew.

Ignoffo, C.M. and Batzer, O.F. (1971). *J. Econ. Ent.* **64**, 850–853.

Ignoffo, C.M. and Garcia, C. (1978). *Environ. Ent.* **7**, 217–218.

Ignoffo, C.M., Puttler, B., Marston, N.L., Hostetter, D.L. and Dickerson, W.A. (1975a). *J. Invertebr. Path.* **25**, 135–137.

Ignoffo, C.M., Hostetter, D.L., Garcia, C. and Pinnell, R.E. (1975b). *Environ. Ent.* **4**, 765–768.

Ignoffo, C.M., Garcia, C. and Hostetter, D.L. (1976a). *Environ. Ent.* **5**, 935–936.

Ignoffo, C.M., Marston, N.L., Hostetter, D.L., Puttler, B. and Bell, J.V. (1976b). *J. Invertebr. Path.* **27**, 191–198.

Ignoffo, C.M., Puttler, B., Hostetter, D.L. and Dickerson, W.A. (1976c). *J. Invertebr. Path.* **28**, 259–262.

Ignoffo, C.M., Hostetter, D.L., Sikorowski, P.P., Sutter, G. and Brooks, W.M. (1977a). *Environ. Ent.* **6**, 411–415.

Ignoffo, C.M., Garcia, C., Hostetter, D.L. and Pinnell, R.E. (1977b). *J. Invertebr. Path.* **29**, 147–152.

Ignoffo, C.M., Garcia, C., Hostetter, D.L. and Pinnell, R.E. (1977c). *J. Econ. Ent.* **70**, 163–164.

Ignoffo, C.M., Hostetter, D.L., Biever, K.D., Garcia, C., Thomas, G.D., Dickerson, W.A. and Pinnell, R.E. (1978). *J. Econ. Ent.* **71**, 165–168.

Ignoffo, C.M., Garcia, C., Hostetter, D.L. and Pinnell, R.E. (1978c). *Environ. Ent.* **7**, 724–727.

Johnson, D.W., Kish, L.P. and Allen, G.E. (1976). *Environ. Ent.* **5**, 964–966.

Kawakami, K. (1960). *Bull. Seric. Exp. Stn.* **16**, 83–99.

Kawakami, K. (1962a). *Bull. Seric. Exp. Stn.* **18**, 133–146.

Kawakami, K. (1962b). *Bull. Seric. Exp. Stn.* **18**, 147–156.

Kawakami, K. (1970). *Acta Serologica* **76**, 58–62.

Kawakami, K. (1973). *Bull. Seric. Exp. Stn.* **25**, 347–370.

Kawakami, K. and Mikuni, T. (1965). *Acta Serologica* **56**, 35–42.

Kawakami, K., Nakazato, Y., Fujimoto, I. and Mikumi, T. (1969). *Acta Serologica* **73**, 52–64.

Kenneth, R. and Olmert, I. (1975). *Isr. J. Ent.,* **10**, 105–112.

King, E.G. and Bell, J.V. (1980). *J. Invertebr. Path.* **31**, 337–340.

Kish, L.P. (1975). "The Biology and Ecology of *Nomuraea rileyi*", 83pp., Ph.D. Diss., Univ. Florida.

Kish, L.P. and Allen, G.E. (1976). *Mycologia* **68**, 436–439.

Kish, L.P. and Allen, G.E. (1978). *Fla. Agric. Exp. Stn. Bull.,* 795.

Kish, L.P., Samson, R.A. and Allen, G.E. (1974). *J. Invertebr. Path.* **24**, 154–158.

Kish, L.P., Greene, G.L. and Allen, G.E. (1976). *Fla. Ent.* **59**, 103–106.

Krassilstschik, I.M. (1888). *Bull. Sci. Fr. Belg.* **19**, 461–472.

Lindenfelser, L.A. and Ciegler, A. (1969). *Dev. Ind. Microbiol.* **10**, 271–278.

Maulblanc, A. (1903). *Bull. Soc. Mycol. Fr.* **19**, 291–296.

Metchnikoff, E. (1879). Commission of Odessa Zemstvo Office. Odessa. 32 pp.

Mikuni, T. and Kawakami, K. (1975). *Jap. J. appl. Ent. Zool.* **19**, 203–207.

Mohamed, K.A., Sikorowski, P. and Bell, J.V. (1977). *J. Invertebr. Path.* **30**, 414–417.

Mohamed, K.A., Sikorowski, P. and Bell, J.V. (1978a). *J. Invertebr. Path.* **31**, 345–352.

Mohamed, K.A., Bell, J.V. and Sikorowski, P. (1978b). *J. econ. Ent.* **71**, 102–104.

Newman, G.G. and Carner, G.R. (1975). *Environ. Ent.* **4**, 231–232.

Petch, T. (1925). VII. *Spicaria. Trans. Brit. Mycol. Soc.* **10**, 183–189.

Phadke, C.H. and Rao, V.G. (1978). *Curr. Sci.* **47**, 511–512.

Phadke, C.H., Rao. V.G. and Pawar, S.K. (1978). *Curr. Sci.* **47**, 476.

Poinar, Jr., G.O., Geest, L. van der, Helle, W. and Wassink, H. (1977). *In* "Tsetse: The Future for Biological Methods in Integrated Control" (Marshall Laird, ed.), pp. 75–92. Ottawa, IDRC.

Puttler, B., Ignoffo, C.M. and Hostetter, D.L. (1976). *J. Invertebr. Path.* **27**, 269–270.

Rao, V.G. and Phadke, C.H. (1977). *Curr, Sci.* **46**, 648–649.

Roach, S.H. (1975). *Environ. Ent.* **4**, 725–728.

Rockwood, L.P. (1950). *J. Econ. Ent.* **45**, 704–707.

Samson, R.A. (1974). *Stud. Mycol. Baarn* **6**, 1–119.

Sawada, K. (1919). Agric. Exp. Sta. Govt. Formosa, p. 606.

Smith, J.W., King, E.G. and Bell, J.V. (1976). *Environ. Ent.* **5**, 224–226.

Snow, F.H. (1891). 21st Ann. Rep. Ent. Soc. Ontario, 93.

Spector, W.S. (1956). *Wright Air Development Center Techn. Rep.* pp. 56–273.

Sprenkel, R.K. and Brooks, W.M. (1975). *J. econ. Ent.* **68**, 847–850.

Sprenkel, R.K. and Brooks, W.M. (1977). *J. Invertebr. Path.* **29**, 262–266.

Sprenkel, R.K., Brooks, W.M. and Van Duyn, J. (1973). "Insect Pest Management Project." Dept. Ent., NC. State Univ., Raleigh, NC. (Unpubl. rep., IPM Project).

Steinhaus, E.A. and Marsh, G.A. (1962). *Hilgardia* **33**, 349–490.

Vandamme, P. and Angelini, A. (1966). *Coton et Fibres Tropicales.* **21**: 333–338.

Wasti, S.S. and Hartman, G.C. (1978). *Appl. Ent. Zool.* **13**, 23–28.

Watson, J.R. (1916). *J. econ. Ent.* **9**, 521–528.

Wolcott, G.N. and Martorell, L.F. (1940). *J. econ. Ent.* **33**, 201–202.

Yen, D.E. (1960). *Pl. Prot. Bull.* **2**, 54–58.

Yen, D.F. and Chu, W.H. (1963). *Mem. Taipei. Natl. Taiwan Univ., Col. Agric.,* **7**, 55–66.

CHAPTER 28

Pest Control by Entomophthorales

N. WILDING

Rothamsted Experimental Station, Harpenden, Hertfordshire, England

I. Introduction

Entomophthora species have long been recognized as able to cause epizootics that decimate arthropod populations. Also, many infect members of only one family or order, a feature commonly deemed an advantage for potential biological control agents. Despite this, there are no fully documented accounts of their successful use in pest control. (See "Note added in proof" to Chapter 7).

Some problems to be overcome before *Entomophthora* species could be used successfully in agriculture were outlined by Roberts (1973): the fungi are difficult to culture; conidia, the usual spore form to develop best *in vitro*, are short lived; resting spores are much longer lived, not produced by all species and not easily induced to germinate; the pathogenicity of the fungi for man and other vertebrates is unknown. In addition the fungi need a water-saturated atmosphere for several hours to infect their hosts and very little is known about the specificity of strains or species, or of the effect of pesticides on fungal activity. This chapter discusses results of research into these and related aspects of *Entomophthora* species that infect insects and mites. Most were published since, or not considered in, Burges and Hussey (1971). Recent reviews of work on *Entomophthora* species that attack aphids were made by Gustafsson (1971), Remaudière (1971), Rabasse (1974) and Delucchi (1976) but there are none concerned specifically with *Entomophthora* spp. attacking a wider range of hosts.

II. Epizootiology

Spread of *Entomophthora* in an insect population is influenced by abiotic factors, essentially weather, and biotic factors, comprising density and distribution of host and fungus and inherent infectivity of the fungus. These factors interact and in the field it is impossible to determine the influence of one in isolation from the others. However, the effect of each of the factors must be estimated so that their relative importance may be assessed.

A. INFLUENCE OF WEATHER ON FUNGUS SPREAD

1. Moisture

Saturated or near saturated air is necessary for many *Entomophthora* species to discharge their conidia (Voronina, 1968; Wilding, 1969, 1970; Newman and Carner, 1975a, b) although *E. muscae* ejected conidia from infected flies at 50% r.h. (Kramer, 1971) and some even at 20% r.h. (Wilding, unpublished observations). Wet air is also essential for conidia to germinate (Yendol, 1968; Newman and Carner, 1975a, b; Shimazu, 1977b). Consequently, infection usually succeeds in the laboratory only in wet air (e.g. Klein and Coppel, 1973; Hartman and Wasti, 1974; Carner, 1976). However, data are few that demonstrate the importance of moisture for spread of the fungi in the field. The mean monthly infection of *Delia (Hylemya) brassicae* by *Strongwellsea castrans* was greater, the more days there were during the month with a mean above 70% r.h. (Nair and McEwen, 1973). Hard (1976) found a smaller ratio of female to male cocoons of *Neodiprion tsugae* the more it rained. He attributed this to deaths of female larvae due to *E. sphaerosperma* which spreads mostly after male larvae have spun cocoons. Infections of pea aphids, *Acyrthosiphon pisum*, by several *Entomophthora* species were positively correlated with the mean rainfall during the preceding 12 days, though the regression coefficients were small (Wilding, 1975). Similarly Berisford and Tsao (1974) found only a weak relationship between daily maximum r.h. and infections of *Delia (Hylemya) platura* by *E. muscae*; and infections of *Delia (Leptohylemyia) coarctata* by *E. muscae* were not related to rainfall during their adult lives (Wilding and Lauckner, 1974).

Some authors consider there are moisture thresholds below which the fungi are inactive in the field. Missonier *et al.* (1970) postulate that a minimum of

8 h/day at >90% r.h. is required to maintain enzootic infection of aphids by *Entomophthora* spp. and that at least 10 h at >90% r.h./day, plus 5 h rain/day for at least three consecutive days are needed to initiate an epizootic. Similarly *D. brassicae* and *Delia (Hylemya) floralis* were infected by *E. muscae* only when the monthly rainfall was at least 20 mm and r.h. was between 64 and 82%: whether these figures were mean or maximum r.h. was not stated (Strazdinya, 1972).

Voronina (1971) defined three zones in Russia, according to their "hydrothermal coefficient". This is presumably the hydrothermal coefficient (GTK) of Selyaninov (see Ventskevich, 1961) given as

$$\frac{\text{total precipitation (mm)} \times 10}{\text{the sum of the mean daily temperatures (}^\circ\text{C)}}$$

at > 10°C. In moist zones, where the summer GTK exceeds 1.4, *Entomophthora* spp. are enzootic in *Ac. pisum* populations and regularly control the aphids to an economically harmless level. In the intermediate zone with a GTK of 1–1.4, the aphid is curbed but still causes moderate damage. Finally, in the arid zone it causes much damage almost every year. While the GTK is not commonly used to express the climate in terms of precipitation and evapotranspiration, its equivalent in more familiar terms (e.g., Thornthwaite, 1948) could be useful to predict the effect of *Entomophthora* on insects in a given area.

2. Temperature

Many authors have studied effects of temperature on stages in the life cycle of *Entomophthora* spp. *in vivo*. Most species eject conidia between 5 and 30°C and conidia germinate between these extremes. However they are ejected earlier and germinate better between 16 and 27°C (Golberg, 1970; Pady *et al.*, 1971; Wilding, 1971a; Newman and Carner, 1975a, b; Shimazu, 1977b). Insects have been experimentally infected within similar temperature extremes, results also being best between 16 and 27°C (Wilding, 1970; Krejzova, 1971b) although larvae of *Galleria mellonella* were infected with *E. destruens* even at 38°C (Krejzova, 1971a).

Roberts and Campbell (1977) state that epizootics in the field usually occur at temperatures optimal in the laboratory for infecting and killing insects. For example, *D. brassicae* and *D. floralis* were infected by *E. muscae* when the air temperature was 15–22°C (Strazdinya, 1972) and a minimum mean temperature of 20°C was required before epizootics occurred in aphid populations (Missonier *et al.*, 1970). However, other authors (listed in Roberts and Yendol, 1971) noted that many flies hibernating in caves and cellars were killed by *Entomophthora* species at mean temperatures < 12°C. These fungi included *E. destruens* which killed most *G. mellonella* at 26°C in the laboratory (Krejzova, 1971a). Attempts to correlate field infections by *Entomophthora* with temperature have failed (Berisford and Tsao, 1974; Wilding, 1975), probably because other more important factors frequently obscure the effect of temperature.

3. Light

More conidia of at least some *Entomophthora* species are ejected in light than in darkness (Voronina, 1968; Callaghan, 1969; Golberg, 1970; Pady *et al.*, 1971; Wilding, 1971a). However, their germination is probably less affected by light, e.g. as many conidia of *E. delphacis* germinated in darkness as in light (Shimazu, 1977b). Similarly, no more conidia of *Basidiobolus ranarum* germinated to form secondary conidia in light than in darkness, although most vegetative germination occurred in light (Callaghan, 1974). Germ tubes and secondary conidiophores from conidia of *E. thaxteriana* and *Conidiobolus coronatus* are positively phototropic (Voronina, 1968; Page and Humber, 1973; Chapter 7, Section II, B, C).

B. INFLUENCE OF HOST DENSITY ON FUNGUS SPREAD

The importance of host density for spread of *Entomophthora* species in field populations of insects is disputed. For example, Missonier *et al.* (1970) stated that the start and development of an epizootic are not directly related to host density — yet Rautapää (1976) in Finland and Suter and Keller (1977) in Switzerland noted that the occurrences of *Entomophthora* species in aphid populations were highly density dependent. Although none of these authors present analyses to substantiate their claims, others have done so. Thus during several seasons, infections of wheat bulb flies (*D. coarctata*) each year for 5 years were much more closely correlated with host density than with any weather factor (Wilding and Lauckner, 1974) and weekly infections of *Ac. pisum* for three seasons were correlated with host density, though the regression coefficients were small (Wilding, 1975). However, the role of host density is probably better shown by simultaneously comparing fungus infections in two or more insect populations of different densities. In this way, Carl (1975) found a greater proportion of thrips (*Thrips tabaci*) infected by *E. parvispora* (as *Entomophthora* sp.) in fields where the thrips population was greater. Similarly, *Entomophthora* sp. spread among aphids allowed to multiply on clover, by using an insecticide to control predators, but was absent in untreated areas (Manglitz and Hill, 1964). By contrast, on a crop of beans, the proportion of infected bean aphids, *Aphis fabae*, in each colony was clearly similar irrespective of colony size (Robert *et al.*, 1973; Dedryver, 1978). However, these aphids aggregate closely whatever the colony size: spread of the fungus may be affected more by the number of colonies per unit area than by their size.

Although the data are few, host density is important at least sometimes for the spread of *Entomophthora*, most probably when the fungus is sparsely distributed.

C. INFLUENCE OF INOCULUM ON FUNGUS SPREAD

1. Inoculum Concentration

The most recent observations on the effect of inoculum concentration on the spread of *Entomophthora* spp. have been on aphids. Missonier *et al.*

(1970) stated that a certain amount of inoculum was required to start an epizootic in populations of *Aphis fabae*. This was 50 to 300 infected aphids in each colony (Robert *et al.*, 1973; Dedryver and Robert, 1975). The proportion of infected aphids in the colonies was less important than the number. Also, the incidence of *Entomophthora* spp. in *Ac. pisum* was correlated with the mean number of conidia in the air above the crop during the preceding days and this factor affected the regression more than any other factor considered (Wilding, 1975).

This evidence suggests that inoculum concentration is important for spread of these fungi at least among aphids. However, there are no recent data on the amount of fungus that remains after an infected insect population has died or dispersed and its effect on the next host generation. That some does remain was shown by Latteur (1977). Soil was sampled from a field which several months earlier was cropped with vetch (*Vicia sativa*) infested with *Ac. pisum*, many of which were infected. When laboratory reared *Ac. pisum* were confined to the surface of the samples for 16 h, some became infected with *E. aphidis* and some with *E. thaxteriana*.

2. Fungus Infectivity

Some *Entomophthora* species are well known to have a wide host range in nature (e.g. Thoizon, 1970) but there is little information on whether such a species comprises different strains, each adapted to a particular host or group of hosts. A species sometimes kills many different but related hosts during an epizootic (e.g. Kushida *et al.*, 1975; Batko and Kmitowa, 1962). Frequently, however, epizootics are either exclusively or largely restricted to one host insect even though other closely related insects are present (e.g. Weiser and Novak, 1964; Selhime and Muma, 1966; Turian and Wuest, 1969; Wuest and Turian, 1971; Remaudière *et al.*, 1976c). Nair and McEwen (1973) noted that in a mixed fly population, adults of the *Delia platura/florilega* complex were infected only with *E. muscae*, and *D. brassicae* only with *S. castrans* – even though *D. brassicae* is a frequent host of *E. muscae* (Harris and Svec, 1966; Coaker and Finch, 1971; Strazdinya, 1972) and *D. platura (Hylemya cilicura)* is a host of *S. castrans* (Strong *et al.*, 1960).

Remaudière (1971) suggested that differences in the incidence of infection, observed in the field, between species of aphid may be caused by host behaviour or microhabitats rather than differences in their inherent susceptibility. This suggestion is supported by the ease with which many *Entomophthora* spp. have been transmitted from one host species to another in the laboratory (Kramer, 1971; Thoizon, 1970; Remaudière *et al.*, 1976a; Krejzova, 1978), sometimes to a species not naturally associated with the fungus (Thoizon, 1967; Shimazu, 1977a). However, differences in susceptibility between host species to some *Entomophthora* strains have been claimed (Golberg, 1970; Lowe and Kennel, 1972; Tyrrell and MacLeod, 1972; Nair and McEwen, 1973). Krejzova (1975) showed that there are different strains of *E. thaxteriana* and *E. virulenta* that differ in their infectivity for different hosts. However, it is difficult to assess the validity of any of these results because in none was the dose of the fungus accurately assessed.

The evidence suggests that although few generalizations can be made about the host range of any strain or species of *Entomophthora,* several species, at least, comprise strains each most infective for one or a few related hosts.

The mechanism that governs the susceptibility of a host is unknown. Conidia of *E. fresenii,* a species infective for *Ap. fabae* but not *Ac. pisum,* germinated on the cuticle of each species and formed an appressorium fixing the germ tube to the cuticle. Later, however, the developing germ tube usually penetrated the cuticle only of *Ap. fabae,* suggesting that the invasion barrier occurred at the host cuticle (Brobyn and Wilding, 1977).

D. FACTORS AFFECTING RESTING SPORE FORMATION

Factors inducing resting spores to form in nature have only recently been studied. Some workers (e.g. Nemoto and Aoki, 1975) note that they form only in autumn. *E. fresenii* formed resting spores in infected *Ap. fabae* either in a short photoperiod, in low temperatures or, more effectively, in both simultaneously (Wilding, 1973a). Others (e.g. Keller, 1976) found resting spores in some infected hosts always. Pickford and Riegart (1964) imply that they form when grasshoppers infected with *E. grylli* die in dry air, whereas conidia form in moist air. Certain *Entomophthora* species seem to respond to differences in the physiological age of their hosts, more resting spores developing with age (MacLeod *et al.,* 1973; Wilding and Lauckner, 1974; Newman and Carner, 1975c).

One cannot, therefore, generalize about the factors that induce resting spores to form. Moreover some species readily produce them *in vitro* (e.g. Tyrrell, 1970): many do not. A fundamental understanding of their controlling factors in nature may facilitate their production *in vitro. E. thaxteriana* formed resting spores *in vitro* more readily in darkness or in low rather than high light intensity (Petrova and Khrameeva, 1972).

E. FACTORS AFFECTING RESTING SPORE GERMINATION

Perhaps inherent with their durability is the difficulty of inducing resting spores of many species to germinate. Most studies suggest that germ tubes that develop from resting spores are not infective but produce one or several infective conidia. For example, some resting spores of *E. fresenii* in *Ap. fabae,* kept on moist soil out of doors during winter, germinated when placed at laboratory temperatures in March and formed anadhesive conidia on slender conidiophores (Wilding, 1971b). Similarly, ca. 5% of resting spores of *E. canadensis* (as *E. aphidis*) collected from aphid cadavers during winter, germinated on water agar at $23°C,$ forming conidia (Tyrrell and MacLeod, 1975). The proportion germinating increased to between 25 and 50% on exposure to light for 14 h or more each day (Wallace *et al.,* 1976). Resting spores of *E. vomitoriae* formed in *Pollenia rudis* during June and July. On moistening 20 of these flies, conidia were produced within 18 h; by 30 h resting spores remained in three, the others containing hyphal bodies only (Newman and Carner, 1975b).

Various factors have induced resting spores produced *in vitro* to germinate. Basova (1972) achieved germination in *E. apiculata* by 1—8 minutes exposure to sonication although further exposure damaged the spores. In *E. virulenta* (= *E. nr. thaxteriana*) germination followed sonication, or high speed blending, and immersion in dilute ethanol (Soper *et al.*, 1975). The proportion of germinating spores was increased by 1—2% Glusalase, and a 16 h photoperiod cf constant darkness (Matanmi and Libby, 1976). All the resting spores of *E. egressa* germinated in a chemically defined medium after 12 h at pH 8.5—9.5 (Nolan *et al.*, 1976). Most spores formed vegetative hyphae. This suggests that the host, which has alkaline gut contents, may be infected by ingesting resting spores. At lower pH, germ conidia usually formed.

III Development of *Entomophthora* for Pest Control

A. SELECTION OF STRAINS

Hartmann and Wasti (1974) and Krejzova (1972, 1975) claim to have increased the infectivity of *in vitro* cultures of entomophthoraceous fungi by serial passage through the living host. Also Krejzova (1977a) stated that the infectivity of *C. coronatus* for three insect hosts increased after similar passage and that the identity of the hosts used for passage did not alter the results. However, it is not certain that the final infectivity exceeded that of the freshly isolated fungus because these authors used only approximate methods to determine infectivity. Sensitive bioassays are needed to detect differences accurately (Wilding, 1976). Only Yendol and Rosario (1972), Wilding (1976) and Papierok and Wilding (1979) have developed adequate techniques but they are laborious.

B. TOXIN PRODUCTION

Little information on *Entomophthora* toxins has appeared since Roberts and Yendol (1971) reviewed the limited work prior to 1969. However Kermarrec and Mauleon (1975) provided further evidence that *C. coronatus* produces a toxin. This killed the ant *Acromyrmex octospinosus* fed on or confined with a culture filtrate of the fungus. Claydon and Grove (1978) obtained 4'-hydroxymethylazoxybenzene-4-carboxylic acid from *E. virulenta* grown *in vitro*. This substance killed *Calliphora vicina* when injected into the haemocoel. It probably also killed flies confined to the surface of cultures of the fungus for 30 min; these died after a few days but were little colonized by the fungus. However, the production of the hydroxy acid has never been demonstrated *in vivo* (Claydon, personal communication). (See also Chapter 23, Section II, A, 10).

C. CULTIVATION AND PRODUCTION

Growth and sporulation *in vitro* are described in Chapter 7, Section V, B. Fungal material for field tests has been produced both *in vitro* and *in vivo*.

Many forest tent caterpillars (*Malacosoma disstria*) were infected by injection with protoplasts produced in cell culture medium (Chapter 7) and distributed in the field in an unsuccessful attempt to control this pest (Tyrrell, 1977). Similarly larvae of *Galleria mellonella* were infected by injection of hyphal bodies of *E. exitialis* isolated from aphids and grown in a liquid medium of unstated composition (Krejzova, 1973a). Conidia developed on these larvae whereas neither conidia nor resting spores had developed *in vitro*. Krejzova suggested that conidia could be produced for aphid control in this way.

However, conidia are usually considered unsuitable for application in insect control because they soon die and cannot be stored (Roberts, 1973). Resting spores are much more suitable and so have attracted most study. Krejzova (1970, 1971c), Egina *et al.* (1972a), Gröner (1975), Latgé (1975a, b) and Soper *et al.* (1975) have developed media suitable for *E. thaxteriana* and *E. virulenta*. Latgé *et al.* (1977) produced 3×10^6 resting spores of *E. virulenta* /ml of an inexpensive medium with commercial grade nutrients in a liquid fermenter.

D. RESISTANCE AND STORAGE

The long term maintenance of *Entomophthora* cultures and the storage of preparations for field use has until recently been difficult. In most collections, strains were maintained by serial subculture. This is unsatisfactory, especially as there is some evidence of infectivity loss after prolonged subculture (Rockwood, 1950; Krejzova, 1971a). Aoki and Tanada (1974) confirmed Prasertphon's (1967) earlier findings that the hyphae of certain species live *in vitro* for several years without oxygen. However the best advance in culture maintenance has been the use of liquid nitrogen. Tyrrell *et al.* (1972) thus preserved protoplasts of *E. egressa*. *In vitro* cultures are so stored at the American Type Culture Collection (Jong, 1978), the Pasteur Institute (G. Remaudière, personal communication) and in my laboratory (Wilding and Best, unpublished observations). Certain fungi have also been preserved cryogenically in host bodies (Remaudière and Michel, 1971).

Some fungi survive in dried host bodies for several months in cool, dry air (Remaudière and Michel, 1971; Kenneth *et al.*, 1972; Wilding, 1973b), but *E. muscae* in *D. coarctata* and *E. vomitoriae* in *P. rudis* died (Wilding, unpublished observations; Newman and Carner, 1975b). Some species may survive in other ways, e.g. some aphids confined to the surface of soil taken from the field in winter became infected with *E. aphidis*. Resting spores of this fungus are almost unknown and its alternative form of survival is unknown (Latteur, 1977).

Conidia are usually regarded as short lived but it has been claimed recently in China (Anon., 1976) that conidia of *E. aphidis* lived for > 100 days. However, in Britain their infectivity for aphids fell after only one week and was lost after 2 weeks on a leaf surface in saturated air at 20°C (Wilding, unpublished observations). Conidia of three species failed to germinate in saturated air after only 9 h at between 50 and 80% r.h. (Yendol, 1968) although, in contrast, conidia of one of these species, *C. coronatus*, took as long as a week to die at 60% r.h. (Kermarrec and Mauleon, 1975), and they remained infective for at least 30 days in saturated air (Stimman, 1968).

The great durability and longevity of resting spores of at least some species has recently been confirmed. Dry resting spores of the *E. thaxteriana* group lived > 1 year (Egina *et al.* 1972b), at least 18 months at − 30°C (Krejzova, 1971d) and 4.5 to 6.5 years at 7 to 15°C (Krejzova, 1973b). However cultures established from resting spores stored for several years were claimed to be slightly less infective for insects than fresh isolates. Resting spores survived boiling for at least 5 min; exposure to 60°C for 4 h; immersion in full strength Ajatin (10% w/v dimethyllaurylbenzylammonium bromide), a disinfectant, for 72 h; 1% solutions of Chlorseptol (another disinfectant, based on chloramin T), HCl, NaOH, and $CuSO_4$; and 0.1% pirimicarb for many hours (Krejzova, 1971d, 1973b, 1977b).

E. SAFETY

The infection of man and other mammals by fungi of the genus *Entomophthora* has never been recorded. In specific safety tests, *E. virulenta* was not toxic when fed to mice (Soper and Bryan, 1974, as *E. nr thaxteriana*; Hartmann and Wasti, 1976).

Two species of *Conidiobolus* infect man and other vertebrates. *C. coronatus* has occasionally been reported as the causative organism of infection of the mouth, nasal mucosae and respiratory tract of mammals, although almost all are isolated cases. Proven infections of man are confined to the tropics and recent records include those of Grateau *et al.* (1974), Onuigbo *et al.* (1975) and Pal *et al.* (1976). It has been isolated from horses as far north as S. Texas (Emmons *et al.*, 1977; Chauhan *et al.*, 1973). *C. coronatus* is a ubiquitous saprophyte in the soil, has occasionally been isolated from arthropods (Samsinakova *et al.* 1974; Remaudière *et al.* 1976b) and can infect many types of insects in experiments (e.g. Getzin and Shanks, 1964). However, despite its ubiquity, it very rarely causes epizootics in field populations of insects, an exception being recorded on the aphid *Lipaphis erysime* (Ramaseshiah, 1967). Mice were not infected with *C. coronatus* by injection, feeding or introducing conidia into wounds (Lowe and Kennel, 1972). However, one of two mice died after inhaling conidia and contained fungal material in its connective tissues, although it could not be determined whether the fungus caused the death. Strains of *C. coronatus* and *C. incongruus* from mammalian hosts, and therefore perhaps more tolerant of mammalian body temperatures than most, were only slightly infective for mice after subcutaneous, intraperitoneal or intratracheal injection (Fromentin, 1976). In such a cosmopolitan species as *C. coronatus* there may be a number of strains with different potentials to attack mammals and insects.

C. incongruus is not known to be entomopathogenic and has been isolated from man only once (Gilbert *et al.*, 1970; King and Jong, 1976). None of the remaining 25 recognized species of *Conidiobolus* is known to attack mammals. *Basidiobolus haptosporus* (= *meristosporus*) has been isolated from man several times (Latgé, 1975c) almost exclusively from the old world tropics. It is not entomopathogenic.

F. COMPATIBILITY WITH PESTICIDES

Fungicides applied to control plant pathogenic fungi will contact *Entomophthora* in the field. In a potato crop, the proportion of *Myzus persicae* infected with *Entomophthora* species was 4-5 times greater in untreated plots than in those treated with mancozeb, captafol or Bordeaux mixture. The aphid population was significantly greatest in the mancozeb and captafol-treated plots (Nanne and Radcliffe, 1971). Similarly, Radcliffe *et al.* (1976) found more spotted alfalfa aphids, *Therioaphis maculata* and *Ac. pisum* and fewer infected with *Entomophthora* in plots treated with mancozeb than in untreated plots. However, the numbers of *Ap. fabae* and their infection with *Entomophthora* on field beans were not affected by weekly applications of mancozeb or captafol (Wilding *et al.,* 1978). Least aphids were infected in plots treated with the insecticide disulfoton (Mellado *et al.,* 1976), possibly because there were fewest aphids in the treated plots, thereby limiting the spread of the fungus, rather than because of a direct effect of the chemical on the fungus. Southall and Sly (1976) obtained slight and rather inconsistent evidence that blight sprays on potatoes favoured aphids, an observation to be interpreted with care because some fungicides directly increase aphid fecundity (Sagenmüller, 1976).

Certain fungicides inhibit *in vitro* growth, sporulation and germination of conidia and resting spores of various *Entomophthora* species (Cadatal and Gabriel, 1970; Soper *et al.,* 1974; Fritz, 1976, 1977; Zimmerman, 1976; Krejzova, 1977b). However, many compounds, antifungal *in vitro,* fail to protect plants from fungal attack in the field (Horsfall, 1956). Similarly, *in vitro* effects of fungicides on *Entomophthora* do not necessarily reflect field effects.

G. CONTROL ATTEMPTS

Entomophthora species have been applied to pests either evenly distributed over the infested crop like insecticides, or as localized inocula dependent on subsequent spread of the fungus.

1. Application as Insecticides

In recent attempts to use *Entomophthora* spp. as insecticides, infective material has been prepared either *in vitro* or *in vivo.* Egina and Tsinovskiy (1972) and Tsinovskiy and Egina (1972) sprayed aphids and spider mites on glasshouse plants with aqueous suspensions of conidia, hyphal bodies and resting spores from *in vitro* cultures of *E. thaxteriana* and *E. sphaerosperma.* More than 95% of the pests died within 24 h, which suggests the involvement of fungal toxins. Similar suspensions of *E. thaxteriana* also killed at least 74% of aphids on apple trees, without affecting the natural predators, Coccinellidae and Chrysopidae (Cinovskis *et al.,* 1974). In China (Anon., 1976), *E. aphidis* was said to have been grown on a cheap bran and water medium from which a suspension of 6×10^6 conidia/ml, sprayed on various aphid species in the field, gave 76–100% mortality. The quickest and best results were claimed when the culture was soaked for 24 h before suspension, perhaps because a toxin diffused into the water. The identity of this fungus needs confirmation since its rate of growth *in vitro* far exceeded that recorded elsewhere for *E. aphidis* and the medium was much simpler than usual (e.g. Gustafsson, 1965).

Field results obtained by distributing fungus from infected insects are contradictory. Kelsey (1965) killed all *Plutella xylostella* (= *maculipennis*) larvae with an aqueous spray of triturated larvae killed by *E. sphaerosperma*. However, Abrahamson and Harper (1973) failed to prevent damage to tree foliage by larvae of *M. disstria* by applying resting spores of *E. megasperma,* collected from dead larvae of the same species the previous season. Furthermore, the fungus was not recovered from larvae in the treated plot.

2. As Localized Inocula.

Attempts to control insects with localized *Entomophthora* inocula have been largely empirical because of ignorance of the factors that affect spread of the fungi. However, such experiments may be a better way of assessing the importance of biotic factors than observing natural epizootics (Tyrrell, 1977).

Remaudière and Michel (1971) tried to control aphids in a peach orchard by placing aphid cadavers in curled leaves, sticking them to leaves and twigs amongst natural aphid colonies and inverting agar cultures of the fungi over aphid colonies. However even though the trees were misted with water, the humidity was too low to allow the fungus to spread: the humidity is probably higher in vegetation near the soil than amongst the branches of trees.

Field introductions by releasing living pests inoculated in the laboratory have usually been successful, but the subsequent spread of the fungus has given inconsistent results. For example, Otvos et al. (1973) released larvae of the eastern hemlock looper, *Lambdina fiscellaria fiscellaria,* injected with protoplasts of *E. egressa.* The disease was transmitted to the natural population but exerted little control. Similarly, Tyrrell (1977) distributed *M. disstria* larvae infected with the same fungus. Even though *E. egressa* is not a natural pathogen of *M. disstria,* it spread but then died out. By contrast, *E. parvispora* controlled large populations of thrips in glasshouses following release of infected thrips (Carl, 1975). Also *E. aphidis* and *E. fresenii* became established in *Ap. fabae* populations after distribution of laboratory inoculated living aphids (Wilding *et al.,* 1978). After two weeks, 47% of adult apterae were infected in treated plots and only 2% in untreated ones. Consequent bean seed yields were twice those of untreated plots, but only half those of insecticide-treated plots.

IV. Discussion and Suggestions for Further Research

Results herein discussed emphasize not only the wide variety of relationships investigated during the past 10 years between *Entomophthora* species and their hosts, but also the range of methods employed. This lack of uniformity hampers generalizations about pest control prospects.

Definitely all *Entomophthora* species require saturated air in which to complete their life cycle. However, more investigations in detail are needed of field conditions required by the fungi to invade their hosts, particularly with localized inocula, to devise the best methods of use. In this context, it is insufficient to observe the behaviour of the fungi in natural conditions in field populations of arthropods — replicated experiments are needed in which host and inoculum density are regulated and the environment manipulated, for example by irrigating and by planting crops at different densities.

Whether broadcast or locally inoculated, it is necessary to find the minimum quantity of fungus required to ensure adequate control. Also, with localized inocula, much more must be known about the pest densities at which the fungi will spread, having consideration of crop damage at those densities.

Experimentation is simpler if abundant fungus is available. Although many resting spores of a few species can be produced, and induced to germinate, some are ineffective in the field. Other fungal forms, such as hyphae in dry host bodies or conidia, should be tried. Accordingly, work on selection of strains suitable for field use, and on their production, formulation and storage, is urgently needed.

Additional needful areas of research include toxic metabolites produced by the fungi and effects of fungicides in the field. To find toxins may not only help to explain the action of some of the fungi *in vivo,* but may lead to the discovery of pesticidal chemicals. Some fungicides clearly affect *Entomophthora* species in the laboratory but existing evidence for their effects in the field is contradictory. More data are needed to substantiate claims that pest numbers are increased by repression of fungal diseases of the pest by chemical applications.

Much is still to be learnt before *Entomophthora* species could be used predictably in pest control. However, their natural regulation of arthropod numbers in certain climatic areas and some experimental evidence encourages the belief that their application could play a useful part in plant protection. It remains to be seen how much this will be limited by climate.

References

Abrahamson, L.P. and Harper, J.D. (1973). Res. Note U.S. Forest Serv. No. SO–157, 1–3.

Anon. (1976). *Acta ent. sin.* **19**, 63–66.

Aoki, J. and Tanada, Y. (1974). *Appl. Ent. Zool.* **9**, 80–86.

Basova, L.P. (1972). Proc. Symp. Path. Microorganisms Pl. Pests, Riga, Latvian S.S.R. 1972, pp. 4–6.

Batko, A. and Kmitowa, K. (1962). *Zesz. probl. Postep. Nauk roln.* **35**, 249–254.

Berisford, Y.C. and Tsao, C.H. (1974). *J. Ga Ent. Soc.* **9**, 104–110.

Brobyn, P.J. and Wilding, N. (1977). *Trans. Br. mycol. Soc.* **69**, 349–366.

Burges, H.D. and Hussey, N.W. (eds) (1971). "Microbial Control of Insects and Mites" Academic Press, London and New York.

Cadatal, T.D. and Gabriel, B.P. (1970). *Philipp. Ent.* **1**, 379–395.

Callaghan, A.A. (1969). *Trans. Br. mycol. Soc.* **53**, 87–97.

Callaghan, A.A. (1974). *Trans. Br. mycol. Soc.* **63**, 13–18.

Carl, K.P. (1975). *Entomophaga* **20**, 381–388.

Carner, G.R. (1976). *J. Invertebr. Path.* **28**, 245–254.

Chauhan, H.V., Sharma, G.L., Kalra, D.S., Malhotra, F.C. and Kapur, M.P. (1973). *Vet. Rec.* **92**, 425–427.

Cinovskis, J., Cudare, Z., Jegina, K., Petrova, V. and Strazdina, A. (1974). *Latv. PSR Zinàt. Akad. Vest.* **7**, 33–36.

Claydon, N. and Grove, J.F. (1978). *J. chem. Soc. Perkin* **1**, 171–173.

Coaker, T.H. and Finch, S. (1971). Rep. natn. Veg. Res. Stn 1970, 23–42.

Dedryver, C. (1978). *Entomophaga* **23**, 137–151.

Dedryver, C. and Robert, Y. (1975). *Annls Phytopath.* **7**, 346.

Delucchi, V.L. (ed.) (1976). *"Myzus persicae* an Aphid of World Importance"*, Int. Biol. Progm. 9, Stud. Biol. Control Part 4. Cambridge Univ. Press.

Egina, K. Ya. and Tsinovskiy, Ya. P. (1972). *In* ["The Pathology of Insects and Mites"] (Ya. P. Tsinovskiy, ed.), pp. 95–110. Zinetne Press, Riga, Latvian S.S.R.

Egina, K. Ya., Tsinovskiy, Ya. P. and Chudare, Z.P. (1972a). *In* ["The Pathology of Insects and Mites"] (Ya. P. Tsinovskiy, ed.), pp. 57–71. Zinetne Press, Riga, Latvian S.S.R.

Egina, K. Ya., Tsinovskiy, Ya. P., Chudare, Z.P., Dobrovolskiy, A.S. and Burgele, D.M. (1972b). Proc. Symp. Path. Microorganisms Pl. Pests, Riga, Latvian S.S.R. 1972, pp. 17–19.

Emmons, C.W., Binford, C.H., Utz, J.P. and Kwon – Chung, K.J. (1977). "Medical Mycology". Lea and Febiger, Philadelphia.

Ferron, P. (1975). Int. Org. Biol. Control, W. Palearctic Reg. Sect., Bull. No. 3, 1–54.

Fritz, R. (1976). *Entomophaga* **21**, 239–249.

Fritz, R. (1977). *Phytiat.-Phytopharm.* **26**, 193–200.

Fromentin, H. (1976). *Bull. Soc. fr. mycol. med.* **5**, 157–160.

Getzin, L.W. and Shanks, C.H. (1964). *J. Invertebr. Path.* **6**, 542–543.

Gilbert, E.F., Khoury, G.H., and Pore, R.S. (1970). *Arch. Path.* **90**, 583–587.

Golberg, A.M. (1970). *Medskaya Parazit.* **39**, 472–478.

Grateau, P., Rigaud, A., Drouhet, E. and Mariat, F. (1974). *Bull. Soc. fr. mycol. med.* **3**, 113–116.

Gröner, A. (1975). *Z. Pflkrankh. PflSchutz* **82**, 22–29.

Gustafsson, M. (1965). *Lantbr Högsk. Annlr.* **31**, 405–457.

Gustafsson, M. (1971). *In* "Microbial Control of Insects and Mites" (H.D. Burges and N.W. Hussey, eds), pp. 375–384. Academic Press, London and New York.

Hard, J.S. (1976). *Can. Ent.* **108**, 485–498.

Harris, C.R. and Svec, H.J. (1966). *J. econ. Ent.* **59**, 569–573.

Hartmann, G.C. and Wasti, S.S. (1974). *Entomophaga* **19**, 353–360.

Hartmann, G.C. and Wasti, S.S. (1976). *Entomophaga* **21**, 377–382.

Horsfall, J.G. (1956). "Principles of Fungicidal Action". Chronica Botanica Company, Waltham, Mass.

Jong, S.C. (1978). *In* "Catalogue of Strains I" (R.L. Gherna, P. Pienta, S.C.

Jong, H. Hsu and P.M. Daggett, eds), 13th edn. Am. Type Cult. Colln, Rockville, Md.

Keller, S. (1976). *Schweiz. landw. Forsch.* **15**, 489–495.

Kelsey, J.M. (1965). *N.Z. Ent.* **36**, 47–49.

Kenneth, R., Wallis, G., Gerson, U. and Plaut, H.N. (1972). *J. Invertebr. Path.* **19**, 366–369.

Kermarrec, A. and Mauleon, H. (1975). *Annls Parasit. hum. comp.* **50**, 351–360.

King, D.S. and Jong, S.C. (1976). *Mycologia* **68**, 181–183.

Klein, M.G. and Coppel, H.C. (1973). *Ann. ent. Soc. Am.* **66**, 1178–1180.

Kramer, J.P. (1971). *J. N.Y. ent. Soc.* **79**, 52–55.

Krejzova, R. (1970). *Ceska Mykol.* **24**, 87–94.

Krejzova, R. (1971a). *Vest. csl. zool. Spol.* **35**, 114–117.

Krejzova, R. (1971b). *Entomophaga* **16**, 221–231.

Krejzova, R. (1971c). *Ceska Mykol.* **25**, 118–124.

Krejzova, R. (1971d). *Ceska Mykol.* **25**, 231–238.

Krejzova, R. (1972). *Vest. csl. zool. Spol.* **36**, 253–255.

Krejzova, R. (1973a). *Vest. csl. zool. Spol.* **37**, 21–22.

Krejzova, R. (1973b). *Ceska Mykol.* **27**, 107–111.

Krejzova, R. (1975). *Vest. csl. zool. Spol.* **39**, 13–22.

Krejzova, R. (1977a). *Vest. csl. zool. Spol.* **41**, 105–113.

Krejzova, R. (1977b). *Anz. Schädlingsk. PflSchutz Umweltschutz* **50**, 83–85.

Krejzova, R. (1978). *Z. angew. Ent.* **85**, 42–52.

Kushida, T., Katagiri, K. and Aoki, J. (1975). *Appl. Ent. Zool.* **10**, 226–230.

Latgé, J.-P. (1975a). *Entomophaga* **20**, 201–207.

Latgé, J.-P. (1975b). *Mycopath. Mycol. appl.* **57**, 53–57.

Latgé, J.-P. (1975c). *Bull. Soc. Mycol. méd.* **4**, 25–28.

Latgé, J.-P., Soper, R.S. and Madore, C.D. (1977). *Biotechnol. Bioengng* **19**, 1269–1284.

Latteur, G. (1977). *C. r. hebd. Séanc. Acad. Sci., Paris* **284**, Sér. D, 2253–2256.

Lowe, R.E. and Kennel, E.W. (1972). *Mosquito News* **32**, 614–620.

MacLeod, D.M., Tyrrell, D., Soper, R.S. and Lyzer, A.J. de (1973). *J. Invertebr. Path.* **22**, 75–79.

Manglitz, G.R. and Hill, R.E. (1964). Res. Bull. Neb. agric. Exp. Stn No. 217, 1–21.

Matanmi, B.A. and Libby, J.B. (1976). *J. Invertebr. Path.* **27**, 279–285.

Mellado, Z., Flores, A.M. and Carillo, L.R. (1976). *Agro Sur* **4**, 111–118.

Missonier, J., Robert, Y. and Thoizon, G. (1970). *Entomophaga* **15**, 169–190.

Nair, K.S.S. and McEwen, F.L. (1973). *J. Invertebr. Path.* **22**, 442–449.

Nanne, H.W. and Radcliffe, E.B. (1971). *J. econ. Ent.* **64**, 1569–1570.

Nemoto, H. and Aoki, J. (1975). *Appl. Ent. Zool.* **10**, 90–95.

Newman, G.G. and Carner, G.R. (1975a). *Env. Ent.* **4**, 615–618.

Newman, G.G. and Carner, G.R. (1975b). *J. Ga Ent. Soc.* **10**, 315–326.

Newman, G.G. and Carner, G.R. (1975c). *J. Invertebr. Path.* **26**, 29–34.

Nolan, R.A., Dunphy, G.B. and MacLeod, D.M. (1976). *Can. J. Bot.* **54**, 1131–1134.

Onuigbo, W.I.B., Gugnani, H.C. and Okafor, B.C. (1975). *J. Lar. Otol.* **89**, 657–662.

Otvos, I.S., MacLeod, D.M. and Tyrrell, D. (1973). *Can. Ent.* **105**, 1435–1441.
Pady, S.M., Kramer, C.L., Long, D.L. and MacBride, T.D. (1971). *Ann. appl. Biol.* **67**, 145–151.
Page, R.M. and Humber, R.A. (1973). *Mycologia* **65**, 335–354.
Pal, A.K., Chopra, S.K., Krishnamurthy, K.V. and Kochar, R.C. (1976). *Indian J. Path. Microbiol.* **19**, 131–134.
Papierok, B. and Wilding, N. (1979). *C. r. hebd. Séanc. Acad. Sci., Paris* Sér. D **288**, 93–95.
Petrova, V.I. and Khrameeva, A.V. (1972). Proc. Symp. Path. Microorganisms Pl. Pests, Riga, Latvian S.S.R. 1972, pp. 27–28.
Pickford, R. and Riegart, P.W. (1964). *Can. Ent.* **96**, 1158–1166.
Prasertphon, S. (1967). *J. Invertebr. Path.* **9**, 140–142.
Rabasse, J.-M. (1974). *Sci. agron. Rennes* 1974, 21–35.
Radcliffe, E.B., Weires, R.W., Stucker, R.E. and Barnes, D.K. (1976). *Environ. Ent.* **5**, 1195–1207.
Ramaseshiah, G. (1967). *J. Invertebr. Path.* **9**, 128–130.
Rautapää, J. (1976). *Annls agric. fenn.* **15**, 272–293.
Remaudière, G. (1971). *Parasitica* **27**, 114–126.
Remaudière, G. and Michel, M.-F. (1971). *Entomophaga* **16**, 75–94.
Remaudière, G., Keller, S., Papierok, B. and Latgé, J.-P. (1976a). *Entomophaga* **21**, 163–177.
Remaudière, G., Latgé, J.-P., Papierok, B. and Coremans Pelseneer, J. (1976b), *C.r. hebd. Séanc. Acad. Sci., Paris* **283**, Sér D, 1065–1068.
Remaudière, G., Papierok, B. and Latgé, J.-P. (1976c). *Méd. Mal. infectieuses* **6**, 418–423.
Robert, Y., Rabasse, J.-M. and Scheltes, P. (1973). *Entomophaga* **18**, 61–75.
Roberts, D.W. (1973). *Ann. N.Y. Acad. Sci.* **217**, 76–84.
Roberts, D.W. and Campbell, A.S. (1977). *Misc. Publs ent. Soc. Am.* **10**, 19–76.
Roberts, D.W. and Yendol, W.G. (1971). *In* "Microbiol Control of Insects and Mites" (H.D. Burges and N.W. Hussey, eds), pp. 125–149. Academic Press, London and New York.
Rockwood, L.P. (1950). *J. econ. Ent.* **43**, 704–707.
Sagenmüller, A. (1976). *Z. angew. Ent.* **82**, 293–300.
Samsinakova, A., Kalalova, S., Daniel, M., Dusbabek, F., Honzakova, E. and Cerny, V. (1974). *Folia Parasit. Praha* **21**, 39–48.
Selhime, A.G. and Muma, M.H. (1966). *Fla Ent.* **49**, 161–168.
Shimazu, M. (1977a). *Appl. Ent. Zool.* **12**, 200–201.
Shimazu, M. (1977b). *Appl. Ent. Zool.* **12**, 260–264.
Soper, R.S. and Bryan, T.A. (1974). *Environ. Ent.* **3**, 346–347.
Soper, R.S., Holbrook, F.R. and Gordon, C.C. (1974). *Env. Ent.* **3**, 560–562.
Soper, R.S., Holbrook, F.R., Majchrowicz, I. and Gordon, C.C. (1975). Life Sci. Agric. Exp. Stn, Univ. Me, Orono. Tech. Bull. No. 76, 1–15.
Southall, D.R. and Sly, J.M.A. (1976). *Pl. Path.* **25**, 89–98.
Stimman, M.W. (1968). *J. econ. Ent.* **61**, 1558–1560.
Strazdinya, A.A. (1972). *In* ["The Pathology of Insects and Mites"] (Ya. P. Tsinovskiy, ed.), pp. 111–121. Zinetne Press, Riga, Latvian S.S.R.

Strong, F.E., Wells, K. and Apple, J.W. (1960). *J. econ. Ent.* **53**, 478–479.
Suter, H. and Keller, S. (1977). *Z. angew. Ent.* **83**, 371–393.
Thoizon, G. (1967). *C.r. hebd. Séanc. Acad. Sci., Paris* **265**, Sér D, 2001–2003.
Thoizon, G. (1970). *Ann. Soc. Ent. Fr. (N.S.)* **6**, 517–562.
Thornthwaite, C.W. (1948). *Geogrl. Rev.* **38**, 55–94.
Tsinovskiy, Ya. P. and Egina, K. Ya. (1972). *In* ["The Pathology of Insects and Mites"] (Ya. P. Tsinovskiy, ed.), pp. 73–94. Zinetne Press, Riga, Latvian S.S.R.
Turian, G. and Wuest, J. (1969). *Mitt. schweiz. ent. Ges.* **42**, 197–201.
Tyrrell, D. (1970). *Bi.-m. Res. Notes* **26**, 12–13.
Tyrrell, D. (1977). *Bi.-m. Res. Notes* **33**, 5.
Tyrrell, D. and MacLeod, D.M. (1972). *J. Invertebr. Path.* **19**, 354–360.
Tyrrell, D. and MacLeod, D.M. (1975). *Can. J. Bot.* **53**, 1188–1191.
Tyrrell, D., Sohi, S. and Welton, M.A. (1972). *Can. J. Microbiol.* **18**, 1967–1968.
Ventskevich, G.Z. (1961). "Agrometeorology" Israel Progm Scient. Translns, Jerusalem.
Voronina, E.G. (1968). *Trudy vses. Inst. Zashch. Rast.* **31**, 394–406 (in Russian).
Voronina, E.G. (1971). *Ent. Rev., Wash.* **50**, 444–453.
Wallace, D.R., MacLeod, D.M., Sullivan, C.R., Tyrrell, D. and DeLyzer, A.J. (1976). *Can. J. Bot.* **54**, 1410–1418.
Weiser, J. and Novak, D. (1964). *Entomophaga Mém. hors. Sér.* No. 2, 149–150.
Wilding, N. (1969). *Trans. Br. mycol. Soc.* **53**, 126–130.
Wilding, N. (1970). Proc. IVth int. Colloq. Insect Path., College Park, Md, 1970, 84–88.
Wilding, N. (1971a). *J. gen. Microbiol.* **69**, 417–422.
Wilding, N. (1971b). *Rep. Rothamsted exp. Stn for 1970,* Part 1, 207.
Wilding, N. (1973a). *Rep. Rothamsted exp. Stn for 1972,* Part 1, 205.
Wilding, N. (1973b). *J. Invertebr. Path.* **21**, 309–311.
Wilding, N. (1975). *Trans. R. ent. Soc. Lond.* **127**, 171–183.
Wilding, N. (1976). Proc. int. Colloq. Invertebr. Path., Kingston, Ont., 1976, 296–300.
Wilding, N. and Lauckner, F.B. (1974). *Ann. appl. Biol.* **76**, 161–170.
Wilding, N., Brobyn, P. and Best, S.K. (1978). *Rep. Rothamsted exp. Stn for 1977,* Part 1, 103.
Wuest, J. and Turian, G. (1971). *C.r. hebd. Séanc. Acad. Sci., Paris* **272**, Sér D, 396–398.
Yendol, W.G. (1968). *J. Invertebr. Path.* **10**, 116–121.
Yendol, W.G. and Rosario, S.B. (1972). *J. econ. Ent.* **65**, 1027–1029.
Zimmerman, G. (1976). *Z. PflKrankh. PflSchutz* **83**, 261–269.

CHAPTER 29

Mosquito Control by the Fungi *Culicinomyces,*
Lagenidium and *Coelomomyces*

B.A. FEDERICI

Division of Biological Control, Department of Entomology,
University of California, Riverside, California, USA

I. Introduction

Mosquitoes are the most important group of insects from the standpoint of human and veterinary medicine. Their ability to vector viral, protozoan, and filarial diseases has made them the target of a variety of biological, chemical and cultural control strategies. Since the advent of DDT in 1939, organic chemical insecticides, particularly chlorinated hydrocarbons, have been the principle means of control. Their extensive use suppressed mosquito populations, thereby reducing the prevalence of mosquito vectored diseases. However, yellow fever, filariasis and especially malaria still prevail over many areas of the world. In tropical Africa alone at least 500 000 children die from malaria every year (Brown, 1973). Furthermore, the development of resistance in many target populations, the high costs of newer insecticides and concern over environmental pollution have aggravated these problems making it imperative to develop alternative methods of mosquito control. During the past decade there has been renewed interest in developing natural enemies for control of mosquitoes (Chapman, 1974). Some of the most significant progress in recent years has come from studies of the fungi, particularly those of the genera *Culicinomyces*, *Lagenidium* and *Coelomomyces*. The purpose of this chapter is to briefly review these advances with emphasis on results relevant to development of these fungi as larvicides.

II. *Culicinomyces clavosporus*

A. GENERAL BIOLOGY AND LIFE CYCLE

Culicinomyces (Class Deuteromycetes) is the most recent addition to fungi with larvicidal potential. The genus consists of two isolates of a facultative parasite of mosquito and related dipteran larvae. The type species, *Cul. clavosporus*, was described from an isolate from a laboratory colony of *Anopheles quadrimaculatus* by Couch *et al.* (1974) in the USA, although an unidentified isolate with similar properties was reported earlier by Sweeney *et al.* (1973) from a colony of *Anopheles amictus hilli* in Australia.

The life cycle is typical of the Deuteromycetes with both isolates reproducing asexually by the formation of conidia. Mosquito infection is initiated by ingestion of conidia and has been studied in *Aedes epactius* (Couch *et al.*, 1974) and *Culex quinquefasciatus* (= *fatigans*) (Sweeney, 1975a). Conidia adhere to the chitinous lining of either the foregut or hindgut. Shortly thereafter they germinate, each forming a germ tube which, over a period of 16–24 h, penetrates through the integument and invades the haemocoel with hyphae and hyphal bodies. Since the oesophagus is the most common site of infection, growth usually starts within the head. Subsequently, hyaline, septate, branched hyphae ramify throughout the haemocoel killing the larva in 2–3 days. Within 48 h of larval death, hyphae pass out through the cuticle, forming branched or unbranched conidiophores which cover the body in 2–6 h and produce conidia.

B. CULTIVATION AND STORAGE

Culicinomyces can be cultured *in vivo* by placing mosquito larvae in water containing dead larvae bearing spores, or *in vitro* on a wide variety of artificial media (Couch *et al.*, 1974). Sweeney (1978a, b) grew the Australian isolate on nutrient agar slopes consisting of 0.3% beef extract, 0.5% peptone and 1.5% agar in 1-litre Roux flasks. Conidia were harvested after 7 days by washing the slopes with sterile distilled water and could be stored for up to a week at 4°C before use. The fungus could also be grown in submerged culture on 0.3% yeast extract and 0.5% peptone in distilled water.

C. EVALUATION OF MICROBIAL CONTROL POTENTIAL

Although studies on the host range must be considered preliminary, both isolates of *Culicinomyces* appear to attack a variety of species. All mosquito species tested to date (Table I) have proven susceptible to infection in all four larval instars, and fungal development in alternative mosquito hosts is similar to that in the original hosts. While bioassay data have not been published for any species, 10^5 conidia/ml killed $> 96\%$ of larvae of instar I of *An. a. hilli* at temperatures varying from 15–27.5°C (Table II). The potential efficacy of *Culicinomyces* is also indicated by Couch et al. (1974) who had extreme difficulty in breeding *An. quadrimaculatus* contaminated by the fungus.

TABLE I

Mosquito species susceptible to isolates of Culicinomyces
(Couch et al., 1974; Sweeney, 1975b, 1978a, b).

Anopheles	Aedes	Culex	
farauti	*epactius*	*erraticus*	*Culiseta*
amictus hilli	*australia*	*quinquefasciatus*	*melanura*
annulipes		(= *fatigans*)	*Psorophora*
punctipennis		*restuans*	*confinnis*
quadrimaculatus		(*pipiens*)	*Uranotaenia*
stephensi		*quinquefasciatus*	*sapphirina*
		territans	

TABLE II

Mortality of Anopheles amictus hilli *larvae exposed to* Culicinomyces *and incubated at different temperatures (Sweeney, 1978a).*

Temperature (°C)	Dosage (conidia ml^{-1})	No. of larvae	Mortality (%)	Corrected mortality (%)
15	10^5	928	97.4	96.8
	0	324	18.2	
20	10^5	944	98.2	98.2
	0	361	1.7	
25	10^5	925	99.4	99.4
	0	279	7.2	
27.5	10^5	972	98.1	98.0
	0	337	4.7	
30	10^5	951	8.4	0.1
	0	314	8.3	

In addition to mosquito larvae, Sweeney (1975b) found larvae of Chironomidae (*Chironomus* sp.) and Ceratopogonidae (*Dasyhelea* sp. and *Bezzia* sp.) susceptible to infection, but not larvae of Psychodidae (*Telmatoscopus albipunctatus*), Trichoptera, Odonata, fresh water shrimp (Atyidae) or fish (*Gambusia* sp.). Couch *et al.* (1974) also found chironomid larvae susceptible to infection. Based on these results, Sweeney (1975b) suggested *Culicinomyces* may infect only species in certain families of the suborder Nematocera.

Sweeney (1978a, b) also studied the effects of temperature and salinity on the infectivity and development of the Australian isolate. The optimum temperature for germination was 27.5°C, although the fungus grew best on nutrient agar at 25°C (Fig. 1). Infectivity was > 95% for larvae of *An. a. hilli* from 15 to 27.5°C, but fell rapidly at 30°C to 0.1% (Table II). In a separate study of infection at 30°C using larvae of *An. a. hilli* and *Cx. fatigans*, dissection of larvae at daily intervals after exposure demonstrated that fungus penetrated the cuticle near the foregut wall, but grew no further. Larvae infected at this temperature were apparently capable of casting off the invading fungus with the larval cuticle at ecdysis. In the studies on salinity, germination decreased directly from 63.3% in freshwater to 2.1% in 200% seawater, with ca. 27% germination at the salinity of seawater. However, larval infection rates, using 10^5 conidia/ml, fell from 100% at 50% seawater to 18.6% at 75% seawater, indicating germination did not result in infection at salinities much above 50% seawater. This was confirmed by dissection of larvae.

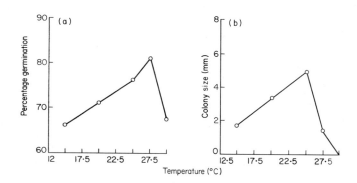

Fig. 1. (a) Percent germination of conidia after 24 h; (b) colony size after 2 weeks incubation on nutrient agar for the Australian isolate of *Culicinomyces*.

In small outdoor trials Sweeney and Panter (1977) introduced their isolate of *Culicinomyces* into rock pools where small numbers of *Ae. rupestris* were breeding. All larvae died within 5 days. Dissections of larval guts indicated death was due to the fungus, so demonstrating that *Culicinomyces* can be effective in natural habitats.

III. *Lagenidium giganteum*

A. GENERAL BIOLOGY AND LIFE CYCLE

Lagenidium giganteum (Class Oomycetes, Order Lagenidiales) is a facultative parasite of mosquito larvae well studied recently because of promising preliminary work which indicates that several isolates have broad mosquito host ranges. Its identification is described in Chapter 8, Section IV, A. Its geographical distribution is wide: USA, UK, Africa, India and Antarctica. Couch (1935) described the type species from a study of two isolates, one from copepods and daphnids in Virginia, and another from mosquito larvae in North Carolina. Much later, using termite wings as bait, Willoughby (1969) reported isolations from mud from England, Uganda, and Antarctica. Glenn and Chapman (1978) documented recently a series of epizootics caused by *L. giganteum* occurring yearly from 1975 to 1978, in larvae of *Cx. territans* breeding in a black gum swamp in Louisiana. In 1976, the mean mortality rate for 16 consecutive weeks was 86%, and in 1975 and 1978 mortalities ranged as high as 100%, although the mean mortality rate for 1975 through 1977 was 61% after initial appearance of the fungus each year. Interestingly, larvae of *Ae. atlanticus, Ae. tormentor, An. crucians, Cx. peccator, Psorophora howardii*, and *Uranotaenia sapphirina* breeding in the same swamp were never found diseased by this strain of *L. giganteum*. Impetus for most of the recent studies, however, arose from a strain from *Culex* sp. in North Carolina (Umphlett and Huang, 1970, 1972) with marked pathogenicity for mosquitoes, particularly culicines.

As a facultative parasite, *L. giganteum* can grow vegetatively either as a parasite of mosquito larvae or as a saprophyte in the aquatic environment, where it apparently prefers a littoral habitat (Willoughby, 1969). The life cycle (Fig. 2) is typical of the genus *Lagenidium*, with both asexual and sexual reproduction (Couch, 1935; Couch and Romney, 1973; Domnas *et al.*, 1974; McCray *et al.*, 1973a, b; Umphlett, 1973; Umphlett and Huang, 1972; Chapter 8 IVA). The parasitic phase is initiated by a laterally biflagellate zoospore, which on contacting a mosquito larva encysts on the cuticle, usually distally on the head capsule, although also on the pharynx, spiracles, anal papillae and posterior abdominal segments. Once encysted, the fungus forms a germ tube which penetrates the cuticle to the haemocoel, apparently aided by a trypsin-like protease (Domnas, personal communication). Later the initial hypha branches into a nonseptate mycelium that soon ramifies the body. After the haemocoel is completely filled with mycelium vegetative growth ceases and reproduction begins. Larval death is most likely the result of starvation (Domnas *et al.* 1974), and occurs at this stage, ca. 72 h after infection. At reproduction, the hyphae form septa, each hyphal segment becoming either an asexual sporangium, an antheridium or oogonium. In asexual reproduction, the most common mode, sporangia become rounded and each forms a discharge tube which passes back out through the cuticle. The protoplasm migrates through the tube into a terminal hyaline vesicle within which it differentiates in 24 h into biflagellate zoospores. The vesicle ruptures and the zoospores seek a mosquito larva or other suitable substrate.

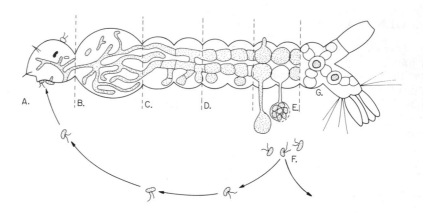

Fig. 2. Major stages of the life cycle and development of *Lagenidium giganteum* growing parasitically in a mosquito larva (schematic). (A) Zoospore encystment, germ tube formation and initial invasion of haemocoel; (B) proliferation of hyphae throughout haemocoel; (C) septa formation; (D) sporangia formation; (E) exit-tube penetration of cuticle, zoospore formation and release; (F) zoospores seek out a mosquito host initiating another parasitic sequence, or encyst and develop saprophytically; (G) oospore formation.

In sexual reproduction, thick-walled oospores (zygotes) form by fusion of antheridia and oogonia from the same or different hyphae. Under favourable conditions, the oospore germinates and grows vegetatively, eventually forming zoospores through asexual reproduction.

The saprophytic cycle is similar to the parasitic cycle except that the zoospore germinates and grows on organic detritus.

B. CULTIVATION AND STORAGE

L. giganteum can be maintained *in vivo* in the laboratory by exposing susceptible larvae to zoospores. The best routine is to place larvae of instars II or III with diseased larval cadavers at maximum zoospore release, which begins ca. 24 h after larval death. Thus McCray *et al.* (1973a) passaged the fungus serially through *Culex* (*pipiens*) *quinquefasciatus* by placing four larvae, dead < 24 h, in 100 ml of tapwater to which 25 3-day old larvae were added 24 h later. Infected cadavers were harvested after 72 h and used for the next passage. This is a good method for providing sporangia, which, after larval death, can either be used within 24 h, or stored prior to zoospore formation. For example, Umphlett and Huang (1972) produced the fungus in larvae of *Cx. restuans* and obtained ca. 16 000 sporangia/larva, yielding ca. 178 640 zoospores (mean, 11/sporangium). McCray *et al.* (1973b) obtained ca. 22 000 sporangia/larva of *Cx. tarsalis*, yielding ca. 240 000 zoospores (mean, 12/sporangium). The best mode of storage was to place cadavers on the day of death in tapwater at 15°C, prior to sporangia formation (McCray *et al.*, 1973a). The fungus retained its ability to form sporangia and produce zoospores for up to 2 weeks at this temperature.

In vitro, L. giganteum grows on a variety of undefined and defined media, either solid or liquid, including 1% beef extract agar (Willoughby, 1969), Difco corn meal agar fortified with 1% dextrose and peptone (Umphlett and Huang, 1972), Cantino peptone–yeast extract–glucose, in agar or broth culture, and Gleason's (1968) defined medium (Domnas *et al.*, 1974). However, while these media provide sufficient nutrients for growth, maximum mycelial size being attained in 5–7 days, they are inadequate for zoospore genesis. This is due to sterol deficiencies and can be overcome by preparing media from oil rich materials, e.g. soy bean or hemp seed extract (Domnas *et al.*, 1974, 1977; Domnas, personal communication). Zoospores were produced by transferring mycelia for 5 days to solid or broth cultures containing a water extract of crushed whole hemp seeds (WHS) or whole soy beans (WSB) at 0.5–1.0 mg protein/ml. Zoospore formation and release were induced by placing the fungus in 0.085 M CaCl$_2$ or 0.0025 M glucose for 24 h. Whereas *L. giganteum* grown on media without exogenous sterol supplements infected few or no larvae, the fungus grown on fortified media killed 95–98% of larvae of *Ae. epactius* and *Cx. quinquefasciatus*. Sistosterol and campesterol were the best individual sterols for enhancement of zoospore production. Ca. 5000 zoospores/ml were produced on the WHS or WSB media and viability decreased to 16% in 16 days storage at 4°C (Domnas, personal communication). Zoospore production could be synchronized by storing mycelia produced on artificial media for a few days at 4°C, and then transferring them directly to water at 28°C. Zoospore release occurred ca. 24 h later. These results are a significant first step to the efficient mass production of zoospores in liquid culture.

C. EVALUATION OF MICROBIAL CONTROL POTENTIAL

1. Laboratory Studies

Although there are conflicting reports about the infectivity of *L. giganteum*

TABLE III

Mosquito species susceptible to isolates of Lagenidium giganteum
*(Couch and Romney, 1973; Glenn and Chapman, 1978;
Umphlett and Huang, 1972; and McCray et al., 1973a)*

Aedes	Culex	Anopheles[a]	
aegypti	quinquefasciatus	punctipennis	Psorophora sp.
(atropalpus) epactius	(= fatigans)	quadrimaculatus	
mediovittatus	nigripalpus	stephensi	Culiseta
nigromaculis	(pipiens)		inornata
polynesiensis	quinquefasciatus		
sollicitans	tarsalis		
taeniorhynchus	territans		
triseriatus			

[a]Susceptible to Couch's isolates (no indication of infection rates given) but completely resistant to Umphlett and Huang's isolate (McCray *et al.*, 1973a).

for anophelines, in general the fungus has a broad host range, particularly against culicines (Table III). Moreover, high mortalities can be obtained against most of these species in the laboratory. When 25 3-day old larvae of *Aedes aegypti, Ae. triseriatus, Ae. mediovittatus, Ae. taeniorhynchus, Ae. sollicitans, Culex quinquefasciatus* (= *fatigans*), *Cx. (p.) quinquefasciatus, Cx. tarsalis,* and *Cx. nigripalpus* were exposed to four diseased cadavers in 100 ml of tapwater for each species, all died (McCray *et al.*, 1973a).

The first three instars of *Cx. restuans* were readily susceptible to infection, but susceptibility decreased with age (Fig. 3). Similar results were obtained with *Ae. aegypti* and *Cx. (p.) quinquefasciatus* (McCray *et al.*, 1973a). Larvae were susceptible to infection during instars I, II, III and the first half of IV, but older larvae, pupae and adults were not. When larvae acquired infection during the first 2 days, they usually died in 40–48 h, while older larvae succumbed in 48–72 h. There was no loss of infectivity when the fungus was cycled through alternative mosquito hosts (McCray *et al.*, 1973a).

Although Couch (1935) described his original isolate as occurring in copepods and daphnids, the more recent isolates have a host range almost exclusively restricted to mosquitoes, the only known exception being the gnat, *Chaoborus astictopus* (Brown and Washino, 1977). McCray *et al.* (1973b) tested Umphlett and Huang's (1970) isolate against *Cyclops* sp., *Daphnia* sp., *Scapholoberis* sp., other unidentified copepods and cladocerans, polycheate worms, dytiscid beetles, and chironomid midges and found none susceptible. Additionally, several vertebrates are insusceptible including fish (*Lebistes* sp. and *Gambusia* sp.), birds (chickens and quail) and rats (Domnas, personal communication).

Effects of pH, salinity and temperature on zoospore production and mosquito infection have received preliminary study by Domnas (personal communication). With *Cx. (p.) quinquefasciatus* as test species, the optimum

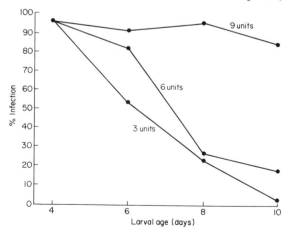

Fig. 3. Effect of age on susceptibility of *Culex restuans* larvae to *Lagenidium giganteum* at three different treatment rates. Each unit equals ca. 178 640 zoospores (Umphlett and Huang, 1972).

pH for the fungus was 6–8 in $0.005\,M$ phosphate, or 5.50–8 with $0.005\,M$ tris maleate as buffer. Little infection was observed at pHs < 4 or > 9: $0.05\,M$ phosphate, borate and acetate repressed zoospore production from pH 3.5–7.0. The fungus tolerated 1.5% NaCl but greater concentrations repressed saprobic growth. Salinities as low as 0.5% almost completely repressed zoospore production.

2. Field Trials

Encouraging results of laboratory studies prompted several preliminary field trials. Umphlett and Huang (1972) introduced 120 diseased cadavers to a 2 × 8-ft pond. After 3 days they collected 30 larvae of *Cx. restuans* of which 13 (43%) were infected. By day 6 no larvae of *Cx. restuans* were present. They also collected 48 anophelines (8% infected) and 4 larvae of *Psorophora* sp., all infected. This study demonstrated that *L. giganteum* could be introduced to a field site and cause considerable deaths of susceptible species.

McCray *et al.* (1973b) conducted trials against *Aedes nigromaculis* and *Cx. tarsalis* in California. The fungus was introduced by spraying (3-gal pressurized spray can, 8002 nozzle, 40 lb pressure) 24 h old triturated diseased cadavers at rates with potential yields of 2.5–3.3 × 10^5 zoospores/ft^2 of water surface. The test sites for *Ae. nigromaculis* were 16-ft^2 sections of an irrigated pasture, while those for *Cx. tarsalis* were 3 × 90 ft seepage ditches adjacent to rice fields. The test against *Ae. nigromaculis* was inconclusive because the larval population was declining rapidly. However, the decline was more rapid in the treated plots and although only three larvae were collected, all were diseased. Interestingly, larvae of *Cx. tarsalis* bred later at this site and sampling suggested that all eventually succumbed to *L. giganteum,* indicating the fungus had cycled at least once at this location. In the trials against *Cx. tarsalis,* three sites were treated. Water conditions at two of the sites were unfavourable for the fungus; at one the pH was 10, while the other had high Cl$^-$ levels. Infection rates at both sites were $< 10\%$. At the third site larvae were eliminated within 5 days of treatment. In the control ditch, larval counts remained similar to pre-treatment levels for up to 17 days post-treatment (Table IV). In follow-up studies, *L. giganteum* was isolated from this site for three consecutive years, 1974–76 (Washino *et al.,* 1975, 1976; Fetter-Lasko and Washino, 1977). While infectivity of the strain isolated in 1974 was low compared to that of the parent strain, passage of the fungus in *Cx. (p.) quinquefasciatus* for several generations restored its activity to original levels. Isolates recovered in 1975 and 1976 were highly infective,

TABLE IV
Mean daily levels of larvae and pupae of Culex tarsalis *before and after treatment on day 0 with* Lagenidium giganteum *(McCray* et al., *1973b)*

Day	−4	−3	−2	−1	0	1	2	3	4	6	17
Control	110	122	102	125	88	72	93	80	123	112	111
Treated	96	96	78	93	89	88	51	36	5	0	0

equivalent to the parent strain. It was concluded that *L. giganteum* was a highly virulent microbial agent for *Cx. tarsalis,* able to survive seasonal droughts and land management practices associated with rice fields.

IV. Coelomomyces

A. GENERAL BIOLOGY AND LIFE CYCLE

The genus *Coelomomyces* (Class Chytridiomycetes, Order Blastocladiales) consists of ca. 40 described species of obligately parasitic, aquatic fungi most often reported from mosquito larvae (see Chapter 8 Section II, A, B for species identification and host list). The genus is widespread and has been found in all continents except Antarctica (Roberts, 1974). Since the 1930s, these fungi have been considered potential biological control agents because of their well documented ability to produce epizootics in larval populations of many mosquito species (Walker, 1938; Couch and Umphlett, 1963; Chapman, 1974; Roberts, 1974). This potential has not been realized or evaluated because, until recently, most attempts to infect larvae in the laboratory and establish *in vivo* cultures for detailed studies failed. The reason for this failure first became apparent with the important discovery by Whisler *et al.* (1974) that *C. psorophorae,* a parasite of *Culiseta inornata,* required a copepod, *Cyclops vernalis,* as an obligate alternate host to complete its life cycle. Five other species of *Coelomomyces* also require an intermediate host, either a copepod or ostracod (Table V). Based on investigations with colonized species of *Coelomomyces,* particularly *C. psorophorae* (Whisler *et al.,* 1975; Whisler, personal communication; Travland,

TABLE V.
Experimentally determined intermediate and definitive hosts for species of
Coelomomyces

Coelomomyces	Intermediate Host	Definitive Host	Reference
C. psorophorae	*Cyclops vernalis*[a]	*Culiseta inornata*	Whisler *et al.,* 1974, 1975
C. punctatus	*Cyclops vernalis*[a]	*Anopheles quadrimaculatus*	Federici, 1975; Federici and Roberts, 1976
C. dodgei	*Cyclops vernalis*[a]	*Anopheles quadrimaculatus*	Federici and Chapman, 1977
C. opifexi	*Trigriopus* sp. near *angulatus*[a]	*Aedes australis*	Pillai *et al.,* 1976
Coelomomyces sp.	*Potomocypris fmaragdina*[b]	*Aedes dorsalis*	Whisler, personal communication; Romney, unpublished observations
C. chironomi	*Heterocypris incongruens*[b]	*Chironomus plumosus*	Weiser, 1976

[a] Crustacea; Copepoda. [b] Crustacea; Ostracoda.

1979), *C. punctatus* (Couch, personal communication; Federici, 1975; Federici and Roberts, 1976) and *C. dodgei* (Federici, 1977a and unpublished observations; Federici and Chapman, 1977), the following tentative life cycle for these species has emerged (Fig. 4).

Mosquito infection is initiated by a posteriorly biflagellate zygote which enters a larva by penetrating through the cuticle. The thallus that develops there is diploid and is known as a sporophyte. Once within the larva the fungus forms an extensive mycelium characterized by coenocytic, dichotomously branching, wall-less hyphae. These hyphae ramify the haemocoel, forming thousands of oval sporangia at their tips as they grow, eventually causing larval death. As the larva putrefies, the sporangia are liberated into the water.

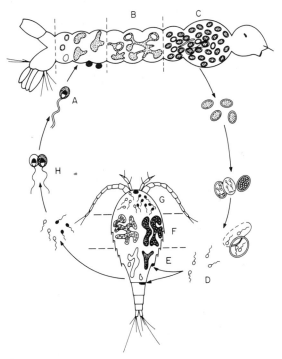

Fig. 4. Generalized life cycle for species of *Coelomomyces*. (A) Biflagellate zygote infects haemocoel of mosquito larva producing hyphagens which later form hyphae; (B, and C) hyphae ramify haemocoel, forming thick-walled resistant sporangia. After liberation from larva, sporangia release meiospores (D) which infect a copepod and produce a gametophyte (E); each gametophyte forms a gametangium (F) that releases gametes of a single mating type (G); opposite mating types may be different colours, as illustrated, or unpigmented. A copepod may contain either, or both types of gametophytes depending on the nature of the infecting meiospore(s). Gametes of opposite mating type fuse inside or outside (H) copepod forming a biflagellate zygote that completes the cycle by infecting another larva (Federici, 1975, 1977a; Whisler *et al.*, 1975).

Meiosis occurs within the sporangia and they subsequently release hundreds of spores, conveniently called meiospores. Each meiospore infects a copepod, selectively encysting and penetrating the cuticle at intersegmental membranes, and produces a haploid gametophyte, a coenocytic wall-less thallus which ramifies the copepod haemocoel much like the sporophyte in the larva except that no sporangia are formed. Instead, the entire gametophyte becomes a gametangium which at maturity cleaves into thousands of uniflagellate gametes. Each gametangium forms gametes of a single mating type. An infected copepod, therefore, may yield gametes of one or both types depending on the number and type of meiospores by which it was infected. After gametogenesis is complete, the fungus kills the copepod and the gametes escape from the carcass. Gametes of opposite mating type fuse, either inside or outside the copepod depending on where they associate, forming a biflagellate zygote which completes the cycle by infecting another mosquito larva.

In *C. dodgei* and *C. punctatus*, gametophytes and gametangia are either orange or amber in colour and produce gametes similar in colour. Experimental studies have demonstrated that these different colours correspond to opposite mating types, and, based on precedents established with related fungi, the orange forms have been designated male and the amber female (Federici, 1977a).

This complex life cycle, with alternation of sporophyte and gametophyte generations, is known as an Euallomyces type of life cycle (Emerson, 1941; Sparrow, 1960) and is remarkably similar to that of other free-living species of fungi of the Order Blastocladiales, particularly species of the genera *Allomyces* and *Blastocladiella*, indicating that the genus *Coelomomyces* bears a close evolutionary relationship to these genera.

B. CULTIVATION AND STORAGE

Most obligate parasites are difficult to establish as *in vitro* cultures, and species of *Coelomomyces* are no exception. Neither the sporophyte nor gametophyte of any species has been established on artificial or defined media. However, Shapiro and Roberts (1976) reported growth of the sporophyte of *C. psorophorae* from *Ae. sollicitans* in a fortified mycoplasma medium, and Castillo and Roberts (1978 personal communication) obtained limited growth and differentiation of both the gametophyte and sporophyte of *C. punctatus* in the same medium.

Lack of methods for *in vitro* cultivation necessitates maintaining these fungi *in vivo*. While this is cumbersome, requiring breeding of mosquitoes and copepods, methods for three species, *C. dodgei*, *C. punctatus*, and *C. psorophorae*, have been devised. Methods have changed considerably since publication (Whisler *et al.*, 1975; Federici and Roberts, 1976; Federici and Chapman, 1977) and those briefly described below are based on current procedures In these systems infection of either copepods or mosquitoes is most readily accomplished by exposing hosts to the fungus during their initial stages of development, when they are most susceptible to infection. Also, once infected, hosts must be given optimal diets and rearing conditions for normal fungal development (Federici, 1980; Whisler, personal communication).

With *C. psorophorae,* after larval death, sporangia from 10 larvae are washed in distilled water, collected and stored on Millipore filters at 5°C until use. For infection a filter is placed in 100 ml of 0.25 strength Whisler's (1965) salts solution along with ca. 200 copepodids of *Cyclops vernalis.* This mixture is held in a petri dish for 7 days at 20°C in diffuse light. Meiospore release and copepod infection occur within the first 2 days, with development and maturation of the gametophytes occurring on days 3 to 7. Copepods are fed nematodes. Ca. 50% of the copepods develop infections. As they begin releasing gametes and zygotes they are combined with 200 instar II larvae of *Culiseta inornata* in 100 ml of half strength salt solution. Normally, all larvae develop patent infections by instar IV.

With *C. dodgei,* a heavily infected instar IV is dissected in 4 ml of distilled water in a 35 × 10-mm petri dish to liberate sporangia. After 48 h, while meiospores are being released, the contents of the dish are combined with 6000–8000 two-day-old copepod nauplii in 1 litre of tapwater to 0.5 litre of culture medium. This mixture is held at 25°C for 2 days to allow optimum infection of the copepods and then transferred to 12 litres of distilled water. Copepods are fed hard-boiled egg yolk and dried brewer's yeast. These cultures produce a total of ca. 1500 infected copepods from day 7–15. Infected copepods, easily differentiated from the others because of the amber or orange colour imparted by the gametophytes, are picked from the surface when needed for mosquito infection. Six of each colour type are placed with 100 instar II larvae of *An. quadrimaculatus* in 50 ml of culture medium. Within 48 h the fungus kills the copepods, forms zygotes, and infects larvae. Ca. 8 days later 25–50% of the larvae develop patent infections during instar IV.

For *C. punctatus,* the *C. dodgei* technique can be used except that × 10 more meiospores must be added to the nauplii to obtain the same rate of copepod infection.

C. EVALUATION OF MICROBIAL CONTROL POTENTIAL

By necessity, most of the recent studies on *Coelomomyces* have investigated life cycles and methods for *in vivo* cultivation, with little emphasis on evaluation of the microbial control potential of these fungi. Still, several aspects of the above studies, some more recent studies and older literature interpreted in the light of the copepods' role in mosquito infection, give relevant information.

1. Laboratory Studies

As a general rule, species of *Coelomomyces* infecting mosquitoes were thought to have a relatively narrow mosquito host range, many being considered species-specific (Couch and Umphlett, 1963). However, recent laboratory studies with all the above colonized species have revealed broader host ranges. For example, *C. dodgei,* originally described as a parasite of *An. crucians,* infects *An. quadrimaculatus* and *An. freeborni; C. punctatus,* originally described from *An.*

quadrimaculatus, infects *An. freeborni* and *An. stephensi* (Federici *et al.,* 1975; Federici, 1977b). Infection of *An. stephensi* is interesting because this species is native to India, indicating these fungi can infect species of mosquitoes from outside their known distribution. Yet, their mosquito host ranges may not be broad, as *C. punctatus* failed to infect *An. albimanus,* as well as *Aedes aegypti, Ae. albopictus, Ae. taeniorhynchus, Culex pipiens* and *Culiseta inornata.* In studies on the host range of *C. psorophorae,* Whisler (personal communication) found his isolate could infect *Ae. aegypti, Ae. epactius, Ae. sierrensis, Ae. triseriatus, Cx. pipiens, Cx. (p.) quinquefasciatus,* and the southern strain of *Culiseta inornata,* but not *Ae. dorsalis, Ae. flavescens, Ae. vexans,* nor any anophelines. These results indicate most species of *Coelomomyces* are probably not host specific, but have relatively restricted host ranges. Possible exceptions may be *C. indicus,* reported occurring naturally in 18 species from three genera (Couch and Umphlett, 1963), and a species of *Coelomomyces* isolated by Romney *et al.* (1971), which has also been observed in the field in larvae of three genera.

Susceptibility of different host life stages to *Coelomomyces* is just beginning to receive study. Although not quantitative, attempts to improve infection rates have shown susceptibility decreases with host development, except that hosts are usually more susceptible immediately after ecdysis rather than a few hours before or after. This may indicate the fungus can penetrate the integument most easily before the cuticle is sclerotized. The most susceptible stages to meiospores and zygotes, respectively, are neonate nauplii and larvae.

In a preliminary semi-quantitative study on infection with *C. dodgei* in larval *An. quadrimaculatus,* Federici and Chapman (1977) found the mean infection rate (= mortality) fell from 37.4% against instar I to 2.4% against instar IV on exposing groups of 100 larvae to 12 infected copepods. Mosquito species may also differ in degree of susceptibility. In comparative studies, *An. freeborni* was almost twice as susceptible to *C. dodgei* (86% vs 42.3%) and *C. punctatus* (90% vs 63%) as was *An. quadrimaculatus* when instar I larvae were exposed as described above.

Little is known about the effects of species of *Coelomomyces* on non-target organisms. However, in the context of mosquito control, the intermediate copepod host most certainly must be considered a non-target organism. Because of the importance of many crustacea in aquatic food chains, the effect of *Coelomomyces* on those organisms must be examined carefully. Interestingly, Nolan *et al.* (1973) found a larva of *Toxorhynchites rutilus septentrionalis,* a predaceous mosquito larva which feeds primarily on other mosquito larvae, infected with *C. macleayae.* Previously, this had been reported only from species of *Aedes.* The predator mosquito must also be considered a non-target organism.

2. Field Trials

No field trials have been undertaken since the discovery of the copepods' role in mosquito infection. Before this, however, Couch (1972) introduced sporangia of *C. punctatus* into ditches in North Carolina along with eggs of *An. quadrimaculatus.* Infection rates in larvae collected 10–15 days later varied from

0–100% (mean 60%). Additionally, Dubitskii (1978) obtained similar results with *C. iliensis* against *Cx. modestus* in the Soviet Union.

V. Assessment of Potential and Future Research

Recent advances in our knowledge of the basic biology of *Cul. clavosporus, L. giganteum* and several species of *Coelomomyces* have been significant and exciting, yet inadequate for critical assessment of the control potential of these fungi. Certainly, none of these fungi has been studied enough or developed to a point where it can be considered for use in mosquito control programs. Yet, in light of emphasis on development of integrated pest management programs that use limited amounts of selective insecticides at critical periods, promising results obtained with these fungi indicate they merit continued and even expanded research and development. While possibly they may suppress mosquito populations over the long term upon introduction to areas where they are absent, it is more likely they will find widespread use only if developed as microbial insecticides, to be applied at will much as *Bacillus thuringiensis* or the registered baculoviruses are today. To attain this goal, efficient techniques must be developed to mass produce, store and distribute infective stages of the best strains of these fungi.

With *Cul. clavosporus,* conidia can be produced on artificial media already. Available isolates of this fungus are effective only at low salinities and moderate temperatures, thereby excluding their use in the tropics or in brackish waters. Nevertheless, their ability to invade and kill a wide range of mosquito species make them attractive candidates for use in temperate zones. *L. giganteum* also has a broad host range and, importantly, grows and produces zoospores on artificial media. The results obtained so far with *Cul. clavosporus* and *L. giganteum* justify further laboratory studies on their target and non-target host ranges, development of methods and media, whether *in vivo* or *in vitro,* for the economic production, formulation, and use of infective stages. The discrepancies reported in pathogenicity and mosquito host range for these fungi, particularly *L. giganteum,* indicate isolates differ significantly from one another. In the future it will be important to accurately identify and characterize known and new isolates, with further emphasis on those with the greatest mosquito control potential. Additionally, there must be carefully designed and executed field trials.

The continued development of species of *Coelomomyces* will most likely proceed slowly until methods are developed for their culture *in vitro*. While this is a major obstacle at present, it must be kept in mind that the characteristics these fungi possess as obligate parasites may make them considerably more useful once developed. Several different species of *Coelomomyces* periodically cause natural epizootics with larval mortalities > 90% (Chapman, 1974). Few other fungi, or parasites of mosquitoes in general, have been reported causing such natural epizootics.

Research on *Coelomomyces* should emphasize colonization of species with broad mosquito host ranges, e.g. *C. indicus*, and the *in vitro* cultivation of gametophytes, the stages which result in production of mosquito-infective zygotes. Furthermore, because several species have a life cycle of the Euallomyces type it is possible the sporophyte in some produces sporangia which yield asexual zoospores that infect mosquitoes directly. These sporangia most likely would be thin-walled. Although thin-walled sporangia have been described from *Coelomomyces* there is no evidence that they produce asexual zoospores. The discovery of such a stage and determination of requirements for *in vitro* culture would be an extremely important advance and so should be pursued diligently. Concurrently, essential studies, e.g. bioassays against mosquito larvae and copepods, factors which influence meiospore infection of copepods, zygote formation, mosquito infection and effects on other non-target organisms should be undertaken. Important information on the effect of *Coelomomyces* on intermediate host populations should also be obtained from studies of natural epizootics.

The extensive and often isolated areas in which mosquitoes breed, as well as the wide range of niches they inhabit, make them particularly difficult to control. Throughout evolution a variety of organisms has evolved which effectively parasitize and kill larvae. Man is now learning to manipulate these pathogens to his advantage. Considering the limited manpower and finance allocated to this work, recent progress has been considerable. The future value of much of the work reported here is difficult to predict in terms of mosquito control. Still, given the present rapid rate of technological development in the biological sciences, it is probable that several of the organisms discussed here, and others, will play a significant role in future integrated mosquito control programs.

References

Brown, A.W.A. (1973). *Bull. ent. Soc. Am.* **19**, 193–196.
Brown, J.K. and Washino, R.K. (1977). *Proc. Pap. Calif. mosq. Control Ass.* **45**, 106.
Chapman, H.C. (1974). *Ann. Rev. Ent.* **19**, 33–59.
Couch, J.N. (1935). *Mycologia* **27**, 376–387.
Couch, J.N. (1972). *Proc. Nat. Acad. Sci. U.S.A.* **69**, 2043–2047.
Couch, J.N. and Romney, S.V. (1973). *Mycologia* **65**, 250–252.
Couch, J.N. and Umphlett, C.J. (1963). *In* "Insect Pathology: An Advanced Treatise" (E.A. Steinhaus, ed). Vol 2, pp. 149–188. Academic Press, New York and London.
Couch, J.N., Romney, S.V. and Rao, B. (1974). *Mycologia* **66**, 374–379.
Domnas, A.J., Giebel, P.E. and McInnis, Jr., T.M. (1974). *J. Invertebr. Path.* **24**, 293–304.
Domnas, A.J., Srebro, J.P. and Hicks, B.F. (1977). *Mycologia* **69**, 875–886.
Dubitskii, A.M. (1978). Biological Control of Bloodsucking Flies in the USSR 267pp. "Nauka" of the Kazakh, SSR.

Emerson, R. (1941). *Lloydia* **4**, 77–144.

Federici, B.A. (1975). *Proc. Pap. Calif. mosq. Control Ass.* **43**, 172–174.

Federici, B.A. (1977a) *Nature Lond.* **267**, 514–515.

Federici, B.A. (1977b). *Proc. Pap. Calif. mosq. Control Ass.* **45**, 107–108.

Federici, B.A. (1980). *Entomophaga* **25**, 209–216.

Federici, B.A. and Chapman, H.C. (1977). *J. Invertebr. Path.* **30**, 288–297.

Federici, B.A. and Roberts, D.W. (1976). *J. Invertebr. Path.* **27**, 333–341.

Federici, B.A., Smedley, G. and Van Leuken, W. (1975). *Annals ent. Soc. Am.* **68**, 669–670.

Fetter-Lasko, J.L. and Washino, R.K. (1977). *Proc. Pap. Calif. mosq. Control Ass.* **45**, 106.

Gleason, F.H. (1968). *Am. J. Bot.* **55**, 1003–1010.

Glenn Jr, F.E. and Chapman, H.C. (1978). *Mosq. News* **38**, 522–524.

McCray, E.M., Umphlett, C.J. and Fay, R.W. (1973a). *Mosq. News* **33**, 54–60.

McCray, E.M., Womeldorf, D.J., Husbands, R.C. and Eliason, D.A. (1973b). *Proc. Pap. Calif. mosq. Control Ass.* **41**, 123–128.

Nolan, R.A., Laird, M., Chapman, H.C. and Glenn, Jr F.E. (1973). *J. Invertebr. Path.* **21**, 172–175.

Pillai, J.S., Wong, T.L. and Dodgshun, T.J. (1976). *J. med. Ent.* **13**, 49–50.

Roberts, D.W. (1970). *Ent. Soc. Am. Misc. Publ.* **7**, 140–154.

Roberts, D.W. (1974). *In* "Le Controle des Moustiques/Mosquito Control" (A. Aubin, A. Belloncik, J.P. Bourassa, E. Lacoursieve, and M. Pellissier, eds), pp. 143–193. Univ. Quebec Press, Montreal.

Romney, S.V., Boreham, M.M. and Nielsen, L.T. (1971). *Utah Mosq. Abatement Ass. Proc.* **24**, 18–19.

Shapiro, M. and Roberts, D.W. (1976). *J. Invertebr. Path.* **27**, 399–402.

Sparrow, F.D. (1960). "Aquatic Phycomycetes" 2nd ed. Univ. Michigan Press, Ann Arbor.

Sweeney, A.W. (1975a). *Aust. J. Zool.* **23**, 49–57.

Sweeney, A.W. (1975b). *Aust. J. Zool.* **23**, 59–64.

Sweeney, A.W. (1978a). *Aust. J. Zool.* **26**, 47–53.

Sweeney, A.W. (1978b). *Aust. J. Zool.* **26**, 55–59.

Sweeney, A.W., Lee, D.J., Panter, C. and Burgess, L.W. (1973). *Search (Sydney)* **4**, 344–345.

Sweeney, A.W. and Panter, C. (1977). *J. med. Ent.* **14**, 495–496.

Travland, L.B. (1979). *J. Invertebr. Path.* **33**, 95–105.

Umphlett, C.J. (1973). *Mycologia* **65**, 970–972.

Umphlett, C.J. and Huang, C.S. (1970). *Bull. Ass. SEast. Biol.* **17**, 68.

Umphlett, C.J. and Huang, C.S. (1972). *J. Invertebr. Path.* **20**, 326–331.

Walker, A.J. (1938). *Ann. trop. Med. Parasitol.* **32**, 231–244.

Washino, R.D., McCray, E.M., Umphlett, C.J. and Lasko, J.F. (1975). *Proc. New Jersey mosq. Control Ass.* **62**, 124.

Washino, R.D., Fetter, J.L., Fukushima, C.K. and Gonot, K. (1976). *Proc. Pap. Calif. mosq. Control Ass.* **44**, 52.

Weiser, J. (1976). *J. Invertebr. Path.* **28**, 273–274.

Whisler, H.C. (1965). *J. Protozool.* **13**, 183–188.

Whisler, H.C. (1979). *In* "Insect-fungus Symbiosis" (L.R. Batra, ed.), pp. 1–32. Allanheld, Osmun, Inc. Montclair, New Jersey.

Whisler, H.C., Zebold, S.L. and Shemanchuk, J.A. (1974). *Nature Lond.* **251**, 715–716.

Whisler, H.C., Zebold, S.L. and Shemanchuk, J.A. (1975). *Proc. Nat. Acad. Sci. U.S.A.* **72**, 693–696.

Willoughby, L.G. (1969). *Trans. Br. mycol. Soc.* **52**, 393–410.

CHAPTER 30

Pest Control by *Nosema locustae*, a Pathogen of Grasshoppers and Crickets

J.E. HENRY and E.A. OMA

Rangeland Insect Laboratory, USDA, Bozeman, Montana, USA

I. Introduction

Reviews by McLaughlin (1971) and Tanada (1976) point out a growing interest in using protozoans, particularly microsporidians, for microbial control of noxious insects. Although some microsporidians are highly pathogenic and could effect short-term control, most species lack such virulence. However many of the less virulent species may suffice when long-term control is acceptable. This depends on whether low pest densities can be tolerated and whether the pathogens can keep the pest at or below the damage threshold.

Grasshoppers are the most serious invertebrate pests on rangelands of western USA (Hewitt, 1977). Forage losses in individual states have been millions of dollars annually, although accurate assessments are impossible because the amount and importance of damage varies with grasshopper species, stage of development, kinds of forage plants, grazing practices, etc. In surveys, grasshopper densities of about 10 or more/m² have been used as the economic thresholds on rangelands (Anonymous, 1969). However, control measures rarely are initiated until the damage is serious, e.g. when densities average 15 or more/m^2. Thus low densities on rangelands are tolerated, so grasshoppers are excellent subjects for long-term microbial control.

For this purpose, the microsporidian *Nosema locustae* has been much studied. Canning (1962b) concluded that it would be ineffective against migratory locusts in Africa and Asia because it cannot be grown *in vitro;* biological races differ in pathogenicity and it grows so slowly that infected locusts continue to cause damage. Although this assessment concerned locusts, which differ economically and behaviourally from grasshoppers, these and other questions also are important with grasshoppers. Extensive research has now indicated that *N. locustae* is potentially useful for long-term control of grasshoppers.

II. Characteristics of Infections

A. HOST AND GEOGRAPHICAL RANGES

Nosema locustae was described by Canning (1953, 1962a) from the African migratory locust, *Locusta migratoria migratorioides,* in cultures in England. It was similar or identical to earlier unnamed organisms in locusts (Goodwin, 1949, 1950, 1952; Goodwin and Srisukh, 1950) and in the grasshoppers *Melanoplus bivittatus, Melanoplus dawsoni,* and *Melanoplus sanguinipes* (= *M. mexicanus*) from Montana (Steinhaus, 1951). Other hosts are *Schistocerca gregaria* and *Dissosteira carolina* (Canning, 1962a) and *Melanoplus brunneri, Aulocara elliotti,* and *Cordillacris occipitalis* from Montana, Wyoming and Canada (Henry, 1969a). A black field cricket (*Gryllus* sp.), a species of pygmy locust (Tetrigidae) and 58 species of grasshoppers were susceptible to infections by *N. locustae* (Henry, 1969a), as was the Mormon cricket, *Anabrus simplex,* but not the eastern house cricket, *Acheta domesticus* (Henry, 1975). Infected grasshoppers were collected in Montana, North Dakota, Minnesota, Oregon, Wyoming, Colorado, Arizona and Idaho (Henry, 1969b). The known geographical range then is North America and England, but it is surely much more extensive, as doubtlessly will be confirmed when other species of grasshoppers from elsewhere are examined. Indeed most species of grasshoppers (Acrididae) and many crickets (Gryllidae) probably are susceptible to infection.

B. DIAGNOSIS OF INFECTIONS

Infections by *N. locustae,* as with most microsporidians, usually are diagnosed by the presence of spores. These are 3.5 to 5.5 μ long by 1.5 to 3.5 μ in diameter (Canning, 1953) with mean measurements of 5.2 μ by 2.8 μ (Henry, 1969b). Triangulate and elongate megaspores, up to 8 μ long, are common. Spores generally are ellipsoidal, occasionally slightly bent or kidney-shaped and refractive to light. Mean lengths of polar filaments extruded by mechanical pressure were 86 μ (max. 145 μ). The spore consists of a chitinous membrane surrounding the sporoplasm within which the polar filament is coiled (Dissanaike and Canning, 1957). The membrane has two separate layers and the polaroplast is about $1.5\,\mu$ large at the anterior end of the sporoplasm (Huger, 1960). The sporoplasm occupies most of the spore's cavity and contains two nuclei. The polar filament is coiled within the outer sporoplasm and attaches to the external membrane anteriorly near the polaroplast.

Canning (1953, 1962a) described uni-, bi-, and tetranucleate schizonts, 2.5 to $6.6\,\mu$ in diameter, and sporonts as uninucleate products of schizogony, each of which subsequently forms a spore, i.e. monosporous development. However we observed binucleate sporonts that undergo karyokinesis to form two daughter binucleate sporoblasts, i.e. disporous, like many species of *Nosema*.

Infections of *N. locustae* are most prominent in the fat body of grasshoppers, (Canning, 1953, 1962b; Henry, 1969b), but they also occur intracellularly in pericardial and neural tissues (Henry, 1969b). Spores have been seen in hind gut and faeces, apparently due to passive movement during normal excretion, and also in gonads, again probably due to passive movement from fat body during histogenesis. In Mormon crickets, infections are restricted to the epithelia of gastric caecae and midgut. Changes in the gross appearance of fat body of grasshoppers after infection by *N. locustae* are diagnostic. It becomes greatly hypertrophied and changes from a translucent bright yellow to an opaque, creamy colour (Canning, 1953, 1962b). In heavy infections fat tissue is almost totally displaced by spores and occasionally there is a pink deposit, termed insectorubin by Goodwin and Srisukh (1950).

C. GROSS PATHOLOGY OF INFECTIONS

After *L. migratoria migratorioides* were inoculated in instar III, $>92\%$ died as nymphs, particularly at moulting (Canning, 1962b). *M. bivittatus*, inoculated in instar III with 5.5×10^3, 5.5×10^4, or 5.5×10^5 spores each became infected and contained 3×10^6 to 5×10^7 spores/mg. However, death rates did not differ significantly from those among uninfected grasshoppers during 25 days of individual rearing (Henry and Oma, 1974b). In contrast, among grasshoppers treated comparably, but reared together for 17 days, infection significantly increased the death rate. Thus infection has a greater effect when the host is simultaneously exposed to other stresses.

Cannibalism, a major cause of death among group-reared grasshoppers, was increased by infection (Henry, 1969b). Also cannibalism occurs most commonly when the victim is moulting, a process lengthened by a weakening due to infection of the fat body, which also causes the deformities – such as twisted wings and legs, and the foregut protruding through the cervical membrane – that are common among infected grasshoppers.

Infection slows development (Henry, 1969b). Instar III *M. bivittatus* inoculated with 10^5 to 10^6 spores generally persisted for prolonged periods, moribund, in the penultimate stage and rarely became adult. Development of *M. sanguinipes* inoculated in instar III was greatly delayed (Fig. 1). Heavily infected grasshoppers fed less than healthy insects, were lethargic and were

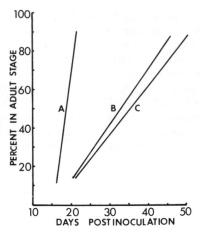

Fig. 1. Development of *Melanoplus sanguinipes* after *per os* inoculation with spores of *Nosema locustae*. A. Untreated ($Y = -211 + 2.87X; r = 0.977$). B. Inoculated with 10^4 spores each ($Y = -45.1 + 2.97X; r = 0.072$). C. Inoculated with 10^6 spores each ($Y = -40.0 + 2.55X; r = 0.987$). 100 grasshoppers/treatment in each of two replicates.

pale in colour. Heavily infected instar V nymphs were robust with extended abdomens due to pressure of the hypertrophied fat body full of spores, which occasionally caused bulging of membranes, particularly the cervical.

D. EFFECTS OF INFECTION ON FECUNDITY

Henry (1969a, 1971) examined 50 000 grasshoppers from both natural enzootic areas and areas where *N. locustae* was applied. Homogenates of even lightly infected females (ca. 10^5 spores each) rarely contained detectable ovarial or egg debris. Thus infections reduced grasshopper fecundity in the field.

In the laboratory, egg production of *Melanoplus differentialis* decreased with increasing percentage infection and spore concentrations in grasshoppers, a result clearly shown despite a certain marring of the test by contaminant infections in grasshoppers of one of the check groups (Table I). *M. sanguinipes* were inoculated at different development stages and paired in small oviposition cages on becoming adult. Few pairs of adults developed from grasshoppers inoculated in instar III and no eggs were laid (Table II). More pairs were derived from grasshoppers inoculated in instar IV and these produced eight egg masses abnormally without pods. Grasshoppers inoculated as instar V nymphs laid a reduced number of eggs normally in pods, in similar numbers irrespective of whether only one sex or both were inoculated. Thus infection in males also depresses fecundity possibly because these males failed to inseminate the uninfected females or because their sperm failed and the females resorbed the eggs. Inoculation of adults also curbed egg production. Thus in both grasshopper species, infections very effectively reduced fecundity. We might also

TABLE I

Effect of infections by Nosema locustae *on fecundity of* Melanoplus differentialis *(Henry, 1969b).*

Inoculation		Infection			Egg pods
Stage of development	No. of spores	No. examined	% infected	Mean no. spores/mg grasshopper	No./♀
Nymph IV	2×10^5	27	96.2	9.3×10^6	0
Nymph IV	2×10^3	27	96.2	4.8×10^6	0.40
Nymph IV	—	30	80.8	7.2×10^6	0.73
Adult	2×10^5	28	51.8	9.7×10^5	2.80
Adult	2×10^3	27	62.9	1.1×10^6	2.73
Adult	—	27	0	—	3.87

TABLE II.

Effect of Nosema locustae *infections on fecundity of* Melanoplus sanguinipes. *Each inoculated grasshopper received 10^5 spores.*

Inoculation			Fecundity			
Stage	Sexes inoculated	No. insects in group	No. of pairs	Mean no. pods/ pair	Mean no. eggs/ pod	Eggs/ ♀
Nymph III	♂, ♀	104	3	0	—	—
Nymph IV	♂, ♀	120	14	0.6	—	—
Nymph V (a)	♂, ♀	40	17	1.5	16.6	20.3
Nymph V (b)	♂ only	40	21	1.4	17.2	23.8
Nymph V (c)	♀ only	40	21	1.2	17.1	20.3
Adult	♂, ♀	25	9	1.9	18.9	35.7
Untreated	none	152	26	3.2	16.5	53.3

expect that infection of the females would reduce the viability of eggs and hatchlings, but this remains to be tested.

III. Natural Occurrence in Grasshoppers and Crickets

Although *N. locustae* has extensive host and geographical ranges, in nature it is generally uncommon (Henry, 1969b). Examinations of thousands of grasshoppers from throughout western USA demonstrated that natural infections among adults total considerably less than 1%. However, enzootic areas were found where *N. locustae* persisted at relatively high levels for periods of several years. In one such area, Henry (1972) found an overall mean of 5.05% infection in 31 570 grasshoppers over five consecutive seasons at 52 permanent sampling

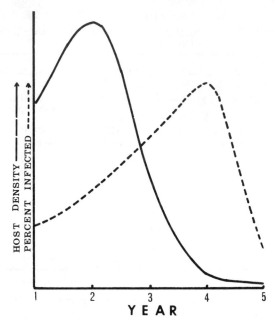

Fig. 2. Scheme of relationship between grasshopper densities and incidence of *Nosema locustae* infections over 5 years. Densities ranged from 34.6/40 sweeps in year 2 to 0.1/40 sweeps in year 5. Infections ranged from 8.1% in year 4 to 1.6% in year 5.

sites over 518 km². Grasshopper density reached a peak then declined, while infection peaked two seasons later, following the typical lag response pattern (Fig. 2) of host—parasite interactions. Although this strongly suggested that *N. locustae* was at least partially responsible for host decline, the actual cause and effect could not be defined because the study was observational not experimental.

An important finding of the study was that infections were commonest in areas with diverse vegetation and grasshopper communities. Usually in these areas, *Oedaleonotes enigma* and *M. bivittatus* were present – both are early summer species and are highly susceptible to *N. locustae* – while *M. sanguinipes* predominated over the whole study area. Usually infections developed first in the early species during early July and most later grasshoppers (up to 97%), mainly *M. sanguinipes*, were infected during late August and early September. Clearly the early infections provided inoculum for the later ones. The initial source of infections was not found, although about 5% of the offspring of moderately infected *M. bivittatus* (ca. 10^7 to 10^9 spores each) also became infected (Henry, unpublished observations). Thus *N. locustae* is transmitted either on or in the eggs, which might account for some infections in early species. Also spores have been seen in grasshopper faeces, another probable source. Finally, Henry (1972) found spores in partly cannibalized bodies. Such horizontal transmission undoubtedly is the principal natural mode of spread of disease during the same season, and such spores could provide inoculum for young grasshoppers the next season.

During 1974 we found spores in Mormon crickets from a cultivated pasture in N.W. Montana. When collected, many were dead or moribund and most at least partially cannibalized: 70% contained spores. These spores were fed to grasshoppers which then became infected. Since we were unable to rear Mormon crickets in the laboratory, spores from these grasshoppers were not fed back to crickets. However, we have since confirmed in a field test that they are susceptible to infection by *N. locustae*.

IV. Results from Applied Studies

A. EXPERIMENTAL FIELD TRIALS

The above information, particularly that infections must be initiated early in the season, was reaffirmed by applied studies. *Melanoplus gladstoni* was curbed much more than two closely related species, *M. sanguinipes* and *Melanoplus infantilis*. This was mainly because it, being a late summer species, was mostly in the instar II stage at treatment, whereas the earlier species were mostly instar IV (Henry, 1971).

Later tests (Henry *et al.*, 1973; Henry and Oma, 1974a) demonstrated that mortalities and infections summed together were maximal when spores were applied while certain main target species, e.g. *M. sanguinipes*, were mostly at instar III. With earlier treatments mortalities were higher and infection lower: *vice versa* with later treatments. This established that ca. 1.6×10^9 to 2.3×10^9 spores on 1.12 to 1.68 kg wheat bran per ha, applied at the proper time, would reduce some prominant species, e.g. *M. sanguinipes* by 50–60% within 4 to 6 weeks, followed by 35 to 40% infection among the survivors. With proper timing this reduction occurred before peak oviposition. Control continued through the ovipositional period by more deaths as well as reduced fecundity.

B. FORMULATION AND APPLICATION OF SPORES

In most of the above studies spores were sprayed onto wheat bran which was then dispersed in test plots by ground equipment. Henry (1971) applied five spore concentrations ranging from 7.75×10^8 to 1.24×10^{10}/ha on 2.24 kg wheat bran/ha. Hydroxymethylcellulose (0.2% w/v of spore suspension) was added as a sticker. There were no differences in mortality or infections between treatments, perhaps because the data on densities were highly variable, possibly as a result of using sweep nets for density sampling.

Later (Henry *et al.*, 1973) spore treatments of 1.5×10^8, 4.65×10^8 and 1.39×10^{10} spores on 1.12 or 4.49 kg/ha were compared on the basis of density reductions and infections after 4 and 6 weeks. Densities were assessed by a pointer technique (Onsager, 1977) in which grasshoppers were counted in

$0.093 \, m^2$ (ft^2) quadrants. Significant differences in densities and infections were due to spore dosages ($P \leqslant 0.01$) and levels of wheat bran ($P \leqslant 0.10$). Thus, wheat bran level was less critical than spore dosage. A standard formulation of 2.47×10^9 spores on 1.68 kg bran/ha was adopted for later studies.

Possibly formulation is the most important factor in the use of *N. locustae* against grasshoppers and crickets due to the relatively low financial return from rangeland, which will not support costly control procedures. During the early part of this century grasshoppers and Mormon crickets were controlled by poisoned baits, usually on wheat bran (Ford and Larrimer, 1921; Whitehead *et al.*, 1937). Later, ultra low volume (ULV) sprays made a significant improvement (Sayer, 1959; Skoog *et al.*, 1965). The area treated per plane load was increased and the costs reduced. Since *N. locustae* must be eaten by the target insect, wheat bran has the advantage of concentrating spores on food particles, but it is bulky and costly to apply by air. However, at least 10 times more spores would be required by ULV to achieve comparable results (Henry *et al.*, 1978). The cost of producing the extra spores for ULV would exceed its cost saving by aerial application, so application of *N. locustae* is cheaper and more efficient on bran.

TABLE III.

Infection among grasshoppers of Melanoplus *spp.* (> 65% *of population) after application of* 2.47×10^9 *spores of* Nosema locustae *on 1.68 kg carrier/ha*

	After 4 weeks[a]		After 6 weeks[a]	
Carrier	No. examined	% infected	No. examined	% infected
Wheat bran	364	12.9 a	213	38.0 a
Rolled barley	332	6.6 ab	173	28.9 ab
Rolled oats	377	4.2 b	210	14.8 bc
Alfalfa meal	283	1.1 b	186	2.2 c
Check	220	2.3 b	210	1.4 c

[a] Analysis of variance conducted as randomized block design with data for 4 and 6 weeks (4 replications each) analysed separately ($P = 0.05$). Mean infections followed by the same letter are not significantly different.

Better and less bulky carriers than wheat bran have been sought. In one test we applied *N. locustae* by ground equipment on three other materials (Table III), readily available at comparable costs and palatable to some grasshoppers (unpublished reports from this laboratory, 1948–1956). Wheat bran was best (Table III). Although rolled barley was not significantly inferior to wheat bran, other factors preclude its use, particularly the cost of aerial application due to narrow swath widths. Wheat bran flakes tend to float and disperse laterally three times as widely as the heavier, rapidly falling, flakes of rolled barley. Applications by ground equipment would not preclude either material since swaths for both would be ca. 9 m wide and the added bulk of the wheat bran would not be so great a problem as by air.

To evaluate the importance of coverage, *N. locustae* was applied by ground equipment in swaths ca. 9 m wide to obtain various percentages of coverage (Henry, unpublished observations). The study area was a pasture of ca. 200 ha, with native mixed grass prairie vegetation interseeded with crested wheat grass, *Agropyron cristatum,* surrounded by cultivated wheat lands. No differences were found between treatments, all of which caused more infection than in untreated plots in *Melanoplus* spp., which comprised ca. 70% of the population (Fig. 3). Therefore, full coverage was not essential with a carrier such as wheat bran. This would be more important with heavier carriers applied aerially.

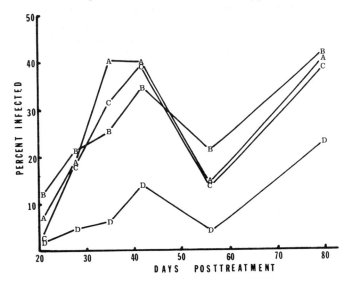

Fig. 3. Infection of *Melanoplus* spp. after field application of 2.47×10^9 spores of *Nosema locustae* on 1.68 kg of wheat bran/ha. Spore coverages were full (A), 50% (B), 25% (C), and check (D), each replicated 3 times.

The infection rate among grasshoppers in untreated plots (Fig. 3) undoubtedly resulted from immigration of grasshoppers from treated plots. This was evident from the lack of infections in grasshoppers about 1.8 km from the study area and also from other studies with small plots (Henry and Oma, 1974a; Henry *et al.,* 1973, 1978). Thus, reliable data can be derived from small plots (4 ha) for a maximum of ca. 42 days (Fig. 3). The sharp drop in percent infection in all treatments, including the untreated check, between days 42 and 56 was caused by immigration of healthy grasshoppers from adjacent croplands at harvest. During the next season infections were common, up to 39.3%, among *Melanoplus* spp. in the study area and the disease had spread over the entire area, so it was impossible to discriminate between treated and untreated plots. Since *N. locustae* is suitable only for the long-term management of grasshoppers and crickets, short-term control must rely on other methods, primarily chemical insecticides. However, the chemicals registered for this in the USA are nonpersistent, so they cannot provide long-term control. Many pathogens are compatible with chemicals (Chapter 38) and *N. locustae* is compatible with low concentrations of malathion, which acted independently with additive effects (Mussgnug and Henry, 1979). The chemical killed within 24 h whereas *N. locustae* took much longer, so the two can be combined to rapidly reduce pest outbreaks to near or below the damage threshold and to provide long-term control below that threshold for several seasons.

C. MASS PRODUCTION AND STORAGE OF SPORES

Spores must be mass produced and stockpiled because outbreaks of grass-hoppers are extensive and the entire infested area must be treated. For efficient use of resources the production system should operate during the entire year, which necessitates prolonged storage of spores without serious loss in viability.

Although obligate pathogens, some microsporidians have been grown in cell cultures (Ishihara, 1969; Sohi and Wilson, 1976; Kurtti and Brooks, 1977). This was considered a major breakthrough in production by Tanada (1976). It may be possible with *N. locustae,* particularly in grasshopper cell lines. However, major developments in cell cultures would be needed to make the method economic (Chapter 16).

Mass production in grasshoppers is relatively easy (Henry 1978). Successive improvements show steady increasing efficiency in the number of spores obtained from each grasshopper from $> 10^9$ spores/*M. bivittatus* to 3×10^9 (Henry, 1971, 1975; Henry *et al.,* 1978). Hatchlings of *M. bivittatus* were reared, 200 to 300 per tube, through the first moult (Cowan, 1966). They were then transferred to large screened cages (ca. $0.24 \, \text{m}^3$, ca. 2000/cage). Instar V nymphs were fed lettuce sprayed with spores (ca. 10^6 spores/cage) for 2 consecutive days and again on the 4th day. This caused $> 99\%$ infection. The food was lettuce, seedlings of balbo rye, and wheat bran. If needed to prevent infections by the amoeba *Malameba locustae,* agar-base diet was given daily with antibiotics, which do not interfere with production of *N. locustae* (Henry and Oma, 1975). At 14 to 20 days postinoculation the grasshoppers were transferred to small rearing vials (ca. $88 \, \text{cm}^3$), 2 adults/vial, to reduce cannibalism. These were fed the agar-base diet in an attached smaller vial along with either wheat bran or crushed dog food (milk bone). Cadavers were stored at $- 10°C$. For processing they were thawed, crushed in a wheat mill, suspended in distilled water, agitated to release spores and then passed three or four times through cheese cloth and cloth screens to remove large tissue fragments. The spores were concentrated and cleaned by differential centrifugation and stock-piled in distilled water at $- 10°C$.

Each cage yielded about 300 to 500 infected adults. Most mortality occurred soon after instar II nymphs were placed in the cages and after inoculation. Spores built up to good numbers in adults, more in females than males (Table IV: in this cage of 405 adults the mean was 3.90×10^9 spores/adult). Estimated costs are ca. U.S. \$0.21/ha, which would be economic against rangeland grass-hoppers. However, with improvements in rearing the spore reproduction per

TABLE IV

Increase of spores in Melanoplus bivittatus *during mass production of* Nosema locustae.

Days after feeding first spores	Mean spores/grasshopper $(\times 10^8)$ in samples of 10	
	Males	Females
20	3.03	5.63
24	18.84	25.31
28	20.98	58.93
32	49.07	111.80

adult could undoubtedly be doubled because some very heavily infected females contained ca. 2×10^{10} spores.

Spores stored at $-10°C$ for 3 years in distilled water were only 10%, and spores stored dry in cadavers for 1 year were only 1%, as effective in field tests as fresh spores (Henry and Oma, 1974a). These studies demonstrate the need to develop effective storage procedures (see also Chapter 35, Section IV, A). For microsporidians that can withstand freezing, which includes most species from terrestrial hosts, Vavra and Maddox (1976) recommended storage in liquid nitrogen since they stored three such species for > 7 years with only 2–3% loss in viability. This procedure is currently being tested with *N. locustae.*

D. RESPONSES OF GRASSHOPPERS TO SPORE APPLICATIONS

Grasshopper control programs aim to simultaneously suppress several to many species. Of 618 species of grasshoppers (Acrididae) in the USA, 26 are economically important and a few of these usually predominate in outbreaks (Hewitt, 1977). The rest are not important, primarily because they do not increase to high densities, and some might be considered beneficial because they feed on undesirable plants. It is impossible in practice to attack only the important species because 20 to 50 species might occur in the outbreak area. In such situations the two to five predominant species usually also are economically important and belong to two or more subfamilies, so as would be expected they differ behaviourally and in their responses to *N. locustae.*

Most often, certain groups, e.g. the Melanoplinae, are more affected by applications of spore–wheat bran formulations than are other groups, such as the Acridinae (Henry 1971; Henry *et al.* 1973; Henry and Oma 1974a). Although Canning (1962b) suggested that such variation might be caused by strain variation of the microorganism. The differences observed in field studies have resulted from variation among species in acceptance of wheat bran. Some omniverous grasshoppers eat grasses, forbs, sedges, organic debris, etc. and some virtually monophagus grasshoppers are restricted to one or two closely related plants (Mulkern *et al.,* 1969). Most Melanoplinae are polyphagous and readily accept wheat bran, whereas the Acridinae are mostly graminivorous and do not. This was particularly evident in the comparison of wheat bran and ULV spray formulations (Henry *et al.*, 1978). The predominant poly-phagous species, mainly *Melanoplus dawsoni* and *M. infantilis*, were curbed only in wheat bran plots, whereas relatively more Acridinae than Melanoplinae were infected in ULV plots. However, Acridinae also are curbed by spore—wheat-bran treatments to some extent (Henry *et al.*, 1973). Possibly reduction of bran-feeding species by *N. locustae* made more Acridinae fall victim to other natural enemies, most of which exhibit wide host ranges. Also Acridinae may eat wheat bran and become infected but, because grassfeeders contain much less fat tissue than polyphagous species and this tissue is destroyed rapidly by *N. locustae*, leading to early deaths, diagnosis of infections in Acridinae is difficult. Anyway, it is important and significant that application of *N. locustae* in wheat bran curbs directly, as well as indirectly, all economically important species of grasshoppers.

E. SAFETY TO NON-TARGET ANIMALS

A number of safety tests with *N. locustae* have been completed against vertebrates, including acute inhalation in rats, acute dermal in guinea pigs, primary skin irritation in rabbits, long-term feeding with rats, and acute LC_{50} treatments of rainbow trout and blue gill sunfish (Henry, 1978). It did not reproduce or accumulate in the tissues of these animals, thus there appears to be little prospect that it would infect warm blooded animals. Honey bees also are immune to *N. locustae* (Menapace *et al.*, 1978), but there is no information about insect parasites or predators of grasshoppers.

V. Assessment of Continued Development

N. locustae curbs grasshoppers, can be produced and applied effectively and efficiently, and can be regarded as safe to use. Thus the major questions involved in its development as a microbial control agent have been answered. However other very important factors, e.g. persistence after application, need assessment. Although *N. locustae* persists at least into a second season more data are needed beyond this. That *N. locustae* remained relatively common in natural enzootics (Henry, 1972) suggests that it will persist until host densities are reduced to levels that will no longer support the parasite. Other factors that need improvement are storage techniques, methods of mass producing great quantities of spores for large-scale treatments, and application techniques, as in other types of bait or carrier formulations. Although most of these problems require only time and research commitment, much of the information could be acquired while using *N. locustae*. Possibly the best approach to its further development would be to initiate one or several large pest management programs in which it would be the principal control procedure. This would have the advantage of acquiring the information under actual use conditions.

References

Anonymous (1969). "Grasshopper Survey. Species Field Guide" Plant Pest Control Div., U.S. Dep. Agric., Agric. Res. Serv. 23 pp.

Canning, E.U. (1953). *Parasitology* **43**, 287–290.
Canning, E.U. (1962a). *J. Invertebr. Path.* **4**, 234–247.
Canning, E.U. (1962b). *J. Invertebr. Path.* **4**, 248–256.
Cowan, F.T. (1966). *In* "Insect Colonization and Mass Production" (C.N. Smith, ed.), pp. 311–321. Academic Press, New York and London.
Dissanaike, A.S. and Canning, E.U. (1957). *Parasitology* **47**, 92–99.
Ford, A.L. and Larrimer, W.H. (1921). *J. econ. Ent.* **14**, 292–299.
Goodwin, T.W. (1949). *Biochem. J.* **45**, 472–479.
Goodwin, T.W. (1950). *Biochem. J.* **47**, 554–562.
Goodwin, T.W. (1952). *Biol. Rev.* **27**, 439–460.
Goodwin, T.W. and Srisukh, S. (1950). *Biochem. J.* **47**, 549–554.
Henry, J.E. (1969a). *Ann. ent. Soc. Am.* **62**, 452–453.
Henry, J.E. (1969b). Ph. D. Thesis, Montana State Univ. 153 pp.
Henry, J.E. (1971). *J. Invertebr. Path.* **18**, 389–394.
Henry, J.E. (1972). *Acrida* **1**, 111–120.
Henry, J.E. (1975). *In* "Proceedings of Advancement in Pesticides". State Dept. Hlth. Environ. Sci., Environ. Sci. Div., Helena, Montana, pp. 16–28.
Henry, J.E. (1978). *In* "Microbial Control of Insect Pests. Future Strategies in Pest Management Systems". (G.E. Allen, C.M. Ignoffo and R.P. Jaques, eds), pp. 195–206. NSF-USDA-Univ. Florida, Gainesville.
Henry, J.E. and Oma, E.A. (1974a). *J. Invertebr. Path.* **23**, 371–377.
Henry, J.E. and Oma, E.A. (1974b). *Acrida* **3**, 223–231.
Henry, J.E. and Oma, E.A. (1975). *Acrida* **4**, 217–226.
Henry, J.E., Tiahrt, K. and Oma, E.A. (1973). *J. Invertebr. Path.* **21**, 263–272.
Henry, J.E., Oma, E.A. and Onsager, J.A. (1978). *J. econ. Ent.* **71**, 629–632.
Hewitt, G.B. (1977). U.S. Dep. Agric., Agric. Res. Ser. Misc. Publ. 1348. 22 pp.
Huger, A. (1960). *J. Insect Path.* **2**, 84–105.
Ishihara, R. (1969). *J. Invertebr. Path.* **14**, 316–320.
Kurtti, T.J. and Brooks, M.A. (1977). *J. Invertebr. Path.* **29**, 126–132.
McLaughlin, R.E. (1971). *In* "Microbial Control of Insects and Mites" (H.D. Burges and N.W. Hussey, eds), pp. 151–172. Academic Press, London and New York.
Menapace, D.M., Sackett, R.R. and Wilson, W.T. (1978). *J. econ. Ent.* **71**, 304–306.
Mulkern, G.B., Pruess, K.P., Hagen, A.F., Campbell, J.B., Knutson, H. and Lambley, J.E. (1969). North Dakota Agric. Exp. Stn Bull. 481, 32 pp.
Mussgnug, G.M. and Henry, J.E. (1979). *Acrida* **8**, 77–81.
Onsager, J.A. (1977). *J. econ. Ent.* **70**, 187–190.
Sayer, H.J. (1959). *Bull. ent. Res.* **50**, 371–386.
Skoog, F.E., Cowan, F.T. and Messenger, K. (1965). *J. econ. Ent.* **58**, 559–565.
Sohi, S.S. and Wilson, G.G. (1976). *Canad. J. Zool.* **54**, 336–342.
Steinhaus, E.A. (1951). *Hilgardia* **20**, 629–678.
Tanada, Y. (1976.) *In* "Comparative Pathobiology, Vol. I, Biology of the Microsporidia" (L.A. Bulla, Jr. and T.C. Cheng, eds.), pp. 247–279. Plenum Press, New York.

Vavra, J. and Maddox, J.V. (1976). *In* "Comparative Pathobiology. Vol.
    Biology of the Microsporidia" (L.A. Bulla, Jr. and T.C. Cheng, eds.
    pp. 281—319. Plenum Press, New York.
Whitehead, F.E., Walton, R.R. and Fenton, F.A. (1937). *J. econ. Ent.* !
    764—768.

CHAPTER 31

Vairimorpha necatrix a Pathogen of Agricultural Pests: Potential for Pest Control

J.V. MADDOX

Section of Economic Entomology, Illinois Natural History Survey,
Urbana, Illinois, USA

W.M. BROOKS

Department of Entomology, North Carolina State University,
Raleigh, North Carolina, USA

and

J.R. FUXA

Department of Entomology, Louisiana State University,
Baton Rouge, Louisiana, USA

I. Introduction

Vairimorpha necatrix, a microsporidium with two distinctly different spore forms, was first reported from *Mythimna* (*Pseudaletia*) *unipuncta* larvae from Hawaii (Tanada and Chang, 1962). Kramer (1965) described it as two separate species, *Nosema necatrix* and *Thelohania diazoma,* in the same host from Illinois. Maddox (1966), Tanabe (1971), Fowler and Reeves (1974), Pilley (1976b), and Maddox and Sprenkel (1978) have demonstrated that *V. necatrix* is a single dimorphic species rather than two separate species, hence its transfer to the genus *Vairimorpha* (Pilley, 1976b). Most workers familiar with this microsporidium have been impressed with its high degree of virulence in a wide range of lepidopterous hosts and hold it in high regard among the entomophilic microsporidia as a candidate microbial control agent.

II. Host—Pathogen Relationships

Of 36 susceptible species of Lepidoptera, 20 are noctuids (Maddox, 1966; Tanabe, 1971; Pilley, 1976a; Kaya, 1977; Maddox and Sprenkel, 1978; Brooks and Fuxa, unpublished observations). Many are major pests, e.g. *M. unipuncta, Heliothis zea, Heliothis virescens, Autographa californica, Trichoplusia ni, Spodoptera frugiperda, Spodoptera ornithogalli, Plathypena scabra, Spodoptera exigua, Pseudoplusia includens, Hyphantria cunea,* and *Ostrinia nubilalis.* Only four lepidopterous species were refractive to infection (Maddox, 1966; Jacques, 1977; Maddox and Sprenkel, 1978), as were all insects tested from several other orders (Maddox and Sprenkel, 1978) including the hymenopterous parasites, *Campoletis sonorensis* and *Cardiochiles nigriceps* (Fuxa and Brooks, 1979b). Thus, *V. necatrix* is primarily, if not exclusively, a pathogen of phytophagous Lepidoptera.

The fat body is the primary infection site. In advanced stages of the disease, its cells enlarge many-fold giving the fat tissue a characteristic lobated appearance (Fig. 1a) in contrast to the normal ribbon-like sheaths of fat tissue in a healthy larva. In this lobated state, it is little more than sacs filled with microsporidian spores (Figs 1b and c). The fat body is the only tissue invaded in some species but in others there are limited additional invasions of the midgut epithelium, Malpighian tubules, labial glands, muscles, epidermis, and haemocytes (Maddox, 1966; Tanabe, 1971; Pilley, 1976b; Brooks, unpublished observations).

Few early external signs of infection are noticeable. However, the abnormally large and white fat body is usually obvious through the integument later, and a dorsal swelling may appear on the last two or three abdominal segments (Pilley, 1976b; Brooks, unpublished observations).

The following features for *M. unipuncta* are probably typical of most noctuid hosts (Maddox, 1966). The lethal period depends on spore dosage, temperature and larval instar when spores are eaten (Table I). Death occurring within 6 days of spore ingestion is usually due to bacterial septicemia rather than microsporidiosis. Bacteria from the midgut enter the haemocoel via polar filament punctures from the extruding spores, and larvae become sluggish about 24 h later, feeding very little before death. Larvae dying from microsporidiosis become sluggish in the late stages of infection and do not respond readily to stimuli. Feeding may be nearly normal at first, decreasing to very little for the last few days before death. Tanabe (1971) infected adults by feeding spores suspended in honey solution and obtained evidence of transovum as well as transovarian transmission. However, larvae surviving infection develop into healthy adults. Thus, *V. necatrix* is not normally transmitted transovarially and

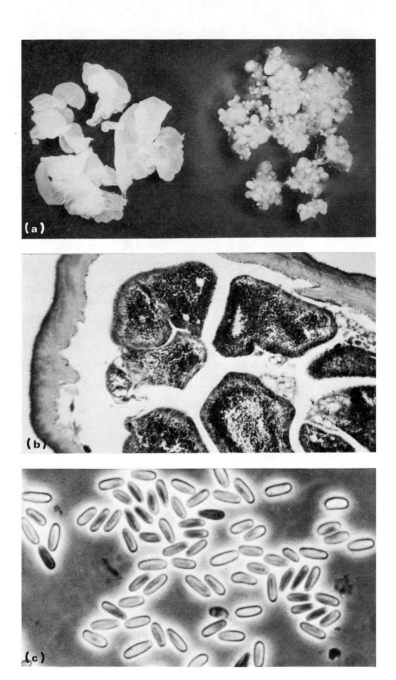

Fig. 1. *Vairimorpha necatrix* in *Heliothis zea*. (a): Gross appearance of fat tissue from healthy (left) and infected (right) larvae. Note cloudy background of the infected tissue due to release of spores from ruptured cells. (b): Section of larval fat body almost completely replaced by spores. Hematoxylin and eosin. (c): Spores from larval fat body, wet-mount, phase contrast microscopy.

TABLE I

Effect of dosage of Vairimorpha necatrix *on mortality and* LT_{50} *for* Mythymna (Pseudaletia) unipuncta *larvae of two ages at 26.7°C*

Instar I			Instar VI		
Spores/ larva	% dead	LT_{50} (days)	Spores/ larva	% dead	LT_{50} (days)
1.1×10^6	100	2.6	4.5×10^6	100	3.7
1.3×10^5	100	3.5	2×10^6	100	5.7
8×10^3	100	4.0	5×10^5	100	13.4
1×10^3	89	10.1	1×10^5	100	13.9
1×10^2	100	9.2	1×10^4	100	11.2
Control 0	10	20.5	Control 0	0	–

the number of spores necessary to infect one-half the population (ID_{50}) equals that necessary to kill one-half the population (LD_{50}) (Maddox, 1966). The ID_{50}s in several different hosts are shown in Table II.

TABLE II

Approximate ID_{50} *(spores/larva) of* Vairimorpha necatrix *for several lepidopteran species*

Host species and stage	ID_{50}
Spilosoma (Diacrisia) virginica – instar I	10
Hyphantria cunea – instar III	54
Mythimna (Pseudaletia) unipuncta – instar I	17
Mythimna unipuncta – instar VI	65
Spodoptera frugiperda – instar I	4
Spodoptera ornithogalli – instar I	5

III. Effect of Environment on Spore Survival

Of all the environmental factors encountered by microsporidian spores in the field, sunlight or its ultraviolet (UV) component is by far the most destructive. Dry spores on various substrates survived only for a few hours when exposed to direct sunlight (Maddox 1977). The half-life of spores exposed to a UV light source was only 2.1 h (Ignoffo *et al.*, 1977). Spores sprayed on bean leaves in the field lost all infectivity between 24 and 78 h (Kaya, 1977). The half-life of unprotected spores sprayed on tobacco, cotton, and soybeans ranged from 0.6 to 2.2 days (Fuxa and Brooks, 1978), although on soybeans some spores remained infective for 7 to 10 days (Gardner *et al.*, 1977; Fuxa and Brooks, 1978). A UV protectant added to spore suspensions increased the half-life to 1.1 days on cotton and 3.7 days on soybeans (Fuxa and Brooks, 1978). Field persistence of protected spores varied from 6 days on beans (Kaya, 1977) to over 14 days on tobacco and soybean (Fuxa and Brooks, 1978). Thus, a UV protectant can prolong spore survival sufficiently to warrant use of *V. necatrix* as a microbial control agent.

Temperature probably plays a minor role in inactivating spores under field conditions, but it has an important effect on stored spores. Dry naked spores of *V. necatrix* were noninfectious after exposure to 40°C for 28 days, 50°C for 10 h, 55°C for 100 min, and 60°C for 40 min (Maddox, 1977). Dry spores formulated in corn meal were highly infective to *H. zea* larvae after 9 months storage at −15°C or 6°C. Spore survival was also high for 18 months at −15°C in host tissues or in a water–glycerol suspension as well as at 6°C in water with antibiotics (Fuxa and Brooks, 1979a). Maddox (1973) found that some spores survived storage in water suspension at 5°C for 2 years, but, compared to fresh spores, infectivity dropped by a factor of 100 after 6 months, 1000 after 1 year and 10 000 after 2 years. Various lyophilization techniques can preserve infectivity of spores for at least 2 years, and spores in a water suspension can be stored in liquid nitrogen for at least 8 years with no loss of viability (Maddox, 1977).

Although little is known about the toxic effect of chemicals on microsporidian spores, various concentrations of such germicides as formaldehyde, sodium hypochlorite, Zephrine chloride, Lysol, and Hyamine 10−X inactivate spores of *V. necatrix,* but not ethyl alcohol at 25% or below, or 0.5% $HgCl_3$ for 5 min (Maddox, 1977; Fuxa and Brooks, 1979a). Some combinations of protozoan spores with insecticides have been evaluated; *V. necatrix* spores survived exposure to most of the insecticides tested (Chapter 38, Section II, D).

IV. Natural Occurrence in Insect Populations

Since *V. necatrix* has a wide host range, it is not surprising that it has been isolated from 14 species of field-collected lepidopterous larvae (Maddox and Sprenkel, 1978; Brooks, unpublished observations). Infection in these natural populations is usually sparse in field crop pests, but high infection rates have been found in some forest insects (Maddox, 1966; Nordin *et al.*, 1972) and in *M. unipuncta* from Hawaiian grasslands (Tanada, 1962, 1964). The reason for this difference is unknown. *V. necatrix* is probably not an important natural regulating factor of field crop pests.

V. Spore Production

Since *V. necatrix,* like all other microsporidia, develops only in living cells, production in living hosts is, at present, the sole practical method of spore production. Many species of noctuid larvae are suitable hosts. The number of spores obtained per infected host depends on the spore inoculum, host species, temperature and age of the host at inoculation and harvest. Maximum spore production of 2×10^{10} spores/g of host was obtained when *H. zea* or *T. ni* larvae were inoculated at 5 days post hatch by placing a drop of spore suspension (1×10^5 spores/ml) on their mouthparts and harvesting after 21 days at 21° or 26°C (Fowler and Reeves, 1975). Similarly, inoculation of instar III *H. zea* larvae by surface treatment of artificial diet, a method perhaps more suited to mass production, produced 1.7×10^{10} spores/larva after 15 days, an increase of 3.3×10^6 times the number of spores inoculated (Fuxa and Brooks, 1979a). Since technology for mass-rearing many noctuid species is currently available and spore production per infected larva is extremely high, mass production of large quantities of *V. necatrix* spores is feasible.

VI. Application Techniques and Formulation

V. necatrix spores must be applied to the underside as well as the top side of plant leaves of agricultural crops. The best time for application is when larvae are young, most susceptible, and can be killed before causing an unacceptable level of damage to the crop (see Section II for LT_{50} s).

Spores in baits or carriers appear to be best for field application (Fuxa and Brooks, 1979b); spores sprayed in water, with and without UV protectants, have generally resulted in mediocre protection of several crops, despite high rates of infection and host mortality (Mistric and Smith, 1973; Jacques, 1977; Maddox, unpublished observations; Fuxa and Brooks, 1979b). Spore formulations plus other pathogens and chemicals have not been field tested but offer possibilities for improved efficacy as do many of the new methods for applying insect pathogens, e.g. aerosols (Falcon *et al.*, 1974).

VII. Evaluation as a Biological Control Agent

The ultimate criterion for evaluating *V. necatrix* as a biological control agent must, of course, be the level of crop protection, but other factors, e.g. infection and mortality, and influence on subsequent pest generations, must also be observed to evaluate fully its potential.

Some early field experiments with noctuid pests of vegetables were not encouraging because crop protection was poor (Maddox unpublished observations). High levels of infection and mortality were observed, but the pests caused unacceptable crop damage before dying. No significant differences in damage were observed between control plots and plots treated with 5×10^{11} spores/acre in water with no UV protectant. This resulted in part from the evaluation procedure. Since protection of vegetables is at least partially a cosmetic problem, a little feeding damage on an ear of corn or a tomato is almost as bad as a large amount. However, Jacques (1977) reported that the proportion of cabbages undamaged or marketable and the weight of marketable heads were significantly higher in treated than in check plots.

In preliminary experiments on tobacco 8×10^{11} spores/acre in water with no UV protectant controlled the tobacco budworm as well as Viron H and Thuricide 90002–A, but not as well as Dipel or the chemical insecticide, methomyl (Mistric and Smith, 1973). More recently, spores sprayed in water with a UV protectant did not produce acceptable crop protection (Fuxa and

Brooks, 1979b). Even dosages of up to 10^{13} spores/acre, or ca. 1000 *H. zea* larval equivalents (LE)/acre, did not reduce damage as well as methomyl, in spite of infection rates up to 63%. However in two tests in 1977, hand applications of a corn meal formulation at a rate of 10^{12} spores/acre reduced budworm damage and gave excellent control that compared favourably with a Dipel—corn meal formulation and was as good or better than methomyl.

In similar tests on soybeans (Fuxa and Brooks, 1979b) *V. necatrix* spores in water with a UV protectant had little effect on *H. zea* density or damage despite infection rates up to 92% at 10^{11} spores/acre and 99% at 10^{13} spores/acre. In these tests 10^{11} spores or more/acre significantly reduced larval densities of the green cloverworm, *P. scabra*. It was not determined if this greater effect on *P. scabra* resulted from *P. scabra* larvae having a lower LD_{50} or if its feeding habits caused it to ingest more spores.

Crop protection with *V. necatrix* will be more or less correlated with target pest mortality; infected larvae that do not die of septicemia while young may live longer and feed more than uninfected larvae, even though none survive to the adult stage.

VIII. Feasibility and Future Work

Unlike most species of microsporidia, the high virulence and wide host range of *V. necatrix* make it an attractive candidate microbial control agent. It can be mass-produced at a reasonable cost and stored for short periods. When formulated with a sunlight protectant, spores survive sufficiently in the field. High rates of infection and mortality were obtained in small field tests but without consistently producing acceptable crop protection. For successful short-term control, relatively large dosages of spores must be eaten by young larvae. Thus the principal immediate need for research lies in the area of formulation and application techniques, particularly baits. Mixtures with other pathogens and stomach poisons should also be studied because the unique action of spores on the host gut could significantly synergize or antagonize any chemical or biological agent passing through or acting on the midgut epithelium. The determination of LD_{50} and LT_{50} for various hosts should aid in the choice of target insects and the design and comparison of laboratory and field tests.

All field tests to date have been on small plots not exceeding 0.1 acres. They have usually included chemical insecticides in nearby and often adjacent plots. As in tests with other microbial insecticides, the chemicals have probably reduced or eliminated the influx of wild parasitoids and predators, thus cancelling one of the main advantages of *V. necatrix* – its safety to beneficial insects. A better evaluation requires larger, more isolated field trials over several years. Inter-actions of parasitoids and predators with *V. necatrix* should also be studied. At least some parasitoids indiscriminately oviposit on healthy and infected armyworm larvae (Melin, 1978). In the latter, although the larval parasitoid is seldom directly infected, some species fail to complete development. Finally, safety evaluations of *V. necatrix* in vertebrates must precede its extensive development as a microbial control agent.

References

Falcon, L.A., Sorenson, A.A. and Akesson, H.B. (1974). *Calif. Agric.* **28**, 11–13.

Fowler, J.L. and Reeves, E.L. (1974). *J. Protozool.* **21**, 538–542.

Fowler, J.L. and Reeves, E.L. (1975). *J. Invertebr. Path.* **25**, 349–353.

Fuxa, J.R. and Brooks, W.M. (1978). *J. econ. Ent.* **71**, 169–172.

Fuxa, J.R. and Brooks, W.M. (1979a). *J. Invertebr. Path.* **33**, 86–94.

Fuxa, J.R. and Brooks, W.M. (1979b). *J. econ. Ent.* **72**, 462–467.

Gardner, W.A., Sutton, R.M. and Noblet, R. (1977). *Environ. Ent.* **6**, 616–618.

Ignoffo, C.M., Hostetter, D.L., Sikorowski, P.P., Sutter, G. and Brooks, W.M. (1977). *Environ. Ent.* **6**, 411–415.

Jacques, R.P. (1977). *J. econ. Ent.* **70**, 111–118.

Kaya, H.K. (1977). *J. Invertebr. Path.* **30**, 192–198.

Kramer, J.P. (1965). *J. Invertebr. Path.* **7**, 117–121.

Maddox, J.V. (1966). Ph.D. Thesis, Univ. Ill, Champaign, Urbana.

Maddox, J.V. (1973). *Misc. Publ. Ent. Soc. Am.* **9**, 99–104.

Maddox, J.V. (1977). *Misc. Publ. Ent. Soc. Am.* **10**, 3–18.

Maddox, J.V. and Sprenkel, R.K. (1978). *Misc. Publ. Ent. Soc. Am.* **11**, 65–84.

Melin, B. (1978). Ph.D. Thesis, Univ. Ill, Champaign, Urbana.

Mistric, W.J. and Smith, F.D. (1973). *J. econ. Ent.* **66**, 979–982.

Nordin, G.L., Rennels, R.G. and Maddox, J.V. (1972). *Environ. Ent.* **1**, 351–354.

Pilley, B.M. (1976a). *J. Invertebr. Path.* **27**, 349–350.

Pilley, B.M. (1976b). *J. Invertebr. Path.* **28**, 177–183.

Tanabe, A.M. (1971). Ph.D. Thesis, Univ. Calif., Berkeley.

Tanada, Y. (1962). *J. Invertebr. Path.* **4**, 495–497.

Tanada, Y. (1964). *Proc. Hawaiian Ent. Soc.* **18**, 435–436.

Tanada, Y. and Chang, G.Y. (1962). *J. Invertebr. Path.* **4**, 129–131.

CHAPTER 32

Nosema fumiferanae, a Natural Pathogen of a Forest Pest: Potential for Pest Management

G.G. WILSON

Forest Pest Management Institute, Canadian Forestry Service, Sault Ste. Marie, Ontario, Canada.

I. Introduction

Other than a few brief reports (Stairs, 1972; McLaughlin, 1973; Weatherston and Retnakaran, 1975) no general review has been presented on the pathogen, *Nosema fumiferanae* (Microsporidia), in the spruce budworm, *Choristoneura fumiferana* (Lepidoptera: Tortricidae). The aim of this chapter is to review available information on the ecology of the pathogen and on its potential in managing a forest pest.

A. THE HOST, *CHORISTONEURA FUMIFERANA*

The spruce budworm is the most destructive defoliating forest insect native to the North American continent. Many millions of acres of fir and spruce timber have been killed in areas without control measures. The larvae feed on various conifers, principally balsam fir (*Abies balsamea*), white spruce (*Picea glauca*) and red spruce (*Picea rubens*).

The pest's life cycle has been fully described by Baker (1972), Bean and Waters (1961) and Prebble and Carolin (1967). Adults emerge in July or August and females lay eggs in masses on the underside of conifer needles. The eggs hatch in about 10 days and each tiny larva spins a silken hibernaculum within which it moults into instar II and spends the winter. The following spring, the larva emerges from the hibernaculum shortly before vegetative conifer buds begin to expand, and mines into old needles, usually two and sometimes three, before boring into expanding buds. There is one generation per year with a total of six instars, the last instar larvae being the largest and most voracious feeders. The later larval stages (IV to VI) eat developing buds, and, if the current growth is depleted, the previous year's foliage, causing severe damage. Eventually, marked defoliation of the tree occurs. In true firs, tree mortality begins after about five successive years of defoliation.

B. THE PATHOGEN, *NOSEMA FUMIFERANAE*

1. *History and Biology*

N. fumiferanae, was probably first reported by Graham (1948), who stated: "A new disease organism possessing characteristics of sporozoan (Protozoa) has been discovered in spruce budworm larvae from three widely separated districts of Port Arthur, Sault Ste. Marie, and Kapuskasing. The internal substance of the affected larvae becomes a thick creamy consistency, without odour." Bird and Whalen (1949) briefly described the pathogen, but the first detailed description was made by Thomson (1955). He reported that although the main site of infection is the midgut cells, other tissues may also become infected. No apparent external symptoms are produced by the disease and there is no hypertrophy of infected cells. Internal symptoms as reported by Graham (1948) were not noted. Thomson found that the parasite goes through the basic cycle of schizogony producing binucleate, tetranucleate and multinucleate schizonts. Sporogony is characterized by binucleate sporonts which undergo nuclear division producing tetranucleate sporonts, each cleaving into two sporoblasts and eventually producing two spores.

2. *Taxonomy*

Recently, the taxonomic characters used to assign microsporidia to the genera *Nosema, Perezia* and *Glugea* have been questioned. The species in the spruce budworm was originally assigned to the genus *Perezia* and later *Glugea* (Weiser, 1961; Thomson, 1960a). Based on the arguments of Sprague and Vernick (1971) and Milner (1972), I now consider the species to be in the genus *Nosema*.

II. Natural Occurrence of *Nosema fumiferanae*

A. HOST RANGE

The host specificity of a microsporidium is an important epizootiological factor. If an obligate pathogen is restricted to one host species, its survival is linked with that of the sole host. A pathogen that can infect more than one host has a greater chance of survival and spread (Thomson 1958a).

Although Thomson (1958a) suggested that *N. fumiferanae* was restricted to the genus *Choristoneura,* more recent observations (Wilson, 1972) indicate that this is not so. In the laboratory, *N. fumiferanae* also infected several species of *Malacosoma* (Lasiocampidae) as well as the mourningcloak butterfly, *Nymphalis antiopa* (Nymphalidae). It must be emphasized that these are laboratory results and it is not known if these insects would support *N. fumiferanae* in the forest. A close relative of the spruce budworm, the jack pine budworm *Choristoneura pinus* is infected by a microsporidium that is very similar to *N. fumiferanae* (Thomson 1959). The two budworms are sympatric, and they may share the same microsporidium or very closely related ones.

B. TRANSMISSION

The ability of a pathogen to disperse throughout the host's environment is of prime importance if that pathogen is to exert any measure of control. Transmission from host to host must be investigated to determine the best means of introducing the pathogen into forest populations for control purposes.

In our laboratory, larvae of the spruce budworm are routinely fed spores of *N. fumiferanae,* and the results show unequivocally that such spores infect healthy insects. In heavy forest infestations of spruce budworm, it is not uncommon to find more than one larva in the same bud, and as the bud is eaten, the larvae contact each other and their by-products. Budworms also regurgitate readily when touched or bothered by predators and entomophagous parasites. In the laboratory, spores in frass and regurgitate are infective for healthy larvae (Wilson, 1972). Thomson (1958a) suggested that in the forest mechanical factors such as wind and rain, may break up infected cadavers and thus spread the spores widely.

Transovarial transmission is the most important means of spreading *N. fumiferanae* throughout budworm populations. Wilson (1972) demonstrated that infected female adults readily transmit *N. fumiferanae* to their offspring; however, this could not be shown for infected males. Schizonts and spores occur within eggs laid by infected females (Thomson 1958a), and overwinter in larvae, which is probably the principal means of persistence in the population.

A number of entomophagous parasites attack the spruce budworm. Larvae of *Apanteles fumiferanae* (Braconidae) and *Glypta fumiferanae* (Ichneumonidae) from infected budworms contained large numbers of *N. fumiferanae* spores in the gut (Thomson, 1958b). Although larval tissues were not invaded, some insects died as pupae.

C. LEVELS OF NATURAL INFECTION

Microsporidia are widespread in the forest affecting all larval instars, pupae and adults of the spruce budworm (Neilson 1963). In over-wintering larvae in the Uxbridge forest of Ontario from 1955 to 1959 Thomson (1960b) reported percentage infections of 36.4, 45.7, 56.1, 69.1 and 81.3 respectively, and stated, "Great reductions in egg numbers in 1959 and the subsequent low larval populations were probably due to direct larval mortality and a reduction in fertility of adults, both caused by microsporidia." Instar V budworms collected from the same area in 1973 to 1977 inclusive had infection levels of 35.9, 43.3, 43.0, 56.0 and 56.2% respectively (unpublished data). Wilson (1973) also demonstrated the increase in *N. fumiferanae* infections as the age of the budworm infestation progressed over 2 or more years. In addition, infections tended to increase throughout any one budworm generation, which means that comparison between and within years should be done at the same time of year in each case.

D. EFFECTS ON HOST

The pathogen retards larval and pupal development, reduces pupal weight and fecundity, and shortens adult life − all effects were more pronounced in females than in males. Some larvae were killed, mostly before instar V (Thomson 1958a). Wilson (1977) substantiated these results, and demonstrated that feeding additional spores of *N. fumiferanae* to naturally infected budworms enhanced the detrimental effects of the pathogen.

III. Factors Influencing Effectiveness as a Control Agent

Studies on factors which influence the controlling effect of *N. fumiferanae* are recent, very limited and have been carried out only in the laboratory.

A. SUNLIGHT AND TEMPERATURE

Rearing the host above 23°C impaired the survival of the microsporidia, as indicated by a significant decrease in the production of spores (23°C, mean spores/larva 47.9×10^5; 28°C, 27.0×10^5). Spores painted on foliage were killed by 5 to 6 h exposure to direct sunlight, or by 3 to 4 h at 30 cm from a 30-W germicidal lamp (Wilson, 1974a).

B. AGE OF HOST

To use a candidate pathogen efficiently in microbial control, one must expose the insects at their most susceptible stage, which should be determined experimentally. In the spruce budworm, the early instars (II and III) are much more susceptible than later instars (Wilson 1974b), but they are difficult to infect in the forest because mining in needles and feeding inside capped buds prevents contact with spores. Although susceptibility decreases with increasing larval age, more larvae can be killed by feeding suitably large dose of spores.

C. OTHER PATHOGENS

There is very little information on the effects of combinations of pathogens on the spruce budworm. Because other microbes have often been applied to populations of spruce budworm that are also naturally infected with *N. fumiferanae*, this subject needs further research.

Smirnoff (1965, 1972) suggested that microsporidian infection in the spruce budworm may improve biological control by *Bacillus thuringiensis*. Recently Wilson (1978) demonstrated that the microsporidium *Pleistophora schubergi* supplemented the effect of natural infection with *N. fumiferanae*. When 5×10^8 *P. schubergi* spores were painted on a $4.7 \, cm^2$-surface of a synthetic diet fed to instar II, 80.7% died compared with 16.5% of larvae bearing only the natural infection. Female pupal weights, adult longevity and fecundity were also significantly reduced by the dual infection.

IV. Forest Application

Only preliminary trials on individual trees have been attempted to determine if microsporidia could be introduced into the host population and persist in the following years (Wilson and Kaupp, 1975, 1976).

A. PRODUCTION AND STORAGE OF SPORES

For spore production instar II spruce budworm larvae were fed on synthetic diet (McMorran, 1965), treated with about 2×10^5 spores/ml of distilled water (Wilson and Kaupp, 1975; Wilson, 1976). Shortly before pupation, spores were harvested by homogenizing larvae in water and straining through two layers of cheesecloth to remove large debris. The spore suspension was concentrated for storage by decanting the supernatant after 2 to 3 days settling at 4°C. Just prior to use, the concentration of pooled, mixed batches was determined by counting with a haemocytometer.

B. FORMULATION AND APPLICATION

White spruce and balsam fir trees were sprayed with 1500 ml of an aqueous formulation containing 25% (v/v) feed grade molasses and 30 g/litre of IMC 90-001 sunlight protectant (Sandoz Wander, Inc., Homestead, Fl, USA), with a packsack-type mist sprayer (KWH Kem San Ltd.) at 2.5×10^{10} and 5.0×10^{10} spores/tree. One group of trees were sprayed when the budworms were in the needle mining stage (II-III instar) and a second group when the buds had opened (IV-V instar). Larval samples were taken 20 and 26 days after spraying. *N. fumiferanae* in smear preparations of larvae was diagnosed with phase contrast optics.

Preliminary results suggest that spraying trees with *N. fumiferanae* spores when budworms are in the needle-mining stage was unsuccessful. Although repeated application at this stage may improve results, it is doubtful that the added expense of time and material would be justified because the pathogen debilitates rather than kills its host. Infection can be increased by later spraying at instars IV and V. Transovarial transmission should infect later generations more severely. Greater spore dosage did not significantly increase infection. The reason for this was unclear, but × 2 differences in dosage rates may have been too small to demonstrate increased infection. Levels of infection were higher in larvae on treated balsam fir than on white spruce, possibly due in part to the earlier opening of balsam fir buds, allowing more spores to reach the insects.

Re-examinations one and two years after spraying indicated that, in this particular population, the levels of infection were advanced by two or three years by the spore spray.

V. Conclusions and Future Research Requirements

We have really only begun to examine the ecology of *N. fumiferanae*. It can be classed as a chronic debilitating agent, affecting host vigour, longevity and fecundity — attributes of more use in the forest than in agriculture where quick kill is usually required.

Many of the research needs for *N. fumiferanae* are similar to those cited by McLaughlin (1973) for all protozoan pathogens. Thorough studies are required to measure the effects of *N. fumiferanae* on forest budworm populations. Although it can be introduced, its effect on budworm numbers has yet to be quantified. Its role in integrated control should be investigated, not only in conjunction with chemicals, but also with other budworm pathogens: other microsporidia, viruses, fungi and bacteria. There is little information on the interaction of these organisms. Further studies are required on spore production, storage and formulations. As yet only budworm larvae are used for propagating *N. fumiferanae*. Possibly a larger insect could be used to provide more spores, or cell culture, but this requires much more study. There are a number of sunlight protectants and sticking agents yet to be tested. Methods of introducing *N. fumiferanae* into the host population, besides spraying, have not been investigated. It has been introduced only into advanced heavy infestations: early introduction at the beginning of an infestation may be more effective.

At present it is difficult to draw definite conclusions on the potential of *N. fumiferanae* in the management of a forest pest. One of its major attributes is that, once applied, it would perpetuate itself in the pest population. Although it may not be effective alone, *N. fumiferanae* might be important in integrated control.

References

Baker, W.L. (1972). U.S. Dep. Agric. Forest Serv. Misc. Publ. No. 1175, pp. 378–381.

Bean, J.L. and Waters, W.E. (1961). U.S. Dep. Agric. Forest Pest Leafl. No. 58, pp. 8.

Bird, F.T. and Whalen, M.M. (1949). Forest Insect Lab. Ann. Rep. No. 2. Sault Ste. Marie, p. 7.

Graham, K. (1948). Can. Dep. Agric. Bimonthly Prog. Rep. No. 4, p. 2.

McLaughlin, R.E. (1973). *Misc. Publ. ent. Soc. Am.* **9**, 95–98.

McMorran, A. (1965). *Can. Ent.* **97**, 58–62.

Milner, R.J. (1972). *J. Invertebr. Path.* **19**, 231–238.

Neilson, M.M. (1963). *In* "The Dynamics of Epidemic Spruce Budworm Populations". (R.F. Morris, ed.), pp. 272–287, Mem. Ent. Soc. Can.

Prebble, M.L. and Carolin, V.M. (1967). *In* "Forest Insects and Diseases of North America" (A.G. Davidson and R.M. Prentice, eds), pp. 75–80. Queen's Printer, Ottawa.

Smirnoff, W.A. (1965). *J. Invertebr. Path.* **7**, 266–269.

Smirnoff, W.A. (1972). *Can. Ent.* **104**, 1153–1159.

Sprague, V. and Vernick, S.H. (1971). *J. Protozool.* **18**, 560–569.

Stairs, G.R. (1972). *Ann. Rev. Ent.* **17**, 355–372.

Thomson, H.M. (1955). *J. Parasitol.* **41**, 1–8.

Thomson, H.M. (1958a). *Can. J. Zool.* **36**, 309–316.

Thomson, H.M. (1958b). *Can. Ent.* **90**, 694–696.

Thomson, H.M. (1959). *Can. J. Zool.* **37**, 117–120.

Thomson, H.M. (1960a). *J. Insect Path.* **2**, 346–385.

Thomson, H.M. (1960b). Can. Dep. Agric. Bi-monthly Prog. Rep. 16, 1.

Weatherston, J. and Retnakaran, A. (1975). *J. environ. Qual.* **4**, 294–303.

Weiser, J. (1961). "Die Mikrosporidien als Parasiten der Insekten". Monograph. *Angew Ent.* **17**, 149 pp.

Wilson, G.G. (1972). Ph.D. Thesis, Cornell Univ., Ithaca, USA, 108 pp.

Wilson, G.G. (1973). Can. For. Serv. Bi-monthly Res. Notes **29**, 35–36.

Wilson, G.G. (1974a). *Can. J. Zool.* **52**, 59–63.

Wilson, G.G. (1974b). *Can. J. Zool.* **52**, 993–996.

Wilson, G.G. (1976). *Can. Ent.* **108**, 383-386.

Wilson, G.G. (1977). *Can. J. Zool.* **55**, 249–250.

Wilson, G.G. (1978). *Can. J. Zool.* **56**, 578–580.

Wilson, G.G. and Kaupp, W.J. (1975). Can. For. Serv. Sault Ste. Marie, Inf. Rep. IP-X-11, 26 pp.

Wilson, G.G. and Kaupp, W.J. (1976). Can. For. Serv. Sault Ste. Marie, Inf. Rep. IP-X-15, 14 pp.

CHAPTER 33

Potential of Nematodes for Pest Control

JEAN R. FINNEY

Research Unit on Vector Pathology (RUVP),
Memorial University of Newfoundland, St. John's, Newfoundland, Canada

I. Introduction

In the recent decade interest in nematodes as biological control agents has grown immensely. Reviews and articles on different aspects have proliferated (Poinar, 1971, 1972; Nickle, 1972a, b, 1973; Webster, 1972, 1973 Benham and Poinar, 1973; Davey and Hominick, 1973; Gordon *et al.* 1973a; Stoffolano, 1973; Gordon and Webster, 1974) and two books have been published — Shephard's (1974) compilation of over 1100 abstracts on insect-nematode associations and Poinar's (1975) Entomogenous Nematodes.

This chapter reviews only those nematodes recently developed or showing potential for biological control of noxious insects. Data on safety is given in Chapter 40, Section IX.

II. Control of Mosquitoes and Blackflies

Petersen and Willis (1972a) contributed greatly to the advancement of biological control of mosquitoes by mass rearing the mosquito mermithid *Romanomermis culicivorax* (as *Reesimermis nielseni*) using larval *Culex pipiens quinquefasciatus* (= *fatigans*) as the live host. The method used is based on the life cycle of this aquatic mermithid. Eggs hatch in water to release infective preparasites, which are allowed to penetrate mosquito larvae where they develop in the haemocoel (Bailey and Gordon, 1973; Gordon *et al.*, 1974). The nematodes emerge, killing their hosts usually at the last larval instar. These postparasites are collected at emergence and placed in trays of damp sand where they mature, mate and lay eggs (Petersen, 1972, 1975b). The stored eggs can be induced to hatch by flooding after 7 weeks. Using this method 10^6 preparasites can be produced for 10¢ U.S. Such availability and the wide host range (Petersen, 1973, 1975a) have led to many recent field trials (Table I). The impact of environmental factors of mosquito habitats, from tree-holes to polluted pools, on the survival and infectivity of the preparasites has been investigated (Table II). Conversely, the possible effect of *R. culicivorax* on nontarget organisms in the environment was examined: it did not harm the series of organisms tested, including vertebrates (Ignoffo *et al.*, 1973, 1974).

Methods of application of the nematodes advanced from hand introduction of postparasites and eggs in sand, which has the advantage of extending the infectivity of the inoculum, to the use of polyethylene squeeze bottles and hand compression sprayers to disseminate preparasites, hatched at the test site. All these methods proved adequate for small field tests. To treat large areas, the viability and infectivity of preparasites were tested when applied by a Simplex low profile aerial spray system equipped with 12 Teejet spray nozzles, similar to ones used for aerial application of larvicides from helicopters (Levy *et al.*, 1976, 1977). Results were encouraging and tests from helicopters continue.

One problem which has loomed greater with larger field trials is nematode transport. Usually aged cultures of adults and eggs in damp sand are sent by air or courier to the field site, where eggs are hatched by addition of water. This is expensive. Sand cultures do not travel well: there may be fungal infections and premature egg hatch due to high humidity or adults and eggs may be crushed. The problem must be solved to make costs competitive with chemical control agents. This is of prime importance to two commercial firms in the USA, Nutrilite Products, Inc. and Fairfax Biological Laboratory, Inc., who produce the nematode for a world market. The Fairfax product "Skeeter Doom" is the first nematode to be marketed, due primarily to the efforts of Nickle (1976a, b).

Advances have been made towards *in vitro* culture, the economical long term solution to mass production of *R. culicivorax*. Roberts and Van Leuken (1973), Sanders *et al.* (1973) and Finney (1976a, b, 1977) have all achieved some success towards this end. The choice of a suitable tissue culture medium continues, aided by work on the mermithid–host interrelationship (Gordon and Webster, 1971, 1972; Gordon *et al.*, 1971, 1973b; Rutherford and Webster, 1976, 1978; Rutherford *et al.*, 1977), as well as studies on the structure and physiology of the trophosome and method of nutrient uptake by parasitic mermithid nematodes from their hosts (Rutherford and Webster, 1974; Poinar and Hess, 1976; Ittycheriah *et al.*, 1977).

The development of *R. culicivorax* as a control agent stimulated exploration for other mermithids better suited to certain mosquito habitats. *Octomyomermis muspratti*, from *Aedes* and *Culex* mosquitoes in Africa (Muspratt, 1945; Obiamiwe and McDonald, 1973) can be maintained in the laboratory (Muspratt, 1965; Petersen, 1977a) in *Culex quinquefasciatus*. Biological studies

TABLE I

Field trials with Romanomermis culicivorax

Site	Treatment[a]	Results (% infection)	Reference
Louisiana, natural	Repeated, 1 000/m²	*Anopheles* spp. 65% instar II, 58% instar III, 33% instar IV. All sites had natural infections once after treatment	Petersen and Willis, 1972b
California, rice fields	418/m² 836/m²	*Anopheles freeborni* 50% 80–85%	Petersen *et al.*, 1972
Louisiana, rice fields	3 261/m² 1 421/m²	*Psorophora confinnis* 94 + % *An. quadrimaculatus* 94 + %	Petersen *et al.*, 1973
Louisiana, natural	836/m² 1 672/m²	*Anopheles* spp. (principally *An. crucians*) 76%, 85%	Petersen and Willis, 1974a
Taiwan, small pool	90 000/m²	*Culex tritaeniorhynchus,* *Cx. t. summorosus,* single infection	Mitchell *et al.*, 1974
Taiwan, paddy field border	370 000 over 95 m perimeter	*Cx. t. summorosus, Cx. rubithoracis,* zero infection	
Taiwan, potholes	11 500/pothole		
Taiwan, fields	116 000/4 m² area	*Cx. t. summorosus* 11%, *Cx. fuscanus* < 1%, *Cx. rubithoracis, An. sinensis,* zero infections	
Taiwan, pools	35 000/m²	*Cx. quinquefasciatus* < 5% infections	
Taiwan	Ground pool 2 000/m² Ground pool 1 000 postparasites 4.5 m²	68% *Cx. quinquefasciatus* Recycled Zero infection *Cx. quinquefasciatus*	Chen Pau-Shu, 1976

TABLE I (*continued*)

Field trials with Romanomermis culicivorax

Site	Treatment[a]	Results (% infection)	Reference
Louisiana, natural	1971 and 1973 1973	1974 *An. crucians* 2–51% 1974 *An. crucians*, 11–85% Establishment, recycling	Petersen and Willis, 1975
Louisiana, natural	Whole cultures (prehatch)	*Aedes atlanticus* 52%; *Ae. tormentor* 59%; *Ps. columbiae* 38%; *Ps. howardii* 57%; *Ps. ferox*, penetration, no development	Petersen and Willis, 1976
Florida, sewage tanks	$1.1 \times 10^5/m^2$	*Cx. quinquefasciatus*, Tank 1 37%, Tank 2, 53%	Levy and Miller, 1977a
Florida, grassy field, potholes, ditches	$3.6 \times 10^3/m^2$	*Ps. ciliata*, *Ae. taeniorhynchus* *Ps. columbiae*, *Cx. nigripalpus*, 96 ± 3% all species	Levy and Miller, 1977b
Manitoba, snow melt	$50\,000/m^2$	*Ae. canadensis*, *Ae. pionips*, 20%, overwintered	Galloway, 1977 Galloway and Brust, 1976a
	$50\,000/m^2$	*Ae. dorsalis*, *Ae. spencerii*, *Ae. communis*, < 1% overwintered	
Manitoba, early spring rain pool	$50\,000/m^2$	*Ae. vexans*, 8–20%	
Tokelau Islands	300/treehole 4 500/166 litre drum 1978	0–14% *Ae. polynesiensis* 11–70% *Ae. aegypti*	Laird, pers. comm.

TABLE I (*continued*)

Field trials with Romanomermis culicivorax

Site	Treatment[a]	Results (% infection)	Reference
Maryland, tractor ruts	3×10^6 overall, 1975	*An. punctipennis/ An. crucians, Ae. vexans, Cx. restuans, Cx. pipiens, Cx. territans,* total infection 30–90%. Overwintered at $-16°C$ and $-19°C$. 1977, 88%; 1978, 30% *Anopheles* spp. infected. *Cx. territans* never infected	Nickle, 1979
Maryland, pond	4.7×10^6 overall, 1975	As above: overwintered 1976, 1977, 1978; summer 1978 25% infected *Anopheles* spp.	
California, natural	706–950/m²	*An. franciscanus* 29% *An. freeborni* 84%	Brown *et al.,* 1977
California, artificial	1 000–25 000/m²	*An. franciscanus* 67–100%, *Cx. tarsalis* 13–62%; *Culiseta inornata* 22–58%	
El Salvador, volcanic lake	2 400–4 800/m² Retreated twice weekly for 6 wks, 1977	*An. albimanus* overall parasitism 58%. Establishment and recycling undetermined	Petersen, Chapman, Willis and Fukuda (Personal communication)

[a] Treatment with preparasites unless otherwise stated.

TABLE II
Effect of environmental factors on activity and infectivity of
Romanomermis culicivorax

Environmental Factor	Results	Reference
Conductivity	No infection in waters 3 000 $\mu\Omega$/cm	Petersen and Willis, 1970
	Survival at 1190−1300 $\mu\Omega$/cm	Levy and Miller, 1977a
	No infection in waters > 4 000 $\mu\Omega$/cm	Brown and Platzer, 1978
pH	Infectivity retained 3.6−8.6	Brown and Platzer, 1978
	Infectivity retained 9.0	Levy and Miller, 1977a
Temperature (temp.)	Preparasites more active > 18° than < 12°C	Petersen and Willis, 1971
	Preparasites optimum infectivity 21−33°C	Brown and Platzer, 1977
	Thermal tolerance of 40°C	Levy and Miller, 1977c
	Rate of parasitism decreased from 30° to 15°C	Kurihara, 1976
	Low temp. prolongs infectivity	Brown and Platzer, 1977
	% Infection by preparasites reduced as temp. reduced	Galloway and Brust, 1977; Brown and Platzer, 1977
	Low temp. prevents infection. Min. 6−8°C	Mitchell *et al.*, 1974
Photoperiod	Mosquito larvae more susceptible to infection in total darkness	Brown and Platzer, 1974
	No significant effect of photo-period on infections	Galloway and Brust, 1977
Cation and anions	Max. tolerated concentration (mM) Na 16; Ca 14; K 13: Cl 24; NO_3 18; PO_4 10; NO_2 9.5; SO_4 8.9; CO_3 2.1	Brown and Platzer, 1978
	No infection in NaCl > 0.04 M	Petersen and Willis, 1970
	Survival when chloride level 244−258 ppm	Levy and Miller, 1977a

are aimed at producing enough material for eventual field application (Petersen, 1977a, b; 1978). This nematode is more tolerant of desiccation, salinity and pollution than *R. culicivorax*. A second species, *Diximermis peterseni* (Nickle, 1972b) (= *Gastromermis* sp.; Petersen and Chapman, 1970; Savage and Petersen, 1971; Chapman *et al.*, 1972), from *Anopheles* only, can be bred in the laboratory and easily established in the field (Petersen and Willis, 1974b; Woodard, 1978). Significantly, it produces high levels of parasitism despite long periods of dry weather. In 1977, four years after the first culture in *Anopheles quadrimaculatus*, the mosquito began to show resistance to infection (Woodard and Fukuda, 1977), because the host colony was restocked only with larvae that escaped infection after exposure to preparasites. Periodic restocking from elsewhere would obviously prevent resistance in the laboratory. This, however, points to resistance as a possible limitation in the use of mermithids artificially established in the field.

Thermal intolerance restricts the use of *R. culicivorax* at low temperatures. However, several other mermithids are being investigated for use in northern climes. Galloway and Brust (1976b) have cultured mermithids from spring hatch—winter diapause snow melt mosquitoes in Manitoba, Canada. A *Romanomermis* sp. has been reared successfully in *Aedes epactius (= atropalpus)*, *Ae. aegypti* and *Cx. quinquefasciatus* at 20°C. Rearing problems and its biology are now being studied. An undescribed *Culicimermis* from adult females of four *Aedes* spp. has accomplished four laboratory generations in *Ae. vexans*. The preparasite infects early larval instars but its development is not complete until after emergence of the adult host, which has distinct advantages for nematode dispersal. The labour required in rearing may restrict its large scale production (Galloway and Brust, 1976b). Another snow melt mosquito parasite *Romanomermis nielseni*, many of whose natural hosts are common to Canada and the USA, was first described from larval, pupal and adult mosquitoes in Wyoming (Tsai and Grundmann, 1969). It was recollected in 1977 and 1978 and the emergent nematodes completed their life cycle through to egg production in the laboratory at 17°C (Finney, 1978), although there is some discrepancy between developmental times reported by Finney (1978), Petersen (1976) and Tsai (1967). Problems hindering further production are lack of a suitable easily reared laboratory host and lack of synchronous egg hatch. Small egg hatches over an extensive period preclude good infection rates in a host, and as the sex of a nematode is partially determined by the number of nematodes in an individual host, this subsequently affects the emergent male:female postparasitic ratios essential to good egg production.

Mermithid infection of blackflies may induce intersex formation or sterility in adults and inevitably causes mortality in larvae and adults on emergence from them. Records of incidence are widespread and high infection rates have been reported (Welch, 1964; Gordon *et al.*, 1973a; Poinar, 1977). In North America recent studies have focused on the control potential of *Neomesomermis flumenalis* (Ebsary and Bennett, 1973, 1974, 1975a, b; Bailey and Gordon, 1977; Bailey *et al.*, 1974; Ezenwa, 1974a, b; Ezenwa and Carter, 1975; Mokry and Finney, 1977). Bailey *et al.* (1977) cultured its free-living stages. Culture of the parasitic phase was hampered by difficulty in rearing the host to maturity. This should be improved at the RUVP by using the new stir-bar larval blackfly rearing system (Colbo and Thompson, 1978). Molloy and Jamnback (1975) tested this nematode in a system simulating the natural habitat of larval black-flies and obtained high rates of parasitism (80%) for first instar *Simulium vittatum*. A field test carried out by the same authors in 1977 indicated high infection rates at the site of application, which decreased further downstream. Molloy and Jamnback (1977) and Bruder (1974) indicated that the use of *N. flumenalis* is limited basically because only early simuliid instars are susceptible, so that regular retreatment would be necessary. The production cost of the nematodes is prohibitive at the moment and is not likely in the near future to compete with insecticides (Molloy and Jamnback, 1977). However, in New-foundland natural populations of mermithids have been experimentally enhanced in certain streams with encouraging results to date (M.H. Colbo, personal communication).

In the absence of an efficient *in vivo* blackfly mermithid rearing system, *in vitro* is the only alternative for mass culture of *N. flumenalis*. Attempts by Myers (personal communication; Finney, 1976b) were limited by the seasonal availability of nematode eggs and preparasites and information on the physiological aspects of the mermithid–simuliid relationship (Condon, 1976; Condon and Gordon, 1977; Gordon *et al.*, 1978).

Mondet and Poinar (1976) and Mondet *et al.* (1976, 1977a, b) have investigated the natural mermithid parasites of *Simulium damnosum* in West Africa. Mondet and his co-workers are now selecting a suitable candidate for mass rearing and eventual release in the field. To this end a blackfly/mermithid rearing system has been built at Bouaké, West Africa, under the auspices of the Organisation de Coordination et de Coopération pour la Lutte contre les Grandes Endémies. In the absence of a naturally occurring simulium mermithid which could be mass produced the feasibility of using *R. culicivorax* for simuliid control was investigated (Finney, 1975). In 1976, Hansen and Hansen found that this nematode would attack early instar *S. damnosum* under simulated field conditions and would initiate development within the host. Further research (Finney and Mokry unpublished observations) indicated that serious blackfly control by *R. culicivorax* was neither a realistic nor an economic possibility.

III. Control of Forest, Soil and Plant Pests

Neoaplectana spp. have been used to control all of these types of pest, primarily because of their wide host range and ease of mass propagation. Recently advances in culture methods have lowered production costs from U.S. $1 to 2¢ per 10^6 infective stage larvae for at least one species (Poinar, 1972; Bedding, 1976). Coarse aspen wood wool coated with homogenized chicken heart, with 30% water added, yields a mean of 10^8 infective *N. carpocapsae* (*agriotos* strain) per 2-litre container. Although long term storage of the in-fective stage of *Neoaplectana* spp. was deemed possible, Jackson (1973) found that the fecundity of *N. glaseri* declined measurably with age when stored for 3 years at $5°C$ in the dark in $1–2$ cm water. This should be investigated further and considered when stock-piling nematodes.

Water must be available in the environment to ensure adequate longevity in the field, a subject well reviewed for the DD-136 strain of *N. carpocapsae* by Moore (1973), Poinar (1971) and Simons and Poinar (1973). Whether sprayed on foliage or crops the nematode must reach a microenvironment with high humidity to survive. Infective stages live for long periods in a variety of inert organic solvents, including paraffin oils, and survive well in spray droplets when 5% high melting point wax is added to the oil. Also, selective rearing for a few generations at high rearing temperatures increased the nematodes' resistance to desiccation (Bedding, 1976). Protection is also needed against ultra-violet light, which soon reduces infectivity and may kill infective stages (Gaugler and Boush, 1977).

Mice fed the DD-136 strain of *N. carpocapsae* and its associated bacterium showed no ill effects (Schmiege, 1963; Poinar, 1972). The temperature range of the nematode (Kaya, 1977) indicated little likelihood of its developing in birds and mammals, although further tests are needed with poikilothermic vertebrates and invertebrates.

One of the largest field trials of *Neoaplectana* sp. in recent years concerns the effects of DD-136 on the total soil fauna of the economically valuable sugar beet fields in Switzerland, Holland and Britain (Edwards, 1977). Preliminary results only are available to date, and these suggest some reduction of certain pest numbers while beneficial predators are not affected (J. Oswald, personal communication).

Attempts were also made to use *Neoaplectana* sp. for the control of *Scolytus scolytus* a vector of the Dutch Elm disease fungus *Ceratocystis ulmi*. Finney (1973), and Finney and Mordue (1976), found that larvae, pupae and adult *S. scolytus* were susceptible to the DD-136 strain of *N. carpocapsae* in the laboratory. In a small field trial treated logs contained significantly more dead nematode-infected insects than the controls (Finney and Walker, 1977). Long term beetle control and nematode survival are still to be evaluated.

One of the greatest advances in forest pest control in the last decade was the integration of *Deladenus siricidicola* (Bedding, 1967, 1968, 1972) into full scale programs against *Sirex* spp. (Hymenoptera), serious pests of *Pinus radiata* in Australia and New Zealand. The potential of this nematode was that it could sterilize adult female woodwasps and that infection rates greater than 90% were recorded (Zondag, 1975). It had two independent life cycles, one parasitic within the host haemocoel and a second associated with the wasp's symbiotic fungus. The mycetophagous phase was utilized for mass culture of the nematode. Potato dextrose agar plates were relatively expensive and time consuming compared with a mass culture method with wheat and water as substrate, which yielded up to 3–10 million nematodes per 500 ml flask, enough to inoculate about 100 m of timber (Bedding and Akhurst, 1974). The nematodes could be stored for several weeks in tap water with an atmosphere of almost pure oxygen at 5–10°C. Transportation was achieved in water.

Several methods for inoculating the nematode into timber were described (Zondag, 1975). The latest and most commonly used involves inoculating *Sirex* infested trees directly with the nematode using a hypodermic syringe. This method can produce over 99% parasitism (Bedding and Akhurst, 1974). Normally logs are infested in this way and transported to locations of high *Sirex* incidence. Field experiments have shown that the nematode can be established in most areas and that it can be widely dispersed by the host. Very high incidences of parasitism, over 75%, occurred in the several field introductions to date (Bedding and Akhurst, 1974; Zondag, 1975). In 1974, Bedding described five new species of *Deladenus* (Neotylenchidae) in woodwasps, now being screened for use in future control programs.

Control of the Colorado beetle, *Leptinotarsa decemlineata*, seemed more realistic after Fedorko and Stanuszek (1971) described *Pristionchus uniformis* (Diplogasteridae) from this host. Two bacteria were extracted from the infective third stage larval nematode by the hanging drop method (Sandner *et al.*, 1972). One is a *Streptococcus* resembling *Streptococcus durans*, the other a rod with the characteristics of *Bacillus subtilis* and *Bacillus pumilis*. *P. uniformis* can be cultured best under xenic conditions on solid media (raw pig or beef liver with 2.5–5% Difco agar) (Fedorko and Stanuszek, 1971). Invasive larvae can be stored in 0.001% formalin at 5°C. *P. uniformis* can develop at temperatures at which the beetle is in anabiosis: 85% of diapausing beetles dug up from fields were dead and contained many nemas of all stages. However, a few invasive larvae in an active (feeding) host do not usually kill it. Spring field applications of 50 000 infective nematodes/m^2 caused a 60% control of the beetle while an autumn application of the same number caused a 97% control. The nematode survived 3 years in the soil (Fedorko, 1974).

Nickle and Grijpma (1974), reviewing lepidopterans parasitized by *Hexamermis* spp., mentioned the mahogany shootborer *Hypsipyla grandella* in Costa Rica, 5–25% of which were killed by *He. albicans*. The parasitized larvae were found in branches 1–2 metres from the ground, most towards the end of the wet season and fewest at the end of the dry season. *Hexamermis* spp. occurred in as many as 50% of Colorado beetles in Southern Poland and may be important in reducing larval and adult populations in the summer (Stanuszek, 1970). Polish studies have been restricted by lack of culture methods. Puttler *et al.* (1973) studied the incidence of *Hexamermis arvalis* in cutworms in the USA. They found 64.5% infection of *Agrotis ipsilon* in the field and discovered that the nematode only attacks overwintering host larvae. Puttler and Thewke (1971) reared field collected *He. arvalis* in the laboratory through to adults which produced viable eggs. If culture methods can be mastered, this nematode has great potential for the control of pests in cultivated crops.

One other nematode which deserves mention in this section is *Steinernema kraussei*. Originally described from sawfly *Cephalcia abietis* larvae from Germany, it has a symbiotic relationship with a strain of *Flavobacterium cytophaga*. It has been cultured *in vitro* on nutrient agar slants at 4–8°C (Weiser, 1976) which will enable further study of its control potential.

IV. Control of Mushroom House and Greenhouse Pests

Howardula husseyi (Tylenchida: Allantonematidae), a naturally occurring obligate parasite, attacks the mushroom phorid fly *Megaselia halterata*, a major pest of cultivated mushrooms in the UK for two decades (Richardson *et al.*, 1977). Both fly and parasite numbers increase from spring onwards; by autumn 60–75% of the flies are parasitized (Riding and Hague, 1974). The flies decline in the winter when parasitism becomes sparse, so if the winter drop in parasitism could be prevented the nematode might exert a continual biological control of the fly (Richardson and Hesling, 1977). Instead of laying eggs, female mushroom flies release larvae or juveniles, which moult and mate in the mushroom compost, then the mated female nematodes become infective. However, these are short-lived and only 10–12% find hosts. A 90% parasitism in the flies would require at least 75×10^6 infective nematodes/tonne compost, a level unlikely to be accepted by growers. Another limit to its use is culture, which is possible only in its natural host and this in itself is a difficult and costly undertaking (Richardson, 1977).

In greenhouses, maggots of the sciarid fly *Bradysia paupera* (Lycoriidae) eat roots and shoots of seedlings. There are two naturally occurring parasites of these pests, *Tripius sciarae* and *Tetradonema plicans*. *Tripius sciarae* reduced greenhouse sciarid populations within 4 weeks of introduction but it lacks a resistant free-living stage and infective female nematodes are viable for only 2 weeks in soil and less in water or on agar (Poinar, 1965). Thus storage would be a major problem.

Tetradonema plicans, originally described from the American greenhouse sciarid *Sciara coprophila*, is a very promising potential control agent in the UK (Hudson, 1972, 1974a, b). The eggs and first stage infective larvae are the only free-living stages and both sexes develop in the haemocoel of the host larvae. Fecundity is high: single females developing in a host produced on average 8 100 eggs/host. Mass production is cheap in the host *B. paupera*, which itself is easily reared. Eggs, the best stage to apply for biocontrol, can be stored at 10°C; some live for one year, but viability dropped to 30% in six months. The eggs in an egg capsule are laid over a two week period and hatch in sequence, a big advantage because pests entering the greenhouse are attacked over a two week period. Single *T. plicans* females in 1–5 day old maggots were lethal but if infected later a few maggots pupated. Of the infected flies emerging, 75% were sterile females, due to reduced ovaries. This is an added advantage in a control agent, allowing manipulation for short term intensive control or longer term moderate results.

V. Use of Nematodes in Integrated Control

Webster (1973) suggested the application of nematodes together with an insect hormonal treatment or another pathogen to debilitate the host and favour parasitism in a variety of ways. Both Altosid 5E (a synthetic insect growth regulator) and *Romanomermis culicivorax* are separately effective in controlling mosquito populations (Schaefer and Wilder, 1973; Petersen, 1973; Mitchell *et al.*, 1974; Petersen and Willis, 1976). Although the development of certain nematodes is affected by juvenile hormone analogues (Davey and Hominick, 1973; Dennis, 1976), Altosid 5E did not interfere with the infectivity of the nematode or its development within *Aedes aegypti*, neither did previous exposure to the chemical affect the subsequent free-living stages of the nematode. Host mortality was considerably increased when a combination of the mermithid and Altosid 5E (5–50 ppb) was used against *Ae. aegypti* in the laboratory, suggesting that an integrated program could prove more effective than either control agent used separately (Finney *et al.*, 1977). Using this system, insects that are pests only as adults could be controlled while immature, before they present a problem. In further work, exposure of preparasitic *R. culicivorax* for 24 h in the laboratory to Abate (1 ppb), malathion (1), Baytex (3), Dursban (1), Dimilin (5) and Altosid (5) did not decrease viability and infectivity of preparasites or viability of resulting postparasites. However, higher concentrations of Baytex (3.5 and 4 ppb) and Dursban (4) significantly lowered preparasite infectivity when compared to controls, even though none of the preparasites died (Levy and Miller, 1977d). Exposure of preparasites to insecticides Abate, dieldrin or gamma-HCH, at concentrations lethal to mosquito larvae, did not adversely affect infectivity of *R. culicivorax* in a subsequent *Cx. quinquefasciatus* assay (Mitchell *et al.*, 1974). Nickle (1979) showed that exposure of *R. culicivorax* to Abate at the concentration recommended for mosquito control killed preparasites; neither Altosid 5E (at the same concentration) or malathion (10 × concentration) had any effect. It was reported that the incidence of unspecified mermithid parasitism of *Simulium venustum* and *Simulium verecundum* in blackfly streams in S. Carolina was not significantly affected by 2% Abate larvicide (Garris and Boblet, 1975).

Hudson (1974c) investigated the effect of routinely used greenhouse pesticides on the sciarid/*T. plicans* system. Of the six tested only two (pirimicarb and benomyl) were not compatible, two (Vydate and dimethirimol) could be integrated into the system when applied at normal concentrations, while gamma-HCH, Parathion and diazinon killed only the sciarid maggots and so exerted a synergistic effect with the nematode on the sciarid population.

Chemicals may be used to make a biological control agent more feasible in a certain area. Control of mosquito larvae by *R. culicivorax* was reduced in sites with dense vegetation or algal mats, which constitutes a problem that might be solved by the use of algacides since those with 1 ppm copper or less have no effect on the infectivity of this nematode (Platzer and Brown, 1976).

Thus, an understanding of the relationship between pesticides and parasites is necessary, not only for integrated control but also to avoid regular insecticide treatment of a site interfering with the natural control exerted by parasites in the field.

Although the association and subsequent use of a nematode with another pathogen has not yet been fully investigated, their occurrence and potential are recorded here. *Paecilomyces farinosus* and *Beauveria bassiana* were isolated from field collected Colorado beetles; infection of *Galleria mellonella* by either one of these fungi in conjunction with *Neoaplectana carpocapsae* increased pathogenicity in experiments oriented primarily to investigate changes in the host (Kamionek *et al.*, 1974a, b). This combination of pathogens caused similar debilitating effects in Colorado beetles in the laboratory (Seryczynska, 1975;

Kamionek, 1976). As yet no field tests have been carried out. The DD-136 nematode plus β-exotoxin from *Bacillus thuringiensis* var *thuringiensis* caused high mortality in leather jackets (Lam and Webster, 1972). Hurpin and Robert (1975) found the mermithid *Melolonthinimermis* (*Pseudomermis*) *hagmeieri* parasitizing the common cockchafer (*Melolontha melolontha*) larvae. Treatment of the soil with preparations of *Bacillus popilliae* or of *Beauveria brongniartii* (= *tenella*) had no clear effect on the pathogenicity of the nematodes, suggesting further investigations are needed of this type of combined use. In addition, Likhovoz (1975) reported that the combined action of *Gastromermis boophthorae* and the microsporidium *Pleistophora simulii* resulted in earlier mortality of a blackfly, *Boophthora erythrocephala*, than either pathogen alone.

As opposed to concurrent application of compatible pathogens for integrated control, it may be possible to capitalize on the ability of some entomophilic nematodes to carry more potent parasites (Gordon and Webster, 1974), e.g. *Neoaplectana carpocapsae,* strain *agriotos,* transmits *Nosema mesnili* and *Pleistophora schubergi* (Veremchuk and Issi, 1970). Both Jackson (1969) and Poinar and Hess (1977) showed that nematodes can sustain other organisms either intra- or extracellularly. Although this potential can be used beneficially for integrated pest control, some caution is needed because it is also possible that mismanagement could aid an invaded vector host, e.g. adult mosquito or simuliid, in the dispersal of an organism pathogenic to man or livestock.

VI. Prospects and Research Requirements

The potential of nematodes as control agents lies in their lethal or sterilizing effect on the host, their ecological adaptation to the host and its environment and man's ability to manipulate these factors. Each nematode prominent as a biological control agent is so because of the ease and economy with which it can be mass produced. The recent acceptance by the US Environmental Protection Agency of *R. culicivorax* as a parasite, not a pesticide, and so outside of their jurisdiction, facilitated its commercialization and encouraged that of others. Commercial production is the key to successful and competitive use of nematodes in a world of chemical insecticides. Species such as *R. culicivorax* and *D. siricidicola*, which have achieved a place in pest control practice, now need continual monitoring.

The accrued knowledge of host and parasite biology, and their physiological and ecological interactions, which have facilitated the use of certain nematodes, should now be sought for those parasites not yet fully exploited. During this review, many associations worthy of further study have been indicated, to which can be added that between *Heterotylenchus autumnalis* and the face flies of cattle. Other approaches to the insect pest problem should not go unmentioned, e.g. Wulker's (1964, 1975) extensive study towards the understanding of mermithid-induced intersexuality and castration, and the utilization of these phenomena for control. In addition, integrated control in all its aspects needs thorough examination. Possibly, in future there will be more demand for nematodes in integrated schemes than for their use alone.

References

Bailey, C.H. and Gordon, R. (1973). *J. Invertebr. Path.* **22**, 435–441.
Bailey, C.H. and Gordon, R. (1977). *Can. J. Zool.* **55**, 148–154.
Bailey, C.H., Gordon, R. and Mokry, J. (1974). *Can. J. Zool.* **52**, 660–661.
Bailey, C.H., Gordon, R. and Mills, C. (1977). *Can. J. Zool.* **55**, 391–397.
Bedding, R.A. (1967). *Nature, Lond.* **214**, 174–175.
Bedding, R.A. (1968). *Nematologica* **14**, 515–525.
Bedding, R.A. (1972). *Nematologica* **18**, 482–493.
Bedding, R.A. (1974). *Nematologica* **20**, 204–225.
Bedding, R.A. (1976). Proc. 1st Int. Colloq. Invertebr. Path. Kingston, Ontario 250–254.
Bedding, R.A. and Akhurst, R.J. (1974). *J. Aust. Ent. Soc.* **13**, 129–135.
Benham, G.S. Jr. and Poinar, G.O. Jr. (1973). *Expl Parasit.* **33**, 248–252.
Brown, B.J. and Platzer, E.G. (1974). *J. Nematol.* **6**, 137.
Brown, B.J. and Platzer, E.G. (1977). *J. Nematol.* **9**, 166–172.
Brown, B.J. and Platzer, E.G. (1978). *J. Nematol.* **10**, 53–61.
Brown, B.J., Platzer, E.G. and Hughes, D.S. (1977). *Mosq. News* **37**, 603–608.
Bruder, K.W. (1974). Ph.D. Thesis, Rutgers Univ., State Univ. New Jersey.
Chapman, H.C., Petersen, J.J. and Fukuda, T. (1972). *Am. J. trop. Med. Hyg.* **21**, 777–781.
Chen, Pau-Shu. (1976). *Bull. Inst. Zool. Acad. Sin.* **15**, 21–28.
Colbo, M.H. and Thompson, B.H. (1978). *Can. J. Zool.* **56**, 507–510.
Condon, W.J. (1976). M.Sc. Thesis, Memorial Univ. Newfoundland, Canada.
Condon, W.J. and Gordon, R. (1977). *J. Invertebr. Path.* **29**, 56–62.
Davey, K.G. and Hominick, W.M. (1973). *Expl Parasit.* **33**, 212–225.
Dennis, R.D. (1976). *Comp. Biochem. Physiol.* **53A**, 53–56.
Ebsary, B.A. and Bennett, G.F. (1973). *Can. J. Zool.* **51**, 637–639.
Ebsary, B.A. and Bennett, G.F. (1974). *Can. J. Zool.* **52**, 65–68.
Ebsary, B.A. and Bennett, G.F. (1975a). *Can. J. Zool.* **53**, 1324–1331.
Ebsary, B.A. and Bennett, G.F. (1975b). *Can. J. Zool.* **53**, 1058–1062.
Edwards, C.A. (1977). *Pedobiologia* **17**, 292–294.
Ezenwa, A.O. (1974a). *Can. J. Zool.* **52**, 557–565.
Ezenwa, A.O. (1974b). *J. Parasit.* **60**, 809–813.
Ezenwa, A.O. and Carter, N.E. (1975). *Environ. Ent.* **4**, 142–144.
Fedorko, A. (1974). *Zesz. probl. Postep. Nauk. roln.* **154**, 413–417.
Fedorko, A. and Stanuszek, S. (1971). *Acta Parasit. Pol.* **19**, 95–112.
Finney, J.R. (1973). *Parasitology* **67**, 1.
Finney, J.R. (1975). *Bull. Wld Hlth Orgn.* **52**, 235.
Finney, J.R. (1976a). Proc. 1st Int. Colloq. Invertebr. Path. Kingston, Ontario, 225–226.
Finney, J.R. (1976b). *J. Nematol.* **8**, 284.
Finney, J.R. (1977). *Nematologica* **23**, 479–480.
Finney, J.R. (1978). *Utah Mosq. Abatement Assoc. Proc.* **31**, 47–48.
Finney, J.R. and Mordue, W. (1976). *Ann. Appl. Biol.* **83**, 311–312.
Finney, J.R. and Walker, C. (1977). *J. Invertebr. Path.* **29**, 7–9.

Finney, J.R., Gordon, R., Condon, W.J. and Rusted, T.N. (1977). *Mosq. News* **37**, 6–11.

Galloway, T.D. (1977). Ph.D. Thesis. Univ. Manitoba, Canada.

Galloway, T.D. and Brust, R.A. (1976a). *Manitoba Entomol.* **10**, 18–25.

Galloway, T.D. and Brust, R.A. (1976b). Proc. 1st Int. Colloq. Invertebr. Path. Kingston, Ontario 227–231.

Galloway, T.D. and Brust, R.A. (1977). *J. Nematol.* **9**, 218–221.

Garris, G.I. and Noblet, R. (1975). *J. Med. Ent.* **12**, 481–482.

Gaugler, R.R. and Boush, G.M. (1977). Abstr. Xth Ann. Meet. Soc. Invertebr. Path. East Lansing 9.

Gordon, R. and Webster, J.M. (1971). *Expl Parasit.* **29**, 66–79.

Gordon, R. and Webster, J.M. (1972). *Parasitology* **64**, 161–172.

Gordon, R. and Webster, J.M. (1974). *Helminth. Abstr. Ser. A.* **43**, 328–349.

Gordon, R., Webster, J.M. and Mead, D.E. (1971). *Can. J. Zool.* **49**, 431–434.

Gordon, R., Ebsary, B.A. and Bennett, G.F. (1973a). *Expl Parasit.* **33**, 226–238.

Gordon, R., Webster, J.M. and Hislop, T.G. (1973b). *Comp. Biochem. Physiol.* **46B**, 575–593.

Gordon, R., Bailey, C.H. and Barber, J.M. (1974). *Can. J. Zool.* **52**, 1293–1302.

Gordon, R., Condon, W.J., Edgar, W.J. and Babie, S.J. (1978). *Parasitology* **77**, 367–374.

Hansen, E.L. and Hansen, J.W. (1976). *Int. Res. Communs System Med. Sci.* **4**, 508.

Hudson, K.E. (1972). *Mushroom Sci.* **8**, 193–197.

Hudson, K.E. (1974a). *J. Invertebr. Path.* **23**, 85–91.

Hudson, K.E. (1974b). *Nematologica* **20**, 455–468.

Hudson, K.E. (1974c). Proc. 3rd Int. Congr. Parasit. Munich. **1**, 1700–1701.

Hurpin, B. and Robert, P.H. (1975). *Annls Soc. ent. Fr. (N.S.)* **11**, 63–72.

Ignoffo, C.M., Biever, K.D., Johnson, W.W., Sanders, H.C., Chapman, H.C., Petersen, J.J. and Woodward, D.B. (1973). *Mosq. News* **33**, 599–602.

Ignoffo, C.M., Petersen, J.J., Chapman, H.C. and Novotny, J.F. (1974). *Mosq. News* **34**, 425–428.

Ittycheriah, P.I., Gordon, R. and Condon, W.J. (1977). *Nematologica* **23**, 165–171.

Jackson, G.J. (1969). *Proc. Helminth. Soc. Wash.* **36**, 188–189.

Jackson, G.J. (1973). *Proc. Helminth. Soc. Wash.* **40**, 74–76.

Kamionek, M. (1976). *Wiad. Parazyt.* **4–5**, 369–377.

Kamionek, M., Sandner, H. and Seryczynska, H. (1974a). *Acta parasit. pol.* **22**, 357–363.

Kamionek, M., Sandner, H. and Seryczynska, H. (1974b). *Bull. Acad. Pol. Sci. Ser. Sci. Biol.* **22**, 253–257.

Kaya, H.K. (1977). *J. Nematol.* **9**, 346–349.

Kurihara, T. (1976). *Jap. J. Parasit.* **25**, 8–16.

Lam, A.B.Q. and Webster, J.M. (1972). *J. Invertebr. Path.* **20**, 141–149.

Levy, R. and Miller, T.W. Jr. (1977a). *Mosq. News* **37**, 410–414.

Levy, R. and Miller, T.W. Jr. (1977b). *Mosq. News* **37**, 483–486.

Levy, R. and Miller, T.W. Jr. (1977c). *J. Nematol.* **9**, 259–260.

Levy, R. and Miller, T.W. Jr. (1977d). *Environ. Ent.* **6**, 447–448.
Levy, R., Murphy, L.J. Jr. and Miller, T.W. Jr. (1976). *Mosq. News* **36**, 498–501.
Levy, R., Cornell, J.A. and Miller, T.W. Jr. (1977). *Mosq. News* **37**, 512–516.
Likhovoz, L.K. (1975). *Medskaya Parazit.* **44**, 230–233.
Mitchell, C.J., Chen, P.S. and Chapman, H.C. (1974). *J. Formosan Med. Assoc.* **73**, 241–254.
Mokry, J.E. and Finney, J.R. (1977). *Can. J. Zool.* **55**, 1370–1372.
Molloy, D. and Jamnback, H. (1975). *Mosq. News* **35**, 337–342.
Molloy, D. and Jamnback, H. (1977). *Mosq. News* **37**, 104–108.
Mondet, B. and Poinar, G.O. Jr. (1976). Proc. 1st Int. Colloq. Invertebr. Path. Kingston, Ontario 232–235.
Mondet, B., Pendriez, B. and Bernadou, J. (1976). *Cah. office Rech. scient. techq. Outre-Mer Sér. Ent. Méd. Parasit.* **14**, 141–149.
Mondet, B., Poinar, G.O. Jr. and Bernadou, J. (1977a). *Can. J. Zool.* **55**, 1275–1283.
Mondet, B., Berl, D. and Bernadou, J. (1977b). *Cah. office Rech. scient. techq. Outre-Mer Sér. Ent. Méd. Parasit.* **15**, 265–269.
Moore, G.E. (1973). *Expl Parasit.* **33**, 207–211.
Muspratt, J. (1945). *J. ent. Soc. Sth. Afr.* **8**, 13–20.
Muspratt, J. (1965). *Bull. Wld. Hlth. Org.* **33**, 140–144.
Nickle, W.R. (1972a). Proc. Ann. Tall Timbers Conf. Ecol. Anim. Control Habitat Mgmt. 145–163.
Nickle, W.R. (1972b). *J. Nematol.* **4**, 113–146.
Nickle, W.R. (1973). *Expl Parasit.* **33**, 303–317.
Nickle, W.R. (1976a). Proc. 1st Int. Colloq. Invertebr. Path. Kingston, Ontario 241–244.
Nickle, W.R. (1976b). *J. Nematol.* **8**, 298.
Nickle, W.R. (1979). *Proc. Helminth. Soc. Wash.* **46**, 21–27.
Nickle, W.R. and Grijpma, P. (1974). *Turrialba* **24**, 222–226.
Obiamiwe, B.A. and McDonald, W.W. (1973). *Ann. trop. Med. Parasit.* **67**, 439–444.
Petersen, J.J. (1972). *J. Nematol.* **4**, 83–87.
Petersen, J.J. (1973). *Expl Parasit.* **33**, 239–247.
Petersen, J.J. (1975a). *J. Nematol.* **7**, 211–214.
Petersen, J.J. (1975b). *J. Nematol.* **7**, 207–210.
Petersen, J.J. (1976). *J. Nematol.* **8**, 273–275.
Petersen, J.J. (1977a). *J. Invertebr. Path.* **30**, 155–159.
Petersen, J.J. (1977b). *J. Nematol.* **9**, 343–346.
Petersen, J.J. (1978). *J. Invertebr. Path.* **31**, 103–105.
Petersen, J.J. and Chapman, H.C. (1970). *Mosq. News* **30**, 420–424.
Petersen, J.J. and Willis, O.R. (1970). *J. econ. Ent.* **63**, 175–178.
Petersen, J.J. and Willis, O.R. (1971). *Mosq. News* **31**, 558–566.
Petersen, J.J. and Willis, O.R. (1972a). *Mosq. News* **32**, 226–230.
Petersen, J.J. and Willis, O.R. (1972b). *Mosq. News* **32**, 312–316.
Petersen, J.J. and Willis, O.R. (1974a). *Mosq. News* **34**, 316–319.

Petersen, J.J. and Willis, O.R. (1974b). *J. Invertebr. Path.* **24**, 20–23.

Petersen, J.J. and Willis, O.R. (1975). *Mosq. News* **35**, 526–532.

Petersen, J.J. and Willis, O.R. (1976). *Mosq. News* **36**, 339–342.

Petersen, J.J., Hoy, J.B. and O'Berg, A.B. (1972). *Calif. Vector Views* **19**, 47–50.

Petersen, J.J., Steelman, C.D. and Willis, O.R. (1973). *Mosq. News* **33**, 573–575.

Platzer, E.G. and Brown, B.J. (1976). Proc. 1st Int. Colloq. Invertebr. Path. Kingston, Ontario 263–271.

Poinar, G.O. Jr. (1965). *Parasitology* **55**, 559–569.

Poinar, G.O. Jr. (1971). *In* "Microbial Control of Insects and Mites" (H.D. Burges and N.W. Hussey, eds), pp. 181–203. Academic Press, London.

Poinar, G.O. Jr. (1972). *Ann. Rev. Ent.* **17**, 103–122.

Poinar, G.O. Jr. (1975). "Entomogenous Nematodes". E.J. Brill, Leiden, The Netherlands. 317 pp.

Poinar, G.O. Jr. (1977). *Bull. World Hlth. Orgn.* **55**, 509–515.

Poinar, G.O. Jr. and Hess, R. (1976). *Int. Res. Communs System* Med. Sci. **4**, 296.

Poinar, G.O. Jr. and Hess, R. (1977). *Nature, Lond.* **266**, 256–257.

Puttler, B. and Thewke, S.E. (1971). *Ann. ent. Soc. Am.* **64**, 1102–1106.

Puttler, B., Sechriest, R.E. and Daugherty, D.M. Jr. (1973). *Environ. Ent.* **2**, 963–964.

Richardson, P.N. (1977). *Rep. Glasshouse Crops Res. Inst.* 1976 108.

Richardson, P.N. and Hesling, J.J. (1977). *Ann. app. Biol.* **86**, 321–327.

Richardson, P.N., Hesling, J.J. and Riding, I.L. (1977). *Nematologica* **23**, 217–231.

Riding, I.L. and Hague, N.G.M. (1974). *Ann. app. Biol.* **78**, 205–211.

Roberts, D.W. and Van Leuken, W. (1973). Abstr. Int. Colloq. Insect Path. Microbiol., Oxford, 91.

Rutherford, T.A. and Webster, J.M. (1974). *J. Parasit.* **60**, 804–808.

Rutherford, T.A. and Webster, J.M. (1976). Proc. 1st Int. Colloq. Invertebr. Path. Kingston, Ontario 272–275.

Rutherford, T.A. and Webster, J.M. (1978). *Can. J. Zool.* **56**, 339–347.

Rutherford, T.A., Webster, J.M. and Barlow, J.S. (1977). *Can. J. Zool.* **55**, 1773–1781.

Sanders, R.D., Stokstad, E.L.R. and Malatesta, C. (1973). *Nematologica* **19**, 567–568.

Sandner, H., Seryczynska, H. and Kamionek, M. (1972). *Bull. Acad. Pol. Sci. Ser. Sci. Biol.* **20**, 567–569.

Savage, K.E. and Petersen, J.J. (1971). *Mosq. News* **31**, 218–219.

Schaefer, C.H. and Wilder, W.H. (1973). *J. econ. Ent.* **66**, 913–916.

Schmiege, D.C. (1963). *J. econ. Ent.* **56**, 427–431.

Seryczynska, H. (1975). *Bull. Acad. Pol. Sci. Ser. Sci. Biol.* **23**, 351–354.

Shephard, M.R.N. (1974). Comm. Inst. Helminthol. Tech. Comm. No. 45.

Simons, W.R. and Poinar, G.O. Jr. (1973). *J. Invertebr. Path.* **22**, 228–230.

Stanuszek, S. (1970). *Zesz. Probl. Postep. Nauk. roln.* **92**, 359–367.

Stoffolano, J.G. Jr. (1973). *Expl Parasit.* **33**, 263–284.

Tsai, Y.H. (1967). Ph.D. Thesis. Univ. Utah, Salt Lake City, Utah. No. 11938.

Tsai, Y.H. and Grundmann, A.W. (1969). *Proc. Helminth. Soc. Wash.* **36**, 61–67.

Veremchuk, G.V. and Issi, I.V. (1970). *Parazitologiya* **4**, 3–7.

Webster, J.M. (1972). *In* "Economic Nematology" (J.M. Webster, ed), pp. 469–496. Academic Press, London and New York.

Webster, J.M. (1973). *Expl Parasit.* **33**, 197–206.

Weiser, R. (1976). Proc. 1st Int. Colloq. Invertebr. Path. Kingston, Ontario 245–250.

Welch, H.E. (1964). *Bull. Wld. Hlth. Org.* **31**, 857–863.

Woodard, D.B. (1978). *Mosq. News* **38**, 80–83.

Woodard, D.B. and Fukuda, T. (1977). *Mosq. News* **37**, 192–195.

Wülker, W. (1964). *Expl Parasitol.* **15**, 561–597.

Wülker, W. (1975). In "Intersexuality in the Animal Kingdom" (R. Reinboth ed.), pp. 121–134. Springer-Verlag, Berlin.

Zondag, R. (1975). Proc. 28th N.Z. Weed and Pest Control Conf. 196–199.

CHAPTER 34

Formulation of Insect Pathogens

T.L. COUCH and C.M. IGNOFFO

*Abbott Laboratories, Chemical and Agricultural Products Division,
North Chicago, Illinois, USA and Biological Control of Insects Laboratory,
U.S. Department of Agriculture, Columbia, Missouri, USA*

I. Introduction

A pesticide formulation is defined as the resultant composition when the candidate pesticide is mixed with anything, including water (van Walkenburg, 1973). Therefore, any combination of an active biocide with a second material is technically a formulation. If this definition were stringently adhered to the scope of this chapter would be too broad and its contents, therefore, too general. Consequently, the discussion will be limited to insect pathogens produced and applied on a large scale and to formulation types and additives which have contributed to pathogen stability, efficacy and acceptance by the end user. No attempt will be made to review all literature on formulations of microbial insecticides. The authors will concentrate on pertinent references published after those described in Angus and Lüthy (1971).

II. Basic Formulation

Angus and Lüthy (1971) mentioned that the development of microbial insecticide formulation closely paralleled that of chemical insecticides. This is true because both chemicals and insect pathogens must be formulated to facilitate mixing and application. For the purpose of this discussion, the definition of the term formulation will be split so that research with basic formulations can be separated from that with tank mix formulations: viz. a basic commercial formulation constitutes the form and contents of the insecticide as supplied by manufacturer to distributor and ultimately to end user: a tank mix formulation is the commercial formulation plus spray vehicle; e.g. water, oil, etc., added by the end user and applicator (Couch, 1978). These two terms are often used interchangeably.

Basic formulations of both insect pathogens and chemicals comprise: (1) liquids (aqueous suspensions, or emulsifiable suspensions), (2) wettable powders, (3) dusts, (4) baits, (5) granules. Since insect pathogens are insoluble living entities, they cannot be formulated as soluble powders.

In commercial development of a basic formulation of an entomopathogen, research concerns maintaining pathogen viability and virulence during the production process and developing a product form which preserves or enhances these properties. To do this, knowledge of the biology of pathogen and target insect is essential. In its initial research, industry considers effects of temperature, humidity and substrate (inert carrier) on the entomopathogen to be the most important. The singular reason for this is that failure to manipulate temperature, humidity and quality (chemical and physical) of the inert carrier can impair viability and virulence of the pathogen. Environmental stability of the pathogen after it reaches the agroecosystem will be discussed in a later section.

Formulation of a pathogen product with an extensive shelf-life (> 18 months) is critical to industrialization. The reason is simple. Unless viable, virulent and stable for a prolonged period, a preparation is simply not commercially economical. Inventory control and manipulation is costly, making extended stability imperative. Formulations of *Bacillus thuringiensis* (*B.t.*) adequately illustrate this point. Since commercialization in the 1950's, the bacterial strain, fermentor yield, product form and stability have been improved, making this microbial an effective, economical alternative to chemical insecticides. In the early years, wettable powder formulations were almost impossible to wet; they distributed poorly in application systems, frequently plugged spray nozzles and provided erratic results (Hall, 1963; Couch, 1978). *B. t.* dust improved efficacy by providing a more even distribution of active ingredient. However, such a specialized formulation is targeted for relatively small markets with little return on investment. The advent of liquid formulations, primarily aqueous concentrates, solved the mixing problem and some spray distribution difficulties. However, because the *B. t.* spore and crystal are suspended in what amounts to a fermentation medium stabilized with fungistatic and bacteriostatic ingredients, biological stability often failed. Containers of liquid stored at ambient temperatures often exploded from gas emitted during vegetation processes or autolysis. Even today, users are cautioned to protect liquids from freezing and temperatures above 90°F (Sandoz label, Thuricide HPC; Abbott label, Dipel LC and Dipel SC).

During the past 5 years more attention has been given to producing agricultural formulations of *B. t.*, the *Heliothis* nuclear polyhedrosis virus (NPV) and certain fungi. New groups of inert additives previously untried with biologicals are being investigated. Angus and Lüthy (1971) summarized a fairly extensive list of these but more detailed listing of others available and descriptions on the merits and uses of inert carriers and emulsifiers were given by Becher (1973) and Polon (1973). Table I lists inert ingredients added to *B. t.* formulations with varying degrees of success — including botanicals (derived from natural sources), minerals, and synthetics. It lists ingredients of candidate wettable powders, dusts, baits and granules, and emulsifiable suspensions.

Additives were selected which significantly improved the desired physical properties of agricultural formulations of *B. t.* The important physical properties considered were ability to flow, wet, disperse and suspend, with little foam and stable physical storage. All these optimize dilution of the product with water to form a homogenous, sprayable suspension (Polon, 1973). Once the desired physical attributes were satisfied, biological stability was studied by accelerated methods to determine adverse effects on the pathogen.

Formulation requirements vary among the different entomopathogens. Formulation research on the fungus, *Hirsutella thompsonii,* revealed — not unexpectedly — that the required moisture content was higher than the relatively low levels in *B. t.* powders. Drying conidia prepared from a semi-solid fermentation had little effect on viability when dried through a series of steps from 80% moisture w/w to 10%. However, infectivity measurements should always be paralleled by bioassay measurements with host species. After drying, virulence and viability of conidia were both eliminated by grinding in a Fitzmill. Because of the moisture requirements and milling problems associated with *H. thompsonii,* research will be required to formulate these conidia in a non-aqueous carrier.

Basic commercial formulations of NPVs and granulosis viruses are generally spray or air dried and diluted with an inert carrier, or freeze dried with a carbohydrate, usually lactose. Freeze drying, however, is generally considered too expensive to use on an industrial scale. Also, liquid suspensions of insect viruses have been kept cold or frozen (Jacques, 1977). However, because commercial formulations of insect pathogens are generally exposed to a range of storage conditions in warehouses, etc., unstabilized, aqueous preparations of virus are generally unsuitable. Moisture levels in the final formulation of insect virus probably resemble those in *B. t.* formulations. Formulation storage requirements are usually more restrictive and to insure an adequate shelf-life refrigeration is suggested, especially for preparations with a high moisture content.

Once prepared and stabilized to preserve viability and virulence, a pathogen can be formulated into a field-strength dust, wettable powder concentrate, granule or bait, emulsifiable aqueous suspension or emulsifiable non-aqueous suspension. This requires experiments with various oils, emulsifiers, extenders and diluents (Table I). After a suitable set of inert carriers has been identified, stability must be studied to insure pathogen—carrier compatibility. Even if a carrier is inert, spreading the pathogen or toxicant over its large surface area could increase the rate of degradation (Polon, 1973) and, particularly with

TABLE I

Inert Carriers Tested as Basic Formulation Components with
Bacillus thuringiensis *(Couch, unpublished observations).*

CLAYS AND DUST DILUENTS	LIQUID VEHICLES	EMULSIFIERS cont.
	Water	Triton GR7M
1095 Marble dust	Preformed oil in water emulsion	Triton X-35
Neosil A	Preformed water in oil emulsion	Plurafac A-24
Silica sand	Mineral oil	Triton-N60
Syloids	Corn oil	AL-1364
Celites	Crude sorbitol	Atlox 848
Pyrax	Aromatic spray oils	Atlox 849
Agsorb	Aliphatic spray oils	Atlox 3404/849
Barden clay	Emulsified cottonseed oil	Witconol H-31A
Kaolin clay		Atplus 448
Continental clay	SUSPENDING AGENTS	
Celite	Bentone-38	BOTANICALS
Al-Sil-Ate	CAB-O-SIL	Citrus pulp
Satintone	SOLOID	Walnut shells
Microcell		Corn cob
Talc	EMULSIFIERS	Corn meal
Attagel	Atplus 300	Wheat bran
Attaclay	AL-1246	Grape pomace
Diluex	Triton X-45	Apple pomace
Emathlite	Triton X-363M	Rice hulls
Lactose	AL-1280	Cracked corn
	AL-1403	

entomopathogens, care must be taken to balance the acidic and alkaline pro-
perties of the carriers as well as the moisture content. As examples, cationic and
anionic surfactants affected thermal death time of *Bacillus megaterium* and
B. subtilis (Rode and Foster, 1960). Atlox, a combination of anionic and non-
ionic surfactants, inhibited *B. t.* fermentation and growth; Span-85, Biofilm, and
Triton B-1956 had no adverse effects (Morris, 1975). Germination of the fungus,
H. thompsonii, was impaired by Triton B-1956 while Miller Nufilm, Chevron
X-77, Plyac, and Spray Oil 435 had no effect at field use rates (McCoy, personal
communication). Thus, many kinds of additives (Table I) can be included in
commercial formulations to improve shelf-life, flowability, anticaking properties
and convenience in handling and mixing. However, all must be thoroughly
checked for subtle effects on the pathogens.

III. Tank Mix Formulations

Dilution of the commercial formulation into a spray for use on a target agroecosystem constitutes production of a tank mix formulation by the end-user.
Most of the tank mix consists of the spray vehicle, usually water, plus additives
to increase persistence and optimize droplet deposition and coverage. These
additives are usually wetting agents which affect the physico-chemical interaction

of toxicant formulation and plant material, i.e. wettability and spreadability (Johnstone, 1973); stickers to improve weathering by forming a film to retard wash off by rain; thickeners to reduce drift and evaporation; humectants to retard desiccation during spraying; botanical additives to stimulate feeding of the host insect; and occasionally enzymes such as chitinase to enhance activity of the pathogen (Smirnoff, 1974). From published literature, Angus and Lüthy (1971) extracted a fairly extensive list of additives which had been combined with *B. t.*, insect viruses, and certain protozoans. Generally because of their similarities, most commercial wetting, sticking and spreading agents are compatible with *B. t.* and virus. Little published information is available on protozoans or fungi. Compatibility of some of these materials with *H. thompsonii* has already been discussed (Section II). *B. t.* and virus producers recommend the addition of these agents when sprays are applied to waxy surfaces such as cabbage and broccoli or when rain-fastness is important.

Thickening agents have also been used to control drift and evaporation of aerial and ground sprays of *B.t.*, fungi and virus. Two commercially available materials Bivert® and the Nalcotrol® have been used with some success (Couch, unpublished observations). Spray deposition and droplet size were increased with Bivert. Effects of Nalcotrol were minimal. These materials form polymers or invert emulsions which reduce evaporation and fine spray droplets lost to drift. Morris (personal communication) used carboxymethylcellulose and Kelzan® (xanthan gum) to thicken sprays of *B. t.* against eastern spruce budworm, *Choristoneura fumiferana,* on spruce-fir forests. Spray deposits and insecticidal activity increased as measured by larval mortality and percent defoliation. Carbon and/or molasses significantly protected spore viability and insecticidal activity of *B. thuringiensis* formulations (Hostetter *et al.*, 1975). In other studies (Morris, 1977, a, b) Cargill Insecticide Base Concentrate (stabilized molasses) and Dowanol TPM increased droplet deposition. Results with molasses were encouraging.

Ignoffo *et al.* (1976) used a commercial adjuvant formulated by Sandoz-Wander, Inc. to reduce evaporation, increase stability in sunlight and increase larval feeding on foliage treated with a mixture of *Heliothis* NPV and water. Evaporation of water from the tank mix formulation of water + virus + adjuvant (20%) was 50% vs 80% from water + virus alone. In an extensive laboratory study, Smith *et al.* (1978) evaluated several tank mix formulations of *Heliothis* NPV against the bollworm *Heliothis zea* using two nozzle-pressure combinations. A wetting agent, oil, polymerizing agent, spreader-penetrant, sunlight protectant, humectant, and thickener were selected as adjuvants to increase coverage and effectiveness. The specific materials used were; Shade® (Sandoz, Inc.), San—285—WP66 (Sandoz, Inc.), a microbial insecticide adjuvant; polyvinyl alcohol (PVA) 99% hydrolysed (Matheson, Colman and Bell, Norwood, OH); Flo—Mo®; Top Oil® (100% mineral oil); and spreader penetrant (Ring-Around Products, Inc., Montgomery, AL); Keltose (Kelco Co., Clark, NJ), a highly refined algin material, thixotrophic at low concentrations; sucrose; and a spreader-binder, Triton CS-7 (Rohm and Haas, Philadelphia, PA). Results indicate that the best combination of ingredients based on bollworm mortality was the PVA–Shade tank mix of NPV. Aqueous suspensions of NPV, mineral oil emulsion and PVA were less effective. The addition of Triton CS-7 wetting agent decreased activity of an aqueous suspension of NPV. Also, sucrose failed to increase insecticidal activity of the PVA + Shade + virus combination. Addition of a polymerizing agent and UV protectant to an aqueous tank mix of NPV provided definite benefits in the laboratory.

Maksymiuk and Neisess (1975) characterized and evaluated the physical properties of 12 experimental, water-based, *B. t.* forest formulations. These were emulsified animal-derived protein (Maywood surfactant), whey, corn oil surfactants (corn oil emulsified with 10% Maypon 4CT), Nu Film BT, a textured vegetable protein, Cargill Insecticide Base concentrate (CIBC), Chevron Spray Sticker, and polyethylene glycol on tank mixes of *B.t.* Maywood, whey and corn oil surfactant thickened and physically stabilized the tank mix. Maywood surfactant increased spread of the formulations on foliage. CIBC was more acidic than polyethylene glycol. None of the additives significantly decreased evaporation or increased sticking properties of the tank mixes but all flowed and atomized readily and produced consistent droplet spectra. Later Neisess (1979) evaluated *B. t.* tank mixes containing water, 25% molasses and 25% Sutro against Douglas-fir tussock moth, *Orgyia pseudotsugata*. He studied 18 sticking agents. Rain negatively affected tank mixes containing Sutro and molasses more than those with only water plus sticker. Carboset depressed *B. t.* activity. Tank mixes with Biofilm, High Tack Fish Glue, Nacrylic X4260 and X4445, Nufilm 17, Plyac, and X-Link 2873 gave the most protection against 2.54 cm of rain. Spray mixtures without stickers lost *ca.* 30% of their activity. Lewis *et al.* (1974) applied *B. t.* formulations at a rate of 8 billion international units (BIU) in 2 gal. of tank mix/acre against the gypsy moth, *Lymantria dispar*. The tank mix formulation contained 25% CIBC and Chevron Spray Sticker. Results were generally favourable.

Smirnoff (1977) combined *B. t.* with Sorbitol, water, Chevron Spray Sticker and chitinase. Tank mixes containing 6.8 BIU/0.5 gal and applied by air at 0.5 gal/acre gave acceptable control of eastern spruce budworm, *Choristoneura fumiferana*. Calibration studies on this tank mix indicated that the sorbitol improved percent spray deposition.

Harper (1974) summarized forest test data with *B. t.* against gypsy moth, spruce budworm, Douglas-fir tussock moth (*Orgyia pseudotsugata*), tent caterpillar (*Malacosoma* spp.), spring cankerworm (*Paleocrita vernata*) and fall cankerworm (*Alsophila pometaria*). Other hardwood and coniferous defoliators are also mentioned. A wide variety of additives to improve spray deposition and weathering were added to aqueous tank mixes. No single spray adjuvant, sticker or spreader was identified to consistently improve insecticidal activity of *B. t.* Often a significantly higher spray deposition can be measured in a particular plot. However, efficacy is rarely significantly better than in a neighbouring stand which received fewer spray droplets.

Application studies with NPVs parallel those for *B. t.* Aerial spray results have been erratic and to achieve predictable results spray deposits must be maximized and uniform coverage insured. Whether this can be accomplished by a "magic bullet" additive to a microbial insecticide tank mix is open for debate. Nozzle arrangement, size, spray platform, type of aircraft, speed, wind currents, humidity, and a host of other climatic and physical factors have similar effects on spray deposition and coverage.

To overcome some of the problems associated with deposition, researchers have tried to attract the target insect from the host plant to the microbial insecticide formulation. Gustatory stimulants, usually of botanical origin, have been tried mostly with the insect viruses. Water extracts of corn (Montoya, et al., 1966) and of corn silks (Allen and Pate, 1966) combined with *Heliothis* NPV were more effective than virus alone in laboratory and field tests. A modified cottonseed oil bait containing sucrose, Dacagin®, hydroxycellulose, glycerin and virus attracted bollworms and reduced the population (McLaughlin et al., 1971). Two consecutive years of small plot field testing of a cottonseed oil bait of the NPV gave control equivalent to that with the chemical standard (Andrews et al., 1975). But, although generally numerically better than conventional virus sprays, the bait applications were not significantly so.

By laboratory feeding, Bell and Kanavel (1975) studied the preferences of neonatal larvae of the pink bollworm, *Pectinophora gossypiella*, to ethanol extracts of the cotton plant, also galactose, raffinose, invert sugar and germinated cottonseed oil. Extracts of all parts of the cotton plant except the leaves, cottonseed oil and sucrose were acceptable to the larvae. A bait derived from these materials with *Autographa* NPV increased the percent virus infected larvae.

Molasses in suspensions of *B. t.* masked the phagodepressant activity of the pathogen (Yendol et al., 1975). Leaf disc consumption by larvae of the gypsy moth, *L. dispar*, equalled that observed in the untreated controls. Corn meal and molasses baits containing *B. t.* have frequently been used to control tobacco budworm, *Heliothis virescens*, on tobacco. These baits have generally been superior to sprays (Abbott Laboratories, unpublished observations). There is at least one such commercial bait used extensively in the USA, manufactured by Soil Serv, Inc., Salinas, California. It is registered for use against insects such as armyworms and cutworms, which are generally not controlled by *B. t.* sprays.

Apple pomace, grape pomace, wheat bran, citrus pulp and almond hulls have also been used in baits with *B. t.* (Abbott Laboratories, unpublished observations). Results have varied but, in general, it is difficult to assess benefit of selective attraction to the bait in field experiments. Pelletized baits are usually better positioned to contact the pest insect and liquid baits generally improve spray deposition. Therefore, positive effects observed on addition of these ingredients may not be due to preferential feeding of larvae. Care must be taken to demonstrate preferential selection of bait over host plant by the target insect before claims are made of benefit of the ingredients.

IV. Shelf-life and Field Stability

A. SHELF-LIFE

A microbial insecticide must be produced, formulated and stabilized so that normal storage conditions do not affect insecticidal properties.

In general, at least 18 months stability under ambient storage conditions is required for servicing the agricultural markets. If the pathogen is to be supplied by contract for application at a specific time, shelf-life is less of a problem and stability for 3 to 6 months may be acceptable.

In the USA, industry must develop degradation profiles for insecticides, including microbial insecticides, as part of the submission for registration to the Environmental Protection Agency. External factors such as temperature and humidity affect *B. t.* preparations (Couch, unpublished observations) and certain NPVs (Ignoffo and Hostetter, 1977). It should be stressed that the effects discussed here are those which have an impact on shelf-life and not field stability.

Formulation work with *B. t.* revealed that a dry dust or wettable powder is more stable than the corresponding aqueous concentrate (Collins and Couch, unpublished observations). As long as the containers were moisture proof, 40°C and 90% r.h. had no effect on viability during a 12-week exposure. Also a non-aqueous concentrate is more stable than an aqueous concentrate (Schmits and Couch, unpublished observations). These differences are especially pronounced at high temperatures and humidities. In several separate studies (Collins and Couch, unpublished observations), the inherent moisture level was a critical factor in the shelf-life of wettable powders. Temperatures of 40°C or less had little effect on activity if moisture content was kept $< 5\%$. Pinnock *et al.* (1977) reviewed environmental effects on *B. t.* and other entomogenous bacteria and concluded from a thorough analysis of the literature that high humidities cause premature germination of spores and autolytic spoilage, a conclusion confirmed by research with new formulations by industry. In contrast, a commercial preparation of *H. thompsonii* (Mycar®) exposed to room temperature, 25°C ± 1°, and ambient humidity lost 99% in viability of conidia within 6 months (McCoy and Couch, 1978). Samples held at 4°C showed no loss of viability after one year.

NPV inclusion bodies of the gypsy moth were stored as air-dried and freeze-dried powders and water suspensions (Lewis and Rollinson, 1978). The effects of elevated (38°C), ambient (17–28°C), refrigerator (4°C) and freezer (− 10°C) temperature were examined. The NPV maintained acceptable activity if stored at 4°C as a suspension, lyophilized powder or frozen, air-dried powder. Suspensions, or powders with moisture content $> 5\%$, degraded slowly at ambient and rapidly at elevated temperatures (Ignoffo, unpublished observations). In general, stability of viral preparations increases as moisture decreases and high temperature accelerates the damaging effects of high moisture content. These observations are fairly typical for NPV and granulosis viruses (Jacques, 1977).

Storage conditions required for Protozoa have been described in detail (Kramer, 1970; Maddox, 1977). In general, *Nosema* spores will not tolerate temperatures $> 50°C$ and are short lived at 10°C. At 5°C some species, e.g. *Octosporea muscaedomesticae* have been kept > 1 year. Only *Nosema locustae* has been applied on a large scale in a commercial type formulation and essentially little is known about the shelf-life of the formulated pathogen on wheat bran. What has been determined is that *N. locustae* spores were most active against grasshoppers when harvested within 4 months of application (Henry and Oma, 1974). Spores stored for one year at $-10°C$ in water lost ca. 90% activity and those in cadavers lost 99%. (See Also Chapter 30, Section IV, C). Longevity of spores in the wheat bran carrier was not examined.

In general, except for *B. t., H. thompsonii* and the *Heliothis* NPV, little is known about shelf-life of entomopathogens formulated for wide scale use in agriculture.

B. FIELD STABILITY

Once a basic formulation with acceptable shelf-life has been developed, emphasis must be shifted to preserving the activity of a microbial insecticide on the substrate to which it is applied. The substrate, usually soil, plant tissue, water or stored grain, influences the effects of environmental factors on activity of the pathogen. These factors are sunlight, temperature, water or humidity, and chemicals. Their effects on the different groups of entomopathogens were reviewed in detail at a symposium in 1974 (*Misc. publ. ent. Soc. Am.* 10(3), June, 1977). Only a few representatives of each type of entomopathogen have been studied. In assessing the above reviews, Ignoffo and Hostetter (1977) observed that most information available on the viruses is from studies of baculoviruses, specifically, NPVs of *Heliothis* spp. and *Trichoplusia ni.* Data on bacteria involve *B. t.* and *B. popilliae*; on protozoans only on *Nosema* spp. and on fungi *Beauveria, Metarhizium, Entomophthora* and *Nomuraea*. Of these, only *Heliothis* NPV, *B. t., B. popilliae* and the fungus *Hirsutella thompsonii* became available to researchers as basic commercial formulations. Only these were applied on a scale large enough to determine effects of environmental factors on season-long population reduction of the target insect, final yield of marketable crop or percent reduction of defoliation.

1. Solar Radiation

Data on field persistence of entomopathogens reveal that sunlight is probably the most destructive environmental factor affecting persistence of microbial insecticides (Ignoffo and Hostetter, 1977). Ignoffo *et al.* (1977) confirmed this by exposing four viruses, a bacterium, a fungus and a protozoan to ultraviolet (UV) light. The UV source provided 0.14 mW/cm^2 (peak 254 nm, range 215–260 nm) and 1.8 mW/cm^2 (peak 365 nm; range 290–400 nm) of short wave and long wave UV energy at the plane of the exposed sample. The half life of all pathogens was $<4 \text{ h}$, ranking: *B. t.* $>$ *Nomuraea rileyi* $>$ entomopox virus $>$ NPV $=$ cytoplasmic polyhedrosis virus $>$ *Vairimorpha* (*Nosema*) *necatrix* $>$ granulosis virus. *B. popilliae* is probably relatively unaffected by sunlight in actual field use since dust formulations are applied and mixed with the soil which protects spores from UV light (Pinnock *et al.*, 1977).

B. t. is very sensitive to UV light. Raun *et al.* (1966) found that exposure of spores to radiation for 24 h drastically reduced activity. Exposures at 48 and 72 h destroyed all insecticidal activity against corn borer, *Ostrinia nubilalis*, and fall armyworm, *Spodoptera frugiperda.* Cantwell and Franklin (1966) exposed *B.t.* spores and crystals to direct sunlight. In 30 min 50% of the spores were inactivated, and in 60 min > 80%. UV light from a germicidal lamp had no effect on the insecticidal activity of purified crystals toward the silk worm, *Bombyx mori* (Cantwell, 1967). Gamma radiation and UV light did not affect insecticidal activity of crystals against *Pieris brassicae* (Burges *et al.*, 1975). In addition to UV, wavelengths in the visible spectrum (near 400 nm) also inactivated *B.t.* spores (Griego and Spence, 1978).

Frye *et al.* (1973), Pinnock *et al.* (1974), Hostetter *et al.* (1975) and Ignoffo and Garcia (1978) all reported the inactivation of various strains of *B.t.* by UV light under field and laboratory conditions. Some studies used spore count alone as a criterion, others larval mortality. Most studies evaluated persistance but not performance as measured by crop protection, increases in yield or quality.

Varying increases in persistence of *B.t.* in sunlight have been obtained with microencapsulation, addition of carbohydrate dyes, proteins, protein hydrolysates, carbon black, nucleic acids, molasses, commercial adjuvants and protectants (Raun and Jackson, 1966; Raun, 1968; Hostetter *et al.* 1975; Krieg, 1975; Morris and Moore, 1975; Ignoffo and Garcia, 1978). Although some materials significantly increased persistence, available data do not conclusively demonstrate that field efficacy is enhanced by addition of UV-protectants, especially in comparison with similar treatments without UV-protectants. Also, since the percentage of protectant by volume or weight often is a significant component of the tank mix, it is difficult, especially with aerial sprays, to determine if a prolonged persistence of the treatment containing a UV-protectant is due to increased droplet size or deposition.

Entomogenous viruses, examined widely in agroecosystems for insect control, like *B.t.*, are very susceptible to UV radiation (review by Jacques, 1977). In general, the half life of unprotected insect viruses both occluded and non-occluded is 24 h direct sunlight (Ignoffo and Hostetter, 1977). Degradation profiles and solutions investigated with viruses closely parallel those considered for *B.t.* Ignoffo and Batzer (1971) tried micro-encapsulation and UV light protectants with *Heliothis* NPV in artificial and natural sunlight. Non-encapsulated virus plus carbon was as effective as encapsulated preparations with carbon. Field and laboratory studies between 1967 and 1970 corroborated the value of carbon with *Heliothis* NPV. Jacques (1968), Bullock (1967), Smirnoff (1972), Young and Yearian (1974), Ignoffo *et al.* (1976), Bull *et al.* (1976), Ignoffo *et al.* (1977) and Smith *et al.* (1978) all demonstrated in laboratory and field the sensitivity of NPV and granulosis viruses to solar radiation. Their solutions to the inactivation problem varied, e.g. addition of blocking and absorbing agents such as carbon black, lignin sulphate, Shade®, and Keltose®, also encapsulation of virus plus UV-screening agents bound with digestible, water soluble polymers. Smirnoff (1972) advocated spraying with *Neodiprion swainei* NPV at sunset to avoid exposure of virus to solar radiation. Generally investigators demonstrated an increase in persistence of virus after addition of a UV-protectant. However, is protected virus superior to unprotected virus when yield and quality are considered and does the addition justify the added cost and formulation complexities? Bull (1978) showed that encapsulated formulations were superior to commercial formulations of *Heliothis* NPV in providing protection from UV light but seedcotton yields were not statistically different. Ignoffo *et al.* (1976) on the other hand demonstrated increased cotton yields (see Chapter 17, Section VII, A). Additional research must be encouraged to determine if the added expense and difficulty in supplementing a basic or tank mix formulation of entomopathogens with a UV-protectant is justified.

2. Temperature, Moisture and Substrate

It is difficult to assess the effects of temperature, moisture, and substrate on an entomopathogen in the field and little such work is published. Temperatures in typical agroecosystems range from ca. 5–50°C and generally do not affect pathogens (Ignoffo and Hostetter, 1977). In recommending use of *B. t.* products, a rule of thumb generally employed is that if the temperatures allow target insects to feed actively the pathogen will be effective.

High humidity or standing water on the substrate generally favour development of epizootics of entomogenous fungi. McCoy (1978) found that *H. thompsonii* growth and infection on the citrus rust mite (*Phyllocoptruta oleivora*) on citrus are greatest on a substrate with free water, although infection also occurs at 90–100% r.h. In the field at 26–27°C it takes <4 h for a spore to penetrate the mite cuticle and ca. 72 h for the total infection process to be completed. In environments not providing the moisture requirements at the microclimatic level, humectants may be required as integral parts of the formulation. These could favour establishment of the fungi and inducement of an epizootic. Roberts and Campbell (1977) extensively reviewed the effects of temperature and humidity on development of insect mycoses in laboratory and field.

Occluded viruses survive for long periods in soil (Jacques, 1977) as does *B. popilliae* which has been recovered 30 years after introduction (Pinnock *et al.*, 1977). Young and Yearian (1974) found that cotton dew played a minor role in the loss of activity of *Heliothis* NPV observed in the field.

V. Conclusions and Suggestions for Future Research

We have attempted to provide a short discourse on factors considered by formulation scientists when attempting to adopt, stabilize and design a recipe for a vehicle to carry an entomopathogen active and intact to target crop and pest insect. No attempt was made to review all the literature addressing the problems involved. The most important literature has been discussed. The predominant finding has been that pitifully little is known about the proprietary expertise developed by commercial formulators for entomopathogens. Some of that known has been presented. Much work has been devoted to observation of isolated environmental effects on the pathogen and pathogen–adjuvant mixtures rather than to resulting crop protection. Thus, future research must evaluate whether the addition of an adjuvant results in more crop protection. In the final analysis industrialization and grower acceptance of a product is based on the economics of its use.

Future impetus should be on development of formulations, basic and tank mix, which insure thorough coverage, including droplet deposition and distribution on the target crop. Conventional spray systems may or may not have to be redesigned to achieve this (Ignoffo, 1972). For pathogens of aquatic insects, the formulations will have to distribute evenly across the aqueous substrate. For insects less susceptible to control with microbial insecticides because of their biology or feeding behaviour, baits or granule formulations may be necessary to insure pest contact with the pathogen.

Finally, the short shelf-life of some pathogens such as certain fungi and protozoans may inhibit commercialization. Efforts should be made to develop specific formulations which can be generated rapidly with minimal handling and manipulation of the pathogen. These can then be applied immediately to insure virulence.

References

Allen, G.E. and Pate, T.L. (1966) *J. Invertebr. Path.* **8**, 129–131.

Andrews, G.L., Harris, F.A., Sikorowski, P.P. and McLaughlin, R.E. (1975). *J. econ. Ent.* **68**, 87–90.

Angus, T.A. and Lüthy, P. (1971). *In* "Microbial Control of Insects and Mites" (H.D. Burges and N.W. Hussey, eds), pp. 623–628. Academic Press, London and New York.

Becher, P. (1973). *In* "Pesticide Formulations" (W.V. van Walkenburg, ed.), pp. 65–92. Marcel Dekker, New York.

Bell, M.R. and Kanavel, R.R. (1975). *J. econ. Ent.* **68**, 389–391.

Bull, D.L. (1978). *Misc. Publ. ent. Soc. Am.* **10**, 11–20.

Bull, D.L., Ridgeway, R.L., House, V.S. and Pryor, N.W. (1976). *J. econ. Ent.* **69**, 731–736.

Bullock, H.R. (1967). *J. Invertebr. Path.* **9**, 434–436.

Burges, H.D., Hillyer, S. and Chanter, D.O. (1975). *J. Invertebr. Path.* **25**, 5–9.

Cantwell, G.E. (1967). *J. Invertebr. Path.* **9**, 138–140.

Cantwell, G.E. and Franklin, B.A. (1966). *J. Invertebr. Path.* **8**, 256–258.

Couch, T.L. (1978). *Misc. Publ. ent. Soc. Am.* **10**, 3–10.

Frye, R.D., Scholl, C.G., Scholz, E.W. and Funke, B.R. (1973). *J. Invertebr. Path.* **22**, 50–54.

Griego, V.M. and Spence, K.D. (1978). *Appl. Environ. Microbiol.* **35**, 906–910.

Hall, I.M. (1963). *In* "Insect Pathology" An Advanced Treatise (E.A. Steinhaus, ed.). Vol. 2, pp. 477–517. Academic Press, New York and London.

Harper, J.D. (1974). Forest Insect Control with *Bacillus thuringiensis:* Survey of Current Knowledge. Univ. Print Serv. Auburn Univ., Auburn, AL. 64 pp.

Henry, J.E. and Oma, E.A. (1974). *J. Invertebr. Path.* **23**, 371–377.

Hostetter, D.L., Ignoffo, C.M. and Kearby, W.H. (1975). *J. Can. ent. Soc.* **48**, 189–193.

Ignoffo, C.M. (1972). *J. Invertebr. Path.* **19**, 1–2.
Ignoffo, C.M. and Batzer, O.F. (1971). *J. econ. Ent.* **64**, 850–853.
Ignoffo, C.M. and Garcia, C. (1978). *Environ. Ent.* **7**, 270–272.
Ignoffo, C.M. and Hostetter, D.L. (1977). *Misc. Publ. ent. Soc. Am.* **10**, 117–119.
Ignoffo, C.M., Hostetter, D.L. and Smith, D.B. (1976). *J. econ. Ent.* **69**, 207–210.
Ignoffo, C.M., Hostetter, D.L., Sikorowski, P.P., Sutter, G. and Brooks, W.M. (1977). *Environ. Ent.* **6**, 411–415.
Jacques, R.P. (1968). *Can. J. Microbiol.* **14**, 1161–1163.
Jacques, R.P. (1977). *Misc. Publ. ent. Soc. Am.* **10**, 99–116.
Johnstone, D.R. (1973). *In* "Pesticide Formulations" (W.V. van Walkenburg, ed.) pp. 343–386. Marcel Dekker, New York.
Kramer, J.P. (1970). *Acta Protozool.* **8**, 217–224.
Krieg, A. (1975). *J. Invertebr. Path.* **25**, 267–268.
Lewis, F.B. and Rollinson, W.D. (1978). *J. econ. Ent.* **71**, 719–722.
Lewis, F.B., Dubois, N.R., Grimble, D., Metterhouse, W. and Quimby, J. (1974). *J. econ. Ent.* **67**, 351–354.
Maddox, J.V. (1977). *Misc. Publ. ent. Soc. Am.* **10**, 3–18.
Maksymiuk, B. and Neisess, J. (1975). *J. econ. Ent.* **68**, 407–410.
McCoy, C.W. (1978). *In* "Microbial Control of Insect Pests: Future Strategies in Pest Management Systems (G.E. Allen, C.M. Ignoffo and R.P. Jacques, eds) pp. 211–219. NSF-USDA. Univ. Florida Workshop, Gainesville.
McCoy, C.W. and Couch, T.L. (1978). *Dev. Ind. Microbiol.* **20**, 89–96.
McLaughlin, R.E., Andrews, G. and Bell, M.R. (1971). *J. Invertebr. Path.* **18**, 304–305.
Montoya, E.L., Ignoffo, C.M. and McGarr, R.L. (1966). *J. Invertebr. Path.* **8**, 320–324.
Morris, O.N. (1975). *J. Invertebr. Path.* **26**, 199–204.
Morris, O.N. (1977a). *Can. Ent.* **109**, 1239–1248.
Morris, O.N. (1977b). *Can. Ent.* **109**, 1319–1323.
Morris, O.N. and Moore, A. (1975). Chem. Control Res. Inst. Rep. cc-X113.
Neisess, J. (1979). U.S. Dep. Agric. For. Scrv. Res. Pap. PNW 254, 6pp.
Pinnock, D.E., Brand, R.J., Jackson, K.L. and Milstead, J.E. (1974). *J. Invertebr. Path.* **23**, 341–346.
Pinnock, D.E., Milstead, J.E., Kirby, M.E. and Nelson, B.J. (1977). *Misc. Publ. ent. Soc. Am.* **10**, 77–97.
Polon, J.A. (1973). *In* "Pesticide Formulations" (W.V. van Walkenburg, ed.). pp. 143–234. Marcel Dekker, New York.
Raun, E.S. (1968). *Crop Soils Mag.* **20**, 16–18.
Raun, E.S. and Jackson, R.D. (1966). *J. econ. Ent.* **59**, 620–622.
Raun, E.S., Sutter, G.R. and Revelo, M.A. (1966). *J. Invertebr. Path.* **8**, 365–375.
Roberts, D.W. and Campbell, A.S. (1977). *Misc. Publ. ent. Soc. Am.* **10**, 19–76.
Rode, L.J. and Foster, J.W. (1960). *Arch. Mikrobiol.* **36**, 67–94.
Smirnoff, W.A. (1972). *J. Invertebr. Path.* **19**, 179–188.

Smirnoff, W.A. (1974). *Can. Ent.* **106**, 429–432.

Smirnoff, W.A. (1977). *Can. Ent.* **109**, 351–358.

Smith, D.B., Hostetter, D.L. and Ignoffo, C.M. (1978). *J. econ. Ent.* **71**, 814–817.

van Walkenburg, W.V. (1973). *In* "Pesticide Formulations" (W.V. van Walkenburg, ed.), pp. 93–112. Marcel Dekker, New York.

Yendol, W.G., Hamlin, R.A. and Rosario, S.B. (1975). *J. econ. Ent.* **68**, 25–27.

Young, S.Y. and Yearian, W.C. (1974). *Environ. Ent.* **3**, 253–255.

CHAPTER 35

Machinery and Factors that Affect the Application of Pathogens

D.B. SMITH

Bioengineering Research Unit, Building T-12, University of Missouri, Columbia, Missouri, USA,

and

L.F. BOUSE

Pest Control Equipment and Methods Research Unit, College Station, Texas, USA

I. Introduction

Research programs on microbial control of insects have been underway and steadily increasing for over two decades. However, probably $<1\%$ of such research deals with application technology. Ignoffo (1970) and Yearian (1978) stated that insect pathogens have been, and will continue to be in the foreseeable future, applied by systems developed for chemical insecticides. Yearian (1978) emphasized "to my knowledge, no commercial equipment has specifically been designed for use with microbial agents. Further, the literature is practically void of references comparing application of microbial agents with existing delivery systems. For example, of the 28 large-scale field tests with *Baculovirus heliothis* (1966–1972) only one compared application methods. The result of the test was not published. One must wonder if advantage has been taken of application systems currently available." While referring to the application of two viruses, Maksymiuk (1975) suggested that their effectiveness would be improved by more work on spray formulations, application equipment and methodology to increase target coverage and to reduce spray drift and possibly dosage. Falcon *et al.* (1974) indicated that most application equipment and procedures used in cotton did not provide optimum doses, good coverage and maximum penetration of the plant canopy with virus applications. Hall (1974) stated that "the dispersion of pathogens of insects has been touched upon by many authors in their treatments of host–pathogen relationships, but the subject has received very little in-depth attention since the inception of modern insect control pathology, with the result that there is a paucity of information in the literature on the varied mechanisms that bring about such movement." Further, *Bacillus thuringiensis* (*B. t.*) sprayable preparations came into their own because of the development of strains with increased toxicity sufficient to make up for poor coverage obtained from spray applications. Falcon (1971) concluded that the application of bacterial pathogens is one of the most important, yet frequently neglected, aspects in field use; also bacterial preparations are commonly applied with the same equipment as for chemical insecticide applications, with little attention being given to the characteristics of spray droplet or dust particle size, deposit, coverage and drift.

Akesson *et al.* (1971) reported, based on first-hand experience and test reports, that several problems had arisen in attempting economic, large-scale application of these materials with available equipment; thus, engineers should re-examine machines and application techniques for this increasingly important area of work.

Taken collectively, these comments show that insufficient effort has been expended on application equipment and technology to ensure the most efficient and effective application of microbial insecticides.

II. Atomization and Deposition Considerations

Many factors affect atomization in agricultural spray systems, e.g. liquid pressure, orifice size, nozzle design, formulation (liquid viscosity, surface tension, concentration of particulate matter), air velocity relative to the nozzle, and nozzle orientation relative to the ambient air stream. As of now, no single relationship will relate each of these characterisitcs to the spray atomization from all spray nozzle designs. In an excellent discussion about the effects of several properties for a given nozzle design, Orr (1966) demonstrated the effect of several important variables on atomization with a fan-type nozzle, thus, the stress imposed on a liquid sheet during its atomization process is illustrated by the following equation:

$$d_s = 722 \left[\frac{Q\sigma}{\Delta P^{3/2} \times \theta} \right]^{1/3} \tag{1}$$

where d_s is surface mean droplet diameter in microns, Q is volumetric flow rate in litres/h, σ is surface tension in dynes/cm, ΔP is pressure drop through the nozzle in kPa, and θ is fan angle in radians. Equation (1) implies that droplet size is increased by increasing volumetric flow rate or interfacial tension or by decreasing either differential pressure drop through the nozzle or fan angle. Even though liquid viscosity does not appear in Eqn. 1, Fraser and Eisenklam (1956) stated that viscosity is undoubtedly the most influential physical property in changing droplet size; a decrease in viscosity increases uniformity and decreases the size of spray droplets. They demonstrate approximately the effect of viscosity on surface mean diameter by:

$$d_s \propto \mu^{0.2} \tag{2}$$

This equation implies that a small increase in viscosity (μ) in a low-viscosity fluid changes droplet size more than does the same increase in a high-viscosity fluid. Dorman (1952) stated that the effect of viscosity on droplet size has not been carefully investigated, but the droplet size appears to be a function of the viscosity raised to a power < 0.1. If so, a small change in viscosity for low-viscosity fluids is even more dramatic than is indicated by Eqn. (2).

Particles in a spray suspension (concentration = 0.5% w/w) had no effect on spray disintegration if they were wetted by the liquid (Dombrowski and Fraser, 1954). In contrast, unwettable particles in suspension had a marked effect on atomization of a spray sheet. In practice, the droplets produced by water-based suspensions were larger than those produced by oil-based suspensions when applied with the same spray equipment under the same operational conditions (Maksymiuk, 1975). Also water-based formulations usually evaporated faster than oil-based formulations.

Deposition of pesticides also depends on several variables. Some of these are droplet size; evaporation rate of the spray formation; air turbulence; temperature; relative humidity; height of release above the target; size, shape, and location of the target; and horizontal wind velocity. Once the insecticide is on target, several detrimental things can happen, e.g. it can be blown off by winds, washed off by rain, react with inactivating chemicals in or on the biological target or degrade by ultraviolet (UV) radiation or temperature. Because these variables interact, the role of individual variables cannot be separated in terms of obtaining a given quantity and quality of spray deposit on a selected target. For example, if the spray formulation is altered, other equipment or meterological conditions or both may have to be selected to obtain a given result. This selection is further complicated when control of a given insect on one crop requires a different quantity and quality of spray coverage than is necessary for control of that same insect on another crop, a factor observed both for row crops and forests.

Feeding habits of various insects differ. Mobility and feeding activity of a given insect species may dictate a different quantity and quality of spray coverage than those required for another insect species. The above and related reasons emphasize the need for cooperative research teams. Such teams can develop the entire control system (e.g., the complete equipment—formulation—biology—meteorology complex) more efficiently than any one discipline acting alone. The effects of equipment, operational and/or meteorological variables on the penetration and deposition of sprays have been widely discussed from several perspectives, e.g. aerial vs ground applications and row crop vs forest-type canopies (Grim and Barry, 1975; Cramer and Boyle, 1976; Threadgill and Smith, 1975; Matthews, 1977; Johnstone and Huntington, 1977).

III. Bacteria

A. AERIAL APPLICATION

Although aerially—applied B. t. has been used to control a variety of foliage feeding insects, relatively little research has been conducted on the effect of aerial—application—equipment parameters on treatment effectiveness. Most such research has been directed toward forestry usage, rather than toward agricultural usage. Forest application factors that influence both research and operational results include droplet size, application height, airspeed, aircraft type, application rate, spray pattern uniformity, swath width, wind and canopy penetration (Harper, 1974). Boving et al. (1971) stated that "to be effective the microbial pathogen (a virus or a bacterium) must be well distributed on the plant material. Good plant coverage is the key to effective control, particularly in forest spraying. The degree of plant coverage is affected by the droplet size and the volume applied, which in turn is affected by the spray equipment used." Basic requirements of equipment for applying microbial insecticides include protection against mechanical or heat damage to the organisms, adequate agitation to maintain uniform formulation concentration, chemical corrosion resistance, and ease of cleaning to permit removal of chemicals toxic to the organisms. These requirements are the same in both forestry and agriculture (Boving et al., 1971).

Harper (1974) concurred with Boving et al. (1971) in concluding that foliage coverage is of extreme importance in controlling forest insects with B. t. Larvae of the Douglas-fir tussock moth (Orgyia pseudotsugata) were more susceptible than those of the gypsy moth or spruce budworm; however, instances of inadequate coverage of host tree foliage reduced effectiveness. Also, in gypsy moth and spruce budworm tests, variation in foliage coverage caused erratic results, and control was generally improved by an increase in the volume of spray applied.

Field studies on the effectiveness of B. t., as in those with chemical insecticides, often involve comparison of various formulations and application rates of active ingredients, without planned treatments to vary the spray coverage. Although these experiments yield valuable information, efforts to relate spray deposit or droplet density (coverage) to insect mortality are usually inconclusive and difficult to interpret (e.g., Klein and Lewis, 1966; Dewey *et al.*, 1973; and McGregor *et al.*, 1976).

One of the most critical, unanswered questions about spray application technology is "What droplet size or size distribution deposits the material most efficiently?" Conclusions drawn by one researcher with specific plants and pests vary from those drawn by others; there may be as many answers to this question as there are combinations of biological and physical parameters. Dipel® (*B. t.*) at 1.12 kg/ha in molasses and water at 18.7 litres/ha produced excellent foliage protection against the Douglas-fir tussock moth (Stelzer *et al.*, 1975). In contrast, the same rate of Dipel in a spray adjuvant without molasses failed to satisfactorily reduce larval densities or protect foliage. Spray deposits collected on Kromekote® cards indicated that the volume median diameter (VMD) of the spray droplets with molasses was considerably larger (276 μm) than that without molasses (160 μm). This larger droplet size resulted in greater spray recovery near sample fir trees with molasses (30%) than without (16%) and in more than twice as much Dipel on foliage at the midcrown of the trees. Evidently the smaller droplets resulted in greater losses due to evaporation and spray drift.

Conclusions by Stelzer *et al.* (1975) as to the effectiveness of various droplet sizes differed from conclusions of other researchers. Barry *et al.* (1977) studied the effect of particle size on impaction of dry–liquid particles on Douglas-fir needles and western spruce budworm larvae; 87% of the particles on fir needles were ≤ 10 μm diameter and 87% of those on larvae were ≤ 15 μm. Also 26% of the particles on needles and 51% of those on larvae were in the 6-to-10 μm size range even though only about 4% of particles collected on silicone-coated glass slides were in that range. Barry and Ekblad (1978) determined the number and size of *B. t.* and chemical insecticide droplets on coniferous foliage after application by a helicopter with a conventional spray boom and flat-fan nozzles. The application rate for the *B. t.* spray was 18.7 litres/ha. Its VMD from Printflex® cards was 350 μm, and 88% to 94% of the droplets that impacted on foliage needles were < 61 μm in diameter, even though only 32% of the droplets on the cards were < 70 μm in diameter. Meteorological conditions during application may have affected the deposition of small particles, although Stelzer *et al.* (1975) reported maximum wind speeds of < 1 m/s, which is within the range reported by Barry *et al.* (1977) and Barry and Ekblad (1978), whose tests were conducted under stable (inversion) and neutral atmospheric conditions. Inversion means that the temperature near the ground is cooler than the temperature above. Neutral conditions means that the air is not trying to rise or settle.

The low percentage of droplets (4%) in the 6-to-10 μm size range obtained on glass slides by Barry *et al.* (1977) and the 32% in the < 70-μm range obtained on Printflex cards by Barry and Ekblad (1978) may be suspect due to the low collection efficiency of this method of sampling small droplets. The collection efficiency of such horizontal flat targets in a 1 m/s horizontal airstream would be about 32–56% for 140 μm droplets, 14–27% for 80 μm droplets, and 1–2% for 20 μm diameter droplets (Miles *et al.*, 1975).

Wedding *et al.* (1978) simulated the aerial spraying of pesticides onto a Douglas-fir canopy in a wind tunnel. Their objectives were to determine: (1) the droplet size range most effective in covering all areas of the foliage; (2) the upper and lower limits of droplet size that reduce waste of pesticide and drift losses while maintaining effective foliage coverage under various wind conditions; and (3) the collection effectiveness of tussock moth larvae placed on the foliage. They concluded that the lower range of particle size to use in effective spraying is limited more by drift considerations and adequate kill mass than by inefficient deposition processes. They suggested a lower droplet-diameter limit of 35 μm and an upper limit ("for efficient deposition, and adequate penetration capability") of 100 μm. Their findings support the field results of Barry *et al.* (1977) and Barry and Ekblad (1978) but are inconsistent with those of Amsden (1962), who maintained that the dynamic catch of droplets down to 100 μm in diameter is about 80% in wind speeds of 2.2–6.7 m/s and that the catch for smaller drops falls off rapidly, so the expected increase in coverage does not occur. Amsden concluded that sprays should comprise droplet sizes from 80 to 120 μm to provide wide swaths and uniform coverage. He further stated that evaporation of sprays having a Sauter mean diameter of 80 to 100 μm can seriously reduce coverage and, sometimes, cause poor biological results. Evidence of the potential for spray loss by evaporation of small droplets of aqueous or other volatile carriers is provided by Eisner *et al.* (1960): 80-μm-diameter water droplets would evaporate completely in 33.7 sec at 20°C and 80% relative humidity (r.h.) or in 9.5 sec at 20°C and 50% r.h. From these data, Amsden (1962) concluded that a 100-μm water droplet would evaporate completely while falling 6.7 m at 20°C and 80% r.h. or 1.8 m at 30°C and 50% r.h. Considerations of deposit efficiency, drift losses, coverage and evaporation indicate the dilemma facing spray equipment personnel who must select the most effective application equipment and methods for the aerial application of bacteria.

Maksymiuk and Orchard (1974) determined the biological swath width, VMD, and mortality of Douglas-fir tussock moth and western spruce budworm larvae for two formulations (Dipel and Thuricide) of *B. t.* applied by a small fixed-wing aircraft. They used a Piper Pawnee aircraft, 8010 (Spraying Systems Co.) flat fan nozzle tips and diaphragm-cutoff nozzle bodies. The nozzles were oriented to direct the spray sheet forward and down 45° to the line of flight. Increasing the spray pressure from 276 to 414 kPa decreased the VMD of Dipel spray from 321 to 266 μm and widened the swath width because of the finer atomization. With the 266-μm VMD spray, the mortality of tussock moth larvae averaged 88% over a swath width of 30.5 m and 78% over a width of 61 m for Dipel applied at a 15-m height with a crosswind of 0.9 m/s. For Thuricide,® the same increase in spray pressure decreased the VMD from 321 to 199 μm. Over 30.5-m and 61-m swath widths, 94% and 87% of the tussock moth larvae died and 100% and 98% of the budworm larvae, respectively. The height was 15 m in a crosswind of 0.2 m/s.

They also sprayed plots in an oak forest to determine droplet size, distribution vertically on tree crowns, horizontally all around the lower portion of tree crowns and on horizontal transects on the forest floor. A Grumman AgCat aircraft with spinning nozzles sprayed at a height of about 14 m in a swath width of 30.5 m. The droplet size was 335 μm VMD for Dipel and 351 μm for Thuricide. No significant differences were found for the droplet size or distribution of spray. However, more spray was consistently deposited on the upper portion of the trees than around their sides. The upper tree crowns collected only about 10% of the 18.7 litres/ha released. The lower crowns, sample plates placed beneath the trees and sample plates placed along transections on the forest floor collected about 8%, 9% and 9% respectively, of the amount released. The authors pointed out the need for more deposit of spray in the forest canopy.

Morris and Armstrong (1973) evaluated aerial applications of B. t. and mixtures of B. t. plus low dosages of an organophosphorous insecticide for the control of spruce budworm on white spruce and balsam fir. A small, fixed-wing agricultural aircraft, with Micronaire® AU 3000 (Britten-Norman) spray units, applied 61-m-wide swaths. At ground level, deposits varied from 13% to 34% of the amount released, based on dye tracer samples, and the droplet density varied from 12 to 37 droplets/cm^2. Estimates of deposits of viable bacterial spores showed no correlation with spray deposits measured by the dye tracer method; thus, spore deposit alone is not an accurate criterion for deposit measurement.

Further evidence of this lack of correlation was provided by Morris and Armstrong (1974). Mixtures of B. t. and an organophosphate insecticide were applied to white spruce and balsam fir trees by a small fixed-wing aircraft with Micronaire AU 3000 spray units, calibrated to deliver 50 to 100-μm-diameter droplets. Swath widths and height of flight above tree tops were both 61 m. Droplet density ranged from 20 to 43 droplets/cm^2 and spray deposits ranged from 12.5% to 40% of the amount released. The deposit of viable spores was not related to spray deposit, and, in one instance, 25% fewer spores reached the ground surface from an application of 30×10^9 IU/ha than from an application of 20×10^9 IU/ha.

Morris (1977) investigated the relationship between size of aerially deposited spray spots on millipore filters and the number of bacterial spores and crystals deposited at ground level. Although there was some variation between individual droplets, the average spore-to-crystal ratio of 1:1, which was the same as for the tank mix, was not affected by spot size. The relationship between spot size and number of spores and crystals/droplet was curvilinear and not affected by spray application rate. All droplets collected at ground level contained some active ingredient. Based on colorimetric analysis of deposits on glass plates, 22% of the volume emitted from the aircraft reached ground level. Data obtained from a commonly used agar plate technique greatly underestimated B. t. deposits when spray droplets were captured directly on the agar. Also, dye tracer methods were found to be more accurate than the plate technique.

B. GROUND APPLICATION

B. t. has been applied with ground equipment as water suspensions, oil—water emulsions, powders extended with clay, and granular formulations. It has been suggested that a strong correlation exists between the latitude of a location and the effectiveness of dusts as opposed to water sprays (National Academy of Science Publication, 1969). Hostetter and Ignoffo (in press) summarize the use of entomopathogens to control cabbage loopers and the wide range of equipment used for their application. Kennedy and Oatman (1976) applied *B. t.* and pirimicarb to broccoli plants with a tractor-mounted, row-crop sprayer with 800067 flat fan nozzles (Spray Systems Co.) that delivered 56 litres/ha at 690 kPa. Pirimicarb is specific against aphids and does not kill caterpillars. They concluded that weekly applications of Dipel at 1.12 kg/ha plus pirimicarb at 0.28 kg/ha during the head formation stage was as effective as methomyl at 0.50 kg/ha in preventing caterpillar damage.

Laboratory performance specifications for *B. t.* on soybeans were studied by Smith *et al.* (1977a). They used a droplet generator to deliver droplets of nearly homogeneous diameters of 90, 180, and 270 μm. Technical application rate, droplet size, deposit density (number of droplets/cm^2), and concentration accounted for 43%, 22%, 16% and 19% respectively of the variation in the insect deaths. When application rate was eliminated as a variable, the droplet size—density—concentration interaction was significant and there were several combinations of these three variables that produced desirable cabbage looper mortalities. In field tests (Smith *et al.*, 1977b), TX-1 (Spraying Systems Co.) nozzles operated at 552 kPa, produced deposits of greater volume and density at the top, middle and bottom of soybean plants than TX-4 nozzles operated at 372 kPa (Fig. 1). However, fewer loopers were killed with the TX-1 nozzles than with the TX-4 nozzles. Since this result did not seem reasonable, damage to the *B. t.* endotoxin crystals was investigated in the laboratory. Bioassays indicated that the crystals were not damaged by nozzle shear or pressures up to 635 kPa, so loss of crystals was examined (Fig. 2). A possible loss of crystals out of spray droplets was indicated when sprayed at 552 kPa versus 138 kPa. The 11% represents a relative loss and possibly not an absolute loss since some crystals could be lost at 138 kPa also. While further verification of the loss phenomenon is required, these results suggest that it may be advantageous to use thickeners but not surfactants. Further study should be conducted to attempt to relate particle loss to physical properties of the spray suspension. Fast (1976) calculated that, for aerial spruce budworm control, up to 97% of the active material was lost from droplets between the time *B. t.* was mixed in the spray tank and the time the droplets landed on their target.

Frye *et al.* (1976) applied *B. t.* on Siberian elm with a cold fogger, a thermal fogger, and a conventional hydraulic sprayer. They stated that "less water is required with foggers, and they were easy to handle when in use. The foggers required less directing of the spray than did the hydraulic sprayer."

Patti and Carner (1974) used a modified sprayer to apply a cottonseed oil bait and a conventional hydraulic sprayer to apply several commercial

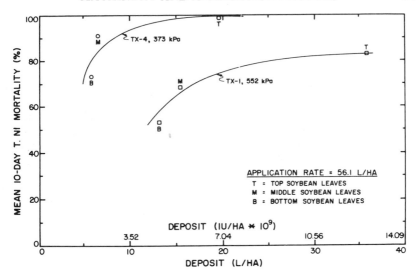

Fig. 1. Effect of two nozzle–pressure combinations on volumetric deposit and on 10-day cabbage looper mortality on soybean leaves with *Bacillus thuringiensis.*

formulations of *B. t.;* Dipel in the cottonseed oil bait killed more caged larvae than other formulations of *B. t.* Stacey *et al.* (1977a) applied the nuclear poly-hedrosis virus (NPV) of *Heliothis* spp. both as baits and sprays and *B. t.* as a bait to control *Heliothis* species on cotton. All yields were significantly lower than those on plots receiving the insecticide standard (toxaphene–methyl parathion). Virus-sprayed plots yielded significantly more cotton than plots treated with dry formulations of virus in cornmeal or pulverized wheat. The virus–cottonseed meal bait produced yields comparable to those with the virus spray. The general conclusion was that dry bait formulations of microbials are not as effective as sprays for insect control, even when gustatory stimulants are added to the baits. Beegle *et al.* (1973) used three solid-cone nozzles per row to apply 905 litres/ha at 414 kPa. One nozzle on each side of the soybean rows was directed upward to improve leaf coverage on the under surfaces. Six days after treatment, there were no significant differences in green cloverworm control obtained with a granulosis virus, *B. t.,* and five out of six chemical insecticides.

Forest defoliators have been controlled by using a wide variety of equipment including hydraulic sprayers, mist blowers, backpack sprayers, dusters, and street pumper type fire engines (Harper, 1974). Many instances of excellent forest insect control were reported with the above diversity of equipment, yet the coverage obtained with the various sprayers was likely not equal. This suggests that the application rate possibly could be reduced if we knew the optimum spray deposit parameters and how to obtain optimum deposists consistently.

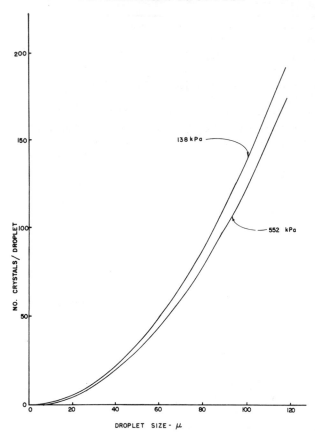

Fig. 2. Possible loss of *Bacillus thuringiensis* crystals from spray droplets caused by increasing the pressure (kPa) through a TX-1 nozzle.

A portable duster was used to apply 2% Dipel dust at 0.45 kg/almond tree to control navel orangeworms (Kellen *et al.*, 1977). For comparison, aqueous Thuricide HPC and Biotrol® XK, diluted 1:32, was sprayed at 1.3 litres/tree with an Ag Tec Spray-All® orchard sprayer. The best control (44.8% damage reduction) was obtained with four applications of Thuricide HPC. Addition of a feeding stimulant (Mo-Bait®) did not increase effectiveness. Although there was no experimental evidence, they suggested that smaller droplets emitted at higher velocities might give more consistent control.

Aerosol applications of microbial insecticides have been studied in California for several years (Falcon *et al.*, 1974). Two cold aerosol foggers, one forming droplets between 10 and 40 μm VMD and the other between 30 and 90 μm produced very wide swaths under temperature-inversion conditions at low wind velocities. The number of viable spores deposited per unit area of cotton leaf decreased rapidly with increasing distance from the atomizer. Fogger equipment (type not specified) provided more uniform deposits from top to bottom of cotton plants than an aerial application. Crude cottonseed oil as a carrier measurably increased deposition of both *B. t.* and NPV and enhanced insect mortality when applied as an aerosol of droplets between 20 and 90 µm VMD (Sorenson, 1977). For an estimated spray droplet size of 40–50 µm VMD, the cumulative percent recovery was 46% for a distance of 304 m downwind.

Burges and Jarrett (1979) studied the application of *B. t.* to tomato crops in greenhouses by thermal fogging equipment. Heat had little or no effect in pulse-jet type fogging machines, but an exhaust-type machine almost completely inactivated both the spores and crystals of *B. t.* Fog applied during windy conditions was blown out of the greenhouse, even though doors and ventilators were closed. Jarrett *et al.* (1978) showed that less *B. t.* was wasted by a controlled droplet applicator (40%) or a thermal fogger (45%) than by a conventional high-volume, hydraulic sprayer (67%) when the *B. t.* was applied on chrysanthemums. However, the vertical distribution provided by the controlled droplets and thermal fogs was not as good as that provided by the high-volume sprays. The controlled droplets and thermal fogs gave poor coverage on the under surfaces of leaves, whereas the high-volume sprays deposited half as much *B. t.* on the lower surfaces as on the upper surfaces. The fog applications were made from above the plants, whereas the controlled droplets and hydraulic sprays were released both from above and below the plants.

C. SPOT RELEASES

Very little research has been done with point or spot releases as opposed to conventional prophylactic dissemination of microbial insecticides. Milky disease bacteria were applied as spot treatments to control grubs of the Japanese beetle, in turf in the eastern United States, with equipment as simple as a rotary hand corn planter modified to inject about 2 g of dry spore powder each time it was tripped (Hall, 1974). An application rate of 1.9 kg of spore powder/ha was adequate when injected every 3.1 m in a checkerboard fashion (Hawley, 1952). These bacteria can spread but Hall (1974) stated that *B. t.* has almost no ability to disperse on its own or with the help of environmental factors. Due to the paucity of reports, additional research on spot releases for selected entomo-pathogens may be needed.

IV. Virus

A. AERIAL APPLICATION

Boving *et al.* (1971) reported that the lack of satisfactory spray equipment has been a major problem in the application of viruses. They evaluated helicopter spray systems for application of virus in an aqueous formulation. A low-volume spray system was modified to overcome excessive heat buildup in the spray pump, which was believed to injure the virus. Flat-fan nozzles applied 9.5 and 19 litres/ha over a 30.5 m, swath width. The VMD was 355 µm for both rates. Four spinning nozzles applied 1.9 litres/ha with a swath width of 30.5 m and a VMD of 106 μm.

A biological evaluation of the above helicopter sprays was made by Maksymiuk et al. (1968a). They used NPV to control the Douglas-fir tussock moth. Rates of 125 × 10⁹ polyhedral inclusion bodies (PIB)/ha were applied to potted trees at both the above low volumes and at 62.5 × 10⁹ PIB/ha when ultra-low volume was used. All three volume rates were equally effective. The swath width could be increased from 30.5 to 61 m for all three volumes without significantly reducing larval mortalities.

Further tests with 125 × 10⁹ PIB/ha applied at the three volume rates confirmed that they were equally effective (1.9 litres/ha, 89% larval mortality; 9.5 litres/ha, 92% and 19 litres/ha, 81%; Maksymiuk et al., 1968b). The authors concluded that coarse atomization (355 μm VMD) was less efficient than fine atomization (106 μm VMD) because, with the coarse spray, more large droplets missed foliage and fell to the ground. Although the recovery of PIB on sample cards at ground level under trees and along a roadway was 5 × greater for the 9.5 litre/ha low volume than for the 1.9 litre/ha ultra-low volume, the mean PIB collected per 50 fir needles was also greater (by 37%). The authors indicated that less fine than coarse spray was recovered due to drift of fine droplets beyond the target area and to limits on the collection efficiency of small droplets. However, they reasoned that as the number of PIB reaching the ground increases (as with the coarse spray), the number deposited on the foliage decreases. In view of the evidence of PIB deposits on fir needles, this reasoning appears questionable. This difficulty in interpretation of data points out the need for improved sampling methods to assess the effect of application parameters on application efficiency; clearly, deposits collected on sampling devices at ground level do not quantify material deposited on tree foliage.

Yendol et al. (1977) reduced gypsy moth egg masses in an oak forest by aerial application of NPV. A fixed-wing aircraft with six spinning nozzles applied 10¹² PIB/ha in 19 litre/ha of finished spray. Post-treatment, the egg mass density in virus-treated plots was ca. 38% of that in untreated check plots, and defoliation was 50%–75%, but untreated checks were almost completely defoliated.

Smirnoff et al. (1976) tested the potential of aerial application of NPV to control the European skipper Thymelicus lineola in a heavily infested timothy crop. An aqueous suspension of 5 × 10⁶ PIB/ml was applied at 19 litres/ha by a fixed-wing aircraft with a Micronair spraying system. Symptoms of viral infection were observed 7 days later and 100% larval mortality after 27 days. At the end of the season, defoliation in the treated crops was 50%, compared to 100% in an untreated check plot.

Specialized mixing and application equipment are needed to handle virus effectively, to keep it in a uniform suspension, and to give optimal droplet size spectra for maximizing target coverage and minimizing spray drift (Maksymiuk, 1975). Use of such equipment can result in reduced field dosage and increased insect control. All spray mixing, loading and application equipment must be cleaned before and after use of microbial agents.

B. GROUND APPLICATION

The amount of research devoted to the application of viruses is less than that for bacteria. Stacey *et al.* (1977b) reported that the mortality of *Heliothis zea* was greatest when larvae were released on plants on which the fruiting structures or lower leaf surfaces had been virus treated. Their data suggest that placement of the virus is an important consideration. Paradoxically, they also reported that nozzle arrangement and orientation did not appear to influence the virus efficacy when directed sprays were used.

Akesson *et al.* (1971) studied downwind deposits from two aerosol foggers. One produced droplets which were ca. 10 μm VMD and the other equipment produced droplets which were from 30 to 90 μm VMD. Flow rates through one machine significantly influenced the size of droplets. Based on downwind spray deposits, they concluded that the aerosol machines could be used to apply viruses if there was a temperature inversion at low wind velocity. A fluorescent dye (type unspecified) added to the spray suspension appeared to damage virus. Tests at Columbia, Missouri (data not reported), indicated that Brilliant Sulfo Flavine fluorescent dye (<0.5%w/w) did not harm *B. t.* or NPV. Yendol *et al.* (1977) applied NPV twice with a truck-mounted mist blower to control the gypsy moth. Egg masses were reduced by 90% in the treated area, compared with a 65% increase in the untreated area.

Stacey *et al.* (1980) compared a mist blower (Span Spray®) with conventional hydraulic equipment using TX-4 nozzles at 207 kPa for the control of *Heliothis* sp. larvae on cotton. Varied spray volumes between 9.4 and 93.5 litres/ha with both machines caused no difference in cotton yields, although the mist blower gave better plant coverage. When 3.0% (w/w) sucrose was added to an NPV (Elcar®) suspension, the resulting cotton yields were comparable to those obtained with an insecticide standard.

Smith *et al.* (1977a) developed laboratory performance specifications for an NPV suspension used to control cotton bollworm larvae on soybeans. They found that the technical application rate, droplet size, deposit density and concentration accounted for 41%, 22%, 30% and 7% respectively, of the variation in bollworm larval mortality. After the effects of technical application rate were eliminated, the droplet size–density and droplet size–concentration interactions were significant. The combination of a small droplet size (90 μm), a high concentration (12.7×10^9 PIB/litre) and a large density (> 35 droplets/cm^2) was superior to other combinations.

The effect of equipment and formulation on the biological–volumetric deposit efficiency and the UV degradability of NPV (Elcar) has been studied (Smith *et al.*, 1978). The bioassay (bollworm)–volumetric deposit data indicated that a 0.003% (w/w) NPV suspension was better when applied with TX-4 nozzles at 373 kPa than when applied with TX-1 nozzles at 552 kPa. Even though no attempts were made to establish whether PIBs were lost from spray droplets during the atomization–transport process, the results suggested that PIBs may have been lost. Aqueous suspensions with 0.5% polyvinyl alcohol (PVA) alone were better than with a 45.3% mineral oil, 0.5% PVA plus 2%

Shade® or 0.01% Triton® CS-7 when applied with TX-1 nozzles at 552 kPa. Also, 0.5% PVA and 0.5% keltose (a refined algin) plus 3% sucrose suspensions were superior to either a 0.5% PVA plus 3% sucrose or a 0.5% keltose suspension applied with TX-1 nozzles at 552 kPa (2% Shade in all formulations). Since each formulation may alter the droplet size distribution produced by a given atomizer operating at a given pressure, these types of tests are really droplet size-formulation tests, as opposed to formulation or droplet-size tests, *per se*. Combination of the bioassay—volumetric deposit with the degradation data, indicated that the PVA and keltose—sucrose suspensions (2% Shade in both) applied with a TX-1 nozzle at 552 kPa were superior to all other equipment—formulation combinations tested.

Chapman and Ignoffo (1972) studied the effect of dosage and spray volume of an NPV on control of *Heliothis* sp. in cotton. Doubling the volume of spray from 94 to 187 litres/ha provided control equal to doubling the dosage from 99 to 198 larval equivalents/ha. A suspension containing 40 larval equivalents of virus, 0.2% pyrethrum and 0.1% piperonyl butoxide was as effective as doubling the dosage or doubling the spray volume per hectare. Bell and Kanavel (1977) used a backpack sprayer fitted with fan-type nozzles (nozzle size and pressure not specified) to apply both aqueous suspensions and a cottonseed oil-flour bait formulation for pink bollworm and *Heliothis sp.* control in cotton. They found that the bait treatment was superior to both the aqueous suspension and an untreated control. These tests represent another example of testing a droplet size—formulation interaction rather than testing formulations or droplet size distributions independently.

V. Assessment of Current Technology

The results obtained at various locations have varied as much as the assessment methods and techniques used. Comparison of results at two or more locations is difficult even when the same insect—crop—microbial insecticide combination was studied. This difficulty is not surprising considering the great number of combinations of application variables involved, e.g. pesticide tank-mix concentration, volumetric application rate, atomizer height above crop, initial droplet sizes and vertical velocities, relative humidity, atmospheric stability, horizontal wind velocity, viscosity and surface tension of spray suspension, solar degradability, and weather (e.g. temperature or rainfall) steadfastness. For sprayers that carry the spray in a stream of air, the list of potentially important variables is further compounded (and possibly confounded) since the machine's air velocity affects droplet size distribution, deposition efficiency on biological targets and, possibly, the loss of particles from spray droplets. With aerial application, the height of flight, aircraft-wake/spray-dispersion interaction, airspeed/nozzle-orientation/spray-atomization interaction, and swath spacing accuracy are additional factors that all affect uniformity of spray deposits and foliage coverage.

Some general trends emerge from the available literature. From the biological perspective, insect mortality is usually enhanced by using droplets of 100 to 150 µm diameter rather than larger ones if the actual technical application rates (units of microbial agent/unit area) are equal. The above statement alludes to two potential problems. The first is that, as droplet sizes decrease, volumetric deposit rate (VDR) may or may not decrease, depending upon several types of variables; e.g. equipment, spray formulation, operating and meteorological conditions. Meteorological variables have greater influence on "drift" than "forced" (e.g. over-the-top or directed) sprays.

The second problem is the loss/or redistribution of particles (e.g. clay, talc, crystals, spores, or PIB) from spray droplets. As stress, e.g. increased pressure or decreased orifice size, on the atomizing sheet is increased, loss/or redistribution of particles from droplets of the same size may increase. Thus, even though the VDR could be held constant by using reduced droplet sizes, the insect mortality could be impaired by particle losses from spray droplets. The particle loss phenomenon suggests that the formulation–equipment interaction may be important in the application of suspensions. Some evidence suggests that increasing the viscosity of a suspension (tank mix) improves its efficacy. Relationships between physical properties of a spray, particle losses/redistribution, and efficacy have not been studied extensively. Knowledge gain about liquid physical properties/particle control/insect efficacy has been slow, partly because physical properties of the suspension and droplet sizes tend to be confounded, so assessment of the effect of each variable on insect control is difficult. Also, the formulation can influence the amount of technical material an insect population will ingest in a given period of time.

Criteria for optimizing "drift" sprays suggest operation during inversion (stable) conditions when the horizontal wind velocity is between 0.6–2.0 m/s. Sufficient data are not available to specify required distances between successive passes for adequate control of a given insect or to specify when to use 30-to-90-μm initial diameter droplets of low volatility.

Laboratory studies have shown that addition of a surfactant may reduce the effectiveness of a formulation, probably because of insect repellency or increased particle loss/redistribution due to reduced surface tension of the suspension.

Limited work has been performed on the use of dusts for agronomic, forest, or point (spot) releases. Reduced deposition efficiency of dusts (relative to sprays) probably accounts for their infrequent use. Dusts and baits usually withstand solar degradation better than commercial tank-mixed spray suspensions. However, differences in such factors as crop yield or tree growth usually have not reflected the improved solar protection. Point releases have been successful in a few instances, predominantly with turf insect control.

The nature of the biological target may influence the level of insect control obtained. Obvious factors that influence VDR are the size, shape and orientation of targets. In addition, significant differences in mortalities of a given insect species have been reported after the spraying of crops or trees of different species that were geometrically similar. Since physical assay and bioassay data were not recorded simultaneously, it is uncertain that reduced mortality was due to differences in plant species, even though such a conclusion seems plausible.

In more than 95% of the microbial application work reported, only the biological responses were discussed. Information on these responses is good since it says, for example, machine-formulation combination "A" is superior to another combination "B". However, it does not answer the question "Why?" That is, did combination "A" produce larger VDR's than combination "B" and, thus, larger insect mortalities? Or did combination "A" produce the same VDR's as "B" but more efficacious deposits? More physical assay and bioassay data for various equipment–formulation–crop combinations are required to guide the development of equipment optimally suited for application of microbial insecticides. Users will not be able to separate equipment and formulation effects in the foreseeable future.

VI. Recommendations

Recommendations listed below are divided into those at the administrative level and others intended to stimulate the progress of scientists and engineers actively involved in research. All of these recommendations are offered as guidelines for optimizing the rate of research progress rather than as a comprehensive list for microbial application research methodology.

A. ADMINISTRATIVE SUPPORT

1. Bring scientists and engineers together into teams with the primary responsibility of developing optimal application systems for microbial insecticides. Desirable disciplines for the inclusion in teams are engineering, entomology, formulation chemistry, meteorology and plant physiology.

2. Provide sufficient equipment, facilities and support personnel to enhance the rate of technical development. Some equipment (e.g., inflight particle size scanners or electron microscopes) is expensive, but crucial, for developing, basic microbial application technology. Newly available equipment now permits researchers to emphasize the in-depth technology aspects of research and to reduce the historical type of "spray-and-look" efforts.

B. APPLICATION RESEARCH

The research areas listed below should speed development of application technology for entomopathogens for insect control.

1. Develop suitable physical assay and bioassay methods and microbial formulations where they are lacking for some insect–crop–entomopathogen combinations.

2. Develop application systems that will provide optimal efficiency in killing insects. The logical order of development is first to determine the optimum formulation for the situation being studied; then, for sprays, determine the best droplet size distribution for this formulation, and finally, select the best volumetric application rate/entomopathogen concentration for this combination. The amount of pathogen should be held constant for all volume-concentration combinations tested.

3. Optimize the application system developed in item 2 by maximizing deposits on plant surfaces (e.g., leaves, stems, needles) under a range of ambient conditions. The "best" system in item 2 may or may not be the same as the one selected in item 3.

(a) Determine the spray droplet or dust-particle-size distribution that will maximize volumetric deposit rates on the targets selected.

(b) Study operating conditions, other than size distribution, that will increase the magnitude and, if possible, the uniformity of the deposits.

4. Select the "optimum" system by collectively considering results from items 2 and 3. This "optimum" system should be compared with other systems historically or intuitively regarded as effective. These tests should be conducted under "field" conditions. Such results may suggest additional work in items 2 or 3 above.

5. Some unique application possibilities should be explored in detail. For example, moths could be attracted to central points in a field, exposed to spray or dust, and then released to spread the entomopathogen over foliage and egg masses; or other critical locations or fields could be spot treated, with dependence on meteorological variables for dispersion of the microbial insecticides.

References

Akesson, N.B., Borgwardt, R. and Yates, W.E. (1971). *Am. Soc. agric. Engrs.*, Paper 71–151, St. Joseph, MI.

Amsden, R.C. (1962). *Agric. Aviation* 4, 88–93.

Barry, J.W. and Ekblad, R.B. (1978). *Trans. ASAE*, 21, 438–441.

Barry, J.W., Ciesla, W.M., Tysowsky, M., Jr. and Ekblad, R.B. (1977). *J. econ. Ent.* 70, 387–388.

Beegle, C.D., Pedigo, L.P., Poston, F.L. and Stone, J.D. (1973). *J. econ. Ent* 66, 1137–1138.

Bell, M.R. and Kanavel, R.F. (1977). *J. econ. Ent.* 70, 625–629.

Boving, P., Maksymiuk, B., Winterfield, R.G., and Orchard, R.D. (1971). *Trans. Am. Soc. agric. Engns.* 14, 48–51.

Burges, H.D. and Jarrett, P. (1979). Proc. 1979 Br. Crop Prot. Conf. Pests Dis. pp. 433–439. British Crop Protection Council, Croydon.

Chapman, A.J. and Ignoffo, C.M. (1972). *J. Invertebr. Path.* 20, 183–186.

Cramer, H.E. and Boyle, D.G. (1976). *In* "Pesticide Spray Application Behavior and Assessment: Workshop Proceedings". U.S. Forest Serv. Tech. Rep. PSW-15/1976.

Dewey, J.E., McGregor, M.D., Marsalis, R.L., Barry, J.W., Williams, C.B. and Ciesla, W.M. (1973). USDA, Forest Serv., State and Private For., Missoula, MT. Rep. No. 74–10.

Dombrowski, N. and Fraser, R.P. (1954). *Phil. Trans.* 247, 101–130.

Dorman, R.G. (1952). *Br. J. Appl. Phys.* 3, 189–192.

Eisner, H.S., Quince, B.W. and Slack, C. (1960). Discussions of the Faraday Society No. 30, 86–95.

Falcon, L.A. (1971). *In* "Microbial Control of Insects and Mites" (H.D. Burges and N.W. Hussey, eds), pp. 67–95. Academic Press, London and New York.

Falcon, L.A., Sorenson, A. and Akesson, N.B. (1974). *Calif. Agric.* 28(4), 11–13.

Fast, P.G. (1976). *Bi-mon. Res. Notes, Can. For. Serv.* 32, 21, and 27.

Fraser, R.P. and Eisenklam, P. (1956). *Trans. Inst. Chem. Eng.* 34, 294–319.

Frye, R.D., McMahon, K.J., and Weinzierl, R.A. (1976). *N.D. Farm Res.* 33, 21–25.

Grim, B.D. and Barry, J.W. (1975). U. S. Army Dugway Proving Ground Rep. No. DPG-FR-C625A, U. S. Army and U. S. Forest Service, Dugway, UT.

Hall, I.M. (1974). *Proc. Summer Inst. Biol. Control Plants, Insects Dis.*, pp. 591–598. University Press of Mississippi.

Harper, J.D. (1974). A report to Abbott Laboratories, N. Chicago, IL. Dec. 1974. 64 pp.

Hawley, I.M. (1952). *In* "Insects–The Yearbook of Agriculture". pp. 394–401. U.S. Dept. Agric.

Hostetter, D.L. and Ignoffo, C.M. (In press). *In* "Monograph on *Trichoplusia ni* (Hübner)." (G.L. Green, ed.). *Fla. Agri. Exp. Stn. Monogr. Ser.* Univ. Fla. Gainesville, FL.

Ignoffo, C.M. (1970). *Proc. Tall Timbers Conf. Ecol. Anim. Control Habitat Mangt.* 2, 41–57.

Jarrett, P., Burges, H.D. and Matthews, G.A. (1978). Proc. Symp. Controlled Drop Application," 75–81. British Crop Protection Council, Croydon.

Johnstone, D.R. and Huntington, K.A. (1977). *Pestic. Sci.* 8, 101–109.

Kellen, W.R., Hunter, D.K., Lindegren, J.E., Hoffmann, D.F. and Collier, S.S. (1977). *J. econ. Ent.* 70, 332–334.

Kennedy, G.G. and Oatman, E.R. (1976). *J. econ. Ent.* 69, 767–772.

Klein, W.H. and Lewis, F.B. (1966). *J. For.* 64, 458–462.

Maksymiuk, B. (1975). *In* "Baculoviruses for Insect Pest Control: Safety Considerations" (Summers, M., Falcon, L.A., and Vail, P., eds) pp. 123–128. American Society for Microbiology, Corvallis, OR.

Maksymiuk, B. and Orchard, R.D. (1974). U. S. Dep. Agric. For. Serv., Res. Paper PNW-183, 13 pp., Portland, OR.

Maksymiuk, B. and Orchard, R.D. (1975). U. S. Dep. Agric. For. Serv., Res. Paper PNW-246. 11 pp., Portland, OR.

Maksymiuk, B., Boving, P.A., Orchard, R.D., and Winterfield, R.G. (1968a). *Prog. Rep., Pacif. NW For. Range Exp. Stn., For. Sci. Lab.*, Corvallis, OR, and *Agric. Eng. Res. Div.*, Forest Grove, OR.

Maksymiuk, B., Boving, P.A., Orchard, R.D. and Winterfield, R.G. (1968b). *Prog. Rep. Pacif. NW For. Range Exp. Stn., For. Sci. Lab.*, Corvallis, OR, and *Agric. Eng. Res. Div.*, 33 pp. Forest Grove, OR.

Matthews, G.A. (1977). *Pestic. Sci.* 8, 96–100.

McGregor, M.D., Hamel, D.R. and Lood, R.C. (1976). U. S. Dep. Agric. For. Serv., State Private For. Missoula, MT Rep. 76–11. 17 pp.

Miles, G.E., Threadgill, E.D., Thompson, J.F. and Williamson, R.E. (1975). *Trans. Am. Soc. agric. Engns.* 18, 74–78.

Morris, O. N. (1977). *Can. Ent.* **109**, 1319–1323.

Morris, O.N. and Armstrong, J.A. (1973). *Chem. Control. Res. Inst.* Ottawa, Ontario, Inf. Rep. CC-X-61. Dec. 1973. 24 pp.

Morris, O.N. and Armstrong, J.A. (1974). *Chem. Control Res. Inst.* Ottawa, Ontario. Rep. CC-X-71. Oct. 1974. 25 pp.

Nat. Acad. Sci. Publ. (1969). *In* "Insect-Pest Management and Control," Vol. 3 : 165–195. Publ. No. 1695.

Orr, C., Jr. (1966). "Particulate Technology". 562 pp. London MacMillan.

Patti, J.H. and Carner, G.R. (1974). *J. econ. Ent.* **67**, 415–418.

Smirnoff, W.A., McNeil, J.N. and Valero, J.R. (1976). *Can. ent.* **108**, 1221–1222.

Smith, D.B., Hostetter, D.L., and Ignoffo, C.M. (1977a). *J. econ. Ent.* **70**, 437–441.

Smith, D.B., Hostetter, D.L. and Ignoffo, C.M. (1977b). *J. econ. Ent.* **70**, 633–637.

Smith, D.B., Hostetter, D.L. and Ignoffo, C.M. (1978). *J. econ. Ent.* **71**, 814–817.

Sorenson, A.A. (1977). Unpubl. Ph.D. Diss. Univ. Calif., Berkley, 158 pp.

Stacey, A.L., Yearian, W.C., and Young, S.Y. (1977a). *Arkansas Farm Res. Bull.* **26**, 3–4.

Stacey, A.L., Young, S.Y. and Yearian, W.C. (1977b). *J. Ga. ent. Soc.* **12**, 167–173.

Stacey, A.L., Yearian, W.C., Young, S.Y., Luttrell, R.G. and Matthews, E.J. (1980). *J. econ. Ent.* **15**, 365–372.

Stelzer, M.J., Neisess, J. and Thompson, C.G. (1975). *J. econ. Ent.* **68**, 269–272.

Threadgill, E.D., and Smith, D.B. (1975). *Trans. Am. Soc. agric. Engns*, **18**, 51–56.

Yearian, W.C. (1978). *In* "Microbial Control of Insect Pests: Future Strategies in Pest Management Systems (Allen, G.E., Ignoffo, C.M. and Jaques, R.P. eds), pp. 100–110. Univ. Florida, Gainsville.

Yendol, W.G., Hedlund, R.B., and Lewis, F.B. (1977). *J. econ. Ent.* **70**, 598–602.

Wedding, J.B., Carncy, T.C., Ekblad, R.B., Montgomery, M.E. and Cermak, J.E. (1978). *Trans. Am. Soc. agric. Engns*, **21**, 253–266.

CHAPTER 36

A Quantitative Approach to the Ecology of the Use of Pathogens for Insect Control

D.E. PINNOCK

Department of Entomology, University of Adelaide,
Waite Agricultural Research Institute, Glen Osmond,
South Australia, Australia and

R.J. BRAND

Department of Biomedical and Environmental Health Sciences,
University of California, Berkeley, California, USA

I. Introduction

Proponents of microbial control of insect pests often emphasize the specificity of pathogens as an advantage of the microbial control approach, implying that the lack of direct effects of the pathogen on parasites and predators contributes to stable, long term suppression of the pest. However, an analysis of operational microbial control programs and strategies shows that very few of these exploit the specificity of the pathogens in a purposeful manner. The reason for this is that most of these programs are based on insufficient information about the pathogens and the biological systems in which they operate, so that the specificity and other attributes of the pathogen cannot be fully exploited. Similarly, many shortcomings of the pathogen, e.g. low field persistence, often are unmeasured or are unknown. As a result, control programs tend to operate with little or no accommodation of the more subtle interactions of the pathogens with other environmental forces. Although this intuitive approach to microbial control has in some cases resulted in operational programs giving reliable suppression of pest populations, the case advanced in this chapter is that a more quantitative and ecological approach to microbial control will improve the reliability and cost-effectiveness of existing programs.

It seems reasonable to hope also that this approach to microbial control perhaps may: (1) indicate fresh avenues to the solution of some of the more intractable pest control problems, particularly those where reliance on chemical control is jeopardized by increasing target insect resistance and by increasing insecticide costs; (2) provide earlier indication of certain pathogens or strategies that will not meet the requirements of a given pest control situation; and (3) result in and eventually be based on a deeper understanding of the ecology and dynamics of insect diseases.

II. Quantitative Modelling of Pathogen Dose and Host Mortality

One of the implicit assumptions in microbial control is that some type of dose—response relationship exists between the pathogen and host insect. Surprisingly, this relationship, which is fundamental to all microbial control trials, is often poorly measured, even in the laboratory. Field trials made without adequate attention to dose—response are usually poorly designed and fail to yield reliable data for the design of subsequent field trials. Such series of field trials provide little cumulative information to further basic understanding of the control process. It is useful to identify individual components of the pest control system.

In the simplest systems, the host—pathogen relationship is unidirectional, or host density independent, i.e. the pathogen population affects that of the host, but not *vice versa*. An example of this type of relationship is that between *Bacillus thuringiensis* and many species of phytophagous larvae of Lepidoptera. In unidirectional systems the important parameters to be quantifield are effective dose of the pathogen to which the target or host insects are exposed, and response of the host population to that dose. Both of these parameters are determined or influenced by various factors operating on the system. For example, the upper limit of the effective dose is usually determined by the initial size of the pathogen population. However, the effective field dose is determined also by the survival of the pathogen and the exposure rate of host to pathogen. The exposure rate is affected by host behaviour, and may be approximated by feeding rate where the pathogen affects the host via the gut, or by rate of contact where the pathogen infects *via* the integument. Some of these factors are strongly influenced by environmental conditions. Among these factors are the survival characteristics of the pathogen and the activity level of the host, which in turn determines its feeding or contact rate and thus its exposure to the pathogen. Thus quantification of important parameters affecting a microbial control attempt is both complex and technically difficult, even with a relatively simple unidirectional system.

Estimation of pathogen population density or dose may be made by various techniques. These include plate counts (e.g. Pinnock *et al.*, 1971; Milner, 1977), haemocytometer counts (Martignoni and Schmid, 1961), fluorescent antibody tracing (Davidson and Pinnock, 1973) and direct bioassay. Burges and Thomson (1971) reviewed and discussed some of these techniques and the determination of their precision. Many problems associated with assessment of control programs using baculovirus preparations arise from paucity of reliable data on effective dose (review; Pinnock, 1975). Wherever possible, estimation of pathogen dose should be verified by bioassay before accepting the estimate as a true index of dose (e.g. Brand *et al.*, 1976).

Although the effects of various single environmental components on pathogen populations have been measured in the laboratory, environmental forces affecting the field persistence of the pathogen, or its infectivity, have been estimated in general by their combined or net effect. Recent reviews of the environmental stability of pathogens were given by Maddox (1977), Roberts and Campbell (1977), Pinnock *et al.* (1977) and Jaques (1977). The general approach has been to estimate the same or similar pathogen populations after varying periods of exposure to environmental forces. Resulting survival or infectivity data may then be incorporated in the estimation of effective dose, and will be most accurate when underlying measurements are made in conditions closely similar to those of the microbial control trial. For example, the species of host plant has a strong influence on the survival of *B. thuringiensis* (Pinnock *et al.*, 1975). Even then, such data allow only for changes in the quantity of viable dose. Where changes in the quality of the pathogen population also occur over time, e.g. with *B. thuringiensis*, this must also be considered. A method accomplishing this was described by Brand *et al.* (1975).

The exposure rate of host insects to pathogen populations effective *per os* may be estimated from the feeding rate, using food consumption, frass production or growth rate as the measured variable. Intuitively, the order given appears to be that of decreasing precision for the estimation of true exposure rate, but no studies could be found giving satisfactory comparative data to illustrate this.

Effective dose in the unidirectional or host—density independent system thus may be determined by:

(i) Initial population size or density of effective pathogens.

(ii) Rate and pattern of decline in numbers of effective pathogens.

(iii) Rate and pattern of change in the quality of the effective pathogens.

(iv) Rate of exposure of the host to the pathogens.

It should be noted that three of these four parameters contain a time element, for they describe rates of change, which may be strongly influenced by environmental conditions.

A time element also enters into a fifth major component of the system, i.e. the response of the host to the effective dose. The type of host response, and the rate of its development, are conditioned by the mode of action of the pathogen, the size and quality of the effective dose, and the duration of time over which this dose was acquired — a reflection of the exposure rate. Combined, these effects are major determinants of this fifth parameter, namely:

(v) Host response.

A microbial control trial described by these five parameters is a dynamic process which may have no biologically distinct endpoint. Usually, e.g. trials with *B. thuringiensis*, the biological end-point occurs at the time of extinction of either the pathogen population or the host population, i.e. when all pathogen—host interaction ceases. In practice, the experimenter will observe the result of the trial after a time interval set by a combination of intrinsic and extrinsic considerations such as the distribution of host response times, or the required crop protection.

The five parameters of the unidirectional system interact as a base for a simple conceptual model of the microbial control trial. In this simple form, the model ignores the effects of heterogeneity of environmental forces and of pathogen and host distribution, both temporally and spatially. Within limits, use of means for parameters (ii) and (iii) can balance some heterogeneity of pathogen quality, and parameter (v) may incorporate heterogeneity of susceptibility or tolerance of host population to pathogen. For both pathogen and host, intraspecific density effects are excluded.

Once trial specific values for each of the parameters are known, it is possible to calibrate a model which will describe the microbial control trial in numerical terms. This model can be used to predict the outcome of later microbial control trials after a pre-set time interval. A predictive model of this type has been made for *B. thuringiensis*, so that a desired host response (in this case, mortality after 4 days) may be preselected within limits by adjustment of the initial pathogen population applied (Pinnock *et al.*, 1978a). Further refinements of this approach are described in Chapter 37. In the present case the probability (P) of target insect mortality after 4 days was estimated by use of the multiple logistic model (Cox 1970) in the form:

$$P = 1/[1 + e^{-(a_0 + a_1 L + a_2 T)}]$$

where L is the effective dose of the pathogen and T the time of its exposure on the treated foliage. For the model describing red-humped caterpillars, *Schizura concinna*, on redbud trees, *Cercis occidentalis*, the fitted parameters are:

$$P = 1/[1 + e^{-(-2.914 + 1.355 L + 0.0304 T)}]$$

where L = effective dose = $\log(r) - 0.363a + a - \log(b_e)$; r = average feeding rate of the *S. concinna* larvae on *C. occidentalis* leaves (Pinnock *et al.*, 1978b), a = the fitted regression parameter representing the initial deposit of *B. thuringiensis* on the leaves (Pinnock *et al.*, 1974), b_e = the fitted regression parameter representing the early segment of the *B. thuringiensis* decay curve on *C. occidentalis* leaves (Pinnock *et al.*, 1974), $T = \pi$, the persistence half-life of *B. thuringiensis* on *C. occidentalis* leaves (Pinnock *et al.*, 1971).

With this model the experimenter can forecast *S. concinna* mortality induced by the pathogen and hence survival. In an integrated pest management program, this approach facilitates the management of *S. concinna* and its two hymenopterous parasites, *Apanteles schizurae* and *Hyposoter fugitivus*, with the pathogen *B. thuringiensis* as a modulator of the *S. concinna* population (Pinnock *et al.*, 1978a). It is emphasized that this model applies only to unidirectional systems — perhaps only to *B. thuringiensis* — and in its present form requires some degree of uniformity of conditions during the 4 days. Thus the differential effects of day and night on the rates of change in parameters (ii), (iii), (iv) and (v) are averaged out, and so accommodated in the model. However, no attempt has been made to accommodate extreme variation, e.g. rain. Therefore rain, especially during the first day of the trial, would nullify the predictive value of the model unless appropriate, calibrated adjustments could be developed and applied.

The *B. thuringiensis* model was constructed to enable the effects of spraying this pathogen to be planned; the preselected number of hosts surviving the spray serving as food or oviposition sites for predators and parasites (Pinnock *et al.*, 1978a). The pathogen can thus be integrated with other biological control agents for the management of the pest complex on redbud trees.

In bidirectional systems, i.e. when the size and/or density of the host population affects the pathogen population and *vice versa*, the problems of quantifying or modelling the two populations may become much more complex. A typical system of this type would have as additional factors the multiplication characteristics of the pathogens within the infected host, and the probability and rate of transmission to a new host insect. These factors would compound to influence the host exposure rate.

A synthesis of all factors affecting the bidirectional system results in a model of the system, often termed an epizootiological model. Selected models, and the fitting of one of them to a *Hydra/Hydramoeba* system were reviewed by Stiven (1973). He described a "community resistance index", $R_{c'}$ for multiple host systems, where

$$R_c = \sum_{i=1}^{s} (p_i/b_i)$$

and p_i = the proportion of numbers or biomass associated with the ith host species, and b_i = the mean growth rate of the pathogen in the ith species. R_c can be determined by bioassay, and will be reflected in the pooled host response as outlined above. He reported that R_c predicted the short term system dynamics under pathogen stress and considered it a realistic measure of community resistance to the pathogen.

The notion of variable resistance to a pathogen (or pathogen-induced stress) of an animal community or species is not new, although this research area has received little attention from workers in insect pathology or microbial control. Martignoni and Schmid (1961) reported that two populations of the California oakworm *Phryganidia californica* had differing susceptibilities to a standardized preparation of nuclear polyhedrosis virus. The two populations also had different degrees of heterogeneity of susceptibility to the virus, and it was suggested that one population was typical of an unselected, i.e. pre-epizootic, population and the other population — less susceptible and less heterogeneous —

was typical of a population recently exposed to selection by a virus epizootic. Such differences in host populations may strongly influence the outcome of a microbial control trial, but usually this type of quantitative data is not obtained.

The fungus *Nomuraea rileyi* may suppress populations of lepidopterous larvae attacking a variety of crops, including soybeans (Ignoffo *et al.*, 1975). The bidirectional *N. rileyi*/lepidopterous host/environment systems are being studied by several workers, and probably there will soon be enough data to make a model for accurate forecasts of epizootics. For example, epizootics of *N. rileyi* in soybean caterpillars were advanced in time by a spray of conidia (Ignoffo *et al.*, 1976), providing much useful data on the effect of the applied dosage of conidia, the field survival rate of the conidia and the host caterpillar populations, e.g. their infection rates and population size required for maintenance of the fungal inoculum.

A simple model of this system predicts the incidence of *N. rileyi* in the velvet bean caterpillar *Anticarsia gemmatalis* on soybean (Kish and Allen, 1978; see Chapter 27 Section VI, A for predictive performance of the model and the ecology). It estimates the mean total daily yield of *N. rileyi* conidia from the estimated mean number of *A. gemmatalis* cadavers/acre multiplied by the mean conidial yield/cadaver. After adjustments to allow for the effects of humidity, rainfall, wind, conidial survival rate and the proportion of conidia not released into the environment, the net total daily yield of conidia is translated into the mean number of conidia/mm^2 of leaf surface. This inoculum density of conidia on bean leaves may be used as an estimate of actual dose and entered into the dose–response curve of Ignoffo *et al.* (1975) for *N. rileyi vs Trichoplusia ni.* Infection rates so indicated predicted field infection rates for *A. gemmatalis* 5 or 6 days later, since the mean period from exposure of the larvae to their death was about 6 days.

Because the incidence of infection depends on both the number of susceptible hosts and their effective exposure rate to the pathogen, the above model comes into effect only above certain threshold densities. A model of the overwintering survival of *N. rileyi* conidia and the initiation of epizootics by this pathogen is lacking, although recent work by Sprenkel and Brooks (1977) and Ignoffo *et al.* (1977) indicates that soon this requirement may be satisfied.

In its present form, the main value of the *N. rileyi* model is to provide short-term predictions of host infection rate so that more informed decisions can be made on intervention with other forms of pest control, mostly chemical insecticides. However, the most recent work suggests the possibility of an extension of this model incorporating host dispersion and pathogen distribution, so that initiation and progression of *N. rileyi* epizootics may be predicted.

Some of the parameters necessary for the calculated integration of a pathogen into a pest management system were outlined in the above section. An equally substantial body of knowledge of the population dynamics of the pest and its ecological interactions, including economic thresholds, is necessary to permit most efficient integration. Much of this information is essentially ecological and will be discussed below.

III. Integration of Pathogen-induced Mortality and Population Dynamics into Pest Control Models

If microbial control is to operate most effectively, it must be applied with knowledge of the ecological framework centred on the target species. This framework, together with inherent developmental patterns, determines the population dynamics of the target species. Often these elements of the pest system exist only as a series of qualitative concepts in the mind of the entomologist. However, with some pest species, attempts have been made to express these elements in mathematical forms. The resulting models have been made so that certain of their components can be used to include the effect of pathogen-induced mortality. In this way more complete, longer-term models of microbial control may be achieved.

An extensive literature attests to the growing interest in, and importance of, mathematical models in population ecology. Streifer (1974) presented an excellent review of past attempts to develop population models, and described in generalized form a population model and submodels of wide applicability. In one of the simplest models of population dynamics, of interest to insect pathologists, the number of female offspring produced by a generation of females is related to survival and fertility rates (Lotka, 1931). In symbols:

R = number of ♀ daughters produced/♀ in the parent generation

$\quad = \Sigma_x l_x m_x$

where

l_x = probability of ♀ survival at age x, and

m_x = number of ♀ progeny produced/♀ at age x.

This equation was used by Weidhaas (1974) to estimate growth rates of mosquito populations by rearranging it to partition total survival into survival of adults and survival of immature stages. The rearranged equation is:

$$R = (\Sigma P_x M_x)S_i$$

where P_x is the probability of survival of the ♀ adults to age x, and S_i is the probability of survival of immatures to the adult stage. This form of the equation better isolates the survival rate for immature stages (S_i).

Since many insect pathogens attack primarily immature stages, the impact on the pest population of a microbial control program, which kills immatures, can be estimated by this simple equation — if the other parameters of the model, i.e. survival of adults and reproductive rate, can be estimated or measured. For microbial control of mosquito larvae, the model may be useful because density dependent regulation of many mosquito populations is mediated through changes in survival of immature stages. In such species, e.g. *Culex pipiens* (Lowe *et al.*, 1973), changes in adult survival or egg-laying pattern contribute little to the recovery potential of the species. Thus the addition of a pathogen, e.g.

Beauveria brongniartii (Pinnock *et al.*, 1973) or *Romanomermis* (Petersen and Willis, 1972), to the larval habitat may in theory suppress the mosquito population for some generations. However, the attractive simplicity of this model is because its survival and fertility parameters are total or net effects. No attempt is made to model the processes which determine these parameters so the model relies on continual updating from field observations if the calculated outcome is to remain true to reality. Without updating the model will have poor predictive value.

More complex population models, with better predictive value, are those that incorporate differences in the quality as well as the quantity of individuals in a population. One obvious quality difference is age. Age-specific population models were developed by Bailey (1931), Lewis (1942), Leslie (1945) and many others. Lewis and Leslie developed the use of matrices to describe the age distribution of a population at successive intervals in time, given the age-specific rates of survival and fertility. True age, or physiological age, e.g. day degrees, may be used as the time scale. A modification of this approach, with physiological age as the time scale, was used by Hughes (1963) to model the population dynamics of the aphid *Brevicoryne brassicae*. The model realistically described the population dynamics of the aphid, including effects of parasites, predators, and the pathogen *Entomophthora aphidis*. Further developments of matrix modelling and its extension to include age-size specific models are described by Streifer (1974).

Stinner *et al.* (1977) described a population model incorporating the effects of intraspecific competition on both interstage and intrastage survival rates, on spatial distribution and on larval growth rate. It described the population dynamics of *Heliothis zea*, in which larval cannibalism occurs and where the spatial distribution of the larvae is an important element of the model. A modification of this approach may be useful for describing the dynamics of pathogen-induced mortality when contagion is an important factor in pathogen transmission.

At present many theoretical ecological models exist, and a few are being adapted for modelling pest insect population dynamics, particularly with respect to parasites and predators, host plant phenology, and climate. The need now is for more data on which to build a submodel including pathogen-induced mortality for incorporation into the general model. When this is accomplished, the pathogens will be part of an integrated pest management system — their most promising role for the future.

IV. Some Economic Aspects of Microbial Control

Economic aspects of development, production and use of microbial insecticides were discussed by Dulmage (1971) and will not be repeated here.

If a pathogen is to be integrated into a multifactorial pest management program, many of the parameters described above must be known or at least approximated. This requires a research and field trial effort, which of course involves a cost. The anticipated benefits of this effort are such that the costs can be justified, although both long-term costs and benefits often exist only as theories or notions rather than definite amounts of money, man-hours and materials. Where the management of a pest is obviously deteriorating because of target insect resistance to chemical insecticides and/or increasing insecticide costs, the introduction of an alternative strategy, e.g. a pathogen, may be justified as the only practical long-term solution. Even so, the optimal diversion of effort to develop the new strategy, while still maintaining adequate crop protection under the old chemical insecticide regime, is not a simple decision. Regev *et al.* (1976) modelled this type of situation and, within the set of assumptions and cost structures used in the model, showed the developmental investment to be made in the new pest management strategy and the optimal path to be taken until the new strategy was available. To the authors' knowledge, this type of optimization analysis has not been applied to any microbial control program.

For most microbial control programs, long-term economic analyses are not readily obtainable. Introductions of specific pathogens into areas where at present they are absent may sometimes be cost-effective. Inundative use of microbial insecticides in highly organized pest management programs can be justified only when environmental or material costs of alternative pest control methods are high. In the control of *Schizura concinna* larvae on highway landscapes, the use of *B. thuringiensis* (Pinnock *et al.*, 1978b) was economic because of the difficulty of repeated insecticide spraying and the highly unstable pest situation that resulted from earlier pest management attempts with non-specific insecticides. Also, the extensive human contact with the pest system required control techniques with minimal health risk. Thus the social and environmental costs of chemical control were very high, so that a non-chemical approach was devised. Even in this type of situation, too much pathogen can result in further destabilization of the pest population and, in extreme cases, causes pest resurgence (Pinnock and Milstead, unpublished observations; Young, 1977). The use of *B. thuringiensis* to manage or modulate the *S. concinna* population in concert with the natural hymenopterous parasites resulted in annual savings in pest control costs of ca. 60% for material and 80% for labour amounting annually to thousands of dollars (Pinnock *et al.*, 1978b). Savings in the high cost of *B. thuringiensis* and other pathogens should be an incentive to the development of the integrated approach, as in many systems now being developed, e.g. the *Nomuraea rileyi/Anticarsia gemmatalis* system on soybeans already described.

V. Conclusions and Suggestions for Future Work

Pathogens are potentially valuable pest management tools, but their effective and reliable use will not be fully realized until enough quantitative and ecological data are available. In general, there appears to be scope for technical improvement of microbial field trials and minimization of error when the outcome of trials is forecast. These issues are discussed in detail in Chapter 37. Even when field trial parameters appear to be adequately quantified, the paucity of fundamental, ecological knowledge about the pathogen may remain a barrier to its effective use. For example, many years of attempts to suppress populations of mosquito larvae with *Coelomomyces* were made with inadequate knowledge of the ecology of the pathogen and, in particular, in ignorance of its alternate host. Research effort should be directed towards the further improvement of microbial control submodels, and toward the gathering of ecological data necessary for the insertion of these submodels into the overall pest management model.

References

Bailey, V.A. (1931). *Q.J. Math.* **2**, 68–77.

Brand, R.J., Pinnock, D.E., Jackson, K.L. and Milstead, J.E. (1975). *J. Invertebr. Path.* **25**, 199–208.

Brand, R.J., Pinnock, D.E., Jackson, K.L. and Milstead, J.E. (1976) *J. Invertebr. Path.* **27**, 141–148.

Burges, H.D. and Thomson, E.M. (1971) *In* "Microbial Control of Insects and Mites" (H.D. Burges and N.W. Hussey, eds), pp. 591–622, Academic Press, London and New York.

Cox, D.R. (1970) "The Analysis of Binary Data" Methuen, London.

Davidson, A. and Pinnock, D.E. (1973). *J. econ. Ent.* **66**, 586–587.

Dulmage, H.T. (1971) *In* "Microbial Control of Insects and Mites" (H.D. Burges and N.W. Hussey, eds), pp. 581–590. Academic Press, London and New York.

Hughes, R.D. (1963). *J. anim. Ecol.* **32**, 393–424.

Ignoffo, C.M., Puttler, B., Hostetter, D.L. and Dickerson, W.A. (1975). *J. Invertebr. Path.* **25**, 135–137.

Ignoffo, C.M., Marston, N.L., Hostetter, D.L. and Puttler, B. (1976). *J. Invertebr. Path.* **27**, 191–198.

Ignoffo, C.M., Giscia, C., Hostetter, D.L. and Pinnell, R.E. (1977). *J. Invertebr. Path.* **29**, 147–152.

Jaques, R.P. (1977). *Misc. Publ. ent. Soc. Am.* **10**, 99–116.

Kish, L.P. and Allen, G.E. (1978). Bull 795 (Tech). Agric. Exp. Stns., Inst. Fd. Agric. Sci., Univ. Fla., Gainesville.

Leslie, P.H. (1945). *Biometrika* **33**, 183–212.

Lewis, E.G. (1942). *Sankhya* **6**, 93–96.

Lotka, A. (1931). *Hum. Biol.* **3**, 459–493.

Lowe, R.E., Ford, H.R., Smittle, B.J. and Weidhaas, D.E. (1973). *Mosq. News* **33**, 221–227.

Maddox, J.V. (1977). *Misc. Publ. ent. Soc. Am.* **10**, 3–18.

Martignoni, M.E. and Schmid, P. (1961). *J. Invertebr. Path.* **3**, 62–74.

Milner, R.J. (1977). *J. Invertebr. Path.* **30**, 283–287.

Petersen, J.J. and Willis, O.R. (1972). WHO/VBC 72, 357.

Pinnock, D.E. (1975). *In* "Baculoviruses for Insect Pest Control: Safety Considerations" (M. Summers, R. Engler, L.A. Falcon and P.V. Vail, eds), pp. 145–154. American Society for Microbiology, Washington, D.C.

Pinnock, D.E., Brand, R.J. and Milstead, J.E. (1971). *J. Invertebr. Path.* **18**, 405–411.

Pinnock, D.E., Garcia, R. and Cubbin, C.M. (1973). *J. Invertebr. Path.* **22**, 143–147.

Pinnock, D.E., Brand, R.J., Jackson, K.L. and Milstead, J.E. (1974). *J. Invertebr. Path.* **23**, 341–346.

Pinnock, D.E., Brand, R.J., Milstead, J.E. and Jackson, K.L. (1975). *J. Invertebr. Path.,* **25**, 209–214.

Pinnock, D.E., Milstead, J.E., Kirby, M.E. and Nelson, B.J. (1977) *Misc. Publ. ent. Soc. Am.* **10**, 77–97.

Pinnock, D.E., Brand, R.J., Milstead, J.E., Kirby, M.E. and Coe, N.F. (1978a) *J. Invertebr. Path.* **31**, 31–36.

Pinnock, D.E., Hagen, K.S., Cassidey, D.V., Brand, R.J., Milstead, J.E. and Tassan, R.L. (1978b) *Calif. Agric.* **37**, 33–34.

Regev, U., Shalit, H., and Gutierrez, A.P. (1976) *In* "Proceedings of a Conference on Pest Management," (J. Norton and C. Holling, eds), pp. xxx–xxx International Institute for Applied Systems Analysis, Luxenberg, Austria.

Roberts, D.W. and Campbell, A.S. (1977). *Misc. Publ. ent. Soc. Am.* **10**, 19–76.

Sprenkel, R.K. and Brooks, W.M. (1977). *J. Invertebr. Path.* **29**, 262–266.

Stinner, R.E., Jones, J.W., Tuttle, C. and Caron R.F. (1977) *Can. Ent.* **109**, 879–890.

Stiven, A.E. (1973) *In* "Current Topics in Comparative Pathobiology (T.C. Cheng, ed.), **2**, 145–212, Academic Press, New York and London.

Streifer, W. (1974). *Adv. ecol. Res.* **8**, 199–266.

Weidhaas, D.E. (1974). *J. econ. Ent.* **67**, 620–624.

Young, L.C. (1977). Ph.D. Thesis, Univ. Calif., Berkeley, Calif., U.S.A.

CHAPTER 37

Application of Biostatistical Modelling to Forecasting the Results of Microbial Control Trials

R.J. BRAND

Department of Biomedical and Environmental Health Sciences,
University of California, Berkeley, California, USA

and

D.E. PINNOCK

Department of Entomology, University of Adelaide,
Waite Agricultural Research Institute, Glen Osmond,
South Australia, Australia

I. Introduction

Biostatistical and related research has produced a large array of models and modelling strategies, research design principles and data analysis methods of use to researchers in the field of microbial pest control. Some of these biostatistical tools have become part of the standard methodology of this field. Obvious examples are probit models used to analyse dose—response relationships, and survival curves used to describe the decline of pest populations in field trials. Currently, there is growing interest in developing procedures for forecasting the outcome of microbial control actions. The purpose of this chapter is to explore the kinds of modelling needed in this area of research.

Bacillus thuringiensis acts rapidly and, in practical pest control, it does not spread from insect to insect. Thus microbial control with this agent seems to be the least complex biologically and a good example to begin a discussion of forecast models.

II. Preliminary Considerations

In this section, a general approach for modelling forecast errors is presented and the possible modelling roles of factors that influence control outcomes are described.

First, consider the context in which microbial pest control occurs. At the time a control action is taken, the target pests exist in a naturally evolving ecosystem. Without intervention, natural evolution would prevail. Typically, a control action has only partial influence on the state of the pest system that emerges after control action has been taken. Therefore, a comprehensive model of a microbial pest control will, in general, involve two integrated submodels. One submodel describes the natural evolution of the pest system; the other describes the supplementary effects of control action *per se.*

The need for explicit consideration of natural forces in a pest system would be particularly important for multi-stage control actions or any type of control action that has a long term effect. Then, almost surely, the natural forces would substantially influence subsequent stages of the system. Certain viruses, nematodes and fungi are all examples of control agents that are likely to persist in a pest system as a communicable disease and have an active effect for a relatively long period of time. By contrast, control actions with *B. thuringiensis* have a relatively short-term effect so this kind of microbial control should be much easier to model. Typically, *B. thuringiensis* has a brief field persistence and completes its active effect in 4 to 7 days for most species of lepidopterous larvae (Brand *et al.*, 1976). A *B. thuringiensis* control action can be viewed as a relatively quick resetting of the state of the system that provides a new set of starting conditions from which the system can evolve according to natural forces.

Therefore, with control by *B. thuringiensis* it seems useful to develop a model to forecast the short-term consequences of a single-stage control action. For this, the most relevant prediction would be the percent reduction of the pest population in each developmental stage. This sort of forecasting would be useful in its own right. In addition, the short-term forecast model could be linked to a natural evolution model to provide longer-term forecasts.

Regardless of the particular method and model used to forecast the outcome of control actions, one important evaluation criterion is the error that occurs so it is useful at the outset to examine the notion of forecast error.

A. FORECAST ERROR

In any one field trial the forecast error is the difference between the forecast and the outcome of the control action. Suppose for now that the forecast is based solely on the log concentration (LC) of the pathogen applied in some standardized amount per unit of the pest environment.

If Y = outcome of control action (percent reduction) and $\hat{P}(LC)$ = forecast (percent reduction) as it depends on LC, then

$$[Y - \hat{P}(LC)] = \text{forecast error}$$

($\hat{P}$ is read as "P hat").

The notion of forecast error can be extended to more than one trial as follows. Assume a specified collection of field trials with the use of the same log concentration, say LC_1, in each trial. Then each trial would have the same forecast, $\hat{P}(LC_1)$, but the actual outcome and therefore the forecast errors would differ from trial to trial. It is the distribution of these forecast errors that measures the performance of the forecast procedure at the particular log concentration equal to LC_1. By considering the forecast error distributions at all relevant log concentrations a comprehensive assessment of the error in the forecast procedure is provided for a specified collection of trials.

A general model for forecast errors in a specified collection of field trials can be represented graphically (Fig. 1). The forecasts $\hat{P}(LC)$ are represented by the dotted line. Distributions of percent reductions for the trials at particular log concentrations, LC_1, LC_2 or LC_3, are also represented in Fig. 1. The means of these distributions are not likely to coincide with the corresponding forecasts. Consequently, the mean outcomes as they depend on LC are represented symbolically by $P(LC)$ as a solid line in Fig. 1.

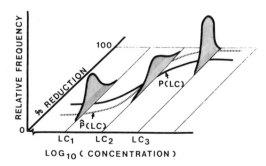

Fig. 1. General model of forecast error for a specified collection of field trials.

At any particular log concentration there are two features of the distribution of control outcomes that are of most interest. First is the difference between the forecast outcome and the mean outcome,

$$B(LC) = \hat{P}(LC) - P(LC) \tag{1}$$

which defines the forecast bias that occurs when the forecast model $\hat{P}(LC)$ is used at log concentration represented by LC.

Secondly the trial outcomes vary about the mean outcome. This variability can be conveniently measured by the standard deviation of the distribution of control outcomes. In symbols,

$$SD_Y(LC) = \sqrt{(V_Y(LC))} = \sqrt{\left(\frac{\sum\limits_{i=1}^{i=N} [Y_i - P(LC)]^2}{N-1} \right)} \tag{2}$$

where Y_i for $i = 1, 2, \ldots N$ are the N outcomes when all trials in a collection of N trials are carried out with log concentration equal to LC and $V_Y(LC)$ represents the variance of the distribution of outcomes. As a general rule, at most 11% of outcomes will occur > 3 standard deviations from the mean, regardless of the shape of the distribution. For a normal (Gaussian) distribution only about 5% of the outcomes will occur > 2 standard deviations from the mean. It is properties like these that make the standard deviation a useful measure of variability. Furthermore, the standard deviation can be interpreted as a measure of the unreliability of the forecast procedure. When $SD_Y(LC)$ is small the procedure is reliable in that outcomes are consistent from trial to trial. Nevertheless, the procedure may be very biased if the consistent outcomes do not coincide with the forecast.

The relationship between forecast error, bias and unreliability can be illustrated by writing

$$[Y - \hat{P}(LC)] = [Y - P(LC)] + [P(LC) - \hat{P}(LC)]$$
$$= [Y - P(LC)] - B(LC). \tag{3}$$

The first term on the right is a deviation-from-mean that stems from unreliability and the second term comes from bias. When the bias and deviation-from-mean are both positive or both negative, they partially cancel to give lower forecast error. When the deviation and bias have opposite signs, they combine to give a larger forecast error.

Implicit in this concept of forecast error is the specification of a particular collection of field trials in which the forecast model will be applied. Furthermore, this view of forecast error involves, in imagination at least, the outcome for each trial for each possible level for LC (assuming for the moment that this characteristic is the sole predictor of the control outcome). It is this concept that makes it possible to use the distribution, mean and standard deviation of control outcomes as a basis for specifying the basic features of forecast error — bias and unreliability.

B. MODELLING ROLES OF FACTORS THAT INFLUENCE CONTROL OUTCOMES

Any specified collection of trials would necessarily exist in some particular geographical region and occur over some particular period of time. In each trial there is a constellation of factors that influence the control outcome. Some of these factors relate to the control action itself and some are inherent in the pest system. Regardless of their origin, each of these factors can be involved in forecast modelling in one of three ways:

(1) A factor can be represented explicitly in the forecast model as a predictor of control outcome. For factors treated in this way a measurement must be devised to actually obtain forecast information. Errors in these measurements are one source of forecast error.

(2) A factor can be implicitly included in the forecast model by restricting the scope of the model to a limited class of B. thuringiensis applications in which the factor does not vary. For example, attention can be focused on a specific insect, host plant, variety of B. thuringiensis, type of application equipment, standardized quantity of control preparation applied per unit pest environment, etc. Then, within this specified class of B. thuringiensis applications, a class-specific forecast model can be developed to relate the explicit predictors to the forecasts for the specified collection of trials. Specialization of the forecast model in this way should substantially reduce forecast error by standardizing major factors that influence control outcomes and that might otherwise vary from trial to trial.

(3) A factor may be relegated to the category of residual effects that do not appear either explicitly or implicitly in the forecast model. Some factors are unknown so there is no possibility of using them. Others may be costly or impractical to use. In any event, to the extent that these remaining factors vary from trial to trial, they influence the control outcome in ways unaccounted for by the forecast procedure. Consequently, these residual factors are another source of forecast error.

Ultimately, the choice of which known factors to include explicitly or implicitly in a forecast model involves a trade-off between the effort required to obtain predictive information, the reasonableness of limiting attention to a particular class of applications and the forecast error that would otherwise occur.

III. Forecast Models Based on Concentration of the Pathogen

The predictors most readily available when a control action is planned are the dimensions of the control action — namely, species or variety of the pathogen, preparation in which it is applied, concentration of pathogen in the control preparation and quantity of the preparation applied per unit of the pest's environment. It would be useful to be able to forecast the outcome of the control action from this information alone. More specifically, if attention were restricted to a class of applications that all used the same variety of B. thuringiensis in the same preparation applied in standardized amounts per unit of the pest environment, then the concentration of the pathogen could be used to predict the outcome of the control action. This sort of forecasting is considered in this section.

It seems unlikely, however, that sufficiently reliable forecasts could be obtained, even within this limited class of applications, unless further restrictions are imposed. For example, attention could also be limited to a particular target insect on a particular host plant under specific kinds of environmental conditions. Focusing on a class of control actions in this way should reduce forecast error without being inconsistent with the sort of specialized considerations that occur in pest management anyway. Some possibilities for developing more generally applicable forecast models are discussed in Section VII.

A. CONSTRUCTING A FORECAST MODEL

Some preliminary data are available to illustrate forecast modelling based on concentration of the control agent for a specified class of control actions. Twelve field trials were conducted in two field studies of six trials each during the summers of 1972 and 1973 near Placerville, California. During these trials there was no rain and days were generally sunny. The mean maximum shade and mean minimum temperatures were 38°C and 17.2°C, respectively. The target pest was *Schizura concinna;* host plant *Cercis occidentalis;* control agent Dipel, a commercial formulation of *B. thuringiensis* H-serotype 3a3b; and the quantity of Dipel per unit leaf area was standardized by spraying to point of run-off with a Hatsuta Blowmic B power sprayer. The outcome measure was the percent reduction in the size of the larval population. Percent reduction was computed for each trial from two single replicate counts (Table I), one made before the treatment and the second 4 days after.

TABLE I

Percent reduction of Schizura concinna *larvae 4 days after application of* Bacillus thuringiensis *on* Cercis occidentalis *foliage in twelve field trials*

Trial	Concentration	Log concentration (LC)	Initial count	Four-day post-treatment count	Percent reduction
1	H_2O Control	-10.00^a	1165	621	46.69
2	0.000025	-4.60	790	399	49.49
3	0.00025	-3.60	1189	596	49.87
4	0.0025	-2.60	1675	410	75.52
5	0.025	-1.60	2060	375	81.80
6	0.25	-0.60	1190	0	100.00
7	H_2O Control	-10.00^a	825	439	46.79
8	0.00125	-2.90	695	413	40.57
9	0.0025	-2.60	1085	494	54.47
10	0.025	-1.60	938	441	52.99
11	0.25	-0.60	836	13	98.44
12	0.5	-0.30	890	10	98.87

[a]This value was used for unweighted least squares calibration of the forecast model.

In these studies there was no attempt either to obtain a random sample from a specified collection of trials or to allocate the various concentrations to the trials in some randomized way. The consequences of these study design features as they relate to sources of forecast bias are discussed later. Nevertheless, these data can be used to illustrate the calibration of a forecast model.

The points plotted in Fig. 2 show the observed percent reduction in the pest population vs $\log_{10}$ (concentration) for the 12 field trials described in Table I. The data suggest a pest reduction asymptote at low concentrations and a pattern of increasing percent reduction with increasing log concentration of *B. thuringiensis* applied. The asymptote at low concentrations derives from two sources. Some insects are dislodged from the host plants by the physical disturbance of spraying. This has been documented for *S. concinna* by pre- and postapplication counts which show substantial insect removal when water is sprayed (Pinnock *et al.*, 1978). Also natural changes that occur in the size of the pest population during the 4-day period between initial and final counts may contribute to the asymptote. Dramatic changes of this type can be anticipated when many larvae are close to pupation at the start of a trial.

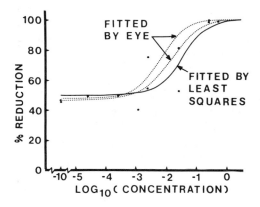

Fig. 2. Percent reduction of *Schizura concinna* larvae 4 days after application of *Bacillus thuringiensis* on *Cercis occidentalis* foliage in twelve field trials, with percent reduction curves fitted by eye and by unweighted least squares.

These qualitative impressions of the percent reduction pattern suggest the following two-part probability model to represent the relationship between percent reduction and $\log_{10}$ (concentration). Let $LC = \log_{10}$ (concentration), $P(LC)$ = percent reduction as it depends on LC, μ = probability scaled from 0 to 100 percent that an insect will be removed by physical–natural forces, and $M(LC)$ = probability scaled from 0 to 1 that a target insect not removed by physical–natural forces will be killed by the pathogen. Then the overall probability of survival equals the probability that the insect is not removed by physical–natural forces times the probability that insects which survive these forces are not killed by the pathogen. This gives the following probability model:

$$100 - P(LC) = (100 - \mu) \times (1 - M(LC)).$$

Rearranging this model we get

$$P(LC) = \mu + (100 - \mu) \times M(LC). \tag{4}$$

The logit model is a good candidate to represent the active effects of the pathogen (Cox, 1970). Then

$$\ln \left[\frac{M(LC)}{1 - M(LC)} \right] = A_0 + A_1 \times LC \tag{5a}$$

or equivalently,

$$M(LC) = \frac{1}{1 + e^{-(A_0 + A_1 \times LC)}} \tag{5b}$$

where A_0 and A_1 are parameters that allow the model to adjust to different empirical percent reduction patterns.

When the parameter A_1 is positive, the logit model gives a family of increasing, symmetric, S-shaped curves virtually identical to the curves provided by the familiar probit model (Cox, 1970). However, the logit model has a more convenient mathematical form than the probit model and is therefore better suited for the modelling extensions considered in later sections.

The model $P(LC)$ has a general structure that is consistent with the percent reduction trend suggested by the data in Fig. 2. When the concentration of the control agent is low or zero so that LC is very negative, then $M(LC) \cong 0$ and $P(LC)$ equals μ which represents the low concentration asymptote for percent reduction. As concentration and therefore LC increases, the model gives percent reductions that increase to 100%. The trend patterns provided by this model are the same as the patterns provided by a standard dose–response model with Abbott's correction for natural mortality (Finney, 1964). However, the associated model of trial-to-trial variations in control outcomes described in Section V is more general than the corresponding variation model typically used in dose–response analysis.

B. CALIBRATING THE FORECAST MODEL

The percent reduction model $P(LC)$ provides a family of curves – one of which can be used to represent the pattern of average outcomes for some particular collection of trials. In practice, data from a sample of trials from the particular collection are available and some procedure is needed to determine the specific values for the model parameters, say $\hat{\mu}$, $\hat{A}_0$ and $\hat{A}_1$, that give a percent reduction trend that best fits the data. This fitted curve then provides an empirically based forecast model, $\hat{P}(LC)$, where

$$\hat{P}(LC) = \hat{\mu} + (100 - \hat{\mu}) \cdot \frac{1}{1 + e^{-(\hat{A}_0 + \hat{A}_1 \times LC)}} \tag{6}$$

This process is termed "calibrating the forecast model"; there are several methods that can be used to do this.

It is possible to visually determine the curve and corresponding parameter values that seem to best represent the percent reduction trend in the data. A systematic approach for calibrating the forecast model by eye is described in Section XA. Two curves independently fitted to the data by eye by the two authors are shown by the dotted lines in Fig. 2. The upper curve has parameter values $\hat{\mu} = 48$, $\hat{A}_0 = 5.64$ and $\hat{A}_1 = 2.56$ and the lower curve has parameter values $\hat{\mu} = 47$, $\hat{A}_0 = 3.84$ and $\hat{A}_1 = 2.08$.

Despite the computational convenience of curve fitting by eye, it has the obvious disadvantage that each person who calibrates the forecast model gets a different result. Consequently a more objective method is desirable. The usual approach is to select some objective criterion of how well the trend curve fits the data and then determine which of the possible values for the model parameters gives the best fitting curve according to that criterion. The most commonly used criterion is unweighted least squares in which the parameter values $\hat{\mu}$, $\hat{A}_0$ and $\hat{A}_1$ are selected to make the sum of squares (SS) given by equation (7) as small as possible.

$$SS = \sum_{i=1}^{n} (Y_i - \hat{P}(LC_i))^2 \qquad (7)$$

where Y_i = observed percent reduction for the i-th of a sample of n field trials and LC_i = log concentration for the i-th trial.

For non-linear models of the type considered here, this method of curve fitting is iterative and best carried out with a computer program using the method described by Draper and Smith (1966, p. 267). The values for μ, A_0, A_1 obtained by fitting the curve by eye provide the starting guess for model fitting by the unweighted least squares method. The fitted curve obtained by this method is shown by the solid line in Fig. 2. The parameter values for this curve are $\hat{\mu} = 50.227$, $\hat{A}_0 = 3.665$ and $\hat{A}_1 = 2.487$.

In effect, the least squares procedure finds the compromise trend line that best represents the curve of mean percent reductions as it appears in just the sample of trials used to calibrate the forecast model. However, the resulting model, $\hat{P}(LC)$, probably will not coincide with $P(LC)$, the model that best represents the curve of mean percent reduction for the full collection of trials under consideration. The calibration sample data may differ in some systematic way from the whole collection from which the sample was taken. Random sampling to obtain the calibration sample and random allocation of concentration levels to trials in the sample would be required to guarantee the minimum of systematic bias. Even under these ideal (although somewhat impractical) conditions, the fitted model $\hat{P}(LC)$ based on the calibration sample will differ from the model $P(LC)$ of the whole because of sampling error.

Further refinements in the calibration procedure must be based on some sort of model that describes how variability in percent reduction outcomes depends on LC. One approach for obtaining such a model is presented in Section V.

On the basis of those results, it may be possible to calibrate the forecast model by the method of weighted least squares. In this method the degree of variability in the percent reduction outcomes at different LC levels is taken into account. Data points that occur at LC levels where variability is low count more heavily in determining the position of the trend line than those that occur where variability is high. The weighted least squares procedure is thereby somewhat more realistic in the way it handles sampling error.

More detailed modelling of the variability and shape of the distributions of percent reduction outcomes shown in Fig. 1 would enable the assessment of sampling error in the calibrated forecast model. For example, a confidence band for $P(LC)$ could be constructed. A pair of forecast curves — an upper curve and a lower curve — can be constructed from the calibration data in such a way that the band between the curves has a high probability of containing $P(LC)$. The width of the confidence band, which reflects the amount of uncertainty due to sampling error, depends on the number of calibration trials. This width then provides a criterion that can be used to decide how many trials are needed in the calibration sample. Research is underway on these extensions of the calibration procedure and will be reported in a future publication.

C. ANALYSIS OF VARIANCE

Discussion in this section has focused primarily on modelling trends in the percent reduction outcomes as they depend on LC. The best fitting trend for the calibration sample provides an empirically based forecast model. However, as shown in Fig. 1, the resulting forecast model would still have been subject to considerable forecast error if it had been used to forecast the outcomes of these 12 calibration trials, so it is of interest to examine the variability in these percent reduction outcomes.

Traditionally, this sort of examination occurs in biostatistical modelling as a complement to the analysis of trends and is called analysis of variance. Analysis of variance is fundamentally a technique by which efforts are made to diagnose the sources of variability in observed data.

In its most common form, analysis of variance in conjunction with trend analysis consists of partitioning the total variation in the outcome data into two parts — one part that stems from the trend in average outcome as it depends on the predictor (LC) and a second part that stems from variation about the average outcome that occurs at each LC. The latter "residual" variation corresponds to forecast unreliability.

Analysis of variance of this type can be carried out either graphically or analytically (Draper and Smith, 1966). Here a graphical approach will be emphasized. The observed percent reductions for the twelve trials ranged from a low of 40.6% to a high of 100% (Table I). The mean ($\bar{Y}$) was 66.3% and the standard deviation (SD_Y) was 23.1%. The variability in these outcomes can be studied graphically by use of a scatter diagram of values ($Y_i - \bar{Y}$) which measure the deviations of particular outcomes (Y_i) from the average outcome $\bar{Y}$. These deviations are plotted in the left-hand side of Fig. 3.

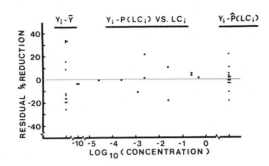

Fig. 3. Deviation of observed percent reductions from mean or forecasting percent reductions in twelve field trials described in Table I.

Some of the variation in these deviations arises from the fact that different concentrations were used in different trials. The variation due to different log concentrations (LC_i) can be removed by considering the "residuals" $(Y_i - \hat{P}(LC_i))$ which are the deviations of the observed outcome Y_i from the predicted outcomes based on the fitted model $\hat{P}(LC_i)$. These "residuals" $(Y_i - \hat{P}(LC_i))$ are plotted vs LC_i in the central portion of Fig. 3. These same "residuals", irrespective of LC_i values, are projected to give the scatter diagram on the right hand side of Fig. 3.

The residuals plotted in the central portion of Fig. 3 suggest that the magnitude of residual variation $SD_Y(LC)$ depends on LC. Therefore, the summary scatter plot on the right side of Fig. 3 represents some sort of mean assessment of residual variation at different LCs. With this limitation in mind, the standard deviation of the residuals is 10.2%. This figure provides a summary measure of the unreliability associated with this forecast procedure.

This analysis is purely descriptive and does not provide for an examination of the effect of sampling error on assessments of residual variation. Nevertheless, this approach does provide a method for assessing forecast unreliability overall, and more specifically as it depends on the predictor (LC). Also, repeated use of this descriptive method could be used to compare the forecast unreliability of different forecast procedures.

The most important question after this analysis is what are the sources of the forecast unreliability ("residual" variation). One likely source, when LC is the only predictor, is the difference in age or developmental stage composition of the target insects in different trials. Forecast methods that use predictive information of this type are described in the next section.

IV. Forecasts Based on the Pathogen and Stage of the Pest

The twelve trials described in Fig. 1 were analysed without regard to the developmental stage of the pest population at the time the control action was taken. If this aspect of the pest population is related to control outcome, and if the age or developmental stage composition of the target insect population differs from trial to trial, then this factor is a source of some of the residual variation in the percent reductions. Fortunately, information about the developmental stages of insects in a particular trial is one of the more readily obtained kinds of predictive information. Consequently, it may be beneficial to remove this factor from residual variation by explicitly incorporating this sort of predictive information into the forecast model.

Some indication of the relevance of stage-adjusted forecasting is provided by data from the second series of six of the twelve trials previously described. In these trials both total and instar-specific insect counts were made. Percent reductions observed by instar and log concentration are plotted in Fig. 4. Only two trials had insects at larval instar I and three trials had insects at instar II. However, instars III, IV and V were each found in all six trials so a better indication of the percent reduction trend with log concentration can be obtained for these instars.

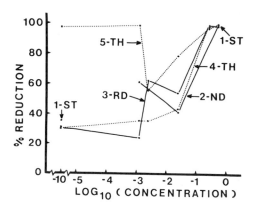

Fig. 4. Percent reduction of *Schizura concinna* larvae 4 days after application of *Bacillus thuringiensis* on *Cercis occidentalis* foliage, by instar.

These sketchy data suggest that instars III and IV show about the same percent reduction trend with LC. The results for instar V show a noticeably different pattern: apparently, in several trials these instar V insects were close to pupation so reduction of their numbers during the 4 days of the trial stems in part from natural development.

Forecasts that incorporate information about the developmental composition of the target insect population require two basic modifications from the forecast method discussed in the previous section. First, calibration data must include pre- and post-treatment insect counts by developmental stage. Then, a percent reduction trend model can be fitted for each part of the pest population to give stage-specific forecast models. Analysis of the stage-specific data from the calibration trials may, however, show that results for some stages are similar so a pooled forecast can be used. It may be important also to assess whether insects in the stage before pupation are in the early or the late part of that stage and treat each part as a separate stage for predictive purposes. Secondly, after calibration at the time of each specific application, the stadial composition of the target population must be determined. For species with synchronous development, the observed current stage of the target population directs the choice of the particular stage-specific forecast model to be used in that application. For species with asynchronous development, the overall forecast can be based on the stage-adjusted forecast method described below.

The specification of a stage adjusted forecast model first requires some definition of symbols. Retain $P(LC)$ as the mean percent reduction in the overall pest population. Let $s =$ stage index that ranges from 1 to S where S represents the number of different stages, $w_s =$ proportion in stage s at the outset of the trial, $P_s(LC) =$ mean percent reduction of stage s insects where log concentration is LC for the specified collection of trials, $\hat{P}_s(LC) =$ stage-specific forecast model, and $\hat{P}(LC) =$ stage-adjusted forecast model. Then the overall mean percent reduction can be expressed as

$$P(LC) = \sum_{s=1}^{s=S} w_s P_s(LC) \tag{8}$$

and the stage-adjusted forecast is

$$\hat{P}(LC) = \sum_{s=1}^{s=S} w_s \hat{P}_s(LC). \tag{9}$$

Thus both the overall mean outcome and the stage-adjusted forecast are weighted means of their stage-specific counterparts with weights provided by the initial stage distribution of the insect population.

The bias in the overall percent reduction is

$$B(LC) = \hat{P}(LC) - P(LC) = \sum_{s=1}^{s=S} w_s B_s(LC) \tag{10}$$

so that $B(LC)$ is the weighted mean of the stage-specific biases, $B_s(LC)$.

Forecast unreliability with stage-adjusted forecasts is described briefly in Section X, B.

V. Forecast Unreliability: Further Analysis of Variance

In the preceding section a stage-adjusted forecast was constructed from stage-specific forecasts. In this section, modelling is used to continue the analysis of variance begun in Section II in order to diagnose the sources of unreliability in these stage-specific forecasts. This can be accomplished by modifying stage-specific forecast models to include elements that specifically represent different sources of variation in the outcome of control actions. It is useful at the outset to identify and describe these sources:

(1) Models are used to forecast according to trial-specific levels of factors used as predictors in the forecast model. However, there are two ways in which this use of a forecast model may be imperfectly implemented. First, a control dimension like log concentration (LC), when used as a predictor, is actually a substitute for the more immediate factor, initial log coverage of the control agent deposited in the pest environment. This varies from trial to trial even when the same log concentration is used in each trial. Secondly, measurement of other kinds of predictors is subject to error that will also vary from trial to trial. In both situations, some degree of misforecasting occurs.

(2) Some factors in the pest control system influence the control outcome but are not explicitly (or implicitly) represented in the forecast model. For example, when log concentration is the only explicit predictor, other factors are: the persistence and pathogenicity of the particular batch of pathogen used in the trial, the susceptibility of the particular insects involved in the trial, immediate environmental conditions, etc. These factors can cause trial-to-trial variation in the mean removal probability of the target insects in the trials. Because the forecast model based only on log concentration has no way of accommodating trial-specific levels of these factors, they are another source of variation in the control outcomes.

(3) In addition to trial-to-trial variation in the (mean) level of these factors, there is also an insect-to-insect variation that occurs in each trial. Also, deposits of the pathogen will be uneven. From a modelling point of view, this means that the probability of removal varies from insect to insect. However, theoretical studies show that probabilistic models with this structure can be represented reasonably well by a simplified model that assumes each insect experiences a removal probability equal to the mean removal probability for the batch of insects involved in the trial (Eisenberg *et al.*, 1963). This simplification will be used in subsequent modelling.

(4) The various factors that influence the control outcome thereby combine to determine a mean removal probability (P) experienced by insects in the trial. Once this removal probability (scaled 0 to 100%) is set by the conditions of the trial, we can imagine that each insect independently undergoes a random experiment that either results in removal, with probability P, or survival with probability 100-P. Under these assumptions, the percent reduction that occurs in a trial with m insects is subject to a source of random variation with standard deviation, $\sqrt{\left(\dfrac{P(100-P)}{m}\right)}$. This derives from binomial variation described in most statistics texts and is the only kind of variation usually incorporated into dose-response models such as the probit model.

(5) Finally, the observed outcome of any trial is based on insect counts that are subject to counting errors. These errors are another source of variation in the observed percent reductions.

The following development provides a way in which the basic stage-specific forecast model given by equations (4) and (5b) can be modified to account for trial-to-trial variation that stems from errors in the measurement of predictors, variation in unobserved forces that influence the outcomes of trials and counting errors.

In any forecast situation the log concentration (LC) is selected to achieve the desired level for the more relevant quantity, the initial log coverage $(LCOV)$ of pathogen in the pest environment. Some preliminary data (Pinnock *et al.*, 1978, Fig. 2) suggest that the mean coverage, $\overline{COV}$, achieved by spraying to the point of run-off is approximately proportional to concentration. That is

$$\overline{COV} \approx K \times \text{concentration.} \tag{11}$$

The proportionality constant, K, reflects units and other details of the coverage and concentration measurements. As a consequence of equation (11) we get the corresponding result,

$$\log(\overline{COV}) \approx \log K + LC. \tag{12}$$

Also, the mean log coverage, $\overline{LCOV}$, is approximately equal to $\log(\overline{COV})$, so

$$\overline{LCOV} \approx \log K + LC. \tag{13}$$

In repeated applications, however, the actual log coverage deviates from the mean because of random influence in the application process. This variation can be represented by expressing the log coverage for the i-th trial in the collection as

$$LCOV_i = \log K + LC_i + D_{LCOV,i} \tag{14}$$

where $D_{LCOV,i}$ = deviation from mean log coverage in the i-th trial.

Suppose further that these deviations have variance V_{LCOV}. This modelling will be incorporated into the forecast model under the assumption that $LCOV_i$ is a better predictor of control outcome than LC_i. Thus LC_i is viewed as a convenient substitute for $LCOV_i$ but it is a substitute that is subject to error with variance, V_{LCOV}.

Also, we can incorporate trial-to-trial variation in the probability of removal due to physical–natural forces. Let

$$U_i = \bar{U} + D_{U,i} \tag{15}$$

where U_i = the physical–natural percent removal probability for the i-th trial, $\bar{U}$ = mean physical–natural removal probability for the specified collection of trials, and $D_{U,i}$ = deviation from mean in the i-th trial.

Suppose further that the deviations have variance V_U. Presumably the variation in U_i stems from physical–natural forces that vary in their intensity from one trial to another.

Other forces may be influencing the insect removal due to the active effect of the control agent. Suppose the combined level of these other relevant forces is represented by X. Then let

$$X_i = \bar{X} + D_{X,i} \tag{16}$$

where X_i represents the combined level of other relevant forces in the i-th trial, $\bar{X}$ = mean level of other forces in the specified collection of trials, and $D_{X,i}$ = deviation from mean in the i-th trial. Suppose further that these deviations have variance V_X.

Finally, suppose counting error — the deviation between observed and true percent reduction — varies from trial to trial. If Y_i = observed percent reduction in the i-th trial, Y_i' = true percent reduction in the i-th trial, and CE_i = counting error in the i-th trial, then

$$Y_i = Y_i' + CE_i. \tag{17}$$

It may be that counting error produces observed results that are consistently low or high relative to true percent reductions. For example, well camouflaged or otherwise elusive insects would tend to be undercounted. Also the variability in counting errors may differ depending on the LC level in the trial since LC influences the number of post-treatment insects left to count and thus LC may be related to the difficulty of counting. These effects can be represented by allowing both the mean counting error and variance in counting errors to depend on LC. Let $\overline{CE}(LC)$ = mean CE as it depends on LC, and $V_{CE}(LC)$ = variance of CE as it depends on LC.

These expressions for values in the i-th trial in a collection of trials can be used to obtain a corresponding expression for the percentage removal probability in the i-th trial. Suppose all trials in the collection are conducted at log concentration equal to LC. Then a two-predictor version of the logit model can be used to give

$$P_i(LC) = \text{percentage removal for the } i\text{-th trial}$$

$$= U_i + (100 - U_i)\left[\frac{1}{1 + e^{-(B_0 + B_1 \, LCOV_i + B_2 \, X_i)}}\right] \tag{18}$$

where new parameters B_0, B_1 and B_2 have been introduced in this model that is extended to explicity relate other factors, X, to percentage removal probability. Then as shown in Section X, C, the following can be deduced.

At the log concentration equal to LC, the mean observed percent reduction is approximately

$$P(LC) \approx \bar{U} + [100 - \bar{U}]\left[\frac{1}{1 + e^{-(B_0 + B_1 \, \log K + B_1 \, LC + B_2 \bar{X})}}\right] + \overline{CE}(LC) \tag{19}$$

If counting error is unbiased so that $\overline{CE}(LC) = 0$, then equation (19) can be directly compared to the earlier trend model given by equations (4) and (5b) to show the following:

(1) the mean physical–natural removal probability, $\bar{U}$, plays the same role as μ;

(2) the coefficient B_1 is the same as A_1; and

(3) $B_0 + B_1 \log K + B_2 \bar{X}$ corresponds to A_0.

Thus in the earlier model, A_0 automatically adjusts to accommodate the proportionality constant that relates coverage to concentration and $\bar{X}$, the mean level of other factors that influence the outcome of control actions.

Also, it can be determined that at a log concentration equal to LC, the standard deviation of the percent reduction, which measures forecast unreliability, is

$$SD_Y(LC) = \sqrt{V_Y(LC)} \tag{20}$$

where the variance $V_Y(LC)$ is approximately

$$V_Y(LC) \approx \frac{P(LC)[100 - P(LC)]}{m} + [1 - M(LC)]^2 V_U$$

$$+ [100 - \bar{U}]^2 [M(LC) [1 - M(LC)]]^2 V_R + V_{CE}(LC). \tag{21}$$

In equation (21) for $V_Y(LC)$ the elements are: $P(LC) =$ mean percentage removal probability (equation (4)), $m =$ number of insects in the trial, $M(LC) =$ removal probability (scaled 0 to 1) due to active effects of the control agent (equation (5b)), $V_U =$ trial-to-trial variance in physical–natural removal probability, $U =$ mean physical–natural removal probability, $V_{CE}(LC) =$ variance of counting error, and

$$V_R = B_1^2 V_{LCOV} + B_2^2 V_X. \tag{22}$$

V_R is a composite source of variance that derives from both coverage variability, V_{LCOV}, and variability of other factors that influence control outcomes, V_X.

Examination of the expression for $V_Y(LC)$ reveals some features that are consistent with biological intuition. The first term stems from binomial variation that depends on the removal probability experienced by insects in the trial. This variation also depends on m, the number of insects in the trial (or the number in a sample selected to monitor the outcome of the control action). By proper choice of m this component of the variation can be controlled.

The second term represents the component of variation that stems from trial-to-trial variability in the physical–natural removal probability. This component is largest when removal due to the pathogen is minimal (i.e., when $M(LC) = 0$). Then the full impact of this source of variability is manifested. By contrast when removal due to the pathogen approaches the upper limit of 1, then the variation in physical–natural forces is overwhelmed by the active effect of the pathogen.

The third term stems from variation in other forces that influence removal due to active effects of the pathogen. This variation is maximized when $M(LC) = 1/2$ and becomes negligible when either no active effect is operating ($M(LC) = 0$) or when the active effect is strong ($M(LC) = 1$) so that variation in log coverage and other factors matter little to the outcome.

Finally, the last source of variation derives from counting error. This variation can be reduced by increasing the number of independent replicate counts used in the trial.

These results show that forecast unreliability depends on the mean percent reduction level and therefore on LC. *They also show that variation due to coverage variability, other factors and trial-to-trial variation in physical– natural forces are inherent effects and cannot be reduced by increasing the number of insects in the trial.*

Although it is beyond the scope of this chapter to go into detail, it appears possible to design experiments to assess the magnitude of the various sources of forecast unreliability in a collection of trials. Unreliability due to variation of physical–natural forces can be assessed directly from a sample of trials with no control agent in the preparation. Variability in the log coverage and variability due to counting error can also be studied directly. Then since the total variation represented by $V_Y(LC)$ can be assessed, it should be possible to deduce the amount of variation represented by V_X that derives from other (unobserved) factors.

VI. Forecast Bias

Forecast error comes partly from forecast unreliability and partly from forecast bias – the difference between the forecast $\hat{P}(LC)$ and mean outcome $P(LC)$ for the specified collection of trials. Recall that the forecast model $\hat{P}(LC)$ is obtained from a two-step process. First some general mathematical form is selected to represent the relationship between the percent reduction and one or more predictive characteristics. This selection, at least at the outset, is an intuitively based choice that draws upon current understanding of the biological process. Then data from a sample of trials are used to calibrate the forecast model.

One possible source of forecast bias comes from use of an inappropriate mathematical form for the forecast model. Every model has structural constraints that limit its ability to flex and represent the actual process under study. For example, a linear forecast model would not have the flexibility to properly represent either a low concentration asymptote for percent reduction or the inherent upper limit of 100% reduction. Bias due to mismodelling of this type typically gives forecasts that are too high at some values of the predictor(s) and too low at other values.

In order to examine the remaining sources of forecast bias it is necessary to pay attention to the collection of field trials for which forecasts are intended. Ideally, the trials in the collection should be identified and listed and a list of plausible log concentration values should also be chosen for study. The ideal calibration sample would be selected at random from the list of trials in the collection and the chosen log concentrations should be randomly assigned to the trials in the sample. These procedures would cause the calibration data to be unbiased relative to the percent reduction trend for the full collection. However, even under these ideal conditions, there is an important source of forecast bias.

The calibration trials represent only a sample of the possible trial and log concentration combinations. Then any parameters fitted from the calibration sample will be subject to sampling error that will cause the fitted parameters to differ from the values of the parameters that would best describe the trend in percent reduction in the full collection of trials. Furthermore, this sampling error persists in the calibrated forecast model for all subsequent applications of the model. Consequently, this discrepancy causes a consistent forecast error (i.e., a forecast bias).

Additional sources of forecast bias can arise when random selection and random assignment are not used to obtain calibration data. The calibration sample gives fitted parameters that are aligned to the mean physical–natural removal probability $\bar{U}$ and the mean level of residual factors, $\bar{X}$, operating when the calibration trials are conducted. If these mean influences are different in the trials to which the forecast model is applied, then the forecasts will be biased accordingly. For example, the forecast model calibrated from trials conducted in one time period can be biased for forecasting the outcome of trials in a subsequent time period if conditions represented by $\bar{U}$ and $\bar{X}$ change.

Forecast procedures with the least trial-specific predictive information would be most subject to bias since there are more potential residual influences whose shift in mean level from calibration to application conditions can cause bias. Models that involve more predictive factors are inherently less subject to forecast bias since they have the potential to adapt to trial-specific levels for all factors that are explicitly used as predictors.

VII. Prospects for Improved Forecast Models

Incorporating additional predictive information into the forecast model can reduce both forecast unreliability and forecast bias. These reductions of forecast error come from the ability of an enlarged model to adapt to trial-to-trial changes in the levels of the added predictors.

The forecast models described so far used two kinds of readily available information – log concentration and stage of the pest. Log concentration can be viewed as a substitute for the effective dose delivered to the pest population, although actually it only relates to the initial deposit of the control agent. Other factors such as persistence and pathogenicity of the control agent and feeding rate and susceptibility of the insects also determine the effective dose delivered to the insect pests. Therefore, the main prospect for improving forecasts lies in getting a better assessment of effective dose in each trial.

A measure of log effective dose (LED) can be constructed in two steps. First, previous work (Brand et $al.$, 1975) has shown that log actual dose (LAD) of $B.$ $thuringiensis$ received through ingestion by insects with a constant feeding rate r can be approximated by

$$LAD \approx LCOV + \log r + \log \pi + K_1 \tag{23}$$

where K_1 is a fixed constant, and π is the persistence half-life of the pathogen in the first few days after application. The log actual dose differs from log effective dose because of the susceptibility of the insects in a particular trial and the potency of the particular batch of pathogen used in the trial. Under the assumptions discussed in Section X, D, the log effective dose is

$$LED \approx LAD + K_2 - \text{LD}_{50} \approx LCOV + \log r + \log \pi - \text{LD}_{50} + K_3 \tag{24}$$

where LD_{50} is measured in logarithmic units, K_2 is explained in Section X, D, and K_3 is another fixed constant. The constant $K_3 = K_1 + K_2$ can be ignored since it is automatically incorporated into the parameters of the dose—response model when

$$LED' = LCOV + \log r + \log \pi - \text{LD}_{50} \tag{25}$$

is used as a predictor.

In principle, LED' (see Section X, D) could be measured to obtain predictive information for each calibration trial and each trial in which the calibrated forecast model is used. The feeding rate could be measured in the few days preceding a trial. (If daily feeding cycles are encountered, equation (25) must be modified to allow for non-constant feeding rate and time of application of the control preparation relative to the feeding cycle). The LD_{50} for the control agent and the specific batch of insects in the trial could be assessed shortly before the start of the trial. Also a pretrial assessment of the persistence half-life (π) could be obtained.

In practice, $LCOV$ can be measured only after control application has occurred so it is not a useful predictor for forecasting. Instead it may be better to revert to use of LC in place of $LCOV$ and accept the forecast unreliability that results from LC being an imprecise substitute for $LCOV$. Also in practice it would be desirable to have a convenient substitute for direct measures of r, LD_{50} and π. For example, it may be possible to relate these characteristics to environmental factors such as temperature, radiation, humidity, etc. and thereby get less precise substitutes that are more easily obtained for use in making trial specific forecasts. Weather forecasts could be used for advance information on these environmental factors. As another intermediate procedure some of the predictors in equation (25) could be ignored altogether. Further research is needed to determine the advantages and disadvantages of the various possible kinds of predictive information that might be used to forecast the outcome of control actions.

VIII. Design of Field Experiments

In the initial research on forecasting outcomes in a particular class of control actions, there are a number of issues to be addressed. How well can outcomes be forecasted with only log concentration as the predictor? How much reduction in forecast error results from the use of stage-adjusted forecasts and by increasing the number of predictors to get a better measure of effective dose delivered to the pest population? How much forecast variation results from sources such as trial-to-trial variation in physical–natural removal and other residual forces that affect control outcomes? How much forecast bias occurs when a forecast model is calibrated from field trial data in one time period and then used to make forecasts later? Field studies are needed to answer these and related questions – preferably in an efficient way since resources are always scarce and there are many classes of control actions to investigate.

There are several broad considerations involved in field study design. First a particular class of control actions must be chosen for study. The field study would thereby be typically focused on a particular target pest, host plant, pathogen, formulation, standardized method of application, geographical region and time period. Then it would be useful to specify in some way the particular collection of trials that are under consideration within the chosen class of trials. Also, the predictors to be explored as candidates for use in forecasting must be identified and appropriate measurement methods devised.

Then the general pattern for a forecast field study must be set. For this, it is helpful to first consider the motivation behind forecasting research. At the outset, this sort of research represents the transition from qualitative to quantitative understanding of microbial pest control processes. Answers are needed to fundamental questions about the forces that operate in a pest control system and the stability of these forces over time. Forecasting research is one mode by which a global, quantitative understanding of these processes can be furthered.

Consequently, the following general features seem reasonable for a forecasting field study conducted primarily for research purposes. Within the specified class of application, a collection of trials over some time period could be identified, at least in general terms. These trials would then be conducted sequentially with the following design features.

(1) Each trial would be conducted with concentration randomly assigned from a spectrum of concentrations designed to range from no active effect to 100% removal. Some non-uniformity in the random assignment may be desirable to give more trials at certain concentrations in order to get data that is most efficient for calibrating the forecast model.

(2) In each trial all predictors that are candidates for use in forecasting control outcome would be measured.

This sort of field study would provide a data base that could be analysed in various ways. For example, the early trials could be used to calibrate a forecast model for application to later trials. Further, this could be done with varying selections from the predictors under study to see how forecast error – both unreliability and bias — depends on the particular predictors involved. Alternatively, a randomly chosen portion of the early trials could be used to forecast outcomes for both the remaining early trials and the later trials. Then forecast bias from sampling error in the estimates of parameters of the forecast model could be distinguished from forecast bias that comes from shifts in the mean level of residual forces that influence control outcomes. Also various sequential calibration strategies could be examined in which the forecast for each trial is based on the forecast model re-calibrated from either the total accumulated data or from the most recent data. From these analyses the feasibility of forecast modelling in practice could be assessed and procedures that might be useful in ongoing pest management identified.

In addition to the broad field study design considerations, there are a number of more specilized design decisions that include the following: (1) How many trials are needed for initial calibration of a forecast model? (2) What concentrations should be used in the calibrating sample in order to effectively estimate the parameters of the forecast model? (3) How many insects at each stage should be included in the trial? (4) How much effort should be invested in measurement of forecast predictors such as persistence half-life, feeding rate and LD_{50}? (5) How many replicate counts should be used to measure the initial and final size of the insect population involved in the trial? Limitations of space and current status of modelling research prevent a full discussion of these issues. However, one issue is considered briefly to indicate how modelling might be used to assist in making these design decisions.

How many insects at each stage should be included in a single trial? If these numbers are large, fewer trials can be conducted for a fixed amount of research effort. Conversely, small numbers involve less effort which is less exhausting so tending to reduce counting error and, more importantly, more trials can be completed. The number (m) in a certain stage has its chief influence through one component of variance in control outcomes. From equation (21) the relevant component is $(P(LC) [100 - P(LC)])/m$ which diminishes as (m) increases. However, trial-to-trial variation due to physical—natural removal and other residual forces occurs even with extremely large trials. This trial-to-trial variation inflates sampling error in fitting the parameters of the forecast model. Then this sampling error becomes a source of forecast bias in subsequent forecasts. Moreover, sampling error can be reduced only by increasing the number of trials used to calibrate the forecast model.

Curbing variation related to trial size follows a pattern of diminishing returns, as can be seen from the following considerations. From equation (21) the standard deviation of the control outcomes (at a particular LC) is

$$SD_Y(LC) = \sqrt{\left(\frac{P(LC) [100 - P(LC)]}{m} + \text{variance from other sources} \right)}$$

For purposes of exploratory calculation, suppose $P(LC)$ equals 50% when variation related to trial size is greatest. Also suppose that $m = 100$ so that standard deviation of variation related strictly to trial size is $\sqrt{\left(\dfrac{50(100-50)}{100}\right)} =$

5%. Finally, suppose variation of other sources is equivalent to variation related to trial size when $m = 100$. Then

$$SD_Y(LC) = \sqrt{((5)^2 + (5)^2)} = 7.1\%$$

For various trial sizes (m) the resulting values for

$$SD_Y(LC) = \sqrt{\left(\frac{50 \times 50}{m} + (5)^2\right)}$$

are

m	25	50	75	100	250	500	750	1000
$SD_Y(LC)\,(\%)$	11.2	8.7	7.6	7.1	5.9	5.4	5.3	5.2

These results show that little reduction in forecast unreliability occurs for trial sizes exceeding, say 250, per stage. For insects with synchronous development this guideline refers to the total size of a trial. Although results will differ somewhat depending on the magnitude of variation from other sources, field studies with many trials of 100–250 insects per stage would be better than a few trials with many more insects per trial.

IX. Discussion

An easy-to-use method of forecast modelling and more involved methods with potentially lower forecast error have been considered. Some more modelling research would be useful even for these limited kinds of forecasting. Calibrating the forecast model can be made more efficient by using a weighted least squares criterion that accounts for the forecast unreliability patterns given by equation (21). As one more example of directions for future work, unreliability in stage-adjusted forecasts, briefly described in Section X, B, should be examined further. And, most importantly, study design modelling should be pursued to provide more specific guidance in planning forecasting field studies.

The appropriate modelling and associated computational work is most easily carried out with a computer. Hopefully, researchers will write programs that can readily accommodate different data file structures; are well documented; are self-contained; are written in a general language, say FORTRAN; and will be offered to other researchers to avoid duplication of effort. This policy will be followed for all programs developed for implementation of the methods presented here.

In this chapter we have restricted attention to the short-term effect of a single-stage control action as measured primarily by percent reduction in the total size of the pest population. Extension to multistage and other kinds of pest control require further research.

For control agents that have a short-term active effect relative to the developmental rate of the target population, multi-stage control can be viewed as a succession of readjustments of the system interspersed by periods of natural evolution. Each phase of natural evolution starts from the state of the systems that existed after the preceding control action. For control agents with a longer term effect, multi-stage control may involve super-positions of successive waves of control effects if later actions occur before the active effects of previous control actions have expired. As the forecast time is increased, the most important outcome to predict may be the damage sustained by the host plants.

With long-term agents, the linkage between the pathogen and the pest population may involve a two-dimensional epidemic process in which the pathogen may migrate or be transmitted from insect to insect and significant replication of the pathogen may occur. This is inherently more complicated than the essentially uni-dimensional effect of *B. thuringiensis* in which the pathogen influences the pest population but the pest population has no important influence on the amount (density) of pathogen in the environment.

Even greater complexities can be envisioned when integrated control is considered in which simultaneous control actions must be represented. It is apparent that many challenges lie ahead in pest control forecasting.

Development and evaluation of forecast procedures will require much field research. Cooperative effort in this venture is also desirable. Persons involved in pest management could identify various classes of *B. thuringiensis* control for which forecasting would be relevant. Appropriate measurement could be identified and standardized data forms devised to facilitate reporting results of trials in each class. These results could be offered to monitoring committees and other interested parties for coordinated analysis and dissemination of the collective results. Names and addresses of chairmen of monitoring committees could be published periodically so that participants in this cooperative research would know where to get a copy of study design guidelines and data reporting forms and where to send results. In the meantime the authors invite communication with others interested in this sort of collaborative effort. Then isolated work that may otherwise go unreported could contribute to the quantitative understanding of microbial pest control processes.

X. Addenda

A. CALIBRATING THE FORECAST MODEL BY EYE

Procedure for fitting the percent reduction curve by eye is based on several structural features of the model given by equations (4) and (5b). At low concentrations, the curve approaches an asymptote represented by μ. At the percent reduction half-way between μ and 100% i.e., at $\dfrac{\mu + 100}{2}$ the slope of the curve is $\dfrac{(100 - \mu)A_1}{4}$. Also, the log concentration (say LC') corresponding to this point is $-A_0/A_1$ (Fig. 5).

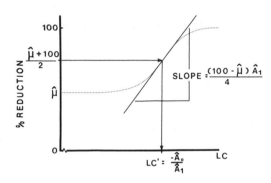

Fig. 5. Structure of the percent reduction trend model.

These features suggest the following procedure for fitting the percent reduction curve by eye: (1) Sketch the S-shaped curve with a low concentration asymptote. (2) Read the percentage reduction asymptote (say $\hat{\mu}$). (3) Draw a horizontal line from $\dfrac{\hat{\mu} + 100}{2}$ to intersect the sketched curve. (4) Draw an extended straight line that seems to best represent the slope of the curve at this intersection. (5) Determine this slope graphically and compute $\hat{A}_1 = \left| \dfrac{4 \cdot \text{slope}}{100 - \hat{\mu}} \right|$.

(6) Drop a line from the intersection and determine the point, LC', on the horizontal scale. (7) Compute $\hat{A}_0 = -\hat{A}_1 \times LC'$.

The resulting values, $\hat{\mu}$, $\hat{A}_0$ and $\hat{A}_1$, then provide the desired estimates of the parameters of the percent reduction curve as fitted by eye. Alternatively, these results can be used as the starting values for the iterative, unweighted least squares estimation procedure.

B. UNRELIABILITY OF STAGE-ADJUSTED FORECASTS

The variability in overall percent reduction in a collection of field trials at log concentration equal to LC can be expressed in terms of the variability and covariability of stage-specific percent reductions. Let $Y =$ overall percent reduction, $s =$ stage index ranging from 1 to S, $Y_s =$ stage-specific percent reduction, and $w_s =$ proportion in stage s. Then

$$Y = \sum_{s=1}^{s=S} w_s Y_s,$$

$$SD_Y(LC) = \sqrt{V_Y(LC)}$$

and from general results for variance of a linear combination (Chiang, 1968, p. 16),

$$V_Y(LC) = \sum_{s=1}^{s=S} w_s^2 V_{Y_s}(LC) + \sum_{s=1}^{s=S} \sum_{\substack{s'=1 \\ s \neq s'}}^{s'=S} w_s w_{s'} C_{Y_s, Y_{s'}}(LC) \qquad (A1)$$

where $C_{Y_s, Y_{s'}}(LC)$ represents the covariance between the percent reduction outcomes for stages s and s'.

The first term in equation (A1) stems from the unreliability in the stage-specific percent reductions. The second term arises from residual factors that simultaneously influence each of the stage-specific percent reductions. For example, wind conditions that may influence physical removal in a particular trial would most likely have some effect on insects at each stage of development. Conditions that prolong or diminish persistence of the control agent would simultaneously influence dosage delivered to insects at each stage. Modelling that extends the results of Section V is needed to further diagnose sources of unreliability in stage-adjusted forecasts.

C. UNRELIABILITY IN STAGE-SPECIFIC FORECASTS

Equation (19) for $P(LC)$ derives from the first order approximation that the mean (expected) value of a function of several random variables is the function of the mean values of those random variables.

Equation (21) for $V_Y(LC)$ derives from the following four considerations:

(1) Counting error (CE) and true percent reduction (Y') are assumed to be statistically independent so the variance of the observed percent reduction (Y) is

$$V_Y(LC) = V_{Y'}(LC) + V_{CE}(LC). \qquad (A2)$$

(2) From the general results for conditional variance (Chiang, 1968, p. 19),

$$V_{Y'} = E[V_{Y'/LCOV, U, X}] + V_{E[Y'/LCOV, U, X]}. \qquad (A3)$$

In equation (A3) notation to indicate the specific log concentration level is omitted for simplification and the usual expected value notation $E[\ ]$ that occurs in statistical writing is used in place of the bar notation used for means in the rest of the paper.

(3) For given values of $LCOV$, U, and X the variance of Y' is obtained by noting it is a binomial random variable times a constant $(1/m)$ where m is the number of insects in the trial. Then using the approximation for the mean of a function of random variables, indicated above, it follows that

$$E[V_{Y'/LCOV, U, X}] \approx \frac{P(LC)[100 - P(LC)]}{m}$$

(4) Finally, from general results for approximating the variance of a function of random variables (Chiang, 1968, p. 18), it follows that

$$V_{E[Y'/LCOV,U,X]} \approx [1 - M(LC)]^2 V_U + [100 - \mu]^2 [M(LC)[1 - M(LC)]] V_R.$$
$$(A4)$$

D. AN EXPRESSION FOR LOG EFFECTIVE DOSE

Suppose the dose-response relationship based on log available dose (LAD) is

$$\ln \left[\frac{M(LAD)}{1 - M(LAD)} \right] = A_0 + A_1 \times LAD \qquad (A5)$$

for the mean potency–susceptibility in a specified collection of trials. Let LD'_{50} represent the 50% lethal dose for this average situation. Now assume that in specific trials the dose–response relationship differs by a translation to a new 50% lethal dose equal to LD_{50}. Then

$$\ln \left[\frac{M(LAD, LD_{50})}{1 - M(LAD, LD_{50})} \right] = A_0 + A_1 (LAD + (LD'_{50} - LD_{50})). \qquad (A6)$$

Then $LAD + (LD'_{50} - LD_{50})$ can be interpreted as the log effective dose (LED). Thus LD'_{50}, corresponding to the mean 50% lethal dose in the collection of trials, corresponds to K_2 in equation (24).

References

Brand, R.J., Pinnock, D.E., Jackson, K.L. and Milstead, J.E. (1975). *J. Invertebr. Path.* **25**, 199–208.

Brand, R.J., Pinnock, D.E., Jackson, K.L. and Milstead, J.E. (1976). *J. Invertebr. Path.* **27**, 141–148.

Chiang, C.L. (1968). "Introduction to Stochastic Processes in Biostatistics". John Wiley, New York.

Cox, D.R. (1970). "The Analysis of Binary Data". Methuen, London.

Draper, N.R. and Smith H. (1966). "Applied Regression Analysis". John Wiley, New York.

Eisenberg, H.B., Geoghagan, R.R.M. and Walsh, J.E. (1963). *Biometrics* **19**, 152–157.

Finney, D.J. (1964). "Probit Analysis: A Statistical Treatment of the Sigmoid Response Curve". Cambridge University Press, Cambridge.

Pinnock, D.E., Brand, R.J., Milstead, J.E., Kirby, M.E. and Coe, N.F. (1978). *J. Invertebr. Path.* **31**, 31–36.

CHAPTER 38

Compatibility of Pathogens with Other Methods of Pest Control and with Different Crops

R.P. JAQUES and O.N. MORRIS

Research Station, Agriculture Canada, Harrow,
Ontario and Canadian Forestry Service, Environment Canada,
Sault Ste Marie, Ontario

I. Introduction

The concept of integrated pest management is based on optimum exploitation of indigenous and introduced mortality factors in maintaining pest species at economically acceptable densities. Insect pathogens can have a major role in pest management but their effective integration into such systems depends on their compatibility with the other components of the system. This chapter concerns compatibility of introduced and indigenous entomopathogens with other factors in the field environment, especially chemical pesticides, other pathogens and parasitic and predaceous arthropods.

Terms describing interactions of mortality factors (synergism, additive effect, etc) have not been used consistently in the literature. To follow definitions by Benz (1971) and common usage, and to avoid confusion in this discussion, joint action of entomopathogens and chemical pesticides will be termed potentiation or supplemental when the combined effect is greater than the sum of the activities of the components acting alone. Additive effect will designate activity similar to the sum of the individual activities whereas antagonism will refer to a combined effect that is less than that of the most active component or of the sum of the individual activities.

II. Compatibility of Entomopathogens and Chemical Pesticides

Compatibility with chemical pesticides is important in the effective usage of entomopathogens in pest management systems. It is essential that pathogens are not inactivated by chemicals in tank mixes of sprays or by deposits of chemicals on the plant. Likewise, the pathogens should not adversely affect activity of the chemical pesticide. Enhanced effectiveness through joint action of pathogens and chemical pesticides, especially insecticides, is particularly interesting because the reduced amount of chemical insecticide required for crop protection would reduce chemical contamination of the field habitat and permit maximum impact of predaceous and parasitic arthropods on pest species.

A moderately accurate estimate of compatibility of chemical pesticides and pathogens may be made by laboratory tests but a final assessment must be made under field conditions. A large number of combinations of chemicals and pathogens may be screened by laboratory tests. Furthermore, laboratory tests can be readily replicated making them more amenable to statistical analysis than are field data. Effects on indigenous pathogens must, of course, be determined by field tests. While numbers of target insects surviving after application of pathogens, or mixtures of pathogens and chemicals, indicate effectiveness, determination of crop protection is a superior criterion on which to assess efficacy.

Laboratory procedures differ with different types of pathogens. Compatibility of mixtures of viruses, protozoans or bacteria with chemicals may be assessed by efficacy against test larvae fed diet or natural food treated with mixtures of the materials. In addition, effects of chemical residues on activity of pathogens may be determined by bioassays of pathogen activity following exposure to the chemical. Growth of bacterial and fungal pathogens on agar or similar nutritive substrates treated with the chemical being tested also indicates compatibility.

A. COMPATIBILITY AND EFFECTIVENESS OF BACTERIA AND PESTICIDES

1. Laboratory Studies on Compatibility

Studies on compatibility of *Bacillus thuringiensis* (*B.t.*) and insecticidal chemicals after a previous review by Benz (1971) are summarized in Table I. Most insecticides were compatible with *B.t.* having little or no effect on spore germination or cell multiplication. Low concentrations of some, e.g. carbamates and organophosphates, either did not affect bacterial growth or improved it,

TABLE I

In vitro tests on the effect of insecticidal chemicals on Bacillus thuringiensis [a]

Effect on spore germination			Effect on cell multiplication		
No effect	Stimulatory	Antagonistic	No effect	Stimulatory	Antagonistic
Acephate[6]	Carbaryl[1]	Carbaryl[2a]	Acephate[7]	Mexacarbate[7]	Aldrin[8]
Carbaryl[7]	Dimethoate[5,7]	Chlordimeform[7]	Carbaryl[3,8]		Chlorpyrifos[7]
Carbofuran[2]	Fenitrothion[6]	Fenitrothion[6]	Carbofuran[8]		DDT[2,8]
Methomyl[1,7]	Gardona[1]	Pyrenone[7]	Chlordimeform[7]		Fenitrothion[7]
Phosmet methyl[2]	Lindane[1]	Stirophos[2]	DDT[4]		Gardona[6]
Trichlorfon[2]	Trichlorfon[1,7]		Diazinon[8]		Heptachlor[8]
			Malathion[8]		Naled[3]
			Methomyl[7]		Parathion[3]
			Parathion[9]		Phosmet[7]
			Phorate[8]		Temephos[7]
			Propoxur[3]		Trichlorfon[7]
					Volaton[7]

[a] Numbers following chemical names indicate references as follows: [1] Altahtawy and Abaless, 1972; [2] Chen et al., 1974; [3] Dougherty et al., 1971; [4] Lebrun, 1976; [5] MacRae and Celo, 1974; [6] Morris, 1975b; [7] Morris, 1977b; [8] Sutter et al., 1971.

whereas others, especially the chlorinated hydrocarbons (DDT, aldrin, heptachlor), inhibited growth. For example, Dougherty *et al.* (1971) assessed growth inhibition on agar plates by the treated-disc procedure and showed that carbaryl and Baygon had no effect on the bacterium whereas the chlorinated hydrocarbons were slightly inhibitory and parathion and naled highly so. Tests on inert surfaces (Chen *et al.*, 1974) are particularly interesting because the test conditions more closely resembled those in the field. Carbaryl and sitophos killed some spores in one formulation of *B.t.* but had little effect on another.

These and other laboratory studies indicate that normal concentrations of most chemical insecticides applied in agriculture and forestry would not significantly inhibit growth of *B.t.* These tests do not describe effects on the endotoxin of *B.t.* It is noteworthy, however, that methyl bromide, a fumigant, in stored grain inactivated the *B.t.* spore but did not appreciably reduce the insecticidal effect of the *B.t.* preparation on the Indian meal moth, *Plodia interpunctella* (McGaughey, 1975).

Extensive tests by Morris (1977b) indicated that emulsifiers and other additives normally used in emulsifiable concentrates and wettable powders are incompatible with the bacteria. Indeed, Morris (1969, 1975b) has shown that several additives, including Atlox and Triton X-100, which are commonly used in chemical pesticide formulations may inhibit *B.t.* spore germination and replication, suggesting that technical grade insecticides without additives are preferable in *B.t.*–insecticide mixtures.

2. *Laboratory Tests on Enhancement of* B. thuringiensis *by Chemicals*

In laboratory bioassays effectiveness of *B.t.* against several insect pests of agricultural crops and forests was enhanced by combination with some chemical insecticides (Table II). Others, e.g. demeton, dimethoate and stirophos, reduced activity of the bacterium. In general, enhancement was greatest among larvae fed low concentrations of the chemicals and low to moderate concentrations of the pathogen.

3. *Field Studies with* B. thuringiensis *and Chemicals*

Field tests against agricultural and forest pests have shown that the amount of chemical insecticide applied can be greatly reduced by integrating entomopathogens into the spray program, without sacrificing crop quality (Table II). Some of the tests did not include applications of the components alone at the rates used in the mixtures and although joint action of the materials could not be assessed properly, there was a strong indication of an additive effect in these and other tests.

(a) Agricultural Insects. The good crop protection by low dosages of chemicals and *B.t.* against agricultural insects is demonstrated by summarized data of tests against *Trichoplusia ni*, *Pieris rapae* and the lesser pest, *Plutella xylostella*, on cabbage in the USA and Canada (Table III). To fully appreciate the significance of these data it must be realized that crop protection in the field is not a linear relationship partly because the relationships of dosage to mortality and

TABLE II

Effect of chemical insecticides on Bacillus thuringiensis in bioassays and field tests

Insecticide	Test, target insect	Effect	Reference
Acephate	Assay, *Orgyia leucostigmata*	Potentiation	Morris, 1975a
	Forest, *Choristoneura fumiferana*	< Additive	Morris and Armstrong, 1975
	Forest, *Choristoneura fumiferana*	Additive	Morris *et al.*, 1975
Binapacryl	Assay, *Plutella xylostella*	Potentiation	Hamilton and Attia, 1977
Carbaryl	Assay, *Heliothis virescens*	Potentiation	Chen *et al.*, 1974
	Field, *Heliothis zea*	Additive[a]	Creighton *et al.*, 1973b
Carbofuran	Assay, *Heliothis virescens*	Additive	Chen *et al.*, 1974
Chlordimeform	Assay, *Plutella xylostella*	Potentiation	Hamilton and Attia, 1977
	Field, *Pieris rapae*	< Additive	Creighton and McFadden, 1975
	Field, *Trichoplusia ni, P. xylostella*	Additive	Creighton and McFadden, 1975
	Field, Lepidopterans on cabbage	Additive[a]	Creighton *et al.*, 1973a;
			Creighton and McFadden, 1974;
			Creighton and McFadden, 1975;
			Jaques and Laing, 1978
	Field, *Keiferia lycopersicella*	Additive	Poe and Everett, 1974
	Field, *Heliothis zea*	Additive[a]	Creighton *et al.*, 1973b
DDT	Assay, *Tenebrio molitor*	Potentiation	Lebrun, 1976
Demeton-S-methyl	Assay, *Plutella xylostella*	Antagonism	Hamilton and Attia, 1977
Dimethoate	Assay, *Plutella xylostella*	Antagonism	Hamilton and Attia, 1977
Fenitrothion	Assay, *Choristoneura fumiferana*	Potentiation	Morris, 1975a; Boucias, 1974
	Assay, *Orgyia leucostigmata*	Potentiation	Morris, 1975a
	Forest, *Choristoneura fumiferana*	Potentiation	Morris and Armstrong, 1975
Methomyl	Assay, *Heliothis virescens*	Additive	Chen *et al.*, 1974

toxaphene and *T. ni* NPV reduced numbers of *T. ni* larvae more than either material alone. More important, the toxaphene–NPV mixtures protected the crop longer than other treatments, indicating a desirable feature of viruses. Combinations of *T. ni* NPV and chemicals at reduced rates gave excellent control of *T. ni* on cabbage in field tests in the USA (Creighton *et al.*, 1970) and in Canada (Jaques, 1977b) but an assessment of additive effect was impossible because plots treated with the insecticides alone at the reduced rates were absent from the tests. Jaques' tests did demonstrate, however, the advantage in combining low concentrations of *T. ni* NPV and *P. rapae* GV with chlordimeform and/or *B.t.* in that plots treated with 1/8 of the recommended dosage of the chemical combined with 1/5 of normal dosage of the viruses yielded 100% marketable crop, being equal or superior to the yield from plots treated with the full dosage of viruses or chemical. Addition of *B.t.* to the mixture increased foliage protection slightly but did not improve protection of the cabbage heads.

C. COMPATIBILITY AND EFFECTIVENESS OF FUNGI AND PESTICIDES

Much of the impact of entomogenous fungi on populations of insects is by naturally occurring fungi rather than by fungi applied for control of specific target insects. Because some of the entomogenous fungi are not readily propagated in the laboratory culture, assessment of the effect of pesticides on these is, by necessity, based on field tests without supporting laboratory data. Comprehensive reviews by Benz (1971) and Roberts and Campbell (1977) show that fungi are adversely affected by many fungicidal and fungistatic agents in commercial use but not by most insecticides. Pesticides that inhibited growth of entomogenous fungi in *in vitro* laboratory tests include the common fungicides benomyl, maneb, zineb, captan, ferbam, chlorothalonil and mancozeb, and the insecticides methyl parathion and phenthoate (Chapter 26, Table III; Chapter 27, Table VI). Benomyl was among the most inhibitory chemicals against *Beaveria bassiana* (Olmert and Kenneth, 1974; Zimmerman, 1975) *Nomuraea rileyi* (Ignoffo *et al.*, 1975), *Verticillium lecanii* (*Cephalosporium aphidicola*) (Wilding, 1972), *Beauveria brongniartii* (*tenella*), *Metarhizium anisopliae*, *Paecilomyces farinosus* (Zimmerman, 1975). Pesticides compatible with fungi in *in vitro* tests included: with *B. bassiana*, copper oxychloride, dinocap and Daconil (Olmert and Kenneth, 1974); with *N. rileyi*, pyroxychlor, acephate, carbaryl, carbofuran, DBCP, DDT, Dimilin, endrin, methoxychlor, bentazone, chlorbromuron, chloroxuron, dalapon-Na, metribuzin, naptalam and trifluran (Ignoffo *et al.*, 1975); and with *Verticillium lecanii*, binapacryl and Daconil (Olmert and Kenneth, 1974). (See also Chapter 26, VI, B, 1 for *Hirsutella.*)

An especially interesting *in vitro* study (Soper *et al.*, 1974) showed that benomyl reduced growth of *Alternaria solani*, a plant pathogen, less than that of four *Entomophthora* species whereas chlorothalonil inhibited *A. solani* but had little effect on *Entomophthora* spp. Benomyl was highly fungistatic to *V. lecanii*, a pathogen of the melon aphid, *Aphis gossypii*, whereas the fungicides triarimol and dimethirimol were much less toxic in laboratory tests

(Wilding, 1972). Dimethirimol had little effect on fungal infection of the aphid in the field indicating the need for careful selection of chemicals for use in pest management systems that involve entomopathogens. (See also Chapter 28, III, F for *Entomophthora*.)

The value in selecting pesticides compatible with entomofungi is well illustrated by the effect of chemical pesticides on infection and mortality of host insects in laboratory bioassay and field tests (Table V). In these tests fungicides generally inhibited entomofungi while most insecticides were compatible or stimulatory. For example, populations of *Psylla mali* on apples in Canada (Jaques and Patterson, 1972) and of aphids on potatoes in USA (Nanne and Radcliffe, 1971) were increased by application of fungicides that inhibited naturally occurring entomofungi. Similarly, application of methidathion to citrus trees increased populations of mites that are often controlled by natural epizootics of the fungus *Hirsutella thompsonii* (Chapter 26, V and VI, B1). In another study benomyl applied alone or mixed with parathion or carbaryl reduced yields of soybeans by suppressing the fungus *N. rileyi* (Chapter 27, V, B), an effective pathogen of *Anticarsia gemmatalis*, a serious pest of soybeans in the USA. Combining insecticides with entomofungi may not only kill more insects than the fungi alone but may also kill them more rapidly as was found with combinations involving *Beauveria* against the Colorado potato beetle and codling moth (See Chapter 24, VI, A).

These limited field data supported by laboratory tests show that entomo-fungi are more frequently adversely affected by pesticides, especially fungicides, than are bacteria or viruses with potentiation by utilization in combination with insecticides seeming to be less probable than for bacteria or viruses.

D. COMPATIBILITY OF PROTOZOA AND NEMATODES WITH PESTICIDES

Compatibility of protozoans and nematodes with chemical pesticides has been studied little (Maddox, 1977). Susceptibility of *Anthonomus grandis* to DDT, malathion, azinphos—methyl and carbaryl increased 5- to 7-fold as infection by the protozoan *Mattesia grandis* progressed, apparently due to degradation of the fat body by the protozoa (Bell and McLaughlin, 1970). Similarly, infection by *Nosema attiorrhynchi* enhanced response of *Brachyrhinus ligustici* to DDT in laboratory studies (Rosiky, 1951). Germination of *Vairimorpha* (*Nosema*) *necatrix* spores was not affected by exposure to Pyrenone but was reduced by malathion, TMPP and TEPP (tetramethyl- and tetraethyl-pyrophosphate) in laboratory tests (Maddox, 1977). Benomyl inhibited various microsporidia in laboratory tests (Hsiao and Hsiao, 1973) and could conceivably suppress them in the field habitat. Low concentrations of malathion are recommended for use with *Nosema locustae* in some circumstances in the field (Chapter 30, IV, B). Entomogenous nematodes tolerated some chemical insecticides but not others (Chapter 33, V).

TABLE V

Effect of chemical pesticides on activity of fungi in laboratory bioassays and field tests

Fungus	Chemical	Test host	Effect	Reference
Beauveria bassiana	Azinphos ethyl, carbaryl	Field, *Leptinotarsa decemlineata*	Compatible, < additive	Fargues, 1975
	DDT	Assay, *L. decemlineata* Assay, *L. decemlineata*	Shorter lethal time Compatible, < additive	Braat-Wildschut, 1966 Fargues, 1972, 1973
Beauveria brongniartii (= *tenella*)	BHC, parathion, trichloronate	Assay, *Melolontha melolontha*	Shorter lethal time, < additive	Ferron, 1970, 1971
Entomophthora sphaerosperma	Captan, ferbam, sulphur	Field, *Psylla mali*	Antagonistic	Jaques and Patterson, 1972
Hirsutella thompsonii	Methidathion	Field, *Phyllocoptruta oleivora*	Antagonistic	McCoy, 1977 Chapter 26, III, B,1
Nomuraea rileyi	Benomyl	Assay, *Trichoplusia ni*	Antagonistic	Ignoffo *et al.*, 1975
	Benomyl, carbaryl, methyl parathion	Field, *Anticarsia gemmatalis*	Antagonistic	Johnson *et al.*, 1976
Verticillium lecanii	Dimethirimol	Field, *Aphis gossypii*	Compatible	Wilding, 1972

III. Compatibility of Entomopathogens with other Biological Agents

The interaction of entomopathogens with other biological agents acting on pest species is a major concern in integration of pathogens into pest management systems. Interference and competition with other pathogens and with parasitic and predaceous arthropods could reduce net effectiveness, perhaps making introduction of a pathogen impractical.

A. COMPATIBILITY WITH OTHER ENTOMOPATHOGENS

Entomopathogens may be compatible, additive through their mutal coexistence, more rarely synergistic or potentiating, or they may compete for the host or antagonize one another (Krieg, 1971; Table VI; Chapter 33).

Synergism of NPV of *Mythimna* (*Pseudaletia*) *unipuncta* by a GV (Table VI) is especially interesting because both viruses occur in field populations of the host. The synergistic factor in the GV capsule that enhanced the infectivity of the NPV was identified as an enzyme (Tanada and Hara, 1975; Tanada et al., 1973, 1975) and later as a phospholipid (Yamamoto and Tanada, 1978) which acts on cell membranes of host tissues. Conversely, interference between two GVs of *C. fumiferana* prevented development of the two viruses in the same cell and when they developed in the same tissue aberrant inclusion bodies were found (Bird, 1976). Although the cytoplasmic polyhedrosis virus (CPV) of *C. fumiferana* interfered with NPV development, sequential combination of the viruses could be advantageous in control (Bird, 1969). (See Chapter 22, III, for interactions of CPV on NPV, GV and *B.t.* and Chapter 32, III,C, for *Nosema*.)

Double infection by different types of pathogens in the laboratory has usually resulted in increased mortality especially when the infections are sequential rather than simultaneous (Table VI). For example, prior infection of *M. melolontha* by entomopox virus (EPV) substantially increased susceptibility to *B. brongniartii*, and NPV-infected *T. ni* larvae were more susceptible to *B.t.* In contrast, simultaneous infection of *Hyphantria cunea* with NPV and a microsporidian and of *C. fumiferana* with NPV and *B.t.* had less than an additive effect on host mortality (see also Chapter 32).

Most field tests on mixtures of entomopathogens have involved *B.t.*–virus combinations (Table VI). Although the results have been variable, most tests to date indicate that introduction of compatible pathogens has considerable potential in pest management. For instance, *B.t.*–virus mixtures were only slightly more effective than the components used alone against *Orgyia pseudotsugata* or *Dendrolimus spectabilis* but were considerably better than either components against *T. ni.*

The technique of applying *B.t.* and viruses separately at strategic intervals (Nef, 1971; Jaques, 1972a, 1973, 1977b; Table VI) warrants further investigation. The components do not compete nor do their effects overlap. The virus becomes established in the field habitat for long-term suppression while the bacterium provides short-term control. It is proposed that entomopathogens and chemical pesticides could be combined in this way.

number of uninfected hosts (Laigo and Tamashiro, 1966). Similarly, the parasite *Hyposoter exiguae* died in *T. ni* larvae exposed to NPV more than 2 days before and up to 4 days after parasitization (Beegle and Oatman, 1975). In other laboratory studies, infection of *Heliothis virescens* with NPV within 2 days after oviposition by *Campoletis sonorensis* substantially reduced survival of the parasite (Irabagon and Brooks, 1974) and emergence of the parasite *Pteromalus* sp. was curtailed by treatment of *P. rapae* larvae with GV or a microsporidian (Laigo and Paschke, 1968). The protozoans *Nosema polyvora* and *M. grandis* adversely affected *A. glomeratus* (Issi and Maslennikova, 1964) and *Bracon mellitor* (Bell and McGovern, 1975) which parasitize *T. ni* and *A. grandis*, respectively. In a USA field study an inverse relationship between incidence of fungal disease and parasitism by hymenopteran species in populations of the bag-worm, *Thyridopteryx ephemeraeformis*, was attributed to different responses of the fungus and the parasites to weather conditions (Berisford and Tsao, 1975).

Joint action of entomopathogens and parasitic arthropods may be advantageous due to transmission of pathogens by parasites: e.g. *Malacosoma disstria* NPV was transmitted by *Sarcophaga aldrichii* (Stairs, 1966), *L. dispar* NPV by *A. melanoscelus* (Raimo, 1975) and *P. rapae* GV by *A. glomeratus* (Kelsey, 1960). Also, *Itoplectis conquisitor* transmitted *Serratia marcescens* among *G. mellonella* (Bucher, 1963) and *Venturia* (*Nemeritis*) *canescens* transmitted *B.t.* among *E. kuehniella* (Vago and Kurstak, 1965). An interesting host–parasite–pathogen relationship exists in *Heliothis* species (Brooks and Cranford, 1972; Brooks, 1973). The parasite *C. sonorensis* was not only susceptible to *N. heliothidis* of its host larva, *H. zea*, but also transmitted the microsporidian to its own progeny. In addition, both *C. sonorensis* and *C. nigriceps* possessed their own transovarially transmitted microsporidia which were not infectious to *Heliothis* larvae.

IV. Compatibility of Entomopathogens with Crop Plants

There is some evidence that the insect's food plant may influence effectiveness of entomopathogens as control agents. The plant may inactivate deposits of the pathogen or influence susceptibility of the insect to the pathogen.

Efficacy of NPV and *B.t.* varies with different species of coniferous and deciduous trees (Poltev and Peshcherskaya, 1967; Smirnoff, 1968; Morris, 1972a; Morris *et al.*, 1974; Cunningham *et al.*, 1975; Pinnock *et al.*, 1975). The effect on *B.t.* may be partly due to bactericidal exudates on the leaf surface, e.g. the terpene components of coniferous and deciduous foliage inhibit growth of *B.t.* (Smirnoff, 1968; Morris, 1972a). Addition of some plant juices to the food of *Archips cerasivoranus* larvae retarded development of the microsporidian *Nosema cerasivoranus* in the larvae, suggesting antagonism (Smirnoff, 1967). Also, the inorganic and organic content of the host insect's food may influence susceptibility to disease (David *et al.*, 1972).

Other evidence that substances produced by plant leaves affect activity of pathogens is largely circumstantial. Deposits of viruses on foliage in the field are rapidly inactivated presumably by ultraviolet light (Jaques, 1977a). However, the partial inactivation of *P. brassicae* GV and *T. ni* NPV on non-wetted leaves of cruciferous plants kept in the dark (David *et al.*, 1971; Jaques, 1972b) suggests that substances of plant origin contribute to virus inactivation. In contrast, *V. necatrix* spores on cruciferous plants retained activity (Chu, 1977).

Significantly deposits of *Heliothis* NPV on cotton lost activity at night when the leaves were wetted with dew (Andrews and Sikorowski, 1973). Dew on cotton leaves was alkaline due to basic ions; prolonged immersion of *Heliothis* NPV in dew caused swelling of the polyhedral bodies. The chemical content of dew differs with the plant; cotton dew was normally more alkaline (pH 8.8) and contained a higher concentration of ions than did dew on soybean leaves (pH 7.8) (Young *et al.*, 1977). *Heliothis* NPV retained activity when held in either dew but, if suspensions were dried and resuspended, the virus resuspended in cotton dew was inactivated whereas that in soybean dew was not. The virus was readily inactivated, however, in highly alkaline dew (pH 9.3) on cotton leaves (McLeod *et al.*, 1977). Composition of dew is undoubtedly influenced to a large extent by deposits from the air but these studies indicate that substances from the plant leaves are involved. (See also Chapter 17, IV,D).

V. Conclusions and Suggested Research Priorities

Several entomopathogens have potential for a major role in pest management systems. This review shows, however, that realization of their full potential depends substantially on their compatibility with factors in the field environment, especially pesticides. Pathogens can enhance the effectiveness of one another and of some chemicals. Conversely, pathogens may reduce the effectiveness of one another, of chemical pesticides, or of other mortality agents such as parasitic arthropods. Therefore, integrated systems involving entomopathogens must be examined carefully, although in general they have much less effect than chemicals on other biological agents. It is evident that while some indication of compatibility can be gained by laboratory procedures, a reliable assessment requires testing in the field environment.

Some attention should be given to selection of fungicidal chemicals compatible with pathogens, especially entomofungi and protozoans, to achieve maximum mortality of pest insects by naturally occurring and introduced pathogens.

The high efficacy of some pathogen—chemical insecticide mixtures suggests their use to reduce chemical contamination of the environment, without sacrificing crop protection. Research to exploit this use of pathogens is warranted. Furthermore, the strategy of separate but strategically timed applications of pathogens and chemicals should be developed.

D. PREINOCULATION OF PLANTS

Another approach is to protect the plant by preinoculating it with an avirulent organism related to the plant pathogen. A serious tomato wilt is caused by *Fusarium oxysporum* f. sp. *lycopersici*. Preinoculation of the roots of tomato seedlings with *Cephalosporium* sp. reduced infection after subsequent inoculation with *Fusarium:* the longer the interval between the two inoculations, the greater the protection (Phillips *et al.*, 1967). *Cephalosporium* may trigger a non-specific occlusion reaction in the vascular system, similar to that which blocks growth and multiplication of the pathogen (*F. oxysporum* f. sp. *cubense*) in banana roots (Beckman and Halmos, 1962).

Crown gall on peach and almond is caused by the soil bacterium *Agrobacterium radiobacter* var *tumefaciens*. In Australia, nearly all the agrobacteria around healthy trees are non-pathogenic, whereas around diseased trees a high proportion are pathogenic. When roots of peach seedlings were experimentally dipped in a suspension of a non-pathogenic strain, wounded, then transferred to soil infested by a virulent strain, no galls developed (New and Kerr, 1972). Three treatments — seed inoculation, root inoculation, or a combination of both — reduced the mean dry weight of gall tissue per plant that developed after growing in naturally infested soil by 78%, 95% and 99% respectively compared with no treatment (Htay and Kerr, 1974). The non-pathogenic strain produces a bacteriocin which inhibits growth of pathogenic agrobacteria. Seed and root inoculation of peach and almond are now widely practised commercially in Australia, giving a spectacular decline in crown gall. Economic control of a root disease is thus achieved by a specific microorganism altering the rhizosphere population.

Considering other tree diseases, there is good evidence that ectomycorrhizal fungi help to protect pines from root infection by *Phytophthora cinnamomi* (Marx, 1972). The mechanisms involved include creation of a mechanical barrier by the fungal sheath, formation of inhibitors by pine roots in response to infection by the ectomycorrhizal fungi and production of antibiotics by these fungi. The many symbiotic fungi differ considerably in these last two respects, and seedling pines could possibly be inoculated with a symbiont selected to confer maximum protection when planting in heavily infested soils. In Australia, where *P. cinnamomi* causes very serious losses in eucalyptus, much work has been done on disease suppression in certain soils (Broadbent and Baker, 1975). Although competition by soil microorganisms undoubtedly plays a part, there is little prospect yet of achieving practical control of this fungus with specific antagonists.

E. TREE STUMP COMPETITORS

Whereas *P. cinnamomi* and related fungi attack small feeder roots, some wood-rotting fungi attack larger roots causing immense losses in tree crops of many kinds. The main sources of infection are stumps or roots of trees already killed by a fungus such as *Armillariella* (*Armillaria*) *mellea*. In California *A. mellea* is controlled in citrus by removing most infected material mechanically and fumigating the soil with carbon disulphide. The fungus then dies out in remaining root debris, an effect partly due to colonization of the treated soil by *Trichoderma viride,* which replaces *A. mellea* in the root fragments (Bliss, 1951) — a good example of biological control integrated with chemical treatment.

In less valuable crops, or in more difficult terrain, stump removal is impracticable, and stumps are treated *in situ* (Rishbeth, 1976). The major sources of infection are stumps of broad-leaved trees, and chemicals used to prevent regrowth on such stumps affect their colonization by fungi. Ammonium sulphamate generally favours species causing rapid wood decay, some of which compete well with *A. mellea.* Effective colonization of birch stumps was obtained by inoculating cut surfaces with basidiospores of *Coriolus versicolor* and then treating them with 40% ammonium sulphamate. In stumps jointly inoculated with *C. versicolor* and *A. mellea,* the amount of wood occupied by the parasite after 4 years was much less than in stumps inoculated with *A. mellea* alone. Such a limitation is virtually certain to restrict the parasite's activity. Similar experiments with wood-rotting fungi (Rayner, 1977; Woodgate-Jones, 1977) have shown that it is not enough to select competitive species only by their effect on *A. mellea in vitro.* One must also consider how the competitor colonizes the roots and whether it is inhibited by chemical treatment of stumps. *Hypholoma fasciculare* and *Phlebia merismoides,* for example, show considerable promise as competitive species on several hardwoods treated with ammonium sulphamate. Such biocontrol would be particularly useful where serious damage is expected, and in forests and other areas where *A. mellea* creates special problems.

Decisions about control measures for *A. mellea* in forests are complicated by the difficulty of assessing economic losses. However, some data are available for another widely distributed root parasite of trees, *Fomes annosus* (= *Heterobasidion annosum*): the annual loss in W. Europe is thought to exceed £ Stg. 30×10^6. The background to biocontrol of this fungus is worth considering in more detail because it is the first extensive application of such a method to a forest disease. *F. annosus* mainly kills conifers, causing butt-rot, the most serious disease in British forests. Typically, air-borne spores colonize the freshly cut surface of a conifer stump, the resulting mycelium occupying part or all of it; the fungus then passes from stump roots to adjacent living ones. It often spreads from tree to tree in this way and increases as stumps are created by successive thinnings. Damage arising during the first rotation of trees may be serious enough, but the fungus persists so long in stumps that it often infects succeeding rotations, causing even greater losses.

In Britain, where most conifers were planted on non-woodland sites free from *F. annosus,* it was highly desirable to exclude the fungus, spore trapping having revealed a risk of stump infection by *F. annosus* everywhere (Rishbeth, 1959). Early attempts at control involved chemical treatment of the newly cut surface, e.g. pine stumps treated with 40% ammonium sulphamate are often colonized by *Trichoderma viride,* competition from which is generally intense enough to exclude *F. annosus.* Another important antagonist, the wood-rotting fungus

For biotherapy, Douglas-fir poles were inoculated with a *Scytalidium* sp. either by pushing impregnated softwood dowels into holes made by removing a sample with an increment hammer, or by forcing impregnated hardwood dowels into timber with a nail gun (Ricard *et al.*, 1969). Standing conifers were similarly inoculated with *Scytalidium* sp. against infection by *F. annosus* by firing metal-tipped birch dowels into the trunks with a rifle (Ricard and Laird, 1968). For inoculating infected fruit trees, wooden dowels (0.8 cm diam. and up to 10 cm long) were impregnated with *T. viride,* pushed into holes drilled into the trunks (at intervals of 10 cm) and sealed with a non-toxic dressing (Corke, 1978).

Residual bark from timber mills is an appropriate medium for pelleting by agricultural equipment (Currier, 1973). With suitable additives pellets, granules of bark or other carriers containing propagules of microorganisms can be designed to break down and release the contents after pre-determined periods. These could be used to increase antagonistic populations in soil, or for implantation in poles and trees. As an alternative to wood dowels or bark pellets (Bio-Innovation AB BINAB, Box 54, S–193 00 Sigtuna, Sweden), spores or a mixture of finely powdered spores and mycelial fragments, can be suspended in liquids for inoculating wounds or incorporated in agar or suitable thickening agents for inoculating trunks.

IV. Commercialization of Microorganisms for use on Edible Crops

An official permit for the sale of biological agents is now almost essential in most countries. The permit may be required by law or because distributors are reluctant to undertake the financial risk involved in marketing a pesticide which has not been approved. Permits are an endorsement of a product by recognized experts and set agreed standards of effectiveness and safety. However, a major problem is that existing rules were designed for new chemical compounds, not microorganisms, so that the granting of a sales permit calls for a rational interpretation of these rules by competent authorities.

A permit was sought for the sale in France (J. Ricard, personal communication), for example, of a preparation for spraying the casing soil of cultivated mushroom beds to control *Verticillium malthousei:* the active ingredient is propagules and metabolites of *Trichoderma* spp. Strains ATCC 20475 and 20476 were incorporated in a wettable powder and marketed under the trade name BINAB T SEPPIC. Although the French authorities required conventional evidence of the effectiveness of the preparation and of its safety for humans, environmental problems were obviated by the ubiquity of *Trichoderma* spp. in nature and their presence, for instance, on grain and vegetables eaten by man. Initial effectivity tests by SEPPIC and Bio-Innovation AB BINAB (de Trogoff and Ricard, 1976) were supplemented by those of L'Institut National de la Recherche Agronomique (INRA) and the Centre Technique du Champignon. The safety data involved acute and chronic (90 day) toxicity tests: no LD_{50} was obtained even at 4 g/kg body weight, indicating less toxicity than that of table salt. Allergenicity tests were negative in both anaphylactic shock induction trials and more conventional intracutaneous injections. Residue tests on mushrooms harvested from trays sprayed with the preparation showed no significant increase in the *Trichoderma* flora on the crop, in spite of increased *Trichoderma* on the casing soil. After review of the data by the Pasteur Institute a use permit was issued, followed by a temporary sales permit on completion of effectiveness tests by INRA. *Trichoderma* is grown on cracked barley kernels in a modification of the koji process (Miall, 1975); the solid substrate allows aerial conidia to form, giving the product a shelf-life of up to 2 years.

V. The Future: Further Research Requirements

It was Aristotle's opinion that "In practical matters the end is not mere speculative knowledge of what is to be done, but rather the doing of it". There is now abundant observational evidence that biocontrol of diseases occurs naturally, but if natural events are to be manipulated for man's benefit, microorganisms must be put where they are needed because, unlike insect predators, they do not seek out their prey. This involves practical problems, but we now have some experience in developing control methods and those most likely to succeed can be visualized. Biocontrol has often succeeded on a limited scale in controlled conditions, under glass or on small field plots, but few examples have been developed to a commercial scale. This may be largely due to the difficulty of providing enough inoculum to spray several acres of crop at an adequate concentration, which requires the development of an appropriate industrial system such as those proved feasible with some insect pathogens, predators and parasites, *P. gigantea* and *Trichoderma* spp. The control of leaf pathogens by spraying with urea is subject to no such restriction, since the raw material is readily available. The development of other indirect methods of control should be actively pursued if they lend themselves readily to exploitation. Whereas large-scale treatment of pruning wounds may at present be partly restricted by the shortage of special tools, development is progressing and equipment can readily be modified. Greater awareness by the horticultural industry of the possibilities of biocontrol of plant diseases would stimulate a demand for supporting industries to provide inoculum and application equipment.

The greatest progress in developing practical techniques for biocontrol of plant pathogens is being made in protecting wounds against infection. Therein lies the best hope for the immediate future. Biocontrol methods should be easier to establish for plant diseases than for insect pests. Some are as good as, or better than chemicals, giving opportunities to replace some chemicals, or at least to integrate the two systems into one program. Accepted methods of biocontrol will act as focal points around which further techniques can be developed as our understanding of mechanisms improves and control of other diseases becomes possible. Industrial processes can be adapted to the commercial production of microorganisms in a form readily acceptable to the horticultural industry, which

I. Introduction

Pathogens are selected for pest control because they are believed to be specific for target organisms. Exposure of many vertebrates to these pathogens has been severe during epizootics of insect disease, e.g. birds and mammals (Lautenschlager and Podgwaite, 1979; Chapter 19, IC2). Man has been exposed to insect virus on food (Heimpel *et al.*, 1973). Yet mention of candidate pesticidal pathogens is absent or rare in the records and researches of human medicine, veterinary science, zoos and breeders of laboratory mammals. Equally there are now extensive data in insect pathology about the specificity of selected pathogens. These facts comprise a vast body of circumstantial evidence that they are harmless.

However, there are possible loopholes in this argument. Mammals are largely unobserved during their exposure to insect pathogens in Nature. There are a few records of natural infections in vertebrates with certain of the candidate pathogens. On certain growth media, some pathogens may produce toxins unrecognized or not produced in insect hosts. Harmful effects may take unexpected forms, e.g. carcinogenicity or teratogenicity. Thus organisms expected to be safe must be tested under strict observation, while giving them every chance to cause such harm as they are capable.

Since safety tests are designed to exaggerate the chances of these pathogens causing harm, their evidence must be subjected to careful hazard assessment. Evidence must be real. Pathogens must not be condemned as biohazards on grounds of remote possibilities or intangible theoretical risks. An example of a remote possibility is a recent ban on use of *Bacillus thuringiensis* against pests in water catchment areas supplying drinking water in W. Germany (Krieg, 1978). This ban was imposed despite the improbability that spores falling to the ground during application to foliage will reach the public water supply, thence the food industry (assessments by Krieg, 1978, and Burges, 1980). Also, even if spores reached food, no harm would be expected because *B. thuringiensis* is registered in W. Germany, the USA and the UK for unlimited application to food crops up to harvest and in the USA to raw agricultural commodities post harvest. Harm due to this bacterium has never been recorded in food or elsewhere (Section XI). An attitude that "one more spore – of any sort – is one too many" is untenable for a biologist. An example of a theoretical risk is the possibility that baculovirus DNA could be incorporated into the mammalian genome, even though tests with baculoviruses have so far failed to show effects or infections in vertebrates or their cell lines.

I believe that a pathogen should be registered as safe when there is reasonable evidence that it is so and in the absence of concrete evidence that it is not. A "no risk" situation does not exist, certainly not with chemical pesticides, and even with biological agents one cannot absolutely prove a negative. Registration of a chemical is essentially a statement of usage in which the risks are acceptable. The same must apply to biological agents. This view is currently shared by M.H. Rogoff, until recently Chief Scientific Officer for pesticide registration in the Environmental Protection Agency (EPA) in the USA (personal communication).

Tests to provide evidence of safety are mandatory for registration. In addition, the frontiers of knowledge must still be advanced by nonmandatory fundamental research. This is now the place for such questions as the possible incorporation of baculovirus DNA into a mammalian genome, which impinges on the poorly understood subject of virus "latency".

The object of this chapter is to evaluate the safety tests applied to various pathogens and the complementary quality control tests. Present problems, attitudes and research will be discussed, leading to future testing systems adequate to provide the necessary assurances about safety, while not so costly as to be self-defeating by delaying or preventing desirable development of microbial alternatives to chemical pesticides. Toxins and the effects of candidate pathogens on invertebrates will be discussed only in general terms because they are well covered in other chapters, with the exception of new data on *B. thuringiensis* var *israelensis* on nontarget fauna which are given in Section IV, C.

II. Hazard and Risk Assessment

In safety testing, it is essential to know what hazards to look for, their routes and timing. The hazards are: (1) infection of man, livestock, useful animals and plants; (2) poisoning by toxins; (3) allergy; (4) carcinogenesis by toxins; and perhaps (5) incorporation of viral DNA into the genomes of non-target organisms. These hazards may arrive by four routes: (1) oral, on food; (2) respiratory, by inhalation; (3) parenteral, in wounds; (4) dermal, on skin and in eyes. They may occur at various times: (1) during production, packaging and storage of pathogens; (2) at application to crops; (3) during crop culture operations, e.g. disbudding, harvest; (4) during post harvest crop storage; (5) consumption of a treated crop; (6) by environmental pollution. Since pathogens multiply in target organisms and some multiply in various substrates as saprophytes, the amount of pathogen in the environment may increase.

Since allergy is a complex subject with a terminology unfamiliar to many, some explanations are included here (Taussing, 1979). Allergies are caused by certain protein and polysaccharide antigens so all types of pathogen are potentially allergenic to man. Early exposures to the antigen cause sensitization and later exposures elicit reactions involving respiratory distress, lachrymation or skin rash. The main route of sensitization is respiratory. Allergies relevant to the safety of biological pesticides are placed in Types I and III in common classifications of allergies into four types.

Type I allergy is an immediate anaphylactic response in sensitized individuals to relatively small amounts of inhaled allergen producing rhinitis (hay fever), lachrymation and asthma. Only some 20% of the population (sensitive or "atopic" individuals) have the genetically determined susceptibility, mediated by the production of reaginic (IgE) antibodies.

Type III allergy is a delayed response to a relatively heavy inhaled dose of allergen, involving intense respiratory distress some 4–8 h after exposure, with shortness of breath, low or high fever, headache and weakness, which resolves fairly rapidly. Susceptibility is general in the population. The allergen (antigen) reaction is with precipitating antibodies (IgG) in and around the alveolar capillaries and elicits a strong polymorphonuclear infiltration resulting in extensive tissue damage. The best known example is farmer's lung.

Allergies are impossible to quantify and reactions differ enormously between different people. Existing tests on laboratory animals are inadequate to predict all effects on humans but must suffice until better tests are available.

III. Evolution of Safety Testing: Safety of Registered Pathogens

A brief look at the historical development of safety testing not only assesses the safety of pathogens already registered, but also provides a deep insight into the problems involved and the difficulties still unresolved. Heimpel (1971) reviewed in detail tests up to 1970.

A. *BACILLUS POPILLIAE*

The first registration of an insect pathogen, *B. popilliae*, occurred about 1950 (Rogoff, 1980). This presented a relatively simple situation where a common sense decision could be made on a few key facts (Heimpel, personal communication). This bacterium did not grow at 37°C (the highest developmental temperature for var *melolontha* is 27°C; Chapter 10, IIIC), precluding systemic infection in man. Spores were applied in small heaps to grassland soil, thus avoiding dispersal in air and residues on food plants. They did not germinate or grow in soil: indeed in Nature increase of the bacterium was known only in larvae and adults of certain scarabeid beetles. Growth *in vitro* occurred only in complex media. No allergies had been encountered among workers preparing experimental bacterial powders from bodies of infected beetle larvae. Spores fed to starlings and chickens had no adverse effects and remained viable without germinating during passage through the avian gut. *B. popilliae* was not related to any vertebrate pathogen. Its mode of action in beetle larvae was gross septicaemia and toxins were not suspected. A much simpler hazard analysis is not likely to be encountered. (Recently, under modern guidelines, more extensive tests, e.g. repeated oral administration to rats and monkeys, gave further convincing evidence of safety: Ignoffo, 1973; Heimpel, personal communication.)

B. *BACILLUS THURINGIENSIS*

The next stage in the evolution of safety testing was the registration of another bacterium, *B. thuringiensis,* in 1960, with full exemption from tolerance for use on food and forage crops (Rogoff, 1980). Information to hand before safety testing indicated a complex situation, posing both practical and theoretical problems. *B. thuringiensis* thrived *in vitro* at 37°C. It was closely related to *B. cereus*, which in turn had biochemical affinities to *B. anthracis*. Thus the practical problem of contamination during production, the possibility of infectivity in mammals and the theoretical question of mutation appeared more real than with *B. popilliae*. The protein crystals of δ endotoxin played the major role in the mode of action in many susceptible insects. The bacterium was to be applied to food plants. It flourished *in vitro* on simple media and so could grow on many liquid human food preparations if held at suitable temperatures.

A dual approach to safety testing was adopted (Fisher and Rosner, 1959; Heimpel, 1971; Ignoffo, 1973). On the one hand, because of the crystal toxin, a series of tests with two commercial products was applied which was the same as at that time required for chemical pesticides. This included acute feeding in rat, chick and man; intraperitoneal injection in mouse, guinea pig, rabbit, swine and chick; inhalation; eye and dermal tests for irritation; dermal tests for allergy; subacute feeding in birds and mammals; and wildlife tests on fish and bees. On the other hand the animals were also examined for infection by the bacteria. In addition, special infectivity tests were included, e.g. bacteria were injected intravenously into mammals and their elimination monitored by plating successive blood samples. To simulate the bacterium growing in human food preparations, extra toxicity and infectivity tests were made with log phase broth cultures.

Attempts were made to induce harmful mutations by serial passage of the bacterium in mammals (Fisher and Rosner, 1959). On the basis of other experiments and theoretical considerations, Steinhaus (1959) considered that the chances of mutation of *B. thuringiensis* to an anthrax-like organism "are exceedingly small", particularly since the modes of action of the two organisms involve distinctly different toxins.

Parallel to safety registration, a code of quality control tests was applied to every production batch (Section X). This ensured the maintenance of pure cultures and guarded against unrecognized contamination with another microorganism, or formation of a mutant which might produce unusual toxins.

Soon after registration of the first two commercial products, it was realized that another toxin, the β exotoxin (= thuringiensin: Chapter 13) was toxic to some mammals parenterally and to fowls even perorally. This problem was overcome by using strains that did not produce β exotoxin, which was banned from commercial products in the USA by Federal Regulation (Commerce Clearing House, 1977).

These tests were at the time a reasonable application of chemical test procedures elaborated to take account of infectivity, with extra tests to look at infectivity in more detail and to look at the possibility of mutations. However, the later realization of potential harm from β exotoxin warned of the need for caution.

inhalation by mice, rats, guinea pigs and rabbits (T.L. Couch, personal communication). Intravenous injection of mice with high levels of viable spores revealed gradual but incomplete elimination of spores from blood and tissues over 5 months, compatible with the mechanical removal of foreign particles (latex particles; Adlersberg et al., 1969) by the mouse immune system, with no indication of infection. Guinea pigs were subjected to a 4-week subacute inhalation study with Dipel powder. Rats were fed ca. 76% of their body weight of Dipel powder over 90 days and others ca. 6.1 times their body weight over 2 years. The HD1 strain of H-type 3a3b did not mutate to an exotoxin-producing form during passage through the gut of young chickens. After application of Dipel in a spruce-fir forest, the health of six species of small rodents was monitored by trapping and dissection. In studies on non-target fauna, Dipel was fed to rainbow trout, blue gills (fish), bobwhite quail and, for 14 days, to chickens. Fish, birds and invertebrates in land and water were observed during forest applications (Couch, personal communication). In the laboratory, it was added to water containing a marine fish *Anguilla anguilla*, the oyster *Crassotrea gigas* ($\times$ 1000 normal rate), numerous invertebrates (Alzieu et al., 1975) and earthworms ($\times$ 100 normal application, Benz and Altwegg, 1975). An unnamed serotype, injected intracardially into the oyster *C. virginica*, was cleared from the body in 6–10 days (Feng, 1966).

WHO has funded tests with the new var *israelensis*, including maximum challenge tests (Section VI, E), before commercialization (J.A. Shadduck, personal communication).

Application of a new product to a different ecosystem may warrant extra tests. Thus the new H-serotype 14 (var *israelensis*) must be applied, for instance, to rivers when used to control black flies (Simuliidae) and so tests were made with non-target river fauna, against which Dejoux (1979) and Colbo and Undeen (in press) found almost total short-term innocuity. At dosages much higher than those effective against mosquito larvae, H-14 had no effect on tested species of Planaria, Hirudinea, Mollusca, Crustacea (Copepoda, Phyllopoda, Cladocera, shrimps), water insects (Odonata, Ephemeroptera, Coleoptera), Acarina, fish and frog larvae (WHO, 1979a; Garcia and Goldberg, 1978). However, there is some toxicity to Chironomidae (WHO, 1979a), Dixidae (Garcia and Goldberg, 1978) and possibly Trichoptera (WHO, 1979a). It is moderately toxic to some Lepidoptera (Vankova et al., 1978) but not to *Ephestia kuehniella* (de Barjac, 1978).

The toxin released by dissolution of the crystalline protoxin of *B. thuringiensis* appears to be specific. It did not affect eight microorganisms, or yeast spheroplasts, but destroyed primary lepidopteran gut epithelial cells and a sensitive moth cell line (Geiser, 1979). An insect predator, the preying mantis, was unharmed by eating moth larvae fed a lethal dose of crystal, which would have been dissolved in the larval gut (Yousten, 1973).

D. RANGE OF TESTS FOR NEW BACULOVIRUS PREPARATIONS

Baculoviruses of insects are a highly specific group of invertebrate pathogens. Many have been subjected to extensive safety tests without record of a single substantiated harmful effect due to virus on vertebrates, including man, insects in an order different to that of the target species, other invertebrates or plants.

In the sub-group A of the Baculoviridae, the NPVs of Lepidoptera and sawflies are a particularly homogeneous group on which most safety tests have been conducted. Three NPVs, those of *Heliothis* spp., the gypsy moth and the sawfly *Neodiprion sertifer*, have undergone long term carcinogenicity tests (Section III, C, E). Bioassay with insects and serological tests failed to show replication of an NPV in test vertebrates, e.g. during exposure of monkeys to *Heliothis* NPV for 26 weeks (Chapter 17, VC). NPV applied to fish disappeared or was inactivated in 24 h (Martignoni, 1978). Over 11 NPVs have been used in a range of short term tests including parenteral routes, 8 including inhalation, 2 intracerebral injection and at least 8 including allergenicity tests. At least 4 NPVs have been used in subacute feeding tests. Infectivity of NPVs has been tested in at least 29 vertebrate species. Ducks were unharmed by 5-day multiple massive dose exposures (Martignoni, 1978). At least 9 NPVs have been applied to cell lines from > 15 vertebrate species. The lines have been grown from > 8 types of tissue. Many cell lines have been challenged with several NPVs. As well as examination for cytopathic effect and visible replication of virus, some lines have been tested for invisible virus proliferation, cytogenetic and chromosomal disturbances (McIntosh, 1975; Miltenburger and David, 1978; Miltenburger, 1980), agglutination of guinea pig erythrocytes, inhibition of replication of Echo-II virus (Chapter 17, VC), Sindbis virus, Encephalomyocarditis virus (Arif and Dobos, 1978) and of Infectious Pancreatic Necrosis virus (Martignoni, 1978; Arif and Dobos, 1978), and presence of virus inside cells using haemadsorption, immunofluorescence and autoradiography techniques (McIntosh and Maramorosch, 1973; Chapter 17, VC). Mammal, fish and bird cell cultures were exposed to virions of the NPV of *Choristoneura fumiferana* (Arif and Dubos, 1979) and of the NPV of *Neodiprion lecontei* (J.C. Cunningham, personal communication); RNA, DNA and protein synthesis was measured in exposed and mock infected cells. Absence of differences indicated that these cells were not infected by these NPVs.

Himeno *et al.* (1967) reported production of polyhedra-like bodies in mammalian amnion cell cultures inoculated with DNA from *Bombyx mori* NPV, but despite extensive efforts these results so far have not been confirmed. A selected strain of *Autographa californica* NPV multiplied in an established moth (*T. ni*) cell line at 37°C, even though the cells cannot multiply at this temperature, but the cell line is readily maintained at it for at least 4 days (McIntosh, 1975). Thus human body temperature does not restrict this virus, nor is cell division a necessary prerequisite for viral multiplication.

The granulosis viruses (GV) of Lepidoptera (Sub-group B of the Baculoviridae), although studied less than the NPVs, appear to have even narrower host ranges than those of the NPVs. GVs have biochemical properties that extensively overlap those of the NPVs so that the two groups cannot be discretely defined biochemically, although they are clearly distinct by morphological criteria. Replication of GV in a cell line — homologous or otherwise — has not yet been achieved, whereas many NPVs will replicate in

C. SUBACUTE TESTS

To test for infectivity, a single massive dose is adequate. If this fails to induce infection, lesser doses applied over 3 months are no more likely to succeed. If really massive dosing is possible in acute tests (Section IV, A) the same applies to the detection of toxins. If a toxin is suspected to be present in small amounts it should be concentrated for further tests, possibly including subacute tests.

D. MAXIMUM TEMPERATURE FOR GROWTH OF A PATHOGEN

Human body temperature is $37°C$ and rectal temperatures of mammals range from $34°$ to $44°C$ (Chapter 27, IV, B1). Thus pathogens unable to grow at $37°C$ are unlikely to cause systemic infection in man and, at $34°C$, in other mammals. Average skin temperature in man is $33°C$, that on feet $28°C$ and fingers $24°C$, varying with the environment (Müller-Kögler, 1967), so maximum temperatures for growth of pathogens are rarely likely to be limiting for dermal infections. These maxima for fungi (reviews: Müller-Kögler, 1967; Roberts and Campbell, 1977) may vary extensively with strain, e.g. $31°$ to $36°C$ for *Verticillium lecanii* (Chapter 25, Table III) and $33°$ to $40°C$ for *Conidiobolus coronatus* (Müller-Kögler, 1967), so strains should be specified in safety tests. Where a pathogen has a limiting temperature of $37°C$ or below, parenteral injections could usefully be made in foot pads rather than centrally in the body.

E. MAXIMUM CHALLENGE EXPERIMENTS

Maximum challenge experiments are designed to maximize opportunities for test pathogens to harm mammals. They involve administration of large quantities of viable pathogens by normal and unusual routes, possibly to the least resistant test animals, e.g. the very old and very young. To avoid complications due to contaminants and additives, pure cultures of test pathogens must be used. Adequate controls, e.g. autoclaved or UV-inactivated material, are essential.

The obvious value of such tests lies in negative results, which strengthen our confidence in the safety of a candidate pathogen. A problem lies in interpretation of positive effects. Carried to the extreme, positive results are meaningless, e.g. one can kill a mouse by injecting any bacterium — if enough are injected — and one would expect some bacteria to increase saprophytically once an animal was overwhelmed. Injection of 1.6×10^9 spores of *B. subtilis* per mouse killed no mice but the minimum lethal dose of a certain strain of *B. cereus* was 4.3×10^8 spores/mouse (Mace, 1977). Is this difference meaningful, bearing in mind that the LD_{50} for the mammalian pathogen *B. anthracis* is < 5 spores/mouse (Wright, 1958) and that injection of 10^6 spores of *B. thuringiensis*/ mouse was selected in the U.S. Federal Regulations as an ample safety margin to reveal contaminants etc. in quality control tests (Commerce Clearing House, 1977)?

The present practice of exposure-related tests, with multiples of acre-doses of the actual commercial product with its contaminants, etc. are probably most meaningful for mandatory tests for registration. Maximum challenge experiments at present are best a subject for research, such as now in progress

(J.A. Shadduck, personal communication). Positive results must be viewed in strict perspective to the experimental conditions.

In acute tests some cursory effects are expected and simple clinical tests, such as temperature, food consumption and body weight, which do not greatly increase the cost of mandatory tests, are useful indications of return to normal. However, the more time consuming clinical chemistry and haematology may do little more than record an expected transient upset although they may be valuable in longer term tests. Shadduck's program investigates their predictive value. Histological examination of macroscopic lesions and of animals that die, also tests to monitor the elimination of candidate pathogens from the mammalian body (at least tests at the end of an observation period) are valuable.

Administration of large doses of test pathogens to weak experimental mammals, e.g. the very old and the very young, means little because it is similar to the use of even larger doses in mature healthy animals, and because the weakened condition may be difficult to reproduce for reliable experimentation. An exception could be to copy a technique already established with mammalian pathogens, e.g. suckling mice may be particularly susceptible to some mammalian viruses.

Use of unusual routes may be suggested by the biology of the candidate pathogen and related species, e.g. injection of Microsporida into the brain because *Nosema* spp. have been found in the brain of rodents (review: Heimpel, 1971).

F. IMMUNOSUPPRESSED ANIMALS

Humans under immunosuppression therapy have been reported more susceptible than normal to diseases and even to microorganisms or viruses not normally pathogenic. Susceptibility to fungal diseases has been increased by use of immunosuppressive drugs or corticosteroids to treat nonmycotic diseases, e.g. a high incidence of aspergillosis was associated with heart transplant patients who had been on corticosteroids (Rickard, 1975). Immunosuppression by malaria (Kaye et al., 1965; Salaman et al., 1969; Stickland et al., 1972) and schistosomiasis (Kayes and Colley, 1979) increases susceptibility to some other diseases. This suggests that immunosuppressed people may be at greater risk from insect pathogens than normal people. Immunosuppression of test animals may offer a more sensitive study of the infective power of candidate pathogens. It may give some information about the mechanisms of resistance of mammals to insect pathogens. It was included in tests on the gypsy moth NPV (F.A. Lewis and J.D. Podgwaite, personal communication). However, the value of the method has not been proved experimentally. There are practical difficulties in that very healthy animals are needed to avoid the effects of their natural pathogens. The dosage of the suppressant may have to be determined to critically narrow limits. A variety of suppressants are available that may act differently, e.g. radiation, neonatal thymectomy, antilymphocyte serum, splenectomy, infection by certain diseases and cortisone treatment. In the final analysis, what is the significance of an infection caused by a high dose of an insect pathogen in an immunosuppressed animal stripped of its normal defence mechanisms? It may have the same significance as giving a higher dose of the test pathogen to a normal animal. Immunosuppressed people may suffer from natural diseases before an insect pathogen would have a chance to act and so worries about the latter may be irrelevant and automatically covered by general precautions that such people would presumably take.

It seems premature to include such poorly defined methods in mandatory test requirements. Instead, basic research to establish their value seems appropriate, as is being funded by WHO (Shadduck, personal communication).

G. TESTS IN CELL CULTURES

A high multiplicity of infection of cells with virus often leads to the production of defective virus. Tests should allow for this possibility.

H. MUTATION HAZARDS

Dangerous mutations may be of two types — mutation to infect a mammal and mutation to produce a toxin harmful in mammals. It is useless to measure the general mutation rate of an agent, because this reveals nothing about dangerous mutants. The most useful test for the detection of an ability to produce infectivity mutations is serial passage of the agent in an environment in which the mutants in question would have a selective advantage over the parent agents and so reveal their presence, i.e. in the mammalian body. Fisher and Rosner (1959) used intravenous passage in mice. Tests for mutation to produce unwanted toxins can be made by serial passage in an animal sensitive to such toxins, e.g. the use of chicks to detect the β exotoxin of *B. thuringiensis* (Section IV, C).

VII. Allergy

The relevant types of allergy are described in Section II. Since the main route of attack by allergens is respiratory, the likelihood of a pathogen causing an allergy largely depends on its ability to penetrate into the lung (Austwick, 1980). Spores or aggregates of spores $> 10\mu$ diam. probably lodge in the nose so large-spored organisms, e.g. *Entomophthora*, are least likely to cause an allergy. Those of $5-10\mu$ impact on the surface of the trachea or bronchi, so *Metarhizium anisopliae* spores may be caught before they reach the bronchioles and may be limited to causing a Type I allergy. Particles of $3-5\mu$ impact on the surface of respiratory and terminal bronchioles and those $< 3\mu$ penetrate to alveoli. Thus fungi such as *Paecilomyces* and *Beauveria*, bacteria, viruses and some microsporidians may cause both Types I and III allergy, although penetration of particles may be reduced by clumping of spores or suspension in spray droplets.

In all safety tests on registered insect pathogens no allergies have been reported. Slight eye irritation from extremely high doses of Gypcheck, which contains gypsy moth NPV, was due to insect remnants (Chapter 18, VI). A number of insects cause allergies, e.g. the tussock moth (Martignoni, 1978) and the mushroom flies *Megaselia agarici* (Kern, 1938; Truitt, 1951) and *M. halterata* (P.N. Richardson, personal communication about personal allergy during laboratory fly rearing). A survey revealed allergic reactions to insects in 113 individuals, including three with anaphylactic shock (Wirtz, 1980).

Reports of reactions during practical use of insect pathogens are limited to numerous reports about high concentrations of *Beauveria bassiana* a non-registered pathogen in Europe and N. America (Westall and Rehnborg, 1964; Samsinakova, 1964: reviews by Müller-Kögler, 1967; Roberts and Yendol, 1971; Heimpel, 1971; Austwick, 1980; Chapter 42, IIC). Symptoms ceased when protective clothing was worn. There is some evidence that sensitization may be dermal rather than respiratory and Austwick (1980) suggests that the symptoms recorded by York (1958) resembled Type III allergy.

Wet spores growing in slime heads on insects, e.g. *Verticillium lecanii,* are not likely to present a hazard, in contrast to the dry dusty spores of *Beauveria* etc. normally distributed by air. Isolations of *Beauveria* were made from small rodent lungs in the wild by MacLeod (1954) and from 15% of 3000 sputum samples from chronic respiratory disease cases by Pore *et al.* (1970), suggesting that these spores penetrate into the lung and survive at least briefly there (Austwick, 1980). Even wet spores on insect bodies dispersed into the air may present a problem since skin test reactivity and precipitins to antigens from *V. lecanii,* suggesting Type I allergy, were found in combine harvester operators (Darke *et al.,* 1976). In contrast, P. Kanagaratnam (personal communication) experienced no reaction to spraying *V. lecanii* blastospores and conidiospores (previously safety tested) in greenhouses without mask, gloves or protective clothing on 13 occasions (mask worn for first two sprays) over 2.5 years, sometimes with accidental splashing into the eyes.

No allergic reactions have been reported to insect pathogenic viruses and bacteria. Reactions to bacteria are not impossible because preparations of other non-pathogenic species, e.g. *Bacillus subtilis,* are allergenic due to their enzymes, and decaying straw debris infected with this organism is believed to cause allergic respiratory problems (Green and Woodcock, 1978).

There are numerous records of the absence of symptoms, including *Beauveria,* probably because lower concentrations and/or non-susceptible people were involved. The answer to the problem lies in the use of protection at sensitizing spore concentrations.

VIII. Safety of Pathogens Not Yet Registered

A. BACTERIA

The safety of unregistered candidate pathogens can be judged to a varying extent from natural records of these candidates or close relatives in vertebrates, safety tests already performed and data for related registered organisms.

For species related to *Bacillus popilliae* I have found no records of infection in vertebrates, and safety data for *B. popilliae* (Section III, A) are remarkably convincing. Unregistered species in this group are highly likely to be safe.

For unregistered varieties of *B. thuringiensis* the conclusion is the same (Sections III, B; IV, C; XI). Its flagellar antigens, the main basis for differentiating varieties, show few cross reactions with flagellar antigens used to subdivide its close relative *B. cereus* (Chapter 3, IV; Gilbert and Parry, 1977).

Some isolates of *B. cereus* produce food poisoning toxins (review: Terranova and Blake, 1978) and others have been isolated from clinical infections in man and mammals, usually associated with stress conditions (e.g. Merck and Burow, 1973; Craig *et al.*, 1974; Feldman and Pearson, 1974; Goullet and Pepin, 1974). Turnbull *et al.* (1979) consider that *B. cereus* may be of clinical importance, not just as an opportunist but as an agent of potentially severe infections in its own right.

There are several reports of *B. sphaericus* pathogenic in man. A case of meningitis and generalized Schwartzman reaction revealed no obvious portal of entry (Allen and Wilkinson, 1969). Two cases of infection were serious (Farrar, 1963) and a batch of infected sausages caused six cases of food poisoning (Elter, 1966). Isolates from these infections were tested in experimental mammals with negative results (Allen and Wilkinson, 1969; Farrar, 1963). A mosquito pathogen, isolate 1593−4, and the mammalian isolates were injected subcutaneously, intraperitoneally and intracerebrally into rats and mice in maximum challenge tests; they were also fed in acute and chronic toxicity studies in rats, mice and guinea pigs (WHO, 1979b). There were no deaths or clinical illnesses and results were negative except that high doses of isolate 1593−4 produced some mild lesions in the brain of rats injected intracerebrally and more severe lesions in the eyes of rabbits injected intraocularly. A large component of the lesions was due to high concentrations of foreign products. Recovery of bacteria from brain and eyes 10−14 days after injection without evidence of replication suggests that the re-isolations were due to prolonged spore survival. Very similar results were obtained with the mammalian isolates. It was concluded that these were probably opportunistic organisms at the time of their isolation from man and that *B. sphaericus* is highly unlikely to pose any hazard to man (WHO, 1979b). The significance of the survival and persistence of spores applied to test animals could be judged in comparison with the persistence of inert materials, e.g. latex particles (Aldersberg *et al.*, 1969).

B. VIRUSES

Baculoviruses are likely to be extremely safe, with no known records in vertebrates or vertebrate cell lines and the results of all safety tests suggesting no risks (Sections III, C; IV, D). Three entomopox viruses were applied to mice and rats intraperitoneally, intracerebrally, intranasally or orally without recorded harm (Ignoffo, 1973).

C. PROTOZOANS

I have found no record of entomopathogenic microsporidia attacking vertebrates despite these being common pathogens of insects in stored food. Heimpel (1971) and WHO (1978a) reviewed *Nosema* and *Thelohania* species in rodents, *Pleistophora* (*Plistophora*) *nyotrophica* in the common toad, common occurrences of species of *Nosema, Glugea* and *Pleistophora* in fresh and salt water fish, and also reports and discussions of microsporidian infections in man. There are distinct serological and morphological differences between microsporidian species found in mammals and in insects (WHO, 1978a). Most entomopathogenic species are reported in a limited range of insect hosts, but this apparent specificity may be mainly due to the scarcity of cross infectivity studies. Smirnoff (1968) found that a sawfly pathogen, *Thelohania pristiphorae,* could thrive in two species of tent caterpillar. Burges *et al.* (1971) found that a beetle pathogen, *Nosema oryzaephili,* attacked five species of beetles and three of moths spanning five insect families. Thus thorough infectivity studies of candidate microsporidia are necessary in mammals. All tests have given negative results with *Nosema locustae* (Chapter 30, IV, E). With *Nosema algerae*, oral, intraperitoneal, subcutaneous, intrafootpad and intracerebral applications by Shadduck (personal communication) have so far given negative results. However, Undeen (1976) obtained transient infections localized in the application site from spores injected into ears, tails and feet of mice. Spore germination tests in blood plasma indicated that spores injected by an infected mosquito bite are unlikely to result in an infection (Undeen, 1976). *N. algerae* can infect pig kidney cell cultures (Undeen, 1975). The limiting factor may be temperature, the maximum for vigorous growth being 28°C and for limited replication 35°C (Undeen, 1975). *Vavrai* (*Pleistophora*) *culicis* does not develop in mosquitoes bred at 35°C (Reynolds, 1970). *Nosema eurytremae* cannot develop above 35°C (J.E. Smith, R.J. Barker and K. Lai, personal communication).

D. FUNGI

Most fungi causing mycoses in man are free-living saprophytes and infect man and animals only when introduced into the body by inhalation or traumatic implantation from their normal saprophytic habitats in various decomposing organic substrates. Optimum temperatures for growth range from 25° to 40°C. Of the two types of fungal disease, the skin infections caused by the specialized group called dermatophytes (ringworm fungi) are contagious, whilst the systemic infections are noncontagious and generally acquired directly from the saprophytic source of the pathogen. One species *Sporothrix schenckii* can infect both animals and plants. Certain pathogenic fungi have been found in pulmonary and systemic lesions in reptiles (Austwick and Keymer, 1980; Austwick, 1980). The commonest species concerned is the soil fungus and insect pathogen *Paecilomyces lilacinus,* mainly found in respiratory infections in which chronic granulomatous and airway plaque lesions both contained sporulating hyphae. *Metarhizium* occurred in more progressive and extensive lesions in Crocodilia, both in miliary granuloma and in surface lesions bearing a copious felt of sporulating mycelium in the lung; it also occurred in associated lesions in liver and kidney. *Beauveria bassiana* has been found in pulmonary mycosis in chill-stressed giant tortoises (Heimpel, 1971) and *Paecilomyces farinosus* in extensive lesions in the lungs of an American alligator (Austwick and Keymer, 1980). Experimentally it has been possible to infect lizards and terrapins with *P. lilacinus* and *M. anisopliae* (Austwick, 1980). Many infections reported from zoos are associated with chill and nutritional stress (Austwick and Keymer, 1980). Whether these infections indicate a real hazard for reptiles if the fungi are used in practical insect control is unknown.

In man, *Beauveria (Tritirachium) shiotae = B. bassiana* was isolated from an abcess in the bile duct (Heimpel, 1971) and species of *Beauveria*, not *B. bassiana*, have also been reported from pulmonary lesions, but these cases are not clearly defined (Austwick, 1980). *P. lilacinus* was reported present in pulmonary lesions from a severe case of pleurisy and there are several records of corneal infection, invariably following penetration of the eye by contaminated material. Many other saprophytic fungi, including *B. bassiana*, can also invade corneal tissue but it seems that prior damage is required for the infection to become established (Austwick, 1980).

No infections have been induced in mammals with the common biocontrol candidates in considerable safety tests. These include short term tests (feeding, inhalation, intravenous and subcutaneous injection) and 90-day subacute inhalation and feeding with *B. bassiana* and *M. anisopliae* in rodents (Westall and Rehnborg, 1964; Müller-Kögler, 1967; Schaerffenberg, 1968; Ignoffo, 1973; Shadduck, personal communication); intraperitoneal (2 years) and subcutaneous injection (2 years), lactation and fertility tests (rats), dusting (3 months; 1.5 years; rats and mice) and skin tests on an atopic human (Westall and Rehnborg 1964); intraocular application of *M. anisopliae* to rabbits (Shadduck, personal communication); intravenous (one set lasting 6 months) and intraperitoneal injections of *P. farinosus* (Ignoffo, 1973; Müller-Kögler, 1965). Short term tests on *Nomuraea rileyi* (Chapter 27, IVB2). *Hirsutella thompsonii* (Chapter 26, VIA4) and *Lagenidium giganteum* (Umphlett and McGray, 1975) gave negative results. *Culicinomyces* sp. conidia were given daily for 6 weeks in drinking water to mice, rats, guinea pigs, cattle, sheep and two species of wild duck. Clinical, pathological, histopathological, haematological, biochemical and serological results were in the normal range and revealed no fungal invasion (Egerton *et al.*, 1978). Risk of infection of mammals with these species seems unlikely. Most records of temperature maxima for growth of *Entomophthora* spp., *M. anisopliae*, *N. rileyi* and *V. lecanii* are $< 37°C$, so ruling out systemic infections: 2/3 records for *Beauveria* spp. are $\geq 37°C$ (review, Roberts and Campbell, 1977; Chapter 25).

Mycoses in horses and humans, some possibly stressed, due to the common soil saprophyte and occasional insect pathogen *Conidiobolus coronatus* (= *Entomophthora coronata*) are reviewed by Heimpel (1971; 11 references) and in Chapter 28, III, E (5 references). Consideration of this species as a candidate for biocontrol ceased at an early date (Müller-Kögler, 1967).

There are no records of species of *Entomophthora* attacking vertebrates and safety tests gave negative results (Chapter 28 IIIE).

Some fungi produce toxins (Chapter 23) and require toxicity tests (Section IV, B).

IX. Nematodes

The well studied nematodes that parasitize insects belong to taxonomic families distinct from families containing species that attack vertebrates. They are unlikely to penetrate the vertebrate body. Some entomophilic nematodes carry bacteria that they normally transmit to the insect, causing septicaemia. The nematode, being unlikely to penetrate and parasitize vertebrates, is not likely to create systemic vertebrate infections. The only hazard appears to be the bacterial content of the nematodes eaten by man, or bacteria from dead insects that foul his food. Although inocula acquired in this way are likely to be small, tests of the ability of the bacteria to infect mammals may be desirable if food crops are treated.

Romanomermis iyengari preparasitic stages were applied to mice, guinea pigs, rabbits and fowls by the intravenous, intraperitoneal, intradermal, subcutaneous and oral routes. All animals stayed healthy. Pathological and histological examination after 6 weeks revealed no lesions suggestive of nematode infection. *Gambusia* fish exposed to a heavy dose for several days remained unaffected (Gajanana *et al.*, 1978). *R. culicivorax* was equally innocuous in similar tests on suckling and adult mice, and normal and immunosuppressed rats (Ignoffo *et al.*, 1974). Mice and rats were not susceptible to *Neoaplectana carpocapsae* (Chapter 33, III; Gaugler and Bousch, 1979).

X. Quality Control

Quality control of production batches is essential to ensure manufacture to the pre-determined standards at which a product has been safety tested. There are two aspects, identity of the pathogen and recognition of contaminating organisms. Registration data should describe both the production methods that ensure quality and the quality checks. For instance, virus production in insects bred on media may involve clean room technique and isolation from operators to minimize contamination from environment and operators.

Identity of the pathogen is ensured by careful maintenance of seed stocks, and also by batch checks during production and after harvest. To prevent loss of virulence seed stocks are usually held inert, e.g. cool or cold stored, freeze-dried, a condition that also facilitates retention of purity. Reports with some baculoviruses suggest that freeze drying may greatly reduce viability. Fermented products can be checked microscopically at each production stage and at harvest.

An example of quality control standards is given by the USDA Forest Service requirements for the NPV of the gypsy moth (Podgwaite and Bruen, 1978) and, copied below, for TM BioControl-1 (NPV of the Douglas-fir tussock moth; Martignoni, 1978):

Activity. The LC_{50} by assay with larvae of strain GL-1 should not exceed 13.066 ng of final product giving a titre of $\geqslant 76.534$ million units$_{GL}$/g.

Purity. Only baculovirus morphotype permitted, determined by darkfield microscopy and electron microscopy. No cases of cytoplasmic polyhedrosis or polyhedrosis caused by unicapsid virus are permitted, nor are cases of death by insecticidal chemicals detected by peroral and intrahaemocoelic injection of a group of larvae at the LC_{50} level. *B. thuringiensis* and its toxins must be absent. The aerobic bacterial count must not exceed 10^9 colonies/g by plate count on trypticase soy agar. Brewer agar is used to detect anaerobic and microaerophilic bacteria.

Physical features. The preparation must pass a screen of 100 meshes/in. (Tyler scale) with no more than 1% residue. A total residue is determined at 104°C.

Safety to vertebrates. No faecal coliform bacteria, or any other bacteria or agents pathogenic for warm blooded vertebrates, are permitted as detected by intraperitoneal injection in mice (observation: 7 days), oral administration to mice (observation: 21 days) and by plate tests. Coliform bacteria (lactose fermentors), typhoid, paratyphoid and dysentry bacteria are detected on formula II Endo agar. Faecal coliforms are detected in EC (*Escherichia coli*) fermentation tubes (44.5°C) after enrichment in lauryl tryptose broth. Pathogenic Entobacteriaceae e.g., *Shigella, Salmonella* (= SS) are detected by spread plates of SS agar.

Sensitivity of detection on agar plates depends on the amount of the preparation plated (0.001 g for the tussock moth NPV: Martignoni, personal communication). For *Trichoderma viride* 0.03 g were plated to detect bacteria and the batch discarded if any were found; 0.06 g were plated for the detection of fungal contaminants; plating whole pellets was a very sensitive test for both types of contaminant (J. Rickard, personal communication). If a pesticidal fungus has a low temperature maximum for growth, many fungal contaminants can be revealed by plating at 2°C above the maximum. Fungal contaminants with similar temperature requirements to those of the pesticidal fungus raise the problem of deciding how many spores to plate to achieve adequate sensitivity. In an analysis of a "worst case example", taking *Aspergillus flavus* producing aflatoxin B_1 as the model worst case, I estimated that the maximum *A. flavus* contaminants likely to be missed when plating 1000 spores could produce aflatoxin that could contaminate, for instance a tomato crop, by far less than the tolerated level in food in countries that set the most sensitive standards (data for aflatoxin from Stoloff, 1977 and Venkitasubramanian, 1977).

At the registration of Elcar (*Heliothis* NPV) a very vigorous limit of 10^7 total bacteria/g was set. This is less per acre of crop than the number of coliform bacteria permitted in 2 gal of unpasteurized milk (Rogoff, 1973).

Insects may be infected by non-occluded viruses, even simultaneously with NPV (Hess *et al.*, 1977; Morris *et al.*, 1979). It is desirable that the healthy and production virus-infected stock should be examined.

XI. Evidence of Safety from Long Term Use of Pathogens

A successful pathogen is used by an increasing number of industrial firms in new products, each subjected to safety tests. *B. thuringiensis* has been registered in products by four different firms, thus giving fourfold replication of safety tests.

After registration of a microbial pesticide the ultimate test of its safety starts, i.e. its continuous use for pest control. Judgements of registration authorities have so far proved sound, since no harm has been reported either from the use of fully registered products or from pathogens given trials clearance. The list of these products and pathogens is impressive.

Of the fully registered products, 82 tons of *B. popilliae* were applied in the USA over a few years after permission for use was granted (Heimpel, 1971) and its use has continued. *B. thuringiensis* has been used worldwide since 1962, and my personal conservative estimate of world commercial application up to 1979 is 7000 tons, much on food crops. In particular, no problems have been reported in the food industry, and it has been absent from cases of food poisoning, a situation not due to failure to identify the organisms because in the UK many *B. cereus*-type isolates from food poisoning cases were examined by a *B. thuringiensis* expert (J.R. Norris, personal communication). In the USA, the NPVs of *Heliothis* spp. and the Douglas-fir tussock moth were registered in 1977 and that of the gypsy moth in 1978. Use of the *Heliothis* NPV totalled 1-2 tons in 1974 (Stanford Research Institute, 1977). Production of the tussock moth NPV will be 1,563 lbs by September 1980 (Martignoni, personal communication 1978). *Trichoderma viride* received a temporary sales permit in about 1978 to control weed fungi in mushrooms beds in France. It has been sold to control tree diseases for much longer. In the UK, registration for this purpose was achieved in 1979. Related strains of *T. viride* have been used industrially to break down large quantities of plant waste and produce a fungal mass rich in protein (Church *et al.*, 1973). Large quantities of enzymes produced by *T. viride* have been exported annually from Japan since 1961 for incorporation in cattle feed in Australia (Toyama, 1969). *Agrobacterium radiobacter* was registered in 1979 in the USA against crown gall of trees.

Experimental use permits existed in the USA in 1979 for *Nosema locustae*, *Hirsutella thompsonii*, *Bacillus sphaericus*, NPVs of sawflies and of *Autographa californica*, and also two fungal weed pathogens *Colleotrichum gloedosporoides* and *Phytophthora citropthora* (Rogoff, personal communication): in the UK; *T. viride* since 1969, *Verticillium lecanii* (1978), NPVs of the sawflies *Gilpinia hercyniae* (1972) and *Neodiprion sertifer* (ca. 1977), NPV of *Mamestra brassicae* (1977), GVs of *Pieris brassicae* (1977), *Cydia pomonella* (1978) and *Lacanobia oleracea* (1979), *Tipula* iridescent virus (1973 for several years): in Germany;

CHAPTER 41

Insect Responses to Microbial Infections

H.G. BOMAN

Department of Microbiology, University of Stockholm,
S-106 91 Stockholm, Sweden

I. Background facts

More than 10^6 species of insects have been described and estimates have indicated the number of individuals to be as high as 10^{18} (Williams, 1960). These figures show that during evolution insects have competed very successfully with other forms of life, including microorganisms. Already on these grounds one could therefore expect insects to have effective mechanisms to counteract microbial infections (Salt, 1970).

The large number of insect species also creates one of the key problems in entomology: to what extent can results obtained with one or a few species be generalized to a larger group of insects? In the case of immunity mechanisms especially only a few species have so far been investigated and safe generalizations are not yet possible. This chapter is therefore a brief review of some reasonably well documented cellular and humoral defence reactions. For a more complete coverage of the earlier literature the reader is referred to an extensive but somewhat uncritical review by Whitcomb *et al.* (1974) or to a symposium volume edited by Maramorosch and Shope (1975).

It is important to emphasize that insects as well as other invertebrates lack specificities have been reported in molluscs and other invertebrates. However, the role of agglutinins in insect defence systems is not yet well documented.

III. Humoral Defence Reactions

Humoral defence reactions (i.e. non-cellular antimicrobial reactions) are often divided into two groups: (A) those pre-existing in the blood at the time of a primary infection, and (B) those which are induced either by a primary infection with viable bacteria or by various vaccines like heat-killed bacteria. However, infections and inducibility are not satisfactory criteria for grouping. Firstly, it is difficult to know if a normal insect is infected or when it last was infected and produced inducible defence factors. Secondly, an inducible factor may also be pre-existing in the haemolymph either at a low level in an active form, or at a significant concentration as an inactive proenzyme. Here, I have chosen to discuss in Section B all factors for which *de novo* synthesis is demonstrated irrespective of whether or not they are pre-existing.

A. PRE-EXISTING DEFENCE FACTORS

The haemolymph of many insects contains the enzyme phenoloxidase, probably mostly as an inactive proenzyme (Nappi, 1975). Phenoloxidase is readily activated by small amounts of proteolytic enzymes, by polysaccharides like zymosan (Pye, 1974), and probably also by injury or alarm reactions not yet clarified. The enzyme is inactivated by phenylthiourea (PTU) which is used routinely during insect surgery and for preventing the haemolymph from darkening (Schneiderman, 1967). Phenoloxidase is generally considered to be of primary importance for encapsulation and nodule formation (Nappi, 1975), and PTU and reduced glutathione have been shown to block the encapsulation of parasite eggs (Brewer and Vinson, 1971).

Kinoshita and Inoue (1977) recently reported that cell-free haemolymph from untreated larvae of *Bombyx mori* contained an antibacterial activity. This finding is contrary to our results with untreated pupae or larvae of *Hyalophora cecropia* or *S. cynthia*. The discrepancy can be explained in at least two ways. Firstly, there could be species differences but this seems unlikely because *B. mori* and the saturniids are rather closely related. Secondly, since Kinoshita and Inoue used pooled haemolymph from more than 2 000 larvae it is possible that some of these animals were infected and therefore carried the inducible antibacterial activity one would expect to find in *B. mori*.

B. INDUCIBLE DEFENCE FACTORS

Convincing studies of inducible antibacterial activities in different insects were early reported by Briggs, Gingrich, Hink and Briggs, Mohrig and Messner and Stephens Chadwick (see reviews by Whitcomb *et al.* 1974 and Chadwick, 1975). The principal difference in the experimental approach between these earlier investigators who chiefly used larvae of *G. mellonella* and our more recent work is our systematic use of saturniid pupae in diapause, a state when the metabolism of the animals is reduced to a few per cent of the normal rate. Such dormant insects can be immunized and if isotopes are injected simultaneously one can obtain a selective labelling of the immune components without much background from other biosynthetic reactions.

We first studied immunity in *Drosophila* (Boman *et al.* 1972) but in order to pursue the biochemistry of the immune response we found it necessary to switch to saturniid pupae in diapause. An early concern was to develop a quantitative assay for the antibacterial activity induced in the haemolymph (Boman *et al.* 1974a). Our main test organism has been *Escherichia coli*, D31, a penicillin and streptomycin resistant mutant. A growing complexity later forced us to use several different test bacteria and to make a further characterization of some of the parameters involved in the assay (Rasmuson and Boman, 1977). With our bacteria-killing assay in combination with injected labelled amino acids we showed that induction of immunity gave rise to an antibacterial activity which increased simultaneously with the synthesis of eight labelled polypeptides (P1—P8), characterized by their electrophoretic mobility in sodium dodecyl sulphate (SDS) containing buffer (Faye *et al.* 1975) (Fig. 3).

Immunity in saturniid pupae could be induced by an injection of either a Gram-negative or a Gram-positive bacterium and in both cases the same pattern of newly synthesized peptides was produced (Fig. 3). It was possible to block the induction with actinomycin D or cycloheximide, experiments which indicated that *de novo* syntheses of RNA and proteins were needed (Faye *et al.*, 1975). The antibacterial activity induced has a rather broad spectrum (Table I and Rasmuson and Boman, 1977) which would indicate that there is a low degree of specificity both in the induction and in the activity produced. However, the induction is fairly specific for viable bacteria which as a first step are phagocytosed but not killed (Faye, 1978). The broad spectrum of the induced killing activity is due to the composite action of at least three different bacteriocidal factors, each one with a given specificity. This was first demonstrated through the use of selective microbial inhibitors (Rasmuson and Boman, 1977; Boman *et al.*, 1978) and later confirmed by separating the factors (Table I).

An early aim of our work has been to fractionate the immune proteins of *H. cecropia* and to reconstitute the antibacterial activity from purified components. The two most abundant proteins, P4 and P5, could be purified by assays for radioactivity and electrophoretic mobility in SDS. P5 has a molecular weight of 96 000 daltons and is composed of four equal size subunits (Pye and Boman, 1977). P4 is a single polypeptide chain of molecular weight 48 000 and it is present both in normal haemolymph and in some tissue (Rasmuson and Boman, 1979). The functions of P4 and P5 are not yet understood. We have also purified P7, a lysozyme-like enzyme, and two bacteriocidal and low molecular weight proteins P9A and P9B, which were previously overlooked (Hultmark *et al.*, 1980).

Lysozymes are a class of very thoroughly studied enzymes which are widely distributed in Nature (Osserman *et al.*, 1974). Initially characterized by their

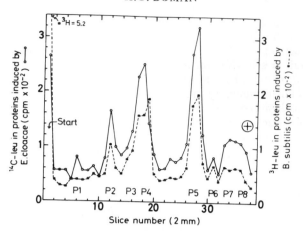

Fig. 3. Immune response in two pupae of *Hyalophora cecropia*, one injected with ³H-leucine and viable *Bacillus subtilis* (■----■), the other with ¹⁴C-leucine and viable *Enterobacter cloacae* (○——○). After 3 days the immune proteins (designated P1–8) were separated by electrophoresis in polyacrylamide gel (7.5%) with 0.1% SDS. Each slice of gel was burnt in an oxidizer which separates the ³H₂O and the ¹⁴CO₂ formed (Faye *et al.*, 1975). Proteins P4, P5 and P7 have been purified as discussed in the text.

ability to lyse suspensions of *Micrococcus luteus* (= *M. lysodeikticus*), lysozymes are more strictly defined as enzymes which hydrolyse $\beta(1 \rightarrow 4)$ bonds between N-acetylmuramic acid and N-acetylglycosamine in alternating polymers of a certain size (Philips, 1974). Using only the *Micrococcus* assays many investigators have demonstrated increased levels of lysozyme after infections in insects (detailed review by Whitcomb *et al.*, 1974). Mohrig and Messner (1968a) claimed that lysozyme was the chief defence agent in insects but more cautious opinions were later expressed by Chadwick (1970) and Kawarabata (1971). The lysozymes from *Galleria mellonella* and *Bombyx mori* were purified by Powning and Davidson (1973) and the *Galleria* enzyme was characterized in detail (Powning and Davidson, 1976).

Egg-white lysozyme can exist as an inactive proenzyme (Palmiter *et al.*, 1977) and increased levels of lysozyme in insect haemolymph could be due either to *de novo* synthesis, release of preformed lysozyme from a reservoir, or activation of an inactive precursor. An induced synthesis of lysozyme in *G. mellonella* was reported by Mohrig and Messner (1968a) but questioned by Kamp (1968) and by Powning and Davidson (1973), who claimed that actinomycin D did not block the increase in lysozyme concentration. We have recently investigated the matter in *H. cecropia* and found that the induction of the bacteriolytic factors in haemolymph was fully blocked by actinomycin D at a concentration of 10 μg/g pupal weight (Hultmark *et al.*, 1980). The most likely explanation for the earlier results of Kamp and of Powning and Davidson is that their concentrations of the inhibitor were too low (0.1 and 1.3 μg/g, respectively).

TABLE I

Bacteriolytic spectra obtained with Hyalophora cecropia haemolymph and two protein fractions[a]

Organism and strain	Normal haemolymph	Immune haemolymph	Pool I (P7 + P9)	Pool II (P9)	Egg-white Lysozyme
Micrococcus luteus	−	+	+	+	+
Bacillus megaterium, Bm 11	−	+	+	(+)	+
Bacillus subtilis, Bs 11	−	(+)	(+)	−	+
Bacillus cereus, Bc 11	−	−	−	−	−
Bacillus thuringiensis, Bt 75	−	−	−	−	−
Staphylococcus aureus, Cowan 1	−	−	−	−	−
Escherichia coli K 12, D 31	−	+	+	(+)	−
Escherichia coli, wild type	−	+	+	+	−
Enterobacter cloacae, β 11	−	+	+	(+)	−
Pseudomonas aeruginosa, OT 97	−	+	−	−	−
Salmonella typhimurium, LT 2	−	+	(+)	−	−
Proteus vulgaris, wild type	−	(+)	−	−	−

[a] Spot test with different bacteria spread in soft agar to give $10^3 - 10^4$ colonies/plate. Amounts applied in 25 µl were for P 7 4 µg, P 9 2 µg, and for egg-white lysozyme 25 µg of protein. Immune and normal haemolymph was 10 µl (containing ~ 75 µg protein). After incubation at 37°C overnight plates were read for zones of growth inhibition, + equals zone of ⩾ 2 mm, (+) < 2 mm, and − no detectable zone. Pools I and II were isolated by chromatography on carboxymethyl sepharose (Rasmuson and Boman, unpubl.). This particular batch of P 9 was probably contaminated with a small amount of P 7 which is responsible for the weak action against *B. megaterium*. Later work has shown P 9 to be two very similar proteins both very active against many bacteria (Hultmark *et al.*, 1980). Note that the spot-test plate method is fast and convenient but it does not distinguish between bactericidal and bacteriostatic effects. Our viable count assay (Boman *et al.*, 1974b) is more sensitive if used with a low concentration of test bacteria (5×10^3 cells/0.1 ml).

One strain, *B.t.* var *wuhanensis*, was isolated from the cotton leaf roller, *Anomis flava*, in 1969 and was shown by electron microscopy to have no flagella. Plated colonies are milky-white with a coarse surface and very irregular margins. It is similar to other *B.t.* though, of course, it has no *H antigen*. The Chinese claim it is distinct from the 18 H-serotypes currently known and that bioassay with houseflies showed it to produce beta exotoxin. In instar III *Pieris rapae* larvae the LD_{50} was 0.129 µg/larva.

B. PRODUCTION OF BACTERIA

Large-scale production on the communes is of two types: − (i) solid fermentation, and (ii) liquid fermentation in tanks. In both, fermentation proceeds in sterile rooms supervised by staff using clean overalls and masks.

The solid medium is usually a mixture of wheat bran and corn meal but soya bean and cotton seed cake may also be used after removal of the oil. After steam treatment the mixture is inoculated with a pure culture provided by the Provincial Institute. After 24 h growth in test tubes the medium reaches the "solid white state" and this is incorporated in a larger bulk of meal in litre bottles. Spore count is assessed after 24 h. This inoculum is added to piles of peanut bran and soyabean meal by mixing either with shovels or hands. The material is dampened, put in shallow trays and, if kept in an airy room, is ready for use in 30 h. On some communes this fermentation is achieved in 2.5−3 days in trays or soil pits.

Fermentation tanks are used at Tahsia commune, Kwantung Province. Electric pumps provide filtered air to aerate the large tanks. The medium within the tanks is a 4% peanut bran, 0.5% yeast, 0.5% silkworm pupal−powder suspension in water. The second stage fermentation is completed in 8 h within two 300-litre tanks and this is used to seed the final fermentation which is completed in two 1500-litre tanks, within 16 h at room temperature. Technicians monitor the process of the fermentation every 2 h day and night. The final product is forced under pressure through a battery of silken filters about 4 m long. The filters are then squeezed in a giant press to remove the water. The solids are scraped from the screens, dried under shelters in the sun, powdered in an electric mill, diluted with chalk and packed into plastic bags.

C. FIELD USE OF BACTERIA

All bacterial products are bioassayed at the communes before field use. *B.t.* is used against paddy borer, army worm, leaf-roller, rice-skipper, *Dendrolimus* spp., tea tussock moth *Euproctis pseudoconspersa*, diamond-backed moth *Plutella xylostella*, and cotton bollworm, *Heliothis armigera*.

It is understood that 2 500 metric tonnes are produced each year. For bollworm control, which is said to be 90% effective, a strain known as B7216 is used. *B.t.* var *wuhanensis* has been used since 1974 to control more than 40 pests in 12 provinces over more than 200 000 ha. The formulation normally contains 18×10^{10} spores/g.

A remarkable technique for applying *B.t.* against *Dendrolimus* on larch in hilly country, demonstrated to us, used mortars to treat the tops of trees in inaccessible sites. A line of mortars, spaced about 10 m apart, is set up along a front about 0.8 km long. The mortars are each about 10 cm diameter and 60 cm long. They are loaded with cartridges containing both the pathogen and an explosive charge at the base, which is ignited. The carton of pathogen is flung for several hundred metres when a second detonator explodes to disperse the contents from about 20 m above the trees. The operation was repeated with the mortars aligned some 30 m lower down the hillside. Apparently, aircraft are little used for spraying as the policy for cultivating all available land leaves no space even for small landing strips.

IV. Insect Pathogenic Viruses

A. RESEARCH ON VIRUSES

Chinese workers appreciate the efficacy of insect viruses as biological control agents. Both laboratory and field research programs are exploring their potential (Chiang and Huffaker, 1976). The Institute of Zoology, Peking, Institute of Entomology, Shanghai, Institute of Microbiology, Wuhan and Institute of Entomology, Kwangchow, were the principal centres of virus research visited. Most practical developmental work on viruses as pesticidal agents by scientific teams at the communes is done in collaboration with these institutes.

The silkworm industry in China is so important that the greatest care is taken to avoid hazard to the silkmoths, *Bombyx mori, Samia (Philosamia) cynthia ricini* and *Antherea pernyi*. This applies particularly to the release of viruses and also to possible pesticide residues on the plant food supplies. Therefore, the virologists in China must consider the potential susceptibility of these three species in any plan to control defoliating caterpillars by virus. They have recently begun to study five nuclear polyhedrosis viruses (NPV) in detail.

1. Spodoptera litura *NPV (Cotton Leafworm)*

This virus was first recorded near Kwangchow (Canton), Kwantung Province, in 1960. Haemolymph containing polyhedra was very infectious when fed to larvae of *S. litura* though no infections were obtained in the oak silkworm *S. cynthia ricini* (Tai, 1973). Hwang and Ding (1975) reported that the LD_{50} in instar III larvae was $10^{5.3}$ polyhedra/ml. The lethal time was influenced by both dosage and temperature, 10^8 polyhedra/ml producing 50% deaths on day 5 at 29–32°C and on day 7 at 25°C. A suspension of polyhedra applied to the upper surfaces of cotton leaves killed all feeding larvae, 2-day old deposits 50%, while 5-day old deposits killed only 10%. However, application to under surfaces of leaves killed 90% of larvae feeding 7 days after treatment and even 50% on day 9. Hwang and Ding concluded that infectivity had lasted longer on the under surfaces of the leaves because the virus had been protected from solar radiation. In the laboratory the *S. litura* NPV did not infect *Agrotis ipsilon, A. segetum,*

TABLE I

Introductions of pathogens since 1970 into invertebrate populations resulting in lasting control. Further instances of examples given by Burges and Hussey (1971) are excluded

Pathogen	Pest	Degree of Success	Place	Authority
GV[a]	Agrotis segetum moth	Residual effect of GV from one year to next: spread 10 m	Pakistan Denmark	Shah et al., 1979 Zethner, 1980
NPV[a]	Choristoneura fumiferana, budworm	Significant NPV and mortality, foliage protection for 3 years	Canada	Howse et al., 1973; Morris et al., 1974; Morris, 1977
NPV[a]	Lymantria dispar, moth	NPV spread from 10 treated ha to 300 ha by the next year, the highest mortality being > 40%	Sardinia	Magnoler, 1974
NPV[a]	Lymantria monarcha, moth	Damage insignificant in treated areas in following year: heavy elsewhere	Denmark Sweden	Zethner, 1976 Eidmann, 1976
NPV[a]	Spodoptera littoralis, moth	Strong evidence of natural spread and persistence. Birds spread virus ca. 0.5 miles	Crete	McKinley, 1980
NPV[a]	Trichoplusia ni, moth	Persistence and spread. Controlled subsequent generations. Respray never necessary. T. ni no longer a pest on cotton	Columbia	Bellotti and Reyes, 1980 Alvaro and Hernan, 1973
NPV, pox virus[b]	Wiseana spp., moths	Pasture management maintains control below economic levels	New Zealand	Crawford and Kalmakoff, 1977
NPV[a]	Neodiprion lecontei, sawfly	Spreads rapidly, often almost eliminating populations in one year	Canada	Chapter 19, VB2

TABLE I (continued)

Introductions of pathogens since 1970 into invertebrate populations resulting in lasting control. Further instances of examples given by Burges and Hussey (1971) are excluded

Pathogen	Pest	Degree of success	Place	Authority
NPV [a]	Neodiprion swainei, sawfly	Sprays gave good control, lasting into following years	Canada	Chapter 19, VB 3
NPV [b]	Cephalcia abietis, sawfly	Natural epizootics curb defoliation by this sawfly with 3-yearly cycles	Austria	Chapter 19, VB 10
Bacillus sphaericus [a]	Mosquito larvae	Recycled through larvae in a drainage ditch for 9 months	Florida	B. Hertlein, personal communication 1979
Aphanomyces astaci, fungus [b]	Crayfish, not a pest	Eradicated from large parts of Europe	Europe	Unestam and Weiss, 1970
Lagenidium giganteum, fungus [a,b]	Mosquitoes	Natural and artificial introductions recycled several years	Louisiana, California	Chapter 29, III
Culicinomyces spp., fungus [a,b]	Mosquitoes	Probably survived 3 years in often dry earthen drain. High initial kills at start of current trials	Australia	Sweeney, 1979
Romanomermis culicivorax, nematode [a]	16 potential mosquito hosts	Significant parasitism, often 90%. Still at many sites in 2nd and 3rd year	Louisiana	Petersen, 1976
Diximermis peterseni, nematode [a]	Mosquito larvae	Released at one site. Still produced 80% infection in 5th year	Louisiana	Petersen, 1976
Hetherorhabditis sp.	Mastotermes darwiniensis, termite	Bait infected workers. Nematode transmitted to successive groups of foragers	Australia	Work of Bedding, D.E.: Pinnock, personal communication

TABLE I (*continued*)

Introductions of pathogens since 1970 into invertebrate populations resulting in lasting control. *Further instances of examples given by Burges and Hussey (1971) are excluded*

Pathogen	Pest	Degree of success	Place	Authority
Dedalenus siricidicola, nematode[a]	*Sirex* spp., wood wasps	Infested logs inoculated with nematodes, which were spread by wasps causing 75% infection in most areas	Australia	Chapter 33, III
Howardula husseyi, nematode[b]	*Megaselia halterata*, mushroom phorid fly	Natural infections reach 60–75% by autumn, decline in winter. Artificial use impracticable	UK	Chapter 33, IV
Riberiroia marini guadeloupensis, trematode[a]	*Biomphalaria glabrata*, snail	Trematode prevents egg laying: 8×10^6 trematode eggs in 15 months added to pond: most snails disappeared	Guadeloupe	Nassi *et al.*, 1979

[a] Artificial introduction. [b] Natural occurrence.

soybean plots than by spraying virus. This seems surprising because it would appear equivalent to killing by spraying an equivalent number of larvae infesting the plants. This approach may produce only modest mortality, e.g. with the NPV of *H. zea* (Gard and Falcon, 1978), which, added to mortality from other causes, may lower crop damage to an economic acceptable level.

Microbial insecticides e.g. *B.t.* may kill only the insects to which they adhere or that eat them. Since, normally, most application methods miss some insects, which survive and breed, these organisms require periodic application. Some pathogens spread, so the initial application reduces an infestation, then spreads in the survivors and their progeny, e.g. the NPVs of sawflies (Chapter 19, V). Such organisms if produced and applied in quantity are regarded, for definition purposes in the analytical Tables II and IV, as both microbial insecticides and introductions, particularly the latter if recycling occurs over many years. Pathogens that naturally and normally curb a pest after it has become numerous can be made effective by application before the pest population reaches economically important levels, e.g. *Nomuraea rileyi* (Chapter 27, VIII) and *Verticillium lecanii* (Chapter 25, IX, X), see review by Ignoffo (1978). In nature the surviving pest inoculum tends to be immobilized in the soil between pest outbreaks (Chapter 19, IC) and this inoculum tends to reach the new outbreak of pests and spreads among them slowly. An application of a pathogen advances this effect, but enough knowledge of the ecosystem must be acquired to time the application effectively. Ideally, prophylactic treatment of a young crop (e.g. with *V. lecanii*) or treatment early in a season (e.g. with *N. rileyi*) can be made if sufficient reliance can be placed on later spread of the pathogen.

Manipulation of the environment is sometimes possible (Section IV; review, Ignoffo, 1978).

III. Effect of Insect Habit on Control Potential of Pathogens

A critical factor in the ecology of a pathogen is its transmission from host to host and this depends largely on the host's habit. Remarkably, almost all pathogens, except nematodes and fungi (excluding yeasts), are transmitted by an insect eating them with its food. In addition, some are transmitted in or on the egg. Insects inhabiting the aerial parts of plants either chew the plant surface, insert their mouthparts through the cuticle and suck the juices, or burrow into the plants.

A. FEEDERS ON PLANT SURFACES

Insects eating plant surfaces, particularly exposed leaves, are invaded by all types of pathogens. These pathogens may be splashed up from soil by rain, blown up by wind, deposited from faeces of insects, birds, etc. or released from infected insect bodies. *Trichoplusia ni*, for example, is attacked by at least one species of all the pathogen groups that attack Lepidoptera, i.e. a very wide range of candidate agents (Table II). Thus two strategies for its control are possible. (1) A microbial insecticide, e.g. *B.t.*, to achieve rapid control. (2) As many pathogens as possible can be introduced into the ecosystem hoping that some will persist, their combined mortalities contributing to biological control. These pathogens are, however, continually inactivated by solar radiation, washed off by rain and carried to the soil on fallen leaves, so they remain effective only while natural replenishment is adequate.

B. SUCKING INSECTS

Sucking insects feed from internal parts of plants free from insect pathogens and so do not ingest pathogens with their food. Consequently, insects such as aphids are attacked almost exclusively by fungi because these pathogens can penetrate the insect cuticle (Table II; Gustafsson, 1971). Fungi normally require high humidity so microbial control of sucking insects is feasible only in humid areas or areas with diurnal or seasonal high humidity, e.g. scales in citrus groves in Florida. Sucking insects acquire pathogens systemically from plants and from vertebrate blood, but these are primary pathogens of the plant or vertebrate and so are not considered for insect control.

C. BORING INSECTS

Boring insects also are protected from infection by eating internal, entomo-pathogen-free parts of plants. Thus larvae of grain weevils suffer no pathogens. The female weevil (*Sitophilus* spp.) drills a small hole in a cereal grain, inserts an egg and plugs the hole with drillings cemented by saliva. From hatching, the larva feeds and pupates entirely within the grain: no other metazoa and probably no microorganisms enter unless the ovipositor is contaminated. The limitation of pathogens is greatest in insects boring solitarily into living, sound plant tissue. Insects, e.g. bark beetles, forming accessible galleries fouled with frass offer food and shelter to a succession of scavaging metazoa, which create openings to the exterior and may carry in pathogens, particularly fungi, which encounter a permanently favourable humid atmosphere. These fungi are worth investigation (Chapter 24, VI C). Some insects, e.g. the codling moth *Cydia pomonella*, lay eggs on plant surfaces where the newly hatched larvae feed, exposed to pathogens, before burrowing to safety inside the plant. Their suitability for microbial control by all types of pathogen is proportional to the time spent on the surface. Thus *B.t.* fails to control codling moth on apple, but succeeds on walnuts where larvae feed on the outer surface for a considerable time before gaining entry [GV is successful on apple, probably because the lethal inoculum is low (Section IXD)]. However, in general, burrowers are poor candidates for microbial control except possibly for control by fungi or by pathogens transmitted transovum, e.g. *Nosema fumiferanae* in spruce budworm (Chapter 32, II B).

D. SOIL INSECTS

Soil insects also feed at the surface of plants, suck from their interiors and bore into them, and so are subject to the above principles. The soil offers protection from airborne spores, from fouling by birds and from other insects.

In contrast, it is an excellent medium for survival of resistant stages, e.g. spores and virus inclusion bodies (Jaques, 1964, 1967), because it is moist and protects them from solar radiation. *Nomuraea rileyi* spores buried in soil had a half life of 80 days (Chapter 27, V A) and this may be true for other fungi. Some fungi give a limited degree of control of scarabaeid larvae (Chapter 24, IV) due to a favourable high humidity in soil but this is probably curbed by the limited dispersal of airborne spores. Some entomopathogenic fungi may also be saprophytic in soil. Amongst bacteria, *Bacillus popilliae* produces extremely dormant spores that do not germinate in soil and can germinate in the gut of a scarabaeid beetle larva only after lengthy vernalization (Chapter 4, II C). Only a proportion can germinate at any one time so extending the effective life of an inoculum, which can be long. At death larvae are large, leaving a large inoculum which is maintained by the preservative quality of soil and absence of disturbance. Spores produced *in vitro* have poor infectivity and survival, probably because they have not acquired this variable and sophisticated dormancy. In contrast, *B.t.* survives relatively poorly in soil, potency falling by about 75% in < 2 weeks (presumably due to denaturation of the crystal). The spore half-life is about 6 weeks in a clay loam at 25°C (Pruett *et al.*, 1980). Equivalent values are 4 weeks and 20 weeks in a more sandy soil (A. West, personal communication). Saleh *et al.* (1969, 1970), Kiselek (1974) and Sekijima *et al.* (1977) obtained spore survival for 3 to 16 months in soil outdoors. *B.t.* can increase saprophytically in soils of very high organic content (Saleh *et al.*, 1970). In most soils spores germinate and die. Thus soil insects are good candidates for microbial control with fungi and specialized bacteria but not with *B.t.* Many organisms have been evaluated for termite control but few were found promising (Table I; Lund, 1971), despite favourable humidity. This suggests that termites have evolved resistance to fungal pathogens even though the fungi would be operating at optimum humidity.

E. AQUATIC INSECTS

Aquatic insects feed on plants by the same three methods as aerial insects, with the same effect on infection by pathogens. In addition, many are filter feeders and these include most pest species. They are attacked by an even wider range of pathogens (Table II; list, Roberts and Strand, 1977) than are the Lepidoptera. Since they ingest inert stages of pathogens and particulate toxins they are susceptible to microbial insecticides. *B.t.* var *israelensis* is a very promising organism probably acting by toxicosis of its crystals (mosquitoes: de Barjac, 1978a; using var *kurstaki*, Panbangred *et al.*, 1979; blackflies: using var *kurstaki*, Lacey *et al.*, 1978). Toxins in the cells of *B. sphaericus* kill by toxicosis and the bacteria multiply after death of the host (Chapter 14, II B1−2, II C3). Spores of *Metarhizium anisopliae* accumulate around the siphon of mosquito larvae, making it less hydrophobic so that the insect cannot remain at the surface to breathe and suffocates. These pathogens present the formulation problem of keeping them suspended in the correct position in the water. This must be solved before their full potential can be realized. For instance,

Goldberg *et al.* (1977) encapsulated *B.t.* var *israelensis* in jojoba oil without reducing insecticidal activity after ingestion. This gives the bacteria both added buoyancy and a hydrophilic surface which holds them just below the water surface without excessive scum formation by wind.

Many aquatic fungal pathogens have motile spores, e.g. *Coelomomyces* (Chapter 29, IV A), or motile infective stages, e.g. mermithid nematodes (Chapter 33), which confer a power to search for prey. A large proportion of epizootics recorded in aquatic pests are caused by these two groups of pathogens, suggesting that a motile stage is very important in the aquatic habitat. However, these infective stages may fall victim as food to non-target fauna. Some fungi have alternate hosts which offer an additional means of dispersal but which may limit success to certain niches where both hosts can thrive. The aqueous habitat largely favours the survival and recycling of pathogens, particularly those with motile stages. It thus favours a strategy of introductions. Application of as many candidate pathogens as possible may together produce a satisfactory cumulative mortality. In complex situations, a strategy of introducing the whole fauna may be successful as with the transfer of whole pond water samples from ponds with successful natural mosquito suppression by *Coelomomyces* in the USSR to ponds lacking the pathogen (A.M. Dubitskij, personal communication).

Some pathogens can grow saprophytically in an aqueous environment, e.g. *Lagenidium giganteum* (Chapter 29, III A). *Bacillus sphaericus* inoculated into the vigorously aerated primary intake tank of open tank sewage systems maintained itself at 10^3 to 10^5 cells/ml. It passed to subsequent settling and percolating tanks and ponds, which are rich breeding sites of *Culex nigripalpus*. Up to 100% control was obtained (mean 80 to 95%), even in ponds past the chlorination tank, because the toxin remained active in the dead cells (B. Hertlein, personal communication).

F. PEST DENSITY AND SPREAD OF PATHOGENS

In insects of all habits, crowding or gregarious behaviour enhances the spread of pathogens by all routes. However, social insects, which normally live crowded, have apparently developed sufficient resistance to prevent pathogens running amok. These apart, insects have a density dependent relationship with pathogens. Successful introduced pathogens must control pests at subeconomic densities. Density dependence can be bypassed by applying pathogens inundatively as insecticides. With introduced pathogens the initial stages of infection in a new insect population may express delayed density dependence, e.g. initiation of infection in spring may depend on pathogen abundance in the previous autumn as with *N. rileyi* (Chapter 27, VI A). This knowledge may be useful in deciding when to use a spring application of a pathogen.

The physiological condition of an insect may affect its susceptibility to a pathogen. For instance, an insect strain with a high gut pH may be less susceptible to baculoviruses because it destroys freed virions (circumstantial evidence; Ripa, 1978) or more susceptible to *B.t.* because it dissolves the endotoxin crystal better. Usually, resistance of larvae increases with age (Section IX, D). Sometimes hibernating insects are more susceptible, reversing this tendency, probably because of changed physiology. Mortality of larvae of *Spilnota occellana* increased in hibernaculae (Legner and Oatman, 1962; Oatman and Legner, 1964). Some insects may hibernate colonially. This may be a good time to apply fungal pathogens when susceptibility is lowered by the insect's physiological state and when the chance of spread of the pathogen is increased by high host density, e.g. application of *Metarhizium* to rhinoceros beetles aggregated in heaps of decaying vegetation (Chapter 20, VIII).

Pathogens with neither a desiccation-resistant nor a motile stage, e.g. non-occluded viruses such as the *Tipula* iridescent virus, are confined to the host and are reliant on trans-ovum transmission or cannibalism (Carter, 1973), which limits their usefulness. They tend to be enzootic rather than epizootic, which probably explains why so few have been discovered and studied.

G. INSECT DEVELOPMENTAL STAGE

The insect stage most vulnerable to pathogens is the larva because it is the main feeding stage. Egg, pupa and non-feeding or sucking adults suffer only the pathogens acquired from the previous stage, except nematodes and fungi which attack through the cuticle. Even so, mortality from disease in these stages is often considerable, viz: bacteria in moth eggs (Lynch *et al.*, 1976) in moth pupae (review, Angus, 1965; Dulmage and Martinez, 1973); *Nosema fumiferanae* in spruce budworm eggs (Chapter 32). *Nosema whitei* (Microsporida) caused small deformed adults of the beetle *Tribolium castaneum* with reduced fecundity (Milner, 1972). *B.t.* had the same effect in the spruce budworm (Smirnoff, 1978).

IV. Influence of Type of Crop and Habitat on Microbial Control

Success of a pest control method depends on its efficacy, cost, current environmental acceptability, labour required and the momentum of research and development. These factors vary with type of crop.

Before considering each type of crop independently, some generalizations can be made about the different ways of using pathogens. Microbial introductions, largely of exogenous organisms, are cheap to apply, usually specific to a single pest, long lasting and — ideally — need no maintenance. The size of any industry producing them is likely to be small. Microbial insecticides require a larger producing industry. They tend to be expensive, specific to one or a few pests in a pest complex (so that other control methods are needed for the rest), but are harmless to beneficial organisms and are thus unlikely to increase pest problems in the long term. Generally, the reverse of all these factors applies to chemical pesticides. Overall cost in the long term depends on the net effect of all factors, e.g. the use of expensive *B.t.* + chitinase against spruce budworm causes no problems in following years, vs chemicals which create a perennial balance between pest and pesticide, and a perennial need for chemicals (Smirnoff, 1978).

Sometimes a pathogen combined with a chemical or another pathogen is beneficial (Chapter 38, II, III A). Attempts to augment endogenous pathogens vary in method and defy generalizations.

Some general principles apply to all methods of using pathogens. Microbial insecticides should be applied near dusk to avoid peak ultra violet (UV) radiation and to coincide with the peak feeding period that occurs at dusk with many pests. A pest should not be eliminated but should be reduced to economic insignificance, so allowing beneficial fauna to persist and exert an effect and — with pathogens able to spread — to maintain a supply of the pathogen in the ecosystem. Where possible, damage thresholds should be measured and applications of pathogens related to them.

A. FOREST

The scale of forest pest control is huge and pollution with chemical pesticides correspondingly great. A food crop is not involved. Natural enemies are of great importance in a long lasting, extensive, stable ecosystem. Cultivation does not remove residual pathogens (Table III) from the pests' environment and the tolerance of damage to trees is high. The saleable product, wood, and recreational facilities are attacked directly by only a few pests. Motivation to control pests is low because forests tend to be national assets and not owned by individuals with high capital investment and income at stake. Even so, widescale use of chemicals in forests has caused considerable social and political upheaval, which has strengthened a search for alternatives. Strategy in these circumstances favours introduction (Table IV) on grounds of low cost and a greater chance of success in this environment than in any other. The cyclic nature of forest pest abundance may necessitate periodic reintroduction or augmentation of agents. Microbial insecticides (Table IV) also have a great chance of success in the forest environment, but their high cost tends to limit them to recreational areas for which their safety makes them well suited. However, the cost of using tussock moth, *Orgyia pseudotsugata,* NPV compares favourably with that of chemicals (M.E. Martignoni, personal communication).

Pathogens, mainly viruses, fungi, microsporidia and nematodes, are prominent in natural control of forest pests (Table III). Introductions of viruses against the main defoliating sawflies have been particularly successful; a nematode has given good control of a wood wasp (Table IV; Chapter 33, III). Microbial insecticides (Table IV) used against Lepidoptera and sawflies are predominantly *B.t.*, which does not persist, and viruses, which are often used initially as microbial insecticides to curb damage, the main objective with sawflies being subsequent viral spread as introductions or augmentations of existing infections.

Often forest pest outbreaks collapse due to disease epizootics before causing economic damage (Table III). Thus, at the start of an outbreak, the forest manager must decide whether to apply control measures. By cumulative studies on pest ecology, we need to approach a situation where this depends less on inspired guesswork. Because of the complexity of forest ecosystems, simulation models (O.N. Morris, personal communication) have been used to co-ordinate

TABLE III

A subjective assessment of the natural control exerted by the best examples of different types
of pathogen on different crops

Pathogen group	Forest	Grass	Arable agriculture	Vegetables	Fruit, nuts,[b] ornamentals	Protected crops[c]	Stored products	Water
Bacillus popilliae	—	+++++	+	+	—	—	—	—
B. thuringiensis	+++[d]	—	—	—	—	—	++	+
B. sphaericus group[a]	—	—	—	—	—	—	+	+++
Baculoviruses	+++++	+++	+++	+++	++++[e]	+	+	++
CPV	+++	++	++	++	++	+	+	++
Other occluded viruses	++	+	+	+	+	—	—	+
Non-occluded viruses	++	+	+	+	+	—	+	++
Hyphomycetes	++++	++++	++++	++++	++++	++	—	+++
Entomophthora spp.	+++	++++	+++++	++++	+++++	+	—	+
Coelomomyces spp.	—	—	—	—	—	—	—	++++
Protozoa	+++	+++	+++	+++	+	—	+	+++
Nematodes	+++	++	+++	++	—	+++	—	+++++

KEY

—	none, virtually none known
+	occasionally enzootic
++	often enzootic
+++	enzootic, cyclic epizootics
++++	enzootic, cyclic epizootics, major natural control
+++++	keeps some species virtually below economic threshold

[a] Includes *B. alvei, B. brevis, B. circulans* complex. [b] Includes coconuts. [c] Includes mushrooms. [d] Rare. [e] Includes the non-occluded virus of citrus mites (Chapter 21).

TABLE IV

A subjective assessment of the efficacy of the best examples of intentional or accidental introductions of pathogens and of microbial insecticides on pests on different crops

Pathogen group	Forest	Grass	Arable agriculture	Vegetables	Fruit, nuts[b] ornamentals	Protected crops[c]	Stored products	Water
Bacillus popilliae	—	x x x x x	—	—	—	—	—	—
	—	+ + + +	+	+	—	—	—	—
B. thuringiensis	x x x x	x x x x	x x x x	x x x x x x	x x x x	x x x x	x x x x	x x
	+ + + a	—	—	—	—	—	+	—
B. sphaericus group[a]	—	—	—	—	—	—	x	x x
	—	—	—	—	—	—	—	+ + + + +
Baculoviruses	x x x x x	x	x x x x x	x x	x x	x	x	—
	+ + + + +	—	+ +	+ + +	+ + + + + e	—	+	—
CPV	x x x	—	—	—	—	—	—	—
	+ + + +	—	—	—	—	—	—	—
Other occluded viruses	x	—	—	—	—	—	—	—
	—	—	—	—	—	—	—	—
Non-occluded viruses	—	—	—	—	—	—	—	—
	—	—	—	—	—	—	—	—
Hyphomycetes	x	x x	x x x	x x x	—	x x	—	x x
	+ + + +	+ + +	+ + +	+ + +	+ + + +	+ + + + +	—	+ +
Entomophthora spp.	—	—	—	—	—	—	—	—
	+ + +	+ + + +	+ + + + +	+ + + +	+ + + + +	+	—	—

TABLE IV (continued)

A subjective assessment of the efficacy of the best examples of intentional or accidental introductions of pathogens and of microbial insecticides on pests on different crops

Pathogen group	Forest	Grass	Arable agriculture	Vegetables	Fruit, nuts[b] ornamentals	Protected crops[c]	Stored products	Water
Coelomomyces spp.	—	—	—	—	—	—	—	—
	—	—	—	—	—	—	—	++
Protozoa	—	××	×	—	—	—	×	×
	+++	++++	+++	—	+	—	+	+
Nematodes	×	—	×	×	—	×	—	×××××
	+++	—	+	+	—	++++	—	++++

KEY

Introductions

—	none or virtually none
+	rare, no economic benefit
++	enzootic, present one year later
+++	enzootic, with epizootic cycles
++++	enzootic, epizootic cycles, major natural control
+++++	keep some species virtually below economic threshold

Microbial Insecticides

—	no promising candidates known
×	candidates with some promise
××	candidates being developed
×××	product in Japan, Brazil, China, USSR
××××	minor sales of registered product(s), sometimes because consumer industry is small
×××××	major use of registered product(s)

a–e See footnotes to Table III.

and synthesize the massive amount of information into a decision-making tool for pest managers. In Canada, for example, investigations on spruce budworm control include manipulation of native and exotic parasites and predators, forest management, pathogens (viruses, bacteria, fungi and microsporidia), pathogen—chemical insecticide combinations, genetic control, pheromones, growth regulators, chemical larvicides and adulticides (Baskerville, 1976). A budworm/forest-system simulation model (Morris, personal communication) involves a budworm survival submodel and a forest response submodel which work in parallel. Major processes of the budworm and forest cycles are linked. Extensive data on budworm, forest and economic interrelations collected over the past 30 years formed the basis for the model construction and validation. It suggests how low-mortality host-specific viruses could be used in a pest management scheme to significantly alter budworm population dynamics in certain stand conditions. This model, despite imperfections, is representative of the type of tool needed in developing integrated forest insect control techniques.

There is considerable scope for the use of microbial agents against tree diseases, e.g. that caused by *Fomes annosus* (Chapter 39, II E).

B. GRASS

Grassland is subject to all the comments made about forests with the following exceptions. An animal food crop is involved. Motivation to control pests is a little higher due to much private ownership but pest damage is unnoticed until severe. Orthoptera, caterpillars and root feeders are the major pests, with fungi, microsporidia, the *B. popilliae* group and occluded viruses dominant in natural control and introductions (Tables III, IV). Introductions are the only economic way to use microbial insecticides. *B.t.* is effective against defoliating caterpillars (McNeil *et al.*, 1977; Henry, 1978) but uneconomic because of low crop value. Strategy is the same as for forest. Wasteland where locusts breed and rangelands seem particularly promising sites. *Wiseana* caterpillars can be maintained below economic levels by spreading and maintaining virus in the pests' microenvironment by regulated grazing, seed oversowing and avoidance of tillage when starting new pastures (Table I).

C. ARABLE AGRICULTURE AND VEGETABLES

Arable crops (cereals, fodder, fibre, seed, sugarcane etc.) and vegetables create much pressure to control pests because they are valuable and privately owned. Pressure is strong against pests attacking the saleable parts of the plant, for which cosmetic appearance is often crucial. Pathogens are most useful against pests attacking the rest of the plant, which can often tolerate great damage without crop loss. Estimation of this damage tolerance is an important aspect of modern pest control strategy, e.g. on cotton (Falcon, 1978). Many of the crops are arranged in small manageable units but multiple ownership of these hampers area-wide policy to reap full benefit by not harming natural enemies. Seasonal cropping, burning off and ploughing remove or bury pathogen reservoirs (Jaques, 1974). Even so some pathogens recur annually, e.g. *Nomuraea rileyi*

(Chapter 27, VI A), and the NPV of *Trichoplusia ni* and GV of *Pieris rapae* splashed from soil to cabbage (Jaques, 1967, 1970a, b, 1972).

Pathogens involved in natural control are mainly fungi, viruses and protozoa (Table III). Strategically there have been some introductions (Tables I and IV). The introduction of *B. popilliae* to waste and pasture land has reduced migration of Japanese beetles to crop land. Some microbial insecticides (Table IV) have their greatest use on food and fibre crops, particularly *B.t.* because it prevents crop damage rapidly. Emphasis has been on lack of harm to beneficial fauna (Falcon, 1978). Some endogenous pathogens can be applied as insecticides and/or seasonal introductions to obtain control earlier than peak season natural control, e.g. *N. rileyi* (Chapter 27, VIII) and the NPV of *T. ni* (Jaques, 1974).

There is limited scope to foster endogenous pathogens by crop management. Thus close row spacings (Jaques, 1978) and plants with a closed rather than an open canopy foster fungi by increasing humidity and protecting spores from solar radiation (Chapter 27, VIII C). Irrigation favours *Entomophthora* on the spotted alfalfa aphid (Hall and Dunn, 1958). However, free water and rain-induced high humidity may stop dispersal of fungi with dusty air-borne spores and rain also washes spores from the atmosphere. Pathogens concentrate at the soil surface, e.g. the NPV of *T. ni* (Jaques, 1969). Thus shallow tillage, or direct drilling of crops coupled with weed killers without ploughing, keeps the pathogens near the surface and hence most able to reach new foliage. Some pathogens survive best in soils of near neutral pH, so pH adjustment may be feasible (review: Jaques, 1978). The nematode *Neoaplectana carpocapsae* was most efficient against root maggots *Delia* (*Hylemya*) *brassicae* in moist soil (Welch, 1962). Interplanted trap plants that encourage pathogens may be feasible.

D. FRUIT, NUTS AND ORNAMENTAL WOODY PLANTS

Fruit and nuts grown on trees, bushes and canes have many similarities to vegetable crops. Motive to control pests on these valuable crops is high, with even greater emphasis on cosmetic appearance of the marketed crop, so pathogens are most useful on the rest of the plant. Although the soil in many orchards is cultivated, the crops are perennial, presenting a stable ecosystem. Natural control of pests by pathogens (Table III) is substantial, with fungi predominating, possibly because of their aerial spore dispersal, since the crop is higher than vegetables above the soil so movement of pathogens without aerial spores from soil to tree is more difficult. However, baculoviruses including the non-occluded virus of the citrus mite are sometimes effective. The role of natural control was demonstrated by resurgence of the green apple sucker in Nova Scotian orchards when fungicides used against apple diseases were changed to compounds lethal to *Entomophthora spaerosperma* (review, Burges and Hussey, 1971; Jaques and Patterson, 1962). Strategy favours introductions with some pathogens, which have been remarkably successful (Table IV; review, Burges and Hussey, 1971) probably due to the stability of the ecosystem, which is similar in this respect to the forest. Some microbial insecticides have been successful, e.g. the GV of the codling moth, *Cydia pomonella*, on apple (Falcon *et al.*, 1968; Huber and Dickler, 1977). Ito *et al.* (1977) proposed simulation models involving the natural enemies of *Adoxophyes orana* and a GV. The simulation strongly suggested that mortality during the 1st generation due to a single early GV spray can reduce the 2nd and 3rd generations and so reduce fruit damage. Other microbial insecticides (Table IV) have succeeded against some pests, but failed against others, largely due to high cosmetic requirements, e.g. *B.t.* failed against *C. pomonella* on apple (Section III, C). Low temperatures in critical spring periods and the leaf rolling habit of many larvae curb the effect of *B.t.* on defoliators. There is little scope for environmental manipulation, although in Israel *Verticillium lecanii* is effective against soft scales on citrus with overhead sprinkler irrigation, but not with low sprinkler or drip irrigation (Soper, 1978).

Cosmetic pressure is less on woody ornamental plants, except those grown for sale, because nothing is harvested. This allows microbial insecticides to obtain partial control calculated to leave enough insects to support a beneficial fauna which continues to exert its regulating effect (Chapter 36, II). Prophylactic treatment of Christmas tree plantations with NPV against sawflies is very effective. There is also scope for the use of microorganisms against tree diseases (Chapter 39).

E. PROTECTED CROPS

In greenhouses all aspects of pest control become exaggerated (Chapter I). Crops are very valuable and cosmetic requirements maximal, so motivation to control pests is great. Integration of methods has proceeded faster than in any other industry, being used in about half the heated acreage in the UK. Post crop practices such as soil steaming against plant pathogens remove residual entomopathogens. Natural control of pests by disease does not exist at acceptable pest populations levels (Table III). There are two strategies for applying pathogens. (1) Rapidly acting microbial insecticides (Table IV), e.g. *B.t.*, which may if necessary be applied frequently due to the high crop value. If the pest is very susceptible with favourable behaviour on a crop conducive to good pesticide coverage, it is as cost-effective as a chemical, e.g. against the tomato moth on tomatoes (Burges and Jarrett, 1979). If one or more of these factors are less favourable it competes only when it safeguards other valuable biological agents, e.g. on chrysanthemums (Burges and Jarrett, unpublished data). (2) Pathogens should be freshly introduced to each new crop (Table IV) to achieve good control, e.g. *V. lecanii* (Chapters 1 and 25). A nematode is promising against sciarid flies (Chapter 33, IV).

Manipulation of the environment and the crop is more feasible than outdoors, but limited in practice, because most factors must be optimized for other reasons, e.g. plant health, or cost of fuel which may prohibit a rise in temperature. Limited increase in humidity is the most useful factor to employ. Jarvis and Slingsby (1977) sprayed plants with water to enhance growth of a hyperparasitic fungus (*Ampelomyces quisqualis*) inoculated on to the plant pathogen *Sphaerotheca fuliginea* that causes powdery mildew on cucumbers at a time when there were no effective fungicides. *V. lecanii* should be applied late in the day to benefit from the increase in humidity which occurs when chrysanthemum plants are covered by black polythene sheets for daylength control to time plant budding (Chapter 1). Overhead irrigation may spread some pathogens, e.g. *V. lecanii*, but it may wash others away, e.g. *B.t.*

On mushrooms, effective mixing of a pathogen into the compost where pests occur is a difficult problem. Nematodes are probably the best candidates because of their motility, but they cannot yet be produced at a low enough cost (Chapter 33, IV).

F. STORED PRODUCTS

Stored materials have high value and, in the raw state, e.g. cereal grains, a moderate cosmetic requirement and some pest tolerance. Cosmetic importance increases vastly and pest tolerance reduces to nil in packaged foods. Thus motive to control pests is high and pathogens are most likely to be of use in raw materials. Pests reach stored material in three ways; (1) prior to storage, e.g. from farm buildings, (2) by flight, (3) from populations breeding in food residues in inaccessible places. Natural control by pathogens occurs long after pest populations have become dense and so is useless, except in residues, where it probably occurs unseen (Table III; Burges and Hurst, 1977). Strategically, stored products can be protected from all routes of attack only by expensive, even, prophylactic, admixture of microbial insecticides (Table IV; Burges, 1973), although peripheral treatment of grain bulks with *B.t.* (the only microbial insecticide on sale for this purpose) is effective against certain moths (McGaughey, 1976). Partial improvement can be obtained by treating inaccessible places with introductions of microbial insecticides (Table IV), e.g. in the first run of grain through a mill after a clean out or by treating crevices in cleaned floors, etc., with a powerful dust blower. This seems more practicable than using pheromone baits (review; Burkolder and Shapas, 1978). Warehouse conditions, dry and shaded from UV radiation, are ideal for spore survival so that infections in residual insect populations survive virtually indefinitely. However, inoculum in all produce except the residues is removed abruptly at unloading. A control procedure could be based on admixture of infected screenings from unloaded products into new produce after killing the pests mechanically in an entoleter machine, which throws powdery materials against its casing by centrifugal force and shatters the bodies of pests. However, monitoring efficacy would be difficult (Burges, 1973). The pests comprise moths (effectively controlled by *B.t.*; Burges, 1964; McGaughey, 1976; some controlled by GV: Hunter *et al.*, 1977), beetle species (most having a specific protozoan) and mites (no known pathogens). Many pathogen species would be needed and a few key pests could not be controlled. The storage environment is manipulated to prevent pests, e.g., by eliminating crevices in buildings and by cooling produce, but not to favour pathogens.

G. AQUATIC PESTS

Most aquatic pests, e.g. mosquitoes, blackflies, tabanids and biting midges, are vectors of mammalian diseases. Vector control is a national and international problem with ponderous drive, poor in underdeveloped countries, but benefiting from funded research, e.g. over 530 mosquito control agencies in the USA and Canada had an overall budget of $69 million (US) in 1975–6. The World Health Organization (WHO) has funded many projects and raised extra finance from 1976 in a Special Program (Chapter 14, II C, Table VII).

A wide variety of pathogens (Table II; list, Roberts and Strand, 1977) exert considerable natural control (Table III). As described in Section III, E, strategies include control by introductions (Table IV) and by microbial insecticides (Table IV; *B.t.* var *israelensis* in the process of industrial development; Goldberg *et al.*, 1977; Goldberg and Margalit, 1977; de Barjac, 1978a, b; Vankova *et al.*, 1978; Weiser and Vankova, 1978; de Barjac and Coz, 1979; Garcia and DesRochers, 1979; Undeen and Berl, 1979; review, WHO, 1979b,d: *B. sphaericus*, Chapter 14: *Metarhizium anisopliae*; Roberts, 1970; Al-Aidroos and Roberts, 1978; review, WHO, 1979a). Aircraft application of *B.t.* var *israelensis* to 12 ha gave 98.8% control of *Culex tarsalis* larvae without a population resurgence as is typical after treatments with organophosphorus larvicides (Schaefer, 1979). In 1979 an advisory group for the WHO Special Programme suggested five priority categories of biocontrol agents, including pathogens, fish and invertebrate predators and parasites. The top priority group contained one organism, *B.t.* var *israelensis* (WHO, 1979c). It is the intention of WHO to review priorities annually as new knowledge accumulates. Aerial application of agents to foliage can be made only during limited periods of suitable inversion weather conditions because small drops must be used to give even cover at a low volume. Application to large water bodies can be made in any conditions suitable for flying because large drops can be used that will fall against upward air currents. By encapsulation in relatively large capsules a foliar canopy can be penetrated. In contrast, application to small water bodies is difficult. Suitable agents for these are transovum transmitted or insect borne pathogens, and invertebrate predators and parasites. Resistant stages are necessary in water bodies that periodically dry up.

V. Role of Organizations in Microbial Control

Many organizations have played a part in developing microbial control and their progress has depended on their motivation, facilities and function.

Industrial production of pathogens for sale and profit is one of the end points of the system. Industry has been attracted by a profit motive, by having spare fermentation capacity, a need to diversify or to find more outlets for raw materials. It is dominated by investment risk and a financial return thereon. Once decisions have been made staff, facilities and plant can be acquired rapidly. This accomplished, the process gains momentum until sales are made or the project dropped. Its research and development are highly directional, tending to minimize time and expense to achieve objectives. Industrial drive has played a vital role in the development of microbial pesticides *and must continue to do so in the future.*

Government organizations are motivated by the views and needs of both consumer industry and public opinion. This is a diffuse and cumbersome process. Vital government roles are research, development of projects where profit is unlikely, and education of consumers through advisory services. Education is essential to support a small producer industry with limited resources to promote its products vs the vast resources of the chemical industry. Governments are the largest source of manpower and finance. Only they can undertake some projects, such as control of vectors, and pests in forests and along highways. They have contracted industry to produce certain pathogens, e.g. the NPV of *Orgyia pseudotsugata*, a practice that should be increased in future.

Universities are motivated by innate interest in research and teaching, and by projects in which they are partially contractors to governments and industry. Their major role is research of the more basic type, e.g. there is great scope for modelling the outcome of pathogen applications in integrated systems on an ecological basis (Chapters 36, 37). Universities also have a place in co-operative programs with government and industry.

International organizations catalyse and co-ordinate co-operation, set standards and finance meetings and research. WHO has financed research on vector control (Section IV, G: Chapter 14, II C).

Chinese organization is unique, being based on local, largely autonomous communes. They choose their own methods and are self-contained in producing and applying microbial agents, and reaping the benefits, aided only by regional centres that supply inocula and advice. Each commune develops its own production methods, doubtlessly aided immensely by the dispersion of university staff and students to work on communes for part of every year (Chapter 42, I, V). They appear to do no safety testing, reasoning that insect pathogens are specific and accepting the results of Western work.

In underdeveloped countries there is little use of microbial agents, apart from commercial sales, and research promoted by international agencies and former colonial powers, e.g. use of rhinoceros beetle NPV. Generally there are no national safety registrations so commerce may find no safety barriers. On the other hand formal approaches of international bodies may be met with a demand for assurances, or objections if a pathogen is not already in use elsewhere. Yet pathogens are ideal for use in these countries because harm is unlikely to result from their misuse. The possibility of their inexpensive local production as "cottage" industries has been mooted, using local materials. Hertlein (1979) has shown that materials such as animal dungs can be used as fermentation media. One can hardly imagine success on the Chinese scale because there is no existing comparable dispersion of scientists to help in the practical use of pathogens. Deep liquid fermentation, so reliant on media sterility and pure culture, would need importation of sophisticated fermenters. Fermentation on solid or semi-solid media (Dulmage and Rhodes, 1971) would be more feasible, being less dependent on absolute sterility or on constant quality of media. Low pressure steam sterilization could be followed by heavy seeding with inocula to depress contaminants, aeration with sterile air, drying with a high input of warm air, finally grinding without excessive heat production. Skilled help would be needed for inocula production and quality control. *In vivo* production of pathogens that can be produced economically only in living insects is more suitable for a cottage industry because diseased or healthy insects could be field collected, or insects could be bred on local plants, e.g. the mosquito pathogen, *Nosema algerae*, can be produced in caterpillars of *Heliothis zea* (Laird, 1977). A major problem would be quality control (Chapter 40, X).

Co-operative programs are rewarding, e.g. that on standardization of *B.t.* (Burges *et al.*, 1967) and the recent program of Dulmage (Chapter 11, I). Local co-operation on individual projects should encompass as wide a team as possible to exploit varied skills. National and international programs bring together locally developed specialized facilities with individual pests or pathogens. A program to investigate the host ranges of individual NPVs and another to investigate GVs would be particularly opportune. Each co-operator could produce enough virus of his pest species in his local insect stock to distribute for bioassay to other members of the program. Use of a single batch of product would ensure comparability of assay results. This could also be improved by standardizing bioassay methods as far as possible, e.g. by using artificial insect rearing media. Aliquots could also be distributed to virologists for virological studies.

Only by such comprehensive programs can sufficient, widely based, results be rapidly accumulated to provide an adequate study of host range and virological relationships within these virus groups. In 1979 a Study Group of the IOBC/WRPS on "Safety of Insect Pathogens" recommended that working groups should be formed (1) on "Characterization and Identification" and (2) for "Bioassay" of baculoviruses. These could form nuclei for international co-operative programs. An essential requirement is an enthusiastic organizer. Other programs to investigate strains of *Beauveria* and of *Verticillium lecanii* would be valuable if sufficiently stable formulations can be produced.

The WHO is sponsoring projects each involving cooperation between a group in an underdeveloped country and one in an advanced country. An important aspect of these is to ensure that the results are politically and socially usable.

Standardization of pathogen products and bioassays by co-operative programs would permit interpretation of bioassay results for practical use. A pathogen that is highly virulent in laboratory assays obviously has good potential for practical control, but the ecology of pathogen and pest, and the nature of the crop, determine the degree to which this potential is realized. These factors are difficult to quantify individually. However, they can be quantified together, in relation to pathogen virulence, by comparing laboratory LC_{50} s with the lowest dosage that gives good field control, thus giving a single value that can be called an efficiency factor (Burges, 1973).

VI. Impact of Changes in Plant Culture and Pest Management

Monoculture is increasing almost universally. It tends towards an unchanging pest complex, enhancing pathogens by allowing them to accumulate in the environment. Predators and parasites may also accumulate, creating an invaluable asset to preserve by avoiding broad spectrum pesticides. However, plant disease organisms also accumulate, which creates a problem, but which also offers more opportunity to use competing and inhibitory organisms (Chapter 39). The spread of monoculture may encourage, or even necessitate, area-wide pest management programs, possibly enforced by law.

Allied to monoculture, minimal tillage techniques maintain near the soil surface, or *in situ* in the region of plant roots, all four types of organisms mentioned above. This requires more research on intensified interactions.

The trend to reduce the industrial labour force makes labour-saving pathogen introductions more attractive. It also makes microbial insecticides, or chemicals, that require frequent application less so. The parallel trend is to employ contract pest control. This offers a chance to train specialist workers in the integration of biocontrol agents (or "biorational" agents, which include pheromones, etc.) and in the attendant observational skills necessary for the employment of sophisticated agents.

The move towards low and ultra-low volume application methods requires extra research with pathogens. *B.t.* can be used in pulse jet type thermal foggers (Burges and Jarrett, 1979) and, probably, most pathogens in cold foggers and ultra-low volume equipment. However, only the upper surfaces of foliage are covered, where pathogens are most sensitive to UV radiation and rain, while some pests are confined to lower surfaces. The significance of this for immobile pathogens must be realized. Use of charged drops from an Electrodyne type machine (Coffee, 1979) should be investigated.

Re-afforestation with equi-aged, unispecies plantations offers a chance of prophylactic pathogen applications to prevent pest outbreaks known to appear at certain plantation ages. The same principle can be used in the seasonal course of annual crops, e.g. with *Nomuraea rileyi* (Chapter 27, VIII B).

In greenhouses, use of "grow bags" and hydroponic systems avoids soil sterilization and destruction of overwintering pests. Pest problems therefore start earlier in the season, sometimes necessitating new control regimes, e.g. the use of *V. lecanii* to limit early whitefly infestations until the hymenopterous parasite *Encarsia formosa*, which is ineffective early in the season, can assume control (Kanagaratnam *et al.*, 1980).

VII. Influence of Microbial Pesticides on Beneficial Organisms

By their more or less specific nature, microbial agents are likely to cause much less harm — if any — to beneficial organisms, food chains and the environment than chemical pesticides, many of which are general poisons. This is clear in Bailey's 1971 review. If they were likely to cause epizootics or epidemics in well known beneficial organisms, they would probably have done so by now, since representative beneficial organisms are normally present in their native environment. Thus in a risk analysis they are likely to be minor risks. Therefore, testing for potential harm should be minimized to beneficial organisms at risk. We now have sufficient knowledge to deduce the most useful tests: they will vary with different pathogens.

Deductive reasoning can start with a consideration of those organisms at risk in the path of pathogen applications. For example, the honey bee (even after extensive tests with NPVs and GVs) is not known to suffer a baculovirus disease, despite being the most pathologically studied insect, neither are any other Apidae (Appendix 2). Thus tests with more NPVs or GVs on bees are unnecessary. In contrast, bees are subject to microsporidiosis, so careful tests should be made with microsporidia. Other considerations concern the candidate pathogen, e.g. aflatoxins are particularly toxic to ducklings, so a variety of vertebrates should be tested with a candidate toxigenic fungus. Some pathogens have alternate hosts, e.g. *Coelomomyces* in copepods or ostracods (Chapter 29, IV A). One would expect such hosts if a candidate pathogen fails to infect the primary host in laboratory experiments.

Monitoring the environment during field trials is probably a better approach than laboratory tests on selected non-target organisms because the range of possible organisms is so vast. In this way the susceptibility of chironomids to *B.t.* var *israelensis* was shown. Later, in laboratory tests, they were less susceptible than many mosquito species and so unlikely to be sufficiently reduced to interrupt important food chains (Garcia and Goldberg, 1977). We now know enough about some candidate pathogens to regard further monitoring as unnecessary, so that tests can be limited to conventional mammalian safety tests (Chapter 40).

B.t. posed an obvious threat to the silkworm industry in Japan, although seven varieties are endemic there, some distributed widely (Aizawa, 1976). All commercial products, including three foreign and five Japanese products, are restricted to experimental use (N. Fujiyoshi, personal communication, 1979). Considerable ecological investigations on likely hazards to the industry are being made and strains of low activity to the silkworm and viable spore-free products are being developed (Aizawa, 1976).

The scope of tests required when introducing a pathogen from one geographical area to another, or from one habitat to another, need be increased only in so far as the pathogen will be in contact with *types* of beneficial organism different from those in its native haunts.

VIII. Commercial Attractiveness of Microbial Insecticides

In a survey, the Stanford Research Institute (1977) compared the attractions of various microbial insecticides to a company newly entering the field. Assuming comparable criteria, the microbials would be worth much less than many chemicals. In these comparisons, in one particular set of criteria, a *B.t.*

product would be worth U.S. $8.8 million, a pheromone for the control of pink bollworm 2.7 million and the *Heliothis* NPV would make a loss of 2.7 million. These absolute values now mean little because they are based on 1975 *assumptions*, but the comparisons are of some value because the assumptions were comparable. However, the survey expected that some products not meeting its commercial feasibility criteria will be promoted to a commercial position. This is most likely when a firm (1) is unaware of the feasibility and risk, (2) is risk orientated, (3) chooses to develop a venture for other than purely economic reasons or (4) if the project is largely in the public sphere. It was considered that, if the risks were known they would be beyond the capacity of a small firm, so large or medium-sized firms are more likely to develop microbials.

Market size and competitiveness are vital. For *B.t.* the maximum attainable market in the USA was 57.9×10^6 lb/annum but the ultimate long term share was more likely to be nearer 8.2×10^6 lb (Stanford Research Institute, 1977). For *Heliothis* NPV, equivalent values were 27.8×10^6 lb (27.8×10^6 acre treatments, assuming 1 lb/acre) and 9.9×10^6 lb. At present the market for *B.t.* appears sound since Sandoz recently opened a new plant at Wasco, California, on June 15th, 1979 (Cooper *et al.*, 1979).

Other important commercial considerations include host spectrum: a wide spectrum increases market potential and provides diversification. It is unlikely that the reduction of specificity possible by selecting wider spectrum microbials will seriously impair beneficial fauna. Good shelf life in ambient conditions, reproducible mechanized technology and modest cost of safety testing and registration requirements are other important features.

IX. Horizons of the Use of Pathogens for Pest Control

A. SHORT TERM AVAILABILITY OF MICROBIAL INSECTICIDES

The Stanford Research Institute (1977) forecast, for the 10 years between 1975 and 1985, that microbial pesticides will not be registered or available in sufficient quantities to dramatically substitute for chemicals, although some are expected to complement their use in selected markets. *B.t.* and some of the baculoviruses have the best chance of commercial success. Since this survey, genuine possibilities of commercial development of *B. sphaericus*, several baculoviruses, *Verticillium lecanii*, *Hirsutella thompsonii*, *Nomuraea rileyi* and *Nosema locustae* have appeared, as well as some bacterial and fungal agents against plant pathogens.

B. PRODUCTIVITY IMPROVEMENT

Productivity will be considered as the biomass of pathogen per amount of gross input.

With *in vivo* production of pathogens the gross input is primarily insect food, space, containers and operator time. Great improvements in host and pathogen production can be made by research on diets, productive strains, large scale production, and timing of both inoculation of the insect and harvest of the pathogen (Milner, 1974; Chapter 17, VI A; Chapter 18, II G). Cell culture offers potential as yet unready for commercialization (Chapter 16, VI).

With *in vitro* production of bacteria and fungi the gross input includes: media; energy for sterilization, aeration and agitation; and equipment and time, all of which have to be optimized. Improvement of media involves juggling with the proportions and quality of a limited number of cheap raw materials and adding small quantities of supplements to suit the organism. Also, the strain of the organism can be adapted to a good medium by serial passage and selection. Bacteriophages must be avoided (Grigorova *et al.*, 1978). Dulmage *et al.* (1978) showed improvements in small-scale productivity of *B.t.* by media modifications. Since technology with this bacterium is now well advanced, improvements of only × 2 or at most × 5 can be expected in commercial fermentation.

For applications where the key factor is the crystal toxin, asporogenic mutants could possibly be improved to divert most of the cell's biomass to crystal production. Since the crystal is an endotoxin, the cell must mature so that the limit is less than the cell's biomass – probably no more than × 3 improvement. Antibiotic producing organisms have been modified to continuously secrete antibiotics into the medium with little increase in cell mass or multiplication so that penicillin cultures now yield more weight of penicillin than weight of fungus (Dulmage *et al.*, 1978). For *B.t.* this would involve the unlikely modification of crystal production to secretion outside the cell.

Production of other bacteria and fungi *in vitro* is at a laboratory or early commercial scale and greater improvements can be expected. Many produce exotoxins, whose production could be increased like that of antibiotics.

C. POTENTIAL OF TOXINS

The value of toxins for pest control depends largely on their potency. Both bacterial and fungal toxins are active *per os* and hence are comparable to stomach poisons with no contact action. The early chemical insecticides were stomach poisons, e.g. lead arsenate, but much less potent than the *B.t.* crystal (Table V), as were the later respiratory inhibitors with contact action, e.g. nicotine. Contact action confers a great advantage over stomach action in pest control and most of the later classes of chemicals combined contact action with neurotoxic effect (Table V). Since the *B.t.* crystal causes muscle paralysis after dissolution in the insect gut it may also have neurotoxic effects, which may explain its greater potency than the respiratory inhibitors. Only some organochlorines, organophosphates and carbamates equal the potency of the crystal. The new synthetic pyrethroids are more potent by ca. × 10, although even these are bettered by × 100 by botulinum toxin, which acts on nerve synapses.

Thus a × 1000 difference is indicated in Table V between the crystal of *B.t.* and botulinum toxin. The toxic entity of the crystal is probably more active than the whole crystal. Lilley (1976) extracted an active molecule from the crystal with increased potency of ca. × 4. The crystal/host insect combination

TABLE V

Comparative potency of toxins from insect pathogens and other poisons. LD_{50} values are approximate lowest acute levels for each class in susceptible hosts, some mammalian

Poison	Class of poison and mode of action	Route of action	Potency $(mg\,kg^{-1})$
Lead arsenate	Respiratory inhibitor	Stomach	50[a]
Cordycepin	3'-deoxyadenosine. Blocks RNA synthesis	Injection	30[e]
Dinitrophenols	Uncouplers of oxidative phosphorylation	Contact	10[a]
B. thuringiensis exotoxin	Inhibition of RNA polymerase	Stomach	6[c]
Oosporein	Dibenzoquinone pigment	Stomach	6[e]
Nicotine, rotenone	Plant derivatives excluding pyrethroids: general respiratory and nervous inhibitors	Contact	5[a]
Bassianolide	Cyclodepsipeptide	Injection	4[e]
Organochlorines	Inhibitors of nervous conduction, action on nerve membrane	Contact, injection	1[a] 0.07[e]
Cyclodienes	Neurotoxic	Contact	0.5[a]
B. thuringiensis exotoxin	Inhibition of RNA polymerase	Injection	0.5[c]
Aflatoxins, B_1 B_2 G_1 G_2; Aspergillus flavus	Hepatoxins, inhibit RNA polymerase, carcinogens	Injection	0.36–3.4[d]
Metarhizium anisopliae	Cyclodepsipeptide	Injection	0.15[e]
B. thuringiensis crystal	Attacks gut epithelium, muscular paralysis	Stomach	0.07[b]

TABLE V (continued)

Comparative potency of toxins from insect pathogens and other poisons. LD_{50} values are approximate lowest acute levels for each class in susceptible hosts, some mammalian

Poison	Class of poison and mode of action	Route of action	Potency $(mg\,kg^{-1})$
Organophosphates, carbamates	Anticholinesterase action in nerve synapses	Contact	0.05[a]
Synthetic pyrethroids	Neurotoxic	Contact	0.007[a]
Botulinum toxin	Neurotoxic at nerve synapses	Stomach	0.00002[a]

[a] Graham-Bryce (1976); [b] Burgerjon and Martouret (1971); [c] Chapter 13, IV A; [d] Carnaghan et al. (1963); [e] Chapter 23, II A.

quoted in Table V, although active, may not have been one of the most active. Thus the real difference between the two active toxins, though substantial, may be much less than x 1000, even so indicating that much further improvement in toxicity of the crystal is possible within the limits of known poisons.

Other toxins from insect pathogens, except the destruxins, fall well short of the potency of the crystal, offering scope for improvement, possibly mainly in those that are neurotoxins.

The greatest potential for improvement in potency lies with strain improvements by genetics and genetic engineering. In 1973, Rogoff stated "... the current [1973] commercial liquid product of *B.t.* contains a potency of 4×10^3 IU/mg... . This high potency was achieved within months after decisions were made to discontinue the use of *B. thuringiensis* var. *thuringiensis* ... and to utilise strains of *B. thuringiensis* var. *alesti*. At that time, reported broth values were only about 200 IU/mg." Substitution of var *kurstaki* for var *alesti* subsequently yielded much higher potencies. These potencies have been selected for, and measured with, *Trichoplusia ni* and *Heliothis virescens,* the key species against which *B. t.* is used in the USA. The current international selection program produced strains only x 1.3 more active vs *T. ni* but x 5 more active vs *H. virescens* (Chapter 11, Table XV). Since ca. 320 naturally occurring isolates were screened we may have found the most potent natural strains against these species and genetic manipulation may be needed for further improvement. The crystal of each strain probably contains several toxic entities of varying potency against different hosts. It may be possible, by genetics, to retain only the most active entity against *T. ni* in greatly increased amount in the crystal. It is likely that some of the toxic factors are carried on plasmids (Chapter 15, II B) so that toxin production could be multiplied by increasing the number of plasmid copies in the cell.

Selection from naturally occurring isolates produced a x 6 improvement in potency against some species (Chapter 11, Table XV) which suggests that all the toxic entities could be crowded into one "superstrain" by genetic manipulation. This would greatly increase the effective host range. This possibility depends on how much of the crystal is carrier protein and how much is toxins that are activated in the insect gut, also on the number of binding sites available on the carrier. Every natural strain is poor against some species. Thus the var *israelensis*, so active against mosquitoes and blackflies, is poor against all Lepidoptera so far tested. This suggests that a universal superstrain may not be possible: rather a limited number of strains will need to be constructed, each aimed at a particular host complex.

Extensive differences are known between strains of other pathogens, particularly *B. sphaericus* (Chapter 14, II A), *Beauveria* spp. and *Metarhizium* spp. (Chapter 24, V A; Al-Aidroos and Roberts, 1978). A strain from one host is often more virulent in another host (compare Table I of Appendix 1 and data in Chapter 11). Thus, at various stages in the practical development of a pathogen, strain selection will often be rewarding.

D. POTENTIAL OF INFECTIVITY

In practical pest control the amount of pathogen used is measured as the dosage (amount applied to a pest's food). The dosage required to kill an insect usually increases greatly with larval instar, e.g. with *B.t.* in the moth *Galleria mellonella* (Hoopingarner and Materu, 1964), *Vairimorpha necatrix* in the moth *Spodoptera exempta* (Pilley, 1976), *Nosema whitei* in the beetle *Tribolium castaneum* (Milner, 1973), a GV in *Pieris brassicae* (C.C. Payne, personal communication), an NPV in *Lymantria dispar* (Doane, 1967), baculoviruses reviewed in ten lepidopterous species by Ignoffo (1966). In terms of dose (amount of pathogen actually eaten by an insect), I know of no pathogen that has achieved the ultimate in infectivity, i.e. one infective spore or inclusion body killing a high proportion of *mature* or nearly mature hosts by the oral or dermal routes. It remains a goal for improved infectivity. The chances of an animal ingesting the requisite small dose of pathogen increase with the amount of food it eats. Pathogens able to increase in the host's gut have the greatest chance of achieving infection at low dosage. Those that must reproduce in or penetrate the gut epithelium, e.g. baculoviruses, must reach the gut wall and their chances of doing so depend on the "gut surface to food volume" ratio, which is obviously most favourable in the youngest host stage, or the smallest host species. The Microsporidia have adapted to this problem by ejecting the infective germplasm from the spore at the tip of a long polar filament.

A dose of one single spore or inclusion body of many pathogens can cause infection *per os* in larval instar I. For instance, some LD_{50} values per instar I larva are one NPV polyhedron in *Heliothis zea* (extrapolated from data of Allen and Ignoffo, 1969, and Ignoffo, 1966); 2 NPV polyhedra in *Operophtera brumata* (Wigley, 1976); 4 (Falcon, 1971) and 5 (Sheppard and Stairs, 1977) GV capsules in the codling moth; 8 GV capsules in *Pieris rapae* ($LD_{15} = 1$ capsule) (Payne, personal communication); 4 to 17 spores of the microsporidian *Vairimorpha necatrix* in four host species (Chapter 31, Table II).

The dose expressed as number of polyhedra/mg larval weight tends to be constant except in the last instar, e.g. 40–90 (Wigley, 1976) and 8–26 (Allen and Ignoffo, 1969). The greatest potency listed above is an LD_{50} of one polyhedron/instar I larva. Greater potency than ca. LD_{90} of one polyhedron/larva I, which has to be applied to a larvae I-sized food meal, would be of little advantage in terms of the practical dosage of microbial insecticide. Lesser LD_{90} dosages would kill older larvae which eat more food, but would allow some instar I to survive, since an instar I-sized meal would often bear no pathogen and since virus is commonly inactivated on leaves in less than the duration of instar I, e.g. 3 days at 24°C to 10 days at 13°C in *Pieris rapae* (Tatchell, 1981).

E. SPREAD AND PERSISTENCE OF PATHOGENS

Spread and persistence are desirable features of pathogens used as microbial insecticides and necessities for those applied as introductions. For instance, a small-spored strain of *V. lecanii*, though producing more conidia per unit of media, failed to spread among aphids in greenhouse experiments in comparison with a large-spored strain (Hall, personal communication). Thus, at least three factors should be measured during strain selection of this species — pathogenicity, productivity of spores and ability to spread. Assays should be developed to measure the latter, which to my knowledge has never been done.

Vertical spread from generation to generation is vital for introduced pathogens. It depends on factors such as transovum transmission and presence of infective pathogen in faecal pellets, on shed wing scales, in insect bodies, in bird droppings and dust. Transovum transmission is efficient because it eliminates the uncertainty of an insect eating resistant stages of a pathogen. A young larva infected by a very virulent pathogen tends to die before it matures, preventing transovum transmission, which applies selection pressure towards attenuation. Most pathogens with a tendency to transovum transmission may already be attenuated to some extent, so it can be speculated that they probably carry virulent genes suppressed by modifier genes. Thus, selection for virulence should produce more virulent strains, which are useful up to the point that they would prevent transovum transmission. At present we are at the stage of asking whether different *species* of pathogen spread among different hosts and in what conditions. We must move on to asking these questions about pathogen *strains*.

F. LONG TERM EFFECTS OF MICROBIAL CONTROL

Beyond the mortality caused by microbial insecticides, sublethally infected survivors are debilitated, delaying population recovery (Smirnoff, 1978). Chemicals debilitate some partially poisoned survivors, but Smirnoff also showed that populations surviving chemical treatment were more vigorous than control populations or the survivors of *B.t.* applications.

With recycling pathogens, changes may occur in the pathogen (Burges, 1971), e.g. better adaptation to the host, attenuation, etc. Changes may also occur in the hosts, e.g. in behaviour and resistance. Changes (possibly attenuation) recorded in the longest-used pathogen, *B. popilliae*, in Connecticut (Chapter 10, III A, IV) should be investigated and steps taken to overcome them.

G. LIMITING PHENOMENA IN MICROBIAL CONTROL

It is important to realize the limitations to microbial control, some more obvious than others. For instance, most fungi do not thrive in dry conditions, *B.t.* does not spread, a motile stage limits a pathogen to an aquatic environment, UV radiation limits pathogen survival, each pathogen has an effective temperature range, a pathogen is to some degree host specific, fungus spores (except resting spores) are short lived, some organisms require particular nutrients for production, certain fungicides kill some fungal insect pathogens.

Most of these limitations are quantitative. For instance, different strains of *V. lecanii* have widely different maximum temperatures for growth (Chapter 25, Table III), nematode infective stages can move about in water films in cabbage heads. While there are obviously ultimate limitations in all these factors, most can be extended by selecting different strains or by adapting a strain. Selection and adaptation should be considered before discarding a pathogen species as impracticable against a certain host. For example, Brassel and Benz (1979) and Brassel (1978) selected a strain of codling moth GV with improved resistance against UV radiation and sunlight. Entomopathogenic fungal strains could be developed with resistance to certain crop protection fungicides.

X. Future Research Requirements

The potential of biocontrol agents is considerable. Throughout this book, a recurrent theme for future development of this potential is the need for *more resources*. Promising agents have already been listed and discussed, so this section will elaborate both on encouraging the use of these agents and on principles. Most can be achieved from available resources by *co-operation*, aimed to assist *industry* to produce the agents. This needs careful *orientation* of objectives. The target must be to *select or create* the best agent for each pest problem, often within the framework of *integrated* control programs.

Since results are proportional to resources, more generous funding is warranted so that government research organizations and universities can advance research to a point where it is financially viable for industry to take the initiative. Further research would then proceed in co-operation with industry. For financially unattractive ventures, it is reasonable for governments to do most of the research and then place production contracts with industry, as with the tussock moth NPV in the USA. Patent rights should be made realistic and alternatives adopted, e.g. regarding safety data as the property of the commercial producer for registration purposes. At present, laws both in the EEC (Vossius, 1979) and the USA (Wegner, 1979a,b) appear to be moving towards the possibility of patenting strains of microorganisms. Priority should be given to speedy registration and waiving of fees, as is at present possible in the EPA of the USA. There is also a need to remove constraints that discourage registration of products designed for small markets. To ease the costly burden of registration, mandatory safety testing requirements must be rationalized asking only for those data required for realistic hazard analysis. Finally, consumer industry must be familiarized with the potential and safety of microbial agents by governmental advisory services.

Most progress can be expected in *"pathogen orientated"* research. For instance, the new *israelensis* variety of *B. thuringiensis* requires concerted effort both to improve production and to develop formulations suitable for application to water. *Cooperative* worldwide testing of products against different disease vectors is necessary in different aquatic habitats. Results, except for commercial production details and patentable information on formulation, should be circulated rapidly. This would provide a body of information to define the ecological systems in which this pathogen would be successful. Baculovirus research has reached a stage where more knowledge is needed about the differences between baculoviruses and about host ranges. This could be achieved most effectively by an international production, exchange, bioassay and virological study program, which would have the additional objectives of detecting strains more active than those native to local pests. International programs with certain fungi, such as *Beauveria* and *Metarhizium*, in which strains are known to vary greatly, could be particularly rewarding.

Cooperation of this sort could also aid projects orientated around individual pests or crops. *Pest orientated* studies require an analysis of existing knowledge to predict the biocontrol agents most likely to be successful (these may not only be pathogens), followed by experimental comparison of the candidate species and strains. *Crop-orientated* projects should involve as many *different cooperators* as possible to include a wide range of skills. Thus knowledge of crop and pest ecology, application technology, etc. can be united from the beginning to enable us to predict and test effects rather than learn by blind and often hard experience. Since some pathogens are relatively specific and others very specific, their main impact will be within *integrated* control programs which must be considered as a whole and must be adaptable to change — alteration of one factor may have many effects. Pathogens must not be regarded as an island in pest control.

Except when a pathogen is used inundatively, like an insecticide, at population levels too low for its spread dispersal is important and the principles of the phenomenon of parasitism govern the results. We must usually surpass the natural performance of a pathogen because we need to maintain pests below economic thresholds, having given them an abundance of food, often of plant varieties selected for their succulence and, unwittingly, for susceptibility to pest attack. This may require a high degree of sophistication. In the wild, the pathogen with the greatest biological success has a high degree of synchrony with the host. To eliminate the host would ultimately spell self destruction to the pathogen. Imbalance in this direction occurs if the organism becomes too pathogenic or is able to spread very effectively. Imbalance in the opposite direction happens if it becomes attenuated or unable to spread sufficiently. Thus we can use pathogens as a blanket correction of pest outbreaks by selecting those most virulent. Having reduced the outbreak, if the most virulent pathogen does not persist, we can select pathogens with intermediate or long term ability to maintain a pest at unimportant levels. Such a pathogen is acting as an introduction. These goals require a search for more pathogens and better strains. Strains must be *selected* for factors such as potency, ability to spread, durability of resistant stages, ability to infect and sporulate at lower humidities (fungi), transovum transmission, etc.

Sometimes a suitable strain may not exist in the wild or there may be room for improvement. This may indicate long term work, such as strain selection or genetic manipulation to *create* a suitable organism. Such programs are likely to be beyond the scope of individual research centres and *cooperative* programs will be imperative. For instance, the existing international program on strain selection of *B. thuringiensis* is leading to a program of cooperation on genetics. Genetic engineering offers further *creative* opportunities, e.g. it may be possible to solve the enigma of *in vitro* spore production of *B. popilliae* by adding characters of the *B.t.* spore, as is being attempted at the USDA laboratory in Beltsville (Faust, 1979). Genetic change may also occur in successful pathogens in the field: the apparent change in *B. popilliae* in Connecticut (Section IX F) should be carefully investigated.

Theoretically, simultaneous use of three agents, with widely differing modes of action (Chapter 41, IV; Goldberg and Ford, 1974), should avoid development of resistance. However, the cost effectiveness of using these agents will need assessment. To justify high cost, it may be possible to select a pest on a high value crop where development of resistance would jeopardize the control program for the whole pest complex, e.g. *Heliothis* spp. on cotton in some areas.

Although many innovative lines of work have been suggested, the main task remains to expand the effort on existing well known agents, the study of new ones and the improvement of production, application and assessment technology. We are entering an era of more sophisticated pest control. However, those involved must understand the sophistication, yet be able to extract the simplest techniques for practical use – a farmer, for instance, prefers simple methods. It is perhaps no accident that the most successful commercial pathogen, and the pathogen accredited top priority for vector control research, is *B.t.* – the simplest to produce and use.

References

Aizawa, K. (1976). *Pl. Prot. Japan, 1976, Agric. Asia*, Special Issue No. 10. pp. 382–387.

Al-Aidroos, K. and Roberts, D.W. (1978). *Can. J. Genet. Cytol.* **20**, 211–219.

Allen, G.E. and Ignoffo, C.M. (1969). *J. Invertebr. Path.* **13**, 378–381.

Alvaro, C.M. and Hernán, A.V. (1973). *Fitotechnia Latinoamericana* **9**, 28–35.

Angus, T.A. (1965). *Proc. ent. Soc. Ont.,* **95**, 133–134.

Bailey, L. (1971). *In* "Microbial Control of Insects and Mites" (H.D. Burges and N.W. Hussey, eds), pp. 491–505. Academic Press, London and New York.

Barjac, H. de. (1978a). *Lille Med.* **286**, 1175–1178.

Barjac, H. de. (1978b). *Entomophaga* **23**, 309–319.

Barjac, H. de and Coz, J. (1979). *Bull. Wld. Hlth. Orgn.* **57**, 139–141.

Baskerville, G. *et al.* (1976). Report of the Task Force for Evaluation of Budworm Alternatives. Dep. Natural Res., Fredericton, New Brunswick, Canada, 210 pp.

Bellotti, A.C. and Reyes, J.A. (1980). Proc. Workshop on Insect Pest Management with Microbial Agents: Recent Achievements, Deficiencies, and Innovations. pp. 20–21. Boyce Thompson Institute, Ithaca.

Brassel, J. (1978). *Bull. Soc. ent. Suisse*, **51**, 155–211.

Brassel, J. and Benz, G. (1979). *J. Invertebr. Path.* **33**, 358–363.

Burgerjon, A. and Martouret, D. (1971). *In* "Microbial Control of Insects and Mites" (H.D. Burges and N.W. Hussey, eds), pp. 305–325. Academic Press, London and New York.

Burges, H.D. (1964). *Mém. hors. Sér. Entomophaga*, No. 2, 323–327.

Burges, H.D. (1971). *In* "Microbial Control of Insects and Mites" (H.D. Burges and N.W. Hussey, eds), pp. 445–457. Academic Press, London and New York.

Burges, H.D. (1973). *Ann. N.Y. Acad. Sci.* **217**, 31–49.

Burges, H.D. and 31 others (1967). *In* "Insect Pathology and Microbial Control". (P.A. van der Laan, ed.), pp. 314–337. NorthHolland Publ. Co., Amsterdam.

Burges, H.D. and Hurst, J.A. (1977). *J. Invertebr. Path.* **30**, 131–139.

Burges, H.D. and Hussey, N.W. (1971). *In* "Microbial Control of Insects and Mites" (H.D. Burges and N.W. Hussey, eds), pp. 687–709. Academic Press, London and New York.

Burges, H.D. and Jarrett, P. (1979). Proc. 1979 Brit. Crop Prot. Conf. – Pests and Diseases Vol. 2, pp. 433–439. Nov. 19–22, 1979. British Crop Protection Council, Brighton.

Burkholder, W.E. and Shapas, T.J. (1978). *In* "Microbial Control of Insect Pests: Future Strategies in Pest Management Systems" (G.E. Allen, C.M. Ignoffo and R.P. Jaques, eds), pp. 236–248, NSF-USDA-Univ. Florida, Gainesville.

Carnaghan, A., Hartley, R.D. and O'Kelley, J. (1963). *Nature, Lond.* **200**, 1101.

Carter, J.B. (1973). *J. Invertebr. Path.* **21**, 136–143.

Coffee, R.A. (1979). Proc. 1979 Brit. Crop Prot. Conf. – Pests and Diseases, Vol. 3, pp. 777–789. Nov. 19–22, 1979. British Crop Protection Council, Brighton.

Cooper, C., Brogan, P. and Tapscott, G., eds. (1979). *Pestic. Toxic Chem. News*, 20th June, 1979, pp. 3–4.

Crawford, A.M. and Kalmakoff, J. (1977). *J. Invertebr. Path.* **29**, 81–87.

Doane, C.C. (1967). *J. Invertebr. Path.* **9**, 376–386.

Dulmage, H.T., Beegle, C.C. and Correa, J. (1978). Production of *Bacillus thuringiensis* δ-endotoxin in submerged culture. 2nd Int. Workshop on *Bacillus thuringiensis*, Sept. 6–8. Darmstadt.

Dulmage, H.T. and Martinez, E. (1973). *J. Invertebr. Path.* **22**, 14–22.

Dulmage, H.T. and Rhodes, R.A. (1971). *In* "Biological Control of Insects and Mites" (H.D. Burges and N.W. Hussey, eds), pp. 507–540. Academic Press, London and New York.

Eidmann, H.H. (1976). *Ambis* **5**, 23–26.

Falcon, L.A. (1971). *In* "Biological Control" (C.B. Huffaker, ed.), pp. 346–364. Plenum Press, New York, London.

Falcon, L.A. (1978). *In* "Microbial Control of Insect Pests: Future Strategies in Pest Management Systems" (G.E. Allen, C.M. Ignoffo and R.P. Jaques, eds), pp. 143–163. NSF-USDA-Univ. Florida, Gainesville.

Falcon, L.A., Kane, W.R. and Bethell, R.S. (1968). *J. econ. Ent.* **61**, 1208–1213.

Faust, R.M. (1979). *Recombinant DNA Tech. Bull.* **2**, 1–8.

Garcia, R. and DesRochers, B. (1979). *Mosq. News*, 541–544.

Garcia, R. and Goldberg, L.J. (1977). *Univ. Calif. Ann. Rep. 1977*, p. 29.

Gard, I.E. and Falcon, L.A. (1978). *In* "Microbial Control of Insect Pests: Future Strategies in Pest Management Systems (G.E. Allen, C.M. Ignoffo and R.P. Jaques, eds), pp. 46–54, NSF-USDA-Univ. Florida, Gainesville.

Goldberg, L.J. and Ford, I. (1974). *Proc. Am. Conf. Calif. Mosq. Control Ass.* **42**, 169–174.

Goldberg, L.J. and Margalit, J. (1977). *Mosq. News* **37**, 355, 358.

Goldberg, L.J., Goldberg, E.M. and Margalit, J. (1977). Unpubl. docum. WHO/VBC/77, 662.

Graham-Bryce, I.J. (1976). *Chem. Ind. 1976*, No. 13, 545–553.

Grigorova, R.T., Markov, K.I., Shivarova, N.I. (1978). *Dokl. Bolg. Akad Nauk.* **31**, 1217–1220.

Gustafsson, M. (1971). *In* "Microbial Control of Insects and Mites" (H.D. Burges and N.W. Hussey, eds), pp. 375–384. Academic Press, London and New York.

Hall, I.M. and Dunn, P.H. (1958). *J. econ. Ent.* **51**, 341–344.

Henry, J.E. (1978). *In* "Microbial Control of Insect Pests: Future Strategies in Pest Management Systems" (G.E. Allen, C.M. Ignoffo and R.P. Jaques, eds), pp. 195–206. NSF-USDA-Univ. Florida, Gainesville.

Hertlein, B.C. (1979). WHO, Unpubl. docum. TDR/BCV/SWG.79/WP.10.

Hoopingarner, R. and Materu, M.E.A. (1964). *J. Insect Path.* **6**, 26–30.

Howse, G.M., Sanders, C.J., Harnden, A.A., Cunningham, J.C., Bird, F.T. and McPhee, J.R. (1973). Can. Forest Serv. Rep. No. O-X-189, 62 pp.

Huber, J. and Dickler, E. (1977). *J. econ. Ent.* **70**, 557–561.

Hunter, D., Collier, S.S. and Hoffman, D.F. (1977). *J. econ. Ent.* **70**, 493–494.

Ignoffo, C.M. (1966). *J. Invertebr. Path.* **8**, 279–282.

Ignoffo, C.M. (1978). *J. Invertebr. Path.* **31**, 1–3.

Ignoffo, C.M., Hostetter, D.L., Biever, K.D., Garcia, C., Thomas, G.D., Dickerson, W. and Pinnell, R. (1978). *J. econ. Ent.* **71**, 165–168.

Ito, Y., Shiga, M., Oho, N. and Nakazawa, H. (1977). *Res. popul. Ecol.* **19**, 33–50.

Jaques, R.P. (1964). *J. Insect Path.* **6**, 251–254.

Jaques, R.P. (1967). *Can. Ent.* **99**, 820–829.

Jaques, R.P. (1969). *J. Invertebr. Path.* **13**, 256–263.

Jaques, R.P. (1970a). *Can. Ent.* **102**, 36–41.

Jaques, R.P. (1970b). *J. Invertebr. Path.* **15**, 328–340.

Jaques, R.P. (1972). *J. econ. Ent.* **65**, 757–760.

Jaques, R.P. (1974). *J. Invertebr. Path.* **23**, 140–152.

Jaques, R.P. (1978). *In* "Microbial Control of Insect Pests: Future Strategies in Pest Management Systems" (G.E. Allen, C.M. Ignoffo and R.P. Jaques, eds), pp. 72–84. NSF-USDA-Univ. Florida, Gainesville.

Jaques, R.P. and Patterson, N.A. (1962). *Can. Ent.* **94**, 818–825.

Jarvis, W.R. and Slingsby, K. (1977). *Plant Disease Reporter* **61**, 728–730.

Kanagaratnam, P., Burges, H.D. and Hall, R.A. (1980). Rep. Glasshouse Crops Res. Inst. 1979, 133–144.

Kiselek, E.V. (1974). *Vestnik Selskokhozyaistvennor Nauki* **5**, 68–71.

Lacey, L.A., Mulla, M.S. and Dulmage, H.T. (1978). *Environ. Ent.* **7**, 583–588.

Laird, M. (1977). Proc. 5th Int. Cong. Protozool., New York, 26 June–2 July, 1977. p. 172–175.

Legner, E.F. and Oatman, E.R. (1962). *J. econ. Ent.* **55**, 677–678.

Lilley, M. (1976). Ph.D. Thesis, Glasgow College of Technology, 185 pp. Glasgow.

Lund, A.E. (1971). *In* "Microbial Control of Insects and Mites" (H.D. Burges and N.W. Hussey, eds), pp. 385–386. Academic Press, London and New York.

Lynch, R.E., Lewis, L.C. and Brindley, T.A. (1976). *J. Invertebr. Path.* 27, 325–331.

Magnoler, A. (1974). *Z. PflKrankh. PflSchutz* 81, 497–511.

McGaughey, W.H. (1976). *Can. Ent.* 108, 105–112.

McKinley, D.J. (1980). Project Rep. 1974–1979. Centre for Overseas Pest Research, London. 16 pp.

McNeil, J.N., Smirnoff, W.A. and Letendre, M. (1977). *Can. Ent.* 109, 37–38.

Milner, R.J. (1972). *J. Invertebr. Path.* 19, 248–255.

Milner, R.J. (1973). *Entomophaga* 18, 305–315.

Milner, R.J. (1974). *J. Invertebr. Path.* 23, 289–296.

Morris, O.N. (1977). *Can. Ent.* 109, 9–14.

Morris, O.N., Armstrong, J.A., Howse, G.M. and Cunningham, J.C. (1974). *Can. Ent.* 106, 813–824.

Nassi, H., Pointier, J-P. and Golvan, Y-J. (1979). *Ann. Parasitol.* 54, 185–192.

Oatman, E.R. and Legner, E.F. (1964). *J. econ. Ent.* 57, 294.

Panbangred, W., Pantuwatana, S. and Bhumiratana, A. (1979). *J. Invertebr. Path.* 33, 340–347.

Petersen, J.J. (1976). Proc. 1st. Int. Colloq. Invertebr. Path. Kingston, Ontario, pp. 236–240.

Pilley, B.M. (1976). Ph.D. Thesis, Univ. London, 180 pp.

Pruett, C.J.H., Burges, H.D. and Wyborn, C.H. (1980). *J. Invertebr. Path.* 35, 168–174.

Ripa, R. (1978). Ph.D. Thesis, Univ. London. 188 pp.

Ripa, R., Tatchell, G.M. and Payne, C.C. (1979). *Proc. Int. Colloq. Invertebr. Path.*, XIth Annual Meet. Soc. Invertebr. Path. Sept. 1978, Prague 165–166.

Roberts, D.W. (1970). *Misc. Publ. ent. Soc. Am.* 7, 140–155.

Roberts, D.W. and Strand, M.A. (1977). Pathogens of Medically Important Arthropods. Bull. Wld. Hlth. Org. Suppl. No. 1 Vol. 55, 419 pp.

Rogoff, M.H. (1973). *Ann. N.Y. Acad. Sci.* 217, 200–210.

Saleh, S.M., Harris, R.F. and Allen, O.N. (1969). *Can. J. Microbiol.* 15, 1101–1104.

Saleh, S.M., Harris, R.F. and Allen, O.N. (1970). *J. Invertebr. Path.* 15, 55–59.

Schaefer, C.H. (1979). WHO, Unpubl. docum. TDR/BVC/SWG.79/WP.09.

Sekijima, Y., Akiba, Y., Ono, K., Aizawa, K. and Fujiyoshi, N. (1977). *Jap. J. appl. Ent. Zool.* 21, 35–40.

Shah, B.H., Zethner, O., Gul, H. and Chaudhry, M.I. (1979). *Entomophaga* 24, 393–401.

Sheppard, R.F. and Stairs, G.R. (1977). *J. Invertebr. Path.* 29, 216–221.

Smirnoff, W.A. (1978). *Forestry Chron.* 54, 309–312.

Soper, R. (1978). *In* "Microbial Control of Insect Pests: Future Strategies in Pest Management Systems" (Allen, G.E., Ignoffo, C.M. and Jaques, R.P. eds), pp. 63–65. NSF-USDA-Univ. Florida, Gainesville.

Stanford Research Institute. (1977). New, Inovative Pesticides: an Evaluation of Incentives and Disincentives for Commercial Development by Industry. Stanford Research Institute, Menlo Park, California. 318 pp.

Sweeney, A.W. (1979). WHO, Unpubl. docum. TDR/BCV/SWG. 79/WP 14.

Tatchell, G.M. (1981). *Entomophaga* **6**, in press.

Undeen, A. and Berl, L. (1979). *Mosq. News* **39**, 742–745.

Unestam, T. and Weiss, D.W. (1970). *J. gen. Microbiol.* **60**, 77–90.

Vankova, J., Weiser, J., Rytir, V. and Hofman, J. (1978). Proc. Int. Colloq. Invertebr. Path., Prague, pp. 219–222.

Vossius, V. (1979). *Biotechnol. Lett.* **1**, 187–192.

Wegner, H.C. (1979a). *Biotechnol. Lett.* **1**, 145–150.

Wegner, H.C. (1979b). *Biotechnol. Lett.* **1**, 193.

Weiser, J. and Vankova, J. (1978). Proc. Int. Colloq. Invertebr. Path. Prague, pp. 243–244.

Welch, H.E. (1962). *Proc. ent. Soc. Ontario* **92**, 11–19.

WHO. (1979a). Biological Control Agent Data Sheet. *Metarhizium anisopliae*. Unpubl. docum. VBC/BCDS/79.06, 8 pp.

WHO. (1979b). Biological Control Agent Data Sheet. *Bacillus thuringiensis* serotype H-14. Unpubl. docum. VBC/BCDS/79.01, 14 pp.

WHO. (1979c). Third Meeting of the Scientific Working Group on Biological Control of Insect Vectors of Diseases, 19–22 Nov. 1979. Unpubl. docum. TDR/BCV–SWG (3)/79.3, 43 pp.

WHO. (1979d). Réunion de Concertation sur l'Évaluation et le Développement du Sérotype 14 de *Bacillus thuringiensis* en Europe. Geneva, 7–8 May, 1979. Unpubl. docum. TDR/BCV/79.05, 10 pp.

Wigley, P.J. (1976). Ph.D. Thesis, Univ. Oxford, Oxford, 185 pp.

Zethner, O. (1976). *Z. Angew. Ent.* **81**, 192–207.

Zethner, O. (1980). *Entomophaga* **25**, 27–35.

Susceptibility of Arthropod Species to *Bacillus thuringiensis*

A. KRIEG and G.A. LANGENBRUCH

Institut für biologische Schädlingsbekämpfung, Darmstadt, Germany

I. Introduction

Bacillus thuringiensis (*B.t.*) is at present the most used pathogen for microbial control of insect pests. Isolates of several H-serotypes (Chapter 3) have been tested in many applications and the data sheets of commercial products contain dosage recommendations for some important pest insects. Data about the sensitivity to *B.t.* of arthropods, including beneficial species, have been published in a wide range of journals and only a few comparative compilations are available (Steinhaus, 1957; Heimpel and Angus, 1958; Isakova, 1958; Krieg, 1961, 1967b; Burgerjon and Biache, 1967b; Takaki, 1975). Thus it is often difficult to find out if an arthropod has already been tested. The following lists, which include data on both susceptible and non-susceptible species, have been compiled for this purpose. Because of the extensive literature and distribution of relevant papers in a great variety of journals, worldwide, the lists of host relationships – although comprehensive – are not complete. Another purpose of these lists is to compare the susceptibilities of the different host species to the different H-serotypes of *B.t.* This is complicated by the fact that not only spores, but also toxins (Chapters 12 and 13) are involved in the pathogenic process. The production of toxins may differ not only in different H-serotypes (Table I), but also in different isolates (see below). Therefore, the species lists are organized according to the quality of *B.t.* preparations. At present, most commercial preparations of *B.t.* contain only spores and parasporal crystals of delta-endotoxin. Because this combination gives a selective microbial control of many lepidopterous pests, this list is by far the longest (Table II). Table III lists data on susceptibilities to the beta-exotoxin (= thermostable exotoxin or "thuringiensin"). This toxin is produced only by some isolates of certain serotypes. Finally Table IV covers *B.t.* preparations that contain spores, crystal toxin and thermostable exotoxin.

II. Comparison of Data

It proved difficult to compare the results of experiments by different workers. Therefore, the degree of susceptibility, indicated in our lists by + or 0 signs, is greatly influenced by the judgement of the authors.* If *B.t.* has been bioassayed several times with the same result against the same host species, only one reference is quoted. A diagonal line (/) denotes a range of efficacy.

Many factors had to be taken into account during these comparisons. There were differences in experimental conditions, e.g. age of larvae, temperature, dosages, application techniques and criteria of measurement (e.g. death, reduction of feeding). Also some results were not comparable if there were differences in the quality or quantity of the active ingredients (e.g. spores and crystals) in the preparations, commercial or otherwise, even based on the same H-serotype of *B.t.* Quantitative differences may arise when the conditions of fermentation processes vary and different media are used (Dulmage, 1970b). Also formulation changes the quality of preparations as a result of additives or formulation conditions. Last, but not least, not all populations and individuals of the same species show the same susceptibility to *B.t.*, especially if infected by other pathogens.

When interpreting results of field trials it is essential either that standardized *B.t.* preparations have been used or that data are available on the yield of crystals (measured as toxicity) and of spores (as viable spore count). Nevertheless, there were variations in efficacy, e.g. due to climatic factors: ultraviolet rays inactivate spores but not the crystal toxin; *B.t.* has little activity below $15°C$ (Fedorinchik, 1964) and increasing temperature ($10°$ to $35°C$) speeds up toxic effects and infection. For example the log of the reciprocal of the time necessary to kill half the larvae ($1/Lt_{50}$) of *Ephestia kuehniella* is proportional to temperature (Matter, 1969). Generally larvae become less susceptible as they become older (McConnell and Cutkomp, 1954; Wiegand, 1963; Afify and Matter, 1970). The mode of application is important and some weakly susceptible species, e.g. *Agrotis segetum* can be controlled only by attractive baits containing a high concentration of *B.t.* (Langenbruch, 1977).

Some reports of field efficiency are contradictory, e.g. those of *Ostrinia nubilalis*. Poor results may be due to the low activity of the preparation, improper application technique, dosage and timing, while in other reports control can be satisfactory (Schubert and Stengel, 1974; Langenbruch, 1976). Only species feeding on open leaves ingest enough *B.t.* for good control. Thus mining species, e.g. *Cydia* (*Laspeyresia*) *pomonella*, show high mortality in laboratory tests when the *B.t.* is mixed in the diet but low mortality in field trials when the apple surface is sprayed. Similarly among stored products pests, mortality is high with species that feed on grain germ and surfaces, e.g. *Plodia interpunctella*, but low with species feeding inside grains, e.g. *Sitotroga cerealella*.

---

* Editorial note: However, this judgement is probably one of the best available, because both authors are widely experienced and Dr. Krieg has studied many aspects of *B.t.* from a very early stage in the development of *B.t.* research.

Wool and textiles can be protected from clothes moth (*Tineola bissielliella*) by *B.t.* (Yamvrias and Angus, 1969). In an advanced program *B.t.* is embedded in foundation wax to control wax moth (*Galleria mellonella*) in beecombs (Burges, 1977).

III. Additional Comments About Host Range

The pathogenicity of *B.t.* to whole taxonomic groups of insects is partially related to certain serotypes. However, different serotypes or groups of isolates may possess different host specifity which is related to the production of parasporal crystals with different toxic qualities (Burges, 1967; Burgerjon and Dulmage, 1977; Chapter 11). Quantitative differences in toxin production, however, will result in isolates of different virulence (Rogoff *et al.*, 1969; Dulmage, 1971; Dubois and Squires, 1971).

With sensitive insects, like Lepidoptera, only such larval stages are killed as can ingest enough spores and crystals of *B.t.* Eggs, pupae and usually adults, however, are unharmed. Injection of spores or vegetative cells of *B.t.* may induce septicemia in all stages of insects. Crystals are innocuous by injection, but exotoxins are more toxic by injection than orally both to larvae and adults.

Many successful field trials have shown that commercial preparations of *B.t.*, containing the spore—crystal complex, effectively control lepidopterous pests in agriculture (Vaňková, 1964; Martouret and Milaire, 1963; Krieg, 1970; Harper 1976) and forestry (Franz and Krieg, 1967; Harper 1974). In general, exotoxin-free preparations are avirulent for Hymenoptera, including sawflies (Heimpel, 1961). Also these preparations do not attack entomophagous insects which may, therefore, help to reduce populations of noxious insects unrestrained by *B.t.* Parasites may be impaired only by the premature death of their host caterpillars induced by *B.t.* The spore—crystal complex of *B.t.* is also harmless for bees and their colonies.

Reports of efficacy of spores and crystals of *B.t.* against Diptera, especially mosquito larvae, vary from inactivity (Kellen and Lewallen, 1960; Rogoff *et al.*, 1969) to striking activity (Lavrentyev *et al.*, 1965; Reeves and Garcia, 1971). This is probably due to differences in the efficacy of different isolates since Hall *et al.* (1977) found both active and inactive isolates within the same II-serotype. Also mosquito species differ widely in susceptibility (Table II) and mortalities were high only at water temperatures near $25°C$ (Lavrentyev *et al.*, 1965; Hall *et al.*, 1977).

Recently a new serotype (H_{14}) of *B.t.* var *israelensis*, has been isolated. Its crystal toxin has a special activity in larvae of mosquitoes like *Culex pipiens* and *Aedes aegypti* (Goldberg and Margalit, 1977; de Barjac, 1978), even at 5 to $8°C$.

Preparations of *B.t.* containing beta-exotoxin are not only effective against Lepidoptera but also against many other arthropods (Table IV). Especially, adults and larvae of the honey bee are attacked by beta-exotoxin, so preparations containing it must not be used on plants in flower. On the other hand, the sensitivity of *Tetranychus urticae* (Krieg, 1968) and other spider mites to beta-exotoxin indicates its use for controlling phytophagous mite pests. To prevent development of fly larvae in manure, preparations of *B.t.* have been used experimentally sometimes as feed additives for cattle (Dunn, 1960) or poultry (Burns *et al.*, 1961). According to Gingrich and Eschle (1971) this effect is caused predominantly by beta-exotoxin. Also Burgerjon (1964) suggested that the effect of some *B.t.* preparations on mosquito larvae may be, in part, due to this exotoxin. A new method of applying *B.t.* is in meat baits to control pharaoh ants, *Monomorium pharaonis*, in hospitals and zoological gardens (Wisniewski, 1975); but the type of toxin involved is still uncertain.

Development of resistance would have an important effect on the use of *B.t.* in pest management. In a 1971 review, Burges concluded that in the laboratory slight resistance to beta-exotoxin had been shown in Diptera (*Musca domestica*). Recently Devriendet and Martouret (1976) as well as Langenbruch and Krieg (1976) found no acquired resistance in larvae of *Plutella xylostella* (= *maculipennis*) after selection by *B.t.* for 10 and 30 generations, respectively. Thus substantial resistance to the spore—crystal complex of *B.t.* in Lepidoptera is not to be expected in the field within a reasonable time.

Recently, concentration-mortality studies with several populations of *Plodia interpunctella* and *Ephestia cautella* demonstrated that the response to the spore—crystal complex of *B.t.* may differ as much as 6- and 10-fold of the LC_{50}. It is suggested that such differences are related, in part, to population vigour (Kinsinger and McGaughey, 1979). Variations in efficacy of *B.t.* in field trials at different places may be caused in some cases by this phenomenon.

TABLE I

Varieties of B. thuringiensis and their ability to produce toxins active per os against insect larvae

Name of variety	Symbol	H-serotype	Endotoxin crystal	β-exotoxin	Isolated from	Found in
thuringiensis	Th	1	+	+ (or 0)	*Ephestia kuehniella*	Europe
finitimus	Fi	2	0[b]	0	*Malacosoma disstria*	N. America
alesti	Al	3a	+	0	*Bombyx mori*	Europe
kurstaki	Ku	3a, 3b	+	0	*Ephestia kuehniella*	Europe
sotto	So	4a, 4b	+	0	*Bombyx mori*	E. Asia
dendrolimus	De	4a, 4b	+	0	*Dendrolimus sibiricus*	Asia
kenyae	Ke	4a, 4c	+	+ (or 0)	*Ephestia cautella*	Africa
galleriae	Ga	5a, 5b	+	0	*Galleria mellonella*	Europe, Asia
canadensis	Ca	5a, 5c	+	0	*Diparopsis sp.*	Africa, N. America
entomocidus	En	6	+	0	*Paralipsa gularis*	N. America
aizawai	Ai	7	+	0	*Ephestia cautella*	E. Asia, Europe
morrisoni	Mo	8a, 8b	+	+ (or 0)	*Galleria mellonella*	Europe
ostriniae	Os	8a, 8c	+	0	*Ostrinia nubilalis*	E. Asia
tolworthi	To	9	+	+ (or 0)	*Plodia interpunctella*	Europe
darmstadiensis	Da	10	+	+ (or 0)	*Galleria mellonella*	Europe
toumanoffi	Tu	11a, 11b	+	0	*Galleria mellonella*	Europe
kyushuensis	Ks	11a, 11c	+	0	*Bombyx mori*	E. Asia
thompsoni	Ts	12	+	0	*Galleria mellonella*	N. America
pakistani	Pa	13	+	0	*Cydia pomonella*	Asia
israelensis	Is	14	+	0	Soil (mosquito breeding place)	Near East
dakota	Dk	15[c]	+	?	soil (crop field)	N. America
indiana	Id	16[c]	+	?	soil (crop field)	N. America
wuhanensis	Wu	—[a]	+	+	*Anomis flava*	E. Asia
fowleri	Fo	—[a]	0[b]	0	water (basin)	N. America

a No flagella. b Parasporal crystals are non-toxic. c *Editorial note*: at present these numbers are confused, see Chapter 3, Table I.

TABLE II

Susceptibility of invertebrates to spores and endotoxin of Bacillus thuringiensis per os. E = Egg; L = Larva; I = Imago; H = Host

Host species	Stage	B.t. var.	Efficiency Lab.	Field	Reference
Arthropoda: Insecta					
COLEOPTERA					
Acanthoscelides obtectus	L + I	Th	(+)		Sander and Cichy, 1967
	L + I	De	0		Sander and Cichy, 1967
Altica ambiens	L	Th	(+)		Hall and Dunn, 1958
Attagenus piceus	L + I	Th	0		Shaikh and Morrison, 1966a
Berosus sp.	I	Is	0		Morawczik and Schnetter, unpublished observations
Coccinella sp.	L + I	Th	0		Steiner, 1960
Coelambus sp.	I	Is	0		Morawczik and Schnetter, unpublished observations
Cryptolestes pusillus	L + I	Ku	0		McGaughey et al., 1975
Dermestes lardarius	L + I	Th	0		Berliner, 1915
Diabrotica longicornis	I	Ga	0		Sutter, 1969
D. undecimpunctata	L	Ga	0		Sutter, 1969
D. virgifera	I	Ga	0		Sutter, 1969
Diaprepes abbreviatus	I	Th		0	Wong et al., 1975
Epilachna varivestis	L + I	Th	0		Shaikh and Morrison, 1966a
Galerucella xanthomelaena	L + I	Th	0		Hall and Dunn, 1958
Gnatocerus cornutus	L + I	Th	0		Berliner, 1915
Gyrinidae gen. sp.	L	Is	0		Weiser and Vaňková, 1978
Hippodamia convergens	L + I	Th		0	Stern, 1961
Hypera brunneipennis	L + I	Th	(+)		Hall, 1957
	L + I	Th		0	Stern, 1961
Laccophilus sp.	I	Is	0		Morawczik and Schnetter, unpublished observations

Species				Reference
Lasioderma serricorne	L+I	Th	0	Thompson and Fletcher, 1972
Leptinotarsa decemlineata	L+I	Th	0	Krieg, 1957b; Drozdowicz, 1964
Melolontha sp.	L+I	Th	0	Krieg, 1957b
Oryctes rhinocerus	L+I	Th	(+)	Steinhaus, 1951b
Oryzaephilus surinamensis	L+I	Ku	0	McGaughey *et al.*, 1975
Popillia japonica	L	Ga	+	Sharpe, 1976
Rhyzopertha dominica	L+I	Th	0	Steinhaus and Bell, 1953
	L+I	Ku	0	McGaughey *et al.*, 1975
Scymnus sp.	L+I	Th	0	Ayyar, 1961
Sitophilus granarius	L+I	Th	0	Berliner, 1915; Shaikh and Morrison, 1966a
	L+I	Th	+/(+)	Steinhaus and Bell, 1953; Sander and Cichy, 1967
	L+I	Ku, De	0	McGaughey *et al.*, 1975
	L+I	De, Ga	0	Sander and Cichy, 1967
S. oryzae	L+I	Th	0	Berliner, 1915
	L+I	Th	(+)	Steinhaus and Bell, 1953
	L+I	Ku	0	McGaughey *et al.*, 1975
Tenebrio molitor	L+I	Th	0	Berliner, 1915
Tenebroides mauritanicus	L+I	Th	0	Shaikh and Morrison, 1966a
Tribolium castaneum	L+I	Ku	0	McGaughey *et al.*, 1975
T. confusum	L+I	Th	0	Steinhaus and Bell, 1953
	L+I	Ku	0	McGaughey *et al.*, 1975
Trogoderma granarium	L+I	Th	0	De and Konar, 1955
DIPTERA				
Aedes aegypti	L	Th	0	Shaikh and Morrison, 1966a
	L	Fi, Al, So, De, Ca, Mo, Da, Tu	0	Hall *et al.*, 1977
	L	Ga, En, Ai	+/0[a]	Hall *et al.*, 1977
	L	Th, Ku, Ke, To	++/0[a]	Hall *et al.*, 1977
	L	Is	++	Goldberg and Margalit, 1977

TABLE II (*Continued*)

Host species	Stage	*B.t.* var	Efficiency Lab.	Efficiency Field	Reference
Aedes cantans	L	Is	+++		Vaňková and Weiser, 1978
Ae. caspius	L	Is	+/++		Sinègre *et al.*, 1979b
Ae. communis	L	Is	+++		Rieger and Schnetter, unpublished observations
Ae. detritus	L	Is	++		Sinègré *et al.*, 1979b
Ae. dorsalis	L	Ku	+		Goldberg, 1978
Ae. nigromaculis	L	Th	0		Kellen and Lewallen, 1960
Ae. punctor	L	Is	+++		Rieger and Schnetter, unpublished observations
Aedes sp.	L	Th, De, Ga	+		Lavrentyev *et al.*, 1965
Ae. stimulans	L	Th	0		Shaikh and Morrison, 1966a
Ae. taeniorhynchus	L	Ku	+		Goldberg, 1978
Ae. tarsalis	L	Ku	+		Goldberg, 1978
Ae. triseriatus	L	Th, Ku, Ke, Ga, En, Ai, To	++/0[a]		Hall *et al.*, 1977
	L	Fi, Al, So, De, Ca, Mo, Da, Tu	0		Hall *et al.*, 1977
Ae. vexans	L	Is	+++		Rieger *et al.*, unpublished observations
Anopheles sergentii	L	Is	+		Goldberg and Margalit, 1977
An. stephensi	L	Is	++		de Barjac, 1978
Anopheles sp.	L	Th, De, Ga	+		Lavrentyev *et al.*, 1965
Bessa fugax	I	Ku	0		Hamed, 1978–1979
Cermoasia auricaudata	H	Ku		(+)[b]	Hamel, 1977
Chaoborus sp.	L	Is	0		Krieg, unpublished observations.
Chironomus plumosus	L	Th, De, Ga	+		Lavrentyev *et al.*, 1965

Species	Stage	Pathogens	Activity	Reference
Chironomus sp.	L	Is	++	Weiser and Vaňková, 1978
C. tummi	L	Is	+++	Rieger and Schnetter, unpublished observations
Chrysomya megacephala	L	Th	0	Sherman *et al.*, 1962
Culex pipiens	L	Th, Fi, Al, So, Ke, Ca, Mo, Tu	0	Hall *et al.*, 1977
	L	Ku	0	Goldberg, 1978
	L	Ku, Ga, Ai	$+/0^a$	Hall *et al.*, 1977;
	L	De	0	Rogoff *et al.*, 1969
	L	En, Da	$(+)/0^a$	Hall *et al.*, 1977
	L	To	+/0	Hall *et al.*, 1977
	L	Is	+++	Goldberg and Margalit, 1977
	L	Is	+++	Rieger and Schnetter, unpublished observations
			+++	
Cu. quinquefasciatus	L	Is	+/+++	Sinègre *et al.*, 1979b
	L	Th, Fi, Al, Ku, So, De, Ke, Ga, Ca, En	0	Hall *et al.*, 1977
	L	Ai, Mo, To, Da	$+/0^a$	Hall *et al.*, 1977
Culex sp.	L	Tu	$+/0^a$	Hall *et al.*, 1977
Cu. tritaeniorhynchus	L	Th, De, Ga	+	Lavrentyev *et al.*, 1965
	L	Ks	++	Ohba and Aizawa, 1979
Cu. tarsalis	L	Th, Ku, Ke, Ga, Ai, To	$++/0^a$	Hall *et al.*, 1977
	L	Fi, So, De, Ca, Da, Tu	0	Hall *et al.*, 1977

TABLE II (Continued)

Host species	Stage	B.t. var	Efficiency Lab.	Efficiency Field	Reference
HOMOPTERA					
Aphis pomi	L + I	Th	0		Oatman, 1965
Kerria lacca	L + I	Ga		0	Malhotra and Choudhary, 1968
Macrosiphum euphorbiae	L + I	Th		0	Shorey and Hall, 1963
Pseudococcus longispinus	L + I	Ku	0		Wysoki et al., 1975
HYMENOPTERA					
Acantholyda nemoralis		Th	0		Glowacka-Pilot, 1968
Ageniaspis fuscicollis		Ga	0		Isakova, 1965
Anagyrus fusciventris		Ku	0	0	Wysoki et al., 1975
Apanteles albipennis		Ga	0		Adashkevich, 1966
A. fumiferanae		Ku		0	Hamel, 1977
A. glomeratus	H	Th	0/(+)[b]		Marchal-Segault, 1975
		Al	0		Biliotti, 1956
		Ga	0		Isakova, 1965
A. melanoscelus		Th	0		Wolliam and Yendol, 1976
A. ruficrus		Ga	0		Adashkevich, 1966
A. sicarius		Ga	0		Adashkevich, 1966
		Th	0	0	Stute, 1963
Apis mellifera	I	Th	0		Wilson, 1962; Krieg, 1964
	I	So, De, Ga	(+)/0[a]		Haragsim and Vaňková, 1968
	I	Th, En	(+)		Haragsim and Vaňková, 1968
	I	Al	(+)		Haragsim and Vaňková, 1968
	L	Th, Al, So, Ga, En	0		Haragsim and Vaňková, 1968
	L	De	(+)		Haragsim and Vaňková, 1968
	I	Ku, Is	0		Pinsdorf and Krieg, unpublished observations
	L + I	Th, Al, So		0	Cantwell et al., 1966

Athalia rosae (as *colibri*)	L	Th	+		Rautapää, 1967
Bracon hebetor	L	Ga	+		Charpentier, 1971
	L	Th	(+)[b]		Tamashiro, 1968
	I	Ku		0	McGaughey, 1979
Cardiochiles nigriceps	L	Ku	0		Dunbar and Johnson, 1975
Coccygomimus (see *Pimpla*)					
Croesus septentrionalis	L	Th	0		Krieg, 1957b
Diadegma armillata	I	Ku	(+)		Hamed, 1978/9
D. cerophaga		Ga	0		Adashkevich, 1966
D. fenestralis		Ga	0		Adashkevich, 1966
Diadromus subtilicornis	L	Ga	0		Adashkevich, 1966
Diprion pini	L	Th	0		Glowacka-Pilot, 1968
Formica polyctena	L + I	Th	0	0	Kneitz, 1966
F. pratensis/nigricans	L + I	Th	0	0	Kneitz, 1966
F. rufa	L + I	Th	0	0	Kneitz, 1966
Gilpinia hercyniae	L	Th	0		Krieg, 1957b
	L	So	0		Angus, 1956b
Glypta fumiferanae		Ku	0	0	Hamel, 1977
Hungariella peregrina	I	Ku	0		Wysoki, 1975
Hyposoter ebeninus		Ga	0		Isakova, 1965
		Al	0		Biliotti, 1965
Leptomastix dactylopii	I	Ku	0		Viggiani and Tranfaglia, 1979
Meteorus versicolor		Ga	0		Isakova, 1965
Microplitis tuberculifera		Ga	0		Isakova, 1965
Monomorium pharaonis	L + I	Th	(+)		Vaňková *et al.*, 1975
	L + I	Th, Al, De, Ga	(+)		Brikman *et al.*, 1967
Nematus (*Pteronidea*) *ribesii*	L	Th	+		Rautapää, 1967
	L	Ga	+		Fedorinchik, 1963
Nemeritis canescens (see *Venturia*)					
Neodiprion abietis	L	So	0		Angus, 1956b
N. banksianae	L	Th	++		Heimpel, 1961

TABLE II (Continued)

Host species	Stage	B.t. var	Efficiency		Reference
			Lab.	Field	
N. banksianae	L	So	0		Angus, 1956
N. lecontei	L	So	(+)		Heimpel, 1961
N. sertifer	L	Th	0		Krieg, 1957b
	L	Th		0	Donaubauer and Schmutzenhofer, 1973
N. swainei	L	So	0		Angus, 1956
	L	So	0		Heimpel, 1961
Phaeogenes hariolus	H	Ku	(+)[b]	(+)[b]	Hamel, 1977
Phanerotoma flavitestacea	H	Th	0		Marchal-Segault, 1975
Phygadeuon trichops	I	Ku	0		Plattner, 1979
Pimpla instigator	I	Th	0		Biache, 1975
P. turionellae	I	Ga	0		Isakova, 1965
	I	Ku	(+)		Bogenschütz, 1979
	I	Ku			Hamed, 1978/1979
Polistes exclamans	L	Th	0		Guthrie et al., 1959
Priophorus tristis	L	Th	+		Rautapää, 1976
Pristiphora abietina	L	Th, Ku	0		Langenbruch and Krieg, unpublished observations
	L	Ku		0	Donaubauer and Schmutzenhofer, 1973
P. erichsonii	L	Th, So	+		Heimpel, 1961
Pteromalus puparum		Ga	0		Isakova, 1965
Telenomus alsophilae		Ku	0		Kaya and Dunbar, 1972
Tetrastichus evynomellae	I	Ku	(+)		Hamed, 1978/79
Theronia atalantae		Ga	0		Isakova, 1965
Thyraeella collaris	I	Ku		0	Hamilton and Attia, 1977
Trichionotus sp.	I	Ku	(+)		Hamed, 1978/79
Trichogramma cacoeciae	I	Th, Ku	0		Hassan and Krieg, 1975

					Reference
T. evanescens	I	Ga		0	Shchepetilnikova *et al.*, 1968a, b
	I	Is		0	Hassan and Krieg, unpublished observations
Trioxys utilis	I	Th		0	Stern, 1961
Venturia (= *Nemeritis*) *canescens*	L	Th		0	Kurstak, 1964
ISOPTERA					
Anacanthotermes sp.	I	Ga	0	0	Kakaliev and Saparliev, 1975
Reticulitermes flavipes	I	Th	0		Smythe and Coppel, 1965
R. hesperus	I	Th	0		Smythe and Coppel, 1965
R. virginicus	I	Th	0		Smythe and Coppel, 1965
Zootermopsis angusticollis	I	Th	0		Smythe and Coppel, 1965
LEPIDOPTERA					
Achaea janata	L	Ga	+++		Govindarajan *et al.*, 1976
Achroia grisella	L	Th	+		Burges and Bailey, 1968
Acleris variana	L	Th	+++		Morris, 1963
	L	Ga	+++	++	Morris, 1963
Acrobasis sp.	L	Ga	++	++	Fedorinchik, 1969
Acrolepia alliella	L	Ku		++	Takaki, 1975
Acrolepiopsis (*Acrolepia*) *assectella*	L	Th, Al	+	+	Goix, 1959
	L	Al	++		Burgerjon and Grison, 1959
Adoxophyes orana	L	Th		++	v.d. Geest and Velterop, 1971
	L	Th	+	+	Takaki, 1975
	L	Ku, Ga	+++	+++	Takaki, 1975
	L	Ku		+++	v.d. Geest and Velterop, 1971
	L	Ai?		(+)	Takaki, 1975
Aegeria (*Sanninoidea*) *exitiosa*	L	Th?		+++	Snapp, 1962
Aglais urticae	L	Th?	+++		Metalnikov and Chorine, 1929
Agriopsis (*Erannis*) *aurantiaria*	L	Th, Ga	++	++	Ceianu *et al.*, 1970
	L	Al			Ceianu *et al.*, 1970
A. (*E.*) *leucophaearia*	L	Th, Ga	++		Ceianu *et al.*, 1970
A. marginaria	L	Al	++	+	Burgerjon and Grison, 1959
Agrotis c-nigrum see *Xestia*					

TABLE II (Continued)

Host species	Stage	B.t. var	Efficiency		Reference
			Lab.	Field	
A. ipsilon	L	Th	++	+	Revelo, 1973
	L	Al	0		Burgerjon and Grison, 1959
A. segetum	L	many	0 to +[a]		Chapter II
	L	Ku	(+)		Langenbruch, 1977
	L	De		++	Sikoura and Tkatsch, 1974
	L	Ga	+		Onishchenko, 1966
	L	Ga		++	Rassoulof et al., 1966
Alabama argillacea	L	Th	+++	+++	Ignoffo et al., 1964
Alsophila aescularia	L	Al	+++		Burgerjon and Grison, 1959
A. pometaria	L	Th, Ga	+++		Larson and Ignoffo, 1971
	L	Th, Ku		+++	Harper, 1974
Amathes c-nigrum see Xestia					
Amorbia essigana	L	Th	+++		Hall and Dunn, 1958
Anadevidia peponis	L	Th	+++	+++	Ayyar, 1961
Anisota senatoria	L	Th, En	+++		Angus, 1956a
	L	Th		+++	Angus and Heimpel, 1959
	L	Ku	++	+++	Kaya, 1974
	L	So	+++	+++	Angus and Heimpel, 1959
	L	En	+++	+++	Angus and Heimpel, 1959
Anomis flava	L	Wu	++	++	Rishbeth, 1978
A. (Cosmophila) sabulifera	L	Th		++	Chatterjee, 1965
Antheraea pernyi	L	Th	++		Burgerjon and Biache, 1967b
	L	Al, So, De, Ga, En	+++		Burgerjon and Biache, 1967b
Anticarsia gemmatalis	L	Ku	+++		Ignoffo et al., 1977
	L	Ku		++	Yearian et al., 1973
Apocheima (Biston) hispidaria	L	Ga	++		Fedorinchik, 1969
A. pilosaria (= pedaria)	L	Al	++		Burgerjon and Grison, 1959

Species					Reference
Aporia crataegi	L	Th	+++	+++	Krieg, unpublished observations
	L	Th?, Ga		+++	Kondrja, 1966a
	L	Ku			Langenbruch, unpublished observations
	L	Ga	+++		Isakova, 1958
	L	Is	+	0	Vaňková and Weiser, 1978
	L	Ku	+++	++	Madsen and Potter, 1977
	L	Th, Ku	++		Pinnock and Milstead, 1978
Archips argyrospilus	L	Th	+++		Smirnoff, 1972
A. cerasivoranus	L	De	++	++	Kudler *et al.*, 1965
A. crataegana	L	Ga	+++	++	Kudler *et al.*, 1959
	L	Th, Ku	++		Fedorinchik, 1969
A. podana	L	Ku	+++		Vaňková, 1973
A. (Cacoecia) rosana	L	Ku	+++	++	Ali Niazee, 1974
	L	De, Ga	++	0	Madsen and Potter, 1977
	L	Ga?	+++		Zukauskiene, 1973
A. xylosteana	L	Ku	+	+	Niemczyk and Bakowski, 1971
Arctia caja	L	Th	++	++	Mihalache *et al.*, 1972
	L	Al, So, De, Ga, En			Burgerjon and Biache, 1967b
	L	Al	++		Burgerjon and Biache, 1967b
Argyrotaenia mariana	L	Th	++	+	Martouret, 1959
A. velutinana	L	Th		++	Jaques, 1961
	L	Ga		+++	McEwen *et al.*, 1960
Ascia (Pieris) monuste orseis	L	Th	++	++	Dolphin *et al.*, 1967
Ascotis selenaria	L	Ku		++	Fugueiredo *et al.*, 1960
Atteva aurea	L	Th	+++		Takaki, 1975
Autographa californica	L	Th	+		Hull and Onuoha, 1962
	L	Th	+		White and Briggs, 1964
A. gamma	L	Th		+	Burges and Jarrett, personal communication

TABLE II (Continued)

Host species	Stage	B.t. var	Efficiency Lab.	Efficiency Field	Reference
A. gamma	L	Ku	++	++	Burges and Jarrett, personal communication
	L	Ga	+		Burges and Jarrett, personal communication
A. nigrisigna	L	Ku, Ai		++	Takaki, 1975
	L	So		+	Takaki, 1975
Autoplusia egena	L	Th	++	++	Genung, 1960
Azochis gripusalis	L	Th	+++	++	Fugueiredo et al., 1960
Bissetia steniella	L	Th		++	Atwal and Paul, 1964
Biston hispidaria see Apocheima					
Boarmia crepuscularia see Ectropis					
Bombyx mori	L	Th	+		Burgerjon and Biache, 1967b
	L	Al, Ga, En	+++		Burgerjon and Biache, 1967b
	L	So	++		Burgerjon and Biache, 1967b
	L	De	0		Burgerjon and Biache, 1967b
	L	Th, De	+		Angus and Norris, 1968
	L	Al, So, En, Ai	+++		Angus and Norris, 1968
	L	Ke, Ga, Mo, To	++		Angus and Norris, 1968
	L	Ks	0		Ohba and Aizawa, 1979
	L	many	0 to +++[a]		Chapter II
Brachionycha sphinx	L	Al	+++		Burgerjon and Grison, 1959
Bucculatrix thurberiella	L	Th	+++		Hall and Dumm, 1958
Bupalus piniarius	L	Ga		++	Aragon, 1964
	L	Th, Ga	+++		Skatulla, 1971
	L	Ga	+++	+++	Fankhänel, 1962
Cacoecia rosana see Archips					

Species					Reference
Cacoecimorpha pronubana	L	Th, Ga	++		Burges, unpublished observations
	L	Ku	++		Burges and Jarrett, personal communication
Cactoblastis cactorum					
Cadra cautella see *Ephestia*					
Caloptilia (Gracillaria) invariabilis	L	Th	+++	++	Huang and Tamashiro, 1966
C. (G.) syringella	L	Th	++		Morris, 1969
C. (G.) theivora	L	Ga	+++		Morris, 1969
Canephora asiatica	L	Ga	++		Fedorinchik, 1969
Carposina niponensis	L	Ku, Ai		+	Takaki, 1975
	L	Ku, Ai		+	Takaki, 1975
	L	Ga		(+)	Takaki, 1975
	L	Th		++	Takaki, 1975
	L	Ku		+	Takaki, 1975
	L	Ga		0	Takaki, 1975
	L	Ai?		(+)	Takaki, 1975
	L	Th		++	Takaki, 1975
Cerapteryx graminis	L	Th		+	Weiser, 1960
Chilo auricilius	L	Th	++		Kalra and Kumar, 1963
	L	Ku			Nayak *et al.*, 1978
C. sacchariphagus indicus	L	Th		+	Kalra and Kumar, 1963
C. suppressalis	L	Th	++		Bounias and Guennelon, 1974
	L	Th, Ai		+	Takaki, 1975
Chloridea = Heliothis					
Choristoneura fumiferana	L	Th, Al	+++		Yamvrias and Angus, 1970
	L	Fi	0		Yamvrias and Angus, 1970
	L	So, Ga, En	++		Yamvrias and Angus, 1970
	L	De, Ke	+		Yamvrias and Angus, 1970
	L	Th, Al, Ku, De, En	+++		Smirnoff, 1965
	L	So	++	++	Smirnoff, 1965
	L	Ku		+++	Smirnoff *et al.*, 1973
	L	Ku			Morris, 1977
C. murinana	L	Th	++		Krieg, 1956

TABLE II (*Continued*)

Host species	Stage	B.t. var	Efficiency		Reference
			Lab.	Field	
C. murinana	L	Th		++	Weiser, 1962
Chrysodeixis (Plusia) chalcites	L	Th		++	Helson, 1965
Clepsis spectrana	L	Ku	++	++	Burges and Jarrett, unpublished observations
Chaphalocrocis medinalis	L	Th, Ga	++	++	Takaki, 1975
	L	Th, Ku	++		Srivastava and Nayak, 1978
	L	Wu	++	++	Rishbeth, 1978
Coleotechnites (Recurvaria) milleri	L	Th	++	++	Struble, 1965
C. nanella	L	Ga?		++	Struble, 1965
	L	Ga		++	Rafal'skiǐ, 1974
Colias eurytheme	L	Th	+++	+++	Steinhaus and Jerrel, 1954
C. lesbia	L	Th	+++	+++	Faldini and Pastrana, 1952
Colotois pennaria	L	Th, Al, Ga	++		Balinski et al., 1969
	L	Al	+++		Biliotti, 1956
	L	Ku		++	Donaubauer and Schmutzenhofer, 1973
Cosmophila sabulifera see Anomis					
Crambus bonifatellus	L	Th	+++		Hall, 1954
C. sperryellus	L	Th	+++	+++	Hall, 1957
Crambus spp.	L	Th		++	Jefferson et al., 1964
Cryptoblabes gnidiella	L	Ku		+++	Wysoki et al., 1975
Cydia funebrana	L	Th		+	Wiackowski and Wiackowska, 1966
	L	Ku		0	Niemczyk, 1975
C. (Grapholitha) molesta	L	Th	+		Roehrich, 1964
	L	Ku		0	Zivanović and Stamenković, 1976
	L	Ai?		0	Takaki, 1975
C. (Laspeyresia) nigricana	L	Th		0	Gould and Legowski, 1965

Species					Reference
C. (L.) pomonella	L	Th	++	++	Roehrich, 1964
	L	Th		++	Jaques, 1961
	L	Ku		++	Niemczyk et al., 1976
	L	Ga		++	Dolphin et al., 1967
Dasychira pudibunda	L	Th	(+)		Krieg, 1957b
Datana integerrima	L	Th	+++		Angus, 1956a
	L	Ku		++	Polles, 1974
D. ministra	L	So	+++		Angus, 1956b
	L	Th	+/++		Angus, 1955
	L	Th	+++	+++	Angus and Heimpel, 1959
	L	So	+++		Angus, 1956b
	L	So	++	+++	Angus and Heimpel, 1959
	L	En	+++		Angus and Heimpel, 1959
Dendrolimus pini	L	Ga	++		Fedorinchik, 1969
D. sibiricus	L	De	+++	+++	Talalaev, 1957, 1959
	L	De		+++	Gukasyan, 1962
	L	Tu		+++	Gukasyan and Rybakova, 1964
	L	Wu	++	++	Rishbeth, 1978
Depressaria marcella	L	Th	+++	++	Celli, 1970
Desmia funeralis	L	Th, Ga		+++	Jensen, 1969
Diachrysia orichalcea	L	Ku		++	Ali Niazee and Jensen, 1973
Diaphania (Margaronia) indica	L	Th?		+++	Basu and Chatterjee, 1969
D. nitidalis	L	Th	+++	++	Ayyar, 1961
Diaphora mendica	L	Ga	+++		Canerday, 1967
Diatraea grandiosella	L	Ga	+++		Isakova, 1958
D. saccharalis	L	Th			Sikorowski et al., 1970
	L	Th		0	Hensley et al., 1961
	L	Ku	+	+	Revelo, 1973
Dichomeris marginella	L	Ga?		+++	Charpentier et al., 1973
Drymonia ruficornis (as chaonia)	L	Th	+++	++	Nordin and Appleby, 1969
	L	Th			Ceianu et al., 1970
	L	Th, Ku, Ga		+++	Pirvescu, 1973

TABLE II (Continued)

Host species	Stage	B.t. var	Efficiency		Reference
			Lab.	Field	
Drymonia sp.	L	Th, Al, Ga	++		Balinski et al., 1969
Dryocampa rubicunda	L	Th	++	+++	Angus and Heimpel, 1959
	L	So	+++		Angus, 1956a
	L	So		+++	Angus and Heimpel, 1959
Earias insulana	L	Th	++	++	Al-Azawi, 1964
	L	Al	+		Burgerjon and Grison, 1959
	L	Ku		0	Hussain and Askari, 1976
Ectropis (Boarmia) crepuscularia	L	Th	+		Morris, 1962
Ennomos subsignarius	L	Ku	+++	+++	Dunbar et al., 1973
Ephestia cautella	L	Th, Ku, Ga,	+++	+++	Kinsinger, McGaughey and Dicke, unpublished observations
		To			Kinsinger, McGaughey and Dicke, unpublished observations
	L	Ai	+		McGaughey, 1979
E. elutella	L	Ku	0 to +++[a]	+++	Chapter II
	L	many	+++		v.d. Laan and Wassink, 1962
E. kuehniella	L	many	0 to +++[a]		Chapter II
	L	Th	+++	++/+++	Berliner, 1911; Vaňková, 1962
	L	Th	++		Jacobs, 1950
	L	Al, De, Ga	+++		Vaňková, 1962
	L	Al	++/+++[a]		Burgerjon and Grison, 1959
	L	En	+		Vaňková, 1962
	L	Pa	0		de Barjac et al., 1977
	L	Is			de Barjac, 1978
Epinotia tsugana	L	Th	++		Morris, 1969
	L	Ga	+++		Morris, 1969
Epiphyas postvittana	L	Ku		++	Buchanan, 1977
Erannis aurantiaria see *Agriopsis*					
E. defoliaria	L	Al	+++		Burgerjon and Grison, 1959

TABLE II (Continued)

Host species	Stage	B.t. var	Efficiency		Reference
			Lab.	Field	
Eutromula (Simaethis) pariana	L	Ga	++	++	Niemczyk and Bakowski, 1971
Euxoa messoria	L	Th, Ga	++	++	Cheng, 1973
	L	Ku	++		Cheng, 1973
Feltia subterranea	L	Th	0		White and Briggs, 1964
Galleria mellonella	L	Th, Ga	+++		Burgerjon and Biache, 1967b
	L	Al	+		Burgerjon and Biache, 1967b
	L	De	0		Burgerjon and Biache, 1967b
	L	En	++/+++[a]		Burgerjon and Biache, 1967b
	L	Al, En	++		Vaňková, 1966
	L	So, De	+		Vaňková, 1966
	L	Ga	+++		Vaňková, 1966
	L	Ku	++	++	Ali *et al.*, 1973
	L	Da	+		Krieg, unpublished observations
	L	Tu	(+)		Krieg, unpublished observations
	L	Ts	(+)		de Barjac and Thompson, 1970
	L	many	0 to +++[a]		Chapter II
Gastropacha quercifolia	L	Th?		++	Panait and Ciortan, 1974
Gnorimoschema lycopersicella see *Keiferia*					
G. operculella see *Phthorimaea*					
Gracillaria spp. see *Caloptilia*					
Grapholitha spp. see *Cydia*					
Hadena (Mamestra) illoba	L	Ai?		0	Takaki, 1975
Halisidota argentata	L	Ga	+		Morris, 1969
H. caryae	L	So	++		Angus and Heimpel, 1959
Harrisina brillians	L	Th, Ku	+++	+	Pinnock *et al.*, 1973
	L	Th		+++	Hall, 1955
Hedya nubiferana	L	Ga	+++		Johansson, 1971

Species	Stage	Pathogen			Reference
Helicoverpa assulta see *Heliothis*					
Heliothis armigera	L	Th	++		Daoust and Roome, 1974
	L	Ku		++	Roome, 1975
	L	Ga	+		Isakova, 1958
	L	Wu	++		Rishbeth, 1978
H. (Helicoverpa) assulta	L	Th		++	Takaki, 1975
	L	Ku		+	Takaki, 1975
Heliothis peltigera	L	Al		+/++	Martouret, 1959
H. virescens	L	Th, Ai, Da	++		Dulmage, 1973
	L	Al, Tu	0		Dulmage, 1973
	L	Th	++		Patti and Carner, 1974
	L	Ku	+++		Patti and Carner, 1974
	L	Ga		++	Mistric and Smith, 1973
H. viriplaca	L	many	0 to +++[a]		Chapter II
H. zea	L	De	++		Charafoutdinof, 1970
	L	Ha, Ga, Ai	+		Rogoff *et al.*, 1969
	L	Fi, Al, So	0		Rogoff *et al.*, 1969
	L	Ku	++		Rogoff *et al.*, 1969
	L	De	(+)		Rogoff *et al.*, 1969
	L	En	0/++[a]		Rogoff *et al.*, 1969
	L	Th	+++		Hall and Dunn, 1958
	L	Th		0	Jaques and Fox, 1960
	L	Th		+++	Shorey and Hall, 1963
	L	Th, Ku, Ga	++/+++		Creighton *et al.*, 1971
	L	Th		++/+++	Tanada, 1956
Hellula undalis					
Hemerocampa see *Orgyia*					
Herpetogramma phaeopteralis	L	Ku		+	Reinert, 1974
Heterocampa guttivitta	L	Th, Ku		+	Wallner, 1971
H. manteo	L	Ku	+++		Ignoffo *et al.*, 1973
Holcocera pulverea	L	Ga	+++		Malhotra and Choudhary, 1968
Homoeosoma electellum	L	Th		+	Carlson, 1967
Homona magnanima	L	Ku		++	Takaki, 1975

TABLE II (*Continued*)

Host species	Stage	B.t. var	Efficiency		Reference
			Lab.	Field	
Hyloicus pinastri	L	Th?	+++		Glowacka-Pilot, 1968
Hypeuryntis coricopa	L	Th?		++	Helson, 1965
Hyphantria cunea	L	Th, De, Ga, En	++		Vaňková, 1962
	L	Al, En	+++		Vaňková, 1962
	L	Th	+++	+++	Krieg and Schmidt, 1962
	L	Ku	+++		Morris, 1972b
	L	Ku, Ai		++	Takaki, 1975
	L	Ku	+++	+++	Zivanović and Stamenković, 1976
	L	So	++		Angus and Heimpel, 1959
	L	Ga		+++	Videnova, 1970
	L	Ga		+	Takaki, 1975
	L	En	+++		Angus and Heimpel, 1959
Hypogymna morio	L	Th, Ga	++		Dobrivojević et al., 1969
	L	Ku	++		Dobrivojević and Injac, 1975
	L	many	0 to +++[a]		Chapter II
Hyponomeuta see *Yponomeuta*					
Itame (Thamnonoma) wauaria	L	Ga	+++	+++	Isakova, 1958
Junonia coenia	L	Th	+		Steinhaus, 1951a
	L	En	++/+++		Steinhaus, 1951a
Kakivoria flavofasciata	L	Ku, Ga, Ai	+		Takaki, 1975
Keiferia (Gnorimoschema) lycopersicella	L	Th		+	Middlekauff et al., 1963
	L	Ku		+	Poe and Everett, 1974
Lacanobia (Polia) oleracea	L	Th	+		Burgerjon and de Barjac, 1960
	L	Ku, Ga		++	Chan Tkho, 1973
	L	Ku	+++	+++	Burges and Jarrett, 1978
	L	Ga	++		Burges, unpublished observations
Lamdina athasaria pellucidaria	L	Ku		++	Sorensen and Barbosa, 1975

L. fiscellaria fiscellaria	L	Th	++/+++	++/+++	Heimpel and Angus, 1959
	L	So	0		Angus, 1956a
L. fiscellaria lugubrosa	L	Th	++/+++	++/+++	Heimpel and Angus, 1959
	L	Ga		+	Harper, 1974
L. fiscellaria somniaria	L	Th	++/+++	++/+++	Heimpel and Angus, 1959
Lampides boeticus	L	Th		+++	Morris, 1962
	L	Ku		+	Takaki, 1975
Laspeyresia see *Cydia*					
Leucoma (Stilpnotia) salicis	L	Th, So	+++		Burgerjon and Biache, 1967b
	L	Al, De, Ga	++		Burgerjon and Biache 1967b
	L	En	+/+++[a]		Burgerjon and Biache, 1967b
	L	Th	+		Morris, 1969
	L	Th		+++	Kudler and Lysenko, 1963
	L	Ku		+++	Donaubauer and Schmutzenhofer, 1973
L. wiltshirei	L	Ga	+++		Morris, 1969
	L	Th, Al, De	++		Alizadeh, 1977
Lithocolletis see *Phyllonorycter*					
Lobesia (= Polychrosis) botrana	L	Th, Ga	+++	++	Roehrich, 1970
	L	Th, Ku		++	Schmid and Antonin, 1977
	L	Ku		0	Viggiani and Tranfaglia, 1975
Loxostege sticticalis	L	Ku	+++		Iacob and Iacob, 1977
	L	Th, Al, Ga, En	++		Burgerjon and Biache, 1967b
Lymantria dispar	L	So, De, En	+		Burgerjon and Biache, 1967b
	L	Th	+/+++		Dubois and Squires, 1971
	L	Al	+++		Dubois and Squires, 1971
	L	De?, En, Mo	++		Dubois and Squires, 1971
	L	Ga	++/+++		Dubois and Squires, 1971
	L	Th?		++	Doane, 1966
	L	Th		++	Ceianu et al., 1970
	L	Th	++		Vaňková, 1973

TABLE II (Continued)

Host species	Stage	B.t. var	Efficiency		Reference
			Lab.	Field	
Lymantria dispar	L	Ku	+++		Vaňková, 1973
	L	Ku	+++	++	Yendol et al., 1973
	L	Ga		++	Doane and Hitchcock, 1964; Ceianu et al., 1970
L. monacha	L	Ai?	0 to +++[a]	(+)	Takaki, 1975
	L	many	++		Chapter II
	L	Th?	++		Glowacka-Pilot, 1968
	L	Th, Al	++		Balinski et al., 1969
	L	Th, De	+++	++	Szmidt and Slizynski, 1965
	L	Ku	++	++	Bejer-Peterson, 1974
	L	Ga	+++		Balinski et al., 1969
	L	Ga	++		Johansson, 1971
Malacosoma americana	L	Th	+++	++	Angus and Heimpel, 1959
	L	Th	++	+++	Jaques, 1961
M. disstria	L	So	+++	+++	Angus and Heimpel, 1959
	L	Th	+++	+++	Heimpel and Angus, 1959
	L	Th	+++	+++	Angus and Heimpel, 1959
	L	Ku	++		Harper, 1974
	L	So		+++	Angus and Heimpel, 1959
	L	Ga?			Morris, 1972a
M. fragilis (= *fragile*)	L	Ga		++	Stelzer, 1965
M. neustria	L	Th, So, En	+		Burgerjon and Biache, 1967b
	L	Al, Ga, En	+++		Burgerjon and Biache, 1967b
	L	De	++		Burgerjon and Biache, 1967b
	L	Th, Al, Ga	+++		Balinski et al., 1969
	L	Th	+		Vaňková, 1973
	L	Th		+++	v.d. Laan and Wassink, 1962
	L	Ku	+++		Vaňková, 1973

Species		Location			Reference
M. neustria	L	Ku		+++	Gürses and Doğanay, 1976
	L	Ga		+++	Shvecova, 1959
	L	Is	+		Vaňková and Weiser, 1978
M. neustria var testacea	L	Ku		+++	Takaki, 1975
M. pluviale	L	Th	++		Morris, 1969
	L	Ga	+++	++	Morris, 1969
Mamestra brassicae	L	Th, Al, So De, Ga, En	(+)		Burgerjon and Biache, 1967b
	L	Th	0		Vaňková, 1973
	L	Ku	++	++	Vaňková, 1973
	L	Ku, So		++	Takaki, 1975
	L	Ga		++	Voskresenskaya, 1977
	L	Ai		(+)	Takaki, 1975
M. illoba see Hadena					
Manduca quinquemaculata	L	Th	+++	+++	Angus, 1955
	L	Th	+++		Rabb *et al.*, 1957
	L	So	+++		Angus, 1956a
M. sexta	L	Th	++	++	Guthrie *et al.*, 1959
	L	So	+++	+++	Guthrie *et al.*, 1959
Margaronia indica see Diaphania					
Maruca testulalis	L	Ga?	+++	+++	Taylor, 1968
Melanolophia imitata	L	Th, Ga	+++	+++	Morris, 1969
	L	Th		++	Morris, 1962
Mesographe forficalis	L	Ga	+++		Isakova, 1958
	L	De, Ga		++	Issi, 1978
	L	Ku	++		Langenbruch, unpublished observations
Mocis repanda	L	Th	++	++	Fugueiredo *et al.*, 1960
Molippa sabina	L	Th	++	++	Fugueiredo *et al.*, 1960
Monema flavescens	L	Ku, Ai		++	Takaki, 1975
	L	Ga		+	Takaki, 1975
Mythimna (Pseudaletia) convecta	L	Ku		0	Hitchcock, 1974

TABLE II (Continued)

Host species	Stage	B. t. var	Efficiency		Reference
			Lab.	Field	
M. loreyimima	L	Ku		0	Hitchock, 1974
M. (Pseudaletia) separata	L	Ku		0	Hitchock, 1974
M. (Pseudaletia) unipuncta	L	Th, So, De, En	(+)		Burgerjon and Biache, 1967b
	L	Al	0		Burgerjon and Biache, 1967b
	L	Ga	+		
Nephantis serinopa	L	Th	++	++	Ayyar, 1961
Noctua pronuba	L	Th, Ga	(+)		Burges and Jarrett, personal communication
	L	Ku	+	+	White and Briggs, 1964
Nomophila noctuella	L	Th	+		Angus, 1956a
Nymphalis antiopa	L	Th	+++		Angus, 1956b
	L	So	+++		Almela Pons et al., 1972
Oiketicus moyanoi	L	Ku	+++		White and Briggs, 1964
Ommatopteryx texana	L	Th	+		Jaques, 1961
Operophtera brumata	L	Th	+++	+++	Ceianu et al., 1970
	L	Th, Ga		++	Biliotti, 1956
	L	Al	++/+++		v. d. Geest, 1971
	L	Ku		++	Okhotnikov and Shpil'chak, 1978
	L	De		++	Isakova, 1958
	L	Ga	+++		Vaňková, 1973
O. fagata	L	Th, Ku	+++		Balinski et al., 1969
Orgyia antiqua	L	Th	+		Balinski et al., 1969
	L	Al, Ga	++		Lipa et al., 1977
	L	Ku, Ai		+++	Niemczyk, 1971
	L	Ga		++	Rossmoore et al., 1970
O. leucostigma	L	Th, Al, So, De, En, Ai	++		Rossmoore et al., 1970
	L	Fi	0		Morris, 1973
O. pseudotsugata	L	Ku	+++		Morris, 1963
	L	Th	++		

O. pseudotsugata	L	Ku	+++		Morris, 1973
	L	Ku		+++	Stelzer *et al.*, 1975
	L	Ga?	++		Morris, 1972a
O. thyellina	L	many	0 to +++[a]		Chapter II
	L	Ku, Ga		+	Takaki, 1975
	L	Ai		++	Takaki, 1975
Oria musculosa	L	Al	0		Burgerjon and Grison, 1959
Orthosia gothica	L	Th, Al, De, En	(+)		Burgerjon and Biache, 1967b
	L	So	0		Burgerjon and Biache, 1967b
	L	Ga, En	+		Burgerjon and Biache, 1967b
Ostrinia nubilalis	L	Th	+++		McConnell and Cutkomp, 1954
	L	many	0 to +++[a]		Chapter II
	L	Th, Ku		++	Schubert and Stengel, 1974
	L	Al		++	Martouret, 1959
Paleacrita vernata	L	Th, Ku		+++	Harper, 1974
Pammene juliana	L	Th	+++		Müller, 1957
Pandemis dumetana	L	Th		++	Balázs Klára, 1966
Panolis flammea	L	Th		++	Glowacka-Pilot and Koehler, 1965
Papilio demoleus	L	Ga	+++		Narayanan *et al.*, 1976
P. philenor	L	Th	+++		Steinhaus, 1957
Paralipsa gularis	L	Th	+++		Steinhaus, 1951a
Paralobesia viteana	L	Ku		++	Biever and Hostetter, 1975
Paramyelois transitella	L	Ku		++	Pinnock and Milstead, 1972
Parnara guttata	L	Th, Ku, Ga, Ai		++	Takaki, 1975
Pectinophora gossypiella	L	Th		+	Graves and Watson, 1970
	L	Ku	+++		Dulmage, 1970a
	L	Ku	++		Kumar and Jayaraj, 1978
Pericallia ricini	L	Th	+		Steinhaus, 1951a
Peridroma saucia	L	Th	++		Ceianu *et al.*, 1970
Phalera bucephala	L	Ga	+++		Isakova, 1958

TABLE II (Continued)

Host species	Stage	B.t. var	Efficiency		Reference
			Lab.	Field	
Phlogophora meticulosa	L	Th	+		Burges and Jarrett, personal communication
	L	Ku	++	++	Burges and Jarrett, personal communication
Phryganidia californica	L	Ga	+		Burges unpublished observations
	L	Th	++		Steinhaus, 1951a
	L	Th, Ga	++	+++	Pinnock and Milstead, 1971
Phthorimaea (= *Gnorimoschema*) *operculella*	L	Al	++		Toumanoff and Grison, 1954
	L	Ku	++		Ali, 1974/76
Phyllonorycter (*Lithocolletis*) *blancardella*	L	Th, Ga	+	+	Jaques, 1965
P. (*L.*) *ringoniella*	L	Ku, Ai?		(+)	Takaki, 1975
	L	Ga		0	Takaki, 1975
Pieris brassicae	L	Th	++		Burgerjon and Biache, 1967b
	L	Al, So, De	+		Burgerjon and Biache, 1967b
	L	Ga	+++		Burgerjon and Biache, 1967b
	L	En	+/+++^{a}		Burgerjon and Biache, 1967b
	L	Th, En, Mo	+++		Galowalia *et al.*, 1973
	L	Al, So, Ai	+		Galowalia *et al.*, 1973
	L	Ga, To	++		Galowalia *et al.*, 1973
	L	Th	+++	+++	Krieg, 1957a
	L	Th, Al	+++	+++	Balinski *et al.*, 1969
	L	Al		+++	Lemoigne *et al.*, 1956
	L	many	0 to +++^{a}		Chapter II
	L	Ga	+++		Isakova, 1958
	L	Ts	+		de Barjac and Thompson, 1970
	L	Pa	+		de Barjac *et al.*, 1977
P. canidia sordida	L	Th	++		Yu-Chen Lee, 1966

Species		Strains			Reference
P. monuste see *Ascia*					
P. rapae	L	Th, Ku, Ga		++	Takaki, 1975
	L	So, Ai		+	Takaki, 1975
	L	Th	+++		Tanada, 1953
	L	Th, En		+++	Jaques and Fox, 1960
	L	Ku, Ga		+++	Jaques, 1972
	L	De	+++		Merdan *et al.*, 1975
	L	Ga	+++		Isakova, 1958
	L	Wu	+++		Hubei Inst., Res. Group, 1976
Plathypena scabra	L	Ku	+++		Ignoffo *et al.*, 1977
Platynota stultana	L	Ku	+++	++	Yearian *et al.*, 1973
Platyptilia carduidactyla	L	Th	+++		Hall and Dunn, 1958
	L	Th	++/+++	++/+++	Tanada and Reiner, 1960
Plodia interpunctella	L	Th, Ku, Ga, Ai, To	+++		Kinsinger, McGaughey and Dicke, unpublished observations
	L	Ku		+++	McGaughey, 1978
	L	many	0 to +++[a]		Chapter II
Plusia chalcites see *Chrysodeixis chalcites*					
Plusia gamma see *Autographa*					
P. nigrisigna see *Autographa*					
P. orichalcea see *Diachrysia*					
P. peponis see *Anadevidia*					
Plutella xylostella as *maculipennis*	L	Th, Al, So, De, En	++		Burgerjon and Biache, 1967b
	L	Ga, En	+++		Burgerjon and Biache, 1967b
	L	Th, Ga		+	Takaki, 1975
	L	Ku, So, Ai		++	Takaki, 1975
	L	Th	++/+++	+++	Tanada, 1956
	L	Th		+++	Creighton and McFadden, 1975
	L	Ku	+++		Langenbruch, unpublished observations

TABLE II (*Continued*)

Host species	Stage	B.t. var	Efficiency Lab.	Efficiency Field	Reference
Plutella xylostella as *macullipennis*	L	Ku		+++	Creighton and McFadden, 1975
	L	Ga		+++	Leskova, 1960
	L	Wu	++	++	Rishbeth, 1978
	L	Is	0		de Barjac, 1978
Polia oleracea see *Lacanobia*					
Polychrosis botrana see *Lobesia*					
Prays citri	L	Th		+++	Giammanco *et al.*, 1966
P. oleae	L	Th		++/+++	Yamvrias, 1972
	L	Ku, Ga		+++	Yamvrias, 1972
Prodenia praefica see *Spodoptera*					
Pseudaletia spp. see *Mythimna*					
Pseudoplusia includens	L	Ku	++		Ignoffo *et al.*, 1977
	L	Ku	++	++	Yearian *et al.*, 1973
	L	Ga	++		Chalfant, 1969
Pygaera anastomosis	L	Th, En	+++		Burgerjon and Biache, 1967b
	L	Al, So, De / Ga, En	++		Burgerjon and Biache, 1967b
Recurvaria milleri see *Coleotechnites*					
Rhyacionia buoliana	L	Th	++		Pointing, 1962
	L	Ga	+++		Morris, 1969
R. frustrana	L	Ku		0	Dupree and Davis, 1975
Sabulodes caberata	L	Th	++		Steinhaus, 1957
Samia cynthia	L	Th, Ga	+		Burgerjon and Biache, 1967b
	L	Al, En	+++		Burgerjon and Biache, 1967b
	L	So	++		Burgerjon and Biache, 1967b
	L	De	0		Burgerjon and Blache, 1967b
Sanninoidea exitiosa see *Aegeria*					
Saturnia pavonia	L	Th	+		Burgerjon and Biache, 1967b

Species		Strain			Reference
Saturnia pavonia	L	Al	+++		Burgerjon and Biache, 1967b
	L	So, Ga, En	++		Burgerjon and Biache, 1967b
	L	De	0		Burgerjon and Biache, 1967b
Schizura concinna	L	Th	+	+++	White and Briggs, 1964
	L	Th, Ku, Ga		+++	Pinnock *et al.*, 1974
Schoenobius bipunctifer	L	Ku	++	++	Nayak *et al.*, 1978
	L	Ga			Kwangsi Kweishien, Biol. Contr. Stat., 1974
Selenephera lunigera	L	De	+++	+++	Firstov, 1965
Sesamia inferens	L	Ku	+		Nayak *et al.*, 1978
Sibine apicalis	L	Ku	+++		Jaramillo Celis *et al.*, 1974
Simaethis pariana see Eutromula					
Sitotroga cerealella	L	Th	(+)	++	Steinhaus and Bell, 1953
	L	Ku	+	+	McGaughey, 1976
Sparganothis pilleriana	L	Th		0	Herfs, 1964
	L	Th			Harranger, 1961
	L	Al			Chaboussou, 1959
Spilonota (Tmetocera) ocellana	L	Th	+++	++	Jaques, 1961
	L	Ga		++	Jaques, 1965
Spilosoma lubricipeda (as *menthastri*)	L	Th, So	+		Burgerjon and Biache, 1967b
	L	Al	+++		Burgerjon and Biache, 1967b
	L	De, Ga, En	++		Burgerjon and Biache, 1967b
S. virginica	L	Th	+		Steinhaus, 1957
Spilosoma sp.	L	Al	++		Burgerjon and Grison, 1959
Spodoptera eridania	L	Al?		++	Creighton *et al.*, 1971
S. exigua	L	Th, De	+		Merdan *et al.*, 1975
	L	Th	++		Hall and Dunn, 1958
	L	Th		++	Shorey and Hall, 1963
	L	Ku	++		Ignoffo *et al.*, 1977
	L	Ku		++	Vail *et al.*, 1972
	L	many	0 to ++[a]		Chapter II
S. frugiperda	L	Th	+	+	Fugueiredo *et al.*, 1960

TABLE II (*Continued*)

Host species	Stage	*B.t.* var	Efficiency Lab.	Efficiency Field	Reference
S. frugiperda	L	Th, Da	+++		Dulmage, 1973
	L	Ku		+++	Creighton *et al.*, 1972
S. littoralis	L	Tu	0		Dulmage, 1973
	L	Th, De	+		Merdan *et al.*, 1975
	L	Ga	++		Altahtawy and Abaless, 1973
S. litura	L	Th	+		Afify and Merdan, 1969
	L	Ku		+	Takaki, 1975
	L	So		0	Takaki, 1975
	L	Ga	0		Narayanan *et al.*, 1976
	L	Ga		(+)	Takaki, 1975
	L	Ai		++	Takaki, 1975
	L	Is	0		de Barjac, 1978
	L	many	0 to ++[a]		Chapter II
S. mauritia	L	Th	++	++	Ayyar, 1961
S. praefica	L	Th	++		Steinhaus, 1951a
	L	Th		+	Middlekauff *et al.*, 1963
	L	En	++/+++		Steinhaus, 1951a
Stilpnotia salicis see Leucoma					
Syllepte derogata	L	Ku	+++	++	Taylor, 1974
S. silicalis	L	Th	++	++	Fugueiredo *et al.*, 1960
Symmerista canicosta	L	Ku		++	Millars *et al.*, 1974
Thamnonoma wauaria see Itame					
Thaumetopoea pityocampa	L	Th	+++	+++	Touzeau, 1971
	L	Al	+++	+/++	Grison and Béguin, 1954
	L	Ku	+++	+++	Videnova *et al.*, 1972
	L	Ga?		+++	Kailidis *et al.*, 1971
	L	Ga	++	++	Videnova *et al.*, 1972
T. processionea	L	Th, So	++	++	Burgerjon and Biache, 1967b

Species	L	Locality			Reference
T. processionea	L	Al, De, Ga, En	+++		Burgerjon and Biache, 1967b
T. wilkinsoni	L	Th	++	++	Moore et al., 1962
Thymelicus lineola	L	Th, Ga	+++	++	Arthur, 1968
	L	Ku	++	+++	McNeil et al., 1977
Thyridopteryx ephemeraeformis	L	Ku	+++	+++	Kearby et al., 1972
Tineola bisselliella	L	Th	++		Yamvrias and Angus, 1969
	L	So			Yamvrias and Angus, 1969
Tmetocera ocellana see *Spilonota*					
Tortrix viridana	L	Th, So, De, En	+++		Burgerjon and Biache, 1967b
	L	Al, Ga	++	+++	Burgerjon and Biache, 1967b
	L	Th, Ga		++	Franz et al., 1967
	L	Al		++	Burgerjon and Klinger, 1959
	L	Ku		+++	Svestka, 1974
	L	De	+++		Okhotnikov and Shpil'chak, 1978
	L	Ga	++		Fankhänel, 1962
	L	Th, Ku, Ga, En			Rogoff et al., 1969
Trichoplusia ni	L	Fi, Al, So, De	0		Rogoff et al., 1969
	L	En, Ai	+		Rogoff et al., 1969
	L	many	0 to +++[a]		Chapter II
	L	Th	++/+++		Tanada, 1956
	L	Th, Ku		++	Creighton and McFadden, 1975
	L	Ku		++	Jaques, 1972
	L	Ga		+	Jaques, 1972
	L	Ga		+++	Creighton et al., 1971
	L	Ts	+		de Barjac and Thompson, 1970
	L	Dk	0		deLucca, personal communication
	L	Id	0		deLucca, personal communication
	L	Th, Ga	(+)		Burges, unpublished observations
Triphaena pronuba see *Noctua*					

TABLE II (Continued)

Host species	Stage	B.t. var.	Efficiency		Reference
			Lab.	Field	
Udea profundalis	L	Th	+		White and Briggs, 1964
U. rubigalis	L	Th	+++		Hall and Dunn, 1958
Vanessa cardui	L	Th	++		Morris, 1969
	L	So	++		Angus and Heimpel, 1959
	L	Ga	+++		Morris, 1969
V. io	L	Th, So	+		Burgerjon and Biache, 1967b
	L	Al, En	+++		Burgerjon and Biache, 1967b
	L	De, Ga, En	++		Burgerjon and Biache, 1967b
Xanthopastis timais	L	Th	++	++	Fugueiredo *et al.*, 1960
Xestia (Amathes) c-nigrum	L	Th, So, Ga, En	+		Burgerjon and Biache, 1967b
Yponomeuta cognatella (= *Y. evonymi*)	L	Al, De	(+)		Burgerjon and Biache, 1967b
	L	Th, Al, So	++		Burgerjon and Biache, 1967b
	L	De, Ga, En	+++		Burgerjon and Biache, 1967b
	L	Th?, Ga	++	+++	Kondrja, 1966a
Y. evonymella	L	Th	++		Hamed, 1978
	L	Ku	+++	+++	Hamed, 1978
Y. mahalebella	L	Al	++		Ceianu *et al.*, 1970
	L	Ga	+++		Ceianu *et al.*, 1970
Y. malinella	L	Th, Ga, En	+++		Burgerjon and Biache, 1967b
	L	Al, So, De	++		Burgerjon and Biache, 1967b
	L	Th		+++	Wildbolz and Staub, 1962
	L	Al		++	Kuchly, 1959
	L	Ku	+++		Vaňková, 1973
	L	De		+++	Rybina, 1966
	L	Ga	+++	+++	Isakova, 1958

Species	Stage	Strain			Reference
Y. malinella	L	Ai		++	Takaki, 1975
Y. padella	L	Th?, Ga	+++	+++	Kondrja, 1966a
	L		+++		Toumanoff, 1955
	L	Al	++		Hamed, 1978
	L	Th	+++		Hamed, 1978
	L	Ku	++		Burgerjon and Biache, 1967b
	L	So, Ga	+++		Burgerjon and Biache, 1967b
	L	En	+++		Karasev, 1968
Y. rorrella	L	Ga		++	Martouret and Auer, 1977
Zeiraphera diniana	L	Th		++	Benz, 1975
	L	Th, Ga?		++	
MALLOPHAGA					
Bovicola bovis		Th	+		Gingrich *et al.*, 1974
B. crassipes		Th	+		Gringrich *et al.*, 1974
		many	0 to ++[a]		Chapter II
B. limbata		Th	+		Gingrich *et al.*, 1974
B. ovis		Th	+		Gingrich *et al.*, 1974
Lipeurus caponis		Th		+	Hoffman and Gingrich, 1968
Menacanthus stramineus		Th		+	Hoffman and Gingrich, 1968
Menopon gallinae		Th		+	Hoffman and Gingrich, 1968
NEUROPTERA					
Chrysopa carnea	L + I	Th	0		Hassan and Krieg, unpublished observations
Chrysopa sp.	L	Th		0	Stern, 1961
ORTHOPTERA					
Blatta orientalis	L + I	Th	0		Brikman *et al.*, 1966
Calliptamus italicus	L + I	Ga	0		Isakova, 1958
Carausius morosus	L + I	Th	0		Pendleton, 1970
Chorthippus dorsatus	L + I	Th(?)	0		Metalnikov and Chorine, 1929
C. pulvinatus	L + I	Th(?)	0		Metalnikov and Chorine, 1929
Diapheromera femorata	L + I	Th	(+)		Ignoffo *et al.*, 1973a
Gryllodes sigillatus	L + I	Th	0		Shaikh and Morrison, 1966b
Periplaneta americana	L + I	Th	0		Shaikh and Morrison, 1966b

TABLE II (Continued)

Host species	Stage	B.t. var	Efficiency		Reference
			Lab.	Field	
Schelfordella tartarica	L + I	Th	(+)		Brikman et al., 1966
Stauroderus biguttulus	L + I	Th(?)	0		Metalnikov and Chorine, 1929
Teleogryllus commodus	L + I	Th	0		Shaikh and Morrison, 1966b
Tenodera aridifolia	L + I	Th	0		Yousten, 1973
THYSANOPTERA					
Frankliniella occidentalis		Th		0	Shorey and Hall, 1963
TRICHOPTERA					
Chaetopteryx sp.	L	Is	0		Weiser and Vaňková, 1978
Hydropsyche pellucida	L	Is	++		Weiser and Vaňková, 1978
Potamophylax rotundipennis	L	Is	+		Weiser and Vaňková, 1978
Arthropoda: Myriopoda					
DIPLOPODA					
Poratophilus pretorianus		Th	0		Fiedler, 1965
P. robustus		Th	0		Fiedler, 1965
Spinotarsus fiedleri		Th	0		Fiedler, 1965
Arthropoda: Arachnoidea					
ACARI					
Anystis agilis	L + I	Th		0	Jaques, 1965
Blattisoccus tarsalis		Ku		0	McGaughey, 1979
Bryobia arborea	L + I	Th		0	Jaques, 1965
Hydacarina sp.		Is	0		Morawczik and Schnetter, unpublished observations
Hydrachna sp.		Is	0		Weiser and Vaňková, 1978
Panonychus ulmi	L + I	Th		0	Oatman, 1965
Pilophorous perplexus	L + I	Th		0	Jaques, 1965
Versates schlechtendali	L + I	Th		0	Jaques, 1965
SCORPIONIDAE					
Buthus occitanus	I	Th	0		Morel, 1974

Arthropoda: Crustacea

Arthemisia salina	Is	0	Sinègre *et al.*, 1979a
Chirocephalus grubei	Is	+	Morawczik and Schnetter, unpublished observations
Cyclops fuscus	Is	0	Sinègre *et al.*, 1979a
Cyclops sp.	Is	(+)	Weiser and Vanková, 1978
	Is, Ku	0	Krieg, unpublished observations
Daphnia magna	Is	0	Sinègre *et al.*, 1979a
D. pulex	Is	0	Morawczik and Schnetter, unpublished observations
Daphnia sp.	Is, Ku	0	Krieg, unpublished observations
Megacyclops sp.	Is	(+)	Weiser and Vanková, 1978
Annelida: Oligochaeta			
Eisenia foetida	Th	+	Heimpel, 1966
	Th	0	Krieg, unpublished observations
Lumbricus terrestris	Th	+	Smirnoff and Heimpel, 1961
	Th	0	White, 1960; Benz and Altwegg, 1975
Tubifex sp.	Is	0	Morawczik and Schnetter, unpublished observations
Annelida: Hirudinea			
Helobdella sp.	Is	0	Weiser and Vanková, 1978
Mollusca: Bivalvia			
Ostrea edulis	Is	0	Sinègre *et al.*, 1979b
Mollusca: Gastropoda			
Aplexa hypnorum	Is	0	Morawczik and Schnetter, unpublished observations
Bithynia tentaculata	Is	0	Morawczik and Schnetter, unpublished observations
Galba palustris	Is	0	Morawczik and Schnetter, unpublished observations

TABLE III (Continued)

Host species	Stage	B.t. var	Efficiency		References
			Lab.	Field	
Orthellia caesarion	L	Th	++		Wasti et al., 1973
Phormia regina	L	Th	++		Wasti et al., 1973
Tipula paludosa	L	Th	++		Lam and Webster, 1972
EPHEMEROPTERA					
Cloeon sp.	L	Th	0		Krieg, unpublished observations
HETEROPTERA					
Notonecta glauca	L	Th	0		Krieg, unpublished observations
Perillus bioculatus	L + I	Th	0		Burgerjon and Biache, 1966
Sigara sp.	L	Th	0		Krieg, unpublished observations
HOMOPTERA					
Myzus persicae	L + I	Th	0		Krieg and Zimmermann, unpublished observations
HYMENOPTERA					
Apis mellifera	L	Th	+		Cantwell et al., 1964b
	I	Th	++		Cantwell et al., 1964b
Athalia rosae (as A. colibri)	L	Th	+/0		Laurent, 1965
Diprion pini	L	Th	++	++	Burgerjon and Biache, 1964
Paravespula vulgaris	I	Th	+		Haragsim and Vaňková, 1973
Pristiphora pallipes	L	Th	++		Burgerjon and de Barjac, 1960, 1964
ISOPTERA					
Reticulitermes flavipes	I	Th	++		Smythe and Coppel, 1965
R. hesperus	I	Th	++		Smythe and Coppel, 1965
R. virginicus	I	Th	++		Smythe and Coppel, 1965
Zootermopsis angusticollis	I	Th	++		Smythe and Coppel, 1965
LEPIDOPTERA					
Achroia grisella	L	Th	0		Heitor, 1962
Agrotis segetum	L	Th	+		Burgerjon and de Barjac, 1960

Species				Reference
Bombyx mori	L	Th	++	Burgerjon and de Barjac, 1960
Ephestia kuehniella	L	Th	++	Yamvrias, 1962
Estigmene acrea	L	Th	0	Mechalas and Beyer, 1963
Galleria mellonella	L + I	Th	++[a]	Ignoffo and Gregory, 1972
	L	Th	0	McConnel and Richards, 1959
	L	Th	+/++[a]	Vaňková, 1966a; Krieg, 1967b
				Burges, 1975
Heliothis zea	L + I	Th	++[a,b]	Ignoffo and Gregory, 1972
H. virescens	L + I	Th	++[a]	Ignoffo and Gregory, 1972
Lacanobia oleracea	L	Th	++[a]	Burgerjon and de Barjac, 1964
Lymantria dispar	L	Th	++[a]	Burgerjon and de Barjac, 1960;
				Burgerjon and Biache, 1967a
Malacosoma neustria	L	Th	++	Burgerjon and de Barjac, 1960, 19 1964
Mamestra brassicae	L	Th	++[a]	Burgerjon and de Barjac, 1960; Burgerjon and Biache, 1967a
	L	Mo	+	De Barjac et al., 1966
	L	To	++	De Barjac et al., 1966
Mythimna (Pseudaletia) unipuncta	L	Th	+	Burgerjon et al., 1964
Ostrinia nubilalis	L	Th	0	Sutter and Raun, 1966
Pectinophora gossypiella	L + I	Th	++[a]	Ignoffo and Gregory, 1972
Peridroma saucia	L	Th	++	Burgerjon and de Barjac, 1960
Pieris brassicae	L	Th	++[a]	Burgerjon and de Barjac, 1960; Burgerjon and Biache, 1967a
	L	Ke, Mo	+	Burgerjon and de Barjac, 1967;
	L	Mo	+	De Barjac et al., 1966
	L	To	++	De Barjac et al., 1966
	L	Da	++	De Barjac and Burgerjon, 1973

TABLE III (Continued)

Host species	Stage	B.t. var	Efficiency		Reference
			Lab.	Field	
P. rapae	L	Th	+		Herfs and Krieg, unpublished observations
Plodia interpunctella	L	Th	0		Cantwell et al., 1964a
Plutella xylostella	L	Th	++[a]		Sicker and Krieg, 1966; Burgerjon and Biache, 1967a
(= P. maculipennis)					
Pseudaletia unipuncta see Mythimna					
Spodoptera (Prodenia) eridania	L	Th	+[b]		Hitchings, 1967
S. exigua	L + I	Th	++[a]		Ignoffo and Gregory, 1972
Trichoplusia ni	L + I	Th	++[a, b]		Ignoffo and Gregory, 1972
Zeiraphera diniana	L	Th	++[a]		Burgerjon and Biache, 1967a;
		Th		++	Benz, 1975
ORTHOPTERA					
Acheta domesticus	E	Th	++		Tremblay et al., 1972
Blatta orientalis	I	Th	0		Krieg and Herfs, 1963
Locusta migratoria	L	Th	++		Burgerjon et al., 1964
Nauphoeta cinerea	L + I	Th	0		Königstedt and Groth, 1972
Periplaneta americana	L + I	Th	0		McConnel and Richards, 1959; Königstedt and Groth, 1972
Schistocerca sp.	L	Th	++		Charles, 1965
Arthropoda: Arachnida					
ACARI					
Panonychus citri	L + I	Th	+++		Hall et al., 1971
Tetranychus pacificus	L + I	Th	+++		Hall et al., 1971
T. urticae (as T. telarius)	L + I	Th	+++		Krieg, 1968
Annelida: Oligochaeta					
Tubifex sp.	L	Th	0		Benz, 1966
Aschelminthes: Nematoda					
Aphelenchus avenae		Th	++		Ignoffo and Dropkin, 1977

Meloidogyne incognita	E	Th	++	Prasad et al., 1972
		Th	++	Ignoffo and Dropkin, 1977
Neoaplectana carpocapsae		Th	0	Lam and Webster, 1972
Panagrellus redivivus		Th	++	Ignoffo and Dropkin, 1977

a Malformation of mouth parts by sublethal doses. b Special inhibitory effect on larval maturation, fecundity and/or adult longevity.

TABLE IV

Susceptibility of arthropods per os to preparations of Bacillus thuringiensis containing spores, endotoxin crystals and (probably) β-exotoxin. E = Egg; L = Larva; I = Imago

Host species	Stage	B.t. var	Efficiency		Reference
			Lab.	Field	
Arthropoda: Insecta					
COLEOPTERA					
Leptinotarsa decemlineata	L$_2$	Th	+++		Lipa, 1976
	L$_3$	Th	+		Lipa, 1976
DIPTERA					
Aedes aegypti	L	Th	+++		Burgerjon, 1964
Haematobia irritans	L	Th	+++		Gingrich, 1965
	L	Th		++	Gingrich and Eschle, 1971
Musca autumnalis	L	Th	++		Gingrich, 1965
	L	Th		++	Hower and Cheng, 1968
M. domestica	L	Th	+++		Dunn, 1960
	L	Th		++	Miller et al., 1971
Stomoxys calcitrans	L	Th	+		Gingrich, 1965
HYMENOPTERA					
Apis mellifera	I	Th	++		Cantwell et al., 1966
ISOPTERA					
Reticulitermes flavipes	I	Th	++		Smythe and Coppel, 1965
R. hesperus	I	Th	++		Smythe and Coppel, 1965
R. virginicus	I	Th	++		Smythe and Coppel, 1965
Zootermopsis angusticollis	I	Th	++		Smythe and Coppel, 1965
LEPIDOPTERA					
Achroia grisella	L	Th	++		Heitor, 1962
Galleria mellonella	L	Th	++		Heitor, 1962
Mamestra brassicae	L	Th	++		Burgerjon, 1964
Spodoptera (Prodenia) eridania	L	Th		++[a]	Hitchings, 1967
Zeiraphera diniana	L	Th		+++	Benz, 1975

Arthropoda: Arachnida

ACARI

Tetranychus urticae (as *telarius*)   L + I   Th   +   Krieg, 1972

a Special inhibitory effect on fecundity and adult longevity.

References

Abreu, J.M. (1974). *Revista Theobroma* **4**, 33–36.
Adashkevich, B.P. (1966). *Zool. Zh.* **45**, 1040–1046.
Afify, A.M. and Matter, M.M. (1970). *Anz. Schädlingsk.* **43**, 97–100.
Afify, A.M. and Merdan, A.I. (1969). *Z. Angew. Ent.* **63**, 263–267.
Al-Azawi, A.F. (1964). *Entomophaga* **9**, 137–145.
Ali, A.U.D.D. (1974/76). *Yearbook Plant Prot. Res., Iraq Minist. Agric. Agrarian Reform* **1**, 54–59.
Ali, A.U.D.D., Abdellatif, M.A., Bakry, N.M. and El-Sawaf, S.K. (1973). *J. apicult. Res.* **12**, 117–123.
Alimdzhanov, R. and Anufrieva, R. (1966). *Zashch. Rast. Vredit. Bolez.* **11**, 27.
Ali Niazee, M.T. (1974). *Can. Ent.* **106**, 393–398.
Ali Niazee, M.T. and Jensen, F.L. (1973). *J. econ. Ent.* **66**, 157–158.
Alizadeh, M.H.S. (1977). *Ent. Phytopath. Appliquées* No. 43, 9–10, 58–65.
Almela Pons, G.R., Domenech, H. and Martini, N.U. (1972). *Rev. Fac. Ciênc. Agrar., (Univ. Nac. Cuyo)* **18**, 55–60.
Altahtawy, M.M. and Abaless, I.M. (1973). *Z. Angew. Ent.* **72**, 299–308.
Angus, T.A. (1955). Ph. D. Thesis, McGill Univ. Montreal.
Angus, T.A. (1956a). *Can. J. Microbiol.* **2**, 111–121.
Angus, T.A. (1956b). *Can. Ent.* **88**, 280–283.
Angus, T.A. and Heimpel, A.M. (1959). *Can. Ent.* **91**, 352–358.
Angus, T.A. and Norris, J.R. (1968). *J. Invertebr. Path.* **11**, 289–295.
Aragon-G., J.H. (1964). *Acta agron. Palmira, Columbia* **14**, 103–224.
Arthur, A.P. (1968). *J. Invertebr. Path.* **10**, 146–150.
Atwal, A.S. and Paul, H.S. (1964). *J. Res. Punjab Agric. Univ.* **1**, 143–148.
Ayyar, G.R. (1961). *Curr. Sci. (Bangalore)* **30**, 29–30.
Balázs Klára, B. (1966). *Folia ent. hung.* **19**, 215–248.
Balinski, I., Ceianu, I. and Mihalache, G. (1969). Commun. Zool. 1st. Conf. natn. ent. (I-a) Bucuresti, 53–63.
Barjac, H. de (1978). *Entomophaga* **23**, 309–319.
Barjac, H. de and Burgerjon, A. (1973). *J. Invertebr. Path* **21**, 325–327.
Barjac, H. de and Thompson, J.V. (1970). *J. Invertebr. Path.* **15**, 141–144.
Barjac, H. de, Cosmao-Dumanoir, V., Shaik, R. and Viviani, A. (1977). *C. r. hebd. Séanc. Acad. Sci., Paris, Sér. D.* **284**, 2051–2053.
Barjac, H. de, Burgerjon, A. and Bonnefoi, A (1966). *J. Invertebr. Path.* **8**, 537–538.
Basu, A.C. and Chatterjee, P.B. (1969). *Indian J. agric. Sci.* **39**, 36–40.
Bejer-Peterson, B. (1974). *Dansk. Skovforen. Tidsskr.* **59**, 59–80.
Benz, G.A. (1966). *Experientia* **22**, 81–82.
Benz, G.A. (1975). *Experientia* **31**, 1288–1290.
Benz, G.A. and Altwegg, A. (1975). *J. Invertebr. Path.* **26**, 125–126.
Berliner, E. (1911). *Z. ges. Getreidew.* **3**, 63–70.
Berliner, E. (1915). *Z. Angew. Ent.* **2**, 29–56.
Biache, G. (1975). *Ann. Soc. Ent. Fr. (N. S.)* **11**, 609–617.
Biever, K.D. and Hostetter, D.L. (1975). *J. econ. Ent.* **68**, 66–68.
Biliotti, E. (1956). *Entomophaga* **1**, 101–103.
Bogenschütz, H. (1979). cited from: Franz *et al. Entomophaga*, in prep.
Bond, R.P.M., Boyce, C.B.C., Rogoff, M.H. and Shieh, T.R. (1971). *In* "Microbial Control of Insects and Mites". (Burges, H.D. and N.W. Hussey, eds), pp. 275–302. Academic Press, London.

Bounias, M. and Guennelon, G. (1974). Centre de Rech. Agronom. du Sud-Est, Inst. natn. Res. Agronomique la Miniére, 8 + (10) pp.
Brikman, L.I., Alekseeva, M.I., Potsheba, T.L., Tokonoshenko, A.P. and Vishnjak, M.J. (1967). *Trudy Zentr. Nauchn. Issled. Dezinf. Inst.* 18, 70–77.
Buchanan, G.A. (1977). *Aust. J. agric. Res.* 28, 125–132.
Burgerjon, A. (1964). *Entomophaga* 2, 227–237.
Burgerjon, A. and Barjac, H. de (1960). *C. r. hebd. Séanc. Acad. Sci. Paris,* 251, 911–912.
Burgerjon, A. and Barjac, H. de (1964). *Entomophaga, Mém. hors Sér.* No. 2, 221–226.
Burgerjon, A. and Barjac, H. de (1967). *J. Invertebr. Path.* 9, 574–577.
Burgerjon, A. and Biache, G..(1964). *J. Insect Path.* 6, 538–541.
Burgerjon, A. and Biache, G. (1966). *Entomophaga* 11, 279–284.
Burgerjon, A. and Biache, G. (1967a). *C. r. hebd. Séanc. Acad. Sci. Paris, Sér. D,* 264, 2423–2425.
Burgerjon, A. and Biache, G. (1967b). *Entomologia exp. appl.* 10, 211–230.
Burgerjon, A. and Dulmage, H.T. (1977). *Entomophaga* 22, 121–129.
Burgerjon, A. and Galichet, P.F. (1965). *J. Invertebr. Path.* 7, 263–264.
Burgerjon, A. and Grison, P. (1959). *Entomophaga* 4, 207–209.
Burgerjon A. and Klinger, K. (1959). *Entomologia exp. appl.* 2, 100–109.
Burgerjon, A. and Martouret, D. (1971). *In* "Microbial Control of Insects and Mites" (H.D. Burges and N.W. Hussey, eds), pp. 305–325. Academic Press, London and New York.
Burgerjon, A. Grison, P. and Kachkouli, A. (1964). *J. Insect Path.* 6, 381–383.
Burges, H.D. (1967). *Nature Lond.* 215, 664–665.
Burges, H.D. (1976). *J. Invertebr. Path.* 28, 217–222.
Burges, H.D. and Bailey, L. (1968). *J. Invertebr. Path.* 11, 184–185.
Burges, H.D. (1977). *Apidologie* 8, 155–168.
Burges, H.D. and Jarrett, P. (1978). *Grower,* 90, 589–595.
Burns, E.C., Wilson, B.H. and Tower, B.A. (1961). *J. econ. Ent.* 54, 913–915.
Canerday, T.D. (1967). *J. econ. Ent.* 60, 1705–1708.
Cantwell, G.E., Heimpel, A.M. and Thompson, M.J. (1964a). *J. Insect Path.* 6, 466–480.
Cantwell, G.E., Knox, D.A., Lehnert, T. and Michael, A.S. (1966). *J. Invertebr. Path.* 8, 228–233.
Cantwell, G.E., Knox, D.A. and Michael, A.S. (1964b). *J. Insect Path.* 6, 532–536.
Carlson, E.C. (1967). *J. econ. Ent.* 60, 1068–1071.
Ceianu, I., Mihalache, G. and Balinski, I. (1970). *Microbiol. (Bucuresti)* (1968), 1, 633–637.
Celli, G. (1970). *Boll. Ist. Ent. Univ. Studi Bologna* (1968–1970), 29, 1–44.
Chaboussou, (1959). ref. by Martouret and Milaire (1963).
Chalfant, R.B. (1969). *J. econ. Ent.* 62, 1343–1344.
Chan Tkho (1973). *Rastit. Zasht. (Sofiya)* 21, 28–31.
Charafoutdinof (1970). ref. by Sikoura, A.J. (1975). *Rev. Zool. Agric. Path. Vég.* 74, 54–60.
Charles, P.J. (1965). Congr. Prot. Cult. Tropic., Marseille 1965. 851–854.
Charpentier, R. (1971). *Meddn. St. VäxtskAnst. (Stockholm)* 15, 139–157.
Charpentier, L.J., Jackson, R.D. and McCormick, W.J. (1973). *J. econ. Ent.* 66, 249–251.
Chatterjee, P.B. (1965). *J. Invertebr. Path.* 7, 512–513.

Cheng, H.H. (1973). *Can. Ent.* **105**, 941–945.
Creighton, C.S. and McFadden, T.L. (1975). *J. econ. Ent.* **68**, 57–60.
Creighton C.S., McFadden, T.L. and Cuthbert, R.B. (1971). *J. econ. Ent.* **64**, 737–739.
Creighton, C.S., McFadden, T.L., Cuthbert, R.B. and Onsager, J.A. (1972). *J. econ. Ent.* **65**, 1399–1402.
Creighton, C.S., McFadden, T.L. and Cuthbert, R.B. (1973). *J. econ. Ent.* **66**, 473–475.
Cruz, P.F.N. (1977). *Revista Theobroma,* **7**, 93–98.
Daoust, R.A. and Roome, R.E. (1974). *J. Invertebr. Path.* **23**, 318–324.
De, R.K. and Konar, G. (1955) *J. econ. Ent.* **48**, 773–774.
Devriendt, M. and Martouret, D. (1976). *Entomophaga* **21**, 189–199.
Doane, C.C. (1966). *J. econ. Ent.* **59**, 618–620.
Doane, C.C. and Hitchcock, S.W. (1964). *Conn. Agric. Exp. Stn. New Haven Bull.* **665**, 5–20.
Dobrivojević, K. and Injac, M. (1975). *Zašt. Bilja (Beograd)* **26**, 365–369.
Dobrivojević, K., Injac, M. and Zabel, A. (1969). *Zašt. Bilja (Beograd)* **20**, 317–324.
Dolphin, R.E., Cleveland, M.L. and Mouzin, T.E. (1967). *Proc. Indiana Acad. Sci.* **76**, 265–269.
Donaubauer, E. and Schmutzenhofer, H. (1973). *Eur. Pl. Prot. Orgn. Bull.* **3**, 111–115.
Dronka, K., Niemczyk, E. and Dadaj, J. (1976). *Roczn. Nauk Roln.* Ser. E **6**, 159–164.
Drozdowicz, A. (1964). *Acta Microbiol. Polon.* **13**, 23–28.
Dubois, N.R. and Squires, A.H. (1971). Proc. 4th Int. Colloq. Insect Path. Maryland 1970, pp. 196–208.
Dulmage, H.T. (1970a). *J. Invertebr. Path.* **15**, 232–239.
Dulmage, H.T. (1970b). *J. Invertebr. Path.* **16**, 385–389.
Dulmage, H.T. (1971). *J. Invertebr. Path.* **18**, 353–358.
Dunbar, J.P. and Johnson, A.W. (1975). *Environ. Ent.* **4**, 352–354.
Dunbar, D.M., Kaya, H.K., Doane, C.C., Anderson, J.F. and Weseloh, R.M. (1973). *Conneticut. Agric. Exp.-Stn. New Haven, Bull.* **735**, 23 pp.
Dupree, M. and Davis, T.S. (1975). *Res. Rep. Agric. Exp. Stn. Univ. Georgia,* No. 203, 10pp.
Dunn, P.H. (1960). *J. Insect Path.* **2**, 13–16.
Faldini, J.D. and Pastrana, J.A. (1952). *Rev. Argent. Agron.* **19**, 154–165.
Fankhänel, H. (1962). *NachrBl. dt. PflSchutzdienst. Berl.* **16**, 121–127.
Fedorinchik, N.S. (1964). *Entomophaga, Mém. hors Sér.* **2**, 51–61.
Fedorinchik, N.S. (1969). *TagBer. dt. Akad. LandwWiss. Berl.* **80**, (3) 595–614.
Feigin, J.M. (1963). *Ann. ent. Soc. Am.* **56**, 878–879.
Fiedler, O.G.H. (1965). *J. ent. Soc. S. Africa* **27**, 219–225.
Firstov, S. (1965). In "Ispol. Mikroorgan. dl. Bor'by s Vred. Nasekom. Les. Vost. Sibiri". (Talalaeva, E.V., ed.), *Izv. biol.-geogr. nauchno.-issled. Inst. Irkutsk gosud. Univ. Zhdanova,* **19**, 71–77.
Franz, J.M. and Krieg, A. (1967). *Gesunde Pfl.* **19**, 2–6.
Franz, J.M., Krieg, A. and Reisch, J. (1967). *NachrBl. dt. PflSchutzdienst. (Braunschweig)* **19**, 36–44.
Fugueiredo, M.B., Coutinho, J.M. and Orlando, A. (1960). *Arquiv. Inst. biol. S. Paulo,* **27**, 77–85.
Galichet, P.F. (1966). *Annls. Zootech.* **15**, 133–145.

Galowalia, M.M.S., Gibson, N.H.E. and Wolf, J. (1973). *J. Invertebr. Path.* **21**, 301–308.

Geest, L.P.S. van der (1971). *Z. Angew. Ent.* **69**, 263–266.

Geest, L.P.S. van der and Velterop, J.H.C. (1971). Proc. 4th Coll. Insect Path., Maryland, 1970. 209–213.

Genung, W.G. (1960). *J. econ. Ent.* **53**, 566–569.

Giammanco, G., Militello, M. and Mineo, G. (1966). *Entomophaga* **11**, 211–212.

Gingrich, R.E. (1965). *J. econ. Ent.* **58**, 363–364.

Gingrich, R.E. and Eschle, J.L. (1966). *J. Invertebr. Path.* **8**, 285–287.

Gingrich, R.E. and Eschle, J.L. (1971). *J. econ. Ent.* **64**, 1183–1188.

Gingrich, R.E., Allan, N. and Hopkins, D.E. (1974). *J. Invertebr. Path.* **23**, 232–236.

Glowacka-Pilot, B. (1968). *Pr. badaw. Inst. badaw. Lesn., Warszawa,* (357) 124–131.

Glowacka-Pilot, B. and Koehler, W. (1965). *Pr. badaw. Inst. badaw. Lesn., Warszawa* (278) 274–278.

Goix (1959). ref. by Martouret and Milaire (1963).

Goldberg, L.J. (1978). Abstr. 2nd Int. Workshop on *Bacillus thuringiensis,* Darmstadt, Sept. 5–9, 18 pp.

Goldberg, L.J. and Margalit, J. (1977). *Mosq. News* **37**, 355–358.

Gould, H.J. and Legowski, T.J. (1965). *Pl. Pathol. Suppl.* **14**, 21–22.

Govindarajan, R., Jayaraj, S. and Narayanan, K. (1976). *Z. Angew. Ent.* **80**, 191–200.

Graves, G.N. and Watson, T.F. (1970). *J. econ. Ent.* **63**, 1828–1830.

Greenwood, E.S. (1964). *N. Z. J. Sci.* **7**, 221–226.

Grison, P. and Béguin, A. (1954). *C. r. Acad. Agric. France* **40**, 413–416.

Guillet, P. and Escaffre, H. (1979a). Unpubl. docum. WHO/VBC 79.730; Geneva, 7 p.

Guillet, P. and Escaffre, H. (1979b). Unpubl. docum. WHO/VBC 79.735; Geneva, 7 pp.

Gukasyan, A.B. (1962). *Zashch. Rast. (Moskva)* **7**, 23–24.

Gukasyan, A.B. and Rybakova, G.M. (1964). *Trudy Sibi. Techn. Inst.* (39), 234–242.

Gürses, A. and Doğanay, Z.Ü. (1976). *Pl Prot. Bull. Ankara* 1976, 190–198.

Guthrie, F.E., Rabb, R.L. and Bowery, T.G. (1959). *J. econ. Ent.* **52**, 798–804.

Hall, I.M. (1954). *Hilgardia* **22**, 536–565.

Hall, I.M. (1955). *J. econ. Ent.* **48**, 675–677.

Hall, I.M. (1957). *In* Steinhaus, E.A., 1957. Mimeogr. Ser. No. 4, Lab., Insect Path. Univ. Calif. Berkeley, 24 pp.

Hall, I.M. and Dunn, P.H. (1958). *J. econ. Ent.* **51**, 296–298.

Hall, I.M., Hunter, D.K. and Arakawa, K.Y. (1971). *J. Invertebr. Path.* **18**, 359–362.

Hall, I.M., Dulmage, H.T. and Arakawa, K.Y. (1972). *J. Invertebr. Path.* **19**, 28–31.

Hall, I.M., Arakawa, K.Y., Dulmage, H.T. and Correa, J.A. (1977). *Mosq. News* **37**, 246–251.

Hamed, A.R. (1978). *Z. angew. Ent.* **85**, 392–412.

Hamed, A.R. (1978/79). *Z. angew. Ent.* **87**, 294–311.

Hamel, D.R. (1977). *Can. Ent.* **109**, 1409–1415.

Hamilton, J.T. and Attia, F.I. (1977). *J. econ. Ent.* **70**, 146–148.

Haragsim, O. and Vaňková, J. (1968). *Ann. Abeille* **11**, 31−40.
Haragsim, O. and Vaňková, J. (1973). *Apidologie* **4**, 87−101.
Harper, J.D. (1974). "Forest Insect Control With *Bacillus thuringiensis*. − Survey of Current Knowledge". Univ. Auburn, Alabama., pp. 64.
Harper, J.D. (1976). Proc. 1st Int. Coll. Invertebr. Path. Kingston (Canada) 69−73.
Harranger (1961). *In* Martouret and Milaire (1963).
Harvey, T.L. and Howell, D.E. (1965). *J. Insect Path.* **7**, 92−100.
Hassan, S. and Krieg, A. (1975). *Z. PflKrankh. PflSchutz.* **82**, 515−521.
Heimpel, A.M. (1961). *J. Insect Path.* **3**, 271−273.
Heimpel, A.M. (1966). *J. Invertebr. Path.* **8**, 295−298.
Heimpel, A.M. and Angus, T.A. (1958). Proc. 10th. Int. Congr. Ent. Montreal 1956, 4, 711−722.
Heimpel, A.M. and Angus, T.A. (1959). *Bienn. Prog. Rep.* **15**, (6) 2.
Heitor, F. (1962). Verh. 11. Ent. Kongr. Wien 1960. **2**, 845−849.
Helson, G.A.H., (1965). *N. Z. J. Agric.* **110**, 101−107.
Hensley, S.D., McCormick, W.J., Long, W.H. and Concienne, E.J. (1961). *J. econ. Ent.* **54**, 1153−1154.
Herfs, W. (1964). *Z. PflKrankh. PflSchutz.* **71**, 332−344.
Herrewege, J. (1970). *Entomophaga* **15**, 209−222.
Hitchings, D.L. (1967). *J. econ. Ent.* **60**, 596−597.
Hitchcock, B.E. (1974). *Cane Growers' Q. Bull.* **37**, 96−97.
Hoffman, R.A. and Gingrich, R.E. (1968). *J. econ. Ent.* **61**, 85−88.
Hower, A. and Cheng, T.H. (1968). *J. econ. Ent.* **61**, 26−31.
Huang, P. (1979). cit. from Franz *et al.*, in prep.
Huang, S.-S. and Tamashiro, M. (1966). *Proc. Hawaii, Ent. Soc.* **19**, 213−221.
Hubei Institut Microbiology, Ent. Res. Grp. (1976). *Acta microbiol. Sin.* **16**, 12−16.
Hull, G. and Onuoha, G.B.I. (1962). *J. Insect Path.* **4**, 357−360.
Hussain, M. and Askari, A. (1976). *J. econ. Ent.* **69**, 343−344.
Iacob, M. and Iacob, N. (1977). *Analele Institutului de Cercetări pentru Protectia Plantelor* **12**, 187−196.
Ignoffo, C.M. and Dropkin, V.H. (1977). *J. Kansas ent. Soc.* **50**, 394−398.
Ignoffo, C.M. and Gregory, B. (1972). *Environ. Ent.* **1**, 269−272.
Ignoffo, C.M., McGarr, R.L. and Martin, D.F. (1964). *J. Insect Path.* **6**, 411−416.
Ignoffo, C.M., Hostetter, D.L. and Kearby, W.H. (1973). *Environ. Ent.* **2**, 807−809.
Ignoffo, C.M., Hostetter, D.L., Pinnell, R.E. and Garcia, C. (1977). *J. econ. Ent.* **70**, 60−63.
Isakova, N.P. (1958). *Ent. Obozr.* **37**, 846−855.
Isakova, N.P. (1965). *Zashch. Rast. Vredit. Bolez. (Moskva)* **10** (3), 51.
Issi, I.V. (1978). *Eur. Pl. Prot. Orgn. conference on good practices in vegetable crop protection.* Kiev (22−26.5.78), papers of Soviet Specialists, S. 6.
Jacobs, S.E. (1950). *Proc. Soc. appl. Bacteriol.* **13**, 83−91.
Jaques, R.P. (1961). *J. Insect Path.* **3**, 167−182.
Jaques, R.P. (1965). *Can. Ent.* **97**, 795−802.
Jaques, R.P. (1972). *J. econ. Ent.* **65**, 757−760.
Jaques, R.P and Fox, C.J.S. (1960). *J. Insect Path.* **2**, 17−23.
Jaramillo Celis, R., Jiménez Lacharme, F. and Hidalgo-Salvatierra, O. (1974). *Turrialba* **24**, 106−107.

Jefferson, R.N., Hall, I.M. and Morishita, F.S. (1964). *J. econ. Ent.* **57**, 150–152.

Jensen, F.L. (1969). *Calif. Agric.* **23** (4), 5–6.

Johansson, K. (1971). *Meddn. St. Växtsk Anst. (Stockholm)* **15** (139), 111–138.

Kailidis, D.-S. *et. al.* (1971). *Dasika Chron. (Athenai)* **13** (152), 7–12.

Kakaliev, K. and Saparliev, K. (1975). *Isv. Akad. Nauk. turkmen. SSR – Ser. Biol. Nauk.* **6**, 39–41.

Kalra, A.N. and Kumar, S. (1963). *Indian J. Sug-Cane Res.* **8**, 75.

Karasev, V.S. (1968). *Zashch. Rast. (Moskva)* **13**, 44.

Kaya, H.K. (1974). *J. econ. Ent.* **67**, 390–392.

Kaya, H.K. and Dunbar, D.M. (1972). *J. econ. Ent.* **65**, 1132–1134.

Kearby, W.H., Hostetter, D.L. and Ignoffo, C.M. (1972). *J. econ. Ent.* **65**, 477–480.

Kellen, W.R. and Lewallen, L.L. (1960). *J. Insect Path.* **2**, 305–309.

Kinsinger, R.A. and McGaughey, W.H. (1979). *J. econ. Ent.* **72**, 346–349.

Kneitz, G. (1966). *Waldhygiene* **6**, 183–187.

König, E. (1975). *Z. Angew. Ent.* **77**, 424–429.

Königstedt, D. and Groth, U. (1972). *Wiss. Z. Univ. Greifswald Mathematisch.-Naturwissenschaftlische. Reihe* **21**, 267–274.

Kondrja, V.S. (1966a). *Trudy moldav. nauchn.-issled. Inst. Sadov., Vinogr. Vinod.* **13**, 101–115.

Kondrja, V.S. (1966b). *Trudy moldav. nauchn.-issled. Inst. Sadov., Vinogr., Vinod.* **13**, 123–130.

Krieg, A. (1956). *Entomophaga* **1**, 98.

Krieg, A. (1957a). *Z. Pfl Krankh. Pfl Schultz.* **64**, 321–327.

Krieg, A. (1957b). ref. by Steinhaus, E.A. (1957). Mimeogr. Ser. No. 4, Lab., Insect Path. Univ. Calif. Berkeley, 24 pp.

Krieg, A. (1961). Mittlg. Biol. Bundesanstalt f. Land-und Forstwirtschaft, Berlin 1961, No. 103, 79 pp.

Krieg, A. (1964). *Anz. Schädlingsk.* **3**, 39–40.

Krieg, A. (1967a). *Anz. Schädlingsk.* **40**, 8–9.

Krieg, A. (1967b). Mittlg. Biol. Bundesanstalt. f. Land- und Forstwirtschaft, Berlin 1967, No. 125, 106 pp.

Krieg, A. (1968). *J. Invertebr. Path.* **12**, 478.

Krieg, A. (1970). *Nachrbl. dt. Pfl Schutzdienst. (Braunschweig)* **22**, 97–103.

Krieg, A. (1971). *J. Invertebr. Path.* **17**, 134–135.

Krieg, A. (1972). *Anz. Schädlingsk.* **45**, 170–171.

Krieg, A. (1973). *Z. Pfl Krankh. Pfl Schutz* **80**, 483–486.

Krieg, A. and Herfs, W. (1963). *Z. Pfl Krankh. Pfl Schutz.* **70**, 11–21.

Krieg, A. and Schmidt, L. (1962). *Nachr Bl. dt. Pfl Schutzdienst. (Braunschweig)* **14**, 177–182.

Kuchly (1959) *In* Martouret and Milaire (1963).

Kudler, J. and Lysenko, O. (1963). *Lesn. Čas. (Bratislava)* **9**, 787–798.

Kudler J., Lysenko, O. and Hochmut, R. (1959). Trans. 1st Int. Conf. Insect. Path. Biol. Control, Praha 1958, 73–79.

Kudler, J., Lysenko, O. and Hochmut R. (1965). *Commun. Inst. For. Čechosl.* 219–226.

Kumar, S. and Jayaraj, S. (1978). *Indian J. Expl. Biol.* **16**, 128–131.

Kurstak, E. (1964). *C. r. hebd. Séanc. Acad. Sci. Paris* **259**, 211–212.

Kwangsi Kweishien (1974). *Acta Ent. Sin.* **17**, 129–134.

Laan, v.d., P.A. and Wassink, H.J.M. (1962). *Entomophaga Mém. hors Sér.* No. 2, 315–322.

Lacey, L.A. and Mulla, M.S. (1977). *J. Invertebr. Path.* 30, 46–49.

Lacey, L.A., Mulla, M.S., Dulmage, H.T. (1978). *Environ. Ent.* 7, 583–588.

Lam, A.B.Q. and Webster, J.M. (1972). *J. Invertebr. Path* 20, 141–149.

Langenbruch, G.A. (1976). *Nachrbl. dt. Pfl Schutzdienst. (Braunschweig)* 28, 148–155.

Langenbruch, G.A. (1977). *Nachr Bl. dt. Pfl Schutzdienst. (Braunschweig)* 29, 133–137.

Langenbruch, G.A. and Krieg, A. (1976). Jahresbericht 1976, Biol. Bundesanstalt f. Land- und Forstwirtschaft (Berlin, Braunschweig), p. 81.

Lappa, N.V. (1964). *Zakhyst Roslyn (Kiev)* 1, 65–72.

Larson, L.V. and Ignoffo, C.M. (1971). *J. econ. Ent.* 64, 1567–1568.

Laurent, J.E. (1965). *Bull. Éc. nat. sup. agron. (Nancy)* 7, 79–91.

Lavrentyev, P.A., Salnikov, V.G. and Anisin, S.D. (1965). *Veterinariya (Moskva)* 42, 107–108.

Lemoigne, M., Bonnefoi, A., Béguin, S., Grison, P., Martouret, D., Schenk, A. and Vago, C. (1956). *Entomophaga* 1, 19–34.

Leskova, A.J. (1960). *Zashch. Rast. (Moskva)* (5), 31–32.

Lipa, J.J. (1976). *Bull. Acad. Polon. Sci., Sci. Biol.* 24, 505–508.

Lipa, J.J., Bakowski, G. and Rychlewka, M. (1977). *Pr. Nauk. Inst. Ochr. Rosl. Poznan,* 19 (1), 183–190.

Madsen, H.F. and Potter, S.A. (1977). *Can. Ent.* 109, 171–174.

Malhotra, C.P. and Choudhary, S.G. (1968). *J. Invertebr. Path.* 11, 429–439.

Marchal-Segault, D. (1975). *Ann. Parasit. (Paris)* 50, 223–232.

Martouret, D. (1959). *Entomophaga* 4, 211–220.

Martouret, D. and Auer, C. (1977). *Entomophaga* 22, 37–44.

Martouret, D. and Milaire, H. (1963). *Phytiat. Phytopharm.* 12, 71–80.

Matter, M.M. (1969). M.Sc. Thesis Dept. Ent. Fac. Sci. Univ. Cairo, 135 pp.

McConnell, E. and Cutkomp, L.K. (1954). *J. econ. Ent.* 47, 1074–1082.

McConnell, E. and Richards, A.G. (1959). *Can. J. Microbiol.* 5, 161–168.

McEwen, F.L., Glass, E.H., Davis, A.C. and Splittstoesser, C.M. (1960). *J. Insect Path.* 2, 152–164.

McGaughey, W.H. (1976). *Can. Ent.* 108, 105–112.

McGaughey, W.H. (1978). *J. econ. Ent.* 71, 835–839.

McGaughey, W.H. (1979). *Newslett. Int. cooperative Program Bacillus thuringiensis* (H.D. Burges ed.) (Mimeogr. in Great Britain) No. 2, p. 4.

McGaughey, W.H., Kinsinger, R.A. and Dicker, E.B. (1975). *Environ. Ent.* 4, 1007–1010.

McNeil, J.N., Smirnoff, W.A. and Letendre, M. (1977). *Can. Ent.* 109, 37–38.

Mechalas, B.J. and Beyer, O. (1963). *Dev. ind. Microbiol.* 4, 141–147.

Merdan, A., Abdel-Rahman, H. and Soliman, A. (1975). *Z. Angew. Ent.* 78, 280–285.

Metalnikov, S. (1937). *C. r. hebd. Séanc. Soc. Biol., Paris* 125, 1020–1023.

Metalnikov, S. and Chorine, V. (1929). *Int. Corn Borer Invest. Sci. Rep.* 2, 60–61.

Middlekauff, W.W., Gonzales, C.Q. and King, R.C. (1963). *J. econ. Ent.* 56, 155–158.

Mihalache, G., Arsenescu, M. and Pirvescu, D. (1972). *Revtă Padŭr.* 87, 362–365.

Millars, I., Hastings, A.R., Ryan, R.O. and Larson, L.V. (1974). U.S. Forest Service, North Central Region, State and Private Forestry Evaluation Rep. S-2-74, 8 pp.

Miller, R.W., Pickens, L.G. and Gordon, C.H. (1971). *J. econ. Ent.* **64**, 902–903.

Mistric, W.J. and Smith, F.D. (1973). *J. econ. Ent.* **66**, 979–982.

Moore, I., Halperin, J. and Navon, A. (1962). *Israel J. agric. Res.* **12**, 167–174.

Morel, G. (1974). *Entomophaga* **19**, 85–95.

Morris, O.N. (1962). *Can. Ent.* **94**, 686–690.

Morris, O.N. (1963). *J. Insect Path.* **5**, 361–367.

Morris, O.N. (1969). *J. Invertebr. Path.* **13**, 285–295.

Morris, O.N. (1972a). *Can. Ent.* **104**, 1419–1425.

Morris, O.N. (1972b). Chem. Control Res. Inst., Ottawa, Ont. Inf. Rep. CC-X-36.

Morris, O.N. (1973). *J. Invertebr. Path.* **22**, 108–114.

Morris, O.N. (1977). *Can. Ent.* **109**, 1239–1248.

Müller, O. (1957). *Z. Angew. Ent.* **41**, 71–111.

Narayanan, K., Jayaraj, S. and Govindarajan, R. (1976). *J. Invertebr. Path.* **28**, 269–270.

Nayak, P., Rao, P.S. and Padmanabhan, S.Y. (1978). *Proc. Indian Acad. Sci.* **87B**, 59–62.

Niemczyk, E. (1971). *Roczn. Nauk Roln.* ser. E, **1**, 7–16.

Niemczyk, E. (1975). *Ochrona Róslin* **(2)**, 17–19.

Niemczyk, E. and Bakowski, G. (1971). *Roczn. Nauk Roln.,* Ser. E **1**, 103–117.

Niemczyk, E., Olszak, R. and Miszczak, M. (1976). *Roczn. Nauk Roln.* Ser. E, **6**, 151–157.

Nordin, G.L. and Appleby, J.E. (1969). *J. econ. Ent.* **62**, 23–24.

Oatman, E.R. (1965). *J. econ. Ent.* **58**, 1144–1147.

Okhotnikov, V.I. and Shpil'chak, M.B. (1978). *Zashch, Rast.* No. 1, 35.

Ohba, M., Aizawa, K. (1979). *J. Invertebr. Path.* **33**, 387–388.

Onishchenko, L. (1966). *Zashch. Rast. (Moskva)* **11**, 25.

Panait, N. and Ciortan, G. (1974). *Analele Institutului de Cercetări pentru Protectia Plantelor* **10**, 265–272.

Patti, J.H. and Carner, G.R. (1974). *J. econ. Ent.* **67**, 415–418.

Pendleton, I.R. (1970). *J. Invertebr. Path.* **15**, 287.

Pigatti, A. and Fugueiredo, M.B. and Orlando, A. (1960). *Biológico (S. Paulo)* **26**, 47–51.

Pinnock, D.E. and Milstead, J.E. (1971). *J. econ. Ent.* **64**, 510–513.

Pinnock, D.E. and Milstead, J.E. (1972). *J. econ. Ent.* **65**, 1747–1749.

Pinnock, D.E. and Milstead, J.E. (1978). *Entomophaga* **23**, 203–206.

Pinnock, D.E. Milstead, J.E., Coe, N.F. and Stegmiller, F. (1973). *J. econ. Ent.* **66**, 194–197.

Pinnock, D.E., Milstead, J.E., Coe, N.F. and Brand, R.J. (1974). *Entomophaga* **19**, 221–227.

Pirvescu, D. (1973). *Revtă Padŭr.* **88**, 557–561.

Poe, S.L. and Everett, P.H. (1974). *J. econ. Ent.* **67**, 671–674.

Pointing, J.P. (1962). *J. Insect Path.* **4**, 484–497.

Polles, S.G. (1974). *J. Georgia Ent. Soc.* **9**, 182–186.

Prasad, S.S.V.S., Tilak, K.V.B.R. and Gollakota, K.G. (1972). *J. Invertebr. Path.* **20**, 377–378.

Prokop'ev, V.N., Yakunin, B.M., Chervyakov, V.D. and Dubitskii, A. (1976). *Parazitologiya* **10**, 222–226.

Rabb, R.L., Steinhaus, E.A. and Guthrie, F.E. (1957). *J. econ. Ent.* **50**, 259–262.

Rafal'skiĭ, A.K. (1974). *Zashch. Rast.* No. **8**, 22.

Rassoulof *et al.* (1966), *In* Sikoura, (1975).

Rautapää, J. (1967). *Ann. agric. Fenn.*, 6, Ser. Anim. Nocentia 25, 103–105.

Reeves, E.L. and Garcia, C. (1971). *Proc. Calif. Mosquito Control Ass.* **39**, 118–120.

Reinert, J.A. (1974). *Fl. Ent.* **57**, 275–279.

Revelo, M.A. (1973). *Rev. Inst. Colomb. Agropecu* 8, 429–490.

Rishbeth, J. (1978). "The Royal Society Delegation on Biological Control to China". The Royal Society Ch/3 (77) London 38 pp.

Roehrich, R. (1964). *J. Insect Path.* **6**, 186–197.

Roehrich, R. (1970). *Rev. Zool. Agric. Path. Vég.* **69**, 74–78.

Rogoff, M.H., Ignoffo, C.M., Singer, S., Gard, I. and Prieto, A.P. (1969). *J. Invertebr. Path.* **14**, 122–129.

Roome, R.E. (1975). *Bull. ent. Res.* **65**, 507–514.

Rossmoore, H.W., Elder, L. and Hoffman, E.A. (1970). *J. Invertebr. Path.* **16**, 102–106.

Rybina, L.M. (1966). In "Biol. metod. bor'b s vredit. sel'sk. lesn. choz. i Karant. sorn." (Rish, M.A., ed.), Izd. Fan, Tashkent, 152–155.

Sander, H. and Cichy, D. (1967). *Ekol. pol. Ser. A.* **15**, 325–333.

Schmid, A. and Antonin, Ph. (1977). *Rev. Suisse Viticult., Arbovicult., Hort.* **9**, 119–126.

Schubert, G. and Stengel, M. (1974). *Rev. Zool. Agric. Path Vég.* **73**, 47–52.

Shaikh, M.U. and Morrison, F.O. (1965). *Ann. ent. Soc. Quebec* **10** (1), 11–12.

Shaikh, M.U. and Morrison, F.O. (1966a). *J. Invertebr. Path.* **8**, 347–350.

Shaikh, M.U. and Morrison, F.O. (1966b). *Ann. Soc. Ent. Québec.* **11**, 120–122.

Sharpe, E.S. (1976). *Proc. 1st Int. Colloq. Invert. Path.* Kingston, Ontario, 418–419.

Shchepetilnikova, V.A., Fedorinchik, N.S. Kolmakova, V.D. and Kapustina, O.V. (1968a). *Trudy vses. nauchno. issl. Inst. Zashch. Rast. (Leningrad)* **31**, 21–62.

Shchepetilnikova, V.A., Kapustina, O.V., Molchanova, V.A. and Shichenkov, P.I. (1968b). *Trudy vses. nauchno.-issl. Inst. Zashch. Rast. (Leningrad)* **31**, 86–98.

Shekar, P.S. and Gopinath, K. (1962). *J. Insect Path.* **4**, 381–391.

Sherman, M., Ross, E. and Komatsu, G.H. (1962). *J. econ. Ent.* **55**, 990–993.

Shorey, H.H. and Hall, I.M. (1963). *J. econ. Ent.* **56**, 813–817.

Shevcova, O.I. (1959). *Zashch. Rast. (Moskva)* **4**, 38.

Sicker, W. and Krieg, A. (1966). *Nachr Bl. dt. Pfl Schutzdienst. (Braunschweig)* **18**, 103–105.

Sikorowski, P. and David, F.M. (1970). *J. Invertebr. Path.* **15**, 131–132.

Sikoura and Tkatsch (1974), *In* Sikoura, A.J. (1975). *Rev. Zool. Agric. Path. Vég.* **74**, 54–60.

Sinègre, G., Gaven, B., Jullien, J.L. (1979a). Unpubl. docum. WHO/VBC 79.742; Geneva, 6 pp.

Sinègre, G., Gaven, B., Jullien, J.L., Crespo, O. (1979b). Unpubl. docum. WHO/VBC 79.743, Geneva, 7 pp.

Skatulla, U. (1971). *Z. Angew. Ent.* **69**, 1–30.

Smirnoff, W.A. (1965). *J. Invertebr. Path.* **7**, 266–269.

Smirnoff, W.A. (1972). *Can. Ent.* **104**, 1153–1159.

Smirnoff, W.A. (1977). *Can. Ent.* **109**, 351–358.
Smirnoff, W.A. and Berlinguet, L. (1966). *J. Invertebr. Path.* **8**, 376–381.
Smirnoff, W.A. and Heimpel, A.M. (1961). *J. Insect Path.* **3**, 403–408.
Smirnoff, W.A., Fettes, J.J. and Desaulniers, R. (1973). Centre de Recherches Forestières des Laurentides Ste-Foy, Quebec, Inf. Rep. Q-X-31.
Smythe, R.V. and Coppel, H.C. (1965). *J. Invertebr. Path.* **7**, 423–426.
Snapp, O.I. (1962). *J. econ. Ent.* **55**, 418–419.
Sorensen, A.J. and Barbosa, P. (1975). *J. econ. Ent.* **68**, 561–562.
Srivastava, R.P. and Nayak, P. (1978). *Z. PflKrankh., PflSchutz.* **85**, 641–644.
Steiner, H. (1960). In Krieg, A., (1961).
Steinhaus, E.A. (1951a). *Hilgardia* **20**, 359–381.
Steinhaus, E.A. (1951b). Pacif. Sci. Board., Mimeogr., 16 pp.
Steinhaus, E.A. (1957). Lab. Insect Path. Dep. Biol. Contr. Univ. Calif., Berkeley. Mimeogr. Ser. 4, 1957, 24 pp.
Steinhaus, E.A. and Bell, C.R. (1953). *J. econ. Ent.* **46**, 582–598.
Steinhaus, E.A. and Jerrel, E.A. (1954). *Hilgardia* **23**, 1–23.
Stelzer, M.J. (1965). *J. Invertebr. Path.* **7**, 122–125.
Stelzer, M.J. Neisess, J. and Thompson, C.G. (1975). *J. econ. Ent.* **68**, 269–272.
Stern, V.M. (1961). *J. econ. Ent.* **54**, 50–55.
Struble, G.R. (1965). *J. econ. Ent.* **58**, 1005–1006.
Stute, K. (1963). *Nachr bl. dt. Pfl. schutzdienist. (Braunschweig)* **15**, 102–104.
Sutter, G.R. (1969). *J. econ. Ent.* **62**, 756–757.
Sutter, G.R. and Raun, E.S. (1966). *J. Invertebr. Path.* **8**, 457–460.
Svestka, M. (1974). *Lesnictvi* **20**, 439–464.
Szmidt, A. and Ślizyński, K. (1965). *Roczn. wyz. Szk. roln. Poznan* **27**, 251–259.
Takaki, S. (1975). *Jap. pestic. Inf.* **25**, 23–26.
Talalaev, E.V. (1957). *Ent. Obozr.* **36**, 845–859.
Talalaev, E.V. (1959). Trans. 1st. Int. Conf. Insect Path., Biol. Contr. Praha 1958, 51–57.
Tamashiro, M. (1968). Proc. Joint U.S.–Japan Seminar Microbiol. Control Insect Pests., U.S.–Japan Committee Sci Cooperation. Panel 8, Fukuoka 1967, 147–153.
Tanada, Y. (1953). *Proc. Hawaii ent. Soc.* **15**, 159–166.
Tanada, Y. (1956). *J. econ. Ent.* **49**, 320–329.
Tanada, Y. and Reiner, C. (1960). *J. Insect Path.* **2**, 230–246.
Taylor, T.A. (1968). *J. Invertebr. Path.* **11**, 386–389.
Taylor, T.A. (1974). *J. econ. Ent.* **67**, 690–691.
Thompson, J.V. and Fletcher, L.W. (1972). *J. Invertebr. Path.* **20**, 341–350.
Toumanoff, C. (1955). *Ann. Inst. Pasteur, Paris,* **88**, 384.
Toumanoff, C. and Grison, P. (1954). *C.r. Acad. Agric. France* **40**, 277–280.
Touzeau, J. (1971). *Phytoma* **23**, 21–27.
Tremblay, F.L.J., Huot, L. and Perron, J.M' (1972). *Entomologia exp. appl.* **15**, 397–398.
Vail, P.V., Soohoo, C.F., Seay, R.S., Killinen, R.G. and Wolf, W.W. (1972). *Environ. Ent.* **1**, 780–785.
Vaňková, J. (1964). *Entomophaga, Mém. hors Sér.* No. 2, 271–291.
Vaňková, J. (1966). *Acta Ent. Bohemoslov.* **63**, 10–16.
Vaňková, J. (1973). *Acta Ent. Bohemoslov.* **70**, 328–333.
Vaňková, J. and Horská, K. (1975). *Acta Ent. Bohemoslov.* **72**, 7–12.

896   A. KRIEG AND G.A. LANGENBRUCH

Vaňková, J. and Weiser, J. (1978). Abstr., Int. Colloq. Invertebr. Path. Praha 1978. p. 118.
Vaňková, J., Vobrázková, E. and Samšiňák, K. (1975). *J. Invertebr. Path.* **26**, 159–163.
Videnova, E. (1970). *Rastit. Zasht. (Sofiya)* **18**, 11–14.
Videnova, E., Tsankow, G. and Chernev, T. (1972). *Gorskostop. Nauka* **9**, 59–65.
Viggiani, G. and Tranfaglia, A. (1975). *Boll., Lab. Ent. Agric. "Filippo silvestri", Portici,* **32**, 140–144.
Viggiani, G. and Tranfaglia, A., (1979). Cited from: Franz *et al.,* in prep.
Voskresenskaya, V.N. (1977). *Zashch. Rast.* No. 4, 54.
Wallner, W.E. (1971). *J. econ. Ent.* **64**, 1487–1490.
Wasti, S.S., Mahadeo, C.R. and Knell, J.D. (1973). *Z. Angew. Ent.* **74**, 157–160.
Weiser, J. (1960). *New Sci., Lond.* **7**, 721–722.
Weiser, J. (1962). *Zesz. probl. Postep. Nauk roln. (Warszawa),* **35**, 79–82.
Weiser, J. and Vaňková, J. (1978). Abstr. 2nd Int. Colloq. Invertebr. Path. Praha, 1978.
White, C.A. (1960). *Bioferm Rep., Wasco (Calif.)* **20**, July 1960.
White, C.A. and Briggs, J.D. (1964). *Entomophaga, Mém. hors Sér.* No. 2, 305–308.
Wiackowski, S.K. and Wiackowska, I. (1966). *Entomophaga* **11**, 261–267.
Wiegand, H. (1963). *Entomophaga* **8**, 35–41.
Wildbolz, T. and Staub, A. (1962). *Schweiz. Z. Obst-u. Weinb.* **71**, 235–240.
Wilson, W.T. (1962). *J. Insect. Path.* **4**, 269–270.
Wilson, B.H. and Burns, E.C. (1968). *J. econ. Ent.* **61**, 1747–1748.
Wisniewski, J. (1975). *Angew. Parasit.* **16**, 43–49.
Wolliam, J.D. and Yendol, W.G. (1976). *J. econ. Ent.* **69**, 113–118.
Wong, T.T.Y., Beavers, J.B., Sutton, R.A. and Norman, P.A. (1975). *J. econ. Ent.* **68**, 119–121.
Wysoki, M., Izhar, Y., Gurevitz, E., Swirski, E. and Greenberg, S. (1975). *Phytoparasitica* **3**, 103–111.
Yamvrias, C. (1962). *Entomophaga* **7**, 101–159.
Yamvrias, C. (1972). *Ann. Inst. Phytopath. Benaki, N.S.* **10**, 256–266.
Yamvrias, C. and Angus, T.A. (1969). *J. Invertebr. Path.* **14**, 423–424.
Yamvrias, C. and Angus, T.A. (1970). *J. Invertebr. Path.* **15**, 92–99.
Yearian, W.C., Livingston, J.M. and Young, S.Y. (1973). Rep. Ser. Arkansas Univ. Agric. Exp. Stn. (Fayetteville) 212, 1–8.
Yendol, W.G., Hamlen, R.A. and Lewis, F.B. (1973). *J. econ. Ent.* **66**, 183–186.
Yousten, A.A. (1973). *J. Invertebr. Path.* **21**, 312–314.
Yu-Chen Lee (1966). *Pl. Prot. Bull. (Nantou, Taiwan)* **8**, 48–53.
Zivanović, V. and Stamenković, S. (1976). *Zast. Bilja (Beograd)* **27** (137–138) 381–387.
Zukauskiene, J. (1973). *Acta ent. Lituanica* **2**, 137–151.

APPENDIX 2

A Catalogue of Viral Diseases of Insects, Mites and Ticks

M.E. MARTIGNONI and P.J. IWAI

Forestry Sciences Laboratory, Pacific Northwest Forest and Range Experiment Station, Forest Service, U. S. Department of Agriculture, Corvallis, Oregon, USA

I. Introduction

This comprehensive Catalogue of insects, mites and ticks reported to have viral diseases is generated from a computer-assisted information system on viral diseases established at the Forestry Sciences Laboratory in 1970 (Martignoni *et al.* 1973). The present Catalogue results from our analysis of the information in 3400 publications (as of this writing). Of these, 733 formed the basis for lists published by Hughes (1957) and by Martignoni and Langston (1960). It should be emphasized that those two lists, as well as the present Catalogue, are not the result of simple title scans, but rather they were generated from a thorough analysis of each article entered in our master file. The methods of input analysis and preparation have been described by Martignoni *et al.* (1973). The techniques of information storage and retrieval are those of the eight FAMULUS program subsystems described by Burton *et al.* (1969). The data base for this Catalogue is preserved in a master file (on magnetic tape and on paper) at our Station. We emphasize that the data base consists only of published host records. Unpublished host records (material stored in virus collections, internal laboratory reports and personal correspondence) do not appear in the Catalogue. This printing of the Catalogue lists 826 host species, each reported to have one or more of 22 viral diseases or disease groups, for a total of 1,271 host—virus records.

Computer routines for the preparation of the present Catalogue were developed in cooperation with the Biometrics Service of our Experiment Station. They sort and list species of arthropods, along with their viral diseases, in four separate printouts:

(1) Species listed alphabetically, by specific names; if one species has one or more subspecies, these are listed, too, in alphabetical sequence. This list serves as a general host index by specific names and disregards arrangement of species by higher taxonomic categories.

(2) Genera listed in alphabetical sequence; species and subspecies listed alphabetically within each genus.

(3) Families listed in alphabetical sequence; genera listed alphabetically within each family, as well as species and subspecies within each genus.

(4) Orders listed in alphabetical sequence; families and each subsequent lower-rank taxon listed alphabetically.

Printout 4 is reproduced here.

Updating these lists requires a minimal effort only, once appropriate information has been retrieved from specialized literature and entered into the FAMULUS master file. New hosts, new disease records for previously listed hosts, and new synonyms are added to our data base several times yearly. Specialists desiring updated records of one or more genera or families of hosts can obtain copies of the pertinent sections of our latest printout on request from the senior author.

Our Catalogue uses the currently accepted scientific names of host species. Recent monographs and several specialists were consulted to determine current and correct specific, generic, and family designations. A search for a particular host would not be considered complete unless synonyms of the generic and specific names were also included in that search. Unfortunately, it would be a Sisyphean task, beyond our scope, to list all synonyms of each host species. We list, however, some of the most common synonyms (those most often found in the literature on viral diseases) along with the accepted scientific names of arthropods concerned.

In most cases, viral diseases have been reported as naturally occurring in their hosts ("natural hosts" or "typical hosts"). In a few cases, however, reports indicate that a disease resulted from inoculation with a virus originally isolated from another host. Thus, some records represent "accidental hosts," i.e. although susceptible, hosts in which the virus is not commonly found. These few instances have not been marked in the present lists; thus it is not possible to identify accidental hosts. Information on host specificity (or host range) of the pathogens is contained in the master file.

A rather disturbing situation, arisen in recent years, forced us to lump several records in the ill-defined category "presumed virosis" (code 17). In our opinion, far too many papers are rushed into print before obtaining sufficient evidence on the viral nature and pathogenicity of "virus-like" particles seen in ultrathin sections. Sometimes, the viral nature of these particles is only conjecture. Faced with the dilemma of ignoring or listing such records, we decided on the latter, hoping that, eventually, the viral nature of the particles would be confirmed in further studies. In the meantime, the reader should consider each code 17 entry with moderate scepticism.

II. A Plea for Assistance

We are grateful to those authors who, for many years, have given us reprints of their publications on viral diseases of arthropods. We trust that their important contribution to our Catalogue will continue. We ask those interested in its maintenance and who may have noticed errors and missing host records to write to us: we would appreciate a reprint of each pertinent publication.

IV. List of Host Insects, Mites and Ticks by Taxonomic Categories

DISEASE NAME	CODE	DISEASE NAME	CODE
Acute paralysis	1	Malaya disease	11
Chronic paralysis	2	Nucleopolyhedrosis	12
CO_2 sensitivity	3	Other nonoccluded-virus disease	13
Crystalline-array virosis	4		
Cytoplasmic polyhedrosis	5	Other occluded -virus disease	14
Densonucleosis	6		
Filamentous-virus disease (see code 22)		Paralysis	15
		Polyhedrosis	16
Flacherie	7	Presumed virosis	17
Gattine	8	Sacbrood	18
Granulosis	9	Spheroidosis	19
Hairless-black syndrome (see code 21)		Watery disintegration	20
		Hairless-black syndrome	21
Iridescent virosis	10	Filamentous-virus disease	22

IV. Alphabetical List of Host Insects, Mites and Ticks by Taxonomic Categories

Acari (Order)
ARGASIDAE
Argas persicus, 17
Ornithodoros lahorensis, 17
O. moubata, 17
O. tartakovskyi, 17
O. tholozani, 17
O. verrucosus, 17

IXODIDAE
Boophilus microplus, 17
Dermacentor marginatus, 17

PHYTOSEIIDAE
Phytoseiulus persimilis, 17

TETRANYCHIDAE
Panonychus citri, 13, 17
P. ulmi, 13
Tetranychus cinnabarinus, 13, 17
T. multisetis, 17
T. telarius see *T. urticae*
T. urticae, 17

Coleoptera (Order)
BUPRESTIDAE
Agrilus suvorovi populneus, 5
Melanophila picta, 5, 16
Trachys auricollis, 10

CERAMBYCIDAE
Batocera lineolata, 12
Stenodryas clavigera, 10
Stenygrinum quadrinotatum see *Stenodryas clavigera*

CHRYSOMELIDAE
Cerotoma trifurcata, 17
Chrysomela vigintipunctata, 10
Microdera vigintipunctata see *Chrysomela vigintipunctata*

COCCINELLIDAE
Coccinella septempunctata bruckii, 10
Epilachna varivestis, 17

CURCULIONIDAE
Anthonomus grandis, 10, 12
Curculio dentipes, 10

DERMESTIDAE
Anthrenus museorum, 12, 16
Dermestes lardarius, 12, 16

GYRINIDAE
Gyrinus natator, 13

LUCANIDAE
Figulus sublaevis, 19
Macrodorcus rectus, 10
M. rubrofemoratus, 10

SCARABAEIDAE
Allomyrina dichotomous, 10
Amphimallon solstitialis, 19
Anomala cuprea, 19
Anoplognathus porosus, 19
Anoxia villosa, 19
Antitrogus morbillosus, 19
Aphodius tasmaniae, 13, 19
Costelytra zealandica, 10
Dasygnathus sp., 19
Demodena boranensis, 19
Dermolepida albohirtum, 19
Geotrupes silvaticus, 19
G. stercorosus, 19
Heteronychus arator, 10, 13
Hoplia sp. 19
Melolontha hippocastani, 20
M. melolontha, 6, 19, 20
Odontria sp., 10
Opogonia sp. 10
Oryctes boas, 11, 20
O. monoceros, 11, 20
O. nasicornis, 11, 20
O. rhinoceros, 11
Othnonius batesi, 19
Pericoptus truncatus, 13
Phyllopertha horticola, 19
Phyllophaga pleei, 19
Rhopaea morbillosa see *Antitrogus morbillosus*
R. verrauxi, 19
Scapanes australis grossepunctatus, 11
Sericesthis pruinosa, 10
Xylotrupes dichotomus, see *Allomyrina dichotomus*

SCOLYTIDAE
Scolytus scolytus, 17

TENEBRIONIDAE
Tenebrio molitor, 10, 17

Diptera (Order)
BIBIONIDAE
Bibio marci, 10

CALLIPHORIDAE
Calliphora sp., 13
C. vomitoria, 10, 12, 16
Phormia sp., 13

CECIDOMYIIDAE
Contarinia tritici, 16
Sitodiplosis mosellana, 16

CERATOPOGONIDAE
Culicoides arboricola, 10
C. cavaticus, 13
C. sp. 10

CHAOBORIDAE
Corethrella appendiculata, 10
C. brakeleyi, 10
Mochlonyx culiciformis see *M. velutinus*
M. velutinus, 10

CHIRONOMIDAE
Camptochironomus tentans, see *Chironomus tentans*
Chironomus attenuatus, 19
C. decorus, 19
C. luridus, 19
C. plumosus, 5, 10
C. tentans, 12, 17, 19
Goeldichironomus holoprasinus, 5, 17, 19

COELOPIDAE
Chaetocoelopa sydneyensis, 13
Coelopa frigida, 17

CULICIDAE
Aedes aegypti, 5, 6, 10, 12, 17, 19
A. albopictus, 10, 11, 13
A. annulipes, 10
A. cantans, 10
A. caspius, 10
A. caspius dorsalis, see *A. dorsalis*
A. cataphylla, 10
A. cinereus, 6
A. detritus, 10
A. dorsalis, 6, 10
A. excrucians, 10

Aedes flavescens, 10
A. fulvus pallens, 10
A. nigromaculis, 12
A. sierrensis, 5, 10
A. sollicitans, 5, 10, 12
A. sticticus, 5, 10
A. stimulans, 10
A. taeniorhynchus, 5, 10, 12, 13
A. thibaulti, 5
A. tormentor, 12
A. triseriatus, 5, 12, 17
A. vexans, 6, 10
Anopheles albimanus, 10, 19
A. bradleyi, 5
A. crucians, 5, 12
A. freeborni, 5
A. quadrimaculatus, 5, 10, 14
A. stephensi, 3, 5, 13
A. subpictus, 17
Culex erraticus, 5
C. peccator, 5, 10
C. pipiens, 3, 5, 6
C. pipiens fatigans see *C. quinquefasciatus*
C. pipiens pipiens see *C. pipiens*
C. pipiens quinquefasciatus see *C. quinquefasciatus*
C. quinquefasciatus, 12
C. restuans, 5, 17
C. salinarius, 5, 10, 12, 17
C. tarsalis, 5, 13, 17
C. territans, 5, 10, 17
Culiseta annulata, 10
C. inornata, 5, 10, 13
C. melanura, 5, 10
C. morsitans, 10
Orthopodomyia signifera, 5, 17
Psorophora confinnis, 5, 10, 12, 17
P. ferox, 5, 10, 12
P. horrida, 10
P. varipes, 10, 12
Uranotaenia sapphirina, 5, 12, 17
Wyeomyia smithii, 12

DROSOPHILIDAE
Drosophila affinis, 3
D. ananassae, 13
D. bifasciata, 17
D. erecta, 13
D. fasciata see *D. melanogaster*
D. hydei, 13

D. immigrans, 3, 13
D. malerkotliana, 13
D. mauritiana, 13
D. melanogaster, 3, 13, 15, 17
D. montium, 13
D. nasuta, 13
D. nebulosa, 13
D. paulistorum, 13, 17
D. pseudoobscura, 17
D. simulans, 13
D. teissieri, 13
D. virilis, 13, 17
D. willistoni, 13, 17
D. yakuba, 13
Zaprionus tuberculatus, 13

MUSCIDAE
Glossina fuscipes fuscipes, 17
G. morsitans centralis, 17
G. morsitans morsitans, 17
G. pallidipes, 17
Musca domestica, 3, 13

SCIARIDAE
Rhynchosciara angelae, 12, 17
R. hollaenderi, 12
R. milleri, 12

SIMULIIDAE
Cnephia mutata, 5
Prosimulium mixtum, 5
P. mixtum fuscum, 5
Simulium ornatum, 10
S. sp., 10
S. tuberosum, 5
S. venustum, 5
S. vittatum, 5, 6

TACHINIDAE
Exorista sorbillans, 10
Ugymyia sericariae, 12

TEPHRITIDAE
Ceratitis capitata, 3, 13
Dacus tryoni, 14

TIPULIDAE
Tipula livida, 10
T. oleracea, 10
T. paludosa, 10, 12, 16

Hemiptera (Order)
APHIDIDAE
Aphis sp. 17

Myzus persicae, 17
Pentalonia nigronervosa, 17
Rhopalosiphum maidis, 17

BELOSTOMATIDAE
Lethocerus columbiae, 10

CICADELLIDAE
Colladonus montanus, 10
Nephotettix cincticeps, 10

DELPHACIDAE
Laodelphax striatella, 10
Tarophagus proserpina, 17

REDUVIIDAE
Panstrongylus megistus, 17

Hymenoptera (Order)
APIDAE
Apis cerana, 10, 13
A. mellifera, 1, 2, 6, 10, 13, 15, 17,
　18, 21, 22
Bombus agrorum, 1
B. hortorum, 1
B. lucorum, 1
B. ruderarius, 1
B. terrestris, 1

ARGIDAE
Arge pectoralis, 12

BRACONIDAE
Apanteles congregatus, 17
A. crassicornis, 17
A. flavipes, 17
A. fumiferanae, 17
A. glomeratus, 17
A. liparidis, 17
A. marginiventris, 17
A. melanoscelus, 17
A. ornigis, 17
A. paleacritae, 17
Cardiochiles nigriceps, 17
Chelonus texanus, 17
Microplitis croceipes, 17
Phanerotoma flavitestacea, 17

DIPRIONIDAE
Diprion hercyniae see *Gilpinia*
　hercyniae
D. nipponica, 12
D. pallida see *Gilpinia pallida*
D. pindrowi see *Gilpinia*
　pindrowi
D. pini, 12

D. polytoma see *Gilpinia*
　polytoma
D. similis, 12
Gilpinia hercyniae, 12
G. pallida, 12
G. pindrowi, 12
G. polytoma, 12
Lophyrus rufus see *Neodiprion sertifer*
Neodiprion abietis, 12, 16
N. americanum see *N. taedae*
　taedae
N. excitans, 12
N. lecontei, 12, 16
N. merkeli, 5
N. mundus, 16
N. nanulus, 16
N. nanulus contortae, 12
N. pratti banksianae, 12
N. pratti pratti, 16
N. sertifer, 12, 16, 17
N. swainei, 12
N. taedae linearis, 12
N. taedae taedae, 12, 16
N. virginiana, 12
Tenthredo sertifera see *Neodiprion*
　sertifer

FORMICIDAE
Formica lugubris, 17
Iridomyrmex itoi, 10
Solenopsis geminata, 17
S. sp., 17

HALICTIDAE
Nomia melanderi, 17

MEGACHILIDAE
Megachile rotundata, 17

PAMPHILIIDAE
Acantholyda nemoralis, 16
Cephalcia abietis, 12, 16
C. alpina see *C. lariciphila*
C. fascipennis, 9
C. issiki, 12
C. lariciphila, 16
Lyda campestris see *Cephalcia*
　abietis
L. hypotrophica see *Cephalcia*
　abietis
L. stellata see *Acantholyda*
　nemoralis
Tenthredo pratensis, see
　Acantholyda nemoralis

SIRICIDAE
Sirex juvencus, 5
S. noctilio, 5
Urocerus gigas gigas, 5
U. tardigradus, 5
Xeris spectrum, 5

TENTHREDINIDAE
Anoplonyx destructor, 5
Cladius viminalis see *Trichiocampus
viminalis*
Nematus olfaciens, 12
Pikonema dimmockii, 12
Pristiphora erichsonii, 12, 16
P. geniculata, 12
Trichiocampus irregularis, 12
T. viminalis, 12

Isoptera (Order)
RHINOTERMITIDAE
Coptotermes lacteus, 17

TERMITIDAE
Nasutitermes exitiosus, 13, 17

TERMOPSIDAE
Porotermes adamsoni, 17

Lepidoptera (Order)
AGARISTIDAE
Phalaenoides glycinae, 13

ANTHELIDAE
Anthela varia, 12
Pterolocera amplicornis, 12

ARCTIIDAE
Amsacta albistriga, 12
A. moorei, 12, 19
A. sp. 12
Apantesis virgo, 16 .
Arctia caja, 5, 12, 16
A. villica, 5, 12
Ardices glatignyi, 12
Cycnia mendica see *Diaphora mendica*
Diacrisia obliqua see *Spilosoma
obliqua*
D. purpurata see *Rhyparia
purpurata*
D. virginica see *Spilosoma
virginica*
Diaphora mendica, 5, 12
Dionychopus amasis, 9, 13
Ecpantheria icasia, 9, 12

Estigmene acrea, 5, 9, 12, 16, 19
Euplagia quadripunctaria, 5
Halisidota argentata, 12
H. caryae, 12
Hyphantria cunea, 5, 9, 12, 16
Hypocrita jacobaeae, 5, 12
Isia isabella, 5
Panaxia dominula, 5, 12
Parasemia plantaginis, 5
Pericallia ricini, 9, 12
Phragmatobia fuliginosa, 5, 9, 16
Rhyparia purpurata, 5, 12
Spilarctia flammeolus, 10
S. imparilis see *Spilarctia
lubricipeda*
S. lubricipeda, 5, 10, 12
S. subcarnea, 5, 12
Spilosoma lubricipeda see *Spilarctia
lubricipeda*
S. lutea, 5
S. menthrastri see *Spilarctia
lubricipeda*
S. obliqua, 9, 12
S. punctaria, 5, 10
S. virginica, 5, 9, 12
Tyria jacobaeae see *Hypocrita
jacobaeae*

BOMBYCIDAE
Bombyx mori, 5, 6, 7, 8, 10, 12, 13,
14, 19
Theophila mandarina, 5, 12

CARPOSINIDAE
Carposina niponensis, 9, 12

COCHYLIDAE
Clysiana ambiguella see *Eupoccilia
ambiguella*
Eupoecilia ambiguella, 16

COLEOPHORIDAE
Coleophora laricella, 12

DANAIDAE
Danaus plexippus, 5, 16, 17

DIOPTIDAE
Phryganidia californica, 12

DREPANIDAE
Drepana lacertinaria, 5

ETHMIIDAE
Ethmia assamensis, 10

GELECHIIDAE
Brachmia macroscopa, 10
Coleotechnites milleri, 9
Gnorimoschema operculella see
  Phthorimaea operculella
Pectinophora gossypiella, 5, 12
Phthorimaea operculella, 9, 12
Recurvaria milleri see *Coleotechnites
  milleri*

GEOMETRIDAE
Abraxas grossulariata, 5, 12
Alsophila pometaria, 5, 9, 12
Amphidasis cognataria, 12
Anaitis plagiata, 5, 12
Anthelia hyperborea, 12
Apocheima pilosaria, 12
Biston betularia, 5, 12
B. hirtaria, 12
B. hispidaria, 12
B. marginata, 16
B. robustum, 12
B. strataria, 12
Boarmia bistortata, 12
Bupalus piniarius, 5, 12, 16
Calospilos miranda, 10
Carecomotis repulsaria, 10
Caripeta divisata, 12
Cleora secundaria, 17
Crocallis elinguaria, 5
Cystidia stratonice stratonice, 10
Ectropis crepuscularia, 16
Ennomos quercaria, 12
E. quercinarias, 12
E. subsignarius, 12
Enypia venata, 12
Erannis defoliaria, 12
E. tiliaria, 5, 12
E. vancouverensis, 12
Eulype hastata see *Rheumaptera
  hastata*
Eupithecia longipalpata, 12
Glena bisulca, 9
Gonodontis arida, 10
Heterolocha aristonaria niphonica, 10
Hibernia defoliaria see *Erannis
  defoliaria*
Lambdina fiscellaria, 12, 16
L. fiscellaria lugubrosa, 12, 16
L. fiscellaria somniaria, 12, 16
Melanolophia imitata, 12

Nepytia canosaria, 16
N. freemani, 12
N. phantasmaria, 12
Nyctobia sp., 16
Oenochroma vinaria, 13
Operophtera bruceata, 5, 12
O. brumata, 5, 12, 16, 17 19
O. fagata, 5
Opisthograptis luteolata, 12
Oporinia autumnata, 5, 12
Ourapteryx sambucaria, 5
Paleacrita vernata, 5, 12
Peribatodes simpliciaria, 12
Pero behrensarius, 12
Phalaena vernata see *Paleacrita
  vernata*
Phigalia pedaria see *Apocheima
  pilosaria*
P. titea, 12
Protoboarmia porcelaria indicataria,
  12
Ptychopoda seriata, 12, 16
Rheumaptera hastata, 9
Sabulodes caberata, 9, 16
Scopula sp., 10
Selenia lunaria, 5
Selidosema suavis, 12
Semiothisa liturata, 5
S. pumila, 13
S. sexmaculata, 9
Sterrha seriata see *Ptychopoda seriata*
Synaxis pallulata, 12

GRACILLARIIDAE
Parectopa geometropis, 10

HELICONIIDAE
Mechanitis veritabilis, 17

HEPIALIDAE
Hepialus lupulinus, 5
Metahepialus xenoctenis, 13
Oncopera alboguttata, 19
Wiseana cervinata, 9, 10, 12, 19
W. signata, 12, 19
W. umbraculata, 9, 12, 19

HESPERIIDAE
Epargyreus clarus, 12
Parnara guttata, 10
Potanthus confucius flava see
  Potanthus flavum

Potanthus flavum, 10
Thymelicus lineola, 12

LASIOCAMPIDAE
Cosmotriche potatoria, 12
Dendrolimus pini, 5, 12, 16
D. punctatus, 5
D. sibiricus, 9
D. spectabilis, 5, 12
D. superans, 5
D. undans, 5, 12
D. undans flaveola, 12
D. yamadai, 5, 12
Entometa apicalis, 13
Eriogaster lanestris, 5
Gastropacha quercifolia, 5
G. quercifolia cerridifolia,
    5, 10, 12
Gonometa podocarpi, 13
G. rufibrunnea, 5
Kunugia yamadai see Dendrolimus
    yamadai
Lasiocampa quercus, 5, 12
L. trifolii, 12
Macrothylacia rubi, 12
Malacosoma alpicola, 12
M. americanum, 5, 12, 16
M. californicum, 12
M. constrictum, 12
M. disstria, 5, 12, 16
M. fragilis, 12
M. lutescens, 12
M. neustria, 5, 12, 16
M. neustria testacea, 5, 12
    16
M. pluviale, 12
Metanastria undans see Dendrolimus
    undans
Pachymetana sp., 13
Pachypasa otus, 12
Selenephera lunigera, 12

LIMACODIDAE
Apoda dentatus, 10
Darna trima, 9, 13
Microleon longipalpis, 10
Narosa conspersa, 17
Natada nararia, 9
Niphadolepis alianta, 17
Parasa consocia, 12
P. lepida, 17
Sibine apicalis, 5

S. fusca, 6
Spatulifimbria castaneiceps, 17
Susica nararia see Natada nararia
Thosea asigna, 13
T. cana, 17
T. cervina, 17
T. recta, 17

LYCAENIDAE
Lycaena phlaeas, 5, 10
Ogyris arbrota, 13

LYMANTRIIDAE
Dasychira abietis, 12
D. basiflava, 12
D. confusa, 12
D. mendosa, 12
D. plagiata, 12
D. pseudabietis, 10, 12
D. pudibunda, 5, 12, 16
D. selenitica, 16
Euproctis chrysorrhoea, 5, 12, 16
E. flava, 10, 12
E. pseudoconspersa, 10, 12
E. similis, 5, 10, 12
E. subflava, 12
E. terminalis, 16
Hemerocampa pseudotsugata see
    Orgyia pseudotsugata
Ivela auripes, 12
Leucoma salicis, 5, 12, 16
Lymantria dispar, 5, 6, 10, 12, 13, 15,
    16, 17, 19
L. dispar japonica, 5, 12
L. fumida fumida, 5, 12
L. incerta, 12
L. mathura aurora, 5, 12
L. monacha, 5, 12, 16
L. obfuscata, 12
Nygmia phaeorrhoea, 16
Ocneria dispar see Lymantria dispar
O. monacha see Lymantria
    monacha
Olene mendosa see Dasychira mendosa
Orgyia anartoides, 12, 13
O. antiqua, 5, 12, 16
O. australis, 12
O. badia, 12
O. gonostigma, 12
O. leucostigma, 5, 12, 16, 17
O. pseudotsugata, 5, 12, 16
O. turbata, 12

Xylena curvimacula, 12
Xylomyges conspicillaris, 5

NOTODONTIDAE
Cerura bifida see *C. hermelina*
C. hermelina, 12, 16
C. vinula, 5
Clostera anachoreta see *Pygaera anachoreta*
C. anastomosis see *Pygaera anastomosis*
Heterocampa guttivitta, 16
Leucodonta bicoloria, 13
Lophopteryx capucina, 5
Melalopha anastomosis see *Pygaera anastomosis*
Nadata gibbosa, 12
Phalera bucephala, 5, 12
Pygaera anachoreta, 10
P. anastomosis, 5, 9, 12, 17
P. anastomosis orientalis, 5, 12
P. anastomosis tristis, 5, 12
Schizura concinna, 5
Semidonta biloba, 12
Stauropus alternus, 16

NYMPHALIDAE
Aglais urticae, 5, 6, 12, 16
Agraulis vanillae, 12
Araschnia levana, 12
Argynnis dia see *Boloria dia*
A. lathonia, 16
A. paphia, 12, 16
Argyreus hyperbius, 10
Asterocampa celtis, 12
Boloria dia, 5
Charaxes jasius, 16
Clossiana dia see *Boloria dia*
Hestina japonica, 10
Inachis io, 5, 10, 12, 16
Junonia coenia, 6, 9, 12, 16
Melitaea didyma, 12
Nymphalis antiopa, 5, 9, 12, 16
N. io, see *Inachis io*
N. polychloros, 12, 16
Polygonia c-album, 5, 12
P. satyrus, 12
Pyrameis atalanta see *Vanessa atalanta*
P. cardui see *Vanessa cardui*
Vanessa atalanta, 12, 16

V. cardui, 5, 10, 12, 16
V. io see *Inachis io*
V. polychloros see *Nymphalis polychloros*
V. prorsa, 12
V. tammeamea, 16
V. urticae see *Aglais urticae*

OECOPHORIDAE
Chimbace fagella see *Diurnea fagella*
Diurnea fagella, 16

OLETHREUTIDAE
Aphania geminata, 10
Argyroploce leucotreta, 5
Eucosma ancyrota, 10
E. griseana see *Zeiraphera diniana*
Exartema appendiceum, 9
Rhyacionia buoliana, 9
R. duplana, 9, 12
Spilonota ocellana, 12
Zeiraphera diniana, 9, 12, 16, 19

PAPILIONIDAE
Graphium sarpedon, 10
Luehdorfia japonica, 12
Papilio anactus, 13
P. machaon, 5
P. machaon hippocrates, 10
P. podalirius, 12

PIERIDAE
Aporia crataegi, 5, 12, 16
Catopsilia pomona, 12
Colias chrysotheme chrysotheme, 16
C. electo, 12, 16
C. eurytheme, 5, 12
C. lesbia, 12
C. philodice, 12, 16
Euchloe cardamines, 5
Gonepteryx rhamni, 5
Neophasia menapia, 12
Pieris brassicae, 5, 9, 10, 13, 17
P. brassicae cheiranthi, 9
P. melete, 10
P. napi, 9
P. rapae, 5, 9, 12, 16
P. rapae crucivora, 5, 9, 12

SCYTHRIDIDAE
Scythris sinensis, 10

SESIIDAE
Bembecia contracta, 5, 12
Paranthrene pernix, 10
Scopelodes contracta see Bembecia
    contracta

SPHINGIDAE
Celerio euphorbiae, 5, 12, 16
C. galii, 12, 16
C. harmuthi, 16
C. kindervateri, 16
C. lineata, 12
C. phileuphorbiae, 16
C. vespertilio, 16
Cephonodes hylas, 10
Deilephila elpenor, 5, 12, 16
Dilina tiliae, 5
Herse convolvuli, 5
Hippotion eson, 12
Hyloicus pinastri, 5, 12
Laothoe populi, 5, 12
Macroglossum pyrrhosticta, 10
Manduca quinquemaculata, 9
M. sexta 9, 12, 17
Pachysphinx modesta, 5
Pergesa elpenor, see Deilephila elpenor
Proserpinus proserpina, 16
Protoparce quinquemaculata see
    Manduca quinquemaculata
P. sexta see Manduca sexta
Psilogramma increta, 10
P. menephron, 9
Smerinthus ocellata, 5, 12, 16
S. ocellata atlanticus, 12, 16
Sphinx ligustri, 5, 12
S. pinastri see Hyloicus pinastri
Theretra japonica, 12
T. nessus, 10
T. oldenlandiae, 10

THAUMETOPOEIDAE
Thaumetopoea pityocampa, 5, 9, 12
T. processionea, 5, 12
T. wilkinsoni, 5, 12

TINEIDAE
Tinea columbariella, 16
T. pellionella, 5, 10, 12
Tineola bisselliella, 5, 12, 16

TORTRICIDAE
Acleris gloverana, 12

A. variana, 12
Adoxophyes fasciata, 5, 9
A. orana, 5, 9, 12
A. reticulana see A. orana
Amelia pallorana, 9
Archippus isshikii, 19
Archips argyrospilus, 9
A. cerasivoranus, 12
A. longicellana, 9
Argyrotaenia velutinana, 9
Cacoecia murinana see Choristoneura
    murinana
Carpocapsa pomonella
    see Cydia
Choristoneura biennis, 19
C. conflictana, 19
C. diversana, 19
C. fumiferana, 5, 9, 12, 13, 19
C. murinana, 9, 12, 16
C. occidentalis, 12
C. pinus, 12
C. rosaceana, 12
Cydia molesta, 9
C. phaseoli see Lathronympha phaseoli
C. pomonella, 9, 16
Epiphyas postvittana, 12
Grapholitha molesta see
    Cydia molesta
Homona coffearla, 10, 16
H. magnanima, 5, 12
Laspeyresia pomonella see
    Cydia
Lathronympha phaseoli, 9
Merophyas divulsana, 12
Pandemis lamprosana, 12
Platynota idaeusalis, 5, 12
Ptycholomoides aeriferana, 12
Sparganothis pettitana, 12
Tortrix loeflingiana, 12, 16
T. viridana, 5, 12, 16

YPONOMEUTIDAE
Argyresthia conjugella, 12
A. cupressella, 9
Plutella maculipennis see P.
    xylostella
P. xylostella, 9, 10, 12
Prays oleae, 16
P. oleellus see Prays oleae
Yponomeuta cognatella, 12
Y. evonymella, 12
Y. malinellus, 12
Y. mayumivorellus, 10
Y. padella, 12

References

Burton, H.D., Russell, R.M. and Yerke, T.B. (1969). USDA For. Serv. Res. Note PSW-193, 6 pp. Pac. Southwest For. and Range Exp. Stn., Berkeley, California.

Hughes, K.M. (1957). *Hilgardia* 26, 597–629.

Martignoni, M.E. and Langston, R.L. (1960). *Hilgardia* 30, 1–40.

Martignoni, M.E., Williams, P. and Reineke, D.E. (1973). *J. Invertebr. Path.* 22, 100–107.

Repository for Data on the Safety of Insect Pathogens

MARSHALL LAIRD

Memorial University of Newfoundland, St. John's, Newfoundland, Canada

During the 1971 Annual Meeting of the Society for Invertebrate Pathology (SIP) in Montpellier, France, a "Working Group on Safety of Microbial Control Agents" was founded. The intention was to provide a forum for frank discussions about the safety regarding health and environment of candidate biological control agents, thus improving communications among all concerned in the government, university and industrial sectors, whether at local, national or international levels.

Membership is open to all interested. The group convenes each year at the SIP's Annual Meeting and its membership list is updated annually, adding more names at each new session. To maintain some continuity between these meetings, and in the hope of enhancing mutual confidences among participants from the above three sectors, the Chairman sends informal circular letters to members about the next annual session's program and the group's Documents Repository. This Repository has been sited at the Research Unit on Vector Pathology, Memorial University of Newfoundland, since 1972. Members have been asked to donate copies of safety-related documents, including unpublished material, which they are prepared to reveal to the membership at large.

Authors have sent copies of papers destined for the regular scientific journals, certain government papers and unpublished industrial reports, etc. As new items are accessioned, titles and availability are indicated by letter to members. On request, individual documents are airmailed until the supply is exhausted. After that, photocopies are sent. One document ("Mammalian toxicity tests of the nuclear polyhedrosis virus of the spruce budworm, *Choristoneura fumiferana*" by Cunningham *et al.*) runs to several hundred pages and is thus somewhat lengthy for frequent photocopying. The original has therefore been loaned to each applicant on the understanding that it will be airmailed back to St. John's within 2 weeks of receipt. It is now a somewhat tattered but impressively well-travelled report in consequence!

Two recent incidents stress the special usefulness of the Repository in making new information available unusually quickly to many of those most closely interested in our subject. Thus, on March 8, 1978, Dr. W.A. Smirnoff of the Laurentian Forest Research Centre of Environment Canada, forwarded a brief typescript about the safety of the baculovirus of *Neodiprion swainei* and *Thymelicus lineola* to non-target vertebrates and invertebrates (including various parasites and predators). A week later, permission was sought to make the information available to Working Group members: permission was received on 7 April.

Similarly, in early March 1978, Dr. A. Krieg of the Institut für Biologische Schädlingsbekämpfung, Darmstadt, offered a short English note summarizing the results of an important paper by Dr. A. Gröner, Dr. J. Huber and himself currently in press. This was gratefully accepted, and the note entitled "Preliminary Safety Tests with Baculoviruses on Mammals" is now available to Group members.

A letter about both new accessions has been airmailed to members as this article is being prepared (early April, 1978). Interested readers of this appendix are urged to join the SIP, if they do not belong already, and to contact the author at the above address so as to receive past and future documentation from the Repository.

At present 19 papers in the repository range from general to very specific. They comprise proceedings of a conference on safety of biological agents, an assessment of regulatory situations in various countries and articles on environmental impact of insect control by microorganisms and on use of viruses against vector insects. There is a scheme for screening and evaluating safety of agents for control of vector insects and a description of laboratory safety at a centre for disease control. For nuclear polyhedrosis viruses (NPV) a production control procedure is described, also tests on mammals with the NPVs of *Choristoneura fumiferana, Neodiprion lecontei* and various other viruses, as well as tests with NPVs on fish and birds, and cytopathogenicity of an NPV in mammalian cells. For bacteria there is a data sheet of safety tests of *Bacillus thuringiensis* and a preliminary report of mutagenicity of its exotoxin. With fungi there are safety tests of *Beauveria bassiana*. For protozoans growth of *Nosema algerae* in tissue culture is described and detection of antibodies against it in mice. The nematodes are represented by papers on susceptibility of various aquatic animals to *Romanomermis culicivorax* and nonsusceptibility of rats to *Neoaplectana carpocapsae*. Some of the papers have since been published in regular scientific journals.

Subject Index

A

Abate, 614
Abbott's formula, 286, 674
Abies balsamea, 381, 596
Ablabesymia monilis, 141
Abraxas grossulariata, 904
Acalitus vaccinii, 502
Acantholyda nemoralis, 902
Acanthopsyche junodi, see *Cryptothelea*
Acanthoscelides obtectus, 467, 842
Acanthotermes, 851
Acari, 19, 264, 748, 876, 899
Hirsutella thompsonii, 500, 506, 509
microsporidia, 176, 178, 180
pathogens, 109, 120, 427–31, 491, 540,
804–5, 839
Acarina, see Acari
Acephate, 697, 699, 702, 704
Acetyle-methyl-carbinol, 40
Achaea janata, 520, 851, 906
Acheta domesticus, 574, 882, 911
Achraea grisella, see *Achroia*
Achroia grisella, 851, 880, 884, 909
Achromobacter, 26, 29
eurydice, 29; *nematophilus*, 26
Acid production, 12
Acineto
*acinetophagus, bosmina, daphniae, pater-
sonii*, 144
Acinetobacter, 27
Acleris
gloverana, 910; *variana*, 851, 910
Acremonium
alternatum, 100; *larvarum*, 102; *zeylanicum*,
102
Acrida turrita, 911
Acrididae, 574, 583–4, 911
Acrobasis, 851
zelleri, 909
Acrobeloides, 155
Acrodontium crateriforme, 102
Acrolepia, see *Acrolepiopsis*
Acrolepiopsis
alliella, 851; *assectella*, 851
Acromyrmex octospinosus, 545
Acronicta aceris, 906
Actebia fennica, 906

Actias
luna, 909; *selene*, 909
Actinomycetes, 8, 30
Actinomycin D, 775–6
Active paralysis virus, 65, 82, 899–911
Activity ratio, 201–2, 204, 206–7, 217
Acyrthosiphon
kondoi, 111; *pisum*, 540–3, 748
Adenine nucleotide, 198
Adjuvant, 507–8
Adoryphorus couloni, 54
Adoxophyes
fasciata, 910; *orana*, 816, 851, 910; *reticulana*, see *orana*
Adris tyrannus amurensis, 906
Aedes, 138–9, 153, 295, 445, 568, 844
aegypti, 449, 453–4, 456
bacteria, 285–6, 289–90, 292
B. thuringiensis, 195, 216, 839, 843,
879, 884
other pathogens, 138, 561–2, 568,
606, 609, 614, 900
albopictus, 138, 290, 292, 411, 453, 568,
900
annulipes, 900
atlanticus, 559, 606
atropalpus, 138
australis, 138, 557, 564
canadensis, 292, 606
cantans, 138, 844, 900
caspius, 844, 900
caspius dorsalis, 138, 900
cataphylla, 900
cinereus, 138, 900
communis, 606, 844
cyprius, 138
detritus, 844, 900
dorsalis, 564, 568, 606, 844, 900
epactius, 138, 449, 453, 456, 556–7,
561, 568, 609
excrucians, 138, 900
flavescens, 568, 901
fulvus pallens, 901
hebrideus, 138
mediovittatus, 561–2
melanimon, 138
multiformis, 138